郭柏灵院士

郭柏灵论文集

第十四卷

Selected Papers of Guo Boling

Volume 14

郭柏灵　著

科　学　出　版　社
北　京

内 容 简 介

郭柏灵论文集第十四卷收集的是郭柏灵先生发表于 2016 年度的主要科研论文,涉及的方程范围宽广,有确定性偏微分方程和随机偏微分方程,研究的问题包括适定性、爆破性、渐近性、孤立波等.

该文集具有较高的学术价值,可供偏微分方程、数学物理、非线性分析、计算数学等方向的科研工作者和研究生参考阅读.

图书在版编目(CIP)数据

郭柏灵论文集·第十四卷/郭柏灵著. —北京:科学出版社,2021.4
ISBN 978-7-03-068124-9

Ⅰ. ①郭… Ⅱ. ①郭… Ⅲ. ①郭柏灵-文集 ②非线性方程-文集
Ⅳ. ①O175-53

中国版本图书馆 CIP 数据核字 (2021) 第 033197 号

责任编辑:李 欣/责任校对:彭珍珍
责任印制:吴兆东/封面设计:陈 敬

科学出版社 出版
北京东黄城根北街 16 号
邮政编码:100717
http://www.sciencep.com

北京建宏印刷有限公司 印刷
科学出版社发行 各地新华书店经销

*

2021 年 4 月第 一 版 开本:720×1000 1/16
2021 年 4 月第一次印刷 印张:39 1/2 插页:2
字数:801 000

定价:298.00 元
(如有印装质量问题,我社负责调换)

致 谢

《郭柏灵论文集》于2006年开始由华南理工大学出版社连续出版了十一卷，论文的收集截止到2012年。从第十二卷开始《郭柏灵论文集》将由科学出版社出版发行，从该卷起将收集郭柏灵院士于2013年以来先后发表的科研成果。在出版过程中科学出版社的同志，特别是陈玉琢老师进行了精心策划和细致编排，付出了辛勤的劳动。再次谨对他们的帮助表示衷心的感谢。

谭绍滨
2018年3月于厦门大学

序

今年是恩师郭柏灵院士 70 寿辰，华南理工大学出版社决定出版《郭柏灵论文集》。郭老师的弟子，也就是我的师兄弟，推举我为文集作序。这使我深感荣幸。我于 1985 年考入北京应用物理与计算数学研究所，师从郭柏灵院士和周毓麟院士。研究生毕业后我留在研究所工作，继续跟随郭老师学习和研究偏微分方程理论。老师严谨的治学作风和对后学的精心培养与殷切期望，给我留下了深刻的印象，同时老师在科研上的刻苦精神也一直深深地印在我的脑海中。

郭老师 1936 年出于福建省龙岩市新罗区龙门镇，1953 年从福建省龙岩市第一中学考入复旦大学数学系，毕业后留校工作。1963 年，郭老师服从祖国的需要，从复旦大学调入北京应用物理与计算数学研究所，从事核武器研制中有关的数学、流体力学问题及其数值方法研究和数值计算工作。他全力以赴地做好了这项工作，为我国核武器的发展作出了积极的贡献。1978 年改革开放以后，他又在非线性发展方程数学理论及其数值方法领域开展研究工作，现为该所研究员、博士生导师，中国科学院院士。迄今他共发表学术论文 300 余篇、专著 9 部，1987 年获国家自然科学奖三等奖，1994 年和 1998 年两度获得国防科工委科技进步奖一等奖，为我国的国防建设与人才培养作出了巨大贡献。

郭老师的研究方向涉及数学的多个领域，其中包括非线性发展方程的数学理论及其数值解、孤立子理论、无穷维动力系统等，其研究工作的主要特点是紧密联系数学物理中提出的各种重要问题。他对力学及物理学等应用学科中出现的许多重要的非线性发展方程进行了系统深入的研究，其中对 Landau-Lifshitz 方程和 Benjamin-Ono 方程的大初值的整体可解性、解的唯一性、正则性、渐近行为以及爆破现象等建立了系统而深刻的数学理论。在无穷维动力系统方面，郭老师研究了一批重要的无穷维动力系统，建立了有关整体吸引子、惯性流形和近似惯性流形的存在性和分形维数精细估计等理论，提出了一种证明强紧吸引子的新方法，并利用离散化等方法进行理论分析和数值计算，展示了吸引子的结构和图像。下面我从这几个方面介绍郭老师的一些学术成就。

Landau-Lifshitz 方程 (又称铁磁链方程) 由于其结构的复杂性，特别是强耦合性和不带阻尼时的强退化性，在 20 世纪 80 年代之前国内外几乎没有从数学上进行理论研究的成果出现。最先进行研究的，当属周毓麟院士和郭老师，他们在 1982 年到 1986 年间，采用 Leray-Schauder 不动点定理、离散方法、Galerkin 方法证明了从一维到高维的各种边值问题整体弱解的存在性，比国外在 1992 年才出现的同类结果早了将近 10 年。

20 世纪 90 年代初期，周毓麟、郭柏灵和谭绍滨，郭柏灵和共敏纯得到了两个在国内外至今影响很大的经典结果。第一，通过差分方法结合粘性消去法，利用十分巧妙的先验估计，证明了一维 Landau-Lifshitz 方程光滑解的存在唯一性，对于一维问题给出了完整的答案，解决了长期悬而未决的难题。第二，系统分析了带阻尼的二维 Landau-Lifshitz 方程弱解的奇性，发现了 Landau-Lifshitz 方程与调和映照热流的联系，其弱解具有与调和映照热流完全相同的奇性。现在，国内外这方面的文章基本上都引用这个结果。调和映照的 Landau-Lifshitz 流的概念，即是源于此项结果。

20 世纪 90 年代中期，郭老师对于 Landau-Lifshitz 方程的长时间性态、Landau-Lifshitz 方程耦合 Maxwell 方程的弱解及光滑解的存在性问题进行了深入的研究，得到了一系列的成果。铁磁链方程的退化性以及缺少相应的线性化方程解的表达式，对研究解的长时间性态带来很大困难。郭老师的一系列成果克服了这些困难，证明了近似惯性流形的存在性、吸引子的存在性，给出了其 Hausdorff 和分形维数的上、下界的精细估计。此外，我们知道，与调和映照热流比较，高维铁磁链方程的研究至今还很不完善。其中最重要的是部分正则性问题，其难点在于单调不等式不成立，导致能量衰减估计方面的困难。另外一个是 Blow-up 解的存在性问题，至今没有解决；而对于调和映照热流来说，这样的问题的研究是比较成熟的。

对于高维问题，20 世纪 90 年代后期至今，郭老师和陈韵梅、丁时进、韩永前、杨士山一道，得出了许多成果，大大地推动了该领域的研究进程。第一，证明了二维问题的能量有限弱解的几乎光滑性及唯一性，这个结果类似于 Freire 关于调和映照热流的结果。第二，得到了高维 Landau-Lifshitz 方程初边值问题的奇点集合的 Hausdorff 维数和测度的估计。第三，得到了三维 Landau-Lifshitz-Maxwell 方程的奇点集合的 Hausdorff 维数和测度的估计。第四，得到了一些高维轴对称问题的整体光滑解和奇性解的精确表达式。郭老师还开创了一些新的研究领域。例如，关于一维非均匀铁磁链方程光滑解的存在唯一性结果后来被其他数学家引用并推广到一般流形上。第五，率先讨论了可压缩铁磁链方程测度值解的存在性。最近，在 Landau-Lifshitz 方程耦合非线性 Maxwell 方程与面，也取得了许多新的进展。

多年来，郭老师还对一大批非线性发展方程解的整体存在唯一性、有限时刻的爆破性、解的渐近性态等开展了广泛而深入的研究，受到国内外同行的广泛关注。研究的模型源于数学物理、水波、流体力学、相变等领域，如含磁场的 Schrödinger 方程组、Zakharov 方程、Schrödinger-Boussinesq 方程组、Schrödinger-KdV 方程组、长短波方程组、Maxwell 方程组、Davey-Stewartson 方程组、Klein-Gordon-Schrödinger

方程组、波动方程、广义 KdV 方程、Kadomtsev-Petviashvili (KP) 方程、Benjamin-Ono 方程、Newton-Boussinesq 方程、Cahn-Hilliard 方程、Ginzburg-Landau 方程等。其中不少耦合方程组都是郭老师得到了第一个结果，开创了研究的先河，对国内外同行的研究产生了深远的影响。

郭老师在无穷维动力系统方面也开展了广泛的研究，取得了丰硕的成果。对耗散非线性发展方程所决定的无穷维动力系统，研究了整体吸引子的存在性、分形维数估计、惯性流形、近似惯性流形、指数吸引子等问题。特别是在研究无界哉上耗散非线性发展方程的强紧整体吸引子存在性时所提出的化弱紧吸引子成为强紧吸引子的重要方法和技巧，颇受同行关注并广为利用。对五次非线性 Ginzburg-Landau 方程，郭老师利用空间离散化方法将无限维问题化为有限维问题，证明了该问题离散吸引子的存在性，并考虑五次 Ginzburg-Landau 方程的定态解、慢周期解、异宿轨道等的结构。利用有限维动力系统的理论和方法，结合数值计算得到具体的分形维数 (不超过 4) 和结构以及走向混沌、湍流的具体过程和图像，这是一种寻求整体吸引子细微结构新的探索和尝试，对其他方程的研究也是富有启发性的。1999 年以来，郭老师集中于近可积耗散的和 Hamilton 无穷维动力系统的结构性研究，利用孤立子理论、奇异摄动理论、Fenichel 纤维理论和无穷维 Melnikov 函数，对于具有小耗散的三次到五次非线性 Schrödinger 方程，证明了同宿轨道的不变性，并在有限维截断下证明了 Smale 马蹄的存在性，目前，正把这一方法应用于具小扰动的 Hamilton 系统的研究上。他对于非牛顿流无穷维动力系统也进行了系统深入的研究，建立了有关的数学理论，并把有关结果写成了专著。以上这些工作得到国际同行们的高度评价，被称为"有重大的国际影响""对无穷维动力系统理论有重要持久的贡献"。最近，郭老师及其合作者又证明了具耗散的 KdV 方程 L^2 整体吸引子的存在性，该结果也是引人注目的。

郭老师不仅自己辛勤地搞科研，还尽心尽力培养了大批的研究生 (硕士生、博士生、博士后)，据不完全统计，有 40 多人。他根据每个人不同的学习基础和特点，给予启发式的具体指导，其中的不少人已成为了该领域的学科带头人，有些人虽然开始时基础较差，经过培养，也得到了很大提高，成为了该方向的业务骨干。

《郭柏灵论文集》按照郭老师在不同时期所从事的研究领域，分成多卷出版。文集中所搜集的都是郭老师正式发表过的学术成果。把这些成果整理成集出版，不仅系统地反映了他的科研成就，更重要的是对于从事这方面学习、研究的学者无疑大有裨益。这本文集的出版得到了多方面的帮助与支持，特别要感谢华南理工大学校长李元元教授、华南理工大学出版社范家巧社长和华南理工大学数学科学院吴敏院长的支持。还要特别感谢华南理工大学的李用声教授、华南师范大学的丁时进教授、北京应用物理与计算数学研究所的苗长兴研究员等人在论文的搜集、选择与校对等工作中付出了辛勤的劳动。感谢华南理工大学出版社的编辑对文集的精心编排工作。

谭绍滨

2005 年 8 月于厦门大学

目录

2016 年

Well-posedness for the Three-dimensional Compressible Liquid Crystal Flows ············ 1

Optimal Decay Rate of the Compressible Quantum Navier-Stokes Equations ············ 34

Mixed-type Soliton Solutions for the N-coupled Higher-order Nonlinear Schrödinger
    Equations in Optical Fibers ································· 47

Global Finite Energy Weak Solution to the Viscous Quantum Navier-Stokes-
    Landau-Lifshitz-Maxwell Model in 2-dimension ································· 59

Global Existence and Long-time Behavior for the Strong Solutions in H^2 to the 3D
    Compressible Nematic Liquid Crystal Flows ································· 84

Global Smooth Solution to a Coupled Schrödinger System in Atomic Bose-Einstein
    Condensates with Two Dimensional Spaces ································· 113

Vanishing Viscosity Limit for the 3D Magnetohydrodynamic System with Generalized
    Navier-slip Boundary Conditions ································· 132

Bright-dark Vector Soliton Solutions for the Coupled Complex Short Pulse Equations in
    Nonlinear Optics ································· 148

Global Existence of Smooth Solutions for the Magnetic Schrödinger Equation Arising
    from Hot Plasma ································· 160

Approximate Solution of Bagley-Torvik Equations with Variable Coefficients and
    Three-point Boundary-value Conditions ································· 192

Blow-up of Smooth Solution to Quantum Hydrodynamic Models in $\mathbf{R}^d$ ··············· 217

Existence and Uniqueness of Global Solutions of the Modified KS-CG Equations for
    Flames Governed by A Sequential Reaction ································· 248

On the Global Well-posedness of the Fractional Klein-Gordon-Schrödinger System
    with Rough Initial Data ································· 271

Global Existence and Semiclassical Limit for Quantum Hydrodynamic Equations with
    Viscosity and Heat Conduction ································· 302

Dynamics for a Generalized Incompressible Navier-Stokes Equations in $\mathbf{R}^2$ ·········· 336

Globally Smooth Solution and Blow-up Phenomenon for a Nonlinearly Coupled

Schrödinger System Atomic Bose-Einstein Condensates ··················· 370

High-order Rogue Wave Solutions for the Coupled Nonlinear Schrödinger Equations-II · 392

A Second-order Finite Difference Method for Two-dimensional Fractional
Percolation Equations ··· 413

Classical Solutions of General Ginzburg-Landau Equations ··················· 440

Implicit Finite Difference Method for Fractional Percolation Equation with Dirichlet and
Fractional Boundary Conditions ··· 461

Coupling Model for Unsteady MHD Flow of Generalized Maxwell Fluid with Radiation
Thermal Transform ··· 479

Global Well-posedness for the Periodic Novikov Equation with Cubic Nonlinearity ···· 500

Quasineutral Limit of the Pressureless Euler-Poisson Equation for Ions ············ 525

Well-posedness and Blow-up Scenario for a New Integrable Four-component System with
Peakon Solutions ··· 561

Darboux Transformation and Classification of Solution for Mixed Coupled Nonlinear
Schrödinger Equations ·· 588

Well-posedness for the Three-dimensional Compressible Liquid Crystal Flows*

Li Xiaoli (李晓莉) and Guo Boling (郭柏灵)

Abstract This paper is concerned with the initial-boundary value problem for the three-dimensional compressible liquid crystal flows. The system consists of the Navier-Stokes equations describing the evolution of a compressible viscous fluid coupled with various kinematic transport equations for the heat flow of harmonic maps into $\mathbb{S}^2$. Assuming the initial density has vacuum and the initial data satisfies a natural compatibility condition, the existence and uniqueness is established for the local strong solution with large initial data and also for the global strong solution with initial data being close to an equilibrium state. The existence result is proved via the local well-posedness and uniform estimates for a proper linearized system with convective terms.

Keywords liquid crystals; compressible; vacuum; strong solution; existence and uniqueness

1 Introduction

In this paper, we establish the well-posedness of a simplified hydrodynamic equation, proposed by Ericksen and Leslie, modeling the flow of nematic liquid crystals formulated in [7]-[9] and [15] in the 1960's. A simplified version of the Ericksen-Leslie model was introduced by Lin [17] and analyzed by Lin and Liu [18]-[20] who used a modified Galerkin approach, and by Shkoller [25] who relied on a contraction mapping argument coupled with appropriate energy estimates. When the Ossen-Frank energy configuration functional reduces to the Dirichlet energy functional, the hydrodynamic

---

* Discrete Contin. Dyn. Syst. Ser. S 9(2016): 1913–1937. DOI:10.3934/dcdss.2016078.

flow equation of liquid crystals in $\mathbb{R}^3$ can be written as follows (see [17]):

$$\rho_t + \nabla \cdot (\rho \mathbf{u}) = 0, \tag{1.1a}$$

$$(\rho \mathbf{u})_t + \nabla \cdot (\rho \mathbf{u} \otimes \mathbf{u}) + \nabla p(\rho) = \mu \Delta \mathbf{u} - \lambda \nabla \cdot \left(\nabla \mathbf{d} \odot \nabla \mathbf{d} - \frac{|\nabla \mathbf{d}|^2}{2} \mathbb{I}_3 \right), \tag{1.1b}$$

$$\mathbf{d}_t + \mathbf{u} \cdot \nabla \mathbf{d} = \theta \left(\Delta \mathbf{d} + |\nabla \mathbf{d}|^2 \mathbf{d} \right), \tag{1.1c}$$

where $\mathbf{u} \in \mathbb{R}^3$ denotes the velocity, $\mathbf{d} \in \mathbb{S}^2$ (the unit sphere in $\mathbb{R}^3$) is the unit-vector field that represents the macroscopic molecular orientations, $p(\rho)$ is the pressure with $p = p(\cdot) \in C^1[0,\infty)$, $p(0) = 0$; and they all depend on the spatial variable $\mathbf{x} = (x_1, x_2, x_3) \in \mathbb{R}^3$ and the time variable $t > 0$. $\sigma = \mu \nabla \mathbf{u} - p \, \mathbb{I}_3$ (the 3×3 identity matrix) is the Cauchy stress tensor given by Stokes' law and $\mu \nabla \mathbf{u}$ stands for the fluid viscosity part of the stress tensor. The term $\lambda \nabla \cdot (\nabla \mathbf{d} \odot \nabla \mathbf{d})$ in the stress tensor represents the anisotropic feature of the system. The parameters μ, λ, θ are positive constants standing for viscosity, the competition between kinetic energy and potential energy, and microscopic elastic relaxation time or the Deborah number for the molecular orientation field, respectively. The symbol $\nabla \mathbf{d} \odot \nabla \mathbf{d}$ denotes a matrix whose (i,j)-th entry is $\partial_{x_i} \mathbf{d} \cdot \partial_{x_j} \mathbf{d}$ for $1 \leqslant i,j \leqslant 3$, and $\nabla \mathbf{d} \odot \nabla \mathbf{d} = (\nabla \mathbf{d})^{\mathrm{T}} \nabla \mathbf{d}$, where $(\nabla \mathbf{d})^{\mathrm{T}}$ denotes the transpose of the matrix $\nabla \mathbf{d}$.

It is well known liquid crystals are states of matter which are capable of flow and in which the molecular arrangement gives rise to a preferred direction. Roughly speaking, the system (1.1) is a coupling between the compressible Navier-Stokes equations and a transport heat flow of harmonic maps into $\mathbb{S}^2$. It is a macroscopic continuum description of the evolution for the liquid crystals of nematic type under the influence of both the flow field $\mathbf{u}$, and the macroscopic description of the microscopic orientation configurations $\mathbf{d}$ of rod-like liquid crystals. As for the nonlinear constraint $\mathbf{d} \in \mathbb{S}^2$, one of the methods used to relax it is to consider a form of penalization, that is, not $|\nabla \mathbf{d}|^2$ in (1.1c), but the Ginzburg-Landau approximation $\frac{1}{\varepsilon^2}(1 - |\mathbf{d}|^2)$ for small ε. There were some similar fundamental results starting from the work in [19], where the density is constant, Lin-Liu proved local existence of the classical solutions the global existence of weak solutions in the two-dimensional and three-dimensional spaces. For the density-dependent case, Liu [22] proved the global existence of weak solutions and classical solutions to the system of incompressible Smectic-A liquid crystals under the general condition of the initial density ρ_0 satisfying $0 < \alpha \leqslant \rho_0 \leqslant \beta$. The global existence of weak solutions in dimension three was established by Liu-Zhang [24] if

$\rho_0 \in L^2(\Omega)$. Later Jiang-Tan [14] pointed out that the condition on initial density can be weaken to belong to $L^\gamma(\Omega)$ for any $\gamma \geqslant \frac{3}{2}$. As for the compressible case, Liu-Liu-Hao [23] established the global existence of strong solutions under the smallness conditions on the initial data in Sobolev spaces in dimension three.

Compared with the Ginzburg-Landau approximation problem, $|\nabla \mathbf{d}|^2$ in (1.1c) brings us some new difficulties. Since the strong solutions of a harmonic map must be blowing up at finite time (see Chang-Ding-Ye [1] for the heat flow of harmonic maps), one cannot expect that there exists a global strong solution to system (1.1) with general initial data. In fact, the global existence of weak solutions to (1.1) with large initial data is an outstanding open problem for high dimensions. By so far, only results in one space dimension have been obtained, for instance, we refer to [5,6]. For the homogeneous case of system (1.1), both the regularity and existence of global weak solutions in dimension two were established by Lin-Lin-Wang [21]. More explicitly, they obtained both interior and boundary regularity theorem for such a flow under smallness conditions, and the existence of global weak solutions that are smooth away from at most finitely many singular times in any bounded smooth domain of $\mathbb{R}^2$. Meanwhile, Hong [10] also showed the global existence of weak solution to this system in two dimensional space. Wang [26] established a global well-posedness theory for the incompressible liquid crystals for rough initial data, provided that $\|\mathbf{u}_0\|_{BMO^{-1}} + [\mathbf{d}_0]_{BMO} \leqslant \varepsilon_0$ for some $\varepsilon_0 > 0$. Assuming that the initial density ρ_0 has a positive bound from below and under smallness conditions on the initial data, Wen-Ding [27] got the global existence and uniqueness of the strong solution to the incompressible density-dependent case in Sobolev spaces in two dimensions, and Li-Wang obtained the result in Sobolev-Besov spaces for three dimensional case in [16]. Concerning the compressible case, Hu-Wu considered the Cauchy problem for the three-dimensional compressible flow of nematic liquid crystals and obtained the existence and uniqueness of the global strong solution in critical Besov spaces provided that the initial data is close to an equilibrium state in a recent work [11]. Local existence of unique strong solutions were proved provided that the initial data $\rho_0, \mathbf{u}_0, \mathbf{d}_0$ are sufficiently regular and satisfy a natural compatibility condition in [12]. A criterion for possible breakdown of such a local strong solution at finite time was given in terms of blow up of the L^∞-norms of ρ and $\nabla \mathbf{d}$. In [13], an alternative blow-up criteria was derived in terms of the L^∞-norms of $\nabla \mathbf{u}$ and $\nabla \mathbf{d}$ in dimension three.

In this paper we consider the initial-boundary value problem of system (1.1) in a bounded smooth domain $\Omega \subset \mathbb{R}^3$, with the following initial-boundary conditions:

$$(\rho, \mathbf{u}, \mathbf{d})\,|_{t=0} = (\rho_0, \mathbf{u}_0, \mathbf{d}_0), \quad \mathbf{x} \in \Omega, \tag{1.2}$$

$$(\mathbf{u}, \mathbf{d})\,|_{\partial\Omega} = (\mathbf{0}, \mathbf{d}_0\,|_{\partial\Omega}), \quad t > 0, \tag{1.3}$$

where $\rho_0 \geqslant 0$, $\mathbf{d}_0 : \Omega \to \mathbb{S}^2$ is given with compatibility. The boundary condition on the velocity implies non-slip on the boundary.

We are interested in the strong solutions to the initial-boundary problem (1.1)-(1.3). In order to obtain strong solutions, we make the following assumptions on the initial data:

$$\rho_0 \in W^{1,q}, \quad \mathbf{u}_0 \in (H_0^1)^3 \cap (H^2)^3, \quad \nabla \mathbf{d}_0 \in (H^2)^9, \tag{1.4}$$

where $3 < q \leqslant 6$. And, $(\rho_0, \mathbf{u}_0, \mathbf{d}_0)$ satisfies a natural compatibility condition

$$-\mu \Delta \mathbf{u}_0 + \nabla p_0 + \lambda \nabla \cdot (\nabla \mathbf{d}_0 \odot \nabla \mathbf{d}_0) = \rho_0^{\frac{1}{2}} \mathbf{g}, \quad \forall\, \mathbf{x} \in \Omega \tag{1.5}$$

with some $(p_0, \mathbf{g}) \in H^1 \times (L^2)^2$. We note here that (1.5) is a compensation to the lack of a positive lower bound of the initial density (see [2]).

Our main result establishes the local well-posedness of the Ericksen-Leslie problem for any regular enough initial data:

Theorem 1.1 *Let the initial data satisfies the regularity condition (1.4) and also the compatibility condition (1.5). There exists a time T^* such that the initial-boundary problem (1.1)-(1.3) has a unique local strong solution $(\rho, \mathbf{u}, \mathbf{d})$ satisfying*

$$\begin{cases} \rho \in C([0,T^*]; W^{1,q}), & \rho_t \in C([0,T^*]; L^q); \\ \mathbf{u} \in C([0,T^*]; H_0^1 \cap H^2) \cap L^2(0,T^*; W^{2,q}), & \mathbf{d} \in C([0,T^*]; H^3); \\ \mathbf{u}_t \in L^2(0,T^*; H_0^1), & \mathbf{d}_t \in C([0,T^*]; H_0^1) \cap L^2(0,T; H^2); \\ \sqrt{\rho}\,\mathbf{u}_t \in C([0,T^*]; L^2), & \mathbf{d}_{tt} \in L^2([0,T^*]; L^2). \end{cases}$$

Furthermore, if we suppose that $(\tilde{\rho}, \tilde{\mathbf{u}}, \tilde{\mathbf{d}})$ is another solution with initial-boundary conditions

$$\begin{cases} (\tilde{\rho}, \tilde{\mathbf{u}}, \tilde{\mathbf{d}})\,|_{t=0} = (\tilde{\rho}_0, \tilde{\mathbf{u}}_0, \tilde{\mathbf{d}}_0), & \mathbf{x} \in \Omega, \\ (\tilde{\mathbf{u}}, \tilde{\mathbf{d}})\,|_{\partial\Omega} = (\mathbf{0}, \tilde{\mathbf{d}}_0), & t > 0, \end{cases}$$

then for any $t \in (0, T^*]$, the quantities

$$\|\rho - \tilde{\rho}\|_{L^q}(t), \quad \|\mathbf{u} - \tilde{\mathbf{u}}\|_{H^1}(t), \quad \|\mathbf{u} - \tilde{\mathbf{u}}\|_{L^2_{T^*}(H^2)}, \quad \|\sqrt{\rho}(\mathbf{u}_t - \tilde{\mathbf{u}}_t)\|_{L^2_{T^*}(L^2)},$$

$$\|\mathbf{d}-\tilde{\mathbf{d}}\|_{H^2}(t), \ \|\mathbf{d}-\tilde{\mathbf{d}}\|_{L^2_{T*}(H^3)}, \ \|\mathbf{d}_t-\tilde{\mathbf{d}}_t\|_{L^2}(t), \ \|\mathbf{d}_t-\tilde{\mathbf{d}}_t)\|_{L^2_{T*}(H^1)}$$

tend to zero as $(\tilde{\rho}_0, \tilde{\mathbf{u}}_0, \tilde{\mathbf{d}}_0) \to (\rho_0, \mathbf{u}_0, \mathbf{d}_0)$.

Moreover, we shall prove the existence of global strong solution for initial data that is close to an equilibrium state $(0, 0, \mathbf{n})$ with a constant vector $\mathbf{n} \in \mathbb{S}^2$. More precisely,

Theorem 1.2 *Let ϱ be a nonnegative constant and $\mathbf{n}$ be a constant unit vector in $\mathbb{R}^3$. Then there exists a suitable positive constant ξ_0 (small) such that if the initial data satisfies further*

$$\max\{\|\rho_0 - \varrho\|_{W^{1,q}}, \ \|\mathbf{u}_0\|_{H^2}, \ \|\mathbf{d}_0 - \mathbf{n}\|_{H^3}, \ \|\mathbf{g}\|_{L^2}^2\} \leq \xi$$

for all $\xi \in (0, \xi_0]$, the problem (1.1)-(1.3) has a unique global strong solution enjoying the regularities as in Theorem 1.1.

Remark 1.1 *We only consider the case $\varrho = 0$ in this paper, because the other case $\varrho > 0$ can be induced to the problem with positive initial density.*

From the viewpoint of partial differential equations, system (1.1) is a highly nonlinear system coupling hyperbolic equations and parabolic equations. It is very challenging to understand and analyze such a system, especially when the density function ρ may vanish or the fluid takes vacuum states (equation (1.1b) becomes a degenerate parabolic-elliptic couples system). Our approach is quite classical. Successive approximation method [3, 4] is employed for two variables, which was used in [23]. It consists in deriving energy estimates without loss of derivatives in sufficiently high order Sobolev spaces for a linearized version of (1.1), and then solve the nonlinear problem through an iterative scheme. There are some difficulties in both steps that will be pointed out along the detailed proofs. Moreover, in order to overcome the difficulty that the initial density has vacuum brings, as usual, the technique is to approximate the nonnegative initial density by a positive initial data.

We organize the rest of this paper as follows. In Sect. 2, we introduce a special linear problem of the original system (1.1)-(1.3) and prove local existence of a strong solution to the linear problem with positive initial density. We also derive a series of uniform *a priori* estimates, which ensure the local strong solution exists when the initial density allows vacuum. In Sect. 3, after constructing a sequence of approximate solutions, a strong solution of (1.1)-(1.3) is obtained. The uniqueness and continuity on the initial data are also proved. Finally, in Sect. 4, our main result on the global existence of a strong solution of (1.1)-(1.3) will be established via iteration and the convergence of the iteration.

2 A linear problem

In this section, let us consider the following auxiliary linear problem:

$$\rho_t + \nabla \cdot (\rho \mathbf{v}) = 0, \tag{2.1a}$$

$$\mathbf{d}_t + \mathbf{v} \cdot \nabla \mathbf{d} = \theta \left(\Delta \mathbf{d} + (\nabla \mathbf{f} : \nabla \mathbf{d}) \, \mathbf{f} \right), \tag{2.1b}$$

$$\rho \mathbf{u}_t + \rho \mathbf{v} \cdot \nabla \mathbf{u} + \nabla p(\rho) = \mu \Delta \mathbf{u} - \lambda \nabla \cdot \left(\nabla \mathbf{d} \odot \nabla \mathbf{d} - \frac{|\nabla \mathbf{d}|^2}{2} \mathbb{I}_3 \right), \tag{2.1c}$$

with $\mathbf{v} \in \mathbb{R}^3$ and $\mathbf{f} \in \mathbb{R}^3$ being given vector functions and enjoying the regularities such that

$$\begin{aligned}
&\mathbf{v} \in C([0,T]; H_0^1 \cap H^2) \cap L^2(0,T; W^{2,q}), \qquad \mathbf{v}_t \in L^2(0,T; H_0^1), \\
&\mathbf{f} \in C([0,T]; H^3), \qquad \mathbf{f}_t \in C([0,T]; H_0^1) \cap L^2(0,T; H^2),
\end{aligned} \tag{2.2}$$

for all $T > 0$.

Conformally to the initial-boundary conditions of the original problem, we suppose

$$\begin{aligned}
(\mathbf{v}, \mathbf{f}) \, |_{t=0} &= (\mathbf{u}_0, \mathbf{d}_0), \quad \mathbf{x} \in \Omega, \\
(\mathbf{v}, \mathbf{f}) \, |_{\partial\Omega} &= (\mathbf{0}, \mathbf{d}_0 \, |_{\partial\Omega}), \quad t \in (0, T).
\end{aligned} \tag{2.3}$$

2.1 Existence of approximate solutions

For each $\delta > 0$, for example, $\delta \in (0,1)$, let $\mathbf{u}_0^\delta$ solve the elliptic boundary value problem:

$$\begin{cases} \mu \Delta \mathbf{u}_0^\delta - \lambda \nabla \cdot \left(\nabla \mathbf{d}_0 \odot \nabla \mathbf{d}_0 - \dfrac{|\nabla \mathbf{d}_0|^2}{2} \mathbb{I}_3 \right) - \nabla p(\rho_0^\delta) = (\rho_0^\delta)^{\frac{1}{2}} \mathbf{g}, \\ \mathbf{u}_0^\delta |_{\partial\Omega} = 0, \end{cases}$$

where $\rho_0^\delta = \rho_0 + \delta$. Then $\mathbf{u}_0^\delta \to \mathbf{u}_0$ in $H_0^1 \cap H^2$ as $\delta \to 0$.

Now for the special linear equations (2.1), we have the following result:

Theorem 2.1   Let $\rho_0, \mathbf{u}_0$ and $\mathbf{d}_0$ satisfy the regularity (1.4) and $(\mathbf{v}, \mathbf{f})$ satisfies (2.2) and (2.3). Then there exists a unique strong solution $(\rho, \mathbf{u}, \mathbf{d})$ to (2.1) with the initial data $(\rho_0^\delta, \mathbf{u}_0^\delta, \mathbf{d}_0)$ and the boundary condition (1.3) such that

$$\rho \in C([0,T]; W^{1,q}), \qquad \rho_t \in C([0,T]; L^q),$$

$$\mathbf{u} \in C([0,T]; H_0^1 \cap H^2) \cap L^2(0,T; W^{2,q}), \quad \mathbf{u}_t \in C([0,T]; L^2) \cap L^2(0,T; H_0^1),$$
$$\mathbf{d} \in C([0,T]; H^3), \quad \mathbf{d}_t \in C([0,T]; H_0^1) \cap L^2(0,T; H^2),$$
$$\mathbf{u}_{tt} \in L^2(0,T; H^{-1}), \quad \mathbf{d}_{tt} \in L^2(0,T; L^2).$$

We will prove Theorem 2.1 through a series of lemmas. To begin with, assume there are three constants c_0, c_1 and c_2 such that

$$c_0 > 1 + \|\rho_0\|_{W^{1,q}} + \|\mathbf{u}_0\|_{H^2} + \|\nabla \mathbf{d}_0\|_{H^2} + \|\mathbf{g}\|_{L^2}^2, \tag{2.4}$$

$$c_1 > \sup_{0 \leqslant t \leqslant T} (\|\mathbf{v}\|_{H_0^1} + \|\mathbf{f}\|_{H^2} + \|\mathbf{f}_t\|_{H_0^1}) + \int_0^T (\|\nabla \mathbf{v}_t\|_{L^2}^2 + \|\mathbf{v}\|_{W^{2,q}}^2 + \|\nabla^2 \mathbf{f}_t\|_{L^2}^2) \, dt, \tag{2.5}$$

$$c_2 > \sup_{0 \leqslant t \leqslant T} (\|\nabla^2 \mathbf{v}\|_{L^2} + \|\nabla^3 \mathbf{f}\|_{L^2}), \tag{2.6}$$

$$c_2 > c_1 > c_0. \tag{2.7}$$

Throughout of the whole paper, sometimes, we make use of $A \lesssim B$ in place of $A \leqslant C_0 B$, where C_0 stands for a "harmless" constant whose exact meaning depends on the context, and $A \approx B$ means that $A \lesssim B$ and $B \lesssim A$.

Lemma 2.1 *Let ρ_0 and $\mathbf{v}$ satisfy (1.4) and (2.2), respectively. Then the system (2.1a) (1.2) has a global unique strong solution such that*

$$\rho \in C([0,T]; W^{1,q}), \quad \rho_t \in C([0,T]; L^q).$$

Moreover, if $\rho_0, \mathbf{v}$ satisfy (2.4)-(2.7), then there exists a time $T_1 = \min\{c_1^{-1}, T\}$ such that for all $t \in [0, T_1]$,

$$\|\rho\|_{W^{1,q}}(t) \lesssim c_0, \quad \|\rho_t\|_{L^q}(t) \lesssim c_0 c_2. \tag{2.8}$$

In particular,

$$\|p\|_{L^q}(t) \lesssim M(c_0), \quad \|\nabla p\|_{L^q}(t) \lesssim M(c_0) c_0, \quad \|p_t\|_{L^q}(t) \lesssim M(c_0) c_0 c_2, \tag{2.9}$$

where $M(c_0) := \sup_{0 \leqslant \cdot \leqslant c_0} (1 + p(\cdot) + p'(\cdot))$.

Proof The existence of the solution follows from the method of characteristics. If we define $V(t, \tau, \mathbf{x})$ to be the solution of

$$\begin{cases} \dfrac{d}{d\tau} V(\tau, t, \mathbf{x}) = \mathbf{v}(\tau, V(\tau, t, \mathbf{x})), \quad \tau \in [0, T], \ \mathbf{x} \in \overline{\Omega}, \\ V(t, t, \mathbf{x}) = \mathbf{x}, \end{cases}$$

(2.1a) can be rewritten as

$$\frac{d}{d\tau}\rho(\tau, V(\tau,t,\mathbf{x})) = -\rho(\tau, V(\tau,t,\mathbf{x}))\nabla \cdot \mathbf{v}(\tau, V(\tau,t,\mathbf{x})),$$

and the explicit formula for ρ is

$$\rho(t,\mathbf{x}) = \rho_0(V(0,t,x))\exp\left(-\int_0^t \nabla \cdot \mathbf{v}(\tau, V(\tau,t,\mathbf{x}))\,d\tau\right). \tag{2.10}$$

Applying the gradient operator ∇ to (2.1a), we have

$$\nabla \rho_t + \mathbf{v} \cdot \nabla(\nabla\rho) + \nabla\mathbf{v}\cdot\nabla\rho + \rho\nabla(\nabla\cdot\mathbf{v}) + \nabla\cdot\mathbf{v}\nabla\rho = 0, \tag{2.11}$$

where $\nabla\mathbf{v}\cdot\nabla\rho = (\nabla\mathbf{v})^T\nabla\rho$.

Multiplying (2.11) by $|\nabla\rho|^{q-2}\nabla\rho$ and integrating over Ω, we obtain

$$\frac{1}{q}\frac{d}{dt}\|\nabla\rho\|_{L^q}^q + \int_\Omega \left(\frac{1}{q}\mathbf{v}\cdot\nabla(|\nabla\rho|^q) + |\nabla\rho|^{q-2}\nabla\rho\cdot(\nabla\mathbf{v}\cdot\nabla\rho)\right.$$
$$\left. + \rho|\nabla\rho|^{q-2}\nabla\rho\cdot\nabla(\nabla\cdot\mathbf{v}) + |\nabla\rho|^q\nabla\cdot\mathbf{v}\right)dx = 0.$$

Bearing in mind that

$$\int_\Omega (\mathbf{v}\cdot\nabla(|\nabla\rho|^q) + |\nabla\rho|^q\nabla\cdot\mathbf{v})\,d\mathbf{x} = \int_\Omega \nabla\cdot(|\nabla\rho|^q\mathbf{v})\,d\mathbf{x} = 0,$$

we find, by Höder's inequality and the imbedding

$$W^{1,q} \hookrightarrow L^\infty, \quad \text{as } 3 < q \leqslant 6, \tag{2.12}$$

$$\frac{1}{q}\frac{d}{dt}\|\nabla\rho\|_{L^q}^q \leqslant \|\nabla\rho\|_{L^q}^{q-1}(\|\rho\|_{L^\infty}\|\nabla(\nabla\cdot\mathbf{v})\|_{L^q} + \|\nabla\rho\|_{L^q}\|\nabla\mathbf{v}\|_{L^\infty}) + \|\nabla\rho\|_{L^q}^q\|\nabla\cdot\mathbf{v}\|_{L^\infty}$$
$$\lesssim \|\nabla\rho\|_{L^q}^{q-1}(\|\nabla\rho\|_{L^q} + \|\rho\|_{L^\infty})\|\mathbf{v}\|_{W^{2,q}},$$

and, by Gronwall's inequality,

$$\|\nabla\rho\|_{L^q}(t) \leqslant \exp\left(\int_0^t \|\mathbf{v}\|_{W^{2,q}}ds\right)\left(\|\nabla\rho_0\|_{L^q} + \int_0^t \|\rho\|_{L^\infty}\|\mathbf{v}\|_{W^{2,q}}ds\right) \tag{2.13}$$

for all $t \leqslant T$.

Using (2.10), (2.13) and choosing $T_1 = \min\{c_1^{-1}, T\}$, we get, for all $t \in [0, T_1]$,

$$\|\rho\|_{W^{1,q}}(t) \lesssim \|\rho_0\|_{W^{1,q}}\left(\|\exp(-\int_0^t \nabla\cdot\mathbf{v}\,d\tau)\|_{L^q} + \exp\left(\left(\int_0^t \|\mathbf{v}\|_{W^{2,q}}^2\,ds\right)^{\frac{1}{2}}\sqrt{t}\right)\right.$$
$$\left.\cdot\left(1 + \int_0^t \|\exp(-\int_0^s \nabla\cdot\mathbf{v}\,d\tau)\|_{L^\infty}\|\mathbf{v}\|_{W^{2,q}}\,ds\right)\right)$$

$$\lesssim c_0 \left(\exp\left(\int_0^t \|\nabla \cdot \mathbf{v}\|_{L^\infty} \, ds \right) + \exp(\sqrt{c_1}t) \right.$$
$$\left. \cdot \left(1 + \int_0^t \exp\left(\int_0^s \|\mathbf{v}\|_{W^{2,q}} d\tau \right) \|\mathbf{v}\|_{W^{2,q}} \, ds \right) \right)$$
$$\lesssim c_0 \left(\exp\left(\left(\int_0^t \|\mathbf{v}\|_{W^{2,q}}^2 \, ds \right)^{\frac{1}{2}} \sqrt{t} \right) \right.$$
$$\left. + \exp(\sqrt{c_1}t) \left(1 + \int_0^t \exp(\sqrt{c_1}s) \|\mathbf{v}\|_{W^{2,q}} \, ds \right) \right)$$
$$\lesssim c_0 \exp(\sqrt{c_1}t) \left(1 + \exp(\sqrt{c_1}t)\sqrt{c_1}t \right) \lesssim c_0,$$

$$\|\rho_t\|_{L^q}(t) \leq \|\mathbf{v}\|_{L^\infty} \|\nabla \rho\|_{L^q} + \|\rho\|_{L^\infty} \|\nabla \mathbf{v}\|_{L^q}$$
$$\lesssim \|\mathbf{v}\|_{H^2} \|\rho\|_{W^{1,q}} \lesssim c_0 c_2,$$

and, by using (2.12), it follows obviously that for all $t \in [0, T_1]$,

$$\|p\|_{L^q}(t) \lesssim M(c_0), \quad \|\nabla p\|_{L^q}(t) \lesssim M(c_0)c_0, \quad \|p_t\|_{L^q}(t) \lesssim M(c_0)c_0 c_2,$$

where

$$M(c_0) := \sup_{0 \leq \cdot \lesssim c_0} (1 + p(\cdot) + p'(\cdot)). \tag{2.14}$$

□

Lemma 2.2 *Let* $\mathbf{v}$, $\mathbf{f}$ *satisfy* (2.2) *and* (2.3). *Then the system* (2.1b),(1.2) *and* (1.3) *has a global unique strong solution such that*

$$\mathbf{d} \in C([0,T]; H^3), \quad \mathbf{d}_t \in C([0,T]; H_0^1) \cap L^2(0,T; H^2), \quad \mathbf{d}_{tt} \in L^2(0,T; L^2).$$

Moreover, if $\mathbf{v}$, $\mathbf{f}$ *satisfy* (2.4)-(2.7), *then there exists a time* $T_3 = \min\{c_0^{-10} c_1^{-18} c_2^{-2}, T\}$ *such that*

$$\sup_{0 \leq t \leq T_3} \left(\|\mathbf{d}\|_{H^1} + c_1^{-4} \|\nabla^2 \mathbf{d}\|_{L^2} + c_0^{\frac{3}{2}} \|\mathbf{d}_t\|_{H_0^1} + c_1^{-7} c_2^{-1} \|\nabla \mathbf{d}\|_{H^2} \right)$$
$$+ \int_0^{T_3} \left(\|\mathbf{d}_t\|_{H^2}^2 + c_0^3 \|\mathbf{d}\|_{H^3}^2 \right) dt \lesssim c_0^3. \tag{2.15}$$

Proof Since (2.1b) is a linear parabolic-type system in $\mathbf{d}$, the existence and uniqueness of $\mathbf{d}$ to the problem (2.1b),(1.2) and (1.3) can be obtained by the standard Faedo-Galerkin method, and also the regularity of $\mathbf{d}$ described in the lemma.

Differentiating (2.1b) with respect to t, multiplying the result by $\mathbf{d}_t$ and then integrating over Ω, we have

$$\frac{1}{2} \frac{d}{dt} \int_\Omega |\mathbf{d}_t|^2 \, d\mathbf{x} + \theta \int_\Omega |\nabla \mathbf{d}_t|^2 \, d\mathbf{x}$$

$$\lesssim \|\mathbf{v}_t\|_{L^6}\|\nabla\mathbf{d}\|_{L^2}\|\mathbf{d}_t\|_{L^3} + \|\mathbf{v}\|_{L^\infty}\|\nabla\mathbf{d}_t\|_{L^2}\|\mathbf{d}_t\|_{L^2} + \|\nabla\mathbf{f}_t\|_{L^6}\|\nabla\mathbf{d}\|_{L^2}\|\mathbf{f}\|_{L^\infty}\|\mathbf{d}_t\|_{L^3}$$
$$+ \|\nabla\mathbf{f}\|_{L^6}\|\nabla\mathbf{d}_t\|_{L^2}\|\mathbf{f}\|_{L^\infty}\|\mathbf{d}_t\|_{L^3} + \|\nabla\mathbf{f}\|_{L^\infty}\|\nabla\mathbf{d}\|_{L^2}\|\mathbf{f}_t\|_{L^6}\|\mathbf{d}_t\|_{L^3}$$
$$:= \sum_{l=1}^{5} I_l. \tag{2.16}$$

Now we estimate the right-hand side of (2.16) term by term.

$$I_1 \lesssim \|\mathbf{v}_t\|_{H^1}\|\nabla\mathbf{d}\|_{L^2}\|\mathbf{d}_t\|_{L^2}^{\frac{1}{2}}\|\mathbf{d}_t\|_{L^6}^{\frac{1}{2}}$$
$$\lesssim \|\nabla\mathbf{v}_t\|_{L^2}\|\nabla\mathbf{d}\|_{L^2}\|\mathbf{d}_t\|_{L^2}^{\frac{1}{2}}\|\nabla\mathbf{d}_t\|_{L^2}^{\frac{1}{2}}$$
$$\leqslant \eta\|\nabla\mathbf{v}_t\|_{L^2}^2\|\nabla\mathbf{d}\|_{L^2}^2 + \eta^{-1}\|\mathbf{d}_t\|_{L^2}^2 + \frac{\theta}{8}\|\nabla\mathbf{d}_t\|_{L^2}^2,$$

where we have used the fact that $\mathbf{d}_t\mid_{\partial\Omega}=\mathbf{0}$.

$$I_2 \leqslant C\|\mathbf{v}\|_{H^2}^2\|\mathbf{d}_t\|_{L^2}^2 + \frac{\theta}{8}\|\nabla\mathbf{d}_t\|_{L^2}^2,$$

$$I_3 + I_5 \leqslant \eta^3(\|\nabla\mathbf{f}_t\|_{H^1}^2\|\mathbf{f}\|_{H^2}^2 + \|\nabla\mathbf{f}_t\|_{L^2}^2\|\mathbf{f}\|_{H^3}^2)\|\nabla\mathbf{d}\|_{L^2}^2 + \eta^{-3}\|\mathbf{d}_t\|_{L^2}^2 + \frac{\theta}{4}\|\nabla\mathbf{d}_t\|_{L^2}^2,$$

$$I_4 \leqslant C\|\nabla\mathbf{f}\|_{L^6}^2\|\mathbf{f}\|_{L^\infty}^2\|\mathbf{d}_t\|_{L^2}^2\|\nabla\mathbf{d}_t\|_{L^2} + \frac{\theta}{8}\|\nabla\mathbf{d}_t\|_{L^2}^2 \leqslant C\|\nabla\mathbf{f}\|_{H^1}^4\|\mathbf{f}\|_{H^2}^4\|\mathbf{d}_t\|_{L^2}^2 + \frac{\theta}{4}\|\nabla\mathbf{d}_t\|_{L^2}^2.$$

Note that the positive constant η will be determined later.

Notice that
$$\frac{d}{dt}\int_\Omega |\nabla\mathbf{d}|^2\,dx = 2\int_\Omega \nabla\mathbf{d}:\nabla\mathbf{d}_t\,dx \leqslant C\|\nabla\mathbf{d}\|_{L^2}^2 + \frac{\theta}{8}\|\nabla\mathbf{d}_t\|_{L^2}^2,$$

and
$$\frac{d}{dt}\int_\Omega |\mathbf{d}-\mathbf{n}|^2\,dx \leqslant \|\mathbf{d}-\mathbf{n}\|_{L^2}^2 + \|\mathbf{d}_t\|_{L^2}^2.$$

Therefore, combining with (2.16), we get

$$\frac{d}{dt}\int_\Omega (|\mathbf{d}_t|^2 + |\mathbf{d}-\mathbf{n}|^2 + |\nabla\mathbf{d}|^2)\,dx + \int_\Omega |\nabla\mathbf{d}_t|^2\,dx$$
$$\lesssim (\|\mathbf{d}_t\|_{L^2}^2 + \|\mathbf{d}-\mathbf{n}\|_{L^2}^2 + \|\nabla\mathbf{d}\|_{L^2}^2)\Big(1 + \eta^{-1} + \eta^{-3} + \eta\|\nabla\mathbf{v}_t\|_{L^2}^2 + \|\mathbf{v}\|_{H^2}^2$$
$$+ \|\nabla\mathbf{f}\|_{H^1}^4\|\mathbf{f}\|_{H^2}^4 + \eta^3(\|\nabla\mathbf{f}_t\|_{H^1}^2\|\mathbf{f}\|_{H^2}^2 + \|\nabla\mathbf{f}_t\|_{L^2}^2\|\mathbf{f}\|_{H^3}^2)\Big). \tag{2.17}$$

It follows from (2.1b) that

$$\|\mathbf{d}_t\|_{L^2}(0) \lesssim \|\Delta\mathbf{d}_0\|_{L^2} + \|\mathbf{u}_0\|_{H^2}\|\nabla\mathbf{d}_0\|_{L^2} + \|\nabla\mathbf{d}_0\|_{H^1}^2\|\mathbf{d}_0\|_{H^1}$$
$$\leqslant \|\Delta\mathbf{d}_0\|_{L^2} + \|\mathbf{u}_0\|_{H^2}\|\nabla\mathbf{d}_0\|_{L^2} + \|\nabla\mathbf{d}_0\|_{H^1}^2(\|\mathbf{d}_0-\mathbf{n}\|_{H^1} + \|\mathbf{n}\|_{H^1})$$
$$\leqslant \|\Delta\mathbf{d}_0\|_{L^2} + \|\mathbf{u}_0\|_{H^2}\|\nabla\mathbf{d}_0\|_{L^2} + \|\nabla\mathbf{d}_0\|_{H^1}^2(\|\nabla\mathbf{d}_0\|_{L^2} + \|\mathbf{n}\|_{H^1})$$
$$\lesssim c_0^3.$$

Hence, by virtue of Gronwall's inequality, we obtain from (2.17) that

$$\int_\Omega (|\mathbf{d}_t|^2 + |\mathbf{d}-\mathbf{n}|^2 + |\nabla \mathbf{d}|^2)\,d\mathbf{x} + \int_0^t \int_\Omega |\nabla \mathbf{d}_s|^2\,d\mathbf{x}ds$$
$$\lesssim c_0^6 \exp\left(\int_0^t (1+\eta^{-1}+\eta^{-3}+\eta\|\nabla \mathbf{v}_t\|_{L^2}^2 + \|\mathbf{v}\|_{H^2}^2 + \eta^3\|\nabla^2 \mathbf{f}_t\|_{L^2}^2 \|\mathbf{f}\|_{H^2}^2\right.$$
$$\left. + \eta^3 \|\nabla \mathbf{f}_t\|_{L^2}^2 \|\mathbf{f}\|_{H^3}^2 + \|\nabla \mathbf{f}\|_{H^1}^4 \|\mathbf{f}\|_{H^2}^4)\,ds\right). \tag{2.18}$$

By taking $\eta = c_1^{-1}$ and $T_2 = \min\{c_1^{-6} c_2^{-2}, T\}$, it follows from (2.18) that

$$\sup_{0\leqslant t\leqslant T_2} \int_\Omega (|\mathbf{d}_t|^2 + |\mathbf{d}-\mathbf{n}|^2 + |\nabla \mathbf{d}|^2)\,d\mathbf{x} + \int_0^{T_2}\int_\Omega |\nabla \mathbf{d}_t|^2\,d\mathbf{x}dt \lesssim c_0^6.$$

From (2.1b), using the elliptic estimates, we get

$$\|\mathbf{d}-\mathbf{n}\|_{H^2} \lesssim \|\mathbf{d}_0-\mathbf{n}\|_{H^2} + \|\mathbf{d}_t\|_{L^2} + \|\mathbf{v}\cdot\nabla \mathbf{d}\|_{L^2} + \|(\nabla \mathbf{f}:\nabla \mathbf{d})\mathbf{f}\|_{L^2}$$
$$\lesssim c_0^3 + \|\mathbf{v}\|_{H_0^1}\|\nabla \mathbf{d}\|_{L^3} + \|\nabla \mathbf{f}\|_{L^6}\|\nabla \mathbf{d}\|_{L^3}\|\mathbf{f}\|_{L^\infty}$$
$$\lesssim c_0^3 + (\|\mathbf{v}\|_{H_0^1} + \|\nabla \mathbf{f}\|_{H^1}\|\mathbf{f}\|_{H^2})\|\nabla \mathbf{d}\|_{L^2}^{\frac{1}{2}}\|\nabla \mathbf{d}\|_{L^6}^{\frac{1}{2}}$$
$$\leqslant C\left(c_0^3 + (\|\mathbf{v}\|_{H_0^1}^2 + \|\mathbf{f}\|_{H^2}^4)\|\nabla \mathbf{d}\|_{L^2}\right) + \frac{1}{2}\|\nabla \mathbf{d}\|_{H^1},$$

which implies $\|\nabla^2 \mathbf{d}\|_{L^2} \lesssim c_0^3 c_1^4$.

Similarly, let us estimate $\|\nabla \mathbf{d}\|_{H^2}$. Since

$$\theta \Delta(\nabla \mathbf{d}) = \nabla \mathbf{d}_t + \nabla(\mathbf{v}\cdot\nabla \mathbf{d}) - \theta \nabla\left((\nabla \mathbf{f}:\nabla \mathbf{d})\mathbf{f}\right), \tag{2.19}$$

then

$$\|\nabla \mathbf{d}\|_{H^2} \lesssim \|\nabla \mathbf{d}_0\|_{H^2} + \|\nabla \mathbf{d}_t\|_{L^2} + \|\nabla \mathbf{v}\|_{L^6}\|\nabla \mathbf{d}\|_{L^3} + \|\mathbf{v}\|_{H^2}\|\nabla^2 \mathbf{d}\|_{L^2}$$
$$+ \|\nabla^2 \mathbf{f}\|_{L^6}\|\nabla \mathbf{d}\|_{L^3}\|\mathbf{f}\|_{L^\infty} + \|\nabla \mathbf{f}\|_{L^6}\|\nabla^2 \mathbf{d}\|_{L^3}\|\mathbf{f}\|_{L^\infty} + \|\nabla \mathbf{f}\|_{L^6}^2 \|\nabla \mathbf{d}\|_{L^6}$$
$$\lesssim c_0 + \|\nabla \mathbf{d}_t\|_{L^2} + (\|\mathbf{v}\|_{H^2} + \|\mathbf{f}\|_{H^2}\|\mathbf{f}\|_{H^3})\|\nabla \mathbf{d}\|_{L^2}^{\frac{1}{2}}\|\nabla \mathbf{d}\|_{L^6}^{\frac{1}{2}} + \|\mathbf{v}\|_{H^2}\|\nabla^2 \mathbf{d}\|_{L^2}$$
$$+ \|\mathbf{f}\|_{H^2}^2 \|\nabla^2 \mathbf{d}\|_{L^6}^{\frac{1}{2}}\|\nabla^2 \mathbf{d}\|_{L^2}^{\frac{1}{2}} + \|\mathbf{f}\|_{H^2}^2\|\nabla \mathbf{d}\|_{L^6}$$
$$\leqslant C\Big(c_0 + \|\nabla \mathbf{d}_t\|_{L^2} + (\|\mathbf{v}\|_{H^2} + \|\mathbf{f}\|_{H^2}\|\mathbf{f}\|_{H^3})\|\nabla \mathbf{d}\|_{L^2}^{\frac{1}{2}}\|\nabla \mathbf{d}\|_{H^1}^{\frac{1}{2}}$$
$$+ \|\mathbf{v}\|_{H^2}\|\nabla^2 \mathbf{d}\|_{L^2} + \|\mathbf{f}\|_{H^2}^4\|\nabla^2 \mathbf{d}\|_{L^2} + \|\mathbf{f}\|_{H^2}^2\|\nabla \mathbf{d}\|_{H^1}\Big) + \frac{1}{2}\|\nabla^2 \mathbf{d}\|_{H^1},$$

and therefore,

$$\|\nabla \mathbf{d}\|_{H^2} \lesssim \|\nabla \mathbf{d}_t\|_{L^2} + c_0^3 c_1^7 c_2. \tag{2.20}$$

Differentiating (2.1b) with respect to time and taking inner product with $\Delta \mathbf{d}_t$, we have

$$\frac{1}{2}\frac{d}{dt}\int_\Omega |\nabla \mathbf{d}_t|^2\, d\mathbf{x} + \theta \int_\Omega |\Delta \mathbf{d}_t|^2\, d\mathbf{x}$$
$$= \int_\Omega (\mathbf{v}_t \cdot \nabla \mathbf{d}) \cdot (\Delta \mathbf{d}_t)\, d\mathbf{x} + \int_\Omega (\mathbf{v} \cdot \nabla \mathbf{d}_t) \cdot (\Delta \mathbf{d}_t)\, d\mathbf{x} - \int_\Omega (\nabla \mathbf{f}_t : \nabla \mathbf{d})\mathbf{f} \cdot (\Delta \mathbf{d}_t)\, d\mathbf{x}$$
$$- \int_\Omega (\nabla \mathbf{f} : \nabla \mathbf{d}_t)\mathbf{f} \cdot (\Delta \mathbf{d}_t)\, d\mathbf{x} - \int_\Omega (\nabla \mathbf{f} : \nabla \mathbf{d})\mathbf{f}_t \cdot (\Delta \mathbf{d}_t)\, d\mathbf{x}$$
$$:= \sum_{l=1}^{5} J_l. \tag{2.21}$$

Here

$$J_1 \leqslant \|\nabla \mathbf{v}_t\|_{L^2}\|\nabla \mathbf{d}\|_{L^6}\|\nabla \mathbf{d}_t\|_{L^3} + \|\mathbf{v}_t\|_{L^3}\|\nabla^2 \mathbf{d}\|_{L^6}\|\nabla \mathbf{d}_t\|_{L^2}$$
$$\leqslant \eta\|\nabla \mathbf{v}_t\|_{L^2}^2 + \eta^{-1}\|\nabla \mathbf{d}\|_{H^1}^2\|\nabla \mathbf{d}_t\|_{L^2}\|\nabla \mathbf{d}_t\|_{H^1} + \eta\|\mathbf{v}_t\|_{L^3}^2\|\nabla \mathbf{d}_t\|_{L^2}^2 + \eta^{-1}\|\nabla^2 \mathbf{d}\|_{L^6}^2$$
$$\leqslant \eta\|\nabla \mathbf{v}_t\|_{L^2}^2 + \eta^{-2}\varepsilon^{-1}c_0^{12}c_1^{16}\|\nabla \mathbf{d}_t\|_{L^2}^2 + \varepsilon\|\nabla \mathbf{d}_t\|_{H^1}^2 + \eta\|\mathbf{v}_t\|_{L^3}^2\|\nabla \mathbf{d}_t\|_{L^2}^2$$
$$+ \eta^{-1}(\|\nabla \mathbf{d}_t\|_{L^2}^2 + c_0^6 c_1^{14} c_2^2),$$

$$J_2 \leqslant \|\mathbf{v}\|_{H^2}\|\nabla \mathbf{d}_t\|_{L^2}\|\Delta \mathbf{d}_t\|_{L^2} \leqslant \varepsilon^{-1}c_2^2\|\nabla \mathbf{d}_t\|_{L^2}^2 + \varepsilon\|\Delta \mathbf{d}_t\|_{L^2}^2,$$

$$J_3 + J_5 \leqslant (\|\nabla \mathbf{f}_t\|_{L^2}\|\mathbf{f}\|_{H^2} + \|\nabla \mathbf{f}\|_{L^3}\|\mathbf{f}_t\|_{L^6})\|\nabla \mathbf{d}\|_{H^2}\|\Delta \mathbf{d}_t\|_{L^2}$$
$$\leqslant \varepsilon^{-1}c_1^4(\|\nabla \mathbf{d}_t\|_{L^2}^2 + c_0^6 c_1^{14} c_2^2) + \varepsilon\|\Delta \mathbf{d}_t\|_{L^2}^2,$$

$$J_4 \leqslant \|\nabla \mathbf{f}\|_{L^\infty}\|\nabla \mathbf{d}_t\|_{L^2}\|\mathbf{f}\|_{L^\infty}\|\Delta \mathbf{d}_t\|_{L^2} \leqslant \varepsilon^{-1}c_1^2(c_1+c_2)^2\|\nabla \mathbf{d}_t\|_{L^2}^2 + \varepsilon\|\Delta \mathbf{d}_t\|_{L^2}^2.$$

Bearing in mind that the elliptic estimate

$$\|\nabla^2 \mathbf{d}_t\|_{L^2} \lesssim \|\Delta \mathbf{d}_t\|_{L^2} + 1 \tag{2.22}$$

while substituting the above estimates into (2.21) and taking ε small enough $\left(< \dfrac{\theta}{4}\right)$, we have

$$\frac{d}{dt}\int_\Omega |\nabla \mathbf{d}_t|^2\, d\mathbf{x} + \int_\Omega |\Delta \mathbf{d}_t|^2\, d\mathbf{x} \lesssim X_\eta(t)\|\nabla \mathbf{d}_t\|_{L^2}^2 + Y_\eta(t), \tag{2.23}$$

where

$$X_\eta(t) = \eta^{-2}c_0^{12}c_1^{16} + \eta\|\nabla \mathbf{v}_t\|_{L^2}^2 + \eta^{-1} + c_1^4 + c_2^2 + c_1^2 c_2^2,$$
$$Y_\eta(t) = \eta\|\nabla \mathbf{v}_t\|_{L^2}^2 + \eta^{-1}c_0^6 c_1^{14} c_2^2 + c_0^6 c_1^{18} c_2^2.$$

Taking $\eta = c_1^{-1}$, we have for all $t \in [0, T_2]$,

$$\int_0^t X_\eta(s)\, ds \lesssim 1 + c_0^{12}c_1^{18}\, t, \quad \int_0^t Y_\eta(s)\, ds \lesssim 1 + c_0^6 c_1^{18} c_2^2\, t.$$

Next we need to estimate $\|\nabla \mathbf{d}_t(0)\|_{L^2}$. From (2.19), we have

$$\|\nabla \mathbf{d}_t(0)\|_{L^2} \lesssim \|\nabla \mathbf{d}_0\|_{H^2} + \|\nabla \mathbf{u}_0\|_{L^3}\|\nabla \mathbf{d}_0\|_{L^6} + \|\mathbf{u}_0\|_{L^3}\|\nabla^2 \mathbf{d}_0\|_{L^6}$$
$$+ \|\nabla^2 \mathbf{d}_0\|_{L^6}\|\nabla \mathbf{d}_0\|_{L^6}\|\mathbf{d}_0\|_{L^6} + \|\nabla \mathbf{d}_0\|_{L^6}^3$$
$$\lesssim c_0^3.$$

Now, applying Gronwall's inequality to (2.23), one has

$$\sup_{0 \leqslant t \leqslant T_3} \int_\Omega |\nabla \mathbf{d}_t|^2 \, d\mathbf{x} + \int_0^{T_3} \|\Delta \mathbf{d}_t\|_{L^2}^2 \, dt \lesssim (c_0^3 + c_0^6 c_1^{18} c_2^2 T_3) \exp(1 + c_0^{12} c_1^{18} T_3)$$
$$\lesssim c_0^3,$$

where $T_3 = \min\{T, c_0^{-10} c_1^{-18} c_2^{-2}\}$. Taking advantage of the elliptic estimate (2.22) again and the fact that $\mathbf{d}_t|_{\partial\Omega} = 0$, we have

$$\int_0^{T_3} \|\mathbf{d}\|_{H^2}^2 dt \lesssim c_0^3.$$

Consequently, coming back to (2.20), we have, for all $0 \leqslant t \leqslant T_3$,

$$\|\nabla \mathbf{d}\|_{H^2}(t) \lesssim c_0^{\frac{3}{2}} + c_0^3 c_1^7 c_2, \quad \int_0^t \|\mathbf{d}\|_{H^3}^2 ds \lesssim 1.$$

All the above estimates can deduce that (2.15) holds. The proof of lemma 2.2 is completed. $\square$

Lemma 2.3 *Under the hypotheses of Theorem 2.1, there exists a global unique solution $\mathbf{u}$ of (2.1c) with the initial data $\mathbf{u}_0^\delta$ and the boundary condition (1.3) such that*

$$\mathbf{u} \in C([0,T]; H_0^1 \cap H^2) \cap L^2(0,T; W^{2,q}), \quad \mathbf{u}_t \in C([0,T]; L^2) \cap L^2(0,T; H_0^1),$$

$$\mathbf{u}_{tt} \in L^2(0,T; H^{-1}).$$

In addition, if $\mathbf{v}$, $\mathbf{f}$ satisfy (2.4)-(2.7), then

$$\sup_{0 \leqslant t \leqslant T_5} \left(M(c_0) c_0^2 \|\mathbf{u}\|_{H_0^1} + M(c_0) c_1^{-10} \|\nabla \mathbf{u}\|_{H^1} + M(c_0) c_0^2 \|\sqrt{\rho} \mathbf{u}_t\|_{L^2} \right)$$
$$+ \int_0^{T_5} (c_0^2 \|\nabla \mathbf{u}_t\|_{L^2}^2 + \|\mathbf{u}\|_{W^{2,q}}^2) \, dt \lesssim M^2(c_0) c_0^2 + c_0^{12}. \tag{2.24}$$

for all $\delta > 0$ small enough, where $T_5 = \min\{T, c_1^{-28} c_2^{-4}\}$.

Proof First, it comes easily from (2.10) that

$$\rho(\mathbf{x},t) \geqslant \delta \exp\left(-\int_0^t \|\nabla \mathbf{v}\|_{W^{1,q}} \, ds\right) \geqslant \underline{\delta},$$

for all $(\mathbf{x},t) \in \overline{\Omega} \times [0,T]$, where $\underline{\delta} > 0$ is a constant.

Now we can rewrite (2.1c) into

$$\mathbf{u}_t + \mathbf{v} \cdot \nabla \mathbf{u} + \rho^{-1} \nabla p(\rho) = \mu \rho^{-1} \Delta \mathbf{u} - \lambda \rho^{-1} \nabla \cdot \left(\nabla \mathbf{d} \odot \nabla \mathbf{d} - \frac{|\nabla \mathbf{d}|^2}{2} \mathbb{I}_3\right). \tag{2.25}$$

Applying the Galerkin method again to (2.25) with the initial data $\mathbf{u}_0^\delta$ and the non-slip boundary condition, we can deduce the existence and regularity of $\mathbf{u}$ described in the lemma.

Next, we prove the estimate (2.24). Differentiating (2.1c) with respect to t, multiplying the result by $\mathbf{u}_t$ and then integrating over Ω, one has

$$\frac{1}{2}\frac{d}{dt}\int_\Omega \rho |\mathbf{u}_t|^2 \, d\mathbf{x} + \mu \int_\Omega |\nabla \mathbf{u}_t|^2 \, d\mathbf{x}$$
$$= \int_\Omega (-\nabla p_t - \rho_t \mathbf{v} \cdot \nabla \mathbf{u} - 2\rho \mathbf{v} \cdot \nabla \mathbf{u}_t - \rho \mathbf{v}_t \cdot \nabla \mathbf{u}) \cdot \mathbf{u}_t \, d\mathbf{x}$$
$$- \lambda \int_\Omega ((\nabla \mathbf{d})^{\mathrm{T}} \Delta \mathbf{d})_t \cdot \mathbf{u}_t \, d\mathbf{x}$$
$$:= \sum_{l=1}^5 K_l. \tag{2.26}$$

Here

$$K_1 = -\int_\Omega \nabla p_t \cdot \mathbf{u}_t \, d\mathbf{x} \leqslant C\|p_t\|_{L^2}^2 + \frac{\mu}{6}\|\nabla \mathbf{u}_t\|_{L^2}^2 \leqslant CM^2(c_0)c_0^2 c_2^2 + \frac{\mu}{6}\|\nabla \mathbf{u}_t\|_{L^2}^2,$$

$$K_2 = -\int_\Omega \rho_t \mathbf{v} \cdot \nabla \mathbf{u} \cdot \mathbf{u}_t \, d\mathbf{x} \leqslant C\|\rho_t\|_{L^q}^2 \|\mathbf{v}\|_{L^{\frac{3q}{q-3}}}^2 \|\nabla \mathbf{u}\|_{L^2}^2 + \frac{\mu}{6}\|\nabla \mathbf{u}_t\|_{L^2}^2$$
$$\leqslant C\|\rho_t\|_{L^q}^2 \|\mathbf{v}\|_{W^{1,q}}^2 \|\nabla \mathbf{u}\|_{L^2}^2 + \frac{\mu}{6}\|\nabla \mathbf{u}_t\|_{L^2}^2 \leqslant Cc_0^2 c_2^4 \|\nabla \mathbf{u}\|_{L^2}^2 + \frac{\mu}{6}\|\nabla \mathbf{u}_t\|_{L^2}^2,$$

$$K_3 = -2\int_\Omega \rho \mathbf{v} \cdot \nabla \mathbf{u}_t \cdot \mathbf{u}_t \, d\mathbf{x} \leqslant C\|\rho\|_{L^\infty}\|\mathbf{v}\|_{L^\infty}^2 \|\sqrt{\rho}\mathbf{u}_t\|_{L^2}^2 + \frac{\mu}{6}\|\nabla \mathbf{u}_t\|_{L^2}^2$$
$$\leqslant Cc_0 c_2^2 \|\sqrt{\rho}\mathbf{u}_t\|_{L^2}^2 + \frac{\mu}{6}\|\nabla \mathbf{u}_t\|_{L^2}^2,$$

$$K_4 = -\int_\Omega \rho \mathbf{v}_t \cdot \nabla \mathbf{u} \cdot \mathbf{u}_t \, d\mathbf{x} \leqslant \|\sqrt{\rho}\mathbf{u}_t\|_{L^2}\|\mathbf{v}_t\|_{L^6}\|\sqrt{\rho}\|_{L^\infty}\|\nabla \mathbf{u}\|_{L^3}$$
$$\lesssim \|\sqrt{\rho}\mathbf{u}_t\|_{L^2}\|\nabla \mathbf{v}_t\|_{L^2}\|\sqrt{\rho}\|_{L^\infty}\|\nabla \mathbf{u}\|_{L^2}^{\frac{1}{2}}\|\nabla \mathbf{u}\|_{H^1}^{\frac{1}{2}}$$
$$\leqslant \eta\|\sqrt{\rho}\mathbf{u}_t\|_{L^2}^2\|\nabla \mathbf{v}_t\|_{L^2}^2 + \eta^{-1}\|\rho\|_{L^\infty}\|\nabla \mathbf{u}\|_{L^2}\|\nabla \mathbf{u}\|_{H^1}$$
$$\leqslant \eta\|\sqrt{\rho}\mathbf{u}_t\|_{L^2}^2\|\nabla \mathbf{v}_t\|_{L^2}^2 + C\eta^{-2}\|\rho\|_{W^{1,q}}^2\|\nabla \mathbf{u}\|_{L^2}^2 + \|\nabla \mathbf{u}\|_{H^1}^2,$$

$$K_5 = -\lambda \int_\Omega ((\nabla \mathbf{d})^T \Delta \mathbf{d})_t \cdot \mathbf{u}_t \, d\mathbf{x}$$

$$\lesssim \|\nabla \mathbf{d}_t\|_{L^2} \|\nabla^2 \mathbf{d}\|_{L^3} \|\mathbf{u}_t\|_{L^6} + \|\nabla \mathbf{d}\|_{L^\infty} \|\nabla \mathbf{d}_t\|_{L^2} \|\nabla \mathbf{u}_t\|_{L^2}$$

$$\leq C \left(\|\nabla \mathbf{d}_t\|_{L^2}^2 \|\nabla^2 \mathbf{d}\|_{L^2} \|\nabla^2 \mathbf{d}\|_{H^1} + \|\nabla \mathbf{d}\|_{H^2}^2 \|\nabla \mathbf{d}_t\|_{L^2}^2 \right) + \frac{\mu}{6} \|\nabla \mathbf{u}_t\|_{L^2}^2$$

$$\leq C c_0^9 c_1^{14} c_2^2 + \frac{\mu}{6} \|\nabla \mathbf{u}_t\|_{L^2}^2.$$

Meanwhile, since

$$\frac{1}{2} \frac{d}{dt} \|\nabla \mathbf{u}\|_{L^2}^2 = \int_\Omega \nabla \mathbf{u} : \nabla \mathbf{u}_t \, d\mathbf{x} \leq C \|\nabla \mathbf{u}\|_{L^2}^2 + \frac{\mu}{6} \|\nabla \mathbf{u}_t\|_{L^2}^2,$$

then, combining with the above estimates and (2.26), we have

$$\frac{d}{dt} \int_\Omega (\rho |\mathbf{u}_t|^2 + |\nabla \mathbf{u}|^2) \, d\mathbf{x} + \int_\Omega |\nabla \mathbf{u}_t|^2 \, d\mathbf{x}$$

$$\lesssim \mathcal{X}_\eta(t)(\|\sqrt{\rho} \mathbf{u}_t\|_{L^2}^2 + \|\nabla \mathbf{u}\|_{L^2}^2) + \mathcal{Y} + \|\nabla \mathbf{u}\|_{H^1}^2, \tag{2.27}$$

where

$$\mathcal{X}_\eta(t) = c_0^2 c_2^4 + \eta \|\nabla \mathbf{v}_t\|_{L^2}^2 + \eta^{-2} c_0^2, \quad \mathcal{Y} = M^2(c_0) c_0^9 c_1^{14} c_2^2.$$

Choosing $\eta = c_1^{-1}$, we have

$$\int_0^t \mathcal{X}_\eta(s) \, ds \lesssim 1 + (c_0^2 c_2^4 + c_0^2 c_1^2) \, t, \quad \forall \, t \in [0, T_3].$$

Multiplying (2.1c) by $\mathbf{u}_t$, integrating the result over Ω and using the Cauchy-Schwarz inequality, we obtain

$$\int_\Omega \rho |\mathbf{u}_t|^2 \, d\mathbf{x} = \int_\Omega \rho^{\frac{1}{2}} \mathbf{u}_t \cdot \left(-\rho^{\frac{1}{2}} \mathbf{v} \cdot \nabla \mathbf{u} + \rho^{-\frac{1}{2}} (\mu \Delta \mathbf{u} - \lambda (\nabla \mathbf{d})^T \Delta \mathbf{d} - \nabla p(\rho)) \right) d\mathbf{x}$$

$$\leq \frac{1}{2} \int_\Omega \rho |\mathbf{u}_t|^2 \, d\mathbf{x} + \frac{1}{2} \int_\Omega \rho |\mathbf{v}|^2 |\nabla \mathbf{u}|^2 + \rho^{-1} |\mu \Delta \mathbf{u} - \lambda (\nabla \mathbf{d})^T \Delta \mathbf{d} - \nabla p(\rho)|^2 \, d\mathbf{x},$$

i.e.,

$$\int_\Omega \rho |\mathbf{u}_t|^2 \, d\mathbf{x} \leq \int_\Omega \rho |\mathbf{v}|^2 |\nabla \mathbf{u}|^2 + \rho^{-1} |\mu \Delta \mathbf{u} - \lambda (\nabla \mathbf{d})^T \Delta \mathbf{d} - \nabla p(\rho)|^2 \, d\mathbf{x},$$

and hence

$$\limsup_{t \to 0+} \int_\Omega \rho |\mathbf{u}_t|^2 \, d\mathbf{x} \leq c_0^5 + \|\mathbf{g}\|_{L^2}^2 \lesssim c_0^5.$$

Here we have used the compatibility condition (1.5).

By virtue of the Gronwall inequality, from (2.27), we have

$$\int_\Omega (\rho |\mathbf{u}_t|^2 + |\nabla \mathbf{u}|^2) \, d\mathbf{x} + \int_0^t \int_\Omega |\nabla \mathbf{u}_t|^2 \, d\mathbf{x} ds$$

$$\lesssim c_0^5 + M^2(c_0) + \int_0^t \|\nabla \mathbf{u}\|_{H^1}^2 \, ds, \tag{2.28}$$

for all $t \in [0, T_4]$, where $T_4 = \min\{T_3, c_0^{-2}c_2^{-4}\}$.

Next, we need to estimate $\|\nabla \mathbf{u}\|_{H^1}^2$. The classical elliptic estimates applied to (2.1c), using equation (2.1b) and the assumption (2.4)-(2.7), it gives rise to

$$\|\nabla \mathbf{u}\|_{H^1} \lesssim \|\rho \mathbf{u}_t\|_{L^2} + \|\rho \mathbf{v} \cdot \nabla \mathbf{u}\|_{L^2} + \|\nabla p\|_{L^2} + \|\nabla \mathbf{u}\|_{L^2} + \|(\nabla \mathbf{d})^T \Delta \mathbf{d}\|_{L^2}$$
$$\lesssim \|\sqrt{\rho}\|_{L^\infty}\|\sqrt{\rho}\mathbf{u}_t\|_{L^2} + \|\rho\|_{L^\infty}\|\mathbf{v}\|_{H_0^1}\|\nabla \mathbf{u}\|_{L^3} + \|\nabla p\|_{L^q} + \|\nabla \mathbf{u}\|_{L^2}$$
$$+ \|\nabla \mathbf{d}\|_{L^3}\|\mathbf{d}_t\|_{L^6} + \|\nabla \mathbf{d}\|_{H^1}^2\|\mathbf{v}\|_{H_0^1} + \|\nabla \mathbf{f}\|_{H^1}\|\nabla \mathbf{d}\|_{H^1}^2\|\mathbf{f}\|_{H^2}$$
$$\leqslant C\left(M(c_0)c_0 + c_0^2 c_1^2(\|\sqrt{\rho}\mathbf{u}_t\|_{L^2} + \|\nabla \mathbf{u}\|_{L^2}) + c_0^6 c_1^{10}\right) + \frac{1}{2}\|\nabla \mathbf{u}\|_{H^1}.$$

Thus, we deduce

$$\|\nabla \mathbf{u}\|_{H^1} \lesssim M(c_0)c_0 + c_0^2 c_1^2(\|\sqrt{\rho}\mathbf{u}_t\|_{L^2} + \|\nabla \mathbf{u}\|_{L^2}) + c_0^6 c_1^{10}. \tag{2.29}$$

Taking (2.29) into (2.28) and using Gronwall's inequality, we have for all $t \in [0, T_4]$,

$$\int_\Omega (\rho|\mathbf{u}_t|^2 + |\nabla \mathbf{u}|^2)\,d\mathbf{x} + \int_0^t \int_\Omega |\nabla \mathbf{u}_t|^2\,d\mathbf{x}ds \lesssim c_0^5 + M^2(c_0), \tag{2.30}$$

and

$$\|\nabla \mathbf{u}\|_{H^1} \lesssim M(c_0)c_0^2 c_1^2 + c_0^6 c_1^{10}.$$

The term $\|\nabla^2 \mathbf{u}\|_{L^q}$ can be estimated by the same way as $\|\nabla \mathbf{u}\|_{H^1}$ above. In fact,

$$\|\nabla^2 \mathbf{u}\|_{L^q} \lesssim \|\rho \mathbf{u}_t\|_{L^q} + \|\rho \mathbf{v} \cdot \nabla \mathbf{u}\|_{L^q} + \|\nabla p\|_{L^q} + \|\nabla \mathbf{u}\|_{L^q} + \|(\nabla \mathbf{d})^T \Delta \mathbf{d}\|_{L^q}$$
$$\lesssim \|\rho\|_{L^\infty}\|\mathbf{u}_t\|_{L^q} + (\|\rho\|_{L^\infty}\|\mathbf{v}\|_{L^\infty} + 1)\|\nabla \mathbf{u}\|_{L^q} + \|\nabla p\|_{L^q} + \|\nabla \mathbf{d}\|_{H^2}^2$$
$$\lesssim c_0\|\nabla \mathbf{u}_t\|_{L^2} + M(c_0)c_0^3 c_1^2 c_2 + c_0^6 c_1^{14} c_2^2. \tag{2.31}$$

Integrating (2.31) over time and using the estimate (2.30), one has for all $t \in [0, T_5]$,

$$\int_0^t \|\nabla^2 \mathbf{u}\|_{L^q}^2 ds \lesssim M^2(c_0)c_0^2 + c_0^{12},$$

where $T_5 = \min\{T, c_1^{-28} c_2^{-4}\}$. $\square$

In conclusion, it is obvious that Lemmas 2.1-2.3 imply Theorem 2.1.

2.2 Local existence of a strong solution to the linear problem (2.1)

Notice that all the estimates obtained in Lemmas 2.1-2.3 are uniform for all small $\delta > 0$, we have the following theorem:

Theorem 2.2 *If the initial data satisfies the regularity assumption (1.4) and the compatibility condition, then there exists a unique strong solution $(\rho, \mathbf{u}, \mathbf{d})$ of (2.1) with the initial-boundary conditions (1.2) (1.3) such that*

$$\rho \in C([0, T_5]; W^{1,q}), \quad \rho_t \in C([0, T_5]; L^q);$$

$$\mathbf{u} \in C([0, T_5]; H_0^1 \cap H^2) \cap L^2(0, T_5; W^{2,q}), \quad \mathbf{u}_t \in L^2(0, T_5; H_0^1), \quad \sqrt{\rho}\mathbf{u}_t \in C([0, T_5]; L^2);$$

$$\mathbf{d} \in C([0, T_5]; H^3), \quad \mathbf{d}_t \in C([0, T_5]; H_0^1) \cap L^2(0, T_5; H^2).$$

Moreover, $(\rho, \mathbf{u}, \mathbf{d})$ also satisfies the inequalities (2.8),(2.9),(2.15) and (2.24).

Proof From Theorem 2.1 and the estimates (2.8),(2.9),(2.15) and (2.24), by virtue of the compactness theorem, there exists a trio $(\rho, \mathbf{u}, \mathbf{d})$ such that

$$(\rho^\delta, \mathbf{u}^\delta, \mathbf{d}^\delta) \to (\rho, \mathbf{u}, \mathbf{d}) \text{ in } L^2(0, T_5; L^r \times W^{1,\alpha} \times H^3), \ \forall\, r \in (1, +\infty), \ \forall\, \alpha \in [2, +\infty),$$

$$p^\delta \xrightarrow{*} \hat{p} \text{ in } L^\infty(0, T_5; W^{1,q}) \text{ as } \delta \to 0.$$

Hence $p = \hat{p}$, a.e..

According to the lower semi-continuity of various norms, the estimates (2.8), (2.15) and (2.24) hold also for $(\rho, \mathbf{u}, \mathbf{d})$. Therefore, $(\rho, \mathbf{u}, \mathbf{d})$ satisfies the system (2.1) almost everywhere on $[0, T_5] \times \Omega$, which means, $(\rho, \mathbf{u}, \mathbf{d})$ is a strong solution to (2.1) with the initial-boundary condition (1.2) (1.3).

We claim the solution $(\rho, \mathbf{u}, \mathbf{d})$ is unique. From Lemma 2.1, ρ is the unique solution to the linear equation (2.1a). Once ρ is computed uniquely from (2.1a), and then (2.1b) and (2.1c) are linear parabolic equations in terms of $\mathbf{d}$ and $\mathbf{u}$, respectively. Uniqueness is obvious.

Next, we prove the time continuity of the solution $(\rho, \mathbf{u}, \mathbf{d})$. Since the solution to the linear equation (2.1a) is unique, then the solution from Lemma 2.1 is the same as from the approximation above, i.e.,

$$\rho \in C([0, T_5]; W^{1,q}).$$

And we can easily obtain

$$\rho_t \in C([0, T_5]; L^q).$$

We know from Lemma 2.2, $\mathbf{d}_t \in L^2(0, T_5; H^2)$, $\mathbf{d} \in L^2(0, T_5; H^3)$, then

$$\mathbf{d}_t \in C([0, T_5]; H^2).$$

Differentiating (2.1b) with respect to both time and space, we have

$$\nabla \mathbf{d}_{tt} + \nabla(\mathbf{v} \cdot \nabla \mathbf{d})_t = \theta \left(\nabla \Delta \mathbf{d}_t + \nabla((\nabla \mathbf{f} : \nabla \mathbf{d})\mathbf{f})_t \right).$$

Combining with the estimate (2.15), we deduce

$$\nabla \mathbf{d}_{tt} \in L^2(0, T_5; H^{-1}).$$

Because $\nabla \mathbf{d}_t \in L^2(0, T_5; H^1)$, we have $\nabla \mathbf{d}_t \in C([0, T_5]; L^2)$, and then

$$\mathbf{d}_t \in C([0, T_5]; H_0^1).$$

Coming back to the linear equation (2.1b) again, we have

$$\Delta \mathbf{d} \in C([0, T_5]; H^1).$$

Similarly, from Lemma 2.3, $\mathbf{u}_t \in L^2(0, T_5; H_0^1)$, $\mathbf{u} \in L^2(0, T_5; W^{2,q})$, then

$$\mathbf{u} \in C([0, T_5]; H_0^1).$$

From the linear equation (2.1c) and the estimates (2.8), (2.15) and (2.24), we obtain $\rho \mathbf{u}_t \in L^2(0, T_5; H^1)$ and $(\rho \mathbf{u}_t)_t \in L^2(0, T_5; H^{-1})$, then

$$\rho \mathbf{u}_t \in C([0, T_5]; L^2).$$

Coming back to the linear equation (2.1c) and taking advantage of the elliptic regularity estimate, we have

$$\mathbf{u} \in C([0, T_5]; H^2).$$

The proof of Theorem 2.2 is now completed. □

3 Local existence in Theorem 1.1

In this section, we prove the local existence and uniqueness of strong solution in Theorem 1.1. The proof will be divided into several steps, including constructing the approximate solutions by iteration, obtaining the uniform estimate, showing the convergence, consistency and uniqueness.

3.1 Construction of approximate solutions and uniform estimates

We initialize the construction of approximate solutions by choosing an initial data $(\mathbf{u}^0(t, \mathbf{x}), \mathbf{d}^0(t, \mathbf{x}))$ of iteration, where $\mathbf{u}^0(t, \mathbf{x})$ satisfies the following heat equation

$$\varphi_t - \Delta \varphi = 0, \quad \varphi|_{t=0} = \mathbf{u}_0(\mathbf{x}), \quad \varphi|_{\partial \Omega} = \mathbf{0},$$

and $\mathbf{d}^0(t, \mathbf{x}) = \mathbf{d}_0(\mathbf{x})$. Set

$$c_1 = M^2(c_0) c_0^2 + c_0^{12}, \quad c_2 = c_1^3, \quad T_5 = \min\{c_1^{-28} c_2^{-4}, T\}.$$

Because (c_1, c_2, T_5) depends only on c_0, we can choose a time $T_*(\leqslant T_5)$ such that (2.5) and (2.6) hold for $\mathbf{u}^0$ and $\mathbf{d}^0$ with T_* instead of T.

For $k = 0, 1, 2, \cdots$, the transport equation (2.1a), the linearized momentum equation (2.1c) and the parabolic equation (2.1b) enable us to define $(\rho^{k+1}(t, \mathbf{x}), \mathbf{u}^{k+1}(t, \mathbf{x}), \mathbf{d}^{k+1}(t, \mathbf{x}))$ inductively (replacing $(\mathbf{v}, \mathbf{f})$ successively by $(\mathbf{u}^k, \mathbf{d}^k)$, $k = 0, 1, 2, \cdots$, and using Theorem 2.2) as the (global, i.e., $[0, T_*] \times \overline{\Omega}$) solution of

$$\begin{cases} \rho_t^{k+1} + \nabla \cdot (\rho^{k+1} \mathbf{u}^k) = 0, \\ \mathbf{d}_t^{k+1} + \mathbf{u}^k \cdot \nabla \mathbf{d}^{k+1} = \theta \left(\Delta \mathbf{d}^{k+1} + (\nabla \mathbf{d}^k : \nabla \mathbf{d}^{k+1}) \mathbf{d}^k \right), \\ \rho^{k+1} \mathbf{u}_t^{k+1} + \rho^{k+1} \mathbf{u}^k \cdot \nabla \mathbf{u}^{k+1} + \nabla p(\rho^{k+1}) \\ \quad = \mu \Delta \mathbf{u}^{k+1} - \lambda \nabla \cdot \left(\nabla \mathbf{d}^{k+1} \odot \nabla \mathbf{d}^{k+1} - \frac{|\nabla \mathbf{d}^{k+1}|^2}{2} \mathbb{I}_3 \right), \end{cases} \quad (3.1)$$

with the initial-boundary conditions:

$$\rho^{k+1}|_{t=0} = \rho_0, \quad \mathbf{u}^{k+1}|_{t=0} = \mathbf{u}_0, \quad \mathbf{d}^{k+1}|_{t=0} = \mathbf{d}_0,,$$

$$\mathbf{u}^{k+1}|_{\partial\Omega} = \mathbf{0}, \quad \mathbf{d}^{k+1}|_{\partial\Omega} = \mathbf{d}_0, \quad |\mathbf{d}_0| = 1.$$

Then we get a solution sequence $\{(\rho^k, \mathbf{u}^k, \mathbf{d}^k)\}_{k \in \mathbb{N}^+}$ and every $(\rho^k, \mathbf{u}^k, \mathbf{d}^k)$ satisfies the following estimates with the same (c_0, c_1, c_2, T_*):

$$\sup_{0 \leqslant t \leqslant T_*} \left(\|\mathbf{u}^k\|_{H_0^1} + \|\mathbf{d}^k\|_{H^1} + \|\mathbf{d}_t^k\|_{H_0^1} + c_1^{-10}(\|\nabla \mathbf{u}^k\|_{H^1} + \|\nabla^2 \mathbf{d}^k\|_{L^2}) \right)$$
$$+ \int_0^{T_*} (\|\nabla \mathbf{u}_t^k\|_{L^2}^2 + \|\mathbf{u}^k\|_{W^{2,q}}^2 + \|\mathbf{d}_t^k\|_{H^2}^2 + \|\mathbf{d}^k\|_{H^3}^2) \, dt$$
$$\lesssim M^2(c_0) c_0^2 + c_0^{12}, \quad (3.2)$$

and

$$\sup_{0 \leqslant t \leqslant T_*} (\|\rho^k\|_{W^{1,q}} + \|\rho_t^k\|_{L^q}) \lesssim c_0 c_2, \quad \sup_{0 \leqslant t \leqslant T_*} \|\sqrt{\rho^k} \mathbf{u}_t^k\|_{L^2} \lesssim c_0^{\frac{5}{2}} + M(c_0), \quad (3.3)$$

$$\sup_{0 \leqslant t \leqslant T_*} (\|p^k\|_{W^{1,q}} + \|p_t^k\|_{L^q}) \lesssim M(c_0) c_0 c_2, \quad \sup_{0 \leqslant t \leqslant T_*} \|\nabla \mathbf{d}^k\|_{H^2} \lesssim c_0^3 c_1^7 c_2. \quad (3.4)$$

3.2 Convergence of the approximate sequence

We claim that $\{(\rho^k, \mathbf{u}^k, \mathbf{d}^k)\}_{k=1}^{\infty}$ is a Cauchy sequence and thus converges. In fact, define

$$\bar{\rho}^{k+1} = \rho^{k+1} - \rho^k, \quad \bar{\mathbf{u}}^{k+1} = \mathbf{u}^{k+1} - \mathbf{u}^k, \quad \bar{\mathbf{d}}^{k+1} = \mathbf{d}^{k+1} - \mathbf{d}^k,$$

and

$$\Phi^{k+1}(t) = \|\bar{\rho}^{k+1}\|_{L^2}^2 + \|\bar{\mathbf{d}}^{k+1}\|_{L^2}^2 + \|\nabla \bar{\mathbf{d}}^{k+1}\|_{L^2}^2 + \|\sqrt{\rho^{k+1}} \bar{\mathbf{u}}^{k+1}\|_{L^2}^2.$$

A straightforward calculation shows that $(\bar\rho^{k+1}, \bar{\mathbf{u}}^{k+1}, \bar{\mathbf{d}}^{k+1})$ satisfies

$$\begin{cases} \bar\rho_t^{k+1} + \nabla\cdot(\bar\rho^{k+1}\mathbf{u}^k) + \nabla\cdot(\rho^k\bar{\mathbf{u}}^k) = 0, \\ \bar{\mathbf{d}}_t^{k+1} - \theta\Delta\bar{\mathbf{d}}^{k+1} = -\bar{\mathbf{u}}^k\cdot\nabla\mathbf{d}^{k+1} - \mathbf{u}^{k-1}\cdot\nabla\bar{\mathbf{d}}^{k+1} + \theta(\nabla\mathbf{d}^k:\nabla\mathbf{d}^{k+1})\,\bar{\mathbf{d}}^k \\ \qquad\qquad + \theta(\nabla\mathbf{d}^k:\nabla\bar{\mathbf{d}}^{k+1} + \nabla\bar{\mathbf{d}}^k:\nabla\mathbf{d}^k)\,\mathbf{d}^{k-1}, \\ \rho^{k+1}\bar{\mathbf{u}}_t^{k+1} + \rho^{k+1}\mathbf{u}^k\cdot\nabla\bar{\mathbf{u}}^{k+1} - \mu\Delta\bar{\mathbf{u}}^{k+1} + \nabla\left(p(\rho^{k+1}) - p(\rho^k)\right) \\ = -\bar\rho^{k+1}(\mathbf{u}^{k-1}\cdot\nabla\mathbf{u}^k + \mathbf{u}_t^k) - \rho^{k+1}\bar{\mathbf{u}}^k\cdot\nabla\mathbf{u}^k \\ \quad -\lambda\nabla\cdot\left(\nabla\bar{\mathbf{d}}^{k+1}\odot\nabla\bar{\mathbf{d}}^{k+1} + \nabla\bar{\mathbf{d}}^{k+1}\odot\nabla\mathbf{d}^k - \dfrac{\nabla\bar{\mathbf{d}}^{k+1}:(\nabla\bar{\mathbf{d}}^{k+1}+\nabla\mathbf{d}^k)}{2}\,\mathbb{I}_3\right). \end{cases} \tag{3.5}$$

From $(3.5)_1$, we have

$$\begin{aligned}\frac{d}{dt}\|\bar\rho^{k+1}\|_{L^2}^2 &\lesssim \|\nabla\mathbf{u}^k\|_{W^{1,q}}\|\bar\rho^{k+1}\|_{L^2}^2 + (\|\nabla\rho^k\|_{L^3} + \|\rho^k\|_{L^\infty})\|\nabla\bar{\mathbf{u}}^k\|_{L^2}\|\bar\rho^{k+1}\|_{L^2} \\ &\leqslant \left(\|\nabla\mathbf{u}^k\|_{W^{1,q}} + \nu^{-1}(\|\nabla\rho^k\|_{L^3}^2 + \|\rho^k\|_{L^\infty}^2)\right)\|\bar\rho^{k+1}\|_{L^2}^2 + \nu\|\nabla\bar{\mathbf{u}}^k\|_{L^2}^2.\end{aligned} \tag{3.6}$$

For $(3.5)_2$, on the one hand, we multiply $(3.5)_2$ by $\bar{\mathbf{d}}^{k+1}$ and integrate over Ω to obtain

$$\begin{aligned}&\frac{d}{dt}\|\bar{\mathbf{d}}^{k+1}\|_{L^2}^2 + \|\nabla\bar{\mathbf{d}}^{k+1}\|_{L^2}^2 \\ &\leqslant C_\nu(\|\nabla\mathbf{d}^{k+1}\|_{L^3}^2 + \|\nabla\mathbf{d}^k\|_{H^1}^2\|\nabla\mathbf{d}^{k+1}\|_{H^1}^2 + \|\nabla\mathbf{d}^k\|_{H^2}^2\|\mathbf{d}^{k-1}\|_{H^2}^2)\|\bar{\mathbf{d}}^{k+1}\|_{L^2}^2 \\ &\quad + (\|\nabla\mathbf{d}^k\|_{H^2}^2\|\mathbf{d}^{k-1}\|_{H^2}^2 + \|\nabla\mathbf{u}^{k-1}\|_{W^{1,q}})\|\bar{\mathbf{d}}^{k+1}\|_{L^2}^2 + \nu(\|\nabla\bar{\mathbf{u}}^k\|_{L^2}^2 + \|\nabla\bar{\mathbf{d}}^k\|_{L^2}^2),\end{aligned} \tag{3.7}$$

where we have used

$$\int_\Omega \mathbf{u}^{k-1}\cdot\nabla\bar{\mathbf{d}}^{k+1}\cdot\bar{\mathbf{d}}^{k+1}\,d\mathbf{x} = \frac{1}{2}\int_\Omega \mathbf{u}^{k-1}\cdot\nabla|\bar{\mathbf{d}}^{k+1}|^2\,d\mathbf{x} = -\frac{1}{2}\int_\Omega |\bar{\mathbf{d}}^{k+1}|^2\nabla\cdot\mathbf{u}^{k-1}\,d\mathbf{x},$$

and

$$\begin{aligned}&\int_\Omega (\nabla\mathbf{d}^k:\nabla\bar{\mathbf{d}}^{k+1})\,\mathbf{d}^{k-1}\cdot\bar{\mathbf{d}}^{k+1}\,d\mathbf{x} \\ &\leqslant \|\nabla\mathbf{d}^k\|_{L^\infty}\|\mathbf{d}^{k-1}\|_{L^\infty}\|\nabla\bar{\mathbf{d}}^{k+1}\|_{L^2}\|\bar{\mathbf{d}}^{k+1}\|_{L^2} \\ &\leqslant C_\theta\|\nabla\mathbf{d}^k\|_{H^2}^2\|\mathbf{d}^{k-1}\|_{H^2}^2\|\bar{\mathbf{d}}^{k+1}\|_{L^2}^2 + \frac{\theta}{2}\|\nabla\bar{\mathbf{d}}^{k+1}\|_{L^2}^2.\end{aligned}$$

On the other hand, we multiply $(3.5)_2$ by $\Delta\bar{\mathbf{d}}^{k+1}$ and integrate over Ω to get

$$\frac{1}{2}\frac{d}{dt}\|\nabla\bar{\mathbf{d}}^{k+1}\|_{L^2}^2 + \theta\|\Delta\bar{\mathbf{d}}^{k+1}\|_{L^2}^2 := \sum_{i=1}^{5} N_i. \tag{3.8}$$

Bearing in mind that $\|\nabla^2\bar{\mathbf{d}}^{k+1}\|_{L^2} \approx \|\Delta\bar{\mathbf{d}}^{k+1}\|_{L^2}$, we estimate the terms N_i ($i = 1, 2, \cdots, 5$) as follows:

$$N_1 = \int_\Omega \bar{\mathbf{u}}^k\cdot\nabla\mathbf{d}^{k+1}\cdot\Delta\bar{\mathbf{d}}^{k+1}\,d\mathbf{x}$$

$$\lesssim \|\nabla \bar{\mathbf{u}}^k\|_{L^2}\|\nabla \mathbf{d}^{k+1}\|_{H^2}\|\nabla \bar{\mathbf{d}}^{k+1}\|_{L^2} + \|\bar{\mathbf{u}}^k\|_{H^1}\|\nabla^2 \mathbf{d}^{k+1}\|_{L^3}\|\nabla \bar{\mathbf{d}}^{k+1}\|_{L^2}$$
$$\leqslant C_\nu(\|\nabla \mathbf{d}^{k+1}\|_{H^2}^2 + \|\nabla^2 \mathbf{d}^{k+1}\|_{L^2}\|\nabla^2 \mathbf{d}^{k+1}\|_{H^1})\|\nabla \bar{\mathbf{d}}^{k+1}\|_{L^2}^2 + \nu\|\nabla \bar{\mathbf{u}}^k\|_{L^2}^2, \quad (3.9)$$

$$N_2 = \int_\Omega \mathbf{u}^{k-1} \cdot \nabla \bar{\mathbf{d}}^{k+1} \cdot \Delta \bar{\mathbf{d}}^{k+1} \, d\mathbf{x} \leqslant C_\theta \|\mathbf{u}^{k-1}\|_{W^{1,q}}^2)\|\nabla \bar{\mathbf{d}}^{k+1}\|_{L^2}^2 + \frac{\theta}{4}\|\Delta \bar{\mathbf{d}}^{k+1}\|_{L^2}^2, \tag{3.10}$$

$$N_3 = -\theta \int_\Omega (\nabla \mathbf{d}^k : \nabla \mathbf{d}^{k+1}) \, \bar{\mathbf{d}}^k \cdot \Delta \bar{\mathbf{d}}^{k+1} \, d\mathbf{x}$$
$$\lesssim (\|\nabla^2 \mathbf{d}^k\|_{H^1}\|\nabla \mathbf{d}^{k+1}\|_{H^1} + \|\nabla \mathbf{d}^k\|_{H^1}\|\nabla^2 \mathbf{d}^{k+1}\|_{H^1})\|\bar{\mathbf{d}}^k\|_{H^1}\|\nabla \bar{\mathbf{d}}^{k+1}\|_{L^2}$$
$$+ \|\nabla \mathbf{d}^k\|_{H^1}\|\nabla \mathbf{d}^{k+1}\|_{H^1}\|\nabla \bar{\mathbf{d}}^k\|_{L^2}^{\frac{1}{2}}\|\Delta \bar{\mathbf{d}}^k\|_{L^2}^{\frac{1}{2}}\|\nabla \bar{\mathbf{d}}^{k+1}\|_{L^2}^{\frac{1}{2}}\|\Delta \bar{\mathbf{d}}^{k+1}\|_{L^2}^{\frac{1}{2}}$$
$$\leqslant \frac{1}{2\nu}(\|\nabla^2 \mathbf{d}^k\|_{H^1}^2\|\nabla \mathbf{d}^{k+1}\|_{H^1}^2 + \|\nabla \mathbf{d}^k\|_{H^1}^2\|\nabla^2 \mathbf{d}^{k+1}\|_{H^1}^2)\|\nabla \bar{\mathbf{d}}^{k+1}\|_{L^2}^2 + \nu\|\nabla \bar{\mathbf{d}}^k\|_{L^2}^2$$
$$+ \frac{1}{8\nu}\|\nabla \mathbf{d}^k\|_{H^1}^2\|\nabla \mathbf{d}^{k+1}\|_{H^1}^2\|\nabla \bar{\mathbf{d}}^{k+1}\|_{L^2}\|\Delta \bar{\mathbf{d}}^{k+1}\|_{L^2} + 2\nu\|\nabla \bar{\mathbf{d}}^k\|_{L^2}\|\Delta \bar{\mathbf{d}}^k\|_{L^2}$$
$$\leqslant \frac{1}{2\nu}(\|\nabla^2 \mathbf{d}^k\|_{H^1}^2\|\nabla \mathbf{d}^{k+1}\|_{H^1}^2 + \|\nabla \mathbf{d}^k\|_{H^1}^2\|\nabla^2 \mathbf{d}^{k+1}\|_{H^1}^2)\|\nabla \bar{\mathbf{d}}^{k+1}\|_{L^2}^2$$
$$+ \frac{\theta}{4}\|\Delta \bar{\mathbf{d}}^{k+1}\|_{L^2}^2$$
$$+ \frac{1}{64\theta\nu^2}\|\nabla \mathbf{d}^k\|_{H^1}^4\|\nabla \mathbf{d}^{k+1}\|_{H^1}^4\|\nabla \bar{\mathbf{d}}^{k+1}\|_{L^2}^2 + \nu(2\|\nabla \bar{\mathbf{d}}^k\|_{L^2}^2 + \|\Delta \bar{\mathbf{d}}^k\|_{L^2}^2), \tag{3.11}$$

$$N_4 = -\theta \int_\Omega (\nabla \mathbf{d}^k : \nabla \bar{\mathbf{d}}^{k+1}) \, \mathbf{d}^{k-1} \cdot \Delta \bar{\mathbf{d}}^{k+1} \, d\mathbf{x}$$
$$\lesssim \|\nabla \mathbf{d}^k\|_{L^\infty}\|\mathbf{d}^{k-1}\|_{L^\infty}\|\nabla \bar{\mathbf{d}}^{k+1}\|_{L^2}\|\Delta \bar{\mathbf{d}}^{k+1}\|_{L^2}$$
$$\leqslant C_\theta \|\nabla \mathbf{d}^k\|_{H^2}^2\|\mathbf{d}^{k-1}\|_{H^2}^2\|\nabla \bar{\mathbf{d}}^{k+1}\|_{L^2}^2 + \frac{\theta}{4}\|\Delta \bar{\mathbf{d}}^{k+1}\|_{L^2}^2, \tag{3.12}$$

$$N_5 = -\theta \int_\Omega (\nabla \bar{\mathbf{d}}^k : \nabla \mathbf{d}^k) \, \mathbf{d}^{k-1} \cdot \Delta \bar{\mathbf{d}}^{k+1} \, d\mathbf{x}$$
$$\lesssim \|\nabla \mathbf{d}^k\|_{H^2}\|\mathbf{d}^{k-1}\|_{H^2}\|\nabla^2 \bar{\mathbf{d}}^k\|_{L^2}\|\nabla \bar{\mathbf{d}}^{k+1}\|_{L^2}$$
$$+ \|\nabla^2 \mathbf{d}^k\|_{H^1}\|\mathbf{d}^{k-1}\|_{H^1}\|\Delta \bar{\mathbf{d}}^k\|_{L^2}\|\nabla \bar{\mathbf{d}}^{k+1}\|_{L^2}$$
$$+ \|\nabla \mathbf{d}^k\|_{H^1}\|\nabla \mathbf{d}^{k-1}\|_{H^1}\|\Delta \bar{\mathbf{d}}^k\|_{L^2}\|\nabla \bar{\mathbf{d}}^{k+1}\|_{L^2}$$
$$\leqslant C_\nu(\|\nabla \mathbf{d}^k\|_{H^2}^2\|\mathbf{d}^{k-1}\|_{H^2}^2 + \|\nabla^2 \mathbf{d}^k\|_{H^1}^2\|\mathbf{d}^{k-1}\|_{H^1}^2$$
$$+ \|\nabla \mathbf{d}^k\|_{H^1}^2\|\nabla \mathbf{d}^{k-1}\|_{H^1}^2)\|\nabla \bar{\mathbf{d}}^{k+1}\|_{L^2}^2 + \nu\|\Delta \bar{\mathbf{d}}^k\|_{L^2}^2. \tag{3.13}$$

Consequently, in view of (3.9)-(3.13) and (3.8), we obtain

$$\frac{d}{dt}\|\nabla \bar{\mathbf{d}}^{k+1}\|_{L^2}^2 + \|\Delta \bar{\mathbf{d}}^{k+1}\|_{L^2}^2$$

$$\begin{aligned}&\lesssim \Big(C_\nu(\|\nabla \mathbf{d}^{k+1}\|_{H^2}^2+\|\nabla^2 \mathbf{d}^{k+1}\|_{H^1}^2)+\|\nabla \mathbf{u}^{k-1}\|_{W^{1,q}}\\&+\|\mathbf{u}^{k-1}\|_{W^{1,q}}^2+\|\nabla \mathbf{d}^k\|_{H^2}^2\|\mathbf{d}^{k-1}\|_{H^2}^2\\&+C_\nu\big(\|\nabla^2 \mathbf{d}^k\|_{H^1}^2\|\nabla \mathbf{d}^{k+1}\|_{H^1}^2+\|\nabla \mathbf{d}^k\|_{H^1}^2\|\nabla^2 \mathbf{d}^{k+1}\|_{H^1}^2+\|\nabla \mathbf{d}^k\|_{H^1}^4\|\nabla \mathbf{d}^{k+1}\|_{H^1}^4\big)\\&+C_\nu\big(\|\nabla \mathbf{d}^k\|_{H^2}^2\|\mathbf{d}^{k-1}\|_{H^2}^2+\|\nabla^2 \mathbf{d}^k\|_{H^1}^2\|\mathbf{d}^{k-1}\|_{H^1}^2\\&+\|\nabla \mathbf{d}^k\|_{H^1}^2\|\nabla \mathbf{d}^{k-1}\|_{H^1}^2\big)\Big)\|\nabla \bar{\mathbf{d}}^{k+1}\|_{L^2}^2\\&+\nu(\|\nabla \bar{\mathbf{u}}^k\|_{L^2}^2+\|\nabla \bar{\mathbf{d}}^k\|_{L^2}^2+\|\Delta \bar{\mathbf{d}}^k\|_{L^2}^2).\end{aligned} \qquad (3.14)$$

Multiplying $(3.5)_3$ by $\bar{\mathbf{u}}^{k+1}$ and integrating over Ω, taking advantage of $(3.1)_1$, we have

$$\begin{aligned}&\frac{1}{2}\frac{d}{dt}\int_\Omega \rho^{k+1}|\bar{\mathbf{u}}^{k+1}|^2\,d\mathbf{x}+\mu\int_\Omega|\nabla \bar{\mathbf{u}}^{k+1}|^2\,d\mathbf{x}\\&=-\int_\Omega \bar{\rho}^{k+1}(\mathbf{u}^{k-1}\cdot\nabla \mathbf{u}^k+\mathbf{u}_t^k)\cdot\bar{\mathbf{u}}^{k+1}\,d\mathbf{x}-\int_\Omega \rho^{k+1}\bar{\mathbf{u}}^k\cdot\nabla \mathbf{u}^k\cdot\bar{\mathbf{u}}^{k+1}\,d\mathbf{x}\\&+\int_\Omega(p(\rho^{k+1})-p(\rho^k))\nabla\cdot\bar{\mathbf{u}}^{k+1}\,d\mathbf{x}\\&+\lambda\int_\Omega\Big(\nabla \mathbf{d}^{k+1}\odot\nabla \bar{\mathbf{d}}^{k+1}+\nabla \bar{\mathbf{d}}^{k+1}\odot\nabla \mathbf{d}^k\\&-\frac{\nabla \bar{\mathbf{d}}^{k+1}:(\nabla \mathbf{d}^{k+1}+\nabla \mathbf{d}^k)}{2}\mathbb{I}_3\Big):\nabla \bar{\mathbf{u}}^{k+1}\,d\mathbf{x}\\&:=\sum_{i=6}^9 N_i,\end{aligned} \qquad (3.15)$$

where

$$\begin{aligned}N_6&\leqslant \|\bar{\rho}^{k+1}\|_{L^2}\|\mathbf{u}^{k-1}\|_{L^6}\|\nabla \mathbf{u}^k\|_{L^6}\|\bar{\mathbf{u}}^{k+1}\|_{L^6}+\|\bar{\rho}^{k+1}\|_{L^2}\|\mathbf{u}_t^k\|_{L^3}\|\bar{\mathbf{u}}^{k+1}\|_{L^6}\\&\lesssim \|\bar{\rho}^{k+1}\|_{L^2}(\|\nabla \mathbf{u}^{k-1}\|_{L^2}\|\nabla \mathbf{u}^k\|_{H^1}+\|\mathbf{u}_t^k\|_{L^2}^{\frac{1}{2}}\|\nabla \mathbf{u}_t^k\|_{L^2}^{\frac{1}{2}})\|\nabla \bar{\mathbf{u}}^{k+1}\|_{L^2}\\&\leqslant C_\mu\|\bar{\rho}^{k+1}\|_{L^2}^2(\|\nabla \mathbf{u}^{k-1}\|_{L^2}^2\|\nabla \mathbf{u}^k\|_{H^1}^2+\|\mathbf{u}_t^k\|_{L^2}\|\nabla \mathbf{u}_t^k\|_{L^2})+\frac{\mu}{4}\|\nabla \bar{\mathbf{u}}^{k+1}\|_{L^2}^2,\end{aligned}$$

$$\begin{aligned}N_7&\leqslant \|\sqrt{\rho^{k+1}}\|_{L^6}\|\bar{\mathbf{u}}^k\|_{L^6}\|\nabla \mathbf{u}^k\|_{L^6}\|\sqrt{\rho^{k+1}}\bar{\mathbf{u}}^{k+1}\|_{L^2}\\&\lesssim \|\sqrt{\rho^{k+1}}\|_{L^6}\|\nabla \bar{\mathbf{u}}^k\|_{L^2}\|\nabla \mathbf{u}^k\|_{H^1}\|\sqrt{\rho^{k+1}}\bar{\mathbf{u}}^{k+1}\|_{L^2}\\&\leqslant C_\nu\|\rho^{k+1}\|_{L^3}\|\nabla \mathbf{u}^k\|_{H^1}^2\|\sqrt{\rho^{k+1}}\bar{\mathbf{u}}^{k+1}\|_{L^2}^2+\nu\|\nabla \bar{\mathbf{u}}^k\|_{L^2}^2,\end{aligned}$$

$$\begin{aligned}N_8&\lesssim \|p(\rho^{k+1})-p(\rho^k)\|_{L^2}\|\nabla \bar{\mathbf{u}}^{k+1}\|_{L^2}\\&\leqslant C_\mu M^2(c_0)\|\bar{\rho}^{k+1}\|_{L^2}^2+\frac{\mu}{4}\|\nabla \bar{\mathbf{u}}^{k+1}\|_{L^2}^2,\end{aligned}$$

$$\begin{aligned}N_9&\lesssim (\|\nabla \mathbf{d}^k\|_{H^2}+\|\nabla \mathbf{d}^{k+1}\|_{H^2})\|\nabla \bar{\mathbf{d}}^{k+1}\|_{L^2}\|\nabla \bar{\mathbf{u}}^{k+1}\|_{L^2}\\&\leqslant C_\mu(\|\nabla \mathbf{d}^k\|_{H^2}^2+\|\nabla \mathbf{d}^{k+1}\|_{H^2}^2)\|\nabla \bar{\mathbf{d}}^{k+1}\|_{L^2}^2+\frac{\mu}{4}\|\nabla \bar{\mathbf{u}}^{k+1}\|_{L^2}^2.\end{aligned}$$

Accordingly, relation (3.15) reduces to

$$\frac{d}{dt}\|\sqrt{\rho^{k+1}}\bar{\mathbf{u}}^{k+1}\|_{L^2}^2 + \|\nabla\bar{\mathbf{u}}^{k+1}\|_{L^2}^2$$
$$\lesssim \left(\|\nabla\mathbf{u}^{k-1}\|_{L^2}^2\|\nabla\mathbf{u}^k\|_{H^1}^2 + \|\mathbf{u}_t^k\|_{L^2}\|\nabla\mathbf{u}_t^k\|_{L^2} + M^2(c_0)\right)\|\bar{\rho}^{k+1}\|_{L^2}^2 + \nu\|\nabla\bar{\mathbf{u}}^k\|_{L^2}^2$$
$$+ C_\nu\|\rho^{k+1}\|_{L^3}\|\nabla\mathbf{u}^k\|_{H^1}^2\|\sqrt{\rho^{k+1}}\bar{\mathbf{u}}^{k+1}\|_{L^2}^2 + (\|\nabla\mathbf{d}^k\|_{H^2}^2 + \|\nabla\mathbf{d}^{k+1}\|_{H^2}^2)\|\nabla\bar{\mathbf{d}}^{k+1}\|_{L^2}^2. \tag{3.16}$$

Finally, by summing (3.6), (3.7), (3.14) and (3.16), it yields to

$$\frac{d}{dt}\Phi^{k+1}(t) + \|\nabla\bar{\mathbf{d}}^{k+1}\|_{L^2}^2 + \|\Delta\bar{\mathbf{d}}^{k+1}\|_{L^2}^2 + \|\nabla\bar{\mathbf{u}}^{k+1}\|_{L^2}^2$$
$$\lesssim A_\nu^k(t)\Phi^{k+1}(t) + \nu(\|\nabla\bar{\mathbf{d}}^k\|_{L^2}^2 + \|\Delta\bar{\mathbf{d}}^k\|_{L^2}^2 + \|\nabla\bar{\mathbf{u}}^k\|_{L^2}^2), \tag{3.17}$$

where

$$A_\nu^k(t) = \|\nabla\mathbf{u}^k\|_{W^{1,q}} + \|\nabla\mathbf{u}^{k-1}\|_{L^2}^2\|\nabla\mathbf{u}^k\|_{H^1}^2 + \|\mathbf{u}_t^k\|_{L^2}\|\nabla\mathbf{u}_t^k\|_{L^2} + \|\nabla\mathbf{d}^k\|_{H^2}^2\|\mathbf{d}^{k-1}\|_{H^2}^2$$
$$+ \|\nabla\mathbf{d}^k\|_{H^2}^2 + \|\nabla\mathbf{d}^{k+1}\|_{H^2}^2 + \|\nabla\mathbf{u}^{k-1}\|_{W^{1,q}} + \|\mathbf{u}^{k-1}\|_{W^{1,q}}^2 + M^2(c_0)$$
$$+ \nu^{-1}\Big(\|\nabla\rho^k\|_{L^3}^2 + \|\rho^k\|_{L^\infty}^2 + \|\rho^{k+1}\|_{L^3}\|\nabla\mathbf{u}^k\|_{H^1}^2 + \|\nabla^2\mathbf{d}^k\|_{H^1}^2\|\nabla\mathbf{d}^{k+1}\|_{H^1}^2$$
$$+ \|\nabla\mathbf{d}^{k+1}\|_{H^2}^2 + \|\nabla\mathbf{d}^k\|_{H^1}^2\|\nabla^2\mathbf{d}^{k+1}\|_{H^1}^2 + \|\nabla\mathbf{d}^k\|_{H^1}^2\|\nabla\mathbf{d}^{k+1}\|_{H^1}^2$$
$$+ \|\nabla\mathbf{d}^k\|_{H^1}^4\|\nabla\mathbf{d}^{k+1}\|_{H^1}^4 + \|\nabla\mathbf{d}^k\|_{H^2}^2\|\mathbf{d}^{k-1}\|_{H^2}^2\Big).$$

Making use of the uniform estimates (3.2)-(3.4), we have

$$\int_0^t A_\nu^k(s)\,ds \lesssim 1 + (1+\nu^{-1})\,t$$

for all $t \in [0, T_*]$.

Now we can apply Gronwall's inequality to (3.17) in order to get the following estimate:

$$\Phi^{k+1}(t) + \int_0^t (\|\nabla\bar{\mathbf{d}}^{k+1}\|_{L^2}^2 + \|\Delta\bar{\mathbf{d}}^{k+1}\|_{L^2}^2 + \|\nabla\bar{\mathbf{u}}^{k+1}\|_{L^2}^2)\,ds$$
$$\leqslant C\nu \exp\left(C + C(1+\nu^{-1})\,t\right) \int_0^t (\|\nabla\bar{\mathbf{d}}^k\|_{L^2}^2 + \|\Delta\bar{\mathbf{d}}^k\|_{L^2}^2 + \|\nabla\bar{\mathbf{u}}^k\|_{L^2}^2)\,ds.$$

Since

$$C\nu\exp\left(C + C(1+\nu^{-1})\,t\right) \leqslant \frac{1}{2}$$

for suitable (small) ν, $T_{**}(\leqslant T_*)$ and all $t \in [0, T_{**}]$, then it is easy to derive that

$$\sum_{k=1}^\infty \sup_{0\leqslant t\leqslant T_{**}} \Phi^{k+1}(t) + \sum_{k=1}^\infty \int_0^{T_{**}} (\|\nabla\bar{\mathbf{d}}^{k+1}\|_{L^2}^2 + \|\Delta\bar{\mathbf{d}}^{k+1}\|_{L^2}^2 + \|\nabla\bar{\mathbf{u}}^{k+1}\|_{L^2}^2)\,dt$$
$$\leqslant C < \infty,$$

and which implies that $\{(\rho^k, \mathbf{u}^k, \mathbf{d}^k)\}_{k=1}^\infty$ is a Cauchy sequence and yields strong convergence:

$$\rho^k \to \rho \text{ in } L^\infty(0, T_{**}; L^2), \quad \mathbf{u}^k \to \mathbf{u} \text{ in } L^2(0, T_{**}; H_0^1),$$

$$\text{and } \mathbf{d}^k \to \mathbf{d} \text{ in } L^\infty(0, T_{**}; H^1) \cap L^2(0, T_{**}; H^2).$$

We remark here that the time of existence T_{**} depends (continuously) on the norms of the data, on the bound for the density, on the domain and on the regularity parameters.

3.3 The limit is a solution

From the argument above, we know the initial-boundary problem (1.1)-(1.3) has a weak solution $(\rho, \mathbf{u}, \mathbf{d})$. Passing to a subsequence of $\{(\rho^k, \mathbf{u}^k, \mathbf{d}^k)\}_{k=1}^\infty$ if necessary the limit for $k \to \infty$, it follows from the estimates (3.2)-(3.4) that $(\rho^k, \mathbf{u}^k, \mathbf{d}^k)$ converges to $(\rho, \mathbf{u}, \mathbf{d})$ in an obvious weak or weak* sense. By using the lower semi-continuity of various norms, we also have $(\rho, \mathbf{u}, \mathbf{d})$ satisfies the regularity estimate:

$$\sup_{0 \leqslant t \leqslant T_{**}} (\|\rho\|_{W^{1,q}} + \|\rho_t\|_{L^q} + \|\mathbf{u}\|_{H_0^1} + \|p\|_{W^{1,q}} + \|p_t\|_{L^q}$$
$$+ \|\mathbf{d}\|_{H^1} + \|\mathbf{d}_t\|_{H_0^1} + \|\nabla \mathbf{u}\|_{H^1} + \|\nabla^2 \mathbf{d}\|_{L^2} + \|\nabla \mathbf{d}\|_{H^2})$$
$$+ \int_0^{T_{**}} (\|\sqrt{\rho}\mathbf{u}_t\|_{L^2}^2 + \|\nabla \mathbf{u}_t\|_{L^2}^2 + \|\mathbf{u}\|_{W^{2,q}}^2 + \|\mathbf{d}_t\|_{H^2}^2 + \|\mathbf{d}\|_{H^3}^2) \, dt$$
$$\leqslant C. \tag{3.18}$$

We claim all those nonlinear terms in (3.1) converge to their corresponding terms in (1.1) almost everywhere in $\Omega \times (0, T_{**})$. Indeed,

$$\|\nabla \cdot (\rho^{k+1} \mathbf{u}^k) - \nabla \cdot (\rho \mathbf{u})\|_{L^2_{T_{**}}(L^{\frac{2q}{q+2}})}$$
$$\leqslant \|\mathbf{u}^k - \mathbf{u}\|_{L^2_{T_{**}}(L^2)} \|\nabla \rho^{k+1}\|_{L^\infty_{T_{**}}(L^q)} + \|\mathbf{u}\|_{L^2_{T_{**}}(L^2)} \|\nabla \rho^{k+1} - \nabla \rho\|_{L^\infty_{T_{**}}(L^q)}$$
$$+ \|\nabla \mathbf{u}^k - \nabla \mathbf{u}\|_{L^2_{T_{**}}(L^2)} \|\rho^{k+1}\|_{L^\infty_{T_{**}}(L^q)} + \|\nabla \mathbf{u}\|_{L^2_{T_{**}}(L^q)} \|\rho^{k+1} - \rho\|_{L^\infty_{T_{**}}(L^2)}$$
$$\lesssim \|\mathbf{u}^k - \mathbf{u}\|_{L^2_{T_{**}}(L^2)} \|\rho^{k+1}\|_{L^\infty_{T_{**}}(W^{1,q})} + \|\mathbf{u}\|_{L^2_{T_{**}}(L^2)} \|\rho^{k+1} - \rho\|_{L^\infty_{T_{**}}(W^{1,q})}$$
$$+ \|\mathbf{u}^k - \mathbf{u}\|_{L^2_{T_{**}}(H^1)} \|\rho^{k+1}\|_{L^\infty_{T_{**}}(L^q)} + \|\mathbf{u}\|_{L^2_{T_{**}}(W^{1,q})} \|\rho^{k+1} - \rho\|_{L^\infty_{T_{**}}(L^2)}$$
$$\to 0 \text{ as } k \to \infty,$$

$$\|\rho^{k+1} \mathbf{u}_t^{k+1} - \rho \mathbf{u}_t\|_{L^2_{T_{**}}(L^{\frac{2q}{q+2}})}$$
$$\leqslant \|\rho^{k+1}\|_{L^\infty_{T_{**}}(L^q)} \|\mathbf{u}_t^{k+1} - \mathbf{u}_t\|_{L^2_{T_{**}}(L^2)} + \|\rho^{k+1} - \rho\|_{L^\infty_{T_{**}}(L^2)} \|\nabla \mathbf{u}_t\|_{L^2_{T_{**}}(L^2)}$$
$$\to 0 \text{ as } k \to \infty,$$

and
$$\|\rho^{k+1}\mathbf{u}^k\cdot\nabla\mathbf{u}^{k+1}-\rho\mathbf{u}\cdot\nabla\mathbf{u}\|_{L^2_{T_{**}}(L^{\frac{3}{2}})}$$
$$\leqslant \|\rho^{k+1}-\rho\|_{L^\infty_{T_{**}}(L^2)}\|\mathbf{u}^k\|_{L^\infty_{T_{**}}(L^\infty)}\|\nabla\mathbf{u}^{k+1}\|_{L^2_{T_{**}}(L^6)}$$
$$+\|\rho\|_{L^\infty_{T_{**}}(L^\infty)}\|\mathbf{u}^k\|_{L^\infty_{T_{**}}(L^6)}\|\nabla\mathbf{u}^{k+1}-\nabla\mathbf{u}\|_{L^2_{T_{**}}(L^2)}$$
$$+\|\rho\|_{L^\infty_{T_{**}}(L^\infty)}\|\mathbf{u}^k-\mathbf{u}\|_{L^2_{T_{**}}(L^2)}\|\nabla\mathbf{u}\|_{L^\infty_{T_{**}}(L^6)}$$
$$\to 0 \text{ as } k\to\infty.$$

Meanwhile,
$$\|\mathbf{u}^k\cdot\nabla\mathbf{d}^{k+1}-\mathbf{u}\cdot\nabla\mathbf{d}\|_{L^2_{T_{**}}(L^{\frac{3}{2}})}$$
$$\leqslant \|\mathbf{u}^k-\mathbf{u}\|_{L^2_{T_{**}}(L^6)}\|\nabla\mathbf{d}^{k+1}\|_{L^\infty_{T_{**}}(L^2)}+\|\mathbf{u}\|_{L^\infty_{T_{**}}(L^2)}\|\nabla\mathbf{d}^{k+1}-\nabla\mathbf{d}\|_{L^2_{T_{**}}(L^6)}$$
$$\lesssim \|\mathbf{u}^k-\mathbf{u}\|_{L^2_{T_{**}}(H^1_0)}\|\mathbf{d}^{k+1}\|_{L^\infty_{T_{**}}(H^1)}+\|\mathbf{u}\|_{L^\infty_{T_{**}}(L^2)}\|\mathbf{d}^{k+1}-\mathbf{d}\|_{L^2_{T_{**}}(H^2)}$$
$$\to 0, \text{ as } k\to\infty,$$

$$\left\|\nabla\cdot\left(\nabla\mathbf{d}^{k+1}\odot\nabla\mathbf{d}^{k+1}-\frac{|\nabla\mathbf{d}^{k+1}|^2}{2}\mathbb{I}_3\right)-\nabla\cdot\left(\nabla\mathbf{d}\odot\nabla\mathbf{d}-\frac{|\nabla\mathbf{d}|^2}{2}\mathbb{I}_3\right)\right\|_{L^2_{T_{**}}(L^{\frac{3}{2}})}$$
$$=\|(\nabla\mathbf{d}^{k+1})^{\mathrm{T}}\Delta\mathbf{d}^{k+1}-(\nabla\mathbf{d})^{\mathrm{T}}\Delta\mathbf{d}\|_{L^2_{T_{**}}(L^{\frac{3}{2}})}$$
$$\leqslant \|\nabla\mathbf{d}^{k+1}-\nabla\mathbf{d}\|_{L^\infty_{T_{**}}(L^2)}\|\Delta\mathbf{d}^{k+1}\|_{L^2_{T_{**}}(L^6)}+\|\Delta\mathbf{d}^{k+1}-\Delta\mathbf{d}\|_{L^2_{T_{**}}(L^2)}\|\nabla\mathbf{d}\|_{L^\infty_{T_{**}}(L^6)}$$
$$\lesssim \|\mathbf{d}^{k+1}-\mathbf{d}\|_{L^\infty_{T_{**}}(H^1)}\|\mathbf{d}^{k+1}\|_{L^2_{T_{**}}(H^3)}+\|\mathbf{d}^{k+1}-\mathbf{d}\|_{L^2_{T_{**}}(H^2)}\|\nabla\mathbf{d}\|_{L^\infty_{T_{**}}(H^1)}$$
$$\to 0, \text{ as } k\to\infty,$$

where we have used the fact that $\nabla\cdot(\nabla\mathbf{d}\odot\nabla\mathbf{d})=\nabla\left(\frac{|\nabla\mathbf{d}|^2}{2}\right)+(\nabla\mathbf{d})^{\mathrm{T}}\Delta\mathbf{d}$, and that the body force term $\nabla\cdot\left(\frac{|\nabla\mathbf{d}|^2}{2}\mathbb{I}_3\right)=\nabla\left(\frac{|\nabla\mathbf{d}|^2}{2}\right)$.

$$\|(\nabla\mathbf{d}^k:\nabla\mathbf{d}^{k+1})\,\mathbf{d}^k-|\nabla\mathbf{d}|^2\mathbf{d}\|_{L^2_{T_{**}}(L^{\frac{3}{2}})}$$
$$\leqslant \|\nabla\mathbf{d}^k-\nabla\mathbf{d}\|_{L^\infty_{T_{**}}(L^2)}\|\nabla\mathbf{d}^{k+1}\|_{L^\infty_{T_{**}}(L^6)}\|\mathbf{d}^k\|_{L^2_{T_{**}}(L^\infty)}$$
$$+\|\nabla\mathbf{d}\|^2_{L^4_{T_{**}}(L^4)}\|\mathbf{d}^k-\mathbf{d}\|_{L^\infty_{T_{**}}(L^6)}$$
$$+\|\nabla\mathbf{d}\|_{L^\infty_{T_{**}}(L^6)}\|\nabla\mathbf{d}^{k+1}-\nabla\mathbf{d}\|_{L^\infty_{T_{**}}(L^2)}\|\mathbf{d}^k\|_{L^2_{T_{**}}(L^\infty)}$$
$$\lesssim \|\mathbf{d}^k-\mathbf{d}\|_{L^\infty_{T_{**}}(H^1)}\|\mathbf{d}^{k+1}\|_{L^\infty_{T_{**}}(H^2)}\|\mathbf{d}^k\|_{L^2_{T_{**}}(H^2)}+\|\nabla\mathbf{d}\|^2_{L^4_{T_{**}}(L^4)}\|\mathbf{d}^k-\mathbf{d}\|_{L^\infty_{T_{**}}(H^1)}$$
$$+\|\mathbf{d}^{k+1}-\mathbf{d}\|_{L^2_{T_{**}}(H^1)}\|\nabla\mathbf{d}\|_{L^\infty_{T_{**}}(H^1)}\|\mathbf{d}^k\|_{L^2_{T_{**}}(H^2)}$$
$$\to 0, \text{ as } k\to\infty,$$

$$\|\nabla p(\rho^{k+1}) - \nabla p(\rho)\|_{L^\infty_{T_{**}}(L^q)}$$
$$\leqslant |p'(\rho^{k+1}) - p'(\rho)|\|\nabla \rho^{k+1}\|_{L^\infty_{T_{**}}(L^q)} + |p'(\rho)|\|\nabla \rho^{k+1} - \nabla \rho\|_{L^\infty_{T_{**}}(L^q)}$$
$$\to 0, \quad \text{as } k \to \infty.$$

Thus, passing to the limit in (3.1) as $k \to \infty$, we conclude that (1.1) holds in $L^2(0, T_{**}; L^{\frac{2q}{q+2}}(\Omega))$, and therefore almost everywhere in $\Omega \times (0, T_{**})$,

Next, we check that $|\mathbf{d}| = 1$ in $\Omega \times (0, T_{**})$. Multiplying the $\mathbf{d}$-system (1.1c) by $\mathbf{d}$, we obtain
$$\frac{1}{2}(|\mathbf{d}|^2)_t + \frac{1}{2}\mathbf{u} \cdot \nabla(|\mathbf{d}|^2) = \Delta \mathbf{d} \cdot \mathbf{d} + |\nabla \mathbf{d}|^2 |\mathbf{d}|^2.$$

Since
$$\Delta(|\mathbf{d}|^2) = 2|\nabla \mathbf{d}|^2 + 2\mathbf{d} \cdot (\Delta \mathbf{d}),$$

then it follows that
$$\frac{1}{2}(|\mathbf{d}|^2)_t + \frac{1}{2}\mathbf{u} \cdot \nabla(|\mathbf{d}|^2) = \frac{1}{2}\Delta(|\mathbf{d}|^2) - |\nabla \mathbf{d}|^2 + |\nabla \mathbf{d}|^2 |\mathbf{d}|^2.$$

Therefore, it is easy to deduce that
$$(|\mathbf{d}|^2 - 1)_t - \Delta(|\mathbf{d}|^2 - 1) + \mathbf{u} \cdot \nabla(|\mathbf{d}|^2 - 1) - 2|\nabla \mathbf{d}|^2(|\mathbf{d}|^2 - 1) = 0. \tag{3.19}$$

Multiplying (3.19) by $(|\mathbf{d}|^2 - 1)$ and then integrating over Ω, using (1.3), we get
$$\frac{d}{dt}\int_\Omega (|\mathbf{d}|^2 - 1)^2 \, d\mathbf{x} \leqslant \int_\Omega \nabla \cdot \mathbf{u} \, (|\mathbf{d}|^2 - 1)^2 \, d\mathbf{x} + 4\int_\Omega |\nabla \mathbf{d}|^2 (|\mathbf{d}|^2 - 1)^2 \, d\mathbf{x}$$
$$\lesssim (\|\nabla \mathbf{u}\|_{L^\infty} + \|\nabla \mathbf{d}\|^2_{L^\infty})\int_\Omega (|\mathbf{d}|^2 - 1)^2 \, d\mathbf{x}. \tag{3.20}$$

Recalling (3.2), we know that $\|\nabla \mathbf{u}\|_{L^\infty} + \|\nabla \mathbf{d}\|^2_{L^\infty} \in L^1(0, T_{**})$. Notice that
$$\int_\Omega (|\mathbf{d}|^2 - 1)^2 \, d\mathbf{x} = 0, \quad \text{at time } t = 0.$$

Thus, using estimate (3.20) together with Grönwall's inequality, it yields $|\mathbf{d}| = 1$ in $\Omega \times (0, T_{**})$.

3.4 Uniqueness and continuity

Let $(\rho_1, \mathbf{u}_1, \mathbf{d}_1)$ and $(\rho_2, \mathbf{u}_2, \mathbf{d}_2)$ be two solutions to (1.1) with the initial-boundary conditions (1.2) (1.3). Denote
$$\bar{\rho} = \rho_1 - \rho_2, \quad \bar{\mathbf{u}} = \mathbf{u}_1 - \mathbf{u}_2, \quad \bar{\mathbf{d}} = \mathbf{d}_1 - \mathbf{d}_2.$$

Note that $(\bar{\rho}, \bar{\mathbf{u}}, \bar{\mathbf{d}})$ satisfies the following system:

$$\begin{cases} \bar{\rho}_t + \nabla \cdot (\bar{\rho}\mathbf{u}_2) + \nabla \cdot (\rho_1 \bar{\mathbf{u}}) = 0, \\ \rho_1 \bar{\mathbf{u}}_t + \rho_1 \bar{\mathbf{u}} \cdot \nabla \mathbf{u}_2 + \rho_1 \mathbf{u}_1 \cdot \nabla \bar{\mathbf{u}} - \mu \Delta \bar{\mathbf{u}} + \nabla \left(p(\rho_1) - p(\rho_2) \right) \\ \quad = -\bar{\rho}(\mathbf{u}_{2t} + \mathbf{u}_2 \cdot \nabla \mathbf{u}_2) - \lambda (\nabla \bar{\mathbf{d}})^T \Delta \mathbf{d}_1 - \lambda (\nabla \mathbf{d}_2)^T \Delta \bar{\mathbf{d}}, \\ \bar{\mathbf{d}}_t - \theta \Delta \bar{\mathbf{d}} = -\mathbf{u}_1 \cdot \nabla \bar{\mathbf{d}} - \bar{\mathbf{u}} \cdot \nabla \mathbf{d}_2 + \theta \left(|\nabla \mathbf{d}_1|^2 \bar{\mathbf{d}} + (\nabla \mathbf{d}_1 + \nabla \mathbf{d}_2) : \nabla \bar{\mathbf{d}} \, \mathbf{d}_2 \right) \end{cases}$$

with the initial-boundary conditions:

$$(\bar{\rho}, \bar{\mathbf{u}}, \bar{\mathbf{d}})|_{t=0} = (0, \mathbf{0}, \mathbf{0}), \quad (\bar{\mathbf{u}}, \bar{\mathbf{d}})|_{\partial \Omega} = (\mathbf{0}, \mathbf{0}).$$

Define a function

$$\Psi(t) = \|\bar{\rho}\|_{L^2}^2 + \|\bar{\mathbf{d}}\|_{L^2}^2 + \|\nabla \bar{\mathbf{d}}\|_{L^2}^2 + \|\sqrt{\rho_1} \bar{\mathbf{u}}\|_{L^2}^2.$$

Repeating the similar argument in subsection 3.2, we get the following estimates:

$$\frac{d}{dt}\|\bar{\rho}\|_{L^2}^2 \lesssim (\|\nabla \mathbf{u}_2\|_{W^{1,q}} + \|\nabla \rho_1\|_{L^3}^2 + \|\rho_1\|_{L^\infty}^2)\|\bar{\rho}\|_{L^2}^2 + \frac{1}{4}\|\nabla \bar{\mathbf{u}}\|_{L^2}^2, \qquad (3.21)$$

$$\frac{d}{dt}\|\sqrt{\rho_1}\bar{\mathbf{u}}\|_{L^2}^2 + \|\nabla \bar{\mathbf{u}}\|_{L^2}^2$$
$$\lesssim \left(\|\rho_1\|_{L^3}\|\nabla \mathbf{u}_1\|_{H^1}^2 + \|\sqrt{\rho_1}\|_{L^\infty}^2 (\|\nabla \mathbf{u}_1\|_{H^1}^2 + \|\nabla \mathbf{u}_2\|_{H^1}^2) \right) \|\sqrt{\rho_1}\bar{\mathbf{u}}\|_{L^2}^2$$
$$+ \left(\|\nabla \mathbf{u}_2\|_{L^2}^2 \|\nabla \mathbf{u}_2\|_{H^1}^2 + \|\mathbf{u}_{2t}\|_{L^2}\|\nabla \mathbf{u}_{2t}\|_{L^2} + M^2(c_0) \right) \|\bar{\rho}\|_{L^2}^2$$
$$+ (\|\nabla \mathbf{d}_1\|_{H^2}^2 + \|\nabla \mathbf{d}_2\|_{H^2}^2) \|\nabla \bar{\mathbf{d}}\|_{L^2}^2, \qquad (3.22)$$

$$\frac{d}{dt}\|\bar{\mathbf{d}}\|_{L^2}^2 + \|\nabla \bar{\mathbf{d}}\|_{L^2}^2$$
$$\lesssim (\|\nabla \mathbf{u}_1\|_{W^{1,q}} + \|\nabla \mathbf{d}_2\|_{L^3}^2 + \|\nabla \mathbf{d}_1\|_{H^2}^2$$
$$+ \|\nabla \mathbf{d}_1 + \nabla \mathbf{d}_2\|_{H^2}^2 \|\mathbf{d}_2\|_{H^2}^2) \|\bar{\mathbf{d}}\|_{L^2}^2 + \frac{1}{4}\|\nabla \bar{\mathbf{u}}\|_{L^2}^2,$$

$$\frac{d}{dt}\|\nabla \bar{\mathbf{d}}\|_{L^2}^2 + \|\Delta \bar{\mathbf{d}}\|_{L^2}^2$$
$$\lesssim (\|\nabla \mathbf{d}_2\|_{H^2}^2 + \|\nabla^2 \mathbf{d}_2\|_{H^1}^2 + \|\mathbf{u}_1\|_{W^{1,q}}^2 + \|\nabla \mathbf{d}_1\|_{H^1}^4$$
$$+ \|\nabla \mathbf{d}_1 + \nabla \mathbf{d}_2\|_{H^2}^2 \|\mathbf{d}_2\|_{H^2}^2) \|\nabla \bar{\mathbf{d}}\|_{L^2}^2 + \frac{1}{4}\|\nabla \bar{\mathbf{u}}\|_{L^2}^2,$$

and

$$\frac{d}{dt}\Psi(t) + \|\nabla \bar{\mathbf{u}}\|_{L^2}^2 + \|\nabla \bar{\mathbf{d}}\|_{L^2}^2 + \|\Delta \bar{\mathbf{d}}\|_{L^2}^2 \lesssim A(t)\Psi(t),$$

where
$$A(t) = \|\nabla \mathbf{u}_1\|_{W^{1,q}} + \|\nabla \mathbf{u}_2\|_{W^{1,q}} + \|\mathbf{u}_1\|_{W^{1,q}}^2 + \|\nabla \rho_1\|_{L^3}^2 + \|\rho_1\|_{L^\infty}^2 + \|\nabla \mathbf{u}_2\|_{L^2}^2 \|\nabla \mathbf{u}_2\|_{H^1}^2$$
$$+ \|\mathbf{u}_{2t}\|_{L^2}\|\nabla \mathbf{u}_{2t}\|_{L^2} + \|\rho_1\|_{L^3}\|\nabla \mathbf{u}_1\|_{H^1}^2 + \|\sqrt{\rho_1}\|_{L^\infty}^2 (\|\nabla \mathbf{u}_1\|_{H^1}^2 + \|\nabla \mathbf{u}_2\|_{H^1}^2)$$
$$+ \|\nabla \mathbf{d}_1\|_{H^1}^4 + M^2(c_0) + (\|\nabla \mathbf{d}_1\|_{H^2}^2 + \|\nabla \mathbf{d}_2\|_{H^2}^2)(1 + \|\mathbf{d}_2\|_{H^2}^2),$$

and obviously, it follows from the estimate (3.18) that $A(t) \in L^1(0, T_{**})$.

Now a straightforward result follows from the Gronwall's inequality, that
$$\Psi(t) + \int_0^t (\|\Delta \bar{\mathbf{d}}\|_{L^2}^2 + \|\nabla \bar{\mathbf{u}}\|_{L^2}^2) \, ds \leqslant 0, \quad t \in [0, T_{**}],$$
and it in turn yields:
$$\bar{\rho} = 0, \quad \bar{\mathbf{u}} = \mathbf{0}, \quad \bar{\mathbf{d}} = \mathbf{0},$$
which implies the property of uniqueness.

Following the argument of uniqueness, we can also easily prove that if $(\rho, \mathbf{u}, \mathbf{d})$ and $(\tilde{\rho}, \tilde{\mathbf{u}}, \tilde{\mathbf{d}})$ are solutions to (1.1) and (1.3) with different initial data $(\rho_0, \mathbf{u}_0, \mathbf{d}_0)$ and $(\tilde{\rho}_0, \tilde{\mathbf{u}}_0, \tilde{\mathbf{d}}_0)$, then for all $t \in [0, T_{**}]$, we have
$$(\|\bar{\bar{\rho}}\|_{L^2}^2 + \|\bar{\bar{\mathbf{d}}}\|_{H^1}^2 + \|\sqrt{\rho}\bar{\bar{\mathbf{u}}}\|_{L^2}^2)(t) + \int_0^t (\|\Delta \bar{\bar{\mathbf{d}}}\|_{L^2}^2 + \|\nabla \bar{\bar{\mathbf{u}}}\|_{L^2}^2) \, ds \to 0 \qquad (3.23)$$
as $(\tilde{\rho}_0, \tilde{\mathbf{u}}_0, \nabla \tilde{\mathbf{d}}_0) \to (\rho_0, \mathbf{u}_0, \nabla \mathbf{d}_0)$ in $W^{1,q} \times (H^2)^3 \times (H^2)^9$, where
$$\bar{\bar{\rho}} = \rho - \tilde{\rho}, \quad \bar{\bar{\mathbf{u}}} = \mathbf{u} - \tilde{\mathbf{u}}, \quad \bar{\bar{\mathbf{d}}} = \mathbf{d} - \tilde{\mathbf{d}}.$$

Since $\mathbf{d}$ and $\tilde{\mathbf{d}}$ satisfy (1.1c), then
$$\bar{\bar{\mathbf{d}}}_t + \bar{\bar{\mathbf{u}}} \cdot \nabla \mathbf{d} + \tilde{\mathbf{u}} \cdot \nabla \bar{\bar{\mathbf{d}}} = \theta(\Delta \bar{\bar{\mathbf{d}}} + |\nabla \mathbf{d}|^2 \bar{\bar{\mathbf{d}}} + (\nabla \mathbf{d} + \nabla \tilde{\mathbf{d}}) : \nabla \bar{\bar{\mathbf{d}}} \, \tilde{\mathbf{d}}). \qquad (3.24)$$

Now we multiply (3.24) by $\Delta \bar{\bar{\mathbf{d}}}_t$ and integrate over Ω while bearing in mind the fact that $\sup_t \|\nabla \mathbf{d}\|_{H^2} \leqslant C$ and $\sup_t \|\nabla \tilde{\mathbf{u}}\|_{H^1} \leqslant C$ (see (3.18)), that the Sobolev imbedding $H^2(\Omega) \subset L^\infty(\Omega)$, and that the boundary conditions, that
$$\frac{d}{dt} \int_\Omega |\Delta \bar{\bar{\mathbf{d}}}|^2 \, d\mathbf{x} + \int_\Omega |\nabla \bar{\bar{\mathbf{d}}}_t|^2 \, d\mathbf{x}$$
$$\lesssim \|\nabla \bar{\bar{\mathbf{u}}}\|_{L^2}^2 + \Big(\|\nabla \tilde{\mathbf{u}}\|_{H^1}^2 + \|\nabla \mathbf{d}\|_{H^1}^4 + \|\nabla \mathbf{d}\|_{L^\infty}^2 \|\nabla^2 \mathbf{d}\|_{L^2}^2$$
$$+ (\|\nabla^2 \mathbf{d}\|_{L^6}^2 + \|\nabla^2 \tilde{\mathbf{d}}\|_{L^6}^2)\|\tilde{\mathbf{d}}\|_{L^6}^2$$
$$+ (\|\nabla \mathbf{d}\|_{L^\infty}^2 + \|\nabla \tilde{\mathbf{d}}\|_{L^\infty}^2)\|\tilde{\mathbf{d}}\|_{L^\infty}^2 + (\|\nabla \mathbf{d}\|_{L^6}^2 + \|\nabla \tilde{\mathbf{d}}\|_{L^6}^2)\|\nabla \tilde{\mathbf{d}}\|_{L^6}^2 \Big) \|\Delta \bar{\bar{\mathbf{d}}}\|_{L^2}^2. \qquad (3.25)$$

Recalling the estimate (3.18), applying Gronwall's inequality to (3.25), making use of the elliptic estimate $\|\bar{\bar{\mathbf{d}}}\|_{H^2} \lesssim \|\Delta \bar{\bar{\mathbf{d}}}\|_{L^2}$ and the convergence (3.23), we get, for all $t \in [0, T_{**}]$,
$$\|\bar{\bar{\mathbf{d}}}\|_{H^2}(t) + \int_0^t \|\nabla \bar{\bar{\mathbf{d}}}_t\|_{L^2}^2 \, ds \to 0 \qquad (3.26)$$
as $(\tilde{\rho}_0, \tilde{\mathbf{u}}_0, \nabla \tilde{\mathbf{d}}_0) \to (\rho_0, \mathbf{u}_0, \nabla \mathbf{d}_0)$ in $W^{1,q} \times (H^2)^3 \times (H^2)^9$.

Similarly, for the momentum conservation equation, we have

$$\rho\bar{\mathbf{u}}_t + \bar{\rho}\tilde{\mathbf{u}}_t + \rho\bar{\mathbf{u}}\cdot\nabla\mathbf{u} + \rho\tilde{\mathbf{u}}\cdot\nabla\bar{\mathbf{u}} + \bar{\rho}\tilde{\mathbf{u}}\cdot\nabla\tilde{\mathbf{u}} + \nabla\left(p(\rho) - p(\tilde{\rho})\right)$$
$$= \mu\Delta\bar{\mathbf{u}} - \lambda(\nabla\bar{\mathbf{d}})^{\mathrm{T}}\Delta\mathbf{d} - \lambda(\nabla\tilde{\mathbf{d}})^{\mathrm{T}}\Delta\bar{\mathbf{d}}. \tag{3.27}$$

On the one hand, similarly from (3.21),(3.22), $\bar{\rho}$ satisfies

$$\frac{d}{dt}\|\bar{\rho}\|_{L^2}^2 \lesssim (\|\nabla\tilde{\mathbf{u}}\|_{W^{1,q}} + \|\nabla\rho\|_{L^3}^2 + \|\rho\|_{L^\infty}^2)\|\bar{\rho}\|_{L^2}^2 + \eta\|\nabla\bar{\mathbf{u}}\|_{L^2}^2$$

for suitable (small) η, and $\bar{\mathbf{u}}$ satisfies

$$\frac{d}{dt}\|\sqrt{\rho}\bar{\mathbf{u}}\|_{L^2}^2 + \|\nabla\bar{\mathbf{u}}\|_{L^2}^2$$
$$\lesssim \left(\|\rho\|_{L^3}\|\nabla\mathbf{u}\|_{H^1}^2 + \|\sqrt{\rho}\|_{L^\infty}^2(\|\nabla\mathbf{u}\|_{H^1}^2 + \|\nabla\tilde{\mathbf{u}}\|_{H^1}^2)\right)\|\sqrt{\rho}\bar{\mathbf{u}}\|_{L^2}^2$$
$$+ \left(\|\nabla\tilde{\mathbf{u}}\|_{L^2}^2\|\nabla\tilde{\mathbf{u}}\|_{H^1}^2 + \|\tilde{\mathbf{u}}_t\|_{L^2}\|\nabla\tilde{\mathbf{u}}_t\|_{L^2} + M^2(c_0)\right)\|\bar{\rho}\|_{L^2}^2$$
$$+ (\|\nabla\mathbf{d}\|_{H^2}^2 + \|\nabla\tilde{\mathbf{d}}\|_{H^2}^2)\|\nabla\bar{\mathbf{d}}\|_{L^2}^2.$$

On the other hand, multiplying (3.27) by $\bar{\mathbf{u}}_t$ and integrating over Ω, we obtain

$$\frac{d}{dt}\|\nabla\bar{\mathbf{u}}\|_{L^2}^2 + \|\sqrt{\rho}\bar{\mathbf{u}}_t\|_{L^2}^2$$
$$\lesssim (\|\rho\|_{L^3}\|\nabla\mathbf{u}\|_{H^1}^2 + \|\sqrt{\rho}\|_{L^\infty}^2\|\nabla\tilde{\mathbf{u}}\|_{H^1}^2)\|\nabla\bar{\mathbf{u}}\|_{L^2}^2 + \|\bar{\mathbf{u}}_t\|_{L^3}(\|\nabla\mathbf{d}\|_{H^2} + \|\nabla^2\tilde{\mathbf{d}}\|_{H^1})\|\nabla\bar{\mathbf{d}}\|_{L^2}$$
$$+ (\|\tilde{\mathbf{u}}_t\|_{L^3} + \|\tilde{\mathbf{u}}\|_{L^6}\|\nabla\tilde{\mathbf{u}}\|_{L^6} + M(c_0))\|\nabla\bar{\mathbf{u}}_t\|_{L^2}\|\bar{\rho}\|_{L^2} + \|\nabla\tilde{\mathbf{d}}\|_{H^2}\|\nabla\bar{\mathbf{u}}_t\|_{L^2}\|\nabla\bar{\mathbf{d}}\|_{L^2}.$$

With (3.23) and (3.26) at hand, evoking the Poincáre inequality and the estimate (3.18), we conclude that

$$\|\bar{\mathbf{u}}\|_{H^1}(t) + \int_0^t \|\sqrt{\rho}\bar{\mathbf{u}}_t\|_{L^2}^2 \, ds \to 0 \tag{3.28}$$

as $(\tilde{\rho}_0, \tilde{\mathbf{u}}_0, \nabla\tilde{\mathbf{d}}_0) \to (\rho_0, \mathbf{u}_0, \nabla\mathbf{d}_0)$ in $W^{1,q} \times (H^2)^3 \times (H^2)^9$.

Finally, it is easy to show that

$$\bar{\rho}_t + \mathbf{u}\cdot\nabla\bar{\rho} + \bar{\rho}\nabla\cdot\mathbf{u} + \nabla\tilde{\rho}\cdot\bar{\mathbf{u}} + \tilde{\rho}\nabla\cdot\bar{\mathbf{u}} = 0. \tag{3.29}$$

Multiplying (3.29) by $q\bar{\rho}^{q-1}$ and integrating over Ω, using the Hölder inequality, we have

$$\frac{d}{dt}\|\bar{\rho}\|_{L^q}^q \lesssim \|\nabla\mathbf{u}\|_{W^{1,q}}\|\bar{\rho}\|_{L^q}^q + \|\nabla\bar{\mathbf{u}}\|_{L^q}\|\nabla\tilde{\rho}\|_{L^q}\|\bar{\rho}\|_{L^q}^{q-1}.$$

Making use of the estimate (3.18) and the convergence (3.28), we can apply Gronwall's inequality to the above inequality to get, for all $t \in [0, T_{**}]$,

$$\|\bar{\rho}\|_{L^q}(t) \to 0 \tag{3.30}$$

as $(\tilde{\rho}_0, \tilde{\mathbf{u}}_0, \nabla \tilde{\mathbf{d}}_0) \to (\rho_0, \mathbf{u}_0, \nabla \mathbf{d}_0)$ in $W^{1,q} \times (H^2)^3 \times (H^2)^9$.

Consequently, together with the equations (1.1b) and (1.1c), we can obtain, for all $t \in [0, T_{**}]$,
$$\begin{cases} \|\bar{\bar{\mathbf{d}}}\|_{L^2}(t) \to 0, & \forall\, t \in [0, T_{**}], \\ \|\bar{\bar{\mathbf{d}}}\|_{L^2_{T_{**}}(H^3)} \to 0, & \|\bar{\mathbf{u}}\|_{L^2_{T_{**}}(H^2)} \to 0 \end{cases} \tag{3.31}$$
as $(\tilde{\rho}_0, \tilde{\mathbf{u}}_0, \nabla \tilde{\mathbf{d}}_0) \to (\rho_0, \mathbf{u}_0, \nabla \mathbf{d}_0)$ in $W^{1,q} \times (H^2)^3 \times (H^2)^9$.

To conclude, (3.23), (3.26), (3.28), (3.30) and (3.31) complete the proof of the continuity in Theorem 1.1.

4 Proof of Theorem 1.2

Suppose that there are two positive constants $\xi\ (< 1)$ and $\tilde{C}$ such that
$$\max\{\|\rho_0\|_{W^{1,q}}, \|\mathbf{u}_0\|_{H^2}, \|\mathbf{d}_0 - \mathbf{n}\|_{H^3}, \|\mathbf{g}\|_{L^2}^2\} < \xi,$$
$$\sup_{0 \leqslant t \leqslant T} (\|\mathbf{v}\|_{H^2} + \|\mathbf{f}\|_{H^3} + \|\mathbf{f}_t\|_{H_0^1}) + \int_0^T (\|\nabla \mathbf{v}_t\|_{L^2}^2 + \|\mathbf{v}\|_{W^{2,q}}^2 + \|\nabla^2 \mathbf{f}_t\|_{L^2}^2)\, dt < \tilde{C}.$$

In this section, we assume the genuine constant C, maybe depending on the constant $M(1)$ which is defined by (2.14).

By Lemma 2.1, there exists a small $\xi_1\ (< 1)$ such that for all $\xi \in (0, \xi_1]$,
$$\|\rho\|_{W^{1,q}} \leqslant C\xi^{\frac{1}{2}},\ \|\rho_t\|_{L^q} \leqslant C\xi^{\frac{1}{3}},\ \|p\|_{W^{1,q}} \leqslant C\xi^{\frac{1}{2}},\ \|p_t\|_{L^q} \leqslant C\xi^{\frac{1}{3}},\ \forall\, t \in [0, T]. \tag{4.1}$$

According to Lemma 2.2, we can find a small $\xi_2\ (< 1)$ such that for all $\xi \in (0, \xi_2]$,
$$\|\mathbf{d}_t\|_{H_0^1}^2(t),\ \|\mathbf{d} - \mathbf{n}\|_{H^2}^2(t),\ \int_0^t \|\mathbf{d} - \mathbf{n}\|_{H^3}^2\, ds \leqslant C\xi^{\frac{1}{2}},\ \forall\, t \in [0, T]. \tag{4.2}$$

Similarly, from Lemma 2.3, a small $\xi_3\ (\leqslant \min\{\xi_1, \xi_2\})$ can be found, satisfying that for all $\xi \in (0, \xi_3]$,
$$\|\mathbf{u}\|_{H^2}^2(t),\ \|\sqrt{\rho}\mathbf{u}_t\|_{L^2}^2(t),\ \int_0^t \|\mathbf{u}_t\|_{H^1}^2\, ds,\ \int_0^t \|\mathbf{u}\|_{W^{2,q}}^2\, ds \leqslant C\xi^{\frac{1}{6}},\ \forall\, t \in [0, T]. \tag{4.3}$$

Making use of the estimates (4.1)-(4.3) and Theorem 2.1, we can obtain the global strong solution of the linear system (2.1), (2.2) with the initial-boundary conditions (1.2)(1.3) provided that
$$\max\{\|\rho_0\|_{W^{1,q}}, \|\mathbf{u}_0\|_{H^2}, \|\mathbf{d}_0 - \mathbf{n}\|_{H^3}, \|\mathbf{g}\|_{L^2}^2\} < \xi,\ \forall\, \xi \in (0, \xi_3].$$

Now let us consider the possibility of the iteration. First, if we choose ξ_3 so small that $C\xi_3 \leq \tilde{C}$, then the process of iteration can be continued for the same ξ_3. Next, we will focus on the convergence of the iteration.

Repeating the same procedure as in Section 3, using the estimates (4.1)-(4.3), we can get

$$H^{k+1}(t) + \int_0^t (\|\nabla \bar{\mathbf{d}}^{k+1}\|_{L^2}^2 + \|\Delta \bar{\mathbf{d}}^{k+1}\|_{L^2}^2 + \|\nabla \bar{\mathbf{u}}^{k+1}\|_{L^2}^2)\, ds$$
$$\leq C\nu \exp\left(C(t + \xi^{\frac{1}{6}}t + \xi^{\frac{1}{6}} + \nu^{-1}\xi^{\frac{1}{2}}t + \nu^{-1}\xi^{\frac{1}{2}})\right)$$
$$\cdot \int_0^t (\|\nabla \bar{\mathbf{d}}^{k}\|_{L^2}^2 + \|\Delta \bar{\mathbf{d}}^{k}\|_{L^2}^2 + \|\nabla \bar{\mathbf{u}}^{k}\|_{L^2}^2)\, ds, \tag{4.4}$$

where

$$H^{k+1}(t) = \|\bar{\rho}^{k+1}\|_{L^2}^2 + \|\bar{\mathbf{d}}^{k+1}\|_{L^2}^2 + \|\nabla \bar{\mathbf{d}}^{k+1}\|_{L^2}^2 + \|\sqrt{\rho^{k+1}}\bar{\mathbf{u}}^{k+1}\|_{L^2}^2,$$

and

$$\bar{\rho}^{k+1} = \rho^{k+1} - \rho^k, \quad \bar{\mathbf{u}}^{k+1} = \mathbf{u}^{k+1} - \mathbf{u}^k, \quad \bar{\mathbf{d}}^{k+1} = \mathbf{d}^{k+1} - \mathbf{d}^k.$$

Now if we choose small constants ν, ξ_0 such that for all $\xi \in (0, \xi_0]$,

$$C\nu \exp\left(C(t + \xi^{\frac{1}{6}}t + \xi^{\frac{1}{6}} + \nu^{-1}\xi^{\frac{1}{2}}t + \nu^{-1}\xi^{\frac{1}{2}})\right) \leq \frac{1}{2},$$

then it is easy to deduce from (4.4) that

$$\sum_{k=1}^{\infty} H^{k+1}(t) + \sum_{k=1}^{\infty} \int_0^T (\|\nabla \bar{\mathbf{d}}^{k+1}\|_{L^2}^2 + \|\Delta \bar{\mathbf{d}}^{k+1}\|_{L^2}^2 + \|\nabla \bar{\mathbf{u}}^{k+1}\|_{L^2}^2)\, ds$$
$$\leq C < \infty.$$

The proof of Theorem 1.2 is completed.

Acknowledgement X. Li's research was supported in part by the National Natural Science Foundation of China under grants 11401036, 11271052 and 11471050, by the China Postdoctoral Science Foundation Funded Project under grant 2013T60085, and by the Fundamental Research for the Central Universities No. 2014RC0901.

References

[1] K. C. Chang, W. Y. Ding, R. Ye. *Finite-time blow-up of the heat flow of harmonic maps from surfaces.* J. Diff. Geom., 36 (1992), 507-515.

[2] H. J. Choe, H. Kim. *Strong solutions of the Navier-Stokes equations for isentropic compressible fluids*. J. Diff. Equations, 190 (2003), 504-523.

[3] Y. Cho, H. J. Choe, H. Kim. *Unique solvability of the initial boundary value prob lems for compressible viscous fluids*. J. Math. Pures Appl., 83 (2004), 243-275.

[4] Y. Cho, H. Kim. *Existence results for viscous polytropic fluids with vacuum*. J. Diff. Equations, 228 (2006), 377-411.

[5] S. Ding, C. Wang, H. Wen. *Weak solution to compressible hydrodynamic flow of liquid crystals in dimension one*. Discrete Conti. Dyna. Sys. Ser. B, 15(2) (2011), 357-371.

[6] S. Ding, J. Lin, C. Wang, H. Wen. *Compressible hydrodynamic flow of liquid crystals in 1-D*. Discrete Conti. Dyna. Sys., 32(2) (2012), 539-563.

[7] J. Ericksen. *Conservation laws for liquid crystals*. Trans. Soc. Rheol., 5 (1961), 22-34.

[8] J. Ericksen. *Equilibrium theory for liquid crystals, in: G. Brown (Ed.)*. Advances in Liquid Crystals, Vol. 2, Academic Press, New York, (1975), 233-398.

[9] J. Ericksen. *Continuum theory of nematic liquid crystals*. Res. Mechanica, 21 (1987), 381-392.

[10] M. Hong. *Global existence of solutions of the simplified Ericksen-Leslie system in dimension two*. Calc. Var. Partial Diff. Equations, 40 (2011), 15-36.

[11] X. Hu, H. Wu. *Global solution to the three-dimensional compressible flow of liquid crystals*. arXiv:1206.2850.

[12] T. Huang, C. Wang, H. Wen. *Strong solutions of the compressible nematic liquid crystal flow*. J. Diff. Equations, 252(3) (2012), 2222-2265.

[13] T. Huang, C. Wang, H. Wen. *Blow up criterion for compressible nematic liquid crystal flows in dimension three*. Arch. Rational Mech. Anal., 204 (2012), 285-311.

[14] F. Jiang, Z. Tan. *Global weak solution to the flow of liquid crystals system*. Math. Meth. Appl. Sci., 32 (2009), 2243-2266.

[15] F. M. Leslie. *Theory of flow phenomena in liquid crystals, in: G. Brown (Ed.)*. Advances in Liquid Crystals, Vol. 4, Academic Press, New York, (1979), 1-81.

[16] X. Li, D. Wang. *Global strong solution to the density-dependent incompressible flow of liquid crystals*. Trans. Amer. Math. Soc., 367 (2015), 2301-2338.

[17] F. H. Lin. *Nonlinear theory of defects in nematic liquid crystal: phase transition and flow phenomena*. Comm. Pure Appl. Math., 42(1989), 789-814.

[18] F. H. Lin. *Existence of solutions for the Ericksen-Leslie system*. Arch. Rat. Mech. Anal., 154 (2000), 135-156.

[19] F. H. Lin, C. Liu. *Nonparabolic dissipative systems modeling the flow of liquid crystals*. Comm. Pure Appl. Math., 48 (1995), 501-537.

[20] F. H. Lin, C. Liu. *Partial regularities of the nonlinear dissipative systems modeling the flow of liquid crystals*. Discrete Conti. Dyna. Sys., 2 (1996), 1-23.

[21] F. H. Lin, J. Lin, C. Wang. *Liquid crystal flows in two dimensions*. Arch. Ration. Mech. Anal., 197 (2010), 297-336.

[22] C. Liu. *Dynamic theory for incompressible smectic-A liquid crystals.* Discrete Conti. Dyna. Sys., 6 (2000), 591-608.

[23] X. Liu, L. Liu, Y. Hao. *Existence of strong solutions for the compressible Ericksen-Leslie model.* arXiv:1106.6140.

[24] X. Liu, Z. Zhang. *Existence of the flow of liquid crystals system.* Chinese Ann. Math., 30A(1) (2009), 1-20.

[25] S. Shkoller. *Well-posedness and global attractors for liquid crystals on Riemannian manifolds.* Comm. Partial Diff. Equations, 27 (2001), 1103-1137.

[26] C. Wang. *Well-posedness for the heat flow of harmonic maps and the liquid crystal flow with rough initial data.* Arch. Ration. Mech. Anal., 200 (2011), 1-19.

[27] H. Wen, S. Ding. *Solutions of incompressible hydrodynamic flow of liquid crystals.* Nonlinear Analysis: Real World Applications, 12 (2011), 1510-1531.

Optimal Decay Rate of the Compressible Quantum Navier-Stokes Equations*

Pu Xueke (蒲学科) and Guo Boling (郭柏灵)

Abstract For quantum fluids governed by the compressible quantum Navier-Stokes equations in $\mathbb{R}^3$ with viscosity and heat conduction, we prove the optimal $L^p - L^q$ decay rates for the classical solutions near constant states. The proof is based on the detailed linearized decay estimates by Fourier analysis of the operators, which is drastically different from the case when quantum effects are absent.

Keywords Compressible quantum Navier-Stokes equations; optimal decay rates; energy estimates

1 Introduction

Let us consider the following classical hydrodynamic equations in $\mathbb{R}^3$ describing the motion of the electrons in plasmas by omitting the electric potential

$$\begin{cases} \dfrac{\partial n}{\partial t} + \operatorname{div} \Pi = 0, & (1.1\text{a}) \\ \dfrac{\partial \Pi}{\partial t} + \operatorname{div}(nu \otimes u - P) = 0, & (1.1\text{b}) \\ \dfrac{\partial W}{\partial t} + \operatorname{div}(uW - Pu + q) = 0, & (1.1\text{c}) \end{cases}$$

where n is the density, $u = (u_1, u_2, u_3)$ is the velocity, $\Pi = (\Pi_1, \Pi_2, \Pi_3)$ and $\Pi_j = nu_j$ is the momentum density, $P = (P_{ij})_{3\times 3}$ is the stress tensor, W is the energy density and $q = -\kappa \nabla T$ is the heat flux and T is the temperature. This system also emerges from descriptions of the motion of the electrons in semiconductor devices, with the electrical potential and the relaxation omitted [4].

In this paper, we consider the following case. The stress tensor is given by

$$P = -nT\mathbb{I} + \frac{\hbar^2 n}{12} \nabla^2 \ln n + \mathbb{S},$$

---

*Ann. of Appl. Math, 2016, 32(3): 275–287.

where $\mathbb{I}$ is the identity matrix and $\mathbb{S}$ is the viscous part of the stress tensor given by

$$\mathbb{S} = \mu(\nabla u + (\nabla u)^T) + \delta(\text{div} u)\mathbb{I},$$

where $\mu > 0$ and δ are the primary coefficients of viscosity and the second coefficients of viscosity, respectively, satisfying $2\mu + 3\delta \geq 0$. The energy density W is given by

$$W = \frac{3}{2}nT + \frac{1}{2}nu^2 - \frac{\hbar^2 n}{24}\Delta \ln n.$$

The quantum correction to the stress tensor was proposed by Ancona and Tiersten [2] on general thermodynamical grounds and derived by Ancona and Iafrate [1] in the Wigner formalism. The quantum correction to the energy density was first derived by Wigner [14]. See also [5]. With these quantum corrections ($\hbar > 0$), the system (2.7) is called the compressible quantum Navier-Stokes (CQNS) equations. When $\hbar = 0$, it reduces to the standard compressible Navier-Stokes (CNS) equations and was studied by Matsumura and Nishida [9] for the existence of smooth small solutions.

Obviously, $(n, u, T) = (1, 0, 1)$ is a solution for (2.7). To show existence of small solutions near $(1, 0, 1)$, we consider $(\rho, u, \theta) = (n - 1, u, T - 1)$ and transform (2.7) into the following quantum hydrodynamic equation

$$\begin{cases} \partial_t \rho + u \cdot \nabla \rho + (1 + \rho)\text{div} u = 0, & (1.2a) \\ \partial_t u - \dfrac{\mu}{\rho + 1}\Delta u - \dfrac{\mu + \delta}{\rho + 1}\nabla \text{div} u = -u \cdot \nabla u - \nabla \theta - \dfrac{\theta + 1}{\rho + 1}\nabla \rho \\ \quad + \dfrac{\hbar^2}{12}\dfrac{\Delta \nabla \rho}{\rho + 1} - \dfrac{\hbar^2}{3}\dfrac{\text{div}(\nabla \sqrt{\rho + 1} \otimes \nabla \sqrt{\rho + 1})}{\rho + 1}, & (1.2b) \\ \partial_t \theta - \dfrac{2\kappa}{3(1 + \rho)}\Delta \theta = -u \cdot \nabla \theta - \dfrac{2}{3}(\theta + 1)\nabla \cdot u \\ \quad + \dfrac{\hbar^2}{36(1 + \rho)}\text{div}((1 + \rho)\Delta u) + \dfrac{2}{3(1 + \rho)}\left\{\dfrac{\mu}{2}|\nabla u + (\nabla u)^T|^2 + \delta(\text{div} u)^2\right\}. & (1.2c) \end{cases}$$

Recently, we obtained the following global existence result of small smooth solutions in [12].

Theorem 1.1 *Let $s \geq 3$ be an integer. Assume that $(\rho_0, u_0, \theta_0) \in X^s$ for $s \geq 3$. There exist positive constants $\hbar_0, \varepsilon_0, C_0 > 0$ and $\nu_0 > 0$ such that if $\hbar \leq \hbar_0$ and*

$$|||(\rho_0, u_0, \theta_0)|||_{X^s} \leq \varepsilon_0,$$

then the Cauchy problem (1.2) has a unique solution (ρ, u, θ) globally in time satisfying

$$|||(\rho, u, \theta)(t)|||_{X^s}^2 + \nu_0 \int_0^t \sum_{k=1}^{s+1} \|\nabla^k(\rho, u, \theta, \hbar\nabla u, \hbar\nabla\rho)(\tau)\|^2 d\tau$$
$$\leqslant C_0 |||(\rho_0, u_0, \theta_0)|||_{X^s}^2, \quad \forall t \geqslant 0.$$

In the above theorem, $\|\cdot\|$ is the L^2-norm and $|||\cdot|||_{X^s}$ is reductively given by

$$|||(\rho, u, \theta)|||_{X^0}^2 := \|(\rho, u, \theta)\|_{L^2}^2 + \|\nabla\rho\|_{L^2}^2 + \|(\hbar\nabla\rho, \hbar\nabla u)\|_{L^2}^2 + \|\hbar^2 \Delta\rho\|_{L^2}^2,$$
$$|||(\rho, u, \theta)|||_{X^s}^2 := |||(\rho, u, \theta)|||_{X^{s-1}}^2 + |||(\nabla^k\rho, \nabla^k u, \nabla^k\theta)|||_{X^0}^2.$$

In this paper, we consider the decay of the solutions constructed in Theorem 1.1 to the constant solution $(\rho, u, \theta) = (0, 0, 0)$. The main result is stated in the following

Theorem 1.2 *Let ε_0 be the constant defined in Theorem 1.1 and $s \geqslant 5$. There exist constants $\varepsilon_1 \in (0, \varepsilon_0)$ and $C > 0$ such that for any $\varepsilon \leqslant \varepsilon_1$, if*

$$|||(\rho_0, u_0, \theta_0)|||_{X^s} \leqslant \varepsilon$$

and $(\rho_0, u_0, \theta_0) \in L^1(\mathbb{R}^3)$, then the solution (ρ, u, θ) in Theorem 1.1 enjoys the estimates

$$|||\nabla(\rho, u, \theta)(t)|||_{X^s} \leqslant C(1+t)^{-5/4}, \quad \forall t \geqslant 0 \tag{1.3}$$

and

$$\|(\rho, u, \theta)(t)\|_{L^p} \leqslant C(1+t)^{-\frac{3}{2}(1-\frac{1}{p})}, \quad 2 \leqslant p \leqslant 6, t \geqslant 0.$$

The decay rate is optimal in the sense that it is consistent with the linear decay rates in Theorem 2.1. To the best of our knowledge, there is no decay estimates for the compressible quantum Navier-Stokes equations (1.2), although there are some decay estimates to related models. For example, Matsumura and Nishida [8] studied the decay for the full Navier-Stokes equations and recently Wang and Tan [13] studied the optimal decay rates for the compressible fluid models of Korteweg type. See also [3, 6, 7, 10, 11] and the references therein to list only a few.

In the next section, we consider the linear decay estimates for (1.2), and in the last section, we consider the nonlinear decay estimates and complete the proof of Theorem 1.2.

2 Linear decay estimates

In this section, we consider the $L_q - L_p$ estimates of solutions to the Cauchy problem to the linearized system in $\mathbb{R}^3$.

2.1 Reformulation

We rewrite (1.2) in the following

$$\begin{cases} \partial_t \rho + \mathrm{div} u = F_1, & \text{(2.1a)} \\ \partial_t u - \mu \Delta u - (\mu + \delta) \nabla \mathrm{div} u + \nabla \theta + \nabla \rho - \dfrac{\hbar^2}{12} \Delta \nabla \rho = F_2, & \text{(2.1b)} \\ \partial_t \theta - \kappa \Delta \theta + \omega \nabla \cdot u - \dfrac{\hbar^2}{36} \mathrm{div} \Delta u = F_3, & \text{(2.1c)} \end{cases}$$

where

$$F_1 = -\mathrm{div}(\rho u),$$

$$F_2 = -u \cdot \nabla u - \frac{\mu \rho}{1+\rho} \Delta u - \frac{(\mu+\delta)\rho}{1+\rho} \nabla \mathrm{div} u - \frac{\theta + \rho}{1+\rho} \nabla \rho$$
$$+ \frac{\hbar^2}{12} \frac{\rho}{1+\rho} \nabla \Delta \rho - \frac{\hbar^2}{3} \frac{\mathrm{div}(\nabla \sqrt{\rho+1} \otimes \nabla \sqrt{\rho+1})}{\rho+1}, \qquad (2.2)$$

$$F_3 = -u \cdot \nabla \theta + \frac{2\kappa}{3} \frac{\rho \Delta \theta}{1+\rho} + \frac{\hbar^2 \nabla \rho \Delta u}{36(1+\rho)}$$
$$+ \frac{2}{3(1+\rho)} \left\{ \frac{\mu}{2} |\nabla u + (\nabla u)^T|^2 + \delta (\mathrm{div} u)^2 \right\}.$$

Take a linear transform of parameters

$$(\rho, u, \theta) = (\tilde{\rho}, \tilde{u}, \sqrt{2/3}\tilde{\theta}),$$

and set

$$\gamma = 1, \quad \omega = \sqrt{2/3}, \quad \tilde{\hbar}^2 = \hbar^2/12, \quad \tilde{\kappa} = 2\kappa/3.$$

By omitting $\tilde{\cdot}$, we rewrite the system (2.1) as

$$\begin{cases} \rho_t + \gamma \nabla \cdot u = F_1, \\ \partial_t u - \mu \Delta u - (\mu+\delta) \nabla \mathrm{div} u + \omega \nabla \theta + \gamma \nabla \rho - 3\hbar^2 \Delta \nabla \rho = F_2, \\ \partial_t \theta - \kappa \Delta \theta + \omega \nabla \cdot u - \dfrac{\hbar^2}{\omega} \mathrm{div} \Delta u = F_3, \end{cases} \qquad (2.3)$$

with initial data (ρ_0, u_0, θ_0).

Let $\mathbb{A}$ to be the following matrix-valued differential operators

$$\mathbb{A} = \begin{pmatrix} 0 & \gamma \mathrm{div} & 0 \\ (\gamma - 3\hbar^2 \Delta)\nabla & -\mu\Delta - (\mu+\delta)\nabla\mathrm{div} & \omega\nabla \\ 0 & \omega\mathrm{div} - \dfrac{\hbar^2}{\omega}\Delta\mathrm{div} & -\kappa\Delta \end{pmatrix}.$$

The system (2.3) can be rewritten in an abstract form

$$\mathbb{U}_t + \mathbb{A}\mathbb{U} = \mathbb{F} \text{ in } [0, \infty) \times \mathbb{R}^3, \quad \mathbb{U}(0) = \mathbb{U}_0 \text{ in } \mathbb{R}^3, \qquad (2.4)$$

where $\mathbb{U} = (\rho, u, \theta)$ and $\mathbb{F} = (F_1, F_2, F_3)$. Then the corresponding semigroup generated by the linear operator $-\mathbb{A}$ is $\mathbb{E}(t) = e^{-t\mathbb{A}}$ for $t \geqslant 0$. Then the problem (2.4) can be rewritten in the integral form

$$\mathbb{U}(t) = \mathbb{E}(t)\mathbb{U}_0 + \int_0^t \mathbb{E}(t-s)\mathbb{F}(\mathbb{U}(s))ds, \quad t \geqslant 0. \tag{2.5}$$

2.2 Linear decay estimates

To consider the decay of the linearized system, we consider the following abstract homogeneous equation

$$\mathbb{U}_t + \mathbb{A}\mathbb{U} = 0 \text{ in } [0, \infty) \times \mathbb{R}^3, \quad \mathbb{U}(0) = \mathbb{F} \text{ in } \mathbb{R}^3, \tag{2.6}$$

where $\mathbb{U} = (\rho, u, \theta)$ and $\mathbb{F} = (f_1, f_2, f_3)$. Note that $\mathbb{F}$ here is different from that in (2.4).

By taking Fourier transform of (2.6) w.r.t. space variable and then solving the ODE, we obtain

$$\mathbb{U}(t) = \mathbb{E}(t)\mathbb{F} = \mathcal{F}^{-1}(e^{-t\widehat{\mathbb{A}}(\xi)}\widehat{\mathbb{F}}(\xi)), \tag{2.7}$$

where $\mathcal{F}^{-1}$ denotes the Fourier inverse transform and $\widehat{\mathbb{A}}(\xi)$ is the 5×5 matrix of the form

$$\widehat{\mathbb{A}}(\xi) = \begin{pmatrix} 0 & -i\gamma\xi_k & 0 \\ -i(\gamma + 3\hbar^2|\xi|^2)\xi_j & \delta_{jk}\mu|\xi|^2 + (\mu+\delta)\xi_j\xi_k & -i\omega\xi_j \\ 0 & -i(\omega^2 + i\hbar^2|\xi|^2)\xi_k/\omega & \kappa|\xi|^2 \end{pmatrix},$$

where $\xi = (\xi_1, \xi_2, \xi_3)^{\mathrm{T}}$, $i = \sqrt{-1}$ and $\delta_{jk} = 0$ when $k \neq j$ and $\delta_{jk} = 1$ when $k = j$.

Theorem 2.1 *Let $\mathbb{E}(t)$ be the solution operator of (2.6) defined by (2.7). Then, we have the decomposition*

$$\mathbb{E}(t)\mathbb{F} = \mathbb{E}_0(t)\mathbb{F} + \mathbb{E}_\infty(t)\mathbb{F},$$

where $\mathbb{E}_0(t)$ and $\mathbb{E}_\infty(t)$ have the following properties:

(1) $\forall m, l \geqslant 0$ integers,

$$\|\partial_t^m \partial_x^l \mathbb{E}_0(t)\mathbb{F}\|_{L^p} \leqslant C(m,l,p,q) t^{-\frac{3}{2}(\frac{1}{q} - \frac{1}{p}) - \frac{m+l}{2}} \|\mathbb{F}\|_{L^q}, \quad \forall t \geqslant 1,$$

where $1 \leqslant q \leqslant 2 \leqslant p \leqslant \infty$, and

$$\|\partial_t^m \partial_x^l \mathbb{E}_0(t)\mathbb{F}\|_{L^p} \leqslant C(m,l,p,q)\|\mathbb{F}\|_{L^q}, \quad 0 < \forall t \leqslant 2,$$

where $1 \leqslant q \leqslant p \leqslant \infty$ and $(p,q) \neq (1,1), (\infty, \infty)$.

(2) Set $(l)^+ = l$ if $l \geq 0$ and 0 if $l < 0$. For any $1 < p < \infty$, there exists $c > 0$ such that $\forall l, n \geq 0$ integers,

$$\left\|\partial_t^m \partial_x^l (\mathbb{I} - \mathbb{P})\Psi_\infty(t)\mathbb{F}\right\|_{L^p} \leq C(m,l,n,p) e^{-ct} t^{-\frac{n}{2}} \|\mathbb{F}\|_{(2m+l+2-n)^+, p},$$
$$\left\|\partial_t^m \partial_x^l \mathbb{P}\Psi_\infty(t)\mathbb{F}\right\|_{L^p} \leq C(m,l,n,p) e^{-ct} t^{-\frac{n}{2}} \|\mathbb{F}\|_{(2m+l+4-n)^+, p},$$
(2.8)

for all $t > 0$.

(3) Let $1 < p < \infty$, then

$$\left\|\partial_t^m \partial_x^l (\mathbb{I} - \mathbb{P})\Psi_\infty(t)\mathbb{F}\right\|_{L^\infty} \leq C(m,l,n,p) e^{-ct} t^{-(\frac{n}{2}+\frac{3}{2p})} \|\mathbb{F}\|_{(2m+l+2-n)^+, p},$$
$$\left\|\partial_t^m \partial_x^l \mathbb{P}\Psi_\infty(t)\mathbb{F}\right\|_{L^\infty} \leq C(m,l,n,p) e^{-ct} t^{-(\frac{n}{2}+\frac{3}{2p})} \|\mathbb{F}\|_{(2m+l+4-n)^+, p},$$
(2.9)

for a suitable constant $c > 0$, for all $t > 0$.

To prove Theorem 2.1, we shall first consider the following stationary problem in $\mathbb{R}^3$ with a complex parameter λ,

$$(\lambda + \mathbb{A})\mathbb{U} = \mathbb{F}, \quad \text{in } \mathbb{R}^3. \tag{2.10}$$

By taking Fourier transform, we obtain

$$\mathbb{U} = \mathcal{F}^{-1}\left\{\left[\lambda + \widehat{\mathbb{A}}(\xi)\right]^{-1} \mathbb{F}\right\},$$

where

$$[\lambda + \widehat{\mathbb{A}}(\xi)]^{-1} = \left\{\det\left[\lambda + \widehat{\mathbb{A}}(\xi)\right]\right\}^{-1} \widetilde{\mathbb{A}}(\lambda, \xi),$$

$$\det\left[\lambda + \widehat{\mathbb{A}}(\xi)\right] = (\lambda + \mu|\xi|^2)^2 F(\lambda, |\xi|),$$

$$F(\lambda, |\xi|) = \lambda^3 + (2\mu + \delta + \kappa)|\xi|^2 \lambda^2 + \left[(2\mu + \delta)\kappa|\xi|^2 + \gamma(\gamma + 3\hbar^2|\xi|^2)\right.$$
$$\left. + (\omega^2 + \hbar^2|\xi|^2)\right]|\xi|^2 \lambda + \gamma(\gamma + 3\hbar^2|\xi|^2)\kappa|\xi|^4,$$

and $\widetilde{\mathbb{A}}(\lambda, \xi) = (\tilde{a}_{k,j}(\lambda, \xi))$ is the 5×5 matrix and the components are

$$\tilde{a}_{1,1} = (\lambda + \mu|\xi|^2)^2 \left\{\lambda^2 + (2\mu + \delta + \kappa)|\xi|^2 \lambda + \left[(2\mu + \delta)\kappa|\xi|^2 + (\omega^2 + \hbar^2|\xi|^2)\right]|\xi|^2\right\},$$

$$\tilde{a}_{5,1} = -(\lambda + \mu|\xi|^2)^2 (\gamma + 3|\xi|^2) (\omega^2 + \hbar^2|\xi|^2) |\xi|^2/\omega,$$

$$\tilde{a}_{1,5} = -\gamma\omega(\lambda + \mu|\xi|^2)^2 |\xi|^2,$$

$$\tilde{a}_{j,1} = -i(\gamma + 3\hbar^2|\xi|^2)(\lambda + \mu|\xi|^2)^2(\lambda + \kappa|\xi|^2)\xi_{j-1},$$

$$\tilde{a}_{1,j} = -i\gamma(\lambda + \mu|\xi|^2)^2(\lambda + \kappa|\xi|^2)\xi_{j-1},$$

$$\tilde{a}_{j,5} = -i\omega\lambda(\lambda + \mu|\xi|^2)^2 \xi_{j-1},$$

$$\tilde{a}_{5,j} = -i(\omega^2 + \hbar^2|\xi|^2)\lambda(\lambda + \mu|\xi|^2)^2 \xi_{j-1}/\omega,$$

$$\tilde{a}_{5,5} = (\lambda + \mu|\xi|^2)^2 \left\{\lambda^2 + (2\mu + \delta)|\xi|^2 \lambda + \gamma(\gamma + 3\hbar^2|\xi|^2)|\xi|^2\right\},$$

$$\tilde{a}_{k,j} = (\lambda + \mu|\xi|^2)\Big\{\lambda(\lambda + \mu|\xi|^2)(\lambda + \kappa|\xi|^2)\delta_{kj} + (\delta_{kj} - \xi_{k-1}\xi_{j-1})\big[(\mu + \delta)\lambda^2$$
$$+ (\beta\kappa|\xi|^2 + (\omega^2 + \hbar^2|\xi|^2) + \gamma(\gamma + 3\hbar^2|\xi|^2))\lambda + \gamma\kappa(\gamma + 3\hbar^2|\xi|^2)|\xi|^2\big]\Big\},$$

where $k, j = 2, 3, 4$.

Lemma 2.1 (1) Let $\{\lambda_j(\xi)\}_{j=1}^5$ be the roots of $\det[\lambda + \widehat{\mathbb{A}}(\xi)] = 0$, where $\lambda_4 = \lambda_5 = -\mu|\xi|^2$. Then for $\lambda_j(\xi), j = 1, 2, 3$, we have the following assertions.

(1.a) There exists a positive constant $r_1 > 0$ such that $\lambda_j(\xi)$ has a Taylor series expansion for $|\xi| < r_1$ as follows

$$\lambda_1(\xi) = \overline{\lambda_2(\xi)} = i(\gamma^2 + \omega^2)^{1/2}|\xi| - \frac{(\gamma^2 + \omega^2)(2\mu + \delta) + \omega^2\kappa}{2(\gamma^2 + \omega^2)}|\xi|^2 + O(|\xi|^3),$$

$$\lambda_3(\xi) = -\frac{\gamma^2\kappa}{\gamma^2 + \omega^2}|\xi|^2 + (\cdots)|\xi|^4 + O(|\xi|^6),$$

where $\lambda_1(\xi)$ and $\overline{\lambda_2(\xi)}$ are complex conjugate and $\lambda_3(\xi)$ is a real number.

(1.b) Similarly, there exists a positive constant $r_2(> r_1)$ such that $\lambda_j(\xi)$ has a Laurent series expansion for $|\xi| > r_2$ as follows. Let $\hbar = o(1)$, then we have

$$\lambda_1(\xi) = -\alpha_{11}|\xi|^2 + \beta_{11}|\xi| + \alpha_{12} + \beta_{12}|\xi|^{-1} + \alpha_{13}|\xi|^{-2} + \cdots,$$
$$\lambda_2(\xi) = -\alpha_{21}|\xi|^2 + \beta_{21}|\xi| + \alpha_{22} + \beta_{22}|\xi|^{-1} + \alpha_{23}|\xi|^{-2} + \cdots, \quad (2.11)$$
$$\lambda_3(\xi) = -\alpha_{31}|\xi|^2 + \alpha_{32} + \alpha_{33}|\xi|^{-2} + \cdots,$$

where $\alpha_{31} > 0$ is real depending only on the parameters and either (i) $\alpha_{11} > 0$ and $\alpha_{21} > 0$ are different real numbers or (ii) α_{11} and α_{21} are complex conjugate and $\Re\alpha_{11} = \Re\alpha_{21} > 0$. In the first case, $\beta_{ij} = 0$ for all $i, j \geqslant 1$.

(2) The matrix exponential has the spectral resolution

$$e^{-t\widehat{\mathbb{A}}(\xi)} = \sum_{j=1}^{4} e^{t\lambda_j(\xi)}\mathbb{P}_j(\xi), \quad \mathbb{P}_j(\xi) = \prod_{i \neq j} \frac{\lambda_i I + \widehat{\mathbb{A}}(\xi)}{\lambda_i - \lambda_j},$$

for all $|\xi| > 0$ except for at most four points of $|\xi| > 0$.

(3) For any $0 < R_1 < R_2 < \infty$, we have

$$\left|e^{-t\widehat{\mathbb{A}}(\xi)}\right| \leqslant C(r_1, r_2)(1 + t)^3 e^{-C_2(r_1, r_2)t}, \quad R_1 \leqslant |\xi| \leqslant R_2,$$

where $C_2 > 0$.

Sketched proof We need only to consider the proof of (1.b). Let $\lambda(\xi)$ be an eigenvalue written in Laurent series for large ξ. The coefficient A before $|\xi|^2$ then should satisfy the following algebraic equation

$$A^3 + (2\mu + \delta + \kappa)A^2 + [(2\mu + \delta)\kappa + (3\hbar^2\gamma + \hbar^2)]A + 3\hbar^2\gamma\kappa = 0. \quad (2.12)$$

It is difficult to give exact expressions for A. But since $\hbar \ll 1$, there is at least one negative solution $A_3 = -\alpha_{31} < 0$. For this, one can show that the eigenvalue has the form $\lambda_3(\xi)$ in (2.11). For the other two roots of (2.12), either A_1 and A_2 are different real numbers, or A_1 and A_2 are complex conjugate (possibly with zero imaginary part). When A_1 and A_2 are different real numbers, the eigenvalues $\lambda_1(\xi)$ and $\lambda_2(\xi)$ both have the form

$$\lambda_j(\xi) = A_j|\xi|^2 + O(1).$$

When A_1 and A_2 are complex conjugate, the eigenvalues $\lambda_1(\xi)$ and $\lambda_2(\xi)$ both have the form

$$\lambda_1(\xi) = \overline{\lambda_2(\xi)} = A_1|\xi|^2 + O(|\xi|).$$

Since $\hbar \ll 1$, the eigenvalues are small perturbations of the eigenvalues in the case when $\hbar = 0$ and hence $\Re A_j < 0$ for $j = 1, 2$. We note that when $\hbar = 0$, the roots of (2.12) can be exactly solved to be $A_1 = -(2\mu + \delta)$, $A_2 = -\kappa$ and $A_3 = 0$ (see [6]). □

Proof of Theorem 2.1 Let $\varphi_0(\xi)$ be a function in $C_0^\infty(\mathbb{R}^3)$ such that $\varphi_0(\xi) = 1$ for $|\xi| \leqslant r_1/2$ and $\varphi_0(\xi) = 0$ for $|\xi| \geqslant r_1$ and set

$$\Psi_0(t)\mathbb{F}(x) = \sum_{j=1}^{4} \mathcal{F}^{-1}\left[e^{t\lambda_j(\xi)}\varphi_0(\xi)\mathbb{P}_j(\xi)\widehat{\mathbb{F}}(\xi)\right](x).$$

Since in the lower mode regime, the eigenvalues as well as the projectors behaves exactly as those in [6], one obtains by the same treatment that

$$\begin{cases} \|\partial_t^m \partial_x^l \Psi_0(t)\mathbb{F}\|_{L^p} \leqslant Ct^{-\frac{3}{2}(\frac{1}{q}-\frac{1}{p})-\frac{m+l}{2}}\|\mathbb{F}\|_{L^q}, & t \geqslant 1,\ 1 \leqslant q \leqslant 2 \leqslant p \leqslant \infty \\ \|\partial_t^m \partial_x^l \Psi_0(t)\mathbb{F}\|_{L^p} \leqslant C\|\mathbb{F}\|_{L^q}, & \begin{cases} 0 < t < 1,\ 1 \leqslant q \leqslant p \leqslant \infty, \\ (p,q) \neq (1,1), (\infty, \infty). \end{cases} \end{cases} \quad (2.13)$$

Next, we consider the high frequency part for large $|\xi|$. Let

$$\Psi_\infty(t)\mathbb{F} = \sum_{j=1}^{4} \mathcal{F}^{-1}\left[e^{t\lambda_j(\xi)}\varphi_\infty(\xi)\mathbb{P}_j(\xi)\widehat{\mathbb{F}}(\xi)\right](x),$$

where φ_∞ is a function in $C^\infty(\mathbb{R}^3)$ such that $\varphi_\infty(\xi) = 1$ for $|\xi| \geqslant 3r_2$ and $\varphi_\infty(\xi) = 0$ for $|\xi| \leqslant 2r_2$. Set

$$[H_l^j(t)f](x) = \sum_{j=1}^{4} \mathcal{F}^{-1}\left[e^{t\lambda_j(\xi)}g_l(\xi)\varphi_\infty(\xi)\hat{f}(\xi)\right](x), \quad j = 1,2,3,4.$$

We write

$$\hat{\rho}(t,\xi) = \sum_{k=1}^{3}\sum_{j=1}^{5} e^{t\lambda_k(\xi)} A_{kj}^1(\xi)\varphi_\infty(\xi)\hat{f}_j(\xi). \quad (2.14)$$

Let us first consider that α_{11} and α_{21} are positive real numbers in (2.11). In this case, we obtain

$$|\xi^\alpha \partial_\xi^\alpha A_{k1}^1(\xi)| \leqslant C(\alpha)|\xi|^2, \quad j=1, \forall k,$$
$$|\xi^\alpha \partial_\xi^\alpha A_{kj}^1(\xi)| \leqslant C(\alpha)|\xi|, \quad j=2,3,4, \forall k,$$
$$|\xi^\alpha \partial_\xi^\alpha A_{k5}^1(\xi)| \leqslant C(\alpha), \quad j=5, \forall k.$$

First, we have

$$\left\|\partial_t^m \partial_x^l \mathcal{F}^{-1}\left[e^{t\lambda_k(\xi)} A_{kj}^1(\xi) \varphi_\infty(\xi) \hat{f}_j(\xi)\right]\right\|_{L^p} \leqslant C(m,l,n,p) t^{-n/2} e^{-ct} \|f_j\|_{(2m+l+2-n)^+, p},$$

for $t > 0$, $1 < p < \infty$, $\forall k \in \{1,2,3\}$ and $\forall j \in \{1,2,3,4,5\}$. For the L^∞-norm, we have

$$\left\|\partial_t^m \partial_x^l \mathcal{F}^{-1}\left[e^{t\lambda_k(\xi)} A_{kj}^1(\xi) \varphi_\infty(\xi) \hat{f}_j(\xi)\right]\right\|_{L^\infty}$$
$$\leqslant C(m,l,n,p) t^{-(\frac{3}{2p}+\frac{n}{2})} e^{-ct} \|f_j\|_{(2m+l+1-n)^+, p},$$

for $t > 0$, $1 < p < \infty$, $\forall k \in \{1,2,3\}$ and $\forall j \in \{1,2,3,4,5\}$. Summarizing these estimates, we have proved the following

$$\begin{aligned}
&\left\|\partial_t^m \partial_x^l (\mathbb{I}-\mathbb{P})\Psi_\infty(t)\mathbb{F}\right\|_{L^p} \leqslant C(m,l,n,p) e^{-ct} t^{-\frac{n}{2}} \|\mathbb{F}\|_{(2m+l+2-n)^+, p}, \\
&\left\|\partial_t^m \partial_x^l (\mathbb{I}-\mathbb{P})\Psi_\infty(t)\mathbb{F}\right\|_{L^\infty} \leqslant C(m,l,n,p) e^{-ct} t^{-(\frac{n}{2}+\frac{3}{2p})} \|\mathbb{F}\|_{(2m+l+2-n)^+, p},
\end{aligned} \quad (2.15)$$

where $(\mathbb{I}-\mathbb{P})\mathbb{F} = (f_1, 0, 0, 0, 0)^T$ and $\mathbb{P}\mathbb{F} = (0, f_2, f_3, f_4, f_5)^T$.

Now, we treat the estimates of $\mathbb{P}\mathbb{U}$. We write

$$\hat{v}_i(t,\xi) = \sum_{k=1}^3 \sum_{j=1}^5 e^{t\lambda_k(\xi)} A_{kj}^{i+1}(\xi) \varphi_\infty(\xi) \hat{f}_j(\xi) + \sum_{j=2}^4 e^{t\lambda_4(\xi)} A_{4j}^{i+1}(\xi) \varphi_\infty(\xi) \hat{f}_j(\xi),$$
$$\hat{\theta}_i(t,\xi) = \sum_{k=1}^3 \sum_{j=1}^5 e^{t\lambda_k(\xi)} A_{kj}^5(\xi) \varphi_\infty(\xi) \hat{f}_j(\xi). \quad (2.16)$$

It can be estimated that

$$|\xi^\alpha \partial_\xi^\alpha A_{kj}^{i+1}(\xi)| \leqslant C(\alpha)|\xi|^3, \quad i=1,2,3, j=1, \forall k;$$
$$|\xi^\alpha \partial_\xi^\alpha A_{kj}^{i+1}(\xi)| \leqslant C(\alpha), \quad i=1,2,3, j=2,3,4,5, \forall k;$$
$$|\xi^\alpha \partial_\xi^\alpha A_{kj}^5(\xi)| \leqslant C(\alpha)|\xi|^3, \quad j=1, \forall k;$$
$$|\xi^\alpha \partial_\xi^\alpha A_{kj}^5(\xi)| \leqslant C(\alpha)|\xi|^4, \quad j=2,3,4, \forall k;$$
$$|\xi^\alpha \partial_\xi^\alpha A_{kj}^5(\xi)| \leqslant C(\alpha)|\xi|, \quad j=5, \forall k.$$

Therefore we have proved the following

$$\left\|\partial_t^m \partial_x^l \mathbb{P}\Psi_\infty(t)\mathbb{F}\right\|_{L^p} \leqslant C(m,l,n,p)e^{-ct}t^{-\frac{n}{2}}\|\mathbb{F}\|_{(2m+l+4-n)^+,p},$$
$$\left\|\partial_t^m \partial_x^l \mathbb{P}\Psi_\infty(t)\mathbb{F}\right\|_{L^\infty} \leqslant C(m,l,n,p)e^{-ct}t^{-(\frac{n}{2}+\frac{3}{2p})}\|\mathbb{F}\|_{(2m+l+4-n)^+,p},$$
(2.17)

where $(\mathbb{I} - \mathbb{P})\mathbb{F} = (f_1, 0, 0, 0, 0)^T$ and $\mathbb{P}\mathbb{F} = (0, f_2, f_3, f_4, f_5)^T$.

Choose $\varphi_M(\xi) \in C_0^\infty$ such that $\varphi_0(\xi) + \varphi_M(\xi) + \varphi_\infty(\xi) = 1$ for all $\xi \in \mathbb{R}^3$ and set

$$\mathbb{E}_0(t)\mathbb{F} = \Psi_0(t)\mathbb{F},$$
$$\mathbb{E}_\infty(t)\mathbb{F} = \Psi_\infty \mathbb{F} + \mathcal{F}^{-1}\left[e^{-t\hat{\mathbb{A}}(\xi)}\varphi_M(\xi)\hat{\mathbb{F}}(\xi)\right].$$

Then it is easy to see that

$$\|\partial_t^m \partial_x^l \mathcal{F}^{-1}\left[e^{-t\hat{\mathbb{A}}(\xi)}\varphi_M(\xi)\hat{\mathbb{F}}(\xi)\right]\|_{L^p} \leqslant C(p,m,l)e^{-ct}\|\mathbb{F}\|_{L^p},$$

for a suitable constant $c > 0$. Combining these estimates, we complete the proof of Theorem 2.1. □

3 Nonlinear decay estimates

In this section, we shall prove two basic inequalities.

Lemma 3.1 *Let $\mathbb{U}$ be the solution to the problem (2.4). Under the assumptions of Theorem 1.2, we have*

$$\|\nabla \mathbb{U}\|_{L^2} \leqslant CK_0(1+t)^{-5/4} + C\varepsilon \int_0^t (1+t-\tau)^{-5/4} \||\nabla \mathbb{U}\||_{X^s} d\tau,$$

where $K_0 = \|(\rho_0, u_0, \theta_0)\|_{H^5 \cap L^1}$ is finite.

Proof From Theorem 2.1 and (2.5), we have

$$\|\nabla \mathbb{U}(t)\| \leqslant CE_0(1+t)^{-\frac{5}{4}}$$
$$+ C \int_0^t (1+t-\tau)^{-\frac{5}{4}}(\|\mathbb{F}(\mathbb{U})(s)\|_{L^1} + \|\mathbb{F}(\mathbb{U})(s)\|_{H^5})d\tau, \quad (3.1)$$

where $\mathbb{F}(\mathbb{U})$ is given in (2.2). It follows from (2.2), Hölder inequality, Theorem 1.1 and Sobolev inequalities that

$$\|\mathbb{F}(\mathbb{U})\|_{L^1} \leqslant C\|(\rho, u)\|\|\nabla(\rho, u)\| + \|(\rho, \theta)\|\|\nabla \rho\| + \hbar^2\|\rho\|\|\nabla \Delta \rho\|$$
$$+ \hbar^2\|\nabla \rho\|\|\nabla^2 \rho\| + \hbar^2\|\nabla \rho\|\|\nabla^2 \rho\|_{L^4}^2 + \|u\|\|\nabla \theta\| + \|\rho\|\|\Delta \theta\|$$
$$+ \hbar^2\|\nabla \rho\|\|\Delta u\| + \|\nabla u\|\|\nabla u\| \leqslant C\varepsilon\||\nabla(\rho, u, \theta)|\|_{X^1} \quad (3.2)$$

and similarly,
$$\|\mathbb{F}(\mathbb{U})(s)\|_{H^5} \leqslant C\varepsilon \||\nabla(\rho,u,\theta)\||_{X^6}. \qquad (3.3)$$

Combining these inequalities together completes the proof. □

Lemma 3.2 *Let $\mathbb{U}$ be the solution to the problem (2.4). Under the assumptions of Theorem 1.2, we have*
$$\frac{dM(t)}{dt} + \nu\||\nabla^2\mathbb{U}(t)\||_{X^s}^2 \leqslant C\varepsilon\|\nabla\mathbb{U}(t)\|^2, \qquad (3.4)$$
where $M(t)$ is equivalent to $\||\nabla\mathbb{U}(t)\||_{X^s}^2$, that is, there exists a positive constant C_2 such that
$$C_2^{-1}\||\nabla\mathbb{U}(t)\||_{X^s}^2 \leqslant M(t) \leqslant C_2\||\nabla\mathbb{U}(t)\||_{X^s}^2, \quad \forall t \geqslant 0. \qquad (3.5)$$

Proof The proof is similar to Proposition 3.2 in [12], and hence omitted here for simplicity. We only note that here, we require $s \geqslant 6$ while $s = 3$ in Proposition 3.2 in [12]. □

Proof of Theorem 1.2 Note that from (3.4), we have
$$\frac{dM(t)}{dt} + DM(t) \leqslant C\|\nabla\mathbb{U}(t)\|^2, \qquad (3.6)$$
for some constant $D > 0$. Define $N(t) = \sup_{0 \leqslant \tau \leqslant t}(1+\tau)^{5/2}M(s)$, then
$$\||\nabla\mathbb{U}(\tau)\||_{X^s} \leqslant C\sqrt{M(\tau)} \leqslant C(1+t)^{-5/4}\sqrt{N(t)}, \quad 0 \leqslant \tau \leqslant t. \qquad (3.7)$$
From Lemma 3.1, we obtain
$$\begin{aligned}\|\nabla\mathbb{U}\|_{L^2} &\leqslant CK_0(1+t)^{-5/4} + C\varepsilon\int_0^t (1+t-\tau)^{-5/4}(1+\tau)^{-5/4}d\tau\sqrt{N(t)}\\ &\leqslant C(1+t)^{-5/4}\left[K_0 + \varepsilon\sqrt{N(t)}\right],\end{aligned} \qquad (3.8)$$
where we have used the following inequality [3]
$$\int_0^t (1+t-\tau)^{-r_1}(1+\tau)^{-r_2}d\tau \leqslant C(r_1,r_2)(1+t)^{-r_2}, \qquad (3.9)$$
where $r_1 > 1$ and $r_2 \in [0, r_1]$ and $C(r_1,r_2) = 2^{r_2+1}/(r_1-1)$. From Lemma 3.2 and Gronwall inequality, we have
$$\begin{aligned}M(t) &\leqslant M(0)e^{-\nu t} + C\int_0^t e^{-\nu(t-\tau)}\|\nabla\mathbb{U}(\tau)\|^2 d\tau\\ &\leqslant M(0)(1+t)^{-5/2} + C\int_0^t (1+t-\tau)^{-5/2}(1+\tau)^{-5/2}d\tau\left[K_0^2 + \varepsilon^2 N(t)\right]\\ &\leqslant C(1+t)^{-5/2}\left[M(0) + K_0^2 + \varepsilon^2 N(t)\right].\end{aligned} \qquad (3.10)$$

By definition of $N(t)$, we obtain

$$N(t) \leqslant C\left[M(0) + K_0^2 + \varepsilon^2 N(t)\right]. \tag{3.11}$$

Choosing ε so small that $\varepsilon \leqslant 1/\sqrt{2C}$, then we have

$$N(t) \leqslant 2C(M(0) + K_0^2).$$

It then follows from definition of $N(t)$ that

$$|||\nabla \mathbb{U}|||_{X^s} \leqslant C\sqrt{M(t)} \leqslant C(1+t)^{-5/4} N(t) \leqslant C(1+t)^{-5/4}. \tag{3.12}$$

In particular,

$$\|\nabla \mathbb{U}\|_{L^2} \leqslant C(1+t)^{-5/4}. \tag{3.13}$$

Next, we consider the L^p-decay of $\mathbb{U}$. Denote $\sigma(p, q; l) = \dfrac{3}{2}\left(\dfrac{1}{q} - \dfrac{1}{p}\right) + \dfrac{l}{2}$. First of all, we have by letting $m = 0$ in Theorem 2.1 that

$$\|\nabla \mathbb{U}(t)\|_{L^p} \leqslant C(1+t)^{-\sigma(p,q;l)} \|\mathbb{U}_0\|_{L^q \cap H^{l+4}}, \quad l = 0, 1, 2, \cdots. \tag{3.14}$$

From (2.4), we have

$$\|\mathbb{U}(t)\|_{L^p} \leqslant CE_0(1+t)^{-\sigma(p,q;0)} + C\int_0^t (1+t-\tau)^{-\sigma(p,q;0)} \|\mathbb{F}(\tau)\|_{L^q \cap H^4} d\tau. \tag{3.15}$$

On the other hand, from (2.2), it is immediate that

$$\|\mathbb{F}(t)\|_{L^q \cap H^4} \leqslant C\varepsilon |||\nabla \mathbb{U}|||_{X^6}. \tag{3.16}$$

It then follows from (3.12) that

$$\begin{aligned}
\|\mathbb{U}(t)\|_{L^p} &\leqslant CE_0(1+t)^{-\sigma(p,q;0)} + C\int_0^t (1+t-\tau)^{-\sigma(p,q;0)} \|\mathbb{F}(\tau)\|_{L^q \cap H^4} d\tau \\
&\leqslant CE_0(1+t)^{-\sigma(p,q;0)} + C\varepsilon \int_0^t (1+t-\tau)^{-\sigma(p,q;0)} (1+\tau)^{-5/4} d\tau \\
&= CE_0(1+t)^{-\sigma(p,q;0)} + C\varepsilon \int_0^t (1+\tau)^{-\sigma(p,q;0)} (1+t-\tau)^{-5/4} d\tau \\
&\leqslant C(1+t)^{-\sigma(p,q;0)}, \tag{3.17}
\end{aligned}$$

where $2 \leqslant p \leqslant 6$ and $1 \leqslant q < 6/5$ to insure application of (3.9). This completes the proof. □

Acknowledgements This work is supported in part by NSFC (11471057), Natural Science Foundation Project of CQ CSTC (cstc2014jcyjA50020) and the Fundamental Research Funds for the Central Universities (Project No. 106112016CDJZR105501). This work was initiated during the visiting of X.P. to Institute of Applied Physics and Computational Mathematics and the first author thanks for their hospitality.

References

[1] M.G. Ancona and G.J. Iafrate. Quantum correction to the equation of state of an electron gas in semiconductor. Phys. Rev. B, 39, (1989)9536-9540.

[2] M.G. Ancona and H.F. Tiersten. Macroscopic physics of the silicon inversion layer. Phys. Rev. B, 35, (1987)7959-7965.

[3] R. Duan, S. Ukai, T. Yang and H. Zhao. Optimal convergence rates for the compressible Navier-Stokes equations with potential forces. Math. Models Meth. Appl. Sci., 17(3), (2007)737-758.

[4] C.L. Gardner. The quantum hydrodynamic model for semiconductor devices. SIAM J. Appl. Math., 54(2), (1994)409-427.

[5] A. Jungel, J.P. Milisic. Full compressible Navier-Stokes equations for quantum fluids: derivation and numerical solution. Kinetic and Related Models, 4(3), (2011)785-807.

[6] T. Kobayashi and Y. Shibata. Decay Estimates of Solutions for the Equations of Motion of Compressible Viscous and Heat-Conductive Gases in an Exterior Domain in $\mathbb{R}^3$. Commun. Math. Phys., 200, 621-659(1999).

[7] H. Li, A. Matsumura and G. Zhang. Optimal Decay Rate of the Compressible Navier-Stokes-Poisson System in $\mathbb{R}^3$. Arch. Rational Mech. Anal., 196, (2010)681-713.

[8] A. Matsumura and T. Nishida. The initial value problem for the equations of motion of compressible viscous and heat-conductive fluids. Proc. Japan Acad. Ser. A, 55, 337-342(1979).

[9] A. Matsumura and T. Nishida. The initial value problems for the equations of motion of viscous and heat-conductive gases. J. Math. Kyoto Univ., 20-1, 67-104(1980).

[10] G. Ponce. Global existence of small solutions to a class of nonlinear evolution equations. Nonl. Anal. TMA., 9, 339-418(1985).

[11] X. Pu and B. Guo. Global existence and convergence rates of smooth solutions for the full compressible MHD equations. Z. Angew. Math. Phys., 64, (2013)519-538.

[12] X. Pu and B. Guo. Global existence and semiclassical limit for quantum hydrodynamic equations with viscosity and heat conduction. arXiv:1504.05304, Submitted.

[13] Y. Wang, Z. Tan. Optimal decay rates for the compressible fluid models of Korteweg type. J. Math. Anal. Appl., 379, (2011)256-271.

[14] E. Wigner. On the quantum correction for thermodynamic equilibrium. Phys. Rev., 40, (1932)749-759.

Mixed-type Soliton Solutions for the N-coupled Higher-order Nonlinear Schrödinger Equations in Optical Fibers*

Guo Boling (郭柏灵) and Wang Yufeng (王玉风)

Abstract Under investigation in this paper are the N-coupled higher-order nonlinear Schrödinger (NLS) equations, which describe the simultaneous N nonlinear wave propagation in a fiber medium with higher order effects. Through the Hirota method, mixed-type one- and two-soliton solutions are obtained. Elastic and inelastic interactions between m-bright-n-dark $(m + n = N)$ solitons are derived through the asymptotic analysis. As an example, the oblique interactions and the bound states of the two-bright-one-dark and one-bright-two-dark solitons for 3-NLS equations are analyzed graphically.

Keywords N-coupled higher-order nonlinear Schrödinger equations; mixed-type soliton solutions; oblique interactions; bound states of solitons; elastic and inelastic interactions

1 Introduction

Optical soliton, which have been theoretically demonstrated [1] and experimentally observed [2] in optical fibers, have been researched intensively in the optical communication system [3, 4]. The interaction between the two components of optical fields can be described by the vector optical solitons, which are as the solutions for the coupled nonlinear Schrödinger (NLS) equations [5]. In contrast with the scalar solitons, vector solitons are more useful in physical applications, such as optical switch, logic gate and data storage [6, 7]. Due to the different profiles in each component, vector solitons may present the bright-bright, dark-dark and mixed-type (bright-dark) forms [8–13]. Shape-preserving (elastic) and shape-changing (inelastic) interactions

---

* Chaos Solitons Fractals, 2016, 93: 246–251. DOI:10.1016/j.chaos.2016.10.015.

have been obtained between the components of the bright-bright solitons [11–13]. In addition, as a direct interaction between two adjacent solitons with the same velocities, bound states of solitons have attracted much more interests [14–16]. Works have been done by certain methods, such as Hirota method [17], traveling-wave transformation method [18], Cole-Hopf transformation method [19] and asymptotic perturbation method [20].

To describe the propagation of optical solitons in the multiple-component of optical fields, the N-coupled NLS typed equations have been focused on [13, 21–23]. The mixed-type vector soliton solutions for the N-coupled NLS typed equations have displayed much more interesting results [13, 22, 23]. In this paper, we will investigate the following N-coupled higher-order NLS equations [24]

$$iq_{j,z} - \frac{k''}{2}q_{j,tt} + \beta \sum_{\rho=1}^{N}|q_\rho|^2 q_j - \frac{ik'''}{6}q_{j,ttt} + i\gamma \left(\sum_{\rho=1}^{N}|q_\rho|^2 q_j\right)_t + i\gamma_s \left(\sum_{\rho=1}^{N}|q_\rho|^2\right)_t q_j = 0,$$
$$j = 1, 2, \cdots, N, \quad \rho = 1, 2, \cdots, N, \quad (1.1)$$

which describe the simultaneous N nonlinear wave propagation in a fiber medium with higher order effects, where N is the positive number. q_j's represent the complex envelope amplitudes, t and z are the time and distance along the direction of propagation, k'' means the group velocity dispersion, β is the phase modulation parameter, k''', γ and γ_s are the higher order dispersion, self-steepening and delayed nonlinear process, respectively.

Dark-soliton solutions for Eqs. (1.1) under the conditions $k'' = \beta$ and $\gamma = -2\gamma_s = k'''$ have been given [24]. When $N = 1$, Eqs. (1.1) will be reduced to the Hirota equation with $\gamma + \gamma_s = 0$ [25]; The Sasa-Satsuma equation will be obtained with $\gamma + 2\gamma_s = 0$ [26]. Works have been done with $N = 2$ [27–30]: The bright and dark soliton solutions have been expressed with $k''\gamma = k'''\beta$ [27]; Ref. [28] has pointed out the mixed-type one- and two-soliton solutions for Eqs. (1.1) with $\gamma + \gamma_s = 0$ and $k''\gamma = k'''\beta$; With $k'' = -1$, $\beta = 1$, $k''' = -6$, $\gamma = 6$ and $\gamma_s = -3$, Painlevé analysis, Lax pair and Darboux-Bäcklund have been carried out [29, 30].

However, to our knowledge, the mixed-type vector soliton (including the vector bound states of solitons) solutions for Eqs. (1.1) with the conditions $\gamma + \gamma_s = 0$ and $k''\gamma = k'''\beta$ have not been obtained yet. The main contexts of this paper will be organized as follows: In Section 2, the bilinear forms, mixed-type one- and two-soliton solutions for Eqs. (1.1) will be obtained. Elastic and inelastic interactions will be studied through the asymptotic analysis. Figures will be plotted to describe the oblique interactions and bound states of the mixed-type vector solitons for the 3-NLS equations in Section 3. Section 4 will be our conclusions.

2 Soliton solutions for Eqs. (1.1)

The Hirota method is a tool for acquiring the analytic solutions for the nonlinear evolution equations (NLEEs) [17]. Once a given NLEE has been bilinearized through certain transformation, via the truncated parameter expansion at different levels, a series of solutions can be obtained. In this section, we will apply this method to construct the m-bright-n-dark ($m + n = N$) soliton solutions for Eqs. (1.1).

Introducing the dependent-variable transformation $q_j = g_j/f$, the bilinear forms for Eqs. (1.1) are given as

$$\left(iD_z - \frac{k''}{2}D_t^2 - \frac{ik'''}{6}D_t^3 + \frac{ik'''}{k''}\lambda D_t + \lambda\right) g_l \cdot f = 0, \quad l = 1, \cdots, m, \quad (2.1a)$$

$$\left(iD_z - \frac{k''}{2}D_t^2 - \frac{ik'''}{6}D_t^3 + \frac{ik'''}{k''}\lambda D_t + \lambda\right) g_\iota \cdot f = 0, \quad \iota = 1, \cdots, n, \quad (2.1b)$$

$$(k'' D_t^2 - 2\lambda) f \cdot f + 2\beta \sum_{j=1}^{N} |g_j|^2 = 0, \quad m + n = N, \quad (2.1c)$$

where g_j's are the complex functions, f is a real one, D_z and D_t are the bilinear derivative operators [17] defined by

$$D_z^{m_1} D_t^{m_2} a \cdot b = \left(\frac{\partial}{\partial z} - \frac{\partial}{\partial z'}\right)^{m_1} \left(\frac{\partial}{\partial t} - \frac{\partial}{\partial t'}\right)^{m_2} a(z,t) b(z',t') \Big|_{z'=z, t'=t}, \quad (2.2)$$

with z' and t' as two formal variables, $a(z,t)$ and $b(z',t')$ being two functions, while m_1 and m_2 are two nonnegative integers.

In order to obtain the m-bright-n-dark soliton solutions for Eqs. (1.1), we will expand g_l, g_ι and f as

$$g_l = \epsilon g_l^{(1)} + \epsilon^3 g_l^{(3)} + \epsilon^5 g_l^{(5)} + \cdots, \quad l = 1, \cdots, m, \quad (2.3a)$$

$$g_\iota = g_\iota^{(0)} \left(1 + \epsilon^2 g_\iota^{(2)} + \epsilon^4 g_\iota^{(4)} + \epsilon^6 g_\iota^{(6)} + \cdots\right), \quad \iota = 1, \cdots, n, \quad (2.3b)$$

$$f = 1 + \epsilon^2 f_2 + \epsilon^4 f_4 + \epsilon^6 f_6 + \cdots, \quad m + n = N, \quad (2.3c)$$

where ϵ is a formal expansion parameter, g_l's ($l = 1, \cdots, m$) and g_ι's ($\iota = 1, \cdots, n$) are the complex functions of z and t, while f_r's ($r = 2, 4, 6, \cdots$) are the real ones.

Without loss of generality, setting $\epsilon = 1$, and substituting Eqs. (2.3) into Bilinear Forms (2.1), we express the mixed-type one- and two-soliton solutions for Eqs. (1.1) respectively, as follows:

$$q_l = \frac{g_l}{f} = \frac{g_l^{(1)}}{1+f_2}, \quad q_\iota = \frac{g_\iota}{f} = \frac{g_\iota^{(0)}\left(1+g_\iota^{(2)}\right)}{1+f_2}, \tag{2.4}$$

and

$$q_l = \frac{g_l}{f} = \frac{g_l^{(1)}+g_l^{(3)}}{1+f_2+f_4}, \quad q_\iota = \frac{g_\iota}{f} = \frac{g_\iota^{(0)}\left(1+g_\iota^{(2)}+g_\iota^{(4)}\right)}{1+f_2+f_4}, \tag{2.5}$$

where the relative parameters are listed in the Appendix.

From Solutions (2.4), the amplitudes for the bright solitons in components q_l are given as $A_l = |\delta_{l,1}|/2\sqrt{\eta_1}$. While the depths for the dark solitons in components q_ι can be expressed as $A_\iota = |\zeta_\iota|\sqrt{2-\xi_1-\xi_1^*}/2$. Solitons in both q_l and q_ι components have the same velocities

$$V = \left[3i\,k''^2(\kappa_1-\kappa_1^*) - k''k'''(\kappa_1^2+\kappa_1^{*2}-\kappa_1\kappa_1^*) + 6\beta\,k'''\sum_{\iota=1}^n|\zeta_\iota|^2\right]/6k''.$$

3 Oblique interactions and bound states of solitons

In order to investigate the interactions between the mixed-type two solitons, we will make the asymptotic analysis on Solutions (2.5).

A. Before the interaction ($z \to -\infty$)

(a) $\theta_1 + \theta_1^* \sim 0$, $\theta_2 + \theta_2^* \sim -\infty$,

$$q_l^{1-} \to \frac{\delta_{l,1}}{2} e^{\frac{\theta_1-\theta_1^*-\ln\eta_1}{2}} \operatorname{sech}\left(\frac{\theta_1+\theta_1^*+\ln\eta_1}{2}\right), \tag{3.1a}$$

$$q_\iota^{1-} \to \zeta_\iota\, e^{ibz}\left[\frac{\xi_1+1}{2} + \frac{\xi_1-1}{2}\tanh\left(\frac{\theta_1+\theta_1^*+\ln\eta_1}{2}\right)\right], \tag{3.1b}$$

(b) $\theta_2 + \theta_2^* \sim 0$, $\theta_1 + \theta_1^* \sim +\infty$,

$$q_l^{2-} \to \frac{\delta_{l,3}}{2\eta_1} e^{\frac{\theta_2-\theta_2^*-\ln\frac{\eta_5}{\eta_1}}{2}} \operatorname{sech}\left(\frac{\theta_2+\theta_2^*+\ln\frac{\eta_5}{\eta_1}}{2}\right), \tag{3.2a}$$

$$q_\iota^{2-} \to \zeta_\iota\, e^{ibz}\left[\frac{\xi_5+\xi_1}{2} + \frac{\xi_5-\xi_1}{2}\tanh\left(\frac{\theta_2+\theta_2^*+\ln\frac{\eta_5}{\eta_1}}{2}\right)\right]. \tag{3.2b}$$

B. After the interaction ($z \to +\infty$)

(a) $\theta_1 + \theta_1^* \sim 0$, $\theta_2 + \theta_2^* \sim +\infty$,

$$q_l^{1+} \to \frac{\delta_{l,4}}{2\eta_4} e^{\frac{\theta_1 - \theta_1^* - \ln\frac{\eta_5}{\eta_4}}{2}} \mathrm{sech}\left(\frac{\theta_1 + \theta_1^* + \ln\frac{\eta_5}{\eta_4}}{2}\right), \qquad (3.3\mathrm{a})$$

$$q_l^{1+} \to \zeta_l e^{ibz} \left[\frac{\xi_5 + \xi_4}{2} + \frac{\xi_5 - \xi_4}{2}\tanh\left(\frac{\theta_1 + \theta_1^* + \ln\frac{\eta_5}{\eta_4}}{2}\right)\right], \qquad (3.3\mathrm{b})$$

(b) $\theta_2 + \theta_2^* \sim 0$, $\theta_1 + \theta_1^* \sim -\infty$,

$$q_l^{2+} \to \frac{\delta_{l,2}}{2} e^{\frac{\theta_2 - \theta_2^* - \ln\eta_4}{2}} \mathrm{sech}\left(\frac{\theta_2 + \theta_2^* + \ln\eta_4}{2}\right), \qquad (3.4\mathrm{a})$$

$$q_l^{2+} \to \zeta_l e^{ibz} \left[\frac{\xi_4 + 1}{2} + \frac{\xi_4 - 1}{2}\tanh\left(\frac{\theta_2 + \theta_2^* + \ln\eta_4}{2}\right)\right], \qquad (3.4\mathrm{b})$$

where $q^{\tau\mp}$ ($\tau = 1, 2$) denotes the status of the τ-th solitons before $(-)$ and after $(+)$ the interactions, respectively.

Comparing Expressions (3.1b) and (3.3b), (3.2b) and (3.4b), we find that the physical characters of q_ι remain unchanged during the interaction, except for some phase shifts. That is, the interactions between the dark solitons are always elastic. However, comparing Expressions (3.1a) and (3.3a), (3.2a) and (3.4a), we find that if $\delta_{l_1,1}/\delta_{l_1,2} = \delta_{l_2,1}/\delta_{l_2,2}$ ($1 \leqslant l_1 < l_2 \leqslant m$), there will be the elastic interaction between two bright solitons. While when $\delta_{l_1,1}/\delta_{l_1,2} \neq \delta_{l_2,1}/\delta_{l_2,2}$ ($1 \leqslant l_1 < l_2 \leqslant m$), inelastic interaction will happen.

To illustrate the interactions between the mixed-type two solitons explicitly, we take $N = 3$ as the example in Eqs. (1.1) for simplify, i.e.,

$$iq_{1,z} - \frac{k''}{2}q_{1,tt} + \beta(|q_1|^2 + |q_2|^2 + |q_3|^2)q_1 - \frac{ik'''}{6}q_{1,ttt}$$
$$+ i\gamma[(|q_1|^2 + |q_2|^2 + |q_3|^2)q_1]_t + i\gamma_s\left(|q_1|^2 + |q_2|^2 + |q_3|^2\right)_t q_1 = 0, \qquad (3.5\mathrm{a})$$

$$iq_{2,z} - \frac{k''}{2}q_{2,tt} + \beta(|q_1|^2 + |q_2|^2 + |q_3|^2)q_2 - \frac{ik'''}{6}q_{2,ttt}$$
$$+ i\gamma[(|q_1|^2 + |q_2|^2 + |q_3|^2)q_2]_t + i\gamma_s\left(|q_1|^2 + |q_2|^2 + |q_3|^2\right)_t q_2 = 0, \qquad (3.5\mathrm{b})$$

$$iq_{3,z} - \frac{k''}{2}q_{3,tt} + \beta(|q_1|^2 + |q_2|^2 + |q_3|^2)q_3 - \frac{ik'''}{6}q_{3,ttt}$$
$$+ i\gamma[(|q_1|^2 + |q_2|^2 + |q_3|^2)q_3]_t + i\gamma_s\left(|q_1|^2 + |q_2|^2 + |q_3|^2\right)_t q_3 = 0. \qquad (3.5\mathrm{c})$$

Case (I) $m = 2$ and $n = 1$.

(i) Oblique interaction

The two-bright-one-dark soliton solutions for Eqs. (3.5) are obtained when we choose $m = 2$ and $n = 1$. Under different conditions, elastic and inelastic oblique interactions happen between the two bright solitons in q_1 and q_2. While for the interaction between the dark solitons in q_3, there are always elastic. When $\delta_{1,1}/\delta_{1,2} = \delta_{2,1}/\delta_{2,2}$, solitons keep their amplitudes, velocities and widths unchanged after the interaction, except for some phase shifts, as seen as in Fig. 1. Otherwise, the head-on-inelastic interaction between the bright solitons S_1 and S_2 occurs with $\delta_{1,1}/\delta_{1,2} \neq \delta_{2,1}/\delta_{2,2}$, as seen in Fig. 2. We can see that after the interaction, the soliton S_1 gets enhanced in q_1 and suppressed in q_2, while the soliton S_2 decreases in q_1 and increases in q_2. The interaction between the two dark solitons in q_3 is elastic. Phase shifts and energy transfers are observed in Fig. 2. The total amount of energy of each soliton in three components is conserved.

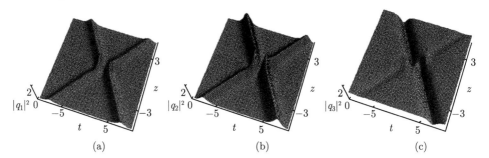

Fig. 1 Elastic oblique interaction between the two-bright-one-dark solitons via Solutions (2.5) with $\kappa_1 = 1$, $\kappa_2 = 1 + 1.5i$, $\delta_{1,1} = 1$, $\delta_{1,2} = 2$, $\delta_{2,1} = 1.5$, $\delta_{2,2} = 3$, $\zeta_1 = 1$, $\nu_1 = 0$, $\nu_2 = 0$, $k'' = -1$, $\beta = 1$ and $k''' = 1$

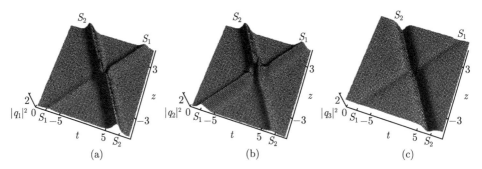

Fig. 2 Inelastic oblique interaction between the two-bright-one-dark solitons via Solutions (2.5) with $\kappa_1 = 1$, $\kappa_2 = 1 + 1.5i$, $\delta_{1,1} = 1$, $\delta_{1,2} = 1$, $\delta_{2,1} = 1$, $\delta_{2,2} = -2.5$, $\zeta_1 = 1$, $\nu_1 = 0$, $\nu_2 = 0$, $k'' = -1$, $\beta = 1$ and $k''' = 1$

(ii) Bound states of the solitons

When the adjacent solitons propagate with the same velocity, periodic attraction and repulsion of the bound states of solitons may occur [8,31]. This kind of complex interaction between solitons have potential applications in optical fiber communications [14–16], such as developing the stable mode-locked fiber lasers and enhancing the bandwidth. The strength of the interaction will be decreasing with the lager soliton separation factor $\Delta\nu = |\nu_2 - \nu_1|$. Fig. 3 shows that the elastic interactions between the bound states of the two bright solitons with $\delta_{1,1}/\delta_{1,2} = \delta_{2,1}/\delta_{2,2}$. When we choose the parameters as $\delta_{1,1}/\delta_{1,2} \neq \delta_{2,1}/\delta_{2,2}$, the redistribution of soliton intensities happened in Fig. 4. The intensity of left soliton increased, while the right soliton get suppressed.

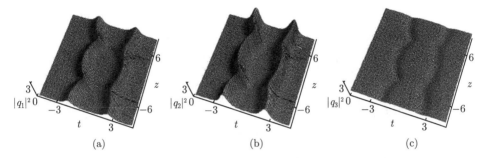

Fig. 3 Bound states of the two-bright-one-dark solitons via Solutions (2.5) with $\kappa_1 = \sqrt{3} + i$, $\kappa_2 = \dfrac{\sqrt{1443}}{20} + \dfrac{21}{20}i$, $\delta_{1,1} = 1$, $\delta_{1,2} = 2$, $\delta_{2,1} = 1.5$, $\delta_{2,2} = 3$, $\zeta_1 = 1$, $\nu_1 = 3$, $\nu_2 = 3$, $k'' = -1$, $\beta = 1$ and $k''' = 1$

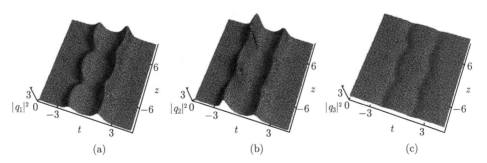

Fig. 4 Bound states of the two-bright-one-dark solitons via Solutions (2.5) with $\kappa_1 = \sqrt{3} + i$, $\kappa_2 = \dfrac{\sqrt{1443}}{20} + \dfrac{21}{20}i$, $\delta_{1,1} = 1$, $\delta_{1,2} = 2$, $\delta_{2,1} = 1$, $\delta_{2,2} = 3.5$, $\zeta_1 = 1$, $\nu_1 = 3$, $\nu_2 = 3$, $k'' = -1$, $\beta = 1$ and $k''' = 1$

Case (II) $m = 1$ and $n = 2$.

(i) Oblique interaction

(ii) Bound states of the solitons

If we choose $m = 1$ and $n = 2$, the one-bright-two-dark soliton solutions for Eqs. (3.5) are obtained. According to Asymptotic Analysis (3.1a)-(3.4b), elastic bright-bright and dark-dark interaction happen in the three components, respectively, which means that either the bright solitons in q_1 or the dark solitons in q_2 and q_3 keep their amplitudes or depths invariant during the interaction. There are no energy transformations, as seen as in Fig. 5. Fig. 6 displays that the bound states of the one-bright-two-dark two solitons. The special kind of interaction which is similar to Fig. 4 cannot occurred in the one-bright-two-dark mode.

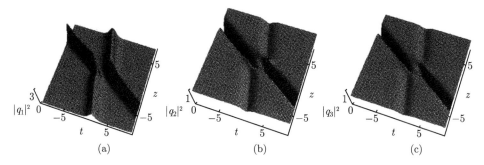

Fig. 5 Elastic oblique interaction between the one-bright-two-dark solitons via Solutions (2.5) with $\kappa_1 = 1$, $\kappa_2 = 1 + i$, $\delta_{1,1} = 1$, $\delta_{1,2} = 1$, $\zeta_1 = 1$, $\zeta_2 = 0.8$, $\nu_1 = 2$, $\nu_2 = 2$, $k'' = -1$, $\beta = 1$ and $k''' = 1$

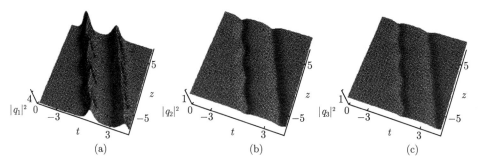

Fig. 6 Bound states of the one-bright-two-dark solitons via Solutions (2.5) with
$$\kappa_1 = \sqrt{3} + i, \kappa_2 = \frac{\sqrt{423}}{10} + 1.1i, \delta_{1,1} = 1, \delta_{1,2} = 2, \zeta_1 = \frac{\sqrt{3}}{2}, \zeta_2 = \frac{\sqrt{2}}{2}, \nu_1 = 2, \nu_2 = 3,$$
$$k'' = -1, \beta = 1 \text{ and } k''' = 1$$

4 Conclusions

In this paper, with the help of Hirota method, we have investigated the N-coupled higher-order NLS equations, i.e., Eqs. (1.1), which describe the simultaneous N nonlinear wave propagation in a fiber medium with higher order effects. The following results should be mentioned:

(i) Mixed-type one- and two-soliton solutions for Eqs. (1.1), i.e., Solutions (2.4) and (2.5), have been explicitly presented based on the Bilinear Forms (2.1).

(ii) Asymptotic Analysis (3.1a)-(3.4b) on Solutions (2.5) have been made. Based on which, figures are illustrated to describe the interactions of the two-bright-one-dark and one-bright-two-dark vector solitons for Eqs. (1.1) with $N = 3$.

(iii) For the case of two-bright-one-dark solitons, the elastic and inelastic oblique interactions between the two bright solitons, and the elastic oblique interaction between the dark solitons have been analyzed, as seen in Figs. 1 and 2, respectively. The elastic and redistribution of the soliton intensities of the bound states have been illustrated in Figs. 3 and 4, respectively.

(iv) For the case of one-bright-two-dark solitons, the elastic oblique interaction [as seen as in Fig. 5] and the bound states of solitons [as seen as in Fig. 6] have been analyzed.

We hope that our study will be helpful to the research and development of optical switch and logic gate in optical fiber communications.

Acknowledgments This work has been supported by the China Postdoctoral Science Foundation under Grant No. 2016M591121.

Appendix

Related parameters in Solutions (2.4) and (2.5) are given as:

$$g_\iota^{(0)} = \zeta_\iota e^{ibz}, \quad g_l^{(1)} = \delta_{l,1} e^{\theta_1} + \delta_{l,2} e^{\theta_2}, \quad g_l^{(3)} = \delta_{l,3} e^{\theta_1+\theta_2+\theta_1^*} + \delta_{l,4} e^{\theta_1+\theta_2+\theta_2^*},$$

$$g_\iota^{(2)} = \xi_1 \eta_1 e^{\theta_1+\theta_1^*} + \xi_2 \eta_2 e^{\theta_1+\theta_2^*} + \xi_3 \eta_3 e^{\theta_2+\theta_1^*} + \xi_4 \eta_4 e^{\theta_2+\theta_2^*}, \quad g_\iota^{(4)} = \xi_5 \eta_5 e^{\theta_1+\theta_1^*+\theta_2+\theta_2^*},$$

$$f_2 = \eta_1 e^{\theta_1+\theta_1^*} + \eta_2 e^{\theta_1+\theta_2^*} + \eta_3 e^{\theta_2+\theta_1^*} + \eta_4 e^{\theta_2+\theta_2^*}, \quad f_4 = \eta_5 e^{\theta_1+\theta_1^*+\theta_2+\theta_2^*},$$

$$b = \lambda = \beta \sum_{\iota=1}^{n} |\zeta_\iota|^2, \quad \theta_\tau = \kappa_\tau t + \omega_\tau z + \nu_\tau,$$

$$\omega_\tau = \frac{1}{6} \kappa_\tau^2 (\kappa_\tau k''' - 3i k'') + \left(i - \kappa_\tau \frac{k'''}{k''}\right) \beta \sum_{\iota=1}^{n} |\zeta_\iota|^2, \quad \tau = 1, 2,$$

$$\xi_1 = \frac{\kappa_1(2i\,k'' + k'''\kappa_1^*)}{\kappa_1^*(-2i\,k'' + k'''\kappa_1)}, \qquad \xi_2 = \frac{\kappa_1(2i\,k'' + k'''\kappa_2^*)}{\kappa_2^*(-2i\,k'' + k'''\kappa_1)},$$

$$\xi_3 = \frac{\kappa_2(2i\,k'' + k'''\kappa_1^*)}{\kappa_1^*(-2i\,k'' + k'''\kappa_2)}, \qquad \xi_4 = \frac{\kappa_2(2i\,k'' + k'''\kappa_2^*)}{\kappa_2^*(-2i\,k'' + k'''\kappa_2)}, \qquad \xi_5 = \xi_1\xi_4,$$

$$\eta_1 = -\frac{\beta\,\kappa_1\kappa_1^*(2k'' + i\,k'''\kappa_1)(2k'' - i\,k'''\kappa_1^*)\sum_{l=1}^{m}|\delta_{l,1}|^2}{k''(\kappa_1+\kappa_1^*)^2\left[\kappa_1\kappa_1^*(2k''+i\,k'''\kappa_1)(2k''-i\,k'''\kappa_1^*) - 4k''\beta\sum_{\iota=1}^{n}|\zeta_\iota|^2\right]},$$

$$\eta_2 = -\frac{\beta\,\kappa_1\kappa_2^*(2k'' + i\,k'''\kappa_1)(2k'' - i\,k'''\kappa_2^*)\sum_{l=1}^{m}\delta_{l,1}\delta_{l,2}^*}{k''(\kappa_1+\kappa_2^*)^2\left[\kappa_1\kappa_2^*(2k''+i\,k'''\kappa_1)(2k''-i\,k'''\kappa_2^*) - 4k''\beta\sum_{\iota=1}^{n}|\zeta_\iota|^2\right]},$$

$$\eta_3 = -\frac{\beta\,\kappa_2\kappa_1^*(2k'' + i\,k'''\kappa_2)(2k'' - i\,k'''\kappa_1^*)\sum_{l=1}^{m}\delta_{l,2}\delta_{l,1}^*}{k''(\kappa_2+\kappa_1^*)^2\left[\kappa_2\kappa_1^*(2k''+i\,k'''\kappa_2)(2k''-i\,k'''\kappa_1^*) - 4k''\beta\sum_{\iota=1}^{n}|\zeta_\iota|^2\right]},$$

$$\eta_4 = -\frac{\beta\,\kappa_2\kappa_2^*(2k'' + i\,k'''\kappa_2)(2k'' - i\,k'''\kappa_2^*)\sum_{l=1}^{m}|\delta_{l,2}|^2}{k''(\kappa_2+\kappa_2^*)^2\left[\kappa_2\kappa_2^*(2k''+i\,k'''\kappa_2)(2k''-i\,k'''\kappa_2^*) - 4k''\beta\sum_{\iota=1}^{n}|\zeta_\iota|^2\right]},$$

$$\delta_{l,3} = \frac{(\kappa_1-\kappa_2)[-2i\,k'' + k'''(\kappa_1-\kappa_1^*)]\,\eta_3\,\delta_{l,1}}{(\kappa_1+\kappa_1^*)[-2i\,k''+k'''(\kappa_1+\kappa_2)]} + \frac{(\kappa_1-\kappa_2)[2i\,k'' + k'''(\kappa_1^*-\kappa_2)]\,\eta_1\,\delta_{l,2}}{(\kappa_2+\kappa_1^*)[-2i\,k''+k'''(\kappa_1+\kappa_2)]},$$

$$\delta_{l,4} = \frac{(\kappa_1-\kappa_2)[-2i\,k'' + k'''(\kappa_1-\kappa_2^*)]\,\eta_4\,\delta_{l,1}}{(\kappa_1+\kappa_2^*)[-2i\,k''+k'''(\kappa_1+\kappa_2)]} + \frac{(\kappa_1-\kappa_2)[2i\,k'' + k'''(\kappa_2^*-\kappa_2)]\,\eta_2\,\delta_{l,2}}{(\kappa_2+\kappa_1^*)[-2i\,k''+k'''(\kappa_1+\kappa_2)]},$$

$$\eta_5 = -\frac{\beta\sum_{l=1}^{m}|\delta_{l,3}|^2}{\eta_1\left\{k''(\kappa_2+\kappa_2^*)^2 + \beta\left[|\xi_1|^2(\xi_4+\xi_4^*) - 2\right]\sum_{\iota=1}^{n}|\zeta_\iota|^2\right\}},$$

while $\delta_{l,1}$'s, $\delta_{l,2}$'s ($l = 1,\cdots,m$), κ_τ's ($\tau = 1,2$) and ζ_ι's ($\iota = 1,\cdots,n$) are all the complex constants, $*$ means the complex conjugate.

References

[1] Hasegawa A and Tappert F. Transmission of stationary nonlinear optical pulses in dispersive dielectric fibers. I. Anomalous dispersion [J]. Appl. Phys. Lett., 1973, 23: 142-144.

[2] Mollenauer L. F, Stolen R. H, Gordon J. P. Experimental observation of picosecond pulse narrowing and solitons in optical fibers [J]. Phys. Rev. Lett., 1980, 45: 1095-1098.

[3] Barnett M. P, Capitani J. F, Zur Gathen J. Von, Gerhard J. Symbolic calculation in chemistry: Selected examples [J]. Int. J. Quantum Chem., 2004, 100: 80-104.

[4] Kivshar Y. S, Luther-Davies B. Dark optical solitons: physics and applications [J]. Phys. Rep., 1998, 298: 81-197.

[5] Manikandan N, Radhakrishnan R, Aravinthan K. Generalized dark-bright vector soliton solution to the mixed coupled nonlinear Schrödinger equations [J]. Phys. Rev. E, 2014, 90: 022902.

[6] Agrawal G. P. Nonlinear Fiber Optics [M]. New York: Academic, 1995.

[7] Kivshar Y. S, Quiroga-Teixeiro M. L. Influence of cross-phase modulation on soliton switching in nonlinear optical fibers [J]. Opt. Lett., 1993, 18: 980-982.

[8] Sheppard A. P, Kivshar Y. S. Polarized dark solitons in isotropic Kerr media [J]. Phys. Rev. E, 1997, 55: 4773-4782.

[9] Ohta Y, Wang D. S, Yang J. K. General N-dark-dark solitons in the coupled nonlinear Schrödinger equations [J]. Stud. Appl. Math., 2011, 127: 345-371.

[10] Kivshar Y. S. Stable vector solitons composed of bright and dark pulses [J]. Opt. Lett., 1992, 17: 1322-1324.

[11] Agalarov A, Zhulego V, Gadzhimuradov T. Bright, dark, and mixed vector soliton solutions of the general coupled nonlinear Schrödinger equations [J]. Phys. Rev. E, 2015, 91: 042909.

[12] Radhakrishnan R, Manikandan N, Aravinthan K. Energy-exchange collisions of dark-bright-bright vector solitons [J]. Phys. Rev. E, 2015, 92: 062913.

[13] Chakraborty S, Nandy S, Barthakur A. Bilinearization of the generalized coupled nonlinear Schrödinger equation with variable coefficients and gain and dark-bright pair soliton solutions [J]. Phys. Rev. E, 2015, 91: 023210.

[14] Tang D. Y, Zhao B, Shen D. Y, Lu C, Man W. S, Tam H. Y. Bound-soliton fiber laser [J]. Phys. Rev. A, 2002, 66: 033806.

[15] Haelterman M, Sheppard A. Bifurcation phenomena and multiple soliton-bound states in isotropic Kerr media [J]. Phys. Rev. E, 1994, 49: 3376-3381.

[16] Li M, Xiao J. H, Jiang Y, Wang M, Tian B. Bound-state dark/antidark solitons for the coupled mixed derivative nonlinear Schrödinger equations in optical fibers [J]. Eur. Phys. J. D, 2012, 66: 297.

[17] Hirota R. The Direct Method in Soliton Theory [M]. Cambridge Univ. Press, Cambridge, 2004.

[18] Yang X. J, Machado J. A. T, Baleanu D, Cattan C. On exact traveling-wave solutions for local fractional Korteweg-de Vries equation [J]. Chaos, 2016, 26: 084312.

[19] Yang X. J, Machado J. A. T, Hristov J. Nonlinear dynamics for local fractional Burgers' equation arising in fractal flow [J]. Nonlinear Dyn., 2016, 84: 3-7.

[20] Yang X. J, Srivastava H. M. Anasymptotic perturbation solution for a linear oscillator of free damped vibrations in fractal medium described by local fractional derivatives [J]. Commun Nonlinear Sci. Numer. Simulat., 2015, 29: 499-504.

[21] Kanna T, Lakshmanan M. Exact soliton solutions of coupled nonlinear Schrödinger equations: Shape-changing collisions, logic gates, and partially coherent solitons [J]. Phys. Rev. E, 2003, 67: 046617.

[22] Jiang Y, Tian B, Liu W. J, Sun K, Li M, Wang P. Soliton interactions and complexes for coupled nonlinear Schrödinger equations [J]. Phys. Rev. E, 2012, 85: 036605.

[23] Li M, Xiao J. H, Liu W. J, Wang P, Qin B, Tian B. Mixed-type vector solitons of the N-coupled mixed derivative nonlinear Schrödinger equations from optical fibers [J]. Phys. Rev. E, 2013, 87: 032914.

[24] Nakkeeran K. Exact dark soliton solutions for a family of N-coupled nonlinear Schrödinger equations in optical fiber media [J]. Phys. Rev. E, 2001, 64: 046611.

[25] Hirota R. Exact envelope-soliton solutions of a nonlinear wave equation [J]. J. Math. Phys., 1973, 14: 805-809.

[26] Sasa N, Satsuma J. New type of soliton solutions for a higher order nonlinear Schrödinger equation [J]. J. Phys. Soc. Jpn., 1991, 60: 409-417.

[27] Radhakrishnan R, Lakshmanan M. Exact soliton solutions to coupled nonlinear Schrödinger equations with higher-order effects [J]. Phys. Rev. E, 1996, 54: 2949-2955.

[28] Jiang Y, Tian B, Liu W. J, Sun K, Wang P. Mixed-type solitons for the coupled higher-order nonlinear Schrödinger equations in multi-mode and birefringent fibers [J]. J. Mod. Opt., 2013, 60: 629-636.

[29] Porsezian K, Sundaram P. S, Mahalingam A, Coupled higher-order nonlinear Schrödinger equations in nonlinear optics: Painlevé analysis and integrability [J]. Phys. Rev. E, 1994, 50: 1543-1547.

[30] Nakkeeran K, Porsezian K, Sundaram P. S, Mahalingam A. Optical solitons in N-coupled higher order nonlinear Schrödinger equations [J]. Phys. Rev. E, 1998, 80: 1425-1428.

[31] Stegeman G. I, Segev M. Optical spatial solitons and their interactions: universality and diversity [J]. Science, 1999, 286: 1518-1523.

Global Finite Energy Weak Solution to the Viscous Quantum Navier-Stokes-Landau-Lifshitz-Maxwell Model in 2-dimension*

Guo Boling (郭柏灵) and Wang Guangwu (王光武)

Abstract In this paper, we prove the global existence of the weak solution to the viscous quantum Navier-Stokes-Landau-Lifshitz-Maxwell equations in two-dimension for large data. The main techniques is the Faedo-Galerkin approximation and weak compactness theory.

Keywords Global finite energy weak solution; viscous quantum Navier-Stokes-Landau-Lifshitz-Maxwell system; Faedo-Galerkin method

1 Introduction

In the studying the dispersive theory of magnetization of ferromagnets, we also consider the viscous quantum of a fluid on motion under the Maxwell electric-magnetic field, i.e. the macroscopic motion of a fluid and the quantum effects and the interactions between electrons in microscopic will be considered similarly.

In this paper we study the viscous quantum Navier-Stokes-Landau-Lifshitz-Maxwell system (QNSLLM) in $(0,T) \times \Omega$:

$$\partial_t \rho + \mathrm{div}(\rho u) = \nu \Delta \rho, \tag{1.1}$$

$$\partial_t(\rho u) + \mathrm{div}(\rho u \otimes u) + \frac{\nabla P}{m} = -\frac{\rho e}{m}(E + u \times B) + \frac{\mu \hbar^2}{2m^2}\rho \nabla \left(\frac{\Delta \sqrt{\rho}}{\sqrt{\rho}}\right)$$

$$+ \nu \Delta(\rho u) - \frac{\rho u}{\tau} - \lambda \nabla \cdot \left(\nabla d \odot \nabla d - \frac{|\nabla d|^2}{2}I\right), \tag{1.2}$$

$$d_t + u \cdot \nabla d + \alpha_1 d \times (d \times (\Delta d + B)) = \alpha_2 d \times (\Delta d + B), \tag{1.3}$$

---
* Ann. of Appl. Math., 2016, 32(2): 111–132.

$$E_t - \nabla \times B = e\rho u, \tag{1.4}$$

$$B_t + \nabla \times E = -\lambda m(d_t + u \cdot \nabla d), \tag{1.5}$$

$$\nabla \cdot B = 0 \quad |d(x,t)| = 1. \tag{1.6}$$

Here We denote by $\Omega \subset \mathbf{R}^2$ the two-dimensional 2D-periodic domain, i.e., $\Omega = \{x = (x_1, x_2) | |x_i| < D; (i = 1, 2)\}$. ρ and u represent the density and the velocity field of the flow respectively. P is the pressure function. We consider the isentropic case, i.e., $P = A\rho^\gamma (A > 0$ is a constant). $d(x,t) : \Omega \to S^1$, the unit sphere in $\mathbf{R}^2$, denotes the magnetization field. E and B represent the electric field and the magnetic field. The physic constants m, e, $\hbar$ are positive and represents the mass, the charge of the particle and Planck constants. ν and μ are positive viscosity constants. The positive constants λ, τ and β represent the competition between kinetic energy and potential energy, the relaxation time of electron and microscopic elastic relaxation time for the molecular orientation field. α_1 is a positive constant, and $\alpha_2 \geq 0$ is Gilbert damping coefficient. $\nabla \cdot$ denotes the divergence operator, and $\nabla d \odot \nabla d$ denotes the 2×2 matrix whose (i,j)-the entry is given by $\nabla_i d \cdot \nabla_j d$ for $1 \leq i, j \leq 2$. The expression $\dfrac{\Delta \sqrt{\rho}}{\sqrt{\rho}}$ can be interpreted as the quantum Bohm potential.

Roughly speaking, the system (1.1)-(1.6) is a coupling between the viscous isentropic quantum Navier-Stokes equations and Landau-Lifshitz-Maxwell equations. This model can be used to describe the dispersive theory of magnetization of ferromagnets with the electromagnetic field.

We call a function $f(x)$ is a 2D-periodic if $f(x + 2De_i) = f(x), (i = 1, 2)$, where (e_1, e_2) forms the unit orthogonal basis of R^2, $D > 0$ is a constant.

For the system (1.1)-(1.6), we impose the following initial conditions

$$\rho|_{t=0} = \rho_0(x), \quad u|_{t=0} = u_0(x), \quad E|_{t=0} = E_0(x), \quad d|_{t=0} = d_0(x), \quad B|_{t=0} = B_0(x), \tag{1.7}$$

which satisfy that

$$\rho_0(x) > 0,$$
$$|d_0(x)| = 1 \quad d_0(x) \in H^2(\Omega), \quad \inf_x d_0^2 > 0, \tag{1.8}$$
$$E_0(x), \quad B_0(x) \in L^2(\Omega).$$

Furthermore, we always assume that $\rho_0(x), u_0(x), d_0(x), E_0(x), B_0(x)$ are 2D-periodic.

Firstly setting $E = B = d = 0$, and using a effective velocity transformation [18] the system (1.1)-(1.6) becomes the isentropic compressible quantum Navier-Stokes equation (IQCNS). Set $\mu = 0$, we get the isentropic compressible Navier-Stokes equation(ICNS). For the Navier-Stokes equations with constant viscosity, the existence of global weak solution of ICNS with large initial data in $\mathbf{R}^n$ was first obtained by P. L. Lions in [22], where $\gamma \geq \dfrac{3n}{n+2}$ for $n = 2$ or 3. Feireisl et al. in [6,7] extended Lions's work to the case $\gamma > \dfrac{3}{2}$ for $n = 3$. For solutions with spherical symmetry, Jiang and Zhang in [15] relaxed the restriction on γ in [22] to the case $\gamma > 1$, and got the global existence of the weak solutions for $N = 2$ or 3. Luo, Xin and Yang [24] proved the existence and regularity of weak solutions with density connecting to vacuum continuously in 1D.

If $\nu = E = B = d = 0$, the system (1.1)-(1.6) is called quantum hydrodynamic model (QHD). In [24] Jüngel and Li proved the existence of local-in-time solutions and global-in-time solutions to (1.1)-(1.6) for the one-dimensional case with Dirichlet and Neumann boundary conditions for the particle density ρ. In [16], the authors have given the local-in-time existence of the solutions for equations (1.1)-(1.6) in the multi-dimensional torus for the irrotational velocity, and they also proved that local-in-time solutions exit globally in time and exponentially converge toward the corresponding steady-state under a "subsonic" type stability condition. In [14], the authors proved the local-in-time and global-in-time existence of the solutions of equations (1.1)-(1.6) in $\mathbf{R}^n$ for rotational fluid.

If $E = B = d = 0$, the system (1.1)-(1.6) becomes the viscous quantum hydrodynamic model (vQHD) (see [10], [17] for the derivation). In [18], Jüngel proved the existence of the global finite energy weak solution of the vQHD.

If $u = E = B = 0$, and ρ is a constant, the system (1.1)-(1.6) is Landau-Lifshitz model(LL). In 1981, a group headed by Zhou and Guo proved the existence of the global weak solutions to the initial value problems and initial boundary value problems for Landau-Lifshitz equations from one dimension to multi-dimensions [27]. Alouges and Soyeur [1] proved similar results by penalty method in 1992. In 1993, Guo and Hong began the studies on two-dimensional Landau-Lifshitz equations, they established in [12] the relations between two-dimensional Landau-Lifshitz equations and harmonic maps and applied the approaches studying harmonic maps to get the global existence and uniqueness of partially regular weak solution. In this aspect, in 2004, Liu [23] proved that the "stationary" weak solutions of higher dimensional Landau-Lifshitz equations are partially regular.

If $d = 0$, the system (1.1)-(1.6) is the viscous quantum Navier-Stokes-Maxwell model(QNSM). Furthermore if $\mu = \nu = 0$, it becomes the Navier-Stokes-Maxwell model(NSM). In [13], Hong etc. got the existence and uniqueness of global spherically symmetric classical solution of the initial boundary value problem to the system of NSM. Germain, Ibrahim and Masmoudi [9] prove the local and global well-posedness of incompressible NSM.

If $u = 0$, and ρ is costant, it obtains the Landau-Lifshitz-Maxwell model(LLM). In [5], Ding, Guo, Li and Zeng obtained the existence of the global weak solution for the LLM for 3-dimensional case. The global existence of the unique smooth solution for the LLM of the ferromagnetic spin chain without disspation in one or two dimensions were established in [26]. Ding and Guo [4] proved the existence of partial regular weak solution to LLM.

In this paper, we are interested in the global existence of finite energy solutions to the initial problem of (1.1)-(1.6) in two-dimensional case. Besides difficulties pointed in [18], the main difficulty is the coupling of velocity, electric field, magnetic field and magnetization field. Therefore, the key in our analysis is to deal with convection term, the quantum term, the electric-magnetic term and the last two terms about the magnetization field in the momentum equation. Inspired by [18], we choose the test function of the form $\rho\phi$ for the momentum equation, the magnetization field equation and the magnetic field equation.

Next we state our main results

Theorem 1.1 (global existence) *Let $T > 0$, $P(\rho) = A\rho^\gamma (\gamma \geqslant 1)$. (1.8) holds. And $\rho_0, u_0, d_0, E_0, B_0$ are 2D-periodic functions. Furthermore $E(\rho_0, u_0, d_0, E_0, B_0)$ (see (4.2) for the definition of $E(\rho, u, d, E, B)$) is finite. There exists a weak solution (ρ, u, d, E, B) to (1.1)-(1.6) with the regularity*

$$\sqrt{\rho} \in L^\infty([0,T]; H^1(\Omega)) \cap L^2([0,T]; H^2(\Omega)), \quad \rho \geqslant 0, \tag{1.9}$$

$$\rho \in H^1([0,T]; L^2(\Omega)) \cap L^\infty([0,T]; L^\gamma(\Omega)) \cap L^2([0,T]; W^{1,3}(\Omega)), \tag{1.10}$$

$$\sqrt{\rho}u \in L^\infty([0,T]; L^2(\Omega)), \quad \rho u \in L^2([0,T]; W^{1,3/2}(\Omega)), \tag{1.11}$$

$$\rho|\nabla u| \in L^2([0,T]; L^2(\Omega)), \tag{1.12}$$

$$E \in L^\infty([0,T]; L^2(\Omega)), \quad B \in L^\infty([0,T]; L^2(\Omega)), \tag{1.13}$$

$$d \in L^2([0,T]; H^2(\Omega)) \cap L^\infty([0,T]; L^2(\Omega)), \tag{1.14}$$

$$\nabla d \in L^4([0,T]; L^4(\Omega)), \tag{1.15}$$

satisfying (1.1) pointwise and for all smooth test functions satisfying $\phi(\cdot, T) = 0$,

$$-\int_\Omega \rho_0^2 u_0 \phi = \int_0^T \int_\Omega \Big(\rho^2 u \cdot \phi_t - \rho^2 \mathrm{div}(u) u \cdot \phi - \nu(\rho u \otimes \nabla \rho) : \nabla \phi$$
$$+ \rho u \otimes \rho u : \nabla \phi + \frac{\gamma}{\gamma+1}\rho^{\gamma+1}\mathrm{div}\phi + \frac{\rho e}{m}(E + u \times B)\cdot \rho\phi$$
$$-\frac{\mu \hbar^2}{2m^2}\Delta\sqrt{\rho}(2\sqrt{\rho}\nabla\rho\cdot\phi + \rho^{3/2}\mathrm{div}\phi) - \nu\nabla(\rho u):(\rho\nabla\phi + 2\nabla\rho\otimes\phi)$$
$$+\lambda\left(\nabla d \odot \nabla d - \frac{|\nabla d|^2}{2}I\right)\cdot\nabla(\rho\phi)\Big)dxdt, \tag{1.16}$$

$$-\int_\Omega d_0 \rho\phi(x,0)dx = \int_0^T \int_\Omega \Big(d\rho\phi_t + \rho u \cdot \nabla d \cdot \phi + \alpha_1 d\times(d\times(\Delta d + B))\cdot\rho\phi$$
$$- \alpha_2 d\times(\Delta d + B)\cdot\rho\phi\Big)dxdt, \tag{1.17}$$

$$-\int_\Omega E_0\cdot\phi(x,0)dx = \int_0^T\int_\Omega (E\cdot\phi_t - B\cdot(\nabla\times\phi) - e\rho u\cdot\phi)dxdt, \tag{1.18}$$

$$-\int_\Omega (B_0 + \lambda m d_0)\cdot\phi(x,0)dx = \int_0^T\int_\Omega ((B + \lambda m d)\cdot\phi_t$$
$$+ E\cdot(\nabla\times\phi) + \lambda m(u\cdot\nabla d)\cdot\phi)dxdt. \tag{1.19}$$

The product $A:B$ means summation over both indices of matrices A and B.

It is easy to see that (1.1)-(1.6) lack compactness which we need to get the estimate of L^2 or H^1-norm for u. To overcome this difficult, we first add the right hand of (1.2) a viscosity term $\delta\Delta u + \delta u$:

$$\partial_t \rho + \mathrm{div}(\rho u) = \nu\Delta\rho, \tag{1.20}$$

$$\partial_t(\rho u) + \mathrm{div}(\rho u \otimes u) + \frac{\nabla P}{m} = -\frac{\rho e}{m}(E + u\times B) + \frac{\mu\hbar^2}{2m^2}\rho\nabla\left(\frac{\Delta\sqrt{\rho}}{\sqrt{\rho}}\right)$$
$$+\nu\Delta(\rho u) - \frac{\rho u}{\tau} - \lambda\nabla\cdot\left(\nabla d\odot\nabla d - \frac{|\nabla d|^2}{2}I\right) + \delta\Delta u - \delta u, \tag{1.21}$$

$$d_t + u\cdot\nabla d + \alpha_1 d\times(d\times(\Delta d + B)) = \alpha_2 d\times(\Delta d + B), \tag{1.22}$$

$$E_t - \nabla\times B = e\rho u, \tag{1.23}$$

$$B_t + \nabla\times E = -\lambda m(d_t + u\cdot\nabla d), \tag{1.24}$$

$$\nabla\cdot B = 0 \quad |d(x,t)| = 1. \tag{1.25}$$

Then we want to send the viscosity constant δ to 0. Finally, we obtain the desired weak solution to the original problem (1.1)-(1.6).

This paper is organized as following. In Section 2, we give some preliminaries which will be used in the following section. We show the local existence of (1.1)-(1.6) in Section 3. In Section 4, we firstly prove the a-priori estimates and the existence of global weak solution to the viscosity system (1.20)-(1.25), then show the limit of $\delta \to 0$. The global-in-time existence weak solutions can be achieved according to the Faedo-Galerkin method and weak compactness techniques.

2 Preliminaries

In this section we first give some notation. In this paper we denote C is the constant dependent of N and δ. $L^p([0,T]; L^q(\Omega))(p,q \geqslant 1)$ is a space whose element is the p-integrable respect to time variable and q-integrable respect to space variable function. $W^{k,p}$ and H^s are the Sobolev spaces. $(H^s)^*$ is the dual space of H^s.

Lemma 2.1 (Gagliardo-Nirenberg inequality) *Let $\Omega \subset \mathbf{R}^d(d \geqslant 1)$ be a bounded open set with $\partial\Omega \in C^{0,1}$, $m \in \mathbf{N}$, $1 \leqslant p,q,r \leqslant \infty$. Then there exists a constant $C > 0$ such that for all $u \in W^{m,p}(\Omega) \cap L^q(\Omega)$,*

$$\|D^\alpha u\|_{L^r(\Omega)} \leqslant C \|u\|_{W^{m,p}(\Omega)}^\theta \|u\|_{L^q(\Omega)}^{1-\theta},$$

where $0 \leqslant |\alpha| \leqslant m-1$, $\theta = |\alpha|/m$, and $|\alpha| - d/r = \theta(m - d/p) - (1-\theta)d/q$. If $m - |\alpha| - d/p \notin \mathbf{N}_0$, then $\theta \in [|\alpha|/m, 1]$ is allowed.

Next we give a weak compactness lemma which will be used in the following section.

Lemma 2.2 (Aubin-Lions lemma) *Assume $X \subset Y \subset Z$ are Banach spaces and $X \hookrightarrow\hookrightarrow Y$. Then the following immbedding are compact:*

$$L^q([0,T];X) \cap \{\varphi : \partial_t \varphi \in L^1([0,T];Z)\} \hookrightarrow\hookrightarrow L^q([0,T];Y), \quad \forall 1 \leqslant q \leqslant \infty, \qquad (2.1)$$

$$L^\infty([0,T];X) \cap \{\varphi : \partial_t \varphi \in L^r([0,T];Z)\} \hookrightarrow\hookrightarrow C([0,T];Y), \quad \forall 1 \leqslant r \leqslant \infty. \qquad (2.2)$$

3 Local existence of solutions

In this section we will show the local existence of solutions to the viscosity system (1.20)-(1.25) by Faedo-Galerkin method. Let $T > 0$, and let $\{\omega_j\}$ be an orthonormal basis of $L^2(\Omega)$ which is also an orthogonal basis of $H^1(\Omega)$. Introduce the finite-dimensional space $X_N = \text{span}\{w_1, \cdots, w_N\}$, $n \in N$. Denote the approximate solutions of the problem (1.21)-(1.25) by $u_N^\delta, d_N^\delta, E_N^\delta, B_N^\delta$ in the following form

$$u_N^\delta(x,t) = \sum_{j=1}^N \lambda_j(t) w_j(x),$$

$$d_N^\delta(x,t) = \sum_{j=1}^N \delta_j(t) w_j(x), \quad E_N(x,t) = \sum_{j=1}^N \eta_j(t) w_j(x),$$

$$B_N^\delta(x,t) = \sum_{j=1}^N \zeta_j(t) w_j(x),$$

where $\beta_j(t), \gamma_j(t), \delta_j(t), \eta_j(t), \zeta_j(t)$ ($j = 1, \cdots, N, N = 1, 2, \cdots$) are 2-dimensional vector-valued functions. For some functions $\lambda_i(t)$, and the norm of v in $C^0([0,T]; X_N)$ can be formulated as

$$\|v\|_{C^0([0,T];X_N)} = \max_{t \in [0,T]} \sum_{i=1}^N |\lambda_i(t)|.$$

As a consequence, v can be bounded in $C^0([0,T]; C^k(\Omega))$ for any $k \in N$, and there exists a constant $C > 0$ depending on k such that

$$\|u\|_{C^0([0,T];C^k(\Omega))} \leq C \|u\|_{C^0([0,T];L^2(\Omega))}. \tag{3.1}$$

The approximate system is defined as follows. Let $\rho \in C^1([0,T]; C^3(\Omega))$ be the classical solution to

$$\rho_t + \text{div}(\rho u) = \nu \Delta \rho, \quad \rho|_{t=0} = \rho_0(x). \tag{3.2}$$

The maximum principle provides the lower and upper bounds ([15], chapter 7.3)

$$\inf_{x \in \Omega} \rho_0(x) \exp\left(-\int_0^t \|\text{div} u\|_{L^\infty(\Omega)} ds\right) \leq \rho(x) \leq \sup_{x \in \Omega} \exp\left(\int_0^t \|\text{div} u\|_{L^\infty} ds\right).$$

Since we assumed that $\rho_0(x) > \bar{\rho} > 0$, $\rho(x)$ is strictly positive. In view of (3.1), for $\|u\|_{C^0([0,T];L^2(\Omega))} \leq C$, there exists constant $\underline{\rho}(C)$ and $\bar{\rho}(C)$ such that

$$0 < \underline{\rho}(C) \leq \rho(x,t) \leq \bar{\rho}(C).$$

We introduce the operator $S_1 : C^0([0,T]; X_N) \to C^0([0,T]; C^3(\Omega))$ by $S_1(u) = \rho$. Since the equation for ρ is linear, S_1 is Lipschitz continuous in the following sense:

$$\|S_1(u_1) - S_1(u_2)\|_{C^0([0,T];C^k(\Omega))} \leq C(N,k) \|u_1 - u_2\|_{C^0([0,T];L^2(\Omega))}. \tag{3.3}$$

Next we wish to solve (1.20)-(1.25) on the space X_N. To this end, for given $\rho = S_1(u)$, we are looking for functions $(u_N^\delta, d_N^\delta, E_N^\delta, B_N^\delta) \in (C^0([0,T]; X_N))^4$ such that

$$-\int_\Omega \rho_0 u_0 \cdot \phi(\cdot, 0) dx = \int_0^T \int_\Omega \Big(\rho u \cdot \phi_t + \rho(u_N \otimes u_N) : \nabla\phi + P(\rho)\mathrm{div}\phi$$
$$- \frac{\rho e}{m}(E + u_N \times B)\phi$$
$$- \frac{\mu \hbar^2}{2m^2} \frac{\Delta\sqrt{\rho}}{\sqrt{\rho}} \mathrm{div}(\rho\phi) - \nu\nabla(\rho u_N) : \nabla\phi$$
$$- \frac{\rho u_N}{\tau}\phi + \lambda(\nabla d \odot \nabla d - \frac{|\nabla d|^2}{2}I)\nabla\phi$$
$$- \delta(\nabla u_N : \nabla\phi + u_N \cdot \phi)\Big) dx dt, \qquad (3.4)$$

$$-\int_\Omega d_0 \phi(x,0) dx = \int_0^T \int_\Omega \Big(d_N \phi_t + u_N \cdot \nabla d_N \cdot \phi + \alpha_1 d_N \times (d_N \times (\Delta d_N + B_N)) \cdot \phi$$
$$- \alpha_2 d_N \times (\Delta d_N + B_N) \cdot \phi\Big) dx dt, \qquad (3.5)$$

$$-\int_\Omega E_0 \cdot \phi(x,0) dx = \int_0^T \int_\Omega (E_N \cdot \phi_t - B_N \cdot (\nabla \times \phi) - e\rho u_N \cdot \phi) dx dt, \qquad (3.6)$$

$$-\int_\Omega (B_0 + \lambda m d_0) \cdot \phi(x,0) dx$$
$$= \int_0^T \int_\Omega ((B_N + \lambda m d_N) \cdot \phi_t + E_N \cdot (\nabla \times \phi) + \lambda m(u_N \cdot \nabla d_N) \cdot \phi) dx dt. \qquad (3.7)$$

for all $\phi \in C^1([0,T]; X_N)$ such that $\phi(\cdot, T) = 0$. Notice that we have added the regularization term $\delta(\Delta u_N - u_N)$. The reason is that we will apply Banach fixed point theorem to prove the local-in-time existence of solutions. The regularization yields the H^1 regularity of u_N needed to conclude the global existence of solutions.

To solve (3.4)-(3.7), we follow ([22], Chapter 7.3.3) and introduce the following family of operators, given a function $\varrho \in L^1(\Omega)$ with $\varrho \geqslant \underline{\varrho} > 0$:

$$M[\varrho] : X_N \to X_N^*, \langle M[\varrho]u, \omega\rangle = \int_\Omega \varrho u \cdot \omega, \quad u, \omega \in X_N.$$

These operators are symmetric and positive definite with the smallest eigenvalue

$$\int_{\|u\|_{L^2(\Omega)}=1} \langle M[\varrho]u, u\rangle = \int_{\|u\|_{L^2(\Omega)}=1} \int_\Omega \varrho |u|^2 dx \geqslant \inf_{x \in \Omega} \varrho(x) \geqslant \underline{\varrho}.$$

Hence, since X_N is finite-dimensional, the operators are invertible with

$$\|M^{-1}[\varrho]\|_{L(X_N^*, X_N)} \leq \underline{\rho}^{-1},$$

where $L(X_N^*, X_N)$ is the set of bounded linear mappings from X_N^* to X_N. Moreover (see [22] Chapter 7.3.3), M^{-1} is Lipschitz continuous in the sense

$$\|M^{-1}[\varrho_1] - M^{-1}[\varrho_2]\|_{L(X_N^*, X_N)} \leq C(N, \underline{\varrho})\|\varrho_1 - \varrho_2\|_{L^1(\Omega)} \tag{3.8}$$

for all $\varrho_1, \varrho_2 \in L^1(\Omega)$ such that $\varrho_1, \varrho_2 \geq \underline{\varrho} > 0$.

Now the integral equation (3.4) can be rephrased as an ordinary differential equation on the finite-dimensional space X_N,

$$\frac{d}{dt}(M[\rho(t)]u_N^\delta(t)) = N[u, d, B, E, u_N^\delta(t)], \quad M[\rho_0]u_N(0) = M[\rho_0]u_0, \tag{3.9}$$

when $\rho = S_1(u)$,

$$\langle N(u, d, B, E, u_N^\delta), \phi \rangle = \int_0^T \int_\Omega \Big(\rho(u \otimes u_N^\delta) : \nabla\phi + P(\rho)\text{div}\phi - \frac{\rho e}{m}(E + u_N^\delta \times B)\phi$$
$$- \frac{\mu\hbar^2}{2m^2} \frac{\Delta\sqrt{\rho}}{\sqrt{\rho}} \text{div}(\rho\phi) - \nu\nabla(\rho u_N^\delta) : \nabla\phi - \frac{\rho u_N^\delta}{\tau}\phi$$
$$+ \lambda\Big(\nabla d \odot \nabla d - \frac{|\nabla d|^2}{2}I\Big)\nabla\phi - \delta(\nabla u_N^\delta : \nabla\phi + u_N^\delta \cdot \phi)\Big) dxdt.$$

For operator $N[u, d, E, B, \cdot]$, defined for every $t \in [0, T]$ as an operator form X_N to X_N^*, is continuous in time. Standard theory for systems of ordinary differential equations then provides the existence of a unique classical solution to (3.9), i.e., for a given $u \in C^0([0, T]; X_N)$, there exists a unique solution $u_N \in C^1([0, T]; X_N)$ to (3.4).

Integrating (3.9) over $(0, t)$ yields the following nonlinear equation:

$$u_N^\delta(t) = M^{-1}S_1(u_N^\delta)(t)\Big(M[\rho_0]u_0 + \int_0^t N[u_N^\delta, u_N^\delta(s)]ds\Big), \quad \text{in } X_N. \tag{3.10}$$

Since the operators S_1 and M are Lipschitz type, (3.10) can be solved by evoking the fixed point theorem of Banach on a short time interval $[0, T']$, where $T' \leq T$, in the space $C^0([0, T]; X_N)$. In fact, we have even $u_N^\delta \in C^1([0, T']; X_N)$. Then we can solve the system (3.5)-(3.7). Thus, there exists a unique local-in-time solution $(\rho_N^\delta, u_N^\delta, d_N^\delta, E_N^\delta, B_N^\delta)$ to (1.20)-(1.25).

4 A priori estimates and global existence

In this section, we will give some a-priori estimates. Using these estimates, we can show that the local-in-time solution $(\rho_N^\delta, u_N^\delta, d_N^\delta, E_N^\delta, B_N^\delta)$ which are proved in Section 2 can be extended globally. In the case of not confusing, we omit the subscript N and superscript δ in this section.

Theorem 4.1 *Set the conditions in Theorem 1.1 holds. Then we have the following energy inequality:*

$$\frac{d}{dt}E(\rho, u, d, E, B)$$
$$+ \int_\Omega \frac{\nu}{m}H''(\rho)|\nabla\rho|^2 + \frac{\mu\nu\hbar^2}{4m^2}\rho|\nabla^2\log\rho|^2 + \nu\rho|\nabla u|^2 + \frac{1}{\tau}\rho|u|^2 + \lambda\alpha_1|d\times(\Delta d+B)|^2$$
$$+ \delta|\nabla u|^2 + \delta|u|^2 dx = 0, \tag{4.1}$$

where

$$E(\rho, u, d, E, B) = \int_\Omega H(\rho) + \frac{\mu\hbar^2}{2m^2}|\nabla\sqrt{\rho}|^2 + \frac{1}{2}\rho|u|^2 + \lambda\frac{1}{2}|\nabla d|^2 + \frac{1}{2m}|E|^2 + \frac{1}{2m}|B|^2 dx, \tag{4.2}$$

here, $H(\rho) = \dfrac{A\rho^\gamma}{\gamma-1}$ for $\gamma > 1$ and $H(\rho) = \rho(\log\rho - 1)$ for $\gamma = 1$.

Proof Multiplying (1.20) by $\dfrac{1}{m}H'(\rho) - \dfrac{|u|^2}{2} - \dfrac{\mu\hbar^2}{2m^2}\dfrac{\Delta\sqrt{\rho}}{\sqrt{\rho}}$, and integrating by parts respect in Ω, we can get

$$\frac{d}{dt}\int_\Omega H(\rho)dx - \int_\Omega \partial_t\rho\frac{|u|^2}{2}dx + \frac{\mu\hbar^2}{2m^2}\frac{d}{dt}\int_\Omega|\nabla\sqrt{\rho}|^2 dx$$
$$+ \frac{1}{m}\int_\Omega \text{div}(\rho u)H'(\rho)dx - \int_\Omega \text{div}(\rho u)\frac{|u|^2}{2}dx$$
$$- \int_\Omega \text{div}(\rho u)\frac{\mu\hbar^2}{2m^2}\frac{\Delta\sqrt{\rho}}{\sqrt{\rho}}dx = -\frac{\nu}{m}\int_\Omega H''(\rho)|\nabla\rho|^2 dx + \nu\int_\Omega \nabla\rho:\nabla u:udx$$
$$- \int_\Omega \frac{\nu\mu\hbar^2}{4m^2}\rho|\nabla^2\log\rho|^2 + \delta|\nabla u|^2 + \delta|u|^2 dx. \tag{4.3}$$

Indeed we have

$$\int_\Omega \partial_t\rho H'(\rho)dx = \frac{d}{dt}\int_\Omega H(\rho)dx,$$
$$-\int_\Omega \partial_t\rho\frac{\mu\hbar^2}{2m^2}\frac{\Delta\sqrt{\rho}}{\sqrt{\rho}}dx = -\frac{\mu\hbar^2}{m^2}\int_\Omega \partial_t(\sqrt{\rho})\Delta\sqrt{\rho}dx = \frac{\mu\hbar^2}{2m^2}\partial_t\int_\Omega|\nabla\sqrt{\rho}|^2 dx,$$
$$\int_\Omega \nu\Delta\rho\frac{1}{m}H'(\rho)dx = -\frac{\nu}{m}\int_\Omega \nabla\rho\nabla H'(\rho)dx = -\frac{\nu}{m}\int_\Omega H''(\rho)|\nabla\rho|^2 dx,$$
$$-\int_\Omega \nu\Delta\rho\frac{|u|^2}{2}dx = \nu\int_\Omega \nabla\rho:\nabla u:udx,$$

and

$$\int_\Omega \frac{\Delta\sqrt{\rho}}{\sqrt{\rho}}\Delta\rho dx = -\int_\Omega \rho\nabla\log\rho\nabla\left(\frac{\Delta\sqrt{\rho}}{\sqrt{\rho}}\right)dx$$
$$= -\frac{1}{2}\int_\Omega \nabla\log\rho\,\text{div}(\rho\nabla^2\log\rho)dx = \frac{1}{2}\int_\Omega \rho|\nabla^2\log\rho|^2 dx.$$

Then multiplying (1.21) by u, and integrating by parts respect in Ω, we have

$$\int_\Omega \partial_t \rho |u|^2 + \rho \partial_t \left(\frac{|u|^2}{2}\right) + \frac{1}{2}\nabla \cdot (\rho u)|u|^2 dx - \frac{1}{m}\int_\Omega H'(\rho)\text{div}(\rho u)dx = -\int_\Omega \frac{\rho e}{m} E \cdot u dx$$

$$-\int_\Omega \frac{\mu \hbar^2}{2m^2}\left(\frac{\Delta\sqrt{\rho}}{\sqrt{\rho}}\right)\text{div}(\rho u)dx - \int_\Omega \nu \nabla \rho : \nabla u : u dx - \int_\Omega \nu \rho |\nabla u|^2 dx - \int_\Omega \frac{1}{\tau}\rho|u|^2 dx$$

$$+\lambda \int_\Omega \nabla d \odot \nabla d : \nabla u dx - \frac{|\nabla d|^2}{2}\nabla u dx \tag{4.4}$$

Indeed we have the following fact:

$$\int_\Omega \partial_t(\rho u) \cdot u + \text{div}(\rho u \otimes u) \cdot u dx$$

$$= \int_\Omega \partial_t \rho |u|^2 + \rho \partial_t \left(\frac{|u|^2}{2}\right) + \nabla \cdot (\rho u)|u|^2 + \rho u \cdot \nabla \left(\frac{|u|^2}{2}\right) dx$$

$$= \int_\Omega \partial_t \rho |u|^2 + \rho \partial_t \left(\frac{|u|^2}{2}\right) + \frac{1}{2}\nabla \cdot (\rho u)|u|^2 dx,$$

$$\int_\Omega \frac{\nabla P}{m}\cdot u dx = \frac{1}{m}\int_\Omega \nabla H \rho u dx = -\frac{1}{m}\int_\Omega H'(\rho)\text{div}(\rho u)dx,$$

$$-\int_\Omega \frac{\rho e}{m} u \times B \cdot u dx = 0,$$

$$\int_\Omega \frac{\mu \hbar^2}{2m^2}\rho \nabla \left(\frac{\Delta\sqrt{\rho}}{\sqrt{\rho}}\right) \cdot u dx = -\int_\Omega \frac{\mu \hbar^2}{2m^2}\left(\frac{\Delta\sqrt{\rho}}{\sqrt{\rho}}\right)\text{div}(\rho u)dx,$$

$$\int_\Omega \nu \Delta(\rho u) \cdot u = -\int_\Omega \nu \nabla(\rho u) \cdot \nabla u = -\int_\Omega \nu \nabla \rho : \nabla u : u dx - \int_\Omega \nu \rho |\nabla u|^2 dx,$$

$$\int_\Omega -\lambda \nabla \cdot \left(\nabla d \odot \nabla d - \frac{|\nabla d|^2}{2}I\right) \cdot u dx = \lambda \int_\Omega \nabla d \odot \nabla d : \nabla u dx - \frac{|\nabla d|^2}{2}\nabla u dx.$$

Multiplying (1.22) by $\Delta d + B$, and integrating by parts respect in Ω, we get

$$-\frac{d}{dt}\int_\Omega \frac{1}{2}|\nabla d|^2 dx - \int_\Omega \nabla d \odot \nabla d : \nabla u dx + \int_\Omega \frac{|\nabla d|^2}{2}\nabla u dx + \int_\Omega d_t B dx$$

$$+ \int_\Omega u \cdot \nabla d \cdot B dx - \alpha_1 \int_\Omega |d \times (\Delta d + B)|^2 dx = 0. \tag{4.5}$$

Here we use the following computation:

$$\int_\Omega d_t \cdot \Delta d dx = -\frac{d}{dt}\int_\Omega \frac{1}{2}|\nabla d|^2 dx,$$

$$\int_\Omega (u \cdot \nabla d) \cdot \Delta d dx = \int_\Omega d_{jj} u^i d_i dx$$

$$= \int_\Omega \left[(d_j u^i d_i)_j - d_i \cdot d_j u^i_j - u^i \left(\frac{|\nabla d|^2}{2}\right)_i\right] dx$$

$$= -\int_\Omega d_i \cdot d_j u_j^i + u_i^i \left(\frac{|\nabla d|^2}{2}\right) dx = -\int_\Omega \nabla d \odot \nabla d : \nabla u dx + \int_\Omega \frac{|\nabla d|^2}{2} \nabla u dx,$$

$$\alpha_1 \int_\Omega d \times (d \times (\Delta d + B)) \cdot (\Delta d + B) dx = -\alpha_1 \int_\Omega |d \times (\Delta d + B)|^2 dx,$$

$$\alpha_2 \int_\Omega d \times (\Delta d + B) \cdot (\Delta d + B) dx = 0.$$

Multiplying (1.23) by $\dfrac{E}{m}$, and integrating by parts respect in Ω, we have

$$\frac{1}{2m}\frac{d}{dt}\int_\Omega |E|^2 dx - \int_\Omega \nabla \times B \cdot \frac{E}{m} dx = \int_\Omega e\rho u \cdot \frac{E}{m} dx. \tag{4.6}$$

Multiplying (1.24) by $\dfrac{B}{m}$, and integrating by parts respect in Ω, we have

$$\frac{1}{2m}\frac{d}{dt}\int_\Omega |B|^2 dx + \int_\Omega \nabla \times E \cdot \frac{B}{m} dx = -\int_\Omega \beta d_t \cdot \frac{B}{m} dx - \int_\Omega \beta u \cdot \nabla d \cdot \frac{B}{m} dx. \tag{4.7}$$

We can easily have

$$\int_\Omega \nabla \times B \cdot \frac{E}{m} dx = \int_\Omega \nabla \times E \cdot \frac{B}{m} dx.$$

Combining (4.3)-(4.7), we can get

$$\frac{d}{dt}\int_\Omega H(\rho) + \frac{\mu\hbar^2}{2m^2}|\nabla\sqrt{\rho}|^2 + \frac{1}{2}\rho|u|^2 + \lambda\frac{1}{2}|\nabla d|^2 + \frac{1}{2m}|E|^2 + \frac{1}{2m}|B|^2 dx$$
$$+ \int_\Omega \frac{\nu}{m} H''(\rho)|\nabla\rho|^2 + \frac{\nu\mu\hbar^2}{4m^2}\rho|\nabla^2 \log\rho|^2 + \nu\rho|\nabla u|^2 + \frac{1}{\tau}\rho|u|^2$$
$$+ \lambda\alpha_1|d \times (\Delta d + B)|^2 + \delta|\nabla u|^2 + \delta|u|^2 dx = 0. \tag{4.8}$$

From Theorem 4.1, by Gronwall inequality, we can easily get the following estimates:

Corollary 4.1 *Set Theorem* 4.1 *holds, we have*

$$\|\sqrt{\rho}\|_{L^\infty([0,T];H^1(\Omega))} \leqslant C, \tag{4.9}$$

$$\|\rho\|_{L^\infty([0,T];L^\gamma(\Omega))} \leqslant C, \tag{4.10}$$

$$\|\sqrt{\rho}u\|_{L^\infty(0,T);L^2(\Omega)} \leqslant C, \tag{4.11}$$

$$\|\sqrt{\rho}\nabla u\|_{L^2([0,T];L^2(\Omega))} \leqslant C, \tag{4.12}$$

$$\|B\|_{L^\infty([0,T];L^2(\Omega))} + \|E\|_{L^\infty([0,T],L^2(\Omega))} \leqslant C, \tag{4.13}$$

$$\|\nabla d\|_{L^\infty([0,T];L^2(\Omega))} \leqslant C, \tag{4.14}$$

$$\|d \times (\Delta d + B)\|_{L^2([0,T];L^2(\Omega))} \leq C, \tag{4.15}$$

$$\|\sqrt{\rho}\nabla^2 \log \rho\|_{L^2([0,T];L^2(\Omega))} \leq C, \tag{4.16}$$

$$\delta\|u\|_{L^2([0,T];H^1(\Omega))} \leq C. \tag{4.17}$$

The energy inequality (4.1) and Corollary 4.1 allows us to conclude some estimates.

Lemma 4.1 *The following uniform estimate holds for some constant $C > 0$,*

$$\|\sqrt{\rho}\|_{L^2([0,T];H^2(\Omega))} + \|\sqrt[4]{\rho}\|_{L^4([0,T];W^{1,4}(\Omega))} \leq C. \tag{4.18}$$

Proof The lemma follow form the energy estimate in Theorem 4.1, the inequality

$$\int_\Omega \rho|\nabla^2 \log \rho|^2 dx \geq \kappa_2 \int_\Omega |\nabla^2 \sqrt{\rho}|^2 dx, \tag{4.19}$$

with κ_2, which is shown in [18], the inequality

$$\int_\Omega \rho|\nabla^2 \log \rho|^2 dx \geq \kappa \int_\Omega |\nabla \sqrt[4]{\rho}|^4 dx, \quad \kappa > 0,$$

which is proved in [18]. □

We able to deduce more regularity from the H^2 bound for $\sqrt{\rho}$.

Lemma 4.2 (space regularity for ρ and ρu) *The following uniform estimates hold for some constant $C > 0$ not depending on N and δ:*

$$\|\rho u\|_{L^2([0,T];W^{1,3/2}(\Omega))} \leq C, \tag{4.20}$$

$$\|\rho\|_{L^2([0,T];W^{2,p}(\Omega))} \leq C, \tag{4.21}$$

$$\|\rho\|_{L^{4\gamma/3+1}([0,T];L^{4\gamma/3+1}(\Omega))} \leq C, \tag{4.22}$$

where $p < 2$.

Proof Since the space $H^2(\Omega)$ embeds continuously into $L^\infty(\Omega)$, showing that $\sqrt{\rho}$ is bounded in $L^2([0,T];L^\infty(\Omega))$. Thus, in view of (4.11)-(4.12) $\rho u = \sqrt{\rho}\sqrt{\rho}u$ is uniformly bounded in $L^2([0,T];L^2(\Omega))$. By (4.9) and (4.18), $\nabla\sqrt{\rho}$ is bounded in $L^2([0,T];L^6(\Omega))$ and $\sqrt{\rho}$ is bounded in $L^\infty([0,T];L^6(\Omega))$. This, together with (4.9), implies that

$$\nabla(\rho u) = 2\nabla\sqrt{\rho} \otimes (\sqrt{\rho}u) + \sqrt{\rho}\nabla u\sqrt{\rho}$$

is uniformly bounded in $L^2([0,T];L^{3/2}(\Omega))$, proving the first claim.

For the second claim, we observe first that, by the Gagliardo-Nirenberg inequality in Lemma 2.1, with $p = 2\gamma/(\gamma + 1)$ and $\theta = 1/2$,

$$\|\nabla\sqrt{\rho}\|^4_{L^4([0,T];L^{2p}(\Omega))} \leqslant C \int_0^T \|\sqrt{\rho}\|^{4\theta}_{H^2(\Omega)} \|\sqrt{\rho}\|^{4(1-\theta)}_{L^{2\gamma}(\Omega)} dt$$

$$\leqslant C\|\sqrt{\rho}\|^{4(1-\theta)}_{L^\infty([0,T];L^{2\gamma}(\Omega))} \int_0^T \|\sqrt{\rho}\|^2_{H^2(\Omega)} dt \leqslant C.$$

Thus, $\sqrt{\rho}$ is bounded in $L^4([0,T]; W^{1,2p}(\Omega))$. Notice that ρ is bounded in $L^\infty([0,T]; H^1(\Omega)) \hookrightarrow L^\infty([0,T]; L^\alpha(\Omega))$ for all $\alpha < \infty$. Then we may replace in the above estimate 2γ by α, obtaining an $L^4([0,T]; W^{1,2p}(\Omega))$ bound for all $p > 2$. The estimate on $\nabla\sqrt{\rho}$ in $L^4([0,T]; L^{2p}(\Omega))$ shows that

$$\nabla^2 \rho = 2(\sqrt{\rho}\nabla^2\sqrt{\rho} + \nabla\sqrt{\rho} \otimes \nabla\sqrt{\rho})$$

is bounded in $L^2([0,T]; L^p(\Omega))$ which proves the second claim.

Finally, the Gagliardo-Nirenberg inequality, with $\theta = 3/(4\gamma + 3)$ and $q = 2(4\gamma + 3)/3$,

$$\|\sqrt{\rho}\|^q_{L^q([0,T];L^q(\Omega))} \leqslant C \int_0^T \|\sqrt{\rho}\|^{q\theta}_{H^2(\Omega)} \|\sqrt{\rho}\|^{q(1-\theta)}_{L^{2\gamma}(\Omega)} dt$$

$$\leqslant C\|\rho\|^{q(1-\theta)}_{L^\infty([0,T];L^\gamma(\Omega))} \int_0^T \|\sqrt{\rho}\|^2_{H^2(\Omega)} dt \leqslant C,$$

shows that ρ is bounded in $L^{q/2}([0,T]; L^{q/2}(\Omega))$. This finishes the proof. □

From (4.16) we can get the estimate about $\|\Delta d\|_{L^2([0,T];H^2(\Omega))}$.

Lemma 4.3

$$\|\Delta d\|_{L^2([0,T];L^2(\Omega))} \leqslant C. \tag{4.23}$$

Proof Since

$$\int_0^T \int_\Omega |d \times (\Delta d + B)|^2 dx dt = \int_0^T \int_\Omega |d \times \Delta d + d \times B|^2 dx dt,$$

and

$$|d \times (\Delta d + B)|^2 = (d \cdot d)((\Delta d + B) \cdot (\Delta d + B)) - (d \cdot (\Delta d + B))^2$$
$$= |\Delta d|^2 + 2(\Delta d \cdot B) + |B|^2 - ((|\nabla d|^2 d)^2 + 2(d \cdot \Delta d)(d \cdot B) + 2|d \cdot B|^2),$$

we have

$$\int_0^T \int_\Omega |\Delta d|^2 dxdt$$

$$\leq \int_0^T \int_\Omega |d \times \Delta d + B|^2 - 2(d \cdot \Delta d) + 2(|\nabla d|^2 d)^2 + 2(d \cdot \Delta d)(d \cdot B)) - |d \times B|^2 dxdt$$

$$\leq C_1 \varepsilon \int_0^T \int_\Omega |\Delta d|^2 dxdt + C_2 \int_0^T \|\nabla d\|_{L^4(\Omega)}^4 dt + C_3$$

$$\leq C_1 \varepsilon \int_0^T \int_\Omega |\Delta d|^2 dxdt + C_4 \int_0^T \int_\Omega |\nabla d|^4 dxdt + C.$$

From [21], we know that if $d : \mathbf{R}^2 \to S^1$, $\nabla d \in H^1(\mathbf{R}^2)$, $\|\nabla d\|_{L^2} \leq C$, $d_2 \geq \varepsilon_0$, then $\|\nabla d\|_{L^4}^4 \leq (1-\delta_0)\|\Delta d\|_{L^2}^2$.

Therefore, if $1 - C_1\varepsilon - C_4(1-\delta_0) > 0$, we can get (4.23).

$\square$

Lemma 4.4(time regularity for ρ and ρu) *The following uniform estimates hold for $s > d/2 + 1$:*

$$\|\partial_t \rho\|_{L^2([0,T];L^{3/2}(\Omega))} \leq C, \qquad (4.24)$$

$$\|\partial_t(\rho u)\|_{L^{4/3}([0,T];(H^s(\Omega))^*)} \leq C. \qquad (4.25)$$

Proof By (4.20)-(4.21), we find that $\partial_t \rho = -\mathrm{div}(\rho u) - \nu \Delta \rho$ is uniformly bounded in $L^2([0,T]; L^{3/2}(\Omega))$, achieving the first claim.

The sequence $\rho u \otimes u$ is bounded in $L^\infty([0,T]; L^1(\Omega))$; hence, $\mathrm{div}(\rho u \otimes u)$ is bounded in $L^\infty([0,T]; (W^{1,\infty}(\Omega))^*)$, and because of the continuous embedding of $H^s(\Omega)$ into $W^{1,\infty}(\Omega)$ for $s > d/2 + 1$, and also in $L^\infty([0,T]; (H^s(\Omega))^*)$. The estimate

$$\int_0^T \int_\Omega \rho \nabla\left(\frac{\Delta\sqrt{\rho}}{\sqrt{\rho}}\right) \cdot \phi dxdt$$

$$= -\int_0^T \int_\Omega \Delta\sqrt{\rho}(2\nabla\sqrt{\rho} \cdot \phi + \sqrt{\rho}\mathrm{div}\phi) dxdt$$

$$\leq \|\Delta\sqrt{\rho}\|_{L^2([0,T];L^2(\Omega))} (2\|\sqrt{\rho}\|_{L^4([0,T];W^{1,3}(\Omega))} \|\phi\|_{L^4([0,T];L^6(\Omega))})$$

$$+ \|\sqrt{\rho}\|_{L^\infty([0,T];L^6(\Omega))} \|\phi\|_{L^2([0,T];W^{1,3}(\Omega))}$$

$$\leq C\|\phi\|_{L^4([0,T];W^{1,3}(\Omega))}$$

for all $\phi \in L^4([0,T]; W^{1,3}(\Omega))$ proves that $\rho \Delta\sqrt{\rho}/\sqrt{\rho}$ is uniformly bounded in $L^{4/3}([0, T]; (W^{1,3}(\Omega))^*) \hookrightarrow L^{4/3}([0,T]; (H^s(\Omega))^*)$. In view of (4.22), ρ^γ is bounded in $L^{4/3}([0, T]; L^{4/3}(\Omega)) \hookrightarrow L^{4/3}([0,T]; (H^s(\Omega))^*)$. Furthermore, by (4.20), $\Delta(\rho u)$ is uniformly bounded in $L^2([0,T];(W^{1,3}(\Omega))^*)$, and by (4.17), $\delta\Delta u$ is bounded in $L^2([0,T];(H^1(\Omega))^*)$. Therefore, using Corollaries 4.1 and 4.3, we get that

$$(\rho u)_t = -\operatorname{div}(\rho u \otimes u) - \nabla P(\rho) + \frac{\rho e}{m}(E + u \times B) + \mu \frac{\hbar^2}{2m^2}\nabla\left(\frac{\nabla\sqrt{\rho}}{\sqrt{\rho}}\right)$$
$$+ \nu\Delta(\rho u) - \frac{\rho u}{\tau} - \lambda\nabla\cdot\left(\nabla d \odot \nabla d - \frac{|\nabla d|^2}{2}I\right)$$

is uniformly bounded in $L^{4/3}([0,T];(H^s(\Omega))^*)$. □

The $L^4([0,T];W^{1,4}(\Omega))$ bound (4.18) on $\sqrt[4]{\rho}$ provides a uniform estimate for $\partial_t\sqrt{\rho}$.

Lemma 4.5 (time regularity for $\sqrt{\rho}$) *The following estimate holds*:

$$\|\partial_t\sqrt{\rho}\|_{L^2([0,T];(H^1(\Omega))^*)} \leqslant C. \tag{4.26}$$

Proof Dividing the mass equation (1.1) by $2\sqrt{\rho}$ gives

$$\partial_t\sqrt{\rho} = -\nabla\sqrt{\rho}\cdot u - \frac{1}{2}\sqrt{\rho}\operatorname{div}u + \nu(\Delta\sqrt{\rho} + 4|\nabla\sqrt[4]{\rho}|^2)$$
$$= -\operatorname{div}(\sqrt{\rho}u) + \frac{1}{2}\sqrt{\rho}\operatorname{div}u + \nu(\Delta\sqrt{\rho} + 4|\nabla\sqrt[4]{\rho}|^2).$$

The first term on the right-hand side is bounded in $L^2([0,T];(H^1(\Omega)^*)$ by (4.11) and (4.12). The remaining terms are uniformly bounded in $L^2([0,T];L^2(\Omega))$; see (4.11), (4.12), (4.17). □

Lemma 4.6 *There holds that*

$$\|d_t\|_{L^\infty([0,T];L^2(\Omega))} \leqslant C. \tag{4.27}$$

Proof Multiplying (1.3) by d_t, and integrating by parts respect to x in Ω, we have

$$\int_\Omega |d_t|^2 dx$$
$$= -\int_\Omega \{u\cdot\nabla d\cdot d_t + \alpha_1 d\times(d\times(\Delta d + B))\cdot d_t - \alpha_2 d\times(\Delta d + B)\cdot d_t\}dxdt$$
$$= -\int_\Omega \{u\cdot\nabla d\cdot d_t + \alpha_1(d\cdot(\Delta d + B))(d\cdot d_t) - \alpha_1(d\cdot d)(\Delta d + B)\cdot d_t$$
$$- \alpha_2(d\times\Delta d)\cdot d_t - \alpha_2(d\times B)\cdot d_t\}dxdt$$
$$= \int_\Omega \{u\cdot\nabla d\cdot d_t + \alpha_1(\Delta d + B)\cdot d_t + \alpha_2(d\times\nabla d_t)\cdot\nabla d + \alpha_2(d\times B)\cdot d_t\}dxdt$$
$$\leqslant \|u\|_{L^4(\Omega)}\|\nabla d\|_{L^4(\Omega)}\|d_t\|_{L^2(\Omega)} + \alpha_1\|\Delta d\|_{L^2(\Omega)}\|d_t\|_{L^2(\Omega)}$$

$$+ (\alpha_1 + \alpha_2)\|B\|_{L^2(\Omega)}\|d_t\|_{L^2(\Omega)} + \alpha_2\|\nabla d\|_{L^2(\Omega)}\|d_t\|_{L^2(\Omega)} - \alpha_1 \frac{d}{dt}\int_\Omega |\nabla d|^2 dx$$

$$\leqslant \frac{1}{2}\|d_t\|^2_{L^2(\Omega)} + C_1\|\Delta d\|^2_{L^2(\Omega)} + C_2\|B\|^2_{L^2(\Omega)} + C_3\|\nabla d\|^2_{L^2(\Omega)}$$

$$+ C_4\|u\|^2_{L^2(\Omega)} - \alpha_1 \frac{d}{dt}\int_\Omega |\nabla d|^2 dx.$$

Here C_1, C_2, C_3, C_4 are constants independent of N. Then integrating by parts respect to t in $[0,T]$, using the Corollary 4.1 and Lemma 4.3 we get (4.27). Thus we complete the proof of this lemma. □

Next we will show the limit of the Fadeo-Galerkin approximated solution. We first perform the limit $N \to \infty$, $\delta > 0$ being fixed. The limit $\delta \to 0$ is carried out in the last part of this section. We consider both limits separately since the weak formulation (1.16)-(1.19) for the viscous quantum NSLLM equations is different from its approximation (3.1) and (3.4)-(3.7).

We conclude from the Aubin-Lions lemma, taking into account the regularity (4.21) and (4.24) for ρ_N, the regularity (4.18) and (4.26) for $\sqrt{\rho_N}$, and the regularity (4.20) and (4.25) for $\rho_N u_N$, that there exist subsequence of ρ_N, $\sqrt{\rho_N}$, and $\rho_N u_N$, which are not related, such that, for some function ρ and J, as $N \to \infty$,

$$\rho_N \to \rho \text{ strongly in } L^2([0,T]; L^\infty(\Omega)),$$
$$\sqrt{\rho_n} \to \sqrt{\rho} \text{ weakly in } L^2([0,T]; H^2(\Omega)),$$
$$\sqrt{\rho_n} \to \sqrt{\rho} \text{ strongly in } L^2([0,T]; H^1(\Omega)),$$
$$\rho_n u_N \to J \text{ strongly in } L^2([0,T]; L^2(\Omega)).$$

Here we have used that the embedding $W^{2,p}(\Omega) \hookrightarrow L^\infty(\Omega) (p > 3/2)$, $H^2(\Omega) \hookrightarrow H^1(\Omega)$, and $W^{1,3/2}(\Omega) \hookrightarrow L^2(\Omega)$ are compact. The estimate (4.17) on u_N provides further the existence of a subsequence (not relabeled) such that, as $N \to \infty$,

$$u_N \rightharpoonup u \text{ weakly in } L^2([0,T]; H^1(\Omega)).$$

Then, since $\rho_n u_N$ converges weakly to ρu in $L^1([0,T]; L^6(\Omega))$, we infer that $J = \rho u$.

We are now in the position to let $N \to \infty$ in the approximate system (3.1) and (3.4)-(3.7) with $\rho = \rho_N$, $u = u_N$, $d = d_N$, $E = E_N$ and $B = B_N$. Clearly, the limit $N \to \infty$ show immediately that n solves

$$\rho_t + \text{div}(\rho u) = \nu \Delta \rho.$$

Next we consider the weak formulation (3.4) term by term. The strong convergence of $\rho_N u_N$ in $L^2([0,T]; L^2(\Omega))$ and the weak convergence of ρ_N in $L^2([0,T]; L^6(\Omega))$ leads to

$$\rho_N u_N \otimes u_N \rightharpoonup \rho u \otimes u \quad \text{weakly in} \quad L^1([0,T]; L^{3/2}(\Omega)).$$

Furthermore, in view of (4.21) (up to a subsequence),

$$\nabla(\rho_n u_N) \rightharpoonup \nabla(\rho u) \quad \text{weakly in} \quad L^2([0,T]; L^{3/2}(\Omega)).$$

The $L^\infty([0,T]; L^\gamma(\Omega))$ bound for ρ_N shows that $\rho^\gamma \rightharpoonup z$ weakly $*$ in $L^\infty([0,T]; L^1(\Omega))$ for some function z, and since $\rho_N^\gamma \to \rho^\gamma$, a.e., $z = \rho^\gamma$. Finally, the above convergence results show that the limit $N \to \infty$ of

$$\int_\Omega \frac{\Delta\sqrt{\rho_N}}{\sqrt{\rho_N}} \operatorname{div}(\rho_N \phi) dx = \int_\Omega \Delta\sqrt{\rho_N}(2\nabla\sqrt{\rho_N} \cdot \phi + \sqrt{\rho_N}\operatorname{div}\phi) dx$$

equals, for sufficiently smooth test functions,

$$\int_\Omega \Delta\sqrt{\rho}(2\nabla\sqrt{\rho} \cdot \phi + \sqrt{\rho}\operatorname{div}\phi) dx.$$

From Lemma 4.3 we know that $\|\nabla d\|_{L^2([0,T];H^2(\Omega))\cap L^4([0,T];L^4(\Omega))\cap L^\infty([0,T];L^2(\Omega))}$ is bounded thus we have

$$\nabla d_N \odot \nabla d_N \rightharpoonup \nabla d \odot \nabla d, \quad \text{weakly in } L^2([0,T]; L^2(\Omega)),$$

$$\frac{|\nabla d_N|^2}{2} I \rightharpoonup \frac{|\nabla d|^2}{2} I, \quad \text{weakly in } L^2([0,T]; L^2(\Omega)).,$$

$$\nabla d_N \odot \nabla d_N \rightharpoonup \nabla d \odot \nabla d, \quad \text{strongly in } L^1([0,T]; L^1(\Omega)),$$

$$\frac{|\nabla d_N|^2}{2} I \rightharpoonup \frac{|\nabla d|^2}{2} I, \quad \text{strongly in } L^1([0,T]; L^1(\Omega)).$$

Since we have $\|E\|_{L^\infty([0,T];L^2(\Omega))} \leqslant C$ and $\|B\|_{L^\infty([0,T];L^2(\Omega))} \leqslant C$, we can get that

$$\frac{\rho_N e}{m} E_N \rightharpoonup \frac{\rho e}{m} E \quad \text{weakly in} \quad L^1([0,T]; L^1(\Omega)),$$

$$\frac{\rho_N e}{m} u_N \times B_N \rightharpoonup \frac{\rho e}{m} u \times B \quad \text{weakly in} \quad L^1([0,T]; L^1(\Omega)).$$

Thus we have shown that (ρ, u, d, E, B) solves $\rho_t + \operatorname{div}(\rho u) = \nu \Delta \rho$ pointwise and for all test function such that the integrals are defined,

$$-\int_\Omega \rho_0 u_0 \cdot \phi(\cdot, 0) dx$$

$$= \int_0^T \int_\Omega \Big(\rho u \cdot \phi_t + \rho(u \otimes u_N) : \nabla\phi + P(\rho)\text{div}\phi - \frac{\rho e}{m}(E + u \times B)\phi$$
$$- \frac{\hbar^2}{2m^2} \frac{\Delta\sqrt{\rho}}{\sqrt{\rho}} \text{div}(\rho\phi) - \nu\nabla(\rho u) : \nabla\phi - \frac{\rho u}{\tau}\phi + \lambda\Big(\nabla d \odot \nabla d - \frac{|\nabla d|^2}{2}I\Big)\nabla\phi$$
$$- \delta(\nabla u : \nabla\phi + u \cdot \phi)\Big) dx dt. \tag{4.28}$$

Similarly, using the a-priori estimates we can show that as $N \to \infty$, the limit of (d_N, E_N, B_N) satisfy:

$$-\int_\Omega d_0\phi(x,0)dx = \int_0^T \int_\Omega \Big(d\phi_t + u \cdot \nabla d \cdot \phi + \alpha_1 d \times (d \times (\Delta d + B)) \cdot \phi$$
$$- \alpha_2 d \times (\Delta d + B) \cdot \phi\Big) dx dt, \tag{4.29}$$

$$-\int_\Omega E_0 \cdot \phi(x,0)dx = \int_0^T \int_\Omega (E \cdot \phi_t - B \cdot (\nabla \times \phi) - e\rho u \cdot \phi) dx dt, \tag{4.30}$$

$$-\int_\Omega (B_0 + \lambda m d_0) \cdot \phi(x,0)dx$$
$$= \int_0^T \int_\Omega ((B + \lambda m d) \cdot \phi_t + E \cdot (\nabla \times \phi) + \lambda m(u \cdot \nabla d) \cdot \phi) dx dt. \tag{4.31}$$

Finally, we will show the limit as $\delta \to 0$. Let $(\rho^\delta, u^\delta, d^\delta, E^\delta, B^\delta)$ be a solution to (3.2),(3.4)-(3.7) with the regularity proved in the previous. By employing the test function $\rho^\delta \phi$ in (3.4)(which is possible as long as the integrals are well defined), we obtain, according to

$$-\int_\Omega \rho_0^2 u_0 \phi = \int_0^T \int_\Omega \Big((\rho^\delta)^2 u^\delta \cdot \phi_t - (\rho^\delta)^2 \text{div}(u^\delta) u^\delta \cdot \phi - \nu(\rho^\delta u^\delta \otimes \nabla\rho^\delta) : \nabla\phi$$
$$+ \rho^\delta u^\delta \otimes \rho^\delta u^\delta : \nabla\phi + \frac{\gamma}{\gamma+1}(\rho^\delta)^{\gamma+1}\text{div}\phi + \frac{\rho^\delta e}{m}(E^\delta + u^\delta \times B^\delta) \cdot \phi$$
$$- 2\varepsilon_0^2 \Delta\sqrt{\rho^\delta}(2\sqrt{\rho^\delta}\nabla\rho^\delta \cdot \phi + (\rho^\delta)^{3/2}\text{div}\phi)$$
$$- \nu\nabla(\rho^\delta u^\delta) : (\rho^\delta \nabla\phi + 2\nabla\rho^\delta \otimes \phi)$$
$$+ \lambda\Big(\nabla d^\delta \odot \nabla d^\delta - \frac{|\nabla d^\delta|^2}{2}I\Big) \cdot \nabla(\rho^\delta \phi))$$
$$- \delta\nabla u^\delta : (\rho^\delta \nabla\phi + \nabla\rho^\delta \otimes \phi) - \delta\rho^\delta u^\delta \cdot \phi\Big) dx dt. \tag{4.32}$$

The Aubin-Lions lemma and the regularity results from the previous allow us to extract subsequences (not relabeled) such that as $\delta \to 0$, for some functions ρ and j,

$$\rho^\delta \to \rho \text{ strongly in } L^2([0,T];W^{1,p}(\Omega)), \quad 3 < p < 6\gamma/(\gamma+3), \tag{4.33}$$

$$\rho^\delta u^\delta \to J \quad \text{strongly in } L^2([0,T]; L^q(\Omega)), \quad 1 \leqslant q < 3, \tag{4.34}$$

$$\sqrt{\rho^\delta} \to \sqrt{\rho} \quad \text{strongly in } L^2([0,T]; L^r(\Omega)), \quad 1 \leqslant r < 6. \tag{4.35}$$

Estimate (4.11)-(4.12) and Fatou lemma yield

$$\int_\Omega \liminf_{\delta \to 0} \frac{|\rho^\delta u^\delta|}{\rho^\delta} dx < \infty.$$

This implies that $J = 0$ in $\rho = 0$. Then, when we define the limit velocity $u := J/\rho$ in $\{\rho \neq 0\}$ and $u := 0$ in $\rho = 0$, we have $J = \rho u$. By (4.11)-(4.12), there exists a subsequence(not relabeled) such that

$$\sqrt{\rho^\delta} u^\delta \rightharpoonup g \quad \text{weakly} * \text{ in } L^\infty([0,T]; L^2(\Omega)) \tag{4.36}$$

for some function g. Hence, since $\sqrt{\rho^\delta}$ converges strongly to $\sqrt{\rho}$ in $L^2([0,T]; L^\infty(\Omega))$, we infer that $\rho^\delta u^\delta = \sqrt{\rho^\delta}(\sqrt{\rho^\delta} u^\delta)$ converges weakly to $\sqrt{\rho} g$ in $L^2([0,T]; L^2(\Omega))$ and $\sqrt{\rho} g = \rho u = J$. In particular, $g = J/\sqrt{\rho}$ in $\{\rho \neq 0\}$.

Now we are able to pass to the limit $\delta \to 0$ in the weak formulation (4.32) term by term. The strong convergence (4.33) and (4.34) imply that

$$(\rho^\delta)^2 u_\delta \to \rho^2 u \quad \text{strongly in } L^1([0,T]; L^q(\Omega)), \quad q < 3,$$

$$\rho^\delta \otimes \nabla \rho^\delta \to \rho u \otimes \nabla \rho \quad \text{strongly in } L^1([0,T]; L^{3/2}(\Omega)).$$

The strong convergence of $\rho^\delta u^\delta$ immediately gives

$$\rho^\delta u^\delta \otimes \rho^\delta u^\delta \to \rho u \otimes \rho u \quad \text{strongly in } L^1([0,T]; L^{q/2}(\Omega)), \quad q < 3.$$

Furthermore, we have

$$\nabla \rho^\delta \to \nabla \rho \text{ strongly in } L^2([0,T]; L^p(\Omega)) \text{ by (4.33)}, \quad p > 3,$$

$$\sqrt{\rho^\delta} \to \sqrt{\rho} \text{ strongly in } L^\infty([0,T]; L^r(\Omega)) \text{ by (4.35) with } r = 2p/(p-2),$$

$$\Delta\sqrt{\rho^\delta} \rightharpoonup \Delta\sqrt{\rho} \text{ weakly in } L^2([0,T]; L^2(\Omega)) \text{ by (4.18)}.$$

It holds that $r < 6$ since we have $p > 3$. This implies that

$$\Delta\sqrt{\rho^\delta}\sqrt{\rho^\delta}\nabla\rho^\delta \rightharpoonup \Delta\sqrt{\rho}\sqrt{\rho}\nabla\rho \quad \text{weakly in } L^1([0,T]; L^1(\Omega)).$$

Since $\nabla\sqrt{\rho^\delta u^\delta}$ converges weakly in $L^2([0,T]; L^{3/2}(\Omega))$(see (4.20)) and $\nabla\rho_\delta$ converges strongly in $L^2([0,T]; L^3(\Omega))$(see (4.33)), we obtain

$$\nabla(\rho^\delta u^\delta) \cdot \rho^\delta \rightharpoonup \nabla(\rho u) \cdot \nabla \rho \quad \text{weakly in } L^1([0,T]; L^1(\Omega)).$$

The almost everywhere convergence of ρ_δ and the $L^{4\gamma/3+1}([0,T]; L^{4\gamma/3+1}(\Omega))$ bound on ρ_δ (see (4.22)), together with the fact that $4\gamma/3 + 1 > \gamma + 1$, proves that

$$(\rho^\delta)^{\gamma+1} \to \rho^{\gamma+1} \text{ strongly in } L^1([0,T]; L^1(\Omega)).$$

Using the estimate (4.17) for $\sqrt{\delta} u_\delta$, we obtain further, for smooth test functions,

$$\delta \int_\Omega \nabla u^\delta : (\rho^\delta \nabla \phi + \nabla \rho^\delta \times \phi) dx$$
$$\leqslant \sqrt{\delta} \|\sqrt{\delta} \nabla u^\delta\|_{L^2([0,T]; L^2(\Omega))} (\|\rho^\delta\|_{L^2([0,T]; L^\infty(\Omega))} \|\phi\|_{L^\infty([0,T]; H^1(\Omega))})$$
$$+ \|\rho^\delta\|_{L^2([0,T]; W^{1,3}(\Omega))} \|\phi\|_{L^\infty([0,T]; L^6(\Omega))} \to 0, \text{ as } \delta \to 0,$$
$$\delta \int_\Omega \rho^\delta u^\delta \cdot \phi dx \leqslant \delta \|\rho^\delta u^\delta\|_{L^2([0,T]; L^3(\Omega))} \|\phi\|_{L^2([0,T]; L^{3/2}(\Omega))} \to 0, \text{ as } \delta \to 0.$$

It remains to show the convergence of $(\rho^\delta)^2 \text{div}(u^\delta) u^\delta$. To this end, we proceed similarly as in [11] and introduce the functions $G_\alpha \in C^\infty([0,\infty))$, $\alpha > 0$, satisfying $G_\alpha(x) = 1$ for $x \geqslant 2\alpha$, $G_\alpha(x) = 0$ for $x \leqslant \alpha$, and $0 \leqslant G_\alpha \leqslant 1$. Then we estimate the low-density part of $(\rho^\delta)^2 \text{div}(u\delta) u^\delta$ by

$$\|(1 - G_\alpha(\rho^\delta))(\rho^\delta)^2 \text{div}(u^\delta) u^\delta\|_{L^1([0,T]; L^1(\Omega))}$$
$$\leqslant \|(1 - G_\alpha(\rho^\delta)) \sqrt{\rho^\delta}\|_{L^\infty([0,T]; L^\infty(\Omega))} \|\sqrt{\rho^\delta} \text{div} u^\delta\|_{L^2([0,T]; L^2(\Omega))} \|\rho^\delta u^\delta\|_{L^2([0,T]; L^2(\Omega))}$$
$$\leqslant C \|(1 - G_\alpha(\rho^\delta)) \sqrt{\rho^\delta}\|_{L^\infty([0,T]; L^\infty(\Omega))} \leqslant C \sqrt{\alpha}, \quad (4.37)$$

where $C > 0$ is independent of δ and α. We write

$$G_\alpha(\rho^\delta) \rho^\delta \text{div} u^\delta = \text{div}(G_\alpha(\rho^\delta) \rho^\delta u^\delta) - \rho^\delta u^\delta \otimes \nabla \rho^\delta \left(G_\alpha'(\rho^\delta) + \frac{G_\alpha(\rho^\delta)}{\rho^\delta} \right). \quad (4.38)$$

As $\delta \to 0$, the first term on the right-hand side converges strongly to $\text{div}(G_\alpha(\rho) \rho u)$ in $L^1([0,T]; (H^1(\Omega))^*)$ since $G_\alpha(\rho^\delta)$ converges strongly to $G_\alpha(\rho)$ in $L^p([0,T]; L^p(\Omega))$ for any $p < \infty$ and $\rho^\delta u^\delta$ converges strongly to ρu in $L^2([0,T]; L^q(\Omega))$ for any $q < 3$. In view of (4.35) and (4.36), we infer the weak* convergence $\rho^\delta u^\delta \rightharpoonup \sqrt{\rho g} = \rho u$ in $L^\infty([0,T]; L^{2r/(r+2)}(\Omega))$ for all $r < 6$. Thus, because of (4.33),

$$\rho^\delta u^\delta \otimes \nabla \rho^\delta \rightharpoonup \rho u \otimes \nabla \rho \text{ weakly in } L^2([0,T]; L^\theta(\Omega)),$$

where $\theta = 2pr/(2p + 2r + pr)$. It is possible to choose $3 < p \leqslant 6\gamma/(\gamma + 3)$ and $r < 6$ such that $\theta > 1$. Then, together with strong convergence of $G_\alpha'(\rho^\delta) + G_\alpha(\rho^\delta)/\rho^\delta$ to $G_\alpha'(\rho) + G_\alpha(\rho)/\rho$ in $L^p([0,T]; L^p(\Omega))$ for any $p < \infty$, the limit $\delta \to 0$ in (4.38) yields the identity

$$G_\alpha(\rho)\rho\mathrm{div}u = \mathrm{div}(G_\alpha(\rho)\rho u) - \rho u \otimes \nabla\rho\left(G'_\alpha(\rho) + \frac{G_\alpha(\rho)}{\rho}\right)$$

in $L^1([0,T];(H^1(\Omega))^*)$. Since $G_\alpha(\rho^\delta)\rho^\delta\mathrm{div}u^\delta$ is bounded in $L^2([0,T];L^2(\Omega))$, we conclude that

$$G_\alpha(\rho^\delta)\rho^\delta\mathrm{div}u^\delta \rightharpoonup G_\alpha(\rho)\rho\mathrm{div}u \quad \text{weakly in } L^2([0,T]\rho;L^2(\Omega)).$$

Moreover, in view of the strong convergence of $\rho^\delta u^\delta$ to ρu in $L^2([0,T];L^q(\Omega))$ for all $q < 3$, we infer that

$$G_\alpha(\rho^\delta)\rho^\delta\mathrm{div}(u^\delta)\rho^\delta u^\delta \rightharpoonup G_\alpha\rho^2\mathrm{div}(u)u \quad \text{weakly in } L^1([0,T];L^{q/2}(\Omega)).$$

We write, for $\phi \in L^\infty([0,T];L^\infty(\Omega))$,

$$\int_\Omega ((\rho^\delta)^2\mathrm{div}(u^\delta)u^\delta - \rho^2\mathrm{div}(u)u) \cdot \phi dx$$
$$= \int_\Omega (G_\alpha(\rho^\delta)(\rho^\delta)^2\mathrm{div}(u^\delta)u^\delta - G_\alpha(\rho)\rho^2\mathrm{div}(u)u) \cdot dx$$
$$+ \int_\Omega (G_\alpha(\rho) - G_\alpha(\rho^\delta))\rho^2\mathrm{div}(u)u \cdot \phi dx$$
$$+ \int_\Omega (1 - G_\alpha(\rho^\delta))((\rho^\delta)^2\mathrm{div}(u^\delta)u^\delta - \rho^2\mathrm{div}(u)u) \cdot \phi dx. \tag{4.39}$$

For fixed $\alpha > 0$, the first integral converges to zero as $\delta \to 0$. Furthermore, the last integral can be estimated by $C\sqrt{\alpha}$ uniformly in δ ((4.37)). For the second integral, we recall that $G_\alpha(\rho^\delta) \to G_\alpha(\rho)$ strongly in $L^p([0,T];L^p(\Omega))$ for all $p < \infty$. Furthermore, by the Gagliardo-Nirenberg inequality, the bounds of ρu in $L^2([0,T];W^{1,3/2}(\Omega))$ and $L^\infty([0,T];L^{3/2}(\Omega))$ imply that $\rho u \in L^{5/2}([0,T];L^{5/2}(\Omega))$. Thus, since $\sqrt{\rho}\mathrm{div}u \in L^2([0,T];L^2(\Omega))$ and $\sqrt{\rho} \in L^q([0,T];L^q(\Omega))$ with $q = 8\gamma/3 + 2$ (see (4.22)),

$$\rho^2\mathrm{div}(u)u = \sqrt{\rho}(\sqrt{\rho}\mathrm{div}u)\rho u \in L^r([0,T];L^r(\Omega)), \quad r = \frac{18\gamma + 21}{20\gamma + 15} > 1.$$

As a consequence, the second integral converges to 0 as $\delta \to 0$. Thus, in the limit $\delta \to 0$, (4.39) can be made arbitrarily small, and hence,

$$(\rho^\delta)^2\mathrm{div}(u^\delta)u^\delta \rightharpoonup \rho^2\mathrm{div}(u)u \quad \text{weakly in } L^1([0,T];L^1(\Omega)).$$

Here we will omit the rest term convergence about d, E, B, you can refer to [11].

We have proved that (ρ, u, d, E, B) solves (1.20)-(1.25) for smooth initial data. Let $(\rho_0, u_0, d_0, E_0, B_0)$ be some finite-energy initial data, i.e. $\rho_0 \geq 0$ and $E(\rho_0, u_0, d_0, E_0, B_0) < \infty$, and let $(\rho_0^\delta, u_0^\delta, d_0^\delta, E_0^\delta, B_0^\delta)$ be smooth approximations satisfying $\rho_0^\delta \geq \delta > 0$ in Ω and $\sqrt{\rho_0^\delta} \to \sqrt{\rho_0}$ strongly in $H^1(\Omega)$ and $\sqrt{\rho_0^\delta} u_0^\delta \to \sqrt{\rho_0} u_0$ strongly in $L^{3/2}(\Omega)$. By the above proof, there exists a weak solution $(\rho^\delta, u^\delta, d^\delta, E^\delta, B^\delta)$ to (1.20)-(1.25) with initial data $(\rho_0^\delta, u_0^\delta, d_0^\delta, E_0^\delta, B_0^\delta)$ satisfying all the above bounds. In particular, $(\rho^\delta, \rho^\delta u^\delta)$ converges strongly in some spaces to $(\rho, \rho u)$ as $\delta \to 0$, and there exist uniform bounds for ρ^δ in $H^1([0,T]; L^{3/2}(\Omega))$ and for $\rho^\delta u^\delta$ in $W^{1,4/3}([0,T]; (H^s(\Omega))^*)$. Thus, up to subsequences, as $\delta \to 0$,

$$\rho_0^\delta = \rho^\delta(\cdot, 0) \rightharpoonup \rho(\cdot, 0) \text{ weakly in } L^{3/2}(\Omega),$$
$$\rho_0^\delta u_0^\delta = (\rho^\delta u^\delta)(\cdot, 0) \rightharpoonup (\rho u)(\cdot, 0) \text{ weakly in } (H^s(\Omega))^*.$$

This shows that $\rho(\cdot, 0) = \rho_0$ and $(\rho u)(\cdot, 0) = \rho_0 u_0$ in the sense of distributions. We conclude the proof of Theorem 1.1.

Note that a priori estimates in Section 4 are independent of D. By using the diagonal method and letting $D \to \infty$, we can obtain the global existence of weak solution to the Cauchy problem of system (1.1)-(1.6) and (1.7). for simplicity, we do not state the theorem here.

In the following work we will show the partial regularity of these weak solution to NSLLM. We will show that If the solution to (1.1)-(1.6) is smooth except finitely many singular point.

References

[1] F. Alouges and A. Soyeur. On global weak solutions for Landau-Lifshitz equations: Existence and nonuniqueness. *Nonlinear Anal. TMA*, **18** (1992), 1071-1084.

[2] D. Bresch, B. Desjardins and C. K. Lin. On some compressible fluid models: Korteweg, lubrication and shallow water systems. *Comm. Partial Differential Equations*, **28** (2003), 1009-1037.

[3] S. Brull and F. Méhats. Derivation of viscous correction terms for the isothermal quantum Euler model. *Z. Angew. Math. Mech.*, **90** (2010), 219-230.

[4] S. J. Ding and B. L. Guo. Existence of partial regularity weak solutions to Landau-Lifshitz-Maxwell equations. *Journal of Differential Equations*, **244(10)** (2008), 2448-2472.

[5] S. J. Ding, B. L. Guo, J. Y. Lin and M. Zeng. Global existence of weak solution for Landau-Lifshitz-Maxwell equtions. *Discrete and Continuous Dynamical Systems*-Series A, **17(4)** (2007), 867-890.

[6] E. Feireisl. *Dynamics of Viscous Compressible Fluids.* Oxford University Press, Oxford, 2004.

[7] E. Feireisl, A. Novotný and H. Petzeltová. On the existence of globally defined weak solutions to the Navier-Stokes equations. *J. Math. Fluid Mech.*, **3**(2001),358-392.

[8] C. Gardner. Numerical simulation of a steady-state electron shock wave in a submicron semiconductor device. *IEEE Trans. El. Dev.*, **38** (1991), 392-398.

[9] P. Germain, S. Ibrahim and N. Masmoudi. Well-posedness of the Navier-Stokes-Maxwell equations. *Proceedings of the Royal Society of Edinburgh*, **144(1)** (2014), 71-86.

[10] M. P. Gualdani and A. Jüngel. Analysis of the viscous quantum hydrodynamic equations for semiconductors. *Eur. J. Appl. Math.*, **15** (2004), 577-595.

[11] B. L. Guo and S. J. Ding. *Landau-Lifshitz equations.* Word Science: Singapore, 2008.

[12] B. L. Guo and M. C. Hong. The Landau-Lifshitz equations of the ferromagnetic spin chain and harmonic maps. *Calc. Var.*, **1** (1993), 311-334.

[13] G. Y. Hong, X. F. Hou, H. Y. Peng and C. J. Zhu. Global spherically symmetric classical solution to the Navier-Stokes-Mawell system with large initial data and vacuum. *Science China Mathematics*, **57(12)** (2014), 2463-2484.

[14] F. M. Huang, H. L. Li, A. Matsumura and S. Odanaka. Well-posedness and stability of multi-dimensional quantum hydrodynamics for semiconductors in R^3. *Series in Contemporary Applied Mathematics CAM 15*, High Education Press, Beijing, 2010.

[15] S. Jiang and P. Zhang. Global sphereically symmetric solutions of the compressible isentropic Navier-Stokes equations. *Comm. Math. Phys.*, **215**(2001), 559-581.

[16] H. L. Li and P. Marcati. Existence and asymptotic behavior of multi-dimensional quantum hydrodynamic model for semiconductors. *Comm. Math. Phys.*, **245** (2004), 215-247.

[17] A. Jüngel. A Steady-state quantum Euler-Poisson system for potential flows. *Commun. Math. Phys.*, **194** (1998), 463-479.

[18] A. Jüngel. Global weak solutions to cmpressible Navier-Stokes equations for quantum fluids. *SIAM J. Appl. Math.*, **42(3)** (2010), 1025-1045.

[19] A. Jüngel. *Quasi-hydrodynamic semiconductor equations.* Birkhäuse, Basel, 2001.

[20] L. D. Landau and E. M. Lifshitz. On the theory of the dispersion of magnetic permeability in ferromagnetic bodies. *Phys. Z. Sovietunion.*, **8** (1935), 153-169.

[21] Z. Lei, D. Li and X. Y. Zhang. Remarks of global wellposedness of liquid crystal flows and heat flow of harmonic maps in two dimensions. *Proceedings of American Mathemathical Society*,**142(11)**(2012),3801-3810.

[22] P. L. Lions. *Mathematical topic in fluid mechanics.* Vol. 2 Compressible models, in; Oxford Lectures Series in Mathematics and its Applications, vol.10, Oxford Science Publications, The Clarendon Press, Oxford University Press, New York, 1998.

[23] X. Liu. Partial regularity for Landau-Lifshitz system of ferromagnetic spin chain. *Calc. Var.*, **20** (2004), 153-173.

[24] T. Luo, Z. P. Xin and T. Yang. Interface behavior of compressible Navier-Stokes equtions with vacuum. *SIAM J. Math. Anal.*, **31**(2000), 1175-1191.

[25] P.M. Markowich, C. Ringhoffer and C. Schmeiser. *Semiconductor equations*. Wien, Springer, 1990.

[26] F. Q. Su and B. L. Guo. The global smooth solution for Landau-Lifshitz-Maxwell equation without dissipaton. *Journal of Partial Differential Equations*, **3** (1998), 193-208.

[27] Y.L. Zhou, H.S. Sun, and B.L. Guo. Existence of weak solution for boundary problems of systems of ferromagnetic chain. *Science in China A*, **27** (1981), 779-811.

Global Existence and Long-time Behavior for the Strong Solutions in H^2 to the 3D Compressible Nematic Liquid Crystal Flows*

Gao Jincheng (高金城), Guo Boling (郭柏灵) and Xi Xiaoyu (席肖玉)

Abstract In this paper, we investigate the global existence and long time behavior of strong solutions for compressible nematic liquid crystal flows in three-dimensional whole space. The global existence of strong solutions is obtained by the standard energy method under the condition that the initial data are close to the constant equilibrium state in H^2-framework. If the initial data in L^1-norm are finite additionally, the optimal time decay rates of strong solutions are established. With the help of Fourier splitting method, one also establishes optimal time decay rates for the higher order spatial derivatives of director.

Keywords compressible nematic liquid crystal flows; global solution; Green function; long-time behavior

1 Introduction

In this paper, we investigate the motion of compressible nematic liquid crystal flows, which are governed by the following simplified version of the Ericksen-Leslie equations

$$\begin{cases} \rho_t + \mathrm{div}(\rho u) = 0, \\ (\rho u)_t + \mathrm{div}(\rho u \otimes u) - \mu \Delta u - (\mu + \nu)\nabla \mathrm{div} u + \nabla P(\rho) = -\gamma \nabla d \cdot \Delta d, \\ d_t + u \cdot \nabla d = \theta(\Delta d + |\nabla d|^2 d), \end{cases} \quad (1.1)$$

where ρ, u, and d stand for the density, velocity and macroscopic average of the nematic liquid crystal orientation field respectively. The pressure $P(\rho)$ is a smooth

---
*Ann. Appl. Math., 2016, 32(4): 331–356.

function in a neighborhood of 1 with $P'(1) = 1$. The constants μ and ν are shear viscosity and the bulk viscosity coefficients of the fluid respectively, that satisfy the physical assumptions

$$\mu > 0, \quad 2\mu + 3\nu \geqslant 0.$$

The positive constants γ and θ present the competition between the kinetic energy and the potential energy, and the microscopic elastic relaxation time for the molecular orientation field, respectively. For simplicity, we set the constant γ and θ to be 1. The symbol $\otimes$ denotes the Kronecker tensor product such that $u \otimes u = (u_i u_j)_{1 \leqslant i, j \leqslant 3}$. To complete system (1.1), the initial data are given by

$$(\rho, u, d)(x, t)|_{t=0} = (\rho_0(x), u_0(x), d_0(x)). \tag{1.2}$$

Furthermore, as the space variable tends to infinity, we assume

$$\lim_{|x| \to \infty} (\rho_0 - 1, u_0, d_0 - w_0)(x) = 0, \tag{1.3}$$

where w_0 is a fixed unit constant vector. The system is a coupling between the compressible Navier-Stokes equations and a transported heat flow of harmonic maps into S^2. Generally speaking, we can not obtain any better results for system (1.1) than those for the compressible Navier-Stokes equations.

The hydrodynamic theory of liquid crystals in the nematic case has been established by Ericksen[1] and Leslie[2] during the period of 1958 through 1968. Since then, the mathematical theory is still progressing and the study of the full Ericksen-Leslie model presents relevant mathematical difficulties. The pioneering work comes from [3–6]. For example, Lin and Liu[5] obtained the global weak and smooth solutions for the Ginzburg-Landau approximation to relax the nonlinear constraint $d \in S^2$. They also discussed the uniqueness and some stability properties of the system. Later, the decay rates for this approximate system were given by Wu[7] in a bounded domain. On the other hand, Dai, Qing and Schonbek[8], Dai and Schonbek[9] established the time decay rates for the Cauchy problem respectively. More precisely, Dai and Schonbek[9] obtained the global existence of solutions in the Sobolev space $H^N(\mathbf{R}^3) \times H^{N+1}(\mathbf{R}^3) (N \geqslant 1)$ only requiring the smallness of $\|u_0\|_{H^1}^2 + \|d_0 - w_0\|_{H^2}^2$, where w_0 is a fixed unit constant vector. If the initial data in L^1-norm are finitely additionally, they also established the following time decay rates

$$\|\nabla^k u(t)\|_{L^2} + \|\nabla^k (d - w_0)(t)\|_{L^2} \leqslant C(1+t)^{-\frac{3+2k}{4}},$$

for $k = 0, 1, 2, \cdots, N$. Recently, Liu and Zhang[10], for the density-dependent model, obtained the global weak solutions in dimension three with the initial density $\rho_0 \in L^2$, which was improved by Jiang and Tan[11] for the case $\rho_0 \in L^\gamma \left(\gamma > \dfrac{3}{2} \right)$. Under the constraint $d \in S^2$, Wen and Ding[12] established the local existence of the strong solutions and obtained the global solutions under the assumptions of small energy and positive initial density, which was improved by Li[13] to be of vacuum. Later, Hong[14] and Lin, Lin and Wang[15] showed independently the global existence of weak solutions in two-dimensional space. Recently, Wang[16] established a global well-posedness theory for rough initial data provided that $\|u_0\|_{\mathrm{BMO}^{-1}} + [d_0]_{BMO} \leqslant \varepsilon_0$ for some $\varepsilon_0 > 0$. Under this condition, Du and Wang[17] obtained arbitrary space-time regularity for the Koch and Tataru type solution (u, d). As a corollary, they also got the decay rates. For more results, the readers can refer to [18–22] and the references therein.

Considering the compressible nematic liquid crystal flows (1.1), Ding et al. [23] gained both the existence and uniqueness of global strong solutions for one dimensional space. And this result about the classical solutions was improved by Ding, Wang and Wen[24] by generalizing the fluids to be of vacuum. For the case of multi-dimensional space, Jiang, Jiang and Wang[25] established the global existence of weak solutions for the initial-boundary problem with large initial energy and without any smallness condition on the initial density and velocity if some component of initial direction field is small. Recently, Lin, Lai and Wang [26] established the existence of global weak solutions in three-dimensional space, provided the initial orientational director field d_0 lies in the hemisphere S_2^+. Local existence of unique strong solutions was proved provided that the initial data were sufficiently regular and satisfied a natural compatibility condition in a recent work [27]. Some blow-up criteria that were derived for the possible breakdown of such local strong solutions at finite time could be found in [28–30]. The local existence and uniqueness of classical solutions to (1.1) were established by Ma in [31]. On one hand, Hu and Wu[32] obtained the existence and uniqueness of global strong solutions in critical Besov spaces provided that the initial data were close to an equilibrium state $(1, 0, \hat{d})$ with a constant vector $\hat{d} \in S^2$; on the other hand, Gao, Tao and Yao[30] attained the global small classical solution in Sobolev spaces $H^m (m \geqslant 3)$ and established decay rates for the compressible nematic liquid crystal flows (1.1). For more results, the readers can refer to [34] for some recent developments of analysis for hydrodynamic flow of nematic liquid crystal flows and references therein.

Recently, Wang and Tan[35] established the global existence of strong solutions and built the time decay rates for the compressible Navier-Stokes equations in H^2-framework (See Matsumura and Nishida[36] in H^3-framework). Precisely, if small initial perturbation belongs to H^2 and initial perturbation in L^1-norm is finite, they built optimal time decay rates as follows

$$\|(\rho-1)(t)\|_{H^{2-k}} + \|u(t)\|_{H^{2-k}} \leq C(1+t)^{-\frac{3+2k}{4}},$$

where $k = 0, 1$. This framework of time convergence rates for compressible flows has been applied to other compressible models, refer to [37–39].

In this paper, motivated by the work [35], we hope to establish the global existence and time decay rates of strong solutions for the compressible nematic liquid crystal flows under the H^2-framework. First, we construct the global existence of strong solutions by the standard energy method under the condition that the initial data are close to the constant equilibrium state $(1, 0, w_0)$ (w_0 is a fixed unit constant vector) in H^2-framework. Second, if the initial data in L^1-norm are finite additionally, the optimal time decay rates of strong solutions are established by the method of Green function. Precisely, we obtain the following time decay rates for all $t \geq 0$

$$\|(\rho-1)(t)\|_{H^{2-k}} + \|u(t)\|_{H^{2-k}} + \|(d-w_0)(t)\|_{H^{3-k}} \leq C(1+t)^{-\frac{3+2k}{4}},$$

where $k = 0, 1$. Although angular momentum equations (1.1)$_3$ are nonlinear parabolic equations, we hope to establish optimal time decay rates for higher order spatial derivatives of director under the condition of small initial perturbation. Motivated by Lemma 3.3, we move the nonlinear terms to the right hand side of (1.1)$_3$ and deal with the nonlinear terms as external force with the property on fast time decay rates. Then, the optimal time decay rates for higher order spatial derivatives of director are built with the help of Fourier splitting method by Schonbek[40]. Finally, we also study the decay rates for the time derivatives of velocity and the mixed space-time derivatives of density and director.

Notation In this paper, we use $H^s(\mathbf{R}^3)(s \in \mathbf{R})$ to denote the usual Sobolev spaces with the norm $\|\cdot\|_{H^s}$ and $L^p(\mathbf{R}^3)(1 \leq p \leq \infty)$ to denote the usual L^p spaces with the norm $\|\cdot\|_{L^p}$. The symbol ∇^l with an integer $l \geq 0$ stands for the usual any spatial derivatives of order l. When l is not a positive integer, ∇^l stands for Λ^l

defined by $\Lambda^l f := \mathscr{F}^{-1}(|\xi|^l \mathscr{F} f)$, where $\mathscr{F}$ is the usual Fourier transform operator ($\mathscr{F}(f) := \hat{f}$) and $\mathscr{F}^{-1}$ is its inverse. We will employ the notation $a \lesssim b$ to mean that $a \leqslant Cb$ for a universal constant $C > 0$ independent of time t. $a \approx b$ means $a \lesssim b$ and $b \lesssim a$. For simplicity, we write $\|(A,B)\|_X := \|A\|_X + \|B\|_X$ and $\int f \, dx := \int_{\mathbf{R}^3} f \, dx$.

Now, we establish the first result concerning the global existence of solutions for the compressible nematic liquid crystal flows (1.1)–(1.3).

Theorem 1.1 *Assume that the initial data $(\rho_0 - 1, u_0, \nabla d_0) \in H^2, |d_0(x)| = 1$ in $\mathbf{R}^3$ and there exists a small constant $\delta_0 > 0$ such that*

$$\|(\rho_0 - 1, u_0, \nabla d_0)\|_{H^2} \leqslant \delta_0, \tag{1.4}$$

then problem (1.1)–(1.3) admits a unique global solution (ρ, u, d) satisfying for all $t \geqslant 0$,

$$\|(\rho-1, u, \nabla d)\|_{H^2}^2 + \int_0^t (\|\nabla \rho\|_{H^1}^2 + \|(\nabla u, \nabla^2 d)\|_{H^2}^2) \, d\tau \leqslant C \|(\rho_0 - 1, u_0, \nabla d_0)\|_{H^2}^2. \tag{1.5}$$

After obtaining the global existence of strong solutions at hand, we investigate the long-time behavior for the density, velocity and direction field.

Theorem 1.2 *Under the assumptions in Theorem 1.1, suppose the initial data $\|d_0 - w_0\|_{L^2}$ and $\|(\rho_0 - 1, u_0, d_0 - w_0)\|_{L^1}$ are finite additionally, then the solution (ρ, u, d) obtained in Theorem 1.1 satisfies for all $t \geqslant 0$,*

$$\begin{aligned} \|\nabla^k(\rho - 1)(t)\|_{H^{2-k}} + \|\nabla^k u(t)\|_{H^{2-k}} &\leqslant C(1+t)^{-\frac{3+2k}{4}}, \\ \|\nabla^l(d - w_0)(t)\|_{L^2} &\leqslant C(1+t)^{-\frac{3+2l}{4}}, \end{aligned} \tag{1.6}$$

where $k = 0, 1$, and $l = 0, 1, 2, 3$.

Remark 1.3 *For any $2 \leqslant p \leqslant 6$, by virtue of Theorem 1.2 and the Sobolev interpolation inequality, we also obtain the following time decay rates:*

$$\begin{aligned} \|(\rho - 1)(t)\|_{L^p} + \|u(t)\|_{L^p} &\leqslant C(1+t)^{-\frac{3}{2}(1-\frac{1}{p})}, \\ \|\nabla^k(d - w_0)(t)\|_{L^p} &\leqslant C(1+t)^{-\frac{3}{2}(1-\frac{1}{p})-\frac{k}{2}}, \end{aligned}$$

where $k = 0, 1, 2$. Furthermore, in the same manner, we also have

$$\begin{aligned} \|(\rho - 1)(t)\|_{L^\infty} + \|u(t)\|_{L^\infty} &\leqslant C(1+t)^{-\frac{5}{4}}, \\ \|\nabla^k(d - w_0)(t)\|_{L^\infty} &\leqslant C(1+t)^{-\frac{3+k}{2}}, \end{aligned}$$

where $k = 0, 1$.

Remark 1.4 Under the assumption of finiteness of $\|d_0 - w_0\|_{L^2}$ in Theorem 1.2, one can obtain the rate of director $d(x,t)$ converging to the constant equilibrium state w_0 in $L^\infty(\mathbf{R}^3)$-norm.

Finally, we also study the convergence rates for time derivatives of velocity and mixed space-time derivatives of density and director.

Theorem 1.5 Under the assumptions in Theorem 1.2, the global solution (ρ, u, d) of problem (1.1)–(1.3) has the following time decay rates for all $t \geqslant 0$,

$$\|\rho_t\|_{H^1} + \|u_t\|_{L^2} \leqslant C(1+t)^{-\frac{5}{4}},$$
$$\|\nabla^k d_t\|_{L^2} \leqslant C(1+t)^{-\frac{7+2k}{4}}, \quad (1.7)$$

where $k = 0, 1$.

This paper is organized as follows. In Section 2, we establish some energy estimates that will play an important role for us to construct the global existence of strong solutions. Then, we close the estimates by the standard continuity argument and the global existence of strong solutions follows immediately. In Section 3, we build the time decay rates by taking the method of Green function and establish optimal time decay rates for the higher order spatial derivatives of director. Finally, we also study the decay rates for the time derivatives of velocity and the mixed space-time derivatives of density and director.

2 Proof of Theorem 1.1

In this section, we construct the global existence of strong solutions for the compressible nematic liquid crystal flows (1.1)–(1.3). By a classical argument (see [36]), the global existence of solutions are obtained by combining the local existence result with a *priori* estimates. Since the local existence and uniqueness of strong solutions were established by Huang, Wang and Wen[27], the global solutions follow in a standard continuity argument after we establish (1.5) a *priori*.

2.1 Energy estimates

Denoting $\varrho = \rho - 1$ and $n = d - w_0$, we rewrite (1.1) in the perturbation form as

$$\begin{cases} \varrho_t + \mathrm{div} u = S_1, \\ u_t - \mu \Delta u - (\mu + \nu)\nabla \mathrm{div} u + \nabla \varrho = S_2, \\ n_t - \Delta n = S_3. \end{cases} \quad (2.1)$$

Here $S_i(i = 1, 2, 3)$ are defined as

$$\begin{cases} S_1 = -\varrho \text{div} u - u \cdot \nabla \varrho, \\ S_2 = -u \cdot \nabla u - h(\varrho)[\mu \Delta u + (\mu + \nu)\nabla \text{div} u] - f(\varrho)\nabla \varrho - g(\varrho)\nabla n \cdot \Delta n, \\ S_3 = -u \cdot \nabla n + |\nabla n|^2 (n + w_0), \end{cases} \quad (2.2)$$

where the three nonlinear functions of ϱ are defined by

$$h(\varrho) := \frac{\varrho}{\varrho + 1}, \quad f(\varrho) := \frac{P'(\varrho + 1)}{\varrho + 1} - 1, \quad g(\varrho) := \frac{1}{\varrho + 1}. \quad (2.3)$$

The associated initial condition is given by

$$(\varrho, u, n)(x, t)|_{t=0} = (\varrho_0, u_0, n_0)(x). \quad (2.4)$$

Assume there exists a small positive constant δ satisfying the following estimate

$$\|(\varrho, u, \nabla n)(t)\|_{H^2} \leqslant \delta, \quad (2.5)$$

for all $t \in [0, T]$. By virtue of (2.5) and Sobolev inequality, it is easy to get

$$\frac{1}{2} \leqslant \varrho + 1 \leqslant \frac{3}{2}.$$

Hence, we immediately have

$$|h(\varrho)|, |f(\varrho)| \leqslant C|\varrho| \quad \text{and} \quad |g^{(k-1)}(\varrho)|, |h^{(k)}(\varrho)|, |f^{(k)}(\varrho)| \leqslant C, \quad \text{for any } k \geqslant 1, \quad (2.6)$$

which can be used frequently to derive a *priori* estimates. The following analytic tool has been proved in Wang and Tan[41]. For simplicity, we only state the results here and omit the proof for brevity.

Lemma 2.1 *Let $2 \leqslant p \leqslant \infty$ and $0 \leqslant m, \alpha \leqslant l$; when $p = \infty$ we require further that $m \leqslant \alpha + 1$ and $l \geqslant \alpha + 2$. Then we have that for any $f \in C_0^\infty(\mathbf{R}^3)$,*

$$\|\nabla^\alpha f\|_{L^p} \lesssim \|\nabla^m f\|_{L^2}^{1-\theta} \|\nabla^l f\|_{L^2}^\theta,$$

where $0 \leqslant \theta \leqslant 1$ and α satisfy

$$\alpha + 3\left(\frac{1}{2} - \frac{1}{p}\right) = m(1 - \theta) + l\theta.$$

Remark 2.2 *If $\|f\|_{H^2} \leqslant M$, then according to Lemma 2.1 we obtain*

$$\|\nabla^\alpha f\|_{L^2} \lesssim \|f\|_{L^2}^{1-\frac{\alpha}{2}} \|\nabla^2 f\|_{L^2}^{\frac{\alpha}{2}} \lesssim M,$$

for any $\alpha \in [0,2]$. Hence, under assumption (2.5), it is easy to obtain

$$\|(\nabla^\alpha \varrho, \nabla^\alpha u, \nabla^\alpha \nabla n)(t)\|_{L^2} \lesssim \delta,$$

for any $\alpha \in [0,2]$.

First of all, we will derive the following energy estimates.

Lemma 2.3 *Under condition (2.5), then for $k = 0, 1$, we have*

$$\frac{d}{dt}\|\nabla^k(\varrho, u, \nabla n)\|_{L^2}^2 + C\|\nabla^{k+1}(u, \nabla n)\|_{L^2}^2 \lesssim \delta \|\nabla^{k+1}\varrho\|_{L^2}^2. \tag{2.7}$$

Proof Taking k-th spatial derivatives to $(2.1)_1$ and $(2.1)_2$ respectively, multiplying the resulting identities by $\nabla^k \varrho$ and $\nabla^k u$ respectively and integrating over $\mathbf{R}^3$ (by parts), it is easy to obtain

$$\frac{1}{2}\frac{d}{dt}\int (|\nabla^k \varrho|^2 + |\nabla^k u|^2)\, dx + \int (\mu|\nabla^{k+1} u|^2 + (\mu+\nu)|\nabla^k \mathrm{div} u|^2)\, dx$$
$$= \int \nabla^k S_1 \nabla^k \varrho\, dx + \int \nabla^k S_2 \nabla^k u\, dx. \tag{2.8}$$

Taking $(k+1)$-th spatial derivatives to $(2.1)_3$, multiplying the resulting identities $\nabla^{k+1} n$ and integrating over $\mathbf{R}^3$ (by parts), we have

$$\frac{1}{2}\frac{d}{dt}\int |\nabla^{k+1} n|^2\, dx + \int |\nabla^{k+2} n|^2\, dx = \int \nabla^{k+1} S_3 \nabla^{k+1} n\, dx. \tag{2.9}$$

Adding (2.8) to (2.9), it follows immediately that

$$\frac{1}{2}\frac{d}{dt}\int (|\nabla^k \varrho|^2 + |\nabla^k u|^2 + |\nabla^{k+1} n|^2)\, dx$$
$$+ \int (\mu|\nabla^{k+1} u|^2 + (\mu+\nu)|\nabla^k \mathrm{div} u|^2 + |\nabla^{k+2} n|^2)\, dx$$
$$= \int \nabla^k S_1 \nabla^k \varrho\, dx + \int \nabla^k S_2 \nabla^k u\, dx + \int \nabla^{k+1} S_3 \nabla^{k+1} n\, dx. \tag{2.10}$$

For the case $k = 0$, the differential identity (2.10) has the following form

$$\frac{1}{2}\frac{d}{dt}\int (|\varrho|^2 + |u|^2 + |\nabla n|^2)\, dx + \int (\mu|\nabla u|^2 + (\mu+\nu)|\mathrm{div} u|^2 + |\nabla^2 n|^2)\, dx$$
$$= \int S_1 \cdot \varrho\, dx + \int S_2 \cdot u\, dx - \int S_3 \cdot \Delta n\, dx = I_1 + I_2 + I_3. \tag{2.11}$$

Applying the Hölder, Sobolev and Young inequalities, it is easy to obtain

$$I_1 \leqslant \|\varrho\|_{L^3}\|\mathrm{div} u\|_{L^2}\|\varrho\|_{L^6} + \|\varrho\|_{L^3}\|\nabla \varrho\|_{L^2}\|u\|_{L^6}$$

$$\lesssim \|\varrho\|_{H^1}\|\nabla u\|_{L^2}\|\nabla\varrho\|_{L^2} + \|\varrho\|_{H^1}\|\nabla\varrho\|_{L^2}\|\nabla u\|_{L^2}$$
$$\lesssim \delta(\|\nabla\varrho\|_{L^2}^2 + \|\nabla u\|_{L^2}^2). \tag{2.12}$$

Integrating by parts and applying (2.6), Hölder, Sobolev and Young inequalities, it arrives at directly

$$-\int h(\varrho)(\mu\Delta u + (\mu+\lambda)\nabla\mathrm{div}u)u dx$$
$$\approx \int (h'(\varrho)\nabla\varrho\cdot u + h(\varrho)\nabla u)\nabla u\, dx$$
$$\lesssim \|\nabla\varrho\|_{L^2}\|u\|_{L^6}\|\nabla u\|_{L^3} + \|\varrho\|_{L^\infty}\|\nabla u\|_{L^2}^2$$
$$\lesssim (\|\varrho\|_{H^2} + \|\nabla u\|_{H^1})(\|\nabla\varrho\|_{L^2}^2 + \|\nabla u\|_{L^2}^2)$$
$$\lesssim \delta(\|\nabla\varrho\|_{L^2}^2 + \|\nabla u\|_{L^2}^2). \tag{2.13}$$

Hence, with the help of (2.6), Hölder, Sobolev and Young inequalities, we deduce

$$I_2 \lesssim (\|u\|_{L^3}\|\nabla u\|_{L^2} + \|\varrho\|_{L^3}\|\nabla\varrho\|_{L^2} + \|\nabla n\|_{L^3}\|\Delta n\|_{L^2})\|u\|_{L^6}$$
$$+ \delta(\|\nabla\varrho\|_{L^2}^2 + \|\nabla u\|_{L^2}^2)$$
$$\lesssim (\|u\|_{H^1}\|\nabla u\|_{L^2} + \|\varrho\|_{H^1}\|\nabla\varrho\|_{L^2} + \|\nabla n\|_{H^1}\|\nabla^2 n\|_{L^2})\|\nabla u\|_{L^2}$$
$$+ \delta(\|\nabla\varrho\|_{L^2}^2 + \|\nabla u\|_{L^2}^2)$$
$$\lesssim \delta(\|\nabla\varrho\|_{L^2}^2 + \|\nabla u\|_{L^2}^2 + \|\nabla^2 n\|_{L^2}^2). \tag{2.14}$$

By virtue of $|d|=1$ (that is, $|n+w_0|=1$), it follows immediately from the Hölder and Sobolev inequalities that

$$I_3 \leqslant (\|u\|_{L^3}\|\nabla n\|_{L^6} + \|\nabla n\|_{L^3}\|\nabla n\|_{L^6})\|\nabla^2 n\|_{L^2}$$
$$\lesssim (\|u\|_{H^1} + \|\nabla n\|_{H^1})\|\nabla^2 n\|_{L^2}^2$$
$$\lesssim \delta\|\nabla^2 n\|_{L^2}^2. \tag{2.15}$$

Substituting (2.12), (2.14) and (2.15) into (2.11) completes the proof of (2.7) for the case of $k=0$. Now, we turn to give the proof of (2.7) for the case of $k=1$. Indeed, taking $k=1$ in (2.10) and integrating by part yield

$$\frac{1}{2}\frac{d}{dt}\int (|\nabla\varrho|^2 + |\nabla u|^2 + |\nabla^2 n|^2)\, dx + \int (\mu|\nabla^2 u|^2 + (\mu+\nu)|\nabla\mathrm{div}u|^2 + |\nabla^3 n|^2)\, dx$$
$$= -\int S_1\Delta\varrho\, dx - \int S_2\Delta u\, dx - \int \nabla S_3\nabla\Delta n\, dx = II_1 + II_2 + II_3. \tag{2.16}$$

Applying Hölder, Sobolev and Young inequalities, we obtain

$$II_1 \leqslant (\|\varrho\|_{L^3}\|\mathrm{div}u\|_{L^6} + \|u\|_{L^3}\|\nabla\varrho\|_{L^6})\|\nabla^2\varrho\|_{L^2}$$
$$\lesssim (\|\varrho\|_{H^1} + \|u\|_{H^1})(\|\nabla^2\varrho\|_{L^2}^2 + \|\nabla^2 u\|_{L^2}^2)$$
$$\lesssim \delta(\|\nabla^2\varrho\|_{L^2}^2 + \|\nabla^2 u\|_{L^2}^2). \tag{2.17}$$

Similarly, it is easy to deduce

$$II_2 \leqslant (\|u\|_{L^3}\|\nabla u\|_{L^6} + \|\varrho\|_{L^\infty}\|\nabla^2 u\|_{L^2})\|\nabla^2 u\|_{L^2}$$
$$+ (\|\varrho\|_{L^3}\|\nabla\varrho\|_{L^6} + \|\nabla n\|_{L^3}\|\Delta n\|_{L^6})\|\nabla^2 u\|_{L^2}$$
$$\lesssim (\|\varrho\|_{H^2} + \|u\|_{H^1} + \|\nabla n\|_{H^1})$$
$$\times (\|\nabla^2\varrho\|_{L^2}^2 + \|\nabla^2 u\|_{L^2}^2 + \|\nabla^3 n\|_{L^2}^2)$$
$$\lesssim \delta(\|\nabla^2\varrho\|_{L^2}^2 + \|\nabla^2 u\|_{L^2}^2 + \|\nabla^3 n\|_{L^2}^2), \tag{2.18}$$

and

$$II_3 \leqslant (\|\nabla u\|_{L^6}\|\nabla n\|_{L^3} + \|u\|_{L^3}\|\nabla^2 n\|_{L^6})\|\nabla^3 n\|_{L^2}$$
$$+ (\|\nabla n\|_{L^3}\|\nabla^2 n\|_{L^6} + \|\nabla n\|_{L^6}^3)\|\nabla^3 n\|_{L^2}$$
$$\lesssim (\|u\|_{H^1} + \|\nabla n\|_{H^1} + \|\nabla^{\frac{3}{2}} n\|_{L^2}^2)(\|\nabla^2 u\|_{L^2}^2 + \|\nabla^3 n\|_{L^2}^2)$$
$$\lesssim \delta(\|\nabla^2 u\|_{L^2}^2 + \|\nabla^3 n\|_{L^2}^2). \tag{2.19}$$

Substituting (2.17)–(2.19) into (2.16), then we complete the proof of (2.7) for the case of $k = 1$.

Next, we derive the second type of energy estimates involving the higher order spatial derivatives of ϱ and u.

Lemma 2.4 *Under condition (2.5), then we have*

$$\frac{d}{dt}\|\nabla^2(\varrho, u, \nabla n)\|_{L^2}^2 + C\|\nabla^3(u, \nabla n)\|_{L^2}^2 \lesssim \delta\|\nabla^2\varrho\|_{L^2}^2. \tag{2.20}$$

Proof Taking 2-th spatial derivatives to $(2.1)_1$ and $(2.1)_2$ respectively, multiplying the resulting identities by $\nabla^2\varrho$ and $\nabla^2 u$ respectively and integrating over $\mathbf{R}^3$ (by parts), we obtain

$$\frac{1}{2}\frac{d}{dt}\int(|\nabla^2\varrho|^2 + |\nabla^2 u|^2)\,dx + \int(\mu|\nabla^3 u|^2 + (\mu+\nu)|\nabla^2\mathrm{div}u|^2)\,dx$$
$$= \int \nabla^2 S_1 \nabla^2\varrho\,dx + \int \nabla^2 S_2 \nabla^2 u\,dx. \tag{2.21}$$

Applying Hölder, Sobolev and Young inequalities, it is easy to obtain

$$-\int \nabla^2(\varrho \mathrm{div} u) \nabla^2 \varrho \, dx$$
$$= -\int (\nabla^2 \varrho \mathrm{div} u + 2\nabla \varrho \nabla \mathrm{div} u + \varrho \nabla^2 \mathrm{div} u) \nabla^2 \varrho \, dx$$
$$\lesssim (\|\nabla u\|_{L^\infty} \|\nabla^2 \varrho\|_{L^2} + \|\nabla \varrho\|_{L^3} \|\nabla^2 u\|_{L^6} + \|\varrho\|_{L^\infty} \|\nabla^3 u\|_{L^2}) \|\nabla^2 \varrho\|_{L^2}$$
$$\lesssim (\|\nabla^2 u\|_{L^2}^{\frac{1}{2}} \|\nabla^3 u\|_{L^2}^{\frac{1}{2}} \|\nabla^2 \varrho\|_{L^2} + \|\nabla \varrho\|_{H^1} \|\nabla^3 u\|_{L^2} + \|\varrho\|_{H^2} \|\nabla^3 u\|_{L^2}) \|\nabla^2 \varrho\|_{L^2}$$
$$\lesssim (\|\nabla^2 u\|_{L^2}^{\frac{1}{2}} \|\nabla^2 \varrho\|_{L^2}^{\frac{1}{2}} + \|\nabla \varrho\|_{H^1} + \|\varrho\|_{H^2})(\|\nabla^2 \varrho\|_{L^2}^2 + \|\nabla^3 u\|_{L^2}^2)$$
$$\lesssim \delta(\|\nabla^2 \varrho\|_{L^2}^2 + \|\nabla^3 u\|_{L^2}^2). \tag{2.22}$$

Integrating by part and applying Hölder, Sobolev and Young inequalities, it arrives at

$$-\int \nabla^2(u \cdot \nabla \varrho) \nabla^2 \varrho \, dx$$
$$= \int \left[-(\nabla^2 u \nabla \varrho + 2\nabla u \nabla^2 \varrho) \nabla^2 \varrho + \frac{1}{2} |\nabla^2 \varrho|^2 \mathrm{div} u \right] dx$$
$$\lesssim (\|\nabla^2 u\|_{L^6} \|\nabla \varrho\|_{L^3} + \|\nabla u\|_{L^\infty} \|\nabla^2 \varrho\|_{L^2}) \|\nabla^2 \varrho\|_{L^2}$$
$$\lesssim \|\nabla \varrho\|_{H^1} \|\nabla^2 \varrho\|_{L^2} \|\nabla^3 u\|_{L^2} + \|\nabla^2 u\|_{L^2}^{\frac{1}{2}} \|\nabla^3 u\|_{L^2}^{\frac{1}{2}} \|\nabla^2 \varrho\|_{L^2}^2$$
$$\lesssim (\|\nabla^2 u\|_{L^2}^{\frac{1}{2}} \|\nabla^2 \varrho\|_{L^2}^{\frac{1}{2}} + \|\nabla \varrho\|_{H^1})(\|\nabla^2 \varrho\|_{L^2}^2 + \|\nabla^3 u\|_{L^2}^2)$$
$$\lesssim \delta(\|\nabla^2 \varrho\|_{L^2}^2 + \|\nabla^3 u\|_{L^2}^2). \tag{2.23}$$

The combination of (2.22) and (2.23) gives rise to

$$\int \nabla^2 S_1 \nabla^2 \varrho \, dx \lesssim \delta(\|\nabla^2 \varrho\|_{L^2}^2 + \|\nabla^3 u\|_{L^2}^2). \tag{2.24}$$

Now, we turn to give the estimate for the second term on the right hand side of (2.21). First of all, by virtue of Hölder and Sobolev inequalities, we have

$$\int \nabla(u \cdot \nabla u) \nabla \Delta u \, dx$$
$$= \int (\nabla u \nabla u + u \nabla^2 u) \nabla \Delta u \, dx$$
$$\leqslant \|\nabla u\|_{L^3} \|\nabla u\|_{L^6} \|\nabla^3 u\|_{L^2} + \|u\|_{L^3} \|\nabla^2 u\|_{L^6} \|\nabla^3 u\|_{L^2}$$
$$\lesssim \|u\|_{L^2}^{\frac{1}{2}} \|\nabla^3 u\|_{L^2}^{\frac{1}{2}} \|\nabla u\|_{L^2}^{\frac{1}{2}} \|\nabla^3 u\|_{L^2}^{\frac{1}{2}} \|\nabla^3 u\|_{L^2} + \|u\|_{H^1} \|\nabla^3 u\|_{L^2}^2$$
$$\lesssim \delta \|\nabla^3 u\|_{L^2}^2. \tag{2.25}$$

In view of (2.6), Hölder and Sobolev inequalities, we have

$$\int \nabla(h(\varrho)(\mu\Delta + (\mu + \lambda)\nabla \mathrm{div} u)) \nabla \Delta u \, dx$$

$$\lesssim (\|\nabla\varrho\|_{L^3}\|\nabla^2 u\|_{L^6} + \|\varrho\|_{L^\infty}\|\nabla^3 u\|_{L^2})\|\nabla^3 u\|_{L^2}$$
$$\lesssim (\|\nabla\varrho\|_{H^1}\|\nabla^3 u\|_{L^2} + \|\varrho\|_{H^2}\|\nabla^3 u\|_{L^2})\|\nabla^3 u\|_{L^2}$$
$$\lesssim \delta\|\nabla^3 u\|_{L^2}^2 \tag{2.26}$$

and

$$\int \nabla(f(\varrho)\nabla\varrho)\nabla\Delta u\,dx$$
$$\lesssim (\|\nabla\varrho\|_{L^4}^2 + \|\varrho\|_{L^\infty}\|\nabla^2\varrho\|_{L^2})\|\nabla^3 u\|_{L^2}$$
$$\lesssim (\|\nabla^{\frac{3}{2}}\varrho\|_{L^2}\|\nabla^2\varrho\|_{L^2} + \|\varrho\|_{H^2}\|\nabla^2\varrho\|_{L^2})\|\nabla^3 u\|_{L^2}$$
$$\lesssim (\|\nabla^{\frac{3}{2}}\varrho\|_{L^2} + \|\varrho\|_{H^2})\|\nabla^2\varrho\|_{L^2}\|\nabla^3 u\|_{L^2}$$
$$\lesssim \delta(\|\nabla^2\varrho\|_{L^2}^2 + \|\nabla^3 u\|_{L^2}^2). \tag{2.27}$$

Similarly, it is easy to deduce

$$\int \nabla(g(\varrho)\nabla n \cdot \Delta n)\nabla\Delta u\,dx$$
$$\lesssim (\|\nabla n\|_{L^\infty}\|\nabla\varrho\|_{L^3}\|\Delta n\|_{L^6} + \|\nabla^2 n\|_{L^4}^2 + \|\nabla n\|_{L^3}\|\nabla^3 n\|_{L^6})\|\nabla^3 u\|_{L^2}$$
$$\lesssim \|\nabla n\|_{H^2}\|\nabla^{\frac{5}{4}}\varrho\|_{L^2}^{\frac{2}{3}}\|\nabla^2\varrho\|_{L^2}^{\frac{1}{3}}\|\nabla n\|_{L^2}^{\frac{1}{3}}\|\nabla^4 n\|_{L^2}^{\frac{2}{3}}\|\nabla^3 u\|_{L^2}$$
$$+ (\|\nabla^{\frac{3}{2}}n\|_{L^2}\|\nabla^4 n\|_{L^2} + \|\nabla n\|_{H^1}\|\nabla^4 n\|_{L^2})\|\nabla^3 u\|_{L^2}$$
$$\lesssim (\|\nabla n\|_{H^2}\|\nabla^{\frac{5}{4}}\varrho\|_{L^2}^{\frac{2}{3}}\|\nabla n\|_{L^2}^{\frac{1}{3}} + \|\nabla^{\frac{3}{2}}n\|_{L^2} + \|\nabla n\|_{H^1})$$
$$\times (\|\nabla^2\varrho\|_{L^2}^2 + \|\nabla^3 u\|_{L^2}^2 + \|\nabla^4 n\|_{L^2}^2)$$
$$\lesssim \delta(\|\nabla^2\varrho\|_{L^2}^2 + \|\nabla^3 u\|_{L^2}^2 + \|\nabla^4 n\|_{L^2}^2). \tag{2.28}$$

Combining (2.25)–(2.27) with (2.28), we deduce

$$\int \nabla^2 S_2 \nabla^2 u\,dx \lesssim \delta(\|\nabla^2\varrho\|_{L^2}^2 + \|\nabla^3 u\|_{L^2}^2 + \|\nabla^4 n\|_{L^2}^2). \tag{2.29}$$

Inserting (2.24) and (2.29) into (2.21), it arrives at immediately

$$\frac{d}{dt}\int (|\nabla^2\varrho|^2 + |\nabla^2 u|^2)\,dx + \int |\nabla^3 u|^2\,dx \lesssim \delta(\|\nabla^2\varrho\|_{L^2}^2 + \|\nabla^4 n\|_{L^2}^2). \tag{2.30}$$

Taking 3-th spatial derivatives to $(2.1)_3$, multiplying the resulting identities by $\nabla^3 n$ and integrating over $\mathbf{R}^3$ (by parts), we obtain

$$\frac{1}{2}\frac{d}{dt}\int |\nabla^3 n|^2\,dx + \int |\nabla^4 n|^2\,dx = \int \nabla^3 S_3 \cdot \nabla^3 n\,dx. \tag{2.31}$$

The application of Hölder, Sobolev and Young inequalities, it is easy to deduce

$$\int \nabla^3 S_3 \cdot \nabla^3 n\, dx$$
$$\lesssim (\|\nabla^2 u\|_{L^6}\|\nabla n\|_{L^3} + \|\nabla u\|_{L^3}\|\nabla^2 n\|_{L^6} + \|u\|_{L^3}\|\nabla^3 n\|_{L^6}$$
$$+ \|\nabla^2 n\|_{L^4}^2 + \|\nabla n\|_{L^3}\|\nabla^3 n\|_{L^6} + \|\nabla n\|_{L^6}^2\|\nabla^2 n\|_{L^6})\|\nabla^4 n\|_{L^2}$$
$$\lesssim (\|\nabla n\|_{H^1}\|\nabla^3 u\|_{L^2} + \|\nabla^{\frac{3}{4}} u\|_{L^2}^{\frac{2}{3}}\|\nabla^2 u\|_{L^2}^{\frac{1}{3}}\|\nabla n\|_{L^2}^{\frac{1}{3}}\|\nabla^4 n\|_{L^2}^{\frac{2}{3}}$$
$$+ \|u\|_{H^1}\|\nabla^4 n\|_{L^2} + \|\nabla^{\frac{3}{2}} n\|_{L^2}\|\nabla^4 n\|_{L^2} + \|\nabla n\|_{H^1}\|\nabla^4 n\|_{L^2}$$
$$+ \|\nabla n\|_{L^2}^{\frac{4}{3}}\|\nabla^4 n\|_{L^6}^{\frac{2}{3}}\|\nabla^{\frac{5}{2}} n\|_{L^2}^{\frac{2}{3}}\|\nabla^4 n\|_{L^2}^{\frac{1}{3}})\|\nabla^4 n\|_{L^2}$$
$$\lesssim \delta(\|\nabla^3 u\|_{L^2}^2 + \|\nabla^4 n\|_{L^2}^2). \qquad (2.32)$$

Substituting (2.32) into (2.31), we have

$$\frac{1}{2}\frac{d}{dt}\int |\nabla^3 n|^2\, dx + \int |\nabla^4 n|^2\, dx \lesssim \delta\|\nabla^3 u\|_{L^2}^2. \qquad (2.33)$$

The combination of (2.30) and (2.33) completes the proof of lemma.

Finally, we will use equations (2.1) to recover the dissipation estimate for ϱ.

Lemma 2.5 *Under condition* (2.5), *then for* $k = 0, 1$, *we have*

$$\frac{d}{dt}\int \nabla^k u \cdot \nabla^{k+1}\varrho\, dx + C\|\nabla^{k+1}\varrho\|_{L^2}^2 \lesssim \|\nabla^{k+1} u\|_{L^2}^2 + \|\nabla^{k+2} u\|_{L^2}^2 + \|\nabla^{k+3} n\|_{L^2}^2. \qquad (2.34)$$

Proof Taking k-th spatial derivatives to the second equation of (2.1), multiplying by $\nabla^{k+1}\varrho$ and integrating over $\mathbf{R}^3$, then we obtain

$$\int \nabla^k u_t \cdot \nabla^{k+1}\varrho\, dx + \int |\nabla^{k+1}\varrho|^2\, dx$$
$$= \int \nabla^k [\mu\Delta u + (\mu+\nu)\nabla\mathrm{div}u]\nabla^{k+1}\varrho\, dx + \int \nabla^k S_2 \nabla^{k+1}\varrho\, dx. \qquad (2.35)$$

In order to deal with $\int \nabla^k u_t \cdot \nabla^{k+1}\varrho\, dx$, following the idea in Guo and Wang [42], we turn the time derivatives of velocity to the density. Then, applying the mass equation $(2.1)_1$, we can transform time derivatives to the spatial derivatives, that is,

$$\int \nabla^k u_t \cdot \nabla^{k+1}\varrho\, dx$$
$$= \frac{d}{dt}\int \nabla^k u \cdot \nabla^{k+1}\varrho\, dx - \int \nabla^k u \cdot \nabla^{k+1}\varrho_t\, dx$$
$$= \frac{d}{dt}\int \nabla^k u \cdot \nabla^{k+1}\varrho\, dx + \int \nabla^k u \cdot \nabla^{k+1}(\mathrm{div}u + \mathrm{div}(\varrho u))\, dx$$

$$= \frac{d}{dt} \int \nabla^k u \cdot \nabla^{k+1} \varrho \, dx - \int \nabla^k \mathrm{div}\, u \cdot \nabla^k (\mathrm{div}\, u + \mathrm{div}(\varrho u)) \, dx$$

$$= \frac{d}{dt} \int \nabla^k u \cdot \nabla^{k+1} \varrho \, dx - \int |\nabla^k \mathrm{div}\, u|^2 \, dx - \int \nabla^k \mathrm{div}\, u \cdot \nabla^k \mathrm{div}(\varrho u) \, dx. \quad (2.36)$$

Substituting (2.36) into (2.35), it is easy to deduce

$$\frac{d}{dt} \int \nabla^k u \cdot \nabla^{k+1} \varrho \, dx + \int |\nabla^{k+1} \varrho|^2 \, dx$$

$$= \int |\nabla^k \mathrm{div}\, u|^2 \, dx + \int \nabla^k \mathrm{div}\, u \cdot \nabla^k \mathrm{div}(\varrho u) \, dx + \int \nabla^k S_2 \nabla^{k+1} \varrho \, dx$$

$$+ \int \nabla^k [\mu \Delta u + (\mu + \nu) \nabla \mathrm{div}\, u] \nabla^{k+1} \varrho \, dx. \quad (2.37)$$

For the case $k = 0$, applying Hölder, Sobolev and Young inequalities, we obtain

$$\int \mathrm{div}\, u \cdot \mathrm{div}(\varrho u) \, dx \lesssim \|\varrho\|_{L^\infty} \|\nabla u\|_{L^2}^2 + \|u\|_{L^3} \|\mathrm{div}\, u\|_{L^6} \|\nabla \varrho\|_{L^2}$$

$$\lesssim (\|\varrho\|_{H^2} + \|u\|_{H^1})(\|\nabla \varrho\|_{L^2}^2 + \|\nabla^2 u\|_{L^2}^2)$$

$$\lesssim \delta(\|\nabla \varrho\|_{L^2}^2 + \|\nabla^2 u\|_{L^2}^2). \quad (2.38)$$

By virtue of Hölder inequality and (2.5), it is easy to deduce

$$\int S_2 \, \nabla \varrho \, dx \lesssim (\|u\|_{L^3} \|\nabla u\|_{L^6} + \|\varrho\|_{L^\infty} \|\nabla^2 u\|_{L^2}) \|\nabla \varrho\|_{L^2}$$

$$+ (\|\varrho\|_{L^\infty} \|\nabla \varrho\|_{L^2} + \|\nabla n\|_{L^3} \|\Delta n\|_{L^6}) \|\nabla \varrho\|_{L^2}$$

$$\lesssim \delta(\|\nabla \varrho\|_{L^2}^2 + \|\nabla^2 u\|_{L^2}^2 + \|\nabla^3 n\|_{L^2}^2) \quad (2.39)$$

and

$$\int [\mu \Delta u + (\mu + \nu) \nabla \mathrm{div}\, u] \nabla \varrho \, dx \leq \frac{1}{2} \|\nabla \varrho\|_{L^2}^2 + \frac{1}{2} \|\nabla^2 u\|_{L^2}^2. \quad (2.40)$$

The combination of (2.38), (2.39) and (2.40) completes the proof of (2.34) for the case of $k = 0$. As for the case $k = 1$, applying Hölder, Sobolev and Young inequalities, we deduce

$$\int \nabla \mathrm{div}\, u \cdot \nabla \mathrm{div}(\varrho u) \, dx$$

$$\lesssim (\|\nabla \varrho\|_{L^3} \|\mathrm{div}\, u\|_{L^6} + \|\varrho\|_{L^\infty} \|\nabla \mathrm{div}\, u\|_{L^2}) \|\nabla^2 u\|_{L^2}$$

$$+ (\|\nabla \varrho\|_{L^3} \|\nabla u\|_{L^6} + \|u\|_{L^\infty} \|\nabla^2 \varrho\|_{L^2}) \|\nabla^2 u\|_{L^2}$$

$$\lesssim \delta(\|\nabla^2 \varrho\|_{L^2}^2 + \|\nabla^2 u\|_{L^2}^2). \quad (2.41)$$

With the help of Hölder inequality and Lemma 2.4, it arrives at

$$\int \nabla S_2 \, \nabla^2 \varrho \, dx \lesssim \delta(\|\nabla^2 \varrho\|_{L^2}^2 + \|\nabla^3 u\|_{L^2}^2 + \|\nabla^4 n\|_{L^2}^2), \quad (2.42)$$

and
$$\int \nabla[\mu\Delta u + (\mu+\nu)\nabla\mathrm{div} u]\nabla^2\varrho\,dx \leqslant \frac{1}{2}\|\nabla^2\varrho\|_{L^2}^2 + \frac{1}{2}\|\nabla^3 u\|_{L^2}^2. \tag{2.43}$$

The combination of (2.41), (2.42) and (2.43) gives rise to the proof of (2.34) for the case of $k=1$.

2.2 Global existence of solutions

In this subsection, we shall combine the energy estimates derived in the previous section to prove the global existence of strong solutions in Theorem 1.1. Summing up (2.7) from $k=l$ ($l=0,1$) to $k=1$, we obtain

$$\frac{d}{dt}\|\nabla^l(\varrho,u,\nabla n)\|_{H^{1-l}}^2 + C\|\nabla^l(\nabla u,\nabla^2 n)\|_{H^{1-l}}^2 \lesssim \delta\|\nabla^{l+1}\varrho\|_{H^{1-l}}^2,$$

which, together with (2.20), arrives at

$$\frac{d}{dt}\|\nabla^l(\varrho,u,\nabla n)\|_{H^{2-l}}^2 + C\|\nabla^{l+1}(u,\nabla n)\|_{H^{2-l}}^2 \leqslant \delta C_1\|\nabla^{l+1}\varrho\|_{H^{1-l}}^2. \tag{2.44}$$

On the other hand, summing (2.34) from $k=l$ ($l=0,1$) to $k=1$, we obtain immediately

$$\frac{d}{dt}\sum_{l\leqslant k\leqslant 1}\int \nabla^k u\cdot\nabla^{k+1}\varrho\,dx + C_2\|\nabla^{l+1}\varrho\|_{H^{1-l}}^2 \leqslant C_3\left(\|\nabla^{l+1}u\|_{H^{2-l}}^2 + \|\nabla^{l+3}n\|_{H^{1-l}}^2\right). \tag{2.45}$$

Multiplying (2.45) by $2\delta C_1/C_2$ and adding the resulting inequality to (2.44), it arrives at

$$\frac{d}{dt}\mathcal{E}_l^2(t) + C_3\left(\|\nabla^{l+1}\varrho\|_{H^{1-l}}^2 + \|\nabla^{l+1}(u,\nabla n)\|_{H^{2-l}}^2\right) \leqslant 0, \tag{2.46}$$

where $\mathcal{E}_l^2(t)$ is defined as

$$\mathcal{E}_l^2(t) = \|\nabla^l(\varrho,u,\nabla n)\|_{H^{2-l}}^2 + \frac{2\delta C_1}{C_2}\sum_{l\leqslant k\leqslant 1}\int \nabla^k u\cdot\nabla^{k+1}\varrho\,dx.$$

By virtue of the smallness of δ, it is easy to obtain

$$C_4^{-1}\|\nabla^l(\varrho,u,\nabla n)\|_{H^{2-l}}^2 \leqslant \mathcal{E}_l^2(t) \leqslant C_4\|\nabla^l(\varrho,u,\nabla n)\|_{H^{2-l}}^2. \tag{2.47}$$

Choosing $l=0$ in (2.46) and integrating over $[0,t]$ yield

$$\|(\varrho,u,\nabla n)(t)\|_{H^2} \leqslant C\|(\varrho_0,u_0,\nabla n_0)\|_{H^2}. \tag{2.48}$$

Since $\|(\varrho, u, \nabla n)(t)\|_{H^2}$ is a continuous function with respect to time (see [27]), there exists a small and positive constant T_0 such that

$$\max_{0 \leqslant t \leqslant T_0} \|(\varrho, u, \nabla n)(t)\|_{H^2} \leqslant 2\|(\varrho_0, u_0, \nabla n_0)\|_{H^2}. \tag{2.49}$$

Choosing

$$\|(\varrho_0, u_0, \nabla n_0)\|_{H^2} \leqslant \min\left\{\frac{\delta}{2}, \frac{\delta}{2C}\right\},$$

which, together with (2.49), gives directly

$$\max_{0 \leqslant t \leqslant T_0} \|(\varrho, u, \nabla n)(t)\|_{H^2} \leqslant \delta.$$

Then, applying estimate (2.48), it is easy to deduce

$$\max_{0 \leqslant t \leqslant T_0} \|(\varrho, u, \nabla n)(t)\|_{H^2} \leqslant \frac{\delta}{2}.$$

Thus, problem (2.1)–(2.4) with the initial data $(\varrho, u, \nabla n)(x, T_0)$ admits a unique solution on $[T_0, 2T_0] \times \mathbf{R}^3$ satisfying the estimate

$$\max_{T_0 \leqslant t \leqslant 2T_0} \|(\varrho, u, \nabla n)(t)\|_{H^2} \leqslant 2\|(\varrho, u, \nabla n)(T_0)\|_{H^2} \leqslant \delta,$$

which, together with (2.48), yields directly

$$\max_{0 \leqslant t \leqslant 2T_0} \|(\varrho, u, \nabla n)(t)\|_{H^2} \leqslant \frac{\delta}{2}.$$

Thus, we can continue the same process for $0 \leqslant t \leqslant nT_0 (n = 1, 2, \cdots)$ and finally get a global solution on $[0, \infty) \times \mathbf{R}^3$. The uniqueness of global strong solutions follows immediately from the uniqueness of local existence of solutions. Choosing $l = 0$ in (2.46), integrating over $[0, t]$ and applying the equivalent relation of (2.47), we obtain

$$\|(\varrho, u, \nabla n)\|_{H^2}^2 + \int_0^t (\|\nabla \varrho\|_{H^1}^2 + \|(\nabla u, \nabla^2 n)\|_{H^2}^2)\, d\tau \leqslant C\|(\varrho_0, u_0, \nabla n_0)\|_{H^2}^2,$$

which completes the proof of Theorem 1.1.

3 Proof of Theorems 1.2 and 1.5

In this section, we will establish the time decay rates for the compressible nematic liquid crystal flows (1.1)–(1.3). First of all, the decay rates are built by the method of the Green function. Secondly, motivated by Lemma 3.3, we enhance the time decay rates for the higher order derivatives of director. Finally, we also establish the convergence rates for the time derivatives of density, velocity and director.

3.1 Decay rates for the nonlinear systems

First of all, let us consider the following linearized systems

$$\begin{cases} \varrho_t + \mathrm{div}\, u = 0, \\ u_t - \mu \Delta u - (\mu + \nu)\nabla \mathrm{div}\, u + \nabla \varrho = 0, \\ n_t - \Delta n = 0, \end{cases} \quad (3.1)$$

with the initial data

$$(\varrho, u, n)|_{t=0} = (\varrho_0, u_0, n_0). \quad (3.2)$$

Obviously, the solution (ϱ, u, n) for the linear problem (3.1)–(3.2) can be expressed as

$$(\varrho, u, n)^{tr} = G(t) * (\varrho_0, u_0, n_0)^{tr}, \quad t \geq 0. \quad (3.3)$$

Here $G(t) := G(x,t)$ is the Green matrix for system (3.1) and the exact expression of the Fourier transform $\hat{G}(\xi, t)$ of Green function $G(x,t)$ as

$$\hat{G}(\xi, t) = \begin{bmatrix} \dfrac{\lambda_+ e^{\lambda_- t} - \lambda_- e^{\lambda_+ t}}{\lambda_+ - \lambda_-} & \dfrac{-i\xi^t(e^{\lambda_+ t} - e^{\lambda_- t})}{\lambda_+ - \lambda_-} & 0 \\ \dfrac{-i\xi(e^{\lambda_+ t} - e^{\lambda_- t})}{\lambda_+ - \lambda_-} & \dfrac{\lambda_+ e^{\lambda_+ t} - \lambda_- e^{\lambda_- t}}{\lambda_+ - \lambda_-} \dfrac{\xi\xi^t}{|\xi|^2} + e^{\lambda_0 t}\left(I_{3\times 3} - \dfrac{\xi\xi^t}{|\xi|^2}\right) & 0 \\ 0 & 0 & e^{\lambda_1 t} I_{3\times 3}, \end{bmatrix}$$

where

$$\lambda_0 = -\mu|\xi|^2, \quad \lambda_1 = -|\xi|^2,$$

$$\lambda_+ = -\left(\mu + \frac{1}{2}\nu\right)|\xi|^2 + i\sqrt{|\xi|^2 - \left(\mu + \frac{1}{2}\nu\right)^2 |\xi|^4},$$

$$\lambda_- = -\left(\mu + \frac{1}{2}\nu\right)|\xi|^2 - i\sqrt{|\xi|^2 - \left(\mu + \frac{1}{2}\nu\right)^2 |\xi|^4}.$$

Since systems (3.1) is an independent coupling of the classical linearized Navier-Stokes equation and heat equation, the representation of Green function $\hat{G}(\xi, t)$ is easy to be verified. Furthermore, we have the following decay rates for system (3.1)–(3.2), refer to [33, 43].

Proposition 3.1 *Assume that (ϱ, u, n) is a solution of the linearized compressible nematic liquid crystal system (3.1)–(3.2) with the initial data $(\varrho_0, u_0, n_0) \in L^1 \cap H^2$, then*

$$\|\nabla^k \varrho\|_{L^2}^2 \leq C \left(\|(\varrho_0, u_0)\|_{L^1}^2 + \|\nabla^k (\varrho_0, u_0)\|_{L^2}^2\right)(1+t)^{-\frac{3}{2}-k},$$

$$\|\nabla^k u\|_{L^2}^2 \leqslant C \left(\|(\varrho_0, u_0)\|_{L^1}^2 + \|\nabla^k(\varrho_0, u_0)\|_{L^2}^2\right)(1+t)^{-\frac{3}{2}-k},$$
$$\|\nabla^k n\|_{L^2}^2 \leqslant C \left(\|n_0\|_{L^1}^2 + \|\nabla^k n_0\|_{L^2}^2\right)(1+t)^{-\frac{3}{2}-k}$$

for $0 \leqslant k \leqslant 2$.

In the sequel, we want to verify some simplified inequalities that play an important role to derive the time decay rates for the compressible nematic liquid crystal flows (2.1)–(2.4). More precisely, we have

$$\|(S_1, S_2, S_3)\|_{L^1} \lesssim \delta(\|\nabla \varrho\|_{L^2} + \|\nabla u\|_{H^1} + \|\nabla n\|_{H^1}),$$
$$\|(S_1, S_2, S_3)\|_{L^2} \lesssim \delta(\|\nabla \varrho\|_{L^2} + \|\nabla u\|_{H^1} + \|\nabla n\|_{H^1}), \qquad (3.4)$$
$$\|\nabla(S_1, S_2, S_3)\|_{L^2} \lesssim \delta(\|\nabla^2 \varrho\|_{L^2} + \|\nabla^2 u\|_{L^2} + \|\nabla^2 n\|_{H^1}) + \|\nabla \varrho\|_{H^1} \|\nabla^3 u\|_{L^2}.$$

Next, we establish decay rates for the compressible nematic liquid crystal flows (2.1)–(2.4).

Lemma 3.2 *Under the assumptions of Theorem 1.2, the global solution (ϱ, u, n) of problem (2.1)–(2.4) satisfies*

$$\|\nabla^k \varrho(t)\|_{H^{2-k}}^2 + \|\nabla^k u(t)\|_{H^{2-k}}^2 + \|\nabla^k n(t)\|_{H^{3-k}}^2 \leqslant C(1+t)^{-\frac{3}{2}-k} \qquad (3.5)$$

for $k = 0, 1$.

Proof First of all, taking $k = 0$ in (2.9), which together with inequality (2.15), we obtain the following inequality immediately

$$\frac{d}{dt} \int |\nabla n|^2 \, dx + \int |\nabla^2 n|^2 \, dx \leqslant 0. \qquad (3.6)$$

Taking $l = 1$ specially in (2.46), it arrives at directly

$$\frac{d}{dt} \mathcal{E}_1^2(t) + C_3 \left(\|\nabla^2 \varrho\|_{L^2}^2 + \|\nabla^2 u\|_{H^1}^2 + \|\nabla^3 n\|_{H^1}^2\right) \leqslant 0,$$

which, together with (3.6), yields directly

$$\frac{d}{dt} \mathcal{F}_1^2(t) + C_4 \left(\|\nabla^2 \varrho\|_{L^2}^2 + \|\nabla^2 u\|_{H^1}^2 + \|\nabla^2 n\|_{H^2}^2\right) \leqslant 0, \qquad (3.7)$$

where $\mathcal{F}_1^2(t)$ is defined as

$$\mathcal{F}_1^2(t) = \|\nabla(\varrho, u)\|_{H^1}^2 + \|\nabla n\|_{H^2}^2 + \frac{2C_1 \delta}{C_2} \int \nabla u \cdot \nabla^2 \varrho \, dx.$$

With the help of Young inequality, it is easy to deduce

$$C_5^{-1} \left(\|\nabla(\varrho, u)\|_{H^1}^2 + \|\nabla n\|_{H^2}^2\right) \leqslant \mathcal{F}_1^2(t) \leqslant C_5 \left(\|\nabla(\varrho, u)\|_{H^1}^2 + \|\nabla n\|_{H^2}^2\right). \qquad (3.8)$$

Adding both hand sides of (3.7) by $\|\nabla(\varrho,u,n)\|_{L^2}^2$ and applying the equivalent relation (3.8), we have
$$\frac{d}{dt}\mathcal{F}_1^2(t) + C\mathcal{F}_1^2(t) \leqslant \|\nabla(\varrho,u,n)\|_{L^2}^2. \tag{3.9}$$

In view of the Gronwall inequality, it follows immediately
$$\mathcal{F}_1^2(t) \leqslant \mathcal{F}_1^2(0)e^{-Ct} + \int_0^t e^{-C(t-\tau)}\|\nabla(\varrho,u,n)\|_{L^2}^2\, d\tau. \tag{3.10}$$

In order to derive the time decay rates for $\mathcal{F}_1^2(t)$, we need to control the term $\|\nabla(\varrho,u,n)\|_{L^2}^2$. In fact, by Duhamel principle, we can represent the solution for system (2.1)–(2.4) as
$$(\varrho,u,n)^{tr}(t) = G(t) * (\varrho_0,u_0,n_0)^{tr} + \int_0^t G(t-s)*(S_1,S_2,S_3)^{tr}(s)ds. \tag{3.11}$$

Denoting
$$F(t) = \sup_{0\leqslant \tau \leqslant t}(1+\tau)^{\frac{5}{2}}(\|\nabla \varrho(\tau)\|_{H^1}^2 + \|\nabla u(\tau)\|_{H^1}^2 + \|\nabla n(\tau)\|_{H^2}^2),$$

by virtue of (3.4), (3.11) and Proposition 3.1, we have
$$\|\nabla(\varrho,u,n)\|_{L^2}$$
$$\leqslant C(1+t)^{-\frac{5}{4}} + C\int_0^t (\|(S_1,S_2,S_3)\|_{L^1} + \|\nabla(S_1,S_2,S_3)\|_{L^2})(1+t-\tau)^{-\frac{5}{4}}\, d\tau$$
$$\leqslant C(1+t)^{-\frac{5}{4}} + C\int_0^t \delta(\|\nabla\varrho\|_{H^1} + \|\nabla u\|_{H^1} + \|\nabla n\|_{H^2})(1+t-\tau)^{-\frac{5}{4}}\, d\tau$$
$$+ C\int_0^t \|\nabla\varrho\|_{H^1}\|\nabla^3 u\|_{L^2}(1+t-\tau)^{-\frac{5}{4}}\, d\tau$$
$$\leqslant C(1+t)^{-\frac{5}{4}} + C\delta\sqrt{F(t)}\int_0^t (1+t-\tau)^{-\frac{5}{4}}(1+\tau)^{-\frac{5}{4}}\, d\tau$$
$$+ C\sqrt{F(t)}\left[\int(1+t-\tau)^{-\frac{5}{2}}(1+\tau)^{-\frac{5}{2}}\, d\tau\right]^{\frac{1}{2}}\left[\int_0^t \|\nabla^3 u(\tau)\|_{L^2}^2\, d\tau\right]^{\frac{1}{2}}$$
$$\leqslant C(1+t)^{-\frac{5}{4}} + C\delta\sqrt{F(t)}(1+t)^{-\frac{5}{4}}$$
$$\leqslant (1+t)^{-\frac{5}{4}}(1+\delta\sqrt{F(t)}),$$

where we have used the fact
$$\int_0^t (1+t-\tau)^{-r}(1+\tau)^{-r}\, d\tau$$
$$= \int_0^{\frac{t}{2}} + \int_{\frac{t}{2}}^t (1+t-\tau)^{-r}(1+\tau)^{-r}\, d\tau$$

$$\leqslant \left(1+\frac{t}{2}\right)^{-r}\int_0^{\frac{t}{2}}(1+\tau)^{-r}\,d\tau + \left(1+\frac{t}{2}\right)^{-r}\int_{\frac{t}{2}}^t (1+t-\tau)^{-r}\,d\tau$$

$$\leqslant (1+t)^{-r},$$

for $r = \dfrac{5}{2}$ and $r = \dfrac{5}{4}$ respectively. Thus, we have the estimate

$$\|\nabla(\varrho, u, n)\|_{L^2}^2 \leqslant C(1+t)^{-\frac{5}{2}}(1+\delta F(t)). \tag{3.12}$$

Inserting (3.12) into (3.10), it follows immediately

$$\begin{aligned}
\mathcal{F}_1^2(t) &\leqslant \mathcal{F}_1^2(0)e^{-Ct} + C\int_0^t e^{-C(t-\tau)}(1+\tau)^{-\frac{5}{2}}(1+\delta F(\tau))\,d\tau \\
&\leqslant \mathcal{F}_1^2(0)e^{-Ct} + C(1+\delta F(t))\int_0^t e^{-C(t-\tau)}(1+\tau)^{-\frac{5}{2}}\,d\tau \\
&\leqslant \mathcal{F}_1^2(0)e^{-Ct} + C(1+\delta F(t))(1+t)^{-\frac{5}{2}} \\
&\leqslant C(1+\delta F(t))(1+t)^{-\frac{5}{2}}, \tag{3.13}
\end{aligned}$$

where we have used the fact

$$\int_0^t e^{-C(t-\tau)}(1+\tau)^{-\frac{5}{2}}\,d\tau$$

$$= \int_0^{\frac{t}{2}} + \int_{\frac{t}{2}}^t e^{-C(t-\tau)}(1+\tau)^{-\frac{5}{2}}\,d\tau$$

$$\leqslant e^{-\frac{C}{2}t}\int_0^{\frac{t}{2}}(1+\tau)^{-\frac{5}{2}}\,d\tau + \left(1+\frac{t}{2}\right)^{-\frac{5}{2}}\int_{\frac{t}{2}}^t e^{-C(t-\tau)}\,d\tau$$

$$\leqslant C(1+t)^{-\frac{5}{2}}.$$

Hence, by virtue of the definition of $F(t)$ and (3.13), it follows immediately

$$F(t) \leqslant C(1+\delta F(t)),$$

which, in view of the smallness of δ, gives

$$F(t) \leqslant C.$$

Therefore, we have the following time decay rates

$$\|\nabla \varrho(t)\|_{H^1}^2 + \|\nabla u(t)\|_{H^1}^2 + \|\nabla n(t)\|_{H^2}^2 \leqslant C(1+t)^{-\frac{5}{2}}. \tag{3.14}$$

On the other hand, by (3.4), (3.11), (3.14) and Proposition 3.1, it is easy to deduce

$$\|(\varrho, u, n)\|_{L^2}^2$$

$$\leqslant C(1+t)^{-\frac{3}{2}} + C\int_0^t \left(\|(S_1,S_2,S_3)\|_{L^1}^2 + \|(S_1,S_2,S_3)\|_{L^2}^2\right)(1+t-\tau)^{-\frac{3}{2}}\,d\tau$$

$$\leqslant C(1+t)^{-\frac{3}{2}} + C\int_0^t \delta\left(\|\nabla\varrho\|_{L^2}^2 + \|\nabla u\|_{H^1}^2 + \|\nabla n\|_{H^1}^2\right)(1+t-\tau)^{-\frac{3}{2}}\,d\tau$$

$$\leqslant C(1+t)^{-\frac{3}{2}} + C\int_0^t (1+t-\tau)^{-\frac{5}{2}}(1+\tau)^{-\frac{3}{2}}\,d\tau$$

$$\leqslant C(1+t)^{-\frac{3}{2}},$$

where we have used the fact

$$\int_0^t (1+t-\tau)^{-\frac{5}{2}}(1+\tau)^{-\frac{3}{2}}\,d\tau \leqslant C(1+t)^{-\frac{3}{2}}.$$

Hence, we have the following decay rates

$$\|(\varrho, u, n)(t)\|_{L^2}^2 \leqslant C(1+t)^{-\frac{3}{2}}. \tag{3.15}$$

Therefore, the combination of (3.14) and (3.15) completes the proof of the lemma.

3.2 Optimal decay rates for the higher order derivatives of director

In this subsection, we will enhance the time decay rates for the higher order spatial derivatives of direction field. This improvement is motivated by the following lemma.

Lemma 3.3 *For some smooth function $F(x,t)$, suppose the smooth function $v(x,t)$ is a solution of heat equation*

$$v_t(x,t) - \Delta v(x,t) = F(x,t), \tag{3.16}$$

for $(x,t) \in \mathbf{R}^3 \times R^+$ with the smooth initial data $v(x,0) = v_0(x)$. If the function $F(x,t)$ and the solution $v(x,t)$ have the time decay rates

$$\|\nabla^k v(t)\|_{L^2}^2 \leqslant C(1+t)^{-(\frac{3}{2}+k)}, \quad \|\nabla^k F(t)\|_{L^2}^2 \leqslant C(1+t)^{-\alpha}, \tag{3.17}$$

where $\alpha \geqslant k + \frac{7}{2}$. Then, we have the following time decay rate for the $(k+1)$-th order of spatial derivatives

$$\|\nabla^{k+1} v(t)\|_{L^2}^2 \leqslant C(1+t)^{-(k+\frac{5}{2})}.$$

Proof Taking $(k+1)$-th spatial derivatives on both hand sides of (3.16), multiplying by $\nabla^{k+1} v$ and integrating over $\mathbf{R}^3$, we obtain

$$\frac{1}{2}\frac{d}{dt}\int |\nabla^{k+1} v|^2 \,dx + \int |\nabla^{k+2} v|^2 \,dx \leqslant \frac{1}{2}\int |\nabla^k F|^2 \,dx + \frac{1}{2}\int |\nabla^{k+2} v|^2 \,dx,$$

which implies
$$\frac{d}{dt}\int |\nabla^{k+1} v|^2\, dx + \int |\nabla^{k+2} v|^2\, dx \leqslant \int |\nabla^k F|^2\, dx. \tag{3.18}$$

For some constant R defined below, denoting the time sphere (see [40])
$$S_0 = \left\{ \xi \in \mathbf{R}^3 \big| |\xi| \leqslant \left(\frac{R}{1+t}\right)^{\frac{1}{2}} \right\},$$

it follows immediately
$$\begin{aligned}
\int_{\mathbf{R}^3} |\nabla^{k+2} v|^2\, dx &\geqslant \int_{\mathbf{R}^3/S_0} |\xi|^{2(k+2)} |\hat{v}|^2\, d\xi \\
&\geqslant \frac{R}{1+t} \int_{\mathbf{R}^3/S_0} |\xi|^{2(k+1)} |\hat{v}|^2\, d\xi \\
&\geqslant \frac{R}{1+t} \int_{\mathbf{R}^3} |\xi|^{2(k+1)} |\hat{v}|^2\, d\xi - \left(\frac{R}{1+t}\right)^2 \int_{S_0} |\xi|^{2k} |\hat{v}|^2\, d\xi,
\end{aligned}$$

or equivalently
$$\int_{\mathbf{R}^3} |\nabla^{k+2} v|^2\, dx \geqslant \frac{R}{1+t} \int_{\mathbf{R}^3} |\nabla^{k+1} v|^2\, dx - \left(\frac{R}{1+t}\right)^2 \int_{\mathbf{R}^3} |\nabla^k v|^2\, dx. \tag{3.19}$$

Choosing $R = k + 3$ and combining inequalities (3.18), (3.19) and the time decay rates (3.17), it arrives at directly
$$\begin{aligned}
\frac{d}{dt}\int |\nabla^{k+1} v|^2\, dx &+ \frac{k+3}{1+t} \int |\nabla^{k+1} v|^2\, dx \\
&\leqslant \left(\frac{k+3}{1+t}\right)^2 \int |\nabla^k v|^2\, dx + \int |\nabla^k F|^2\, dx \\
&\leqslant C(1+t)^{-(k+\frac{7}{2})}.
\end{aligned} \tag{3.20}$$

Multiplying (3.20) by $(1+t)^{k+3}$ and integrating over $[0,t]$, we have
$$\|\nabla^{k+1} v(t)\|_{L^2}^2 \leqslant (1+t)^{-k-3} \left[\|\nabla^{k+1} v_0\|_{L^2}^2 + C(1+t)^{\frac{1}{2}}\right],$$

which implies the time decay rates
$$\|\nabla^{k+1} v(t)\|_{L^2}^2 \leqslant C(1+t)^{-(k+\frac{5}{2})}.$$

Therefore, we complete the proof of the lemma.

Motivated by Lemma 3.3, we will improve the time decay rates for the second and third order derivatives of director.

Lemma 3.4 *Under the assumptions in Theorem 1.2, the global solution (ϱ, u, n) for problem (2.1)–(2.4) satisfies*

$$\|\nabla^k n(t)\|_{L^2}^2 \leqslant C(1+t)^{-\frac{3}{2}-k}, \tag{3.21}$$

where $k = 2, 3$.

Proof Taking $k = 1$ in (2.9), it follows immediately

$$\frac{1}{2}\frac{d}{dt}\int |\nabla^2 n|^2\, dx + \int |\nabla^3 n|^2\, dx = \int \nabla[u\cdot\nabla n - |\nabla n|^2(n+w_0)]\nabla^3 n\, dx. \tag{3.22}$$

In view of (2.19), we have

$$-\int \nabla[|\nabla n|^2(n+w_0)]\nabla^3 n\, dx \lesssim \delta\|\nabla^3 n\|_{L^2}^2. \tag{3.23}$$

By virtue of (3.5), Hölder, Sobolev and Young inequalities, it arrives at

$$\int \nabla(u\cdot\nabla n)\nabla^3 n\, dx$$
$$\lesssim \|\nabla u\|_{L^3}\|\nabla n\|_{L^6}\|\nabla^3 n\|_{L^2} + \|u\|_{L^3}\|\nabla^2 n\|_{L^6}\|\nabla^3 n\|_{L^2}$$
$$\lesssim \|\nabla u\|_{H^1}^2\|\nabla^2 n\|_{L^2}^2 + (\varepsilon+\delta)\|\nabla^3 n\|_{L^2}^2$$
$$\lesssim (1+t)^{-\frac{5}{2}}(1+t)^{-\frac{5}{2}} + (\varepsilon+\delta)\|\nabla^3 n\|_{L^2}^2$$
$$\lesssim (1+t)^{-5} + (\varepsilon+\delta)\|\nabla^3 n\|_{L^2}^2. \tag{3.24}$$

Inserting (3.23) and (3.24) into (3.22) and applying the smallness of ε and δ, we have

$$\frac{d}{dt}\int |\nabla^2 n|^2\, dx + \int |\nabla^3 n|^2\, dx \leqslant C(1+t)^{-5}. \tag{3.25}$$

On the other hand, from inequality (2.31), we have

$$\frac{1}{2}\frac{d}{dt}\int |\nabla^3 n|^2\, dx + \int |\nabla^4 n|^2\, dx = \int \nabla^2[u\cdot\nabla n - |\nabla n|^2(n+w_0)]\nabla^4 n\, dx. \tag{3.26}$$

By virtue of Hölder, Sobolev and Young inequalities, we obtain

$$\int \nabla^2(u\cdot\nabla n)\nabla^4 n\, dx$$
$$\lesssim \|\nabla^2 u\|_{L^2}\|\nabla n\|_{L^\infty}\|\nabla^4 n\|_{L^2} + \|\nabla u\|_{L^3}\|\nabla^2 n\|_{L^6}\|\nabla^4 n\|_{L^2}$$
$$\quad + \|u\|_{L^3}\|\nabla^3 n\|_{L^6}\|\nabla^4 n\|_{L^2}$$
$$\lesssim \|\nabla^2 u\|_{L^2}^2\|\nabla^2 n\|_{L^2}\|\nabla^3 n\|_{L^2} + \|\nabla u\|_{H^1}^2\|\nabla^3 n\|_{L^2}^2$$
$$\quad + (\varepsilon+\delta)\|\nabla^4 n\|_{L^2}^2$$

$$\lesssim \|\nabla u\|_{H^1}^2 \|\nabla^2 n\|_{H^1}^2 + (\varepsilon + \delta)\|\nabla^4 n\|_{L^2}^2. \tag{3.27}$$

Following from the idea of inequality (2.32), we have

$$-\int \nabla^2 [|\nabla n|^2 (n+w_0)] \nabla^4 n \, dx \lesssim \delta \|\nabla^4 n\|_{L^2}^2. \tag{3.28}$$

Inserting (3.27) and (3.28) into (3.26) and applying the smallness of ε and δ, it arrives at immediately

$$\frac{d}{dt} \int |\nabla^3 n|^2 \, dx + \int |\nabla^4 n|^2 \, dx \lesssim \|\nabla u\|_{H^1}^2 \|\nabla^2 n\|_{H^1}^2. \tag{3.29}$$

Combining (3.25) and (3.29) and applying the time decay rates (3.5), we get

$$\frac{d}{dt} \int (|\nabla^2 n|^2 + |\nabla^3 n|^2) \, dx + \int (|\nabla^3 n|^2 + |\nabla^4 n|^2) \, dx \leqslant C(1+t)^{-5}. \tag{3.30}$$

Similar to the analysis of inequality (3.19), it follows immediately

$$\int |\nabla^3 n|^2 \, dx \geqslant \frac{4}{1+t} \int |\nabla^2 n|^2 \, dx - \left(\frac{4}{1+t}\right)^2 \int |\nabla n|^2 \, dx, \tag{3.31}$$

and

$$\int |\nabla^4 n|^2 \, dx \geqslant \frac{5}{1+t} \int |\nabla^3 n|^2 \, dx - \left(\frac{5}{1+t}\right)^2 \int |\nabla^2 n|^2 \, dx. \tag{3.32}$$

The combination of (3.30), (3.31) and (3.32) yields directly

$$\frac{d}{dt} \int (|\nabla^2 n|^2 + |\nabla^3 n|^2) \, dx + \frac{4}{1+t} \int (|\nabla^2 n|^2 + |\nabla^3 n|^2) \, dx$$
$$\leqslant \frac{25}{(1+t)^2} \int (|\nabla n|^2 + |\nabla^2 n|^2) \, dx + C(1+t)^{-5}$$
$$\leqslant C(1+t)^{-\frac{9}{2}}, \tag{3.33}$$

where we have used the convergence rates (3.5). Multiplying (3.33) by $(1+t)^4$, we obtain

$$\frac{d}{dt}\left[(1+t)^4 (\|\nabla^2 n\|_{L^2}^2 + \|\nabla^3 n\|_{L^2}^2)\right] \leqslant C(1+t)^{-\frac{1}{2}}. \tag{3.34}$$

Integrating (3.34) over $[0, t]$, we have the following decay rate

$$\|\nabla^2 n(t)\|_{L^2}^2 + \|\nabla^3 n(t)\|_{L^2}^2 \leqslant C(1+t)^{-\frac{7}{2}}. \tag{3.35}$$

On the other hand, applying the convergence rates (3.5), (3.35) and inequality (3.29), it arrives at

$$\frac{d}{dt} \int |\nabla^3 n|^2 \, dx + \int |\nabla^4 n|^2 \, dx \lesssim (1+t)^{-\frac{5}{2}} (1+t)^{-\frac{7}{2}} \lesssim (1+t)^{-6},$$

which, together with (3.32) and (3.35), yields

$$\frac{d}{dt}\int |\nabla^3 n|^2\,dx + \frac{5}{1+t}\int |\nabla^3 n|^2\,dx \leqslant C(1+t)^{-\frac{11}{2}}. \qquad (3.36)$$

Multiplying (3.36) by $(1+t)^5$ and integrating over $[0,t]$, it follows immediately

$$\|\nabla^3 n(t)\|_{L^2}^2 \leqslant C(1+t)^{-\frac{9}{2}}.$$

Therefore, we complete the proof of the lemma.

Proof of Theorem 1.2 With the help of Lemmas 3.2 and 3.4, we complete the proof of Theorem 1.2.

Remark 3.5 *In order to obtain the rate of $d(x,t)$ converging to w_0, we suppose the finiteness of $\|d_0 - w_0\|_{L^2}$ in Theorem 1.2 additionally. Then, the density and velocity (ρ,u) enjoy the same decay rate with the director field $d(x,t) - w_0$. However, (ρ,u) will have the same decay rate with $\nabla(d(x,t) - w_0)$ without the assumption of finiteness of $\|d_0 - w_0\|_{L^2}$.*

3.3 Decay rates for the mixed space-time derivatives of density and velocity

In this subsection, we will establish the decay rates for the time derivatives of velocity and the mixed space-time derivatives of density and director.

Lemma 3.6 *Under the assumptions in Theorem 1.2, the global solution (ϱ, u, n) of problem (2.1)-(2.4) satisfies*

$$\|\varrho_t(t)\|_{H^1} + \|u_t(t)\|_{L^2} \leqslant C(1+t)^{-\frac{5}{4}},$$

$$\|\nabla^k n_t(t)\|_{L^2} \leqslant C(1+t)^{-\frac{7+2k}{4}},$$

for $k = 0, 1$.

Proof By virtue of equation $(2.1)_1$ and the convergence rates (1.6), we have

$$\|\varrho_t\|_{L^2} = \|\mathrm{div}\,u + \varrho\,\mathrm{div}\,u + u\cdot\nabla\varrho\|_{L^2}$$
$$\leqslant \|\mathrm{div}\,u\|_{L^2} + \|\varrho\|_{L^\infty}\|\mathrm{div}\,u\|_{L^2} + \|\nabla\varrho\|_{L^3}\|u\|_{L^6}$$
$$\leqslant C(1+t)^{-\frac{5}{4}}.$$

Similarly, it follows immediately

$$\|\nabla\varrho_t\|_{L^2} = \|\nabla\mathrm{div}\,u + \nabla\varrho\,\mathrm{div}\,u + \varrho\nabla\mathrm{div}\,u + \nabla u\cdot\nabla\varrho + u\cdot\nabla^2\varrho\|_{L^2}$$

$$\lesssim \|\nabla \mathrm{div} u\|_{L^2} + \|\nabla \varrho\|_{L^3}\|\nabla u\|_{L^6} + \|u\|_{L^\infty}\|\nabla^2 \varrho\|_{L^2}$$
$$\lesssim \|\nabla^2 u\|_{L^2} + \|\nabla^2 \varrho\|_{L^2}$$
$$\leqslant C(1+t)^{-\frac{5}{4}},$$

and

$$\|u_t\|_{L^2} \lesssim \|\mu \Delta u + (\mu+\nu)\nabla \mathrm{div} u\|_{L^2} + \|\nabla \varrho\|_{L^2} + \|u\|_{L^3}\|\nabla u\|_{L^6}$$
$$+ \|h(\varrho)\|_{L^\infty}\|\mu \Delta u + (\mu+\nu)\nabla \mathrm{div} u\|_{L^2} + \|f(\varrho)\|_{L^\infty}\|\nabla \varrho\|_{L^2}$$
$$+ \|g(\varrho)\|_{L^\infty}\|\nabla n\|_{L^\infty}\|\nabla^2 n\|_{L^2}$$
$$\lesssim \|\nabla^2 u\|_{L^2} + \|\nabla \varrho\|_{L^2} + \|\nabla^2 n\|_{L^2}$$
$$\lesssim (1+t)^{-\frac{5}{4}} + (1+t)^{-\frac{7}{4}}$$
$$\leqslant C(1+t)^{-\frac{5}{4}}.$$

By virtue of $(2.1)_3$, (1.6), Hölder and Sobolev inequalities, we obtain

$$\|n_t\|_{L^2} = \|-u\cdot \nabla n + \Delta n + |\nabla n|^2 (n+w_0)\|_{L^2}$$
$$\lesssim \|u\|_{L^3}\|\nabla n\|_{L^6} + \|\Delta n\|_{L^2} + \|\nabla n\|_{L^3}\|\nabla n\|_{L^6}$$
$$\lesssim \|u\|_{H^1}\|\nabla^2 n\|_{L^2} + \|\nabla^2 n\|_{L^2} + \|\nabla n\|_{H^1}\|\nabla^2 n\|_{L^2}$$
$$\leqslant C(1+t)^{-\frac{7}{4}}.$$

In the same manner, it arrives at directly

$$\|\nabla n_t\|_{L^2} = \|\nabla(-u\cdot \nabla n + \Delta n + |\nabla n|^2(n+w_0))\|_{L^2}$$
$$\lesssim \|\nabla u\|_{L^3}\|\nabla n\|_{L^6} + \|u\|_{L^3}\|\nabla^2 n\|_{L^6} + \|\nabla \Delta n\|_{L^2}$$
$$+ \|\nabla n\|_{L^3}\|\nabla^2 n\|_{L^6} + \|\nabla n\|_{L^6}^3$$
$$\lesssim \|\nabla u\|_{H^1}\|\nabla^2 n\|_{L^2} + \|\nabla^3 n\|_{L^2} + \|\nabla^2 n\|_{L^2}^3$$
$$\lesssim (1+t)^{-\frac{5}{4}}(1+t)^{-\frac{7}{4}} + (1+t)^{-\frac{9}{4}} + (1+t)^{-\frac{21}{4}}$$
$$\leqslant C(1+t)^{-\frac{9}{4}}.$$

Therefore, we complete the proof of the lemma.

Proof of Theorem 1.5 With the help of Lemma 3.6, we complete the proof of Theorem 1.5.

References

[1] Ericksen J. Hydrostatic theory of liquid crystals [J]. Arch. Rational Mech. Anal., 1962, 9: 371-378.

[2] Leslie F. Some constitutive equations for liquid crystals [J]. Arch. Rational Mech. Anal., 1968, 28: 265-283.

[3] Hardt R, Kinderlehrer D, Lin F. Existence and partial regularity of static liquid configurations [J]. Commun. Math. Phys., 1986, 105: 547-570.

[4] Lin F. Nonlinear theory of defects in nematic liquid crystals: Phsse transition and flow phenomena [J]. Commun. Pure Appl. Math., 1989, 42: 789-814.

[5] Lin F, Liu C. Nonparabolic dissipative systems modeling the flow of liquid crystals [J]. Commun. Pure Appl. Math., 1995, 48: 501-537.

[6] Lin F, Liu C. Partial regularity of the dynamic system modeling the flow of liquid crystals [J]. Discrete Contin.Dyn.Syst., 1996, 2: 1-22.

[7] Wu H. Long-time behavior for nonlinear hydrodynamic system modeling the namatic liquid crystal flows [J]. Discrete Contin. Dyn. Syst., 2010, 26: 379-396.

[8] Dai M, Qing J, Schonbek M. Asymptotic behavior of solutions to the liquid crystals systems in $\mathbf{R}^3$ [J]. Comm. Partial Differential Equations, 2012, 37: 2138-2164.

[9] Dai M, Schonbek M. Asymptotic behavior of solutions to the liquid crystal system in $H^m(\mathbf{R}^3)$ [J]. SIAM J. Math. Anal., 2014, 46: 3131-3150.

[10] Liu X, Zhang Z. Existence of the flow of liquid crystals system [J]. Chinese Ann. Math., 2009, 30A: 1-20.

[11] Jiang F, Tan Z. Global weak solution to the flow of liquid crystals system [J]. Math. Methods Appl. Sci., 2009, 32: 2243-2266.

[12] Wen H, Ding S. Solution of incompressible hydrodynamic flow of liquid crystals [J]. Nonlinear Anal. Real World Appl., 2011, 12: 1510-1531.

[13] Li J. Global strong solutions to incompressible nematic liquid crystal flow [J]. arXiv:1211.5864v1.

[14] Hong M. Global existence of solutions of the simplified Ericksen-Leslie system in dimension two [J]. Calc.Var.Partial Differential Equations, 2011, 40: 15-36.

[15] Lin F, Lin J, Wang C. Liquid crystal flows in two dimensions [J]. Arch. Ration. Mech. Anal., 2010, 197: 297-336.

[16] Wang C. Well-posedness for the heat flow of harmonic maps and the liquid crystal flow with rough initial data [J]. Arch. Ration. Mech. Anal., 2011, 200: 1-19.

[17] Du Y, Wang K. Space-time regularity of the Koch and Tataru solutions to the liquid crystal equations [J]. SIAM J. Math. Anal., 2013, 45: 3838-3853.

[18] Du Y, Wang K. Regularity of the solutions to the liquid crystal equations with small rough data [J]. J. Differential Equations, 2014, 256: 65-81.

[19] Li J. Global strong and weak solutions to nematic liquid crystal flow in two dimensions [J]. Nonlinear Anal., 2014, 99: 80-94.

[20] Li X, Wang D. Global solution to the incompressible flow of liquid crystals [J]. J. Differential Equations, 2012, 252: 745-767.

[21] Li X, Wang D. Global strong solution to the density-dependent incompressible flow of liquid crystal [J]. Trans. Amer. Math. Soc., 2015, 367(4): 2301-2338.

[22] Hao Y, Liu X. The existence and blow-up criterion of liquid crystals system in critical Besov space [J]. Commun. Pure Appl. Anal., 2014, 13(1): 225-236.

[23] Ding S, Lin J, Wang C, Wen H. Compressible Hydrodynamic Flow of Liquid Crystals in 1-D [J]. Discrete Contin. Dyn. Syst., 2012, 32: 539-563.

[24] Ding S, Wang C, Wen H. Weak solution to comprssible hydrodynamic flow of liquid crystals in dimension one [J]. Discrete Contin. Dyn. Syst., 2011, 15: 357-371.

[25] Jiang F, Jiang S, Wang D. On multi-dimensional compressible flows of nematic liquid crystals with large initial energy in a bounded domain [J]. J. Funct. Anal., 2013, 265: 3369-3397.

[26] Lin J, Lai B, Wang C. Global finite energy weak solutions to the compressible nematic liquid crystal flow in dimension three [J]. SIAM J. Math. Anal., 2015, 47(4): 2952-2983.

[27] Huang T, Wang C, H. Y. Wen, Strong solutions of the compressible nematic liquid crystal [J]. J. Differential Equations, 2012, 252: 2222-2265.

[28] T. Huang, Wang C, Wen H. Blow up criterion for compressible nematic liquid crystal flows in dimension three [J]. Arch. Rational Mech. Anal., 2012, 204: 285-311.

[29] Huang X, Wang Y. A Serrin criterion for compressible nematic liquid crystal flows [J]. Math. Meth. Appl. Sci., 2013, 36: 1363-1375.

[30] Gao J, Tao Q, Yao Z. A blowup criterion for the compressible nematic liquid crystal flows in dimension two [J]. J. Math. Anal. Appl., 2014, 415: 33-52.

[31] Ma S. Classical solutions for the compressible liquid crystal flows with nonnegative initial densities [J]. J. Math. Anal. Appl., 2013, 397: 595-618.

[32] Hu X, Wu H. Global solution to the three-dimensional compressible flow of liquid crystals [J]. SIAM J. Math. Anal., 2013, 45: 2678-2699.

[33] Gao J, Tao Q, Yao Z. Long-time behavior of solution for the compressible nematic liquid crystal flows in $\mathbf{R}^3$ [J]. J. Differential Equations, 2016, 261: 2334-2383.

[34] Lin F, Wang C. Recent developments of analysis for hydrodynamic flow of nematic liquid crystals [J]. Phil. Trans. R. Soc. A, 2014, 372: 20130361.

[35] Wang Y, Tan Z. Global existence and optimal decay rate for the strong solutions in H^2 to the compressible Navier-Stokes equations [J]. Appl. Math. Lett., 2011, 24: 1778-1784.

[36] Matsumura A, Nishida T. The initial value problems for the equations of motion of viscous and heat-conductive gases [J]. J. Math. Kyoto Univ., 1980, 20: 67-104.

[37] Hu X, Wu G. Global existence and optimal decay rates for three-dimensional compressible viscoelastic flows [J]. SIAM J. Math. Anal., 2013, 45: 2815-2833.

[38] Wang W, Wang W. Decay rates of the compressible Navier-Stokes-Korteweg equations with potential forces [J]. Discrete Contin. Dyn. Syst., 2015, 35: 513-536.

[39] Wang W. Large time behavior of solutions to the compressible Navier-Stokes equations with potential force [J]. J. Math. Anal. Appl., 2015, 423: 1448-1468.

[40] Schonbek M. L^2 decay for weak solutions of the Navier-Stokes equations [J]. Arch. Rational Mech. Anal., 1985, 88: 209-222.

[41] Tan Z, Wang Y. Global solution and large-time behavior of the 3D compressible Euler equations with damping [J]. J. Differential Equations, 2013, 254: 1686-1704.

[42] Guo Y, Wang Y. Decay of dissipative equations and negative Sobolev spaces [J]. Comm. Partial Differential Equations, 2012, 37: 2165-2208.

[43] Duan R, Liu H, Ukai S, Yang T. Optimal L^p-L^q convergence rates for the compressible Navier-Stokes equations with potential force [J]. J.Differential Equations, 2007, 238: 220-233.

Global Smooth Solution to a Coupled Schrödinger System in Atomic Bose-Einstein Condensates with Two Dimensional Spaces*

Guo Boling (郭柏灵) and Li Qiaoxin (李巧欣)

Abstract In this paper, we obtain the global smooth solution of a nonlinear Schrödinger equations in atomic Bose-Einstein condensates with two dimensional spaces. By using the Galërkin method and *a priori* estimates, we establish the global existence and uniqueness of the smooth solution.

Keywords Schrödinger equations; the Galërkin method; *a priori* estimates; global smooth solution

1 Introduction

In this paper we consider the following nonlinear Schrödinger equations [13]

$$\begin{cases} i\hbar u_t = \left(-\dfrac{\hbar^2}{2M}\Delta + \lambda_u|u|^2 + \lambda|v|^2\right)u + \sqrt{2}\alpha \bar{u}v, \\ i\hbar v_t = \left(-\dfrac{\hbar^2}{4M}\Delta + \varepsilon + \lambda_v|v|^2 + \lambda|u|^2\right)v + \dfrac{\alpha}{\sqrt{2}}u^2, \end{cases} \quad (1.1)$$

with the initial condition and periodic boundary condition

$$u(x,0) = u_0(x), \quad v(x,0) = v_0(x), \quad x \in \Omega, \quad (1.2)$$

$$u(x+2\pi e_i, t) = u(x,t), \quad v(x+2\pi e_i, t) = v(x,t), \quad x \in \Omega, \quad i = 1,2, \quad (1.3)$$

where $\Omega = (-\pi, \pi) \times (-\pi, \pi) \subseteq \mathbb{R}^2$, $t > 0$ and $e_1 = (1,0), e_2 = (0,1)$ is an orthonormal basis of $\mathbb{R}^2$. In the system (1.1), $i = \sqrt{-1}$ is the imaginary unit, $\hbar$ is the Planck constant, $M > 0$ is the mass of a single atom, $\lambda_u, \lambda_v, \lambda$ represent the strengths of the

---
* Front. Math. China, 2016, 11(6): 1515–1532.

atom-atom, molecule-molecule and atom-molecule interactions, respectively and ϵ, α are any real constants.

The observation of Feshbach resonances in the inter-particle interactions of a dilute Bose-Einstein condensate of Na-atoms by Ketterle's group at MIT [13] was an anticipated event. The significance of this experimental breakthrough appears of singular importance as its consequences are far-reaching in two subfields of physics: (i) Atomic and Molecular Physics.(ii) BEC-physics. Experimental advances in achieving and observing Bose-Einstein condensation (BEC) in trapped neutral atomic vapors have spurred great excitement in the atomic physics community and renewed the interest in studying the collective dynamics of macroscopic ensembles of atoms occupying the same one-particle quantum state [3]. The properties of a BEC are usually well modeled by a nonlinear Schrödinger equation (NLSE) for the macroscopic wave function known as the Gross-Pitaevskii equation (GPE) [14].

Many authors have considered the Gross-Pitaevskii equation with two dimensions. In [12], Sadhan K. Adhikari obtained the numerical solution of the time-independent nonlinear Gross-Pitaevskii equation in two dimensions. In [4], Bao Weizhu and Tang Weijun propose a new numerical method to compute the ground state solution of Bose-Einstein condensation by directly minimizing the energy functional through the finite element discretization. They begin with the 3d Gross-Pitaevskii equation, make it dimensionless to obtain a three-parameter model, show how to approximately reduce it to a 2d GPE and a 1d GPE in certain limits. Tingchun Wang and Xiaofei Zhao [6] also studied this problem. They proposed and analyzed the finite difference methods for solving in two dimensions. Further discussion can be found in Refs [1, 2, 11].

However, to our knowledge, the TGPE with two dimensions has not yet been fully studied. In this paper, we consider the system (1.1) replacing the usual Gross-Pitaevskii equation that describes the time evolution of the dilute single condensate system as a mathematical model in nonlinear partial differential equations. And our aim is to obtain the global smooth solution of system (1.1) in two dimensional spaces. The main difficulty is to establish certain delicate *a priori* estimates that govern our strategy to prove the existence of the smooth solution.

In [7]- [10], Guo Boling studied a class of systems of nonlinear Schrödinger equations for the initial value problem and the periodic boundary value problem. In this paper, we prove the existence and uniqueness of the global solution to the periodic boundary value problem for the nonlinearly coupled Schrödinger system (1.1) by using the Faedo-Galëkin method.

Before starting the main results, we review the notations and the calculus inequalities used in this paper.

Denote $L^p = L^p(\Omega)$ be the Banach space endowed with the norm $\|\cdot\|_{L^p}$, when $p = 2$, $L^2(\Omega)$ denote the Hilbert space with the usual scalar product $(\cdot, \cdot)$. Denote $H^m(\Omega), m = 1, 2, \cdots$ be the Sobolev space of complex-valued functions

$$H^m(\Omega) = \{u \in L^2(\Omega) : D^\alpha u \in L^2(\Omega), 0 \leqslant |\alpha| \leqslant m\},$$

whose inner product is given by $(u, v)_m = \sum_{0 \leqslant |\alpha| \leqslant m}(D^\alpha u, D^\alpha v)$ with the induced norm

$$\|u\|_{H^m} = \left(\int_\Omega \sum_{0 \leqslant |\alpha| \leqslant m} |D^\alpha u|^2 dx\right)^{\frac{1}{2}}.$$

To simplify the notation in this paper, we shall denote by $\int U(x)dx$ the integration $\int_\Omega U(x)dx$, C is a generic constant and may assume different values in different formulates.

The following auxiliary lemmas will be needed.

Lemma 1.1 *Assuming $u \in H^1(\mathbb{R}^2)$, then we have*

$$\frac{1}{2}\|u\|_4^4 \leqslant \left(\frac{\|u\|_2^2}{\|Q\|_2^2}\right)\|\nabla u\|_2^2, \tag{1.4}$$

where Q is the solution of the equation $\Delta Q - Q = -|Q|^2 Q$.

$$\|\nabla u\|_4^4 \leqslant C_{12}^4 \|\Delta u\|_2^2 \|\nabla u\|_2^2, \tag{1.5}$$

where

$$C_{12}^4 = \frac{2}{\|\varphi\|_2^2}, \quad \Delta\varphi - \varphi + \varphi^3 = 0.$$

Lemma 1.2 *Let $u \in W^{k,p}(\mathbb{R}^n) \cap W^{s,q}(\mathbb{R}^n)$, $k, s > 0$, $p > 1$, $q \geqslant 1$ and $kp = d < sq$, then we have*

$$\|u\|_\infty \leqslant C\|u\|_{W^{k,p}}\left(1 + \ln\left(1 + \frac{\|u\|_{W^{s,q}}}{\|u\|_{W^{k,p}}}\right)\right)^{1-\frac{1}{p}}, \tag{1.6}$$

where C is a constant depending only on k, s, p, q, n.

In particularly, when $n = 2, k = 1, s = 2$ and $p = q = 2$, for $u \in H^2(\mathbb{R}^2)$ and $\|u\|_{H^1} \leqslant K$, we have

$$\|u\|_\infty \leqslant C(1 + \ln(1 + \|u\|_{H^2}))^{\frac{1}{2}}, \tag{1.7}$$

where C depends on K.

Lemma 1.3 (the Gronwall inequality) *Let c be a constant, and $b(t), u(t)$ be nonnegative continuous functions in the interval $[0, T]$ satisfying*

$$u(t) \leqslant c + \int_0^t b(\tau) u(\tau) d\tau, \quad t \in [0, T].$$

Then $u(t)$ satisfies the estimate

$$u(t) \leqslant c \exp\left(\int_0^t b(\tau) d\tau\right) \tag{1.8}$$

for $t \in [0, T]$.

Our main results are:

Theorem 1.1 *Let $u_0(x) \in H_{per}^m(\Omega), v_0(x) \in H_{per}^m(\Omega)$, and $m > 1$. Then $\forall T > 0$, the system (1.1)–(1.3) has a uniquely global smooth solution*

$$(u, v) \in L^\infty\left([0, T); H_{per}^m(\Omega)\right)^2. \tag{1.9}$$

Theorem 1.2 *Let $u_0(x) \in H^m(\mathbb{R}^2), v_0(x) \in H^m(\mathbb{R}^2)$, and $m > 1$. The system (1.1)–(1.2) has a uniquely global smooth solution*

$$(u, v) \in L_{loc}^\infty\left([0, \infty); H^m(\mathbb{R}^2)\right)^2. \tag{1.10}$$

2 *A Priori* Estimates

In this section, we give the demonstration of *a priori* estimates that guarantee the existence of the global smooth solution of the system (1.1)–(1.3).

Lemma 2.1 *If $(u_0, v_0) \in L^2(\Omega)$, then for the solution (u, v) of problem (1.1), we have*

$$\sup_{0 \leqslant t < \infty} \left(\|u(x,t)\|_2^2 + 2\|v(x,t)\|_2^2\right) \equiv \left(\|u_0(x)\|_2^2 + 2\|v_0(x)\|_2^2\right). \tag{2.11}$$

The proof of this lemma is the same to Lemma 2.1 in [1]. So we omit it here.

Lemma 2.2 *Let $(u_0, v_0) \in H_{per}^1(\Omega)$, (u, v) be the solution of the system (1.1), and $\|u_0\|_2^2 + 2\|v_0\|_2^2 < \min\left\{\dfrac{\hbar^2 \|Q\|_2^2}{2|M|(|\lambda_u| + |\lambda| + \sqrt{2}|\alpha|)}, \dfrac{\hbar^2 \|Q\|_2^2}{4|M|(|\lambda_v| + |\lambda|)}\right\}$, then we have*

$$\sup_{0 \leqslant t < \infty} \left(\|u(\cdot, t)\|_{H_{per}^1} + \|v(\cdot, t)\|_{H_{per}^1}\right) \leqslant C, \tag{2.12}$$

where C is a constant depending only on $\|u_0\|_{H_{per}^1}, \|v_0\|_{H_{per}^1}$.

Proof The inner product is taken to the first equation of the system (1.1) with $\bar{u}_t$ and the second equation with $\bar{v}_t$, and then integrating and taking the real part of the resulting equations, we get

$$\begin{cases} 0 = \dfrac{\hbar^2}{4M}\dfrac{d}{dt}\int |\nabla u|^2 dx + \dfrac{\lambda_u}{4}\dfrac{d}{dt}\int |u|^4 dx + \lambda Re\int |v|^2 u\bar{u}_t dx + \sqrt{2}\alpha Re\int \bar{u}v\bar{u}_t dx \\ 0 = \dfrac{\hbar^2}{8M}\dfrac{d}{dt}\int |\nabla v|^2 dx + \dfrac{\varepsilon}{2}\dfrac{d}{dt}\int |v|^2 dx + \dfrac{\lambda_v}{4}\dfrac{d}{dt}\int |v|^4 dx + \lambda Re\int |u|^2 v\bar{v}_t dx \\ \quad + \dfrac{\alpha}{\sqrt{2}}Re\int u^2 \bar{v}_t dx. \end{cases} \quad (2.13)$$

Summing up the two equations of the system (2.13), we have

$$\dfrac{\hbar^2}{4M}\dfrac{d}{dt}\left(\int |\nabla u|^2 dx + \dfrac{1}{2}\int |\nabla v|^2 dx\right) + \dfrac{1}{4}\dfrac{d}{dt}\left(\lambda_u \int |u|^4 dx + \lambda_v \int |v|^4 dx\right)$$
$$+ \dfrac{\varepsilon}{2}\dfrac{d}{dt}\int |v|^2 dx + \dfrac{\lambda}{2}\dfrac{d}{dt}\int |u|^2|v|^2 dx + \dfrac{\alpha}{\sqrt{2}}Re\dfrac{d}{dt}\int u^2 \bar{v} dx = 0.$$

Let

$$\text{I} := \dfrac{\hbar^2}{4M}\left(\int |\nabla u|^2 dx + \dfrac{1}{2}\int |\nabla v|^2 dx\right), \quad \text{II} := \dfrac{1}{4}\left(\lambda_u \int |u|^4 dx + \lambda_v \int |v|^4 dx\right),$$

$$\text{III} := \dfrac{\lambda}{2}\int |u|^2|v|^2 dx, \quad \text{IV} := \dfrac{\varepsilon}{2}\int |v|^2 dx, \quad \text{V} := \dfrac{\alpha}{\sqrt{2}}Re\int u^2 \bar{v} dx.$$

Then

$$E(t) = \text{I} + \text{II} + \text{III} + \text{IV} + \text{V} \equiv E(0). \quad (2.14)$$

Applying Lemma 1.1, we have

$$\|u\|_4^4 \leqslant \dfrac{2\|u\|_2^2}{\|Q\|_2^2}\|\nabla u\|_2^2, \quad \|v\|_4^4 \leqslant \dfrac{2\|v\|_2^2}{\|Q\|_2^2}\|\nabla v\|_2^2. \quad (2.15)$$

Therefore

$$|\text{II}| + |\text{III}| \leqslant \left(\dfrac{|\lambda_u| + |\lambda|}{4}\right)\|u\|_4^4 + \left(\dfrac{|\lambda_v| + |\lambda|}{4}\right)\|v\|_4^4$$
$$\leqslant \left(\dfrac{|\lambda_u| + |\lambda|}{2}\right)\dfrac{\|u\|_2^2}{\|Q\|_2^2}\|\nabla u\|_2^2 + \left(\dfrac{|\lambda_v| + |\lambda|}{2}\right)\dfrac{\|v\|_2^2}{\|Q\|_2^2}\|\nabla v\|_2^2. \quad (2.16)$$

For the term V, using Hölder inequality, we have

$$\text{V} = \dfrac{\alpha}{\sqrt{2}}Re\int u^2 \bar{v} dx \leqslant \dfrac{|\alpha|}{\sqrt{2}}\int |u|^2|v|dx$$
$$\leqslant \dfrac{|\alpha|}{2\sqrt{2}}(\|v\|_2^2 + \|u\|_4^4) \leqslant \dfrac{|\alpha|}{2\sqrt{2}}\|v\|_2^2 + \dfrac{|\alpha|}{\sqrt{2}}\dfrac{\|u\|_2^2}{\|Q\|_2^2}\|\nabla u\|_2^2. \quad (2.17)$$

Combining (2.16) and (2.17), we have

$$|\text{II}| + |\text{III}| + |\text{IV}| + |\text{V}|$$
$$\leqslant \left(\frac{|\lambda_u| + |\lambda| + \sqrt{2}|\alpha|}{2}\right)\frac{\|u\|_2^2}{\|Q\|_2^2}\|\nabla u\|_2^2 + \left(\frac{|\lambda_v| + |\lambda|}{2}\right)\frac{\|v\|_2^2}{\|Q\|_2^2}\|\nabla v\|_2^2$$
$$+ \left(\frac{\sqrt{2}|\alpha|}{4} + \frac{|\epsilon|}{2}\right)\|v\|_2^2$$
$$\leqslant \left(\frac{|\lambda_u| + |\lambda| + \sqrt{2}|\alpha|}{2}\right)\frac{\|u_0\|_2^2 + 2\|v_0\|_2^2}{\|Q\|_2^2}\|\nabla u\|_2^2 + \left(\frac{|\lambda_v| + |\lambda|}{2}\right)\frac{\|u_0\|_2^2 + 2\|v_0\|_2^2}{\|Q\|_2^2}\|\nabla v\|_2^2$$
$$+ \left(\frac{\sqrt{2}|\alpha|}{4} + \frac{|\epsilon|}{2}\right)\|v\|_2^2.$$

By (2.15), we have

$$\int |\nabla u|^2 dx + \frac{1}{2}\int |\nabla v|^2 dx \equiv \frac{4M}{\hbar^2}E(0) - \frac{4M}{\hbar^2}(\text{II} + \text{III} + \text{IV} + \text{V})$$
$$\leqslant \frac{4M}{\hbar^2}E(0) + \frac{4|M|}{\hbar^2}(|\text{II}| + |\text{III}| + |\text{IV}| + |\text{V}|).$$

So when $\|u_0\|_2^2 + 2\|v_0\|_2^2 < \min\left\{\dfrac{\hbar^2\|Q\|_2^2}{2|M|(|\lambda_u| + |\lambda| + \sqrt{2}|\alpha|)}, \dfrac{\hbar^2\|Q\|_2^2}{4|M|(|\lambda_v| + |\lambda|)}\right\}$,

$$2\|\nabla u\|_2^2 + \|\nabla v\|_2^2 \leqslant C,$$

where C is a constant depending only on $\|u_0\|_{H^1}, \|v_0\|_{H^1}$.

This completes the proof of Lemma 2.2.

Lemma 2.3 *Let $(u_0, v_0) \in H^2(\Omega)$, (u, v) be the solution of the system (1.1), we have*

$$\sup_{0\leqslant t<\infty}(\|u(\cdot, t)\|_{H^2} + \|v(\cdot, t)\|_{H^2}) \leqslant C, \tag{2.18}$$

where the constant C depends on $\|u_0\|_{H^2}, \|v_0\|_{H^2}$.

Proof Making the inner product of $\Delta^2 \overline{u}$ with the first equation of the system (1.1) and $\Delta^2 \overline{v}$ with the second equation and integrating the resulting equations with respect to x on Ω, and then taking the imaginary part of the resulting equations, we obtain

$$\frac{\hbar}{2}\frac{d}{dt}\|\Delta u\|_2^2 = \lambda_u Im \int \left(u^2(\Delta\overline{u})^2 + 4u|\nabla u|^2\Delta\overline{u} + 2\overline{u}(\nabla u \cdot \nabla u)\Delta\overline{u}\right) dx$$
$$+ \lambda Im \int \left(u\overline{v}\Delta v\Delta\overline{u} + vu\Delta\overline{v}\Delta\overline{u} + 2|\nabla v|^2 u\Delta\overline{u} + 2(\nabla|v|^2 \cdot \nabla u)\Delta\overline{u}\right) dx$$
$$+ \sqrt{2}\alpha Im \int \left(v(\Delta\overline{u})^2 + \overline{u}\Delta v\Delta\overline{u} + 2(\nabla\overline{u} \cdot \nabla v)\Delta\overline{u}\right) dx,$$

$$\frac{\hbar}{2}\frac{d}{dt}\|\Delta v\|_2^2 = \lambda_v Im \int \left(v^2(\Delta\bar{v})^2 + 4v|\nabla v|^2\Delta\bar{v} + 2\bar{v}(\nabla v \cdot \nabla v)\Delta\bar{v}\right)dx$$

$$+ \lambda Im \int \left(\bar{u}v\Delta u\Delta\bar{v} + uv\Delta\bar{u}\Delta\bar{v} + 2v|\nabla u|^2\Delta\bar{v} + 2(\nabla|u|^2 \cdot \nabla v)\Delta\bar{v}\right)dx$$

$$+ \frac{2\alpha}{\sqrt{2}} Im \int (u\Delta u + \nabla u \cdot \nabla u)\Delta\bar{v}dx.$$

Denote $\dfrac{\hbar}{2}\dfrac{d}{dt}\|\Delta u\|_2^2 = \text{I} + \text{II} + \text{III}$, where

$$\text{I} \equiv \lambda_u Im \int \left(u^2(\Delta\bar{u})^2 + 4u|\nabla u|^2\Delta\bar{u} + 2\bar{u}(\nabla u \cdot \nabla u)\Delta\bar{u}\right)dx,$$

$$\text{II} \equiv \lambda Im \int \left(u\Delta v\bar{v}\Delta\bar{u} + vu\Delta\bar{v}\Delta\bar{u} + 2|\nabla v|^2 u\Delta\bar{u} + 2(\nabla|v|^2 \cdot \nabla u)\Delta\bar{u}\right)dx,$$

$$\text{III} \equiv \sqrt{2}\alpha Im \int \left(v(\Delta\bar{u})^2 + \bar{u}\Delta v\Delta\bar{u} + 2(\nabla\bar{u} \cdot \nabla v)\Delta\bar{u}\right)dx.$$

And denote $\dfrac{\hbar}{2}\dfrac{d}{dt}\|\Delta v\|_2^2 = \text{IV} + \text{V} + \text{VI}$, where

$$\text{IV} \equiv \lambda_v Im \int \left(v^2(\Delta\bar{v})^2 + 4v|\nabla v|^2\Delta\bar{v} + 2\bar{v}(\nabla v \cdot \nabla v)\Delta\bar{v}\right)dx,$$

$$\text{V} \equiv \lambda Im \int \left(\bar{u}v\Delta u\Delta\bar{v} + uv\Delta\bar{u}\Delta\bar{v} + 2v|\nabla u|^2\Delta\bar{v} + 2(\nabla|u|^2 \cdot \nabla v)\Delta\bar{v}\right)dx,$$

$$\text{VI} \equiv \frac{2\alpha}{\sqrt{2}} Im \int (u\Delta u + \nabla u \cdot \nabla u)\Delta\bar{v}dx.$$

Firstly, we estimate the term I.

By using Sobolev embedding theorem, we have

$$\int u^2 \Delta\bar{u}^2 dx \leqslant \|u\|_\infty^2 \|\Delta u\|_2^2. \tag{2.19}$$

Applying the Hölder's inequality and the inequality (1.5)

$$\int_\Omega |\nabla u|^2 u\Delta\bar{u}dx \leqslant \|u\|_\infty \|\nabla u\|_4^2 \|\Delta u\|_2$$

$$\leqslant C\|u\|_\infty \|\Delta u\|_2 \|\nabla u\|_2 \|\Delta u\|_2$$

$$\leqslant C(\|u\|_\infty^2 + 1)\|\Delta u\|_2^2. \tag{2.20}$$

$$\int \bar{u}(\nabla u \cdot \nabla u)\Delta\bar{u}dx \leqslant \int |\nabla u|^2 |\Delta u||u|dx$$

$$\leqslant \|u\|_\infty \int |\nabla u|^2 |\Delta u|dx,$$

noticing the inequality (2.20), we can get

$$\int \bar{u}(\nabla u \cdot \nabla u)\Delta \bar{u} dx \leqslant C\left(1 + \|u\|_\infty^2\right)\|\Delta u\|_2^2. \tag{2.21}$$

Combining the inequalities (2.19), (2.20) and (2.21),

$$|\mathrm{I}| \leqslant C(1 + \|u\|_\infty^2)(\|\Delta u\|_2^2 + 1). \tag{2.22}$$

Next, we deal with the term II.

Applying the Sobolev embedding theorem and the Hölder inequality

$$\begin{aligned}\int vu\Delta \bar{v}\Delta \bar{u} dx &\leqslant \frac{\|u\|_\infty^2 + \|v\|_\infty^2}{4}(\|\Delta u\|_2^2 + \|\Delta v\|_2^2) \\ &\leqslant C\left(1 + \|u\|_\infty^2 + \|v\|_\infty^2\right)\left(\|\Delta u\|_2^2 + \|\Delta v\|_2^2 + 1\right),\end{aligned} \tag{2.23}$$

$$\begin{aligned}\int \bar{v}u\Delta v\Delta \bar{u} dx &\leqslant \frac{\|u\|_\infty^2 + \|v\|_\infty^2}{4}(\|\Delta u\|_2^2 + \|\Delta v\|_2^2) \\ &\leqslant C\left(1 + \|u\|_\infty^2 + \|v\|_\infty^2\right)\left(\|\Delta u\|_2^2 + \|\Delta v\|_2^2 + 1\right).\end{aligned} \tag{2.24}$$

Using Hölder's inequality and the inequality (1.5)

$$\begin{aligned}\int u|\nabla v|^2 \Delta \bar{u} dx &\leqslant \|u\|_\infty \|\Delta u\|_2 \|\nabla v\|_4^2 \\ &\leqslant C\|u\|_\infty \|\Delta u\|_2 \|\Delta v\|_2 \|\nabla v\|_2 \\ &\leqslant C\|u\|_\infty(\|\Delta u\|_2^2 + \|\Delta v\|_2^2 + 1) \\ &\leqslant C(1 + \|u\|_\infty^2)(\|\Delta u\|_2^2 + \|\Delta v\|_2^2 + 1),\end{aligned} \tag{2.25}$$

$$\begin{aligned}\int (\nabla |v|^2 \cdot \nabla u)\Delta \bar{u} dx &\leqslant \int |\nabla |v|^2||\nabla u||\Delta u| dx \\ &\leqslant C\|v\|_\infty \int |\nabla v||\nabla u||\Delta u| dx \\ &\leqslant C_1 \|v\|_\infty \|\nabla v\|_4 \|\nabla u\|_4 \|\Delta u\|_2 \\ &\leqslant C_2 \|v\|_\infty \|\Delta v\|_2^{\frac{1}{2}} \|\Delta u\|_2^{\frac{3}{2}} \\ &\leqslant C(1 + \|v\|_\infty^2)(\|\Delta v\|_2^2 + \|\Delta u\|_2^2 + 1).\end{aligned} \tag{2.26}$$

Using the inequalities (2.23)–(2.26), we get

$$|\mathrm{II}| \leqslant C(1 + \|u\|_\infty^2 + \|v\|_\infty^2)(\|\Delta v\|_2^2 + \|\Delta u\|_2^2 + 1). \tag{2.27}$$

For the estimate of the term III, we only need the inequality

$$\int (\nabla \bar{u} \cdot \nabla v) \Delta \bar{u} dx \leqslant \int |\nabla u||\nabla v||\Delta u| dx \leqslant \|\nabla u\|_4 \|\nabla v\|_4 \|\Delta u\|_2$$

$$\leqslant C(\|\Delta v\|_2^2 + \|\Delta u\|_2^2 + 1).$$

So we get

$$|\text{III}| \leqslant C(1 + \|u\|_\infty^2 + \|v\|_\infty^2)(\|\Delta v\|_2^2 + \|\Delta u\|_2^2 + 1). \tag{2.28}$$

Comparing the term I with the term IV, we can immediately get the estimate

$$|\text{IV}| \leqslant C(1 + \|v\|_\infty^2))(\|\Delta v\|_2^2 + 1). \tag{2.29}$$

And comparing the term II with the term V,

$$|\text{V}| \leqslant C(1 + \|u\|_\infty^2 + \|v\|_\infty^2))(\|\Delta v\|_2^2 + \|\Delta u\|_2^2 + 1). \tag{2.30}$$

For the term VI, using Hölder inequality

$$\int u \Delta u \Delta \bar{v} dx \leqslant C \|u\|_\infty (\|\Delta u\|_2^2 + \|\Delta v\|_2^2) \leqslant C(1 + \|u\|_\infty^2)(\|\Delta v\|_2^2 + \|\Delta u\|_2^2 + 1),$$

$$\int (\nabla u \cdot \nabla u) \Delta \bar{v} dx \leqslant C_1 \|\nabla u\|_4^2 \|\Delta v\|_2 \leqslant C_2(\|\Delta u\|_2^2 + \|\Delta v\|_2^2 + 1).$$

So we have

$$|\text{VI}| \leqslant C(1 + \|u\|_\infty^2 + \|v\|_\infty^2)(\|\Delta u\|_2^2 + \|\Delta v\|_2^2 + 1). \tag{2.31}$$

Combining the estimates (2.22), (2.27) and (2.28)

$$\frac{d}{dt} \|\Delta u\|_2^2 \leqslant C(1 + \|u\|_\infty^2 + \|v\|_\infty^2)(\|\Delta v\|_2^2 + \|\Delta u\|_2^2 + 1).$$

And applying (2.27), (2.30) and (2.31), we can obtain the following estimate

$$\frac{d}{dt} \|\Delta v\|_2^2 \leqslant C(1 + \|u\|_\infty^2 + \|v\|_\infty^2)(\|\Delta v\|_2^2 + \|\Delta u\|_2^2 + 1)).$$

Therefore

$$\frac{d}{dt}(\|\Delta u\|_2^2 + \|\Delta v\|_2^2 + 1)$$
$$\leqslant C(1 + \|u\|_\infty^2 + \|v\|_\infty^2)(\|\Delta v\|_2^2 + \|\Delta u\|_2^2 + 1)$$
$$\leqslant C\left(1 + \ln(1 + \|\Delta u\|_2^2 + \|\Delta v\|_2^2)\right)(\|\Delta u\|_2^2 + \|\Delta v\|_2^2 + 1).$$

Using the Gronwall inequality, we get

$$\|\Delta u\|_2^2 + \|\Delta v\|_2^2 \leqslant C,$$

where C is a constant depending only on $\|u_0\|_{H^2}, \|v_0\|_{H^2}$.

This completes the proof of Lemma 2.3.

Lemma 2.4 Let $m \geq 0$ be any integer number, $u_0 \in H^m_{per}(\Omega), v_0 \in H^m_{per}(\Omega)$, and (u,v) be the solution of the system (1.1), then we have

$$\sup_{0 \leq t \leq T} (2\|u(\cdot, t)\|_{H^m} + \|v(\cdot, t)\|_{H^m}) \leq C, \quad \forall T > 0, \tag{2.32}$$

where the constant C depends only on T and $\|u_0\|_{H^m}, \|v_0\|_{H^m}$.

Proof This lemma is proved by mathematical induction as follows. When $m = 0, 1, 2$, according to Lemmas 2.1–2.3, the inequality (2.32) is held.

Suppose that (2.32) is valid for $m \leq k$. We will prove that (2.32) holds for $m = k+1$.

Making the inner product of $D^{2(k+1)}\overline{u}$ with the first equation of the system (1.1) and $D^{2(k+1)}\overline{v}$ with the second equation, and then integrating and taking the imaginary part of the resulting equations, we get

$$\frac{\hbar}{2}\frac{d}{dt}\|D^{k+1}u\|_2^2 = \lambda_u Im \int D^{k+1}(|u|^2 u) \cdot D^{k+1}\overline{u}dx + \lambda Im \int D^{k+1}(|v|^2 u) \cdot D^{k+1}\overline{u}dx$$
$$+ \sqrt{2}\alpha Im \int D^{k+1}(\overline{u}v) \cdot D^{k+1}\overline{u}dx := \mathrm{I} + \mathrm{II} + \mathrm{III},$$

$$\frac{\hbar}{2}\frac{d}{dt}\|D^{k+1}v\|_2^2 = \lambda_v Im \int D^{k+1}(|v|^2 v) \cdot D^{k+1}\overline{v}dx + \lambda Im \int D^{k+1}(|u|^2 v) \cdot D^{k+1}\overline{v}dx$$
$$+ \frac{\alpha}{\sqrt{2}} Im \int D^{k+1}(u^2) \cdot D^{k+1}\overline{v}dx := \mathrm{IV} + \mathrm{V} + \mathrm{VI}.$$

By the normal computation, we can obtain

$$\int D^{k+1}(|u|^2 u) \cdot D^{k+1}\overline{u}dx$$
$$= \int u D^{k+1}|u|^2 \cdot D^{k+1}\overline{u} + C_1 \int D^k|u|^2 \cdot Du \cdot D^{k+1}\overline{u}dx$$
$$+ \cdots + C_k \int D|u|^2 \cdot D^k u \cdot D^{k+1}\overline{u}dx + \int |u|^2 D^{k+1}u \cdot D^{k+1}\overline{u}dx. \tag{2.33}$$

Then we estimate every term of the right hand of the identity (2.33).

For the first term, we have

$$\int u D^{k+1}|u|^2 \cdot D^{k+1}\overline{u}dx$$
$$= \int |u|^2 |D^{k+1}u|^2 dx + C_1 \int u D^k u \cdot D\overline{u} \cdot D^{k+1}\overline{u}dx$$
$$+ \cdots + C_k \int u Du \cdot D^k\overline{u} \cdot D^{k+1}\overline{u}dx + \int u^2 D^{k+1}\overline{u} \cdot D^{k+1}\overline{u}dx.$$

By the induction assumption and Sobolev's embedding theorem, we have

$$\int u D^{k+1}|u|^2 \cdot D^{k+1}\bar{u} dx$$

$$\leqslant \|u\|_\infty^2 \|D^{k+1}u\|_2^2 + \frac{C_1}{2}\|Du\|_\infty \|u\|_\infty (\|D^k u\|_2^2 + \|D^{k+1}u\|_2^2)$$

$$+ \cdots + \frac{C_k}{2}\|Du\|_\infty \|u\|_\infty (\|D^k u\|_2^2 + \|D^{k+1}u\|_2^2) + \|u\|_\infty^2 \|D^{k+1}u\|_2^2$$

$$\leqslant C_0 \|D^{k+1}u\|_2^2 + C.$$

For the rest terms of the identity (2.33), using the same computation and induction, the following estimates can be obtained

$$\int (D^k |u|^2) \cdot Du \cdot D^{k+1}\bar{u} dx \leqslant C_1 \|D^{k+1}u\|_2^2 + C_2,$$

$$\int (D^{k-1}|u|^2) \cdot D^2 u \cdot D^{k+1}\bar{u} dx \leqslant C_1 \|D^{k+1}u\|_2^2 + C_2,$$

$$\cdots\cdots$$

$$\int (D|u|^2) \cdot D^k u \cdot D^{k+1}\bar{u} dx \leqslant C_1 \|D^{k+1}u\|_2^2 + C_2,$$

$$\int |u|^2 D^{k+1}u \cdot D^{k+1}\bar{u} dx \leqslant C \|D^{k+1}u\|_2^2.$$

Combining the above estimates, the term I can be bounded by

$$\lambda_u Im \int D^{k+1}(|u|^2 u) \cdot D^{k+1}\bar{u} dx \leqslant C_1 \|D^{k+1}u\|_2^2 + C_2. \tag{2.34}$$

Comparing the term I with the term IV, we get the inequality

$$\lambda_v Im \int D^{k+1}(|v|^2 v) \cdot D^{k+1}\bar{v} dx \leqslant C_1 \|D^{k+1}v\|_2^2 + C_2. \tag{2.35}$$

Applying Sobolev's embedding theorem and the induction computation, we have

$$\int (D^{k+1}\bar{u}v) D^{k+1}\bar{u} dx$$

$$= \int v D^{k+1}\bar{u} \cdot D^{k+1}\bar{u} dx + C_1 \int D^k \bar{u} \cdot Dv \cdot D^{k+1}\bar{u} dx$$

$$+ \cdots + C_k \int D\bar{u} \cdot D^k v \cdot D^{k+1}\bar{u} dx + \int \bar{u} D^{k+1}v \cdot D^{k+1}\bar{u} dx$$

$$\leqslant \|v\|_\infty \|D^{k+1}u\|_2^2 + \frac{C_1\|Dv\|_\infty}{2}(\|D^k u\|_2^2 + \|D^{k+1}u\|_2^2)$$

$$+ \frac{C_2\|D^2 v\|_\infty}{2}(\|D^{k-1}u\|_2^2 + \|D^{k+1}u\|_2^2)$$

$$+ \cdots + \frac{C_k\|Du\|_\infty}{2}(\|D^k v\|_2^2 + \|D^{k+1}u\|_2^2) + \frac{\|u\|_\infty}{2}(\|D^{k+1}v\|_2^2 + \|D^{k+1}u\|_2^2)$$

$$\leqslant C_1 \|D^{k+1}u\|_2^2 + C_2 \|D^{k+1}v\|_2^2 + C_3.$$

So we have the estimate of the term III,

$$\sqrt{2}\alpha Im \int D^{k+1}(\overline{u}v) \cdot D^{k+1}\overline{u}dx \leqslant C_1\|D^{k+1}u\|_2^2 + C_2\|D^{k+1}v\|_2^2 + C_3. \qquad (2.36)$$

For the term VI, by direct computation, we have

$$\int D^{k+1}u^2 \cdot D^{k+1}\overline{u}dx$$
$$= \int uD^{k+1}u \cdot D^{k+1}\overline{u}dx + C_1 \int D^k u \cdot Du \cdot D^{k+1}\overline{u}dx$$
$$+ \cdots + C_k \int Du \cdot D^k u \cdot D^{k+1}\overline{u}dx + \int uD^{k+1}u \cdot D^{k+1}\overline{u}dx$$
$$\leqslant \|u\|_\infty \|D^{k+1}u\|_2^2 + \frac{C_1\|Du\|_\infty}{2}(\|D^k u\|_2^2 + \|D^{k+1}u\|_2^2) + \frac{C_2\|D^2 u\|_\infty}{2}(\|D^{k-1}u\|_2^2$$
$$+ \|D^{k+1}u\|_2^2) + \cdots + \frac{C_k\|Du\|_\infty}{2}(\|D^k u\|_2^2 + \|D^{k+1}u\|_2^2)$$
$$+ \frac{\|u\|_\infty}{2}(\|D^{k+1}u\|_2^2 + \|D^{k+1}u\|_2^2)$$
$$\leqslant C_1\|D^{k+1}u\|_2^2 + C_2.$$

Therefore

$$|\text{VI}| \leqslant C_1\|D^{k+1}u\|_2^2 + C_2. \qquad (2.37)$$

Combining (2.34)–(2.37), we have

$$\frac{d}{dt}(\|D^{k+1}u\|_2^2 + \|D^{k+1}v\|_2^2) \leqslant C(\|D^{k+1}u\|_2^2 + \|D^{k+1}v\|_2^2 + 1).$$

Using the Gronwall inequality, we can get that

$$\|D^{k+1}u\|_2^2 + \|D^{k+1}v\|_2^2 \leqslant C,$$

where the constant C depends only on T and $\|u_0\|_{H^m}, \|v_0\|_{H^m}$.

This completes the proof of Lemma 2.4.

3 The Global Existence of Smooth Solution

In this section, we prove the existence of global smooth solution to the problem (1.1)–(1.3) by using Galërkin-Fourier method. We need the following lemmas.

Lemma 3.1 Let B_0, B and B_1 be three Banach spaces. Assume that $B_0 \subset B \subset B_1$ and B_i, $i = 0, 1$ are reflective. Suppose also that B_0 is compactly embedded in B. Let

$$W = \left\{ v | v \in L^{p_0}(0, T; B_0), v' = \frac{dv}{dt} \in L^{p_1}(0, T; B_1) \right\},$$

where T is finite and $1 < p_i < \infty, i = 0, 1$. W is equipped with the norm

$$\|v\|_{L^{p_0}(0,T;B_0)} + \|v'\|_{L^{p_1}(0,T;B_1)}.$$

Then W is compactly embedded in $L^{p_0}(0,T;B)$.

Lemma 3.2 Suppose that Q is a bounded domain in $R_x^n \times R_t, g_\mu, g \in L^q(Q)$ $(1 < q < \infty)$ and $\|g_\mu\|_{L^q(Q)} \leqslant C$. Furthermore, suppose that

$$g_\mu \to g \text{ a.e. in } Q.$$

Then

$$g_\mu \rightharpoonup g \text{ weakly in } L^q(Q).$$

Lemma 3.3 X is a Banach space. Suppose that $g \in L^p(0,T;X), \frac{\partial g}{\partial x} \in L^p(0,T;X)$ $(1 \leqslant p \leqslant \infty)$. Then $g \in C([0,T], X)$ (after possibly being redefined on a set of measure zero).

Proof of Theorem 1.1 We prove Theorem 1.1 by the following three steps.

Step 1. Constructing the approximate solutions by the Galerkin-Fourier method.

Let $\{\omega_j(x)\}(j = 1, 2, \cdots)$ be a complete orthonormal basis of eigenfunctions for the periodic boundary problem $-\Delta u = \lambda u$ in Ω. For every integer m, we are looking for an approximate solution of the system (1.1) of the form

$$u_m(t) = \sum_{j=1}^{m} \xi_{jm}(t)\omega_j, \quad v_m(t) = \sum_{j=1}^{m} \mu_{jm}(t)\omega_j,$$

where ξ_{jm}, μ_{jm} satisfy the following nonlinear equations

$$\begin{cases} \left(-i\hbar u_{mt} + \left(\frac{\hbar^2}{2M} - \Delta + \lambda_u|u_m|^2 + \lambda|v_m|^2\right)u_m + \sqrt{2}\alpha\bar{u}_m v_m, \omega_j\right) = 0, \\ \left(-i\hbar v_{mt} + \left(\frac{\hbar^2}{4M} - \Delta + \epsilon + \lambda_v|v_m|^2 + \lambda|u_m|^2\right)v_m + \frac{\alpha}{\sqrt{2}}u_m^2, \omega_j\right) = 0, \end{cases} \quad (3.38)$$

the nonlinear equations (3.38) satisfy the following initial-value conditions

$$\begin{cases} u_m(0) = u_{0m}, \ u_{0m} = \sum_{i=1}^{m} f_{im}\omega_i \to u_0 \text{ in } H_{per}^m(\Omega) \text{ as } m \to \infty, \\ v_m(0) = v_{0m}, \ v_{0m} = \sum_{i=1}^{m} g_{im}\omega_i \to v_0 \text{ in } H_{per}^m(\Omega) \text{ as } m \to \infty. \end{cases} \quad (3.39)$$

Then (3.38) becomes the system of nonlinear ODE subject to the initial condition (3.39). According to standard existence theory for nonlinear ordinary differential equations, there exists a unique solution for a.e. $0 \leqslant t \leqslant t_m$. By a priori estimates we obtain that $t_m = T$.

Step 2. A priori estimates.

As the proof of Lemmas 2.1, 2.2, we have

$$(u_m, v_m) \in L^\infty(0, T; H^1_{per}(\Omega))^2. \tag{3.40}$$

$\forall (\varphi, \phi) \in H^1_{per}(\Omega) \times H^1_{per}(\Omega)$, we have

$$\begin{cases} \left(-i\hbar u_{mt} + \left(\dfrac{\hbar^2}{2M} - \Delta + \lambda_u |u_m|^2 + \lambda |v_m|^2 \right) u_m + \sqrt{2}\alpha \bar{u}_m v_m, \varphi \right) = 0, \\ \left(-i\hbar v_{mt} + \left(\dfrac{\hbar^2}{4M} - \Delta + \epsilon + \lambda_v |v_m|^2 + \lambda |u_m|^2 \right) v_m + \dfrac{\alpha}{\sqrt{2}} u_m^2, \phi \right) = 0. \end{cases} \tag{3.41}$$

So

$$\begin{aligned} &|(u_{mt}, \varphi)| \\ &\leqslant C_1 |(Du_m, D\varphi)| + C_2 \left| (|u_m|^2 u_m, \varphi) \right| + C_3 \left| (|v_m|^2 u_m, \varphi) \right| \\ &\quad + C_4 |(\bar{u}_m v_m, \varphi)| \\ &\leqslant C_1 \|Du_m\| \|D\varphi\| + C_2 \|u_m\|_4^3 \|\varphi\|_4 + C_3 \|v_m\|_4^2 \|u_m\|_4 \|\varphi\|_4 \\ &\quad + C_4 \|u_m\|_3 \|v_m\|_3 \|\varphi\|_3, \end{aligned} \tag{3.42}$$

$$\begin{aligned} &|(v_{mt}, \phi)| \\ &\leqslant C_1 |(Dv_m, D\phi)| + \epsilon |(v_m, \phi)| + C_2 \left| (|v_m|^2 v_m, \phi) \right| \\ &\quad + C_3 |(|u_m|^2 v_m, \phi)| + C_4 |(u_m^2, \phi)| \\ &\leqslant C_1 \|Dv_m\| \|D\phi\| + \epsilon \|v_m\|_2 \|\phi\|_2 + C_2 \|v_m\|_4^3 \|\phi\|_4 \\ &\quad + C_3 \|u_m\|_4^2 \|v_m\|_4 \|\phi\|_4 + C_4 \|u_m\|_4^2 \|\phi\|_2. \end{aligned} \tag{3.43}$$

Using the Sobolev embedding theorem, we have

$$\|\varphi\|_4 \leqslant C\|D\varphi\| + C_1, \quad \|\varphi\|_3 \leqslant C_2\|D\varphi\| + C_3, \quad \|\phi\|_4 \leqslant C_4\|D\phi\| + C_5.$$

So by (3.42) and (3.43), we get

$$|(u_{mt}, \varphi)| \leqslant C\|D\varphi\| + C_1, \quad |(v_{mt}, \phi)| \leqslant C_2\|D\phi\| + C_3, \quad \forall \varphi, \phi \in H^1_{per}(\Omega).$$

Therefore

$$(u_{mt}, v_{mt}) \in L^\infty(0, T; H^{-1}_{per}(\Omega))^2. \tag{3.44}$$

Step 3. Passaging to the limit.

By applying (3.40) and (3.44), we deduce that there exists a subsequence u_μ from u_m, v_k from v_m such that

$$u_\mu \rightharpoonup u \text{ *-weakly in } L^\infty(0,T;H^1_{per}(\Omega)),$$
$$u_{\mu t} \rightharpoonup u_t \text{ *-weakly in } L^\infty(0,T;H^{-1}_{per}). \tag{3.45}$$

$$v_k \rightharpoonup v \text{ *-weakly in } L^\infty(0,T;H^1_{per}(\Omega)),$$
$$v_{kt} \rightharpoonup v_t \text{ *-weakly in } L^\infty(0,T;H^{-1}_{per}(\Omega)). \tag{3.46}$$

By (3.40), we have

$$(u_m, v_m) \text{ is bounded in } L^2(0,T;H^1_{per}(\Omega))^2. \tag{3.47}$$

By (3.44), we have

$$(u_{mt}, v_{mt}) \text{ is bounded in } L^2(0,T;H^{-1}_{per}(\Omega))^2. \tag{3.48}$$

Define

$$W = \{v | v \in L^2(0,T;H^1_{per}(\Omega)), v_t \in L^2(0,T;H^{-1}_{per}(\Omega))\}$$

We equip W with the norm:

$$\|v\|_W = \|v\|_{L^2(0,T;H^1_{per}(\Omega))} + \|v_t\|_{L^2(0,T;H^{-1}_{per}(\Omega))}.$$

Since $H^1_{per}(\Omega)$ is compactly embedded in $L^2(\Omega)$, by Lemma 3.1 we have that W is compactly embedded in $L^2(0,T;L^2(\Omega))$. By (3.40) and (3.44), $u_m, v_m \in W$. Then, there exists the subsequence u_μ, v_k (not rebelled) which satisfies

$$u_\mu \to u, \quad v_k \to v \text{ strongly in } L^2(0,T;L^2(\Omega)) \text{ and a. e.}. \tag{3.49}$$

By using (3.40), (3.49) and Lemma 3.2, we have

$$|u_\mu|^2 u_\mu \rightharpoonup |u|^2 u \text{ *-weakly in } L^\infty(0,T;L^{\frac{4}{3}}(\Omega)), \tag{3.50}$$

$$|v_k|^2 v_k \rightharpoonup |v|^2 v \text{ *-weakly in } L^\infty(0,T;L^{\frac{4}{3}}(\Omega)). \tag{3.51}$$

Fixing j, we get

$$\begin{cases} \left(-i\hbar u_{mt} + \left(\frac{\hbar^2}{2M}-\Delta+\lambda_u|u_m|^2+\lambda|v_m|^2\right)u_m + \sqrt{2}\alpha\bar{u}_m v_m, \omega_j\right) = 0, \\ \left(-i\hbar v_{mt} + \left(\frac{\hbar^2}{4M}-\Delta+\epsilon+\lambda_v|v_m|^2+\lambda|u_m|^2\right)v_m + \frac{\alpha}{\sqrt{2}}u_m^2, \omega_j\right) = 0. \end{cases} \tag{3.52}$$

By applying (3.45), (3.46), (3.50) and (3.51), we deduce that there exists a subsequence u_μ from u_m, v_k from v_m such that

$$(-\Delta u_\mu, \omega_j) \rightharpoonup (-\Delta u, \omega_j) \text{ *-weakly in } L^\infty(0,T),$$

$$(u_{\mu t}, \omega_j) \rightharpoonup (u_t, \omega_j) \text{ *-weakly in } L^\infty(0,T),$$

$$((\lambda_u |u_\mu|^2 + \lambda |v_\mu|^2)u_\mu, \omega_j) \rightharpoonup ((\lambda_u |u|^2 + \lambda |v|^2)u, \omega_j) \text{ *-weakly in } L^\infty(0,T),$$

$$(\overline{u}_\mu v_\mu, \omega_j) \rightharpoonup (\overline{u} v, \omega_j) \text{ *-weakly in } L^\infty(0,T),$$

$$(-\Delta v_\mu, \omega_j) \rightharpoonup (-\Delta v, \omega_j) \text{ *-weakly in } L^\infty(0,T),$$

$$(v_{\mu t}, \omega_j) \rightharpoonup (v_t, \omega_j) \text{ *-weakly in } L^\infty(0,T),$$

$$(v_\mu, \omega_j) \rightharpoonup (v, \omega_j) \text{ *-weakly in } L^\infty(0,T),$$

$$((\lambda_v |v_\mu|^2 + \lambda |u_\mu|^2)v_\mu, \omega_j) \rightharpoonup ((\lambda_v |v|^2 + \lambda |u|^2)v, \omega_j) \text{ *-weakly in } L^\infty(0,T),$$

$$(u_\mu^2, \omega_j) \rightharpoonup (u^2, \omega_j) \text{ *-weakly in } L^\infty(0,T).$$

Then from (3.17), we have

$$\begin{cases} \left(-i\hbar u_t + \left(\dfrac{\hbar^2}{2M} - \Delta + \lambda_u |u|^2 + \lambda |v|^2\right)u + \sqrt{2}\alpha \overline{u} v, \omega_j\right) = 0, \\ \left(-i\hbar v_t + \left(\dfrac{\hbar^2}{4M} - \Delta + \epsilon + \lambda_v |v|^2 + \lambda |u|^2\right)v + \dfrac{\alpha}{\sqrt{2}} u^2, \omega_j\right) = 0, \end{cases} \quad (3.53)$$

the above equalities hold for any fixed j. By the density of the basis $\omega_j, (j \in Z)$, we have:

$$\begin{cases} \left(-i\hbar u_t + \left(\dfrac{\hbar^2}{2M} - \Delta + \lambda_u |u|^2 + \lambda |v|^2\right)u + \sqrt{2}\alpha \overline{u} v, h\right) = 0, & \forall h \in H^1_{per}(\Omega), \\ \left(-i\hbar v_t + \left(\dfrac{\hbar^2}{4M} - \Delta + \epsilon + \lambda_v |v|^2 + \lambda |u|^2\right)v + \dfrac{\alpha}{\sqrt{2}} u^2, g\right) = 0, & \forall g \in H^1_{per}(\Omega). \end{cases} \quad (3.54)$$

Hence (u,v) satisfies the system (1.1). By (3.3), (3.7) and Lemma 3.3, we obtain that

$$u_\mu \in C(0,T; H^{-1}_{per}(\Omega)), \quad v_k \in C(0,T; H^{-1}_{per}(\Omega)).$$

Then

$$u_\mu(0) \rightharpoonup u(0) \text{ weakly in } H^{-1}_{per}(\Omega), \quad v_k(0) \rightharpoonup v(0) \text{ weakly in } H^{-1}_{per}(\Omega).$$

But from (3.2), we have

$$u_\mu(0) \rightharpoonup u_0 \text{ weakly in } H^1_{per}(\Omega), \quad v_k(0) \rightharpoonup v_0 \text{ weakly in } H^1_{per}(\Omega).$$

Therefore, $u(0) = u_0$, $v(0) = v_0$.

Remark 3.1 Using a priori estimates of Lemma 2.3–Lemma 2.5 and Theorem 1.1, we can get the global smooth solution of (1.1)–(1.3) satisfying

$$(u,v) \in L^\infty([0,T]; H^m_{per}(\Omega))^2.$$

Finally, we prove the uniqueness of the solution to the system (1.1)–(1.3) in the following.

Let $(u_1, v_1), (u_2, v_2)$ be two solutions which satisfy the system (1.1)–(1.3), then $(s = u_1 - u_2, m = v_1 - v_2)$ satisfies

$$\begin{cases} i\hbar s_t = -\dfrac{\hbar^2 \nabla^2}{2M} s + \lambda_u(|u_1|^2 u_1 - |u_2|^2 u_2) + \lambda(|v_1|^2 u_1 - |v_2|^2 u_2) \\ \qquad + \sqrt{2}\alpha(\bar{u}_1 v_1 - \bar{u}_2 v_2), \\ i\hbar m_t = -\dfrac{\hbar^2 \nabla^2}{4M} m + \epsilon m + \lambda_v(|v_1|^2 v_1 - |v_2|^2 v_2) + \lambda(|u_1|^2 v_1 - |u_2|^2 v_2) \\ \qquad + \dfrac{\alpha}{\sqrt{2}}(u_1^2 - u_2^2), \end{cases} \quad (3.55)$$

$$s(0) = 0, \quad m(0) = 0.$$

Taking the inner product of the first equation of the system (3.28) with $\bar{s}$ and the second equation with $\bar{m}$, considering the imaginary part of the resulting equations, we obtain:

$$\frac{\hbar}{2}\frac{d}{dt}\|s\|_2^2 = \lambda_u Im \int (|u_1|^2 u_1 - |u_2|^2 u_2)\bar{s}dx + \lambda Im \int (|v_1|^2 u_1 - |v_2|^2 u_2)\bar{s}dx$$
$$+ \sqrt{2}\alpha Im \int (\bar{u}_1 v_1 - \bar{u}_2 v_2)\bar{s}dx,$$

$$\frac{\hbar}{2}\frac{d}{dt}\|m\|_2^2 = \lambda_v Im \int (|v_1|^2 v_1 - |v_2|^2 v_2)\bar{m}dx + \lambda Im \int (|u_1|^2 v_1 - |u_2|^2 v_2)\bar{m}dx$$
$$+ \frac{\alpha}{\sqrt{2}} Im \int (u_1^2 - u_2^2)\bar{m}dx.$$

But

$$\lambda_u Im \int (|u_1|^2 u_1 - |u_2|^2 u_2)\bar{s}dx$$

$$\leqslant C\int |(|u_1|^2 s\bar{s} + (|u_1|^2 - |u_2|^2)u_2\bar{s})|dx$$

$$\leqslant C\|u_1\|_\infty^2\|s\|_2^2 + C\int(|u_1|^2 - |u_2|^2)u_2\bar{s}dx$$

$$\leqslant C\|u_1\|_\infty^2\|s\|_2^2 + C\|u_2\|_\infty\|(|u_1|^2 - |u_2|^2)\|_2\|s\|_2$$

$$\leqslant C\|s\|_2^2.$$

And

$$\lambda Im \int (|v_1|^2 u_1 - |v_2|^2 u_2)\bar{s}dx$$

$$\leqslant C\int |(|v_1|^2 s\bar{s} + (|v_1|^2 - |v_2|^2)u_2\bar{s})|dx$$

$$\leqslant C\|v_1\|_\infty^2\|s\|_2^2 + C\int(|v_1|^2 - |v_2|^2)u_2\bar{s}dx$$

$$\leqslant C\|v_1\|_\infty^2\|s\|_2^2 + C\|u_2\|_\infty\|(|v_1|^2 - |v_2|^2)\|_2\|s\|_2$$

$$\leqslant C(\|s\|_2^2 + \|m\|_2^2),$$

$$\int (\bar{u}_1 v_1 - \bar{u}_2 v_2)\bar{s}dx$$

$$= \int (\bar{u}_1 v_1 - \bar{u}_1 v_2 + \bar{u}_1 v_2 - \bar{u}_2 v_2)\bar{s}dx$$

$$\leqslant C_1 \int (m\bar{s} + |s|^2)dx \leqslant C_2(\|m\|_2^2 + \|s\|_2^2),$$

$$\int (u_1^2 - u_2^2)\bar{m}dx$$

$$= \int ((u_1^2 - u_1 u_2) + (u_1 u_2 - u_2^2))\bar{m}dx$$

$$\leqslant C_1 \int s\bar{m}dx \leqslant C_2(\|s\|_2^2 + \|m\|_2^2).$$

By the above inequalities, one can easily check that

$$\frac{d}{dt}(\|s\|_2^2 + \|m\|_2^2) \leqslant C(\|s\|_2^2 + \|m\|_2^2).$$

Applying the Gronwall inequality, we get $s = 0, m = 0$. Thus the uniqueness is obtained.

Remark 3.2 All the above estimates are unconcerned with the period π and only depend on the norm of initial data. Therefore, by using the *a priori* estimates of the solution to the system (1.1)–(1.3) for π, as in Ref. [15], we derive the global smooth solution as $\pi \to \infty$. So that, Theorem 1.2 is obtained.

References

[1] Adhikari S K. Numerical study of the spherically sysmmetric Gross-Pitaevskii equation in two space dimensions. *Journal. Phys. Rev. E*, 2000, 65, 2937-2944.

[2] Adhikari S K, Muruganandam P. Bose-Einstein condensation dynamics from the numerical solution of the Gross-Pitaevskii equation. *Journal. Phys. B.*, 2002, 35.

[3] Bao W Z, Jaksch D, Markowich P A. Numerical solution of the Gross-Pitaevskii equation for Boes-Einstein condensation. *Journal of Computational Physics.*, 2003, 187, 318-342.

[4] Bao W Z and Tang W J. Ground-state solution of Bose-Einstein condensate by directly minimizing the energy functiona. *Journal of Computational Physics.*, 2003, 187, 230-254.

[5] Guo B L. The global solution for some systems of nonlinear Schrödinger equations. *Proc. of DD-1 Symposium*, 1980, 3, 1227-1246.

[6] Gross E P. Structure of a quantized vortex in boson systems. *Nuovo Cimento*, 1961, 20, 454-477.

[7] Guo B L. The initial and periodic value problem of one class nonlinear Schrödinger equations describing excitons in molecular crystals. *Acta Mathematica Scientia*, 1982, 2(3), 269-276.

[8] Guo B L. The initial value problems and periodic boundary value problem of one class of higher order multi-dimensional nonlinear Schrödinger equations. *Chinese Science Bulletin*, 1982, 6, 324-327.

[9] Guo B L. Nonlinear Evolution Equations. *Shanghai Scientific and Technological Education Publishing House*, Shanghai, 1985 (in Chinese).

[10] Lee M D, Morgan S A, Davis M J and Burnett K. Energy-dependent scattering and the Gross-Pitaevskii equation in two-dimensional Bose-Einstein condensates. *Journal. Phys. Rev. A.*, 2002, 65.

[11] Pérez-Garca V M, Michinel H J, Cirac I, Lewenstein M, Zoller P. Dynamics of Bose-Einstein condensates: Variational solutions of the Gross-Pitaevskii equations. *Journal. Phys. Rev. A.*, 1997, 56.

[12] Sadhan K and Adhikari. Numerical solution of the two-dimensional Gross-Pitaevskii equation for trapped interacting atoms. *Physics Letters A*, 2000, 265, 91-96.

[13] Timmermans E, Tommasini P, Hussein M and Kerman A. Feshbach resonances in atomic Bose-Einstein condensates. *Physics Reports*, 1999, 315, 199-230.

[14] Wang T C and Zhao X F. Optimal l^∞ error estimates of finite difference methods for the coupled Gross-Pitaevskii equations in high dimensions. *Science China Mathematics.*, 2014, 57, doi: 10.1007/s11425-014-4773-7.

[15] Zhou Y and Guo B L. Periodic boundary problem and initial value problem for the generalized Korteweg-de Vries systems of higher order. *Acta Math. Sinica*, 1984, 27, 154-176(in Chinese).

Vanishing Viscosity Limit for the 3D Magnetohydrodynamic System with Generalized Navier-slip Boundary Conditions*

Guo Boling (郭柏灵) and Wang Guangwu (王光武)

Abstract In this paper, we investigate the vanishing viscosity limit problem for the 3-dimensional (3D) incompressible magnetohydrodynamic (MHD) system in a general bounded smooth domain of $\mathbf{R}^3$ with the generalized Navier-slip boundary conditions. We also obtain rates of convergence of the solution of viscous MHD to the corresponding ideal MHD.

Keywords viscous incompressible magnetohydrodynamic; general Navier-slip boundary conditions; vanishing viscosity limit

1 Introduction

Let $\Omega \subset \mathbf{R}^3$ be a bounded smooth domain, we consider the initial and boundary problem (IBVP) for the system of viscous magnetohydrodynamic (MHD) equations

$$\partial_t v - \nu \Delta v + (\nabla \times v) \times v + H \times (\nabla \times H) + \nabla p = 0, \tag{1.1}$$

$$\nabla \cdot v = 0, \tag{1.2}$$

$$\partial_t H - \mu \Delta H + v \cdot \nabla H - H \cdot \nabla v = 0, \tag{1.3}$$

$$\nabla \cdot H = 0, \tag{1.4}$$

with the following general Navier-slip boundary condtions

$$v \cdot n = 0, \quad n \times (\nabla \times v) = [\beta v]_\tau, \tag{1.5}$$

$$H \cdot n = 0, \quad n \times (\nabla \times H) = [\beta H]_\tau. \tag{1.6}$$

---

* Math. Method Appl. Sci., 2016, 39(15): 4526–4534.

Here $\nabla\cdot$ and $\nabla\times$ denotes the div and curl operators. n is the outward normal vector. β are given smooth symmetric tensor on the boundary, τ is the unit tangential vector of Ω, $[\cdot]_\tau$ represents the tangential component.

The corresponding ideal MHD system are the following

$$\partial_t v^0 + (\nabla \times v^0) \times v^0 + H^0 \times (\nabla \times H^0) + \nabla p = 0, \tag{1.7}$$

$$\nabla \cdot v^0 = 0, \tag{1.8}$$

$$\partial_t H^0 + v^0 \cdot \nabla H^0 - H^0 \cdot \nabla v^0 = 0, \tag{1.9}$$

$$\nabla \cdot H^0 = 0, \tag{1.10}$$

with slip boundary conditions

$$v^0 \cdot n = 0, \quad H^0 \cdot n = 0. \tag{1.11}$$

The vanishing viscosity limit problem for the Navier-Stokes equations has been well studied whether the domain is bounded or unbounded. For Navier-Stokes equation in the bounded domain(see [2], [3], [5], [11], [12], [14]-[18], [21]-[23], [25]), Various boundary conditions have been proposed. The most common boundary condition is the classical non-slip boundary condition, $u = 0$ on $\partial\Omega$, which gives rise to the phenomenon of strong boundary layers is general as formally derived by Prandtl. Other commonly used boundary conditions are Navier-slip boundary conditions, in which there is a stagnant layer of fluid close to the wall allowing a fluid to slip, and the slip velocity is proportional to the shear stress, i.e.,

$$u \cdot n = 0, \quad (S(u)n)_\tau = \alpha_\tau u_\tau, \quad \text{on } \partial\Omega, \tag{1.12}$$

where $S(u) = \nabla u + (\nabla u)^T$ is the shear stress, τ is the unit tangential vector of $\partial\Omega$. Such slip boundary conditions which was first introduced by Navier in [13] are used to in the large eddy simulations of turbulent flows, which seek to compute the large eddies of a turbulent flow accurately while neglecting small flow structure. In [12], the author proved the uniform bounded of the solution to Navier-Stokes equation and the vanishing viscosity limit.

It is noticed that

$$(2(S(u)n) - (\nabla \times u) \times n)_\tau = GD(u)_\tau, \quad \text{on } \partial\Omega, \tag{1.13}$$

where $GD(u) = -2S(n)u$. Hence the generalized slip condition (1.12) are equivalent to the following generalized vorticity-slip condition:

$$u \cdot n = 0, \quad n \times (\nabla \times u) = [Bu]_\tau, \quad \text{on } \partial\Omega, \tag{1.14}$$

with B a given smooth symmetric tensor on the boundary.

In [21], Xiao and Xin got the vanishing viscosity limit of Navier-Stokes equation with a new slip boundary condtion:

$$u \cdot n = 0 \quad (\nabla \times u) \cdot n = 0, \quad n \times (\Delta u) = 0, \quad \text{on } \partial\Omega. \tag{1.15}$$

On the other hand, the research about invicid limit of viscous MHD is less than the Navier-Stokes equation. The viscous MHD system in the whole space or with non-slip boundary conditions have been studied extensively and there is a large literature on various topics concerning the MHD system such as the well-posedness in various functional spaces(see [1], [4], [7]- [9], [19], [20], [24]). However, very little is known about the MHD system with a slip boundary condition. The solvability of (1.1)-(1.6) is far from being obvious due to the compatibility issues of the nonlinear term with the slip boundary conditions. So far Xiao, Xin and Wu [24] investigate the vanishing viscosity limit problem of 3D MHD in a flat boundary bounded domain with homogeneous slip boundary conditions:

$$v \cdot n = 0, \quad n \times (\nabla \times v) = 0, \tag{1.16}$$

$$H \cdot n = 0, \quad n \times (\nabla \times H) = 0. \tag{1.17}$$

In this paper, we follow the approach of [22] and [24] and formulate the boundary value problem in a suitable functional setting so that the Stokes operator is well-behaved. In these functional setting, the nonlinear terms naturally fall into desired functional spaces. These facts allow us to establish the existence and regularity of solutions through the Galerkin approximation and appropriate a-priori bounds.

With this well-posedness theory at our disposal, we pursue the vanishing viscosity limit of (1.1)-(1.6). The issue of vanishing viscosity limits of the Navier-Stokes equations and the viscous MHD equations is classical and of fundamental importance in fluid dynamics and turbulence theory. When a non-slip boundary condition is imposed, the vanishing viscosity limit of the MHD equations is not well understood due to the formation of turbulent boundary layer. Mathematically, one difficulty is due to the mismatch between the boundary condition for the viscous MHD system and that for it potential limit, the ideal MHD.

As pointed out [22], the key in studying the vanishing viscosity limit is to control the vorticity created at the boundary. Thus to obtain a uniform convergence of solutions of (1.1)-(1.6) of that of the ideal problem (1.7)-(1.11), one needs to obtain some uniform estimates on vorticity. Our approach here is motivated by the ideal introduced in [22], [24] to study the same problem for the Navier-Stokes equations.

The major results are organized into four sections. Section 2 contains several notations and results to be used in the subsequent sections. Section 3 established the existence of global weak solution through the method of Galerkin approximation. Strong solutions are studied in Section 4 for general domains. The vanishing viscosity limit results are presented in Section 5.

2 Preliminaries

Throughout the rest of this paper, $\Omega \subset \mathbf{R}^3$ denotes a simply connected smooth domain. $H^s(\Omega)$ with $s \geq 0$ denotes the standard Sobolev spaces and $H^{-s}(\Omega)$ with $s \geq 0$ denotes the dual of $H_0^s(\Omega)$ (the closur of $C_0^\infty(\Omega)$ in $H^s(\Omega)$). For notational convenience, Ω may be omitted when we write these spaces without confusion.

The following lemmas allow as us to control the H^s-norm of a vector-valued function u by its H^{s-1}-norms of $\nabla \times u$ and $\nabla \cdot u$, together with the $H^{s-\frac{1}{2}}(\partial\Omega)$-norm of $u \cdot n$.

Lemma 2.1 *Let $s \geq 0$ be an integer. Let $u \in H^s$ be a vector-valued function. Then*

$$||u||_s \leq C(||\nabla \times u||_{s-1} + ||\nabla \cdot u|| + |n \cdot u|_{s-\frac{1}{2}} + ||u||_{s-1}). \tag{2.1}$$

Lemma 2.2 *Let $s \geq 0$ be an integer. Let $u \in H^s(\Omega)$. Then*

$$||u||_s \leq C(||\nabla \times u||_{s-1} + ||\nabla \cdot u|| + |u \times n|_{s-\frac{1}{2}} + ||u||_{s-1}). \tag{2.2}$$

Let $X = \{u \in L^2; \nabla \cdot u = 0, u \cdot n = 0\}$ be the Hilbert space with the L^2 inner product, and let

$$V = H^1 \cap X \subset X,$$
$$W = \{u \in H^2; n \times (\nabla \times u) = [\beta u]_\tau, \text{ on } \partial\Omega\}.$$

It is well know that for any $v \in V$, one has $||v||_1 \leq c||\nabla \times v||$. For any $u \in W$ and $v \in V$

$$(-\Delta u, v) = (\nabla \times (\nabla \times u), v) = \int_{\partial\Omega} n \times u \cdot v dx + \int_\Omega (\nabla \times u) \cdot (\nabla \times v) dx$$
$$= \int_{\partial\Omega} \beta u \cdot v dx + \int_\Omega (\nabla \times u) \cdot (\nabla \times v) dx.$$

Therefore, $-\Delta$ can be extended to the closure of W in V. The extended operator is denotes by A and its domain by $D(A)$. Obviously,
$$W \subseteq D(A) \subset V.$$
The following lemma states that A is well-behaved in these functional setting.

Lemma 2.3 *The Stokes operator $A = -\Delta$ with $D(A) = W \subset V$ satisfying*
$$(Au, v) = \int_{\partial \Omega} \beta u \cdot u dx + \int_{\Omega} (\nabla \times u) \cdot (\nabla \times u) dx$$
is a self-adjoint and positive operator, with its inverse being compact. Consequently, its countable eigenvalues can be listed as
$$0 < \lambda_1 \leqslant \lambda_2 \leqslant \cdots \to \infty$$
and the corresponding eigenvectors $\{e_j\} \subset W \cap C^\infty$ make an orthogonal complete basis of X.

For notational convenience, we still write $-\Delta$ for A. Now, we consider the nonlinear terms in these functional settings. For $v, H \in C^\infty \cap W$, define
$$B_1(v, H) = (\nabla \times v) \times v + H \times (\nabla \times H) + \nabla p,$$
where p satisfies
$$\Delta p = \nabla \cdot ((\nabla \times v) \times v + H \times (\nabla \times H)),$$
$$\nabla p \cdot n = ((\nabla \times v) \times v + H \times (\nabla \times H)) \cdot n,$$
and
$$B_2(v, H) = v \cdot \nabla H - H \cdot \nabla v.$$
Thus we can easily get that $B_1(v, H) \in X$ and $B_2(v, H) \in X$.

3 The weak solutions

This section establishes the global existence of weak solutions to the MHD system (1.1)-(1.6). The approach is the Galerkin approximation following the argument of Constantin and Foias [4]. Here as in the next section, we consider a general smooth bounded simply connected domain in $\mathbf{R}^3$.

Definition 3.1 (v, H) is called a weak solution of (1.1)-(1.6) with the initial data $(v_0, H_0) \in X$ on the time interval $[0, T)$ if $(v, H) \in L^2(0, T; V) \cap C_w(0, T; X)$ satisfies $(v', H') \in L^1(0, T; V^*)$ and

$$(v', \phi) + \nu(\omega_v, \nabla \times \phi) + \nu \int_{\partial \Omega} \beta v \cdot \phi ds + (\omega_v \times v, \phi) + (H \times \omega_H, \phi) = 0,$$

$$(H', \phi) + \nu(\omega_H, \nabla \times \phi) + \mu \int_{\partial \Omega} \beta H \cdot \phi ds + (v \cdot \nabla H - H \cdot \nabla v, \phi) = 0,$$

for all $\phi \in V$ and for a.e. $t \in [0, T)$, and

$$v(0) = v_0, \quad H(0) = H_0,$$

where $\omega_v = \nabla \times v$ and $\omega_H = \nabla \times H$.

The major result of this section is the global existence of a weak solution.

Theorem 3.1 Let $(v_0, H_0) \in X$. Let $T > 0$. Then there exists at least one weak solution (v, H) of (1.1)-(1.6) on $[0, T)$ which satisfies the energy inequality

$$\frac{d}{dt}(||v||^2 + ||H||^2) + 2(\nu||\nabla \times v||^2 + \mu||\nabla \times H||^2) + 2(\nu|v|^2_{L^2(\partial \Omega)} + \mu|H|^2_{L^2(\partial \Omega)}) \leqslant 0 \quad (3.1)$$

in the sense of distribution.

Proof We start with a sequence of approximate functions $(v^{(m)}, H^{(m)})$,

$$v^{(m)}(t) = \sum_{j=1}^m v_j(t)e_j, \quad H^{(m)}(t) = \sum_{j=1}^m H_j(t)e_j,$$

where v_j and H_j for $j = 1, \cdots, m$ solve the following ordinary differential equations

$$v'_j(t) + \nu \lambda_j v_j(t) + g^1_j(U) = 0, \quad (3.2)$$

$$H'_j(t) + \nu \lambda_j H_j(t) + g^2_j(U) = 0, \quad (3.3)$$

$$v_j(0) = (v_0, e_j), \quad H_j(0) = (H_0, e_j), \quad (3.4)$$

with $U = (v_1, v_2, \cdots, v_m, H_1, \cdots, H_m)$ and

$$g^1_j(U) = (B_1(v^{(m)}, H^{(m)}), e_j), \quad g^2_j(U) = (B_2(v^{(m)}, H^{(m)}), e_j).$$

Since $g^k_j(U)$ are Lipshitz in U, (3.2)-(3.4) is locally well posed, say on $[0, T)$. Consequently, for any $t \in [0, T)$, $(v^{(m)}, H^{(m)})$ solves the following system of equations

$$(v^{(m)}(t))' + \nu \lambda_j v^{(m)}(t) + P_m B_1(v^{(m)}, H^{(m)}) = 0, \quad (3.5)$$

$$(H^{(m)}(t))' + \nu\lambda_j H^{(m)}(t) + P_m B_2(v^{(m)}, H^{(m)}) = 0, \tag{3.6}$$

$$v^{(m)}(0) = P_m v_0, \quad H^{(m)}(0) = P_m H_0, \tag{3.7}$$

where P_m denotes the projection of X onto the space spanned by $\{e_j\}_1^m$.

Noting that
$$(P_m B_1(v^{(m)}, H^{(m)}), v^{(m)}) = \int_\Omega (H^{(m)} \times (\nabla \times H^{(m)})) \cdot v^{(m)} dx,$$
$$(P_m B_2(v^{(m)}, H^{(m)}), H^{(m)}) = \int_\Omega H^{(m)} \cdot \nabla \times (H^{(m)} \times v^{(m)}) dx$$
$$= \int_\Omega \nabla \times H^{(m)} \cdot (H^{(m)} \times v^{(m)}) dx,$$

and taking the inner products $((3.5), v^{(m)})$ and $((3.6), H^{(m)})$, adding them up, we obtain

$$\frac{d}{dt}(\|v^{(m)}\|^2 + \|H^{(m)}\|^2) + 2(\nu\|\nabla \times v^{(m)}\|^2$$
$$+ \mu\|\nabla \times H^{(m)}\|^2) + 2(\nu|v^{(m)}|_{L^2(\partial\Omega)}^2 + \mu|H^{(m)}|_{L^2(\partial\Omega)})^2 = 0. \tag{3.8}$$

Therefore,
$$(v^{(m)}, H^{(m)}) \text{ is bounded in } L^\infty(0, T; X), \text{ uniform for } m, \tag{3.9}$$
$$(\nabla \times v^{(m)}, \nabla \times H^{(m)}) \text{ is bounded in } L^2(0, T; X), \text{ uniform for } m. \tag{3.10}$$

Note that for $\phi \in V$, we have
$$|(-\Delta v^{(m)}, \phi)| = |(\nabla \times v^{(m)}, \nabla \times \phi) + \int_{\partial\Omega} \beta v^{(m)} \cdot \phi ds|.$$

Therefore, $\{-\Delta v^{(m)}\}$ is bounded in $L^2(0, T; V^*)$. Similarly, $\{-\Delta H^{(m)}\}$ is bounded in $L^2(0, T; V^*)$.

For the nonlinear terms, since $H^1 \hookrightarrow L^6$ and $\|u\|_{L^3} \leq c\|\nabla u\|_{H^1}^{\frac{1}{2}}\|u\|_{L^2}^{\frac{1}{2}}$, we have, for any $\phi \in V$,

$$|(P_m B_1(v^{(m)}, H^{(m)}), \phi)| = |(B_1(v^{(m)}, H^{(m)}), \phi_m)|$$
$$\leq \|v\|_{L^3}\|\nabla \times v\|_{L^2}\|\phi_m\|_{L^6} + \|H\|_{L^3}\|\nabla \times H\|_{L^2}\|\phi_m\|_{L^6}$$
$$\leq C(\|v^{(m)}\|^{\frac{1}{2}}\|v^{(m)}\|_1^{\frac{3}{2}} + \|H^{(m)}\|^{\frac{1}{2}}\|H^{(m)}\|_1^{\frac{3}{2}})\|\phi_m\|_1,$$

where $\phi_m = P_m \phi$. Because of the uniform bound for $\|v^{(m)}\|$ is (3.9) and the bound for $\|v^{(m)}\|_1$ in (3.10), we obtain $\{B_1(v^{(m)}, H^{(m)})\}$ is bounded in $L^{\frac{4}{3}}(0, T; V^*)$. Similarly, $\{B_2(v^{(m)}, H^{(m)})\}$ is bounded in $L^{\frac{4}{3}}(0, T; V^*)$. Therefore, $((v^{(m)})', (H^{(m)})')$ is bounded in $L^{\frac{4}{3}}(0, T; V^*)$.

The rest of the proof is similar to the arguments in Constatin and Foias [4] and thus further details are omitted. This completes the proof of Theorem 3.1. □

4 The strong solutions

In this section we study the local well-posedness of the strong solution of (1.1)-(1.6) corresponding to an initial data $(v_0, H_0) \in V$ and its higher regularities.

Let $(v_0, H_0) \in V$ and let $(v^{(m)}, H^{(m)})$ be the Galerkin approximation constructed in the previous section.

Firstly, we note that

$$(v^{(m)})' - \nu \Delta v^{(m)} + \sum g_j^1 \times e_j = 0, \tag{4.1}$$

$$(H^{(m)})' - \mu \Delta H^{(m)} + \sum g_j^2 \times e_j = 0, \tag{4.2}$$

$$v^{(m)}(0) = P_m v_0, \quad H^{(m)}(0) = P_m H_0, \tag{4.3}$$

where we recall g_j^1 satisfies $\sum_{j=1}^{m} g_j^1 e_j = P_m B_1(v^{(m)}, H^{(m)})$. Taking the inner product $((4.1), -\Delta v^{(m)}) + ((4.2), -\Delta H^{(m)})$, and noting that

$$-\int_\Omega (v^{(m)})' \cdot (\Delta v^{(m)}) dx = \frac{1}{2}\frac{d}{dt}\left(\int_{\partial\Omega} \beta v^{(m)} \cdot v^{(m)} ds + \int_\Omega (\nabla \times v^{(m)}) \cdot (\nabla \times v^{(m)}) dx\right),$$

we obtain

$$\frac{d}{dt}(\|\nabla \times v^{(m)}\|^2 + \|\nabla \times v^{(m)}\|^2 + \int_{\partial\Omega} \beta v^{(m)} \cdot v^{(m)} + \beta H^{(m)} \cdot H^{(m)} ds)$$
$$+ 2(\nu\|\Delta v^{(m)}\|^2 + \mu\|\Delta H^{(m)}\|^2) + 2((B_1(v^{(m)}, H^{(m)}), -\Delta v^{(m)}))$$
$$+ (B_2(v^{(m)}, H^{(m)}), -\Delta v^{(m)})) = 0. \tag{4.4}$$

Applying the Agmon inequality

$$\|\phi\|_{L^\infty} \leqslant \|\phi\|_1^{\frac{1}{2}} \|\phi_2\|_2^{\frac{1}{2}}, \quad \forall \phi \in H^2,$$

we find

$$(B_1(v^{(m)}, H^{(m)}), -\Delta v^{(m)}) \leqslant C(\|\nabla \times v^{(m)}\|^{\frac{3}{2}} + \|\nabla \times H^{(m)}\|^{\frac{3}{2}})(\|\Delta v^{(m)}\|^{\frac{1}{2}}$$
$$+ \|\Delta H^{(m)}\|^{\frac{1}{2}})\|\Delta v^{(m)}\|,$$
$$(B_2(v^{(m)}, H^{(m)}), -\Delta H^{(m)}) \leqslant C(\|\nabla \times v^{(m)}\|^{\frac{3}{2}} + \|\nabla \times H^{(m)}\|^{\frac{3}{2}})(\|\Delta v^{(m)}\|^{\frac{1}{2}}$$
$$+ \|\Delta H^{(m)}\|^{\frac{1}{2}})\|\Delta H^{(m)}\|,$$

and

$$\frac{d}{dt}(\|\nabla \times v^{(m)}\|^2 + \|\nabla \times v^{(m)}\|^2 + \int_{\partial\Omega} \beta v^{(m)} \cdot v^{(m)}$$

$$+ \beta H^{(m)} \cdot H^{(m)} ds) + 2(\nu ||\Delta v^{(m)}||^2 + \mu ||\Delta H^{(m)}||^2)$$
$$\leqslant C(||\nabla \times v^{(m)}|| + ||\nabla \times v^{(m)}||)^6, \tag{4.5}$$

where C depends on ν and μ. Comparing with the ordinary differential equation

$$\frac{d}{dt}y = Cy^3, \tag{4.6}$$

we find that there is time $T_0 > 0$ such that, for any fixed $T \in (0, T_0)$,

$$(v^{(m)}, H^{(m)}) \text{ is bounded in } L^\infty(0, T; H^1),$$
$$(v^{(m)}, H^{(m)}) \text{ is bounded in } L^2(0, T; H^2).$$

Note that $||P_m(v \times u)|| \leqslant ||v \times u|| \leqslant C||v||||u||_{L^\infty}$, it follows that

$$\{(v^{(m)})'\}, \{(H^{(m)})'\} \text{ is bounded in } L^2(0, T; L^2). \tag{4.7}$$

The standard compactness results allow us to find a subsequence of $(v^{(m)}, H^{(m)})$ (still denoted by $(v^{(m)}, H^{(m)})$) and (v^0, H^0) such that

$$v^{(m)} \to v^0, \quad H^{(m)} \to H^0 \text{ in } L^\infty(0, T; H^1) \text{ weak-star},$$
$$v^{(m)} \to v^0, \quad H^{(m)} \to H^0 \text{ in } L^2(0, T; H^2) \text{ weakly},$$
$$v^{(m)} \to v^0, \quad H^{(m)} \to H^0 \text{ in } L^2(0, T; H^1) \text{ strongly}.$$

Passing to the limit, we find the weak solution obtained in the previous section may be chosen such that $(v^0, H^0) \in L^\infty(0, T; H^1) \cap L^2(0, T; H^2)$. From (4.7), we find $(v^0)', (H^0)' \in L^2(0, T; L^2)$ and $v^0, H^0 \in C(0, T; H^1)$. We call such solution a strong solution. To show that strong solutions are unique, we consider two strong solution (v_1, H_1) and (v_2, H_2). Then their difference $\tilde{v} = v_1 - v_2$, $\tilde{H} = H_1 - H_2$, satisfies

$$\tilde{v} - \nu \Delta \tilde{v} + B_1(v_1, H_1) - B_1(v_2, H_2) = 0, \tag{4.8}$$
$$\tilde{H} - \mu \Delta \tilde{H} + B_2(v_1, H_1) - B_2(v_2, H_2) = 0. \tag{4.9}$$

Taking the inner products $((4.8), \tilde{v}) + ((4.9), \tilde{H})$, we find

$$\frac{d}{dt}(||\tilde{v}||^2 + ||\tilde{H}||^2) \leqslant g(t)(||\tilde{v}||^2 + ||\tilde{H}||^2)$$

on $[0, T]$ for some positive integrable function $g(t)$. Then, $v_1 = v_2$, $H_1 = H_2$ follows from $\tilde{v}(0) = 0, \tilde{H}(0) = 0$ and the Gronwall inequality. From the standard extension method of time evolution, we can conclude

Theorem 4.1 Let $(v_0, H_0) \in W$. Then there is $T^* > 0$ depending on ν, μ and the H^1-norm of (v_0, H_0) only such that (1.1)-(1.6) has a strong solution (v, H) on $[0, T^*)$ satisfying

$$v, H \in L^2(0, T; W) \cap C(0, T; V),$$
$$v', H' \in L^2(0, T; X)$$

for any $T \in (0, T^*)$. In addition, the energy equation

$$\frac{d}{dt}(||\nabla \times v||^2 + ||\nabla \times v||^2 + \int_{\partial\Omega} \beta v \cdot v + \beta H \cdot H ds)$$
$$+ 2(\nu ||\Delta v||^2 + \mu ||\Delta H||^2) + 2((B_1(v, H), -\Delta v) + (B_2(v, H), -\Delta v)) = 0. \quad (4.10)$$

5 The vanishing viscosity limit

We now turn to the purpose of this paper to establish the converge with a rate for the solution of viscous MHD to the solution of ideal MHD.

Theorem 5.1 Let $(v_0, H_0) \in W$. Then there is a $T_0 > 0$ depending on $||(v_0, H_0)||_{H^2}$ only such that the strong solution $v = v(\nu, \mu), H = H(\nu, \mu)$ of the MHD system (1.1)-(1.6) with the initial data (v_0, H_0) obeys the following uniform bound

$$||v||_{H^1} + ||H||_{H^1} \leq C, \quad for \ t \in [0, T_0],$$

where C is a constant independent of ν and μ.

Proof According to Theorem 4.1, we only need to estimate the last two term in the left hand of (4.10). Since $H^1 \hookrightarrow L^6$ and $||u||_{L^3} \leq ||u||_{H^1}^{\frac{1}{2}} ||u||_{L^2}^{\frac{1}{2}}$, we can get that

$$|(B_1(v, H), -\Delta v)| \leq C(||\nabla \times v||^{\frac{3}{2}} + ||\nabla \times H||^{\frac{3}{2}}),$$

and

$$|(B_2(v, H), -\Delta v)| \leq C(||\nabla \times v||^{\frac{3}{2}} + ||\nabla \times H||^{\frac{3}{2}}).$$

Then comparing the ordinary differential equation

$$y'(t) = Cy(t)^{\frac{3}{2}},$$
$$y(0) = ||\nabla \times v_0||^2 + ||\nabla \times H_0||^2,$$

we can get $T(\nu, \mu) \geq T_0$ for all $\nu, \mu > 0$. This completes the proof of Theorem 5.1. $\square$

Theorem 5.2 Let $(v_0, H_0) \in W$. Let $T_0 > 0$ and let $v = v(\nu, \mu)$, $H = H(\nu, \mu)$ be the corresponding strong solution of the MHD system (1.1)-(1.6) on $[0, T_0]$. Then, as $\nu, \mu \to 0$, (v, H) converges to the unique solution (v^0, H^0) of the ideal MHD system with the same initial data (1.7)-(1.11) in the sense

$$(v(\nu,\mu), H(\nu,\mu)) \to (v^0, H^0) \text{ in } L^2(0,T;V); \tag{5.1}$$

$$(v(\nu,\mu), H(\nu,\mu)) \to (v^0, H^0) \text{ in } C(0,T;X), \tag{5.2}$$

for any $1 \leq q < \infty$ and $T > T_0$.

Proof It follows from Theorem 4.1, we know that $T > T_0$ and

$$v(\nu,\mu), H(\nu,\mu) \text{ is uniformly bounded in } L^2(0,T;W) \cap C(0,T;V),$$

$$v'(\nu,\mu), H'(\nu,\mu) \text{ is uniformly bounded in } L^2(0,T;X),$$

for all $\nu, \mu > 0$. By the standard compactness result, there is a subsequence ν_n, μ_n of ν, μ and vector function v^0, H^0 such that

$$(v(\nu_n,\mu_n), H(\nu_n,\mu_n)) \to (v^0, H^0) \text{ in } L^2(0,T;V),$$

$$(v(\nu_n,\mu_n), H(\nu_n,\mu_n)) \to (v^0, H^0) \text{ in } C(0,T;X),$$

for any $1 \leq q < \infty$, as $\nu_n, \mu_n \to 0$. Passing to limit, we find the limit (v^0, H^0) solves the following limit equations

$$\partial_t v^0 + (\nabla \times v^0) \times v^0 + H \times (\nabla \times H^0) + \nabla p = 0, \tag{5.3}$$

$$\nabla \cdot v^0 = 0, \tag{5.4}$$

$$\partial_t H^0 + v^0 \cdot \nabla H^0 - H^0 \cdot \nabla v^0 = 0, \tag{5.5}$$

$$\nabla \cdot H^0 = 0, \tag{5.6}$$

with slip boundary conditions

$$v^0 \cdot n = 0, \quad n \times (\nabla \times v^0) = 0, \quad H^0 \cdot n = 0, \quad n \times (\nabla \times H^0) = 0 \tag{5.7}$$

and p satisfying

$$\Delta p = \nabla \cdot ((\nabla \times v^0) \times v^0 + H^0 \times (\nabla \times H^0)),$$

$$\nabla p \cdot n = 0.$$

As in the proof of the uniqueness of the strong solutions of the MHD system in the previous section, we can show that (v^0, H^0) is unique. We then show the convergence of whole sequence. □

Finally, we present the convergence rate.

Theorem 5.3 *Let $(v_0, H_0) \in W$ satisfy the assumptions stated in Theorem 4.1 and (v^0, H^0) be the solution of the ideal MHD (1.7)-(1.11) on $[0, T_0)$ with $(v(0) = v_0, H(0) = H_0)$ and $(v, H) = (u(\nu, \mu), H(\nu, \mu))$ be the solution of the viscous MHD (1.1)-(1.5) with $(v(0) = v_0, H(0) = H_0)$. Then*

$$||u(\nu,\mu) - u||^2 + ||H(\nu,\mu) - H||^2 + \nu \int_0^{T_0} ||u(\nu,\mu) - u||_1^2$$
$$+ ||H(\nu,\mu) - H||_1^2 dt \leqslant C(T_0)(\nu^{2-s} + \mu^{2-s}), \text{ on } [0, T_0], \quad (5.8)$$

where $s > 0$.

Proof Denote by $\tilde{v} = v(\nu, \mu) - v^0$, $\tilde{H} = H(\nu, \mu) - H^0$. We find that

$$\partial_t \tilde{v} - \nu \Delta \tilde{v} + B_1(v, H) - B_1(v^0, H^0) = \nu \Delta v^0, \quad (5.9)$$

$$\nabla \cdot \tilde{v} = 0, \quad (5.10)$$

$$\partial_t \tilde{H} - \nu \Delta \tilde{H} + B_2(v, H) - B_2(v^0, H^0) = \nu \Delta H^0, \quad (5.11)$$

$$\nabla \cdot \tilde{H} = 0, \quad (5.12)$$

$$\tilde{v} \cdot n = 0, \quad \tilde{H} \cdot n = 0. \quad (5.13)$$

Taking the L^2 inner product of (5.9) with $\tilde{v}$ and (5.11) with $\tilde{H}$, integrating by parts, one can get

$$\frac{d}{dt}(||\tilde{v}||^2 + ||\tilde{H}||^2) + 2(\nu||\nabla \times \tilde{v}||^2 + \mu||\nabla \times \tilde{H}||^2) + 2\left(\nu \int_{\partial\Omega} n \times (\nabla \times \tilde{v}) \cdot \tilde{v} ds\right.$$
$$\left. + \mu \int_{\partial\Omega} n \times (\nabla \times \tilde{H}) \cdot \tilde{H}\right) ds$$
$$+ 2(B_1(v, H) - B_1(v^0, H^0), \tilde{v}) + 2(B_2(v, H) - B_2(v^0, H^0), \tilde{H})$$
$$= 0. \quad (5.14)$$

We denote

$$B_0 = \nu \int_{\partial\Omega} n \times (\nabla \times \tilde{v}) \cdot \tilde{v} ds + \mu \int_{\partial\Omega} n \times (\nabla \times \tilde{H}).$$

Then we have

$$B_0 = \nu \int_{\partial\Omega} (\beta(v - v^0) + \beta v^0 - n \times (\nabla \times v^0))(v - v^0) ds$$
$$+ \mu \int_{\partial\Omega} (\beta(H - H^0) + \beta H^0 - n \times (\nabla \times H^0))(H - H^0) ds$$

$$\leqslant C\int_{\partial\Omega}(|v-v^0|^2+|v-v^0|+|H-H^0|^2+|H-H^0|)$$
$$\leqslant C(||v-v^0||||\nabla\times v-\nabla\times v^0||+||v-v^0||_{L^1(\partial\Omega)}$$
$$+||v-v^0||||\nabla\times v-\nabla\times v^0||+||v-v^0||_{L^1(\partial\Omega)}).$$

By using the trace theorem that

$$||v-v^0||_{L^1(\partial\Omega)}\leqslant C||v-v^0|_{L^q(\partial\Omega)}\leqslant C||v-v^0||_{H^s(\Omega)}, \quad (5.15)$$
$$||H-H^0||_{L^1(\partial\Omega)}\leqslant C||H-H^0|_{L^q(\partial\Omega)}\leqslant C||H-H^0||_{H^s(\Omega)}, \quad (5.16)$$

for any $q>1$ and any $s>0$.

By interpolation, we have

$$||v-v^0||_{H^s(\Omega)}\leqslant C||v-v^0||^{(1-s)}||v-v^0||_1^s\leqslant C||v-v^0||^{1-s}||\nabla\times v-\nabla\times v^0||^s, \quad (5.17)$$

$$||H-H^0||_{H^s(\Omega)}\leqslant C||H-H^0||^{(1-s)}||H-H^0||_1^s$$
$$\leqslant C||H-H^0||^{1-s}||\nabla\times H-\nabla\times H^0||^s. \quad (5.18)$$

Note that

$$\nu||v-v^0||^{1-s}||\nabla\times v-\nabla\times v^0||^s\leqslant\alpha\nu||\nabla\times v-\nabla\times v^0||^2+C_\alpha(||v-v^0||^2+\nu^{2-s}), \quad (5.19)$$

$$\mu||H-H^0||^{1-s}||\nabla\times H-\nabla\times H^0||^s$$
$$\leqslant\beta\mu||\nabla\times H-\nabla\times H^0||^2+C_\beta(||H-H^0||^2+\mu^{2-s}), \quad (5.20)$$

for any $s\in(0,1)$ and

$$\nu||v-v^0||||\nabla\times v-\nabla\times v^0||\leqslant\nu^2||\nabla\times v-\nabla\times v^0||^2+||v-v^0||^2, \quad (5.21)$$
$$\mu||H-H^0||||\nabla\times H-\nabla\times H^0||\leqslant\mu^2||\nabla\times H-\nabla\times H^0||^2+||H-H^0||^2. \quad (5.22)$$

Then we have

$$B_0\leqslant 2\alpha\nu||\nabla\times v-\nabla\times v^0||^2+C_\alpha||v-v^0||^2+\nu^{2-s}$$
$$+2\beta\nu||\nabla\times H-\nabla\times H^0||^2+C_\beta||H-H^0||^2+\mu^{2-s},$$

for any $s\in(0,1)$ and ν,μ small enough.

Since $H^1\hookrightarrow L^6$ and $||v||_{L^3}\leqslant||v||_{L^2}^{\frac{1}{2}}||v||_{H^1}^{\frac{1}{2}}$, we have that

$$\left|\int_\Omega((\nabla\times v)\times v,v-v^0)-((\nabla\times v^0)\times v^0,v-v^0)dx\right|$$

$$= \left| \int_\Omega ((\nabla \times v) \times v, v - v^0) - ((\nabla \times v^0) \times v, v - v^0) \right.$$
$$\left. + ((\nabla \times v^0) \times v, v - v^0) - ((\nabla \times v^0) \times v^0, v - v^0) dx \right|$$
$$\leqslant \left| \int_\Omega ((\nabla \times (v - v^0)) \times v, v - v^0) + ((\nabla \times v^0) \times (v - v^0), v - v^0) dx \right|$$
$$\leqslant \|\nabla \times (v - v^0)\|_{L^2} \|v\|_{L^6} \|v - v^0\|_{L^3} + \|\nabla \times v^0\|_{L^2} \|v - v^0\|_{L^3} \|v - v^0\|_{L^6}$$
$$\leqslant \|\nabla \times (v - v^0)\|_{L^2}^{\frac{3}{2}} \|v - v^0\|_{L^2}^{\frac{1}{2}} \|v\|_{H^1} + \|\nabla \times v^0\|_{L^2} \|v - v^0\|_{L^2}^{\frac{1}{2}} \|\nabla \times (v - v^0)\|_{L^2}^{\frac{3}{2}}$$
$$\leqslant \|v\|_{H^1} \|\nabla \times (v - v^0)\|_{L^2}^2 + \|v\|_{H^1} \|v - v^0\|_{L^2}^2 + \|\nabla \times v^0\|_{L^2} \|\nabla \times (v - v^0)\|_{L^2}^2$$
$$+ \|\nabla \times v^0\|_{L^2} \|v - v^0\|_{L^2}^2,$$

and
$$\left| \int_\Omega (H \times (\nabla \times H), v - v^0) - (H^0 \times (\nabla \times H^0), v - v^0) dx \right|$$
$$\leqslant \|\nabla \times H\|_{L^2}^2 \|\nabla \times (v - v^0)\|_{L^2} + \|\nabla \times (H - H^0)\|^2 \|v - v^0\|_{L^2}$$
$$+ \|\nabla \times (H - H^0)\|_{L^2}^2 \|v - v^0\|_{L^2} + \|\nabla \times H\|_{L^2}^2 \|\nabla \times (v - v^0)\|_{L^2}.$$

Therefore, we can obtain that
$$\left| \int_0^{T_0} (B_1(v, H) - B_1(v^0, H^0), v - v^0) dt \right|$$
$$\leqslant c(\|v - v^0\|_{L^2}^2 + \|H - H^0\|_{L^2}^2 + \int_0^{T_0} \|\nabla \times (v - v^0)\|_{L^2}^2 + \|\nabla \times (H - H^0)\|_{L^2}^2 dt).$$

Similarly, we can get
$$\left| \int_0^{T_0} (B_2(v, H) - B_2(v^0, H^0), H - H^0) dt \right|$$
$$\leqslant c \left(\|v - v^0\|_{L^2}^2 + \|H - H^0\|_{L^2}^2 + \int_0^{T_0} \|\nabla \times (v - v^0)\|_{L^2}^2 + \|\nabla \times (H - H^0)\|_{L^2}^2 dt \right).$$

Note also that
$$|\nu(\Delta v, v - v^0)| \leqslant \|v - v^0\|^2 + c\nu^2, \quad |\mu(\Delta H, H - H^0)| \leqslant \|H - H^0\|^2 + c\mu^2.$$

By the Gronwall inequality we can complete the proof. □

References

[1] Agapito R, Schonbek M. Non-uniform decay of MHD equations with and without magnetic diffusion. *Comm. Partial Differential Equations*, 2007, **32**: 1791-1821.

[2] Beirão da Veiga H, Crispo F. Concerning the $W^{k,p}$-inviscid limit for 3D flows under a slip boundary condition. *J. Math. Fluid Mech.*, 2011, **13**: 117-135.

[3] Beirão da Veiga H, Crispo F. Sharp inviscid limit results under Navier type boundary conditions. An L^p theory. *J. Math. Fluid Mech.*, 2010, **12**:397-411.

[4] Biskamp D. *Nonlinear Magnetohydrodynamics*. Cambridge, UK: Cambridge University Press, 1993.

[5] Constantin P, Foias C. *Navier-Stokes equations*. Chicago Lectures in Mathematics. Chicago, IL: University of Chicago Press, 1988.

[6] Constantin P, Wu J H. Inviscid limit for vortex patches. *Nonlinearity*, 1995, **8**: 735-742.

[7] Duvaut G, Lions J L. Inéquation en thermoélasticite et magnétohydrodynamique. *Arch. Ration. Mech. Anal.*, 1972, **46**: 241-279.

[8] He C, Xin Z P. Partial regularity of suitable weak solutions to the incompressible magnetohydrodynamic equations. *J. Funct. Anal.*, 2005, **227**:113-152.

[9] He C, Xin Z P. On the regularity of weak solutions to the magnetohydrodynamic equations. *J. Diff. Eqns.*, 2005, **213**:235-254.

[10] Kato T, Lai C. Nonlinear evolution equations and the Euler flow. *J. Funct. Anal.*, 1984, **56**:15-28.

[11] Masmoudi N. Remarks about the inviscid limit of the Navier-Stokes system. *Commun. Math. Phys.*, 2007, **270(3)**:777-788.

[12] Masmoudi N, Rousset F. Uniform regularity for the Navier-Stokes Equation with Navier boundary conditon. *Arch. Rational Mech. Anal.*, 2012, **203**:529-575.

[13] Navier C L, H. L M. Sur les lois de l'équilibre et du mouvement des corps élastiques. *Mem. Acad. R. Sci. Inst. France*, 1827, **6**:369.

[14] Sammartino M, Caflisch R. Zero viscosity limit for analytic solutions of the Navier-Stokes equation on a halfspace. I. Existence for Euler and Prandtl equations. *Comm. Math. Phys.*, 1998, **192**:433-461.

[15] Sammartino M, Caflisch R. Zero viscosity limit for analytic solutions of the Navier-Stokes equation on a halfspace. II. Construction of the Navier-Stokes solution. *Comm. Math. Phys.*, 1998, **192**:463-491.

[16] Wang L, Xin Z P, Zang A B. Vanishing viscous limits for 3D Navier-Stokes equations with a Navierslip boundary condition. *J. Math. Fluid Mech.*, 2012, **14(4)**:791-825.

[17] Wang X, Wang Y, Xin Z P. Boundary layers in incompressible Navier-Stokes equations with Navier boundary conditions for the vanishing viscosity limit. *Commun. Math. Sci.*, 2010, **8(4)**:965-998.

[18] Wang Y, Xin Z P. Zero-viscosity limit of the linearized compressible Navier-Stokes equations with highly oscillatory forces in the half-plane. *SIAM J. Math. Anal.*, 2005, **37**:1256-1298.

[19] Wu J H. Regularity criteria for the generalized MHD equations. *Comm. Partial Differential Equations*, 2008, **33**:285-306.

[20] Wu J H. Viscous and inviscid magnetohydrodynamics equations. *J. Anal. Math.*, 1997, **73**:251-265.

[21] Xiao Y L, Xin Z P. A new boundary condition for the 3D Navier-Stokes equation and the vanishing viscosity limit. *J. Math. Phys.*, 2012, **53**:115617.

[22] Xiao Y L, Xin Z P. On the vanishing viscosity limit for the 3D Navier-Stokes equations with a slip boundary condition. *Comm. Pure Appl. Math.*, 2007, **LX**:1027-1055.

[23] Xiao Y L, Xin Z P. On the invisid limit of the 3D Navier-Stokes equations with generalized Navier-slip boundary conditions. *Commun. Math. Stat.*, 2013, **1**:259-279.

[24] Xiao Y L, Xin Z P, Wu J H. Vanishing viscosity limit for the 3D magnetohydrodynamic system with a slip boundary condition. *J. Funct. Anal.*, 2009, **257**:3375-3394.

[25] Xin Z P, Yanagisawa T. Zero-viscosity limit of the linearized Navier-Stokes equations for a compressible viscous fluid in the half-plane. *Comm. Pure Appl. Math.*, 1999, **52**:479-541.

Bright-dark Vector Soliton Solutions for the Coupled Complex Short Pulse Equations in Nonlinear Optics

Guo Boling (郭柏灵), and Wang Yufeng (王玉风)

Abstract Under investigation in this paper are the coupled complex short pulse equations, which describe the propagation of ultra-short optical pulses in cubic nonlinear media. Through the Hirota method, bright-dark one- and two-soliton solutions are obtained. Interactions between two bright or two dark solitons are verified to be elastic through the asymptotic analysis. With different parameter conditions, the oblique interactions, bound states of solitons and parallel solitons are analyzed.

Keywords coupled complex short pulse equations; bright-dark vector soliton solutions; oblique interactions; bound states of solitons

1 Introduction

The propagations of picosecond optical pulses in the single-mode nonlinear media are usually described by the cubic nonlinear Schrödinger (NLS) equation [1–3], but to the ultra-fast signal transmissions, the cubic NLS equation is no longer valid [3]. On the one hand, people study the subpicosecond and femtosecond pulses by incorporating higher terms in the cubic NLS equation [4–6], including the higher-order dispersion, self-steepening and Raman scattering. On the other hand, people try to derive some new equations besides the NLS equation to describe the ultra-short pulses under certain assumptions [7–12].

The short pulse equation which is derived under the assumption that the center of pulse is far from the nearest resonance frequency of the material's susceptibility reads as [12]

$$u_{XT} = u + \frac{1}{6}(u^3)_{XX}, \qquad (1.1)$$

which describes the propagation of ultra-short optical pulses in cubic nonlinear media, where the real function u of T and X stands for the magnitude of the electric field [12], the subscripts mean the partial differentiations. Eq. (1.1) is proved to be completely integrable [13, 14], which owns the Lax pair in the Wadati-Konno-Ichikawa type. Solitary solutions for Eq. (1.1) have been derived [15]. Periodic solutions, multi-loop soliton and multi-breather solutions for Eq. (1.1) have been constructed by a hodograph transformation which converts Eq. (1.1) into the sine-Gordon equation [16, 17]. The integrable discretization and the geometric interpretation of Eq. (1.1) have been given in Refs. [18, 19].

Advantages have been verified during the process of dealing with the NLS equation that the complex function can describe the propagation of optical pulse along the optical fibers more properly than the real one [20]. The complex function contains the amplitude and phase, which are two characters of an optical pulse envelop [20]. In addition, the complex functions pave the ways to study the soliton interaction [20]. The complex form of Eq. (1.1) reads as [21]

$$q_{xt} + q + \frac{1}{2}(|q|^2 q_x)_x = 0, \qquad (1.2)$$

where q is the complex function of x and t, which represents the magnitude of the electric field. The derivation of Eq. (1.2) from the Maxwell equation has been pointed in Ref. [21]. Integrability and multi-bright-soliton solutions for Eq. (1.2) have been given [21]. The link between the complex coupled dispersionless equation and Eq. (1.2) has been carried out [22].

Contrast to the single-component equation, the multi-component ones own the inter-component and inner-component nonlinearity terms [3, 23]. Owing to the different polarization of each component may has, vector soliton solutions for the multi-component equations have been displayed in the bright-bright, bright-dark and dark-dark forms [24–28]. Vector solitons have shown to be applicable in the optical switch and logic gate [3, 29], since multiple kinds of interactions between the bright vector solitons have been experimentally and theoretically performed, including the shape-changing and shape-preserving interactions [24, 25].

To describe the propagation of optical pulses in birefringence fibers, Ref. [21] derived the coupled complex short pulse equations [21]

$$q_{1,xt} + q_1 + \frac{1}{2}\left[(|q_1|^2 + |q_2|^2)q_{1,x}\right]_x = 0, \tag{1.3a}$$

$$q_{2,xt} + q_2 + \frac{1}{2}\left[(|q_1|^2 + |q_2|^2)q_{2,x}\right]_x = 0, \tag{1.3b}$$

where q_1 and q_2 are the complex functions of x and t, which indicate the magnitudes of the electric fields. The Lax pair, conservation laws and bright soliton solutions in pfaffians have been displayed [21]. The multi-bright-soliton, multi-breather and higher-order rogue wave solutions for Eqs. (1.3) have been obtained through Darboux transformation [30]. More on solitons can be seen in Refs. [31–33].

However, to our knowledge, the bright-dark vector soliton solutions and their interactions for Eqs. (1.3) have not been put forward. Motivated by the above, in this paper, we will study the bright-dark vector soliton solutions and their interactions for Eqs. (1.3) via the Hirota method. In Section 2, the bright-dark one- and two-soliton solutions for Eqs. (1.3) will be obtained based on the bilinear forms under scale transformations. Asymptotic analysis will be performed on the two-soliton solutions in Section 3. Oblique interaction, bound states of solitons and parallel solitons will be analyzed in Section 4. Section 5 will be our conclusions.

2 Soliton solutions for Eqs. (1.3)

The Hirota method is a directive and effective way to deal with the solutions for the nonlinear evolution equations (NLEEs) [34]. Soliton solutions will be obtained through the truncated parameter expansion at different levels based on the bilinear forms, which can be obtained through certain transformations [34].

With the dependent variable transformations

$$q_1 = \frac{g_1}{f} e^{i(y-s)}, \quad q_2 = \frac{g_2}{f} e^{i(y-s)}, \tag{2.1}$$

and the scale transformations

$$x = \lambda y + 2\lambda s - 2(\ln f)_s, \quad t = -s, \tag{2.2}$$

the bilinear forms for Eqs. (1.3) will be written as

$$(D_y D_s + iD_s - iD_y - \lambda - 1)g_1 \cdot f = 0, \tag{2.3a}$$

$$(D_yD_s + iD_s - iD_y - \lambda - 1)g_2 \cdot f = 0, \qquad (2.3b)$$

$$(D_s^2 - 2\lambda)f \cdot f = \frac{1}{2}(|g_1|^2 + |g_2|^2), \qquad (2.3c)$$

where g_1 and g_2 are both complex functions, f is a real one, λ is a constant to be determined, D_y and D_s are the bilinear derivative operators [34] defined by

$$D_y^{m_1} D_s^{m_2} a \cdot b = \left(\frac{\partial}{\partial y} - \frac{\partial}{\partial y'}\right)^{m_1} \left(\frac{\partial}{\partial s} - \frac{\partial}{\partial s'}\right)^{m_2} a(y,s)b(y',s')\bigg|_{y'=y, s'=s}, \qquad (2.4)$$

with y, s, y' and s' being formal variables, $a(y,s)$ and $b(y',s')$ being two functions, while m_1 and m_2 are two nonnegative integers.

Proof of Bilinear Forms (2.3) Dividing the Bilinear Forms (2.3) by f^2, we get

$$\left(\frac{g_1}{f}\right)_{ys} + \frac{g_1}{f} \cdot 2(\ln f)_{ys} + i\left(\frac{g_1}{f}\right)_s - i\left(\frac{g_1}{f}\right)_y = (\lambda + 1)\frac{g_1}{f}, \qquad (2.5a)$$

$$\left(\frac{g_2}{f}\right)_{ys} + \frac{g_2}{f} \cdot 2(\ln f)_{ys} + i\left(\frac{g_2}{f}\right)_s - i\left(\frac{g_2}{f}\right)_y = (\lambda + 1)\frac{g_2}{f}, \qquad (2.5b)$$

$$2(\ln f)_{ss} - 2\lambda = \frac{1}{2}(|q_1|^2 + |q_2|^2). \qquad (2.5c)$$

From the Scale Transformations (2.2), we have

$$\frac{\partial x}{\partial s} = 2\lambda - 2(\ln f)_{ss}, \qquad \frac{\partial x}{\partial y} = \lambda - 2(\ln f)_{sy}, \qquad (2.6)$$

i.e.,

$$\partial_y = \varrho^{-1}\partial_x, \qquad \varrho^{-1} = \lambda - 2(\ln f)_{sy},$$

$$\partial_s = [2\lambda - 2(\ln f)_{ss}]\partial_x - \partial_t = -\frac{1}{2}(|q_1|^2 + |q_2|^2)\partial_x - \partial_t.$$

Via Transformations (2.1), Eq. (2.5a) can be rewritten as

$$\left[q_1 e^{-i(y-s)}\right]_{ys} = [-2(\ln f)_{ys} + i\partial_y - i\partial_s + \lambda + 1] q_1 e^{-i(y-s)}, \qquad (2.7)$$

which means

$$q_{1,ys} = [\lambda - 2(\ln f)_{ys}] q_1, \qquad (2.8)$$

$$\varrho^{-1}\partial_x \left[-\partial_t - \frac{1}{2}(|q_1|^2 + |q_2|^2)\partial_x\right] q_1 = \varrho^{-1} q_1, \qquad (2.9)$$

$$q_{1,xt} + \frac{1}{2}\left[(|q_1|^2 + |q_2|^2)q_{1,x}\right]_x + q_1 = 0. \qquad (2.10)$$

It is obviously that Eq. (2.10) is exactly equivalent to Eq. (1.3a). In the similar way, Eq. (1.3b) can be derived from Eq. (2.5b). Therefore, Eqs. (1.3) can be reduced from Eqs. (2.5). Proof of Bilinear Forms (2.3) is completed.

To obtain the bright-dark soliton solutions for Eqs. (1.3), we truncate g_1, g_2 and f at the formal expansion parameter ϵ as follows:

$$g_1 = \epsilon g_1^{(1)} + \epsilon^3 g_1^{(3)} + \epsilon^5 g_1^{(5)} + \cdots, \tag{2.11a}$$

$$g_2 = g_2^{(0)} \left(1 + \epsilon^2 g_2^{(2)} + \epsilon^4 g_2^{(4)} + \epsilon^6 g_2^{(6)} + \cdots \right), \tag{2.11b}$$

$$f = 1 + \epsilon^2 f_2 + \epsilon^4 f_4 + \epsilon^6 f_6 + \cdots, \tag{2.11c}$$

$g_1^{(l)}$'s ($l = 1, 3, 5, \cdots$) and $g_2^{(\iota)}$'s ($\iota = 0, 2, 4, \cdots$) are the complex functions of y and s, while f_r's ($r = 2, 4, 6, \cdots$) are the real ones, which will be determined later.

Terminating the expansions as

$$g_1 = \epsilon g_1^{(1)} = \epsilon \delta_1 e^{\theta_1},$$

$$g_2 = g_2^{(0)}\left(1 + \epsilon^2 g_2^{(2)}\right) = \rho e^{ibs}\left(1 + \epsilon^2 \xi_1 \eta_1 e^{\theta_1 + \theta_1^*}\right),$$

$$f = 1 + \epsilon^2 f_2 = 1 + \epsilon^2 \eta_1 e^{\theta_1 + \theta_1^*},$$

we express the one-soliton solutions for Eqs. (1.3) as

$$q_1 = \frac{1}{2}\delta_1 e^{i(y-s)} e^{\frac{\theta_1 - \theta_1^* - \ln \eta_1}{2}} \operatorname{sech}\left(\frac{\theta_1 + \theta_1^* + \ln \eta_1}{2}\right), \tag{2.12a}$$

$$q_2 = \rho e^{ibs} e^{i(y-s)}\left[\frac{1+\xi_1}{2} - \frac{1-\xi_1}{2}\tanh\left(\frac{\theta_1 + \theta_1^* + \ln \eta_1}{2}\right)\right], \tag{2.12b}$$

$$x = \lambda y + 2\lambda s - (\omega_1 + \omega_1^*)\left[1 + \tanh\left(\frac{\theta_1 + \theta_1^* + \ln \eta_1}{2}\right)\right], \quad t = -s, \tag{2.12c}$$

with $\theta_1 = k_1 y + \omega_1 s + \nu_1$, k_1, ν_1, δ_1 and ρ being all the complex constants, while λ, ω_1, b, ξ_1 and η_1 are listed in Appendix.

Terminating g_1, g_2 and f as $g_1 = \epsilon g_1^{(1)} + \epsilon^3 g_1^{(3)}$, $g_2 = g_2^{(0)}\left(1 + \epsilon^2 g_2^{(2)} + \epsilon^4 g_2^{(4)}\right)$ and $f = 1 + \epsilon^2 f_2 + \epsilon^4 f_4$, we get the bright-dark two-soliton solutions for Eqs. (1.3),

$$q_1 = \frac{g_1^{(1)} + g_1^{(3)}}{1 + f_2 + f_4} e^{i(y-s)}, \quad q_2 = \frac{g_2^{(0)}\left(1 + g_2^{(2)} + g_2^{(4)}\right)}{1 + f_2 + f_4} e^{i(y-s)},$$

$$x = \lambda y + 2\lambda s - 2\ln(1 + f_2 + f_4)_s, \quad t = -s, \tag{2.13}$$

where

$$g_1^{(1)} = \delta_1 e^{\theta_1} + \delta_2 e^{\theta_2}, \quad g_1^{(3)} = \delta_3 e^{\theta_1 + \theta_2 + \theta_1^*} + \delta_4 e^{\theta_1 + \theta_2 + \theta_2^*},$$

$$g_2^{(2)} = \xi_1\eta_1 e^{\theta_1+\theta_1^*} + \xi_2\eta_2 e^{\theta_1+\theta_2^*} + \xi_3\eta_3 e^{\theta_2+\theta_1^*} + \xi_4\eta_4 e^{\theta_2+\theta_2^*},$$

$$f_2 = \eta_1 e^{\theta_1+\theta_1^*} + \eta_2 e^{\theta_1+\theta_2^*} + \eta_3 e^{\theta_2+\theta_1^*} + \eta_4 e^{\theta_2+\theta_2^*},$$

$$g_2^{(4)} = \xi_5\eta_5 e^{\theta_1+\theta_1^*+\theta_2+\theta_2^*}, \qquad f_4 = \eta_5 e^{\theta_1+\theta_1^*+\theta_2+\theta_2^*},$$

with $\theta_2 = k_2 y + \omega_2 s + \nu_2$, k_2, ν_2 and δ_2 being all the complex constants, while the relevant coefficients will be shown in Appendix.

3 Asymptotic analysis

In this part, we will make the asymptotic analysis on Solutions (2.13) to investigate the interactions between the bright-dark two solitons.

A. Before the interaction ($s \to -\infty$),

(a) $\theta_1 + \theta_1^* \sim 0$, $\theta_2 + \theta_2^* \sim -\infty$,

$$q_1 \to S_1^{1-} = \frac{\delta_1}{2} e^{\frac{\theta_1-\theta_1^*-\ln\eta_1}{2}} e^{i(y-s)} \operatorname{sech}\left(\frac{\theta_1+\theta_1^*+\ln\eta_1}{2}\right), \tag{3.1a}$$

$$q_2 \to S_2^{1-} = \rho e^{ibs} e^{i(y-s)} \left[\frac{\xi_1+1}{2} + \frac{\xi_1-1}{2}\tanh\left(\frac{\theta_1+\theta_1^*+\ln\eta_1}{2}\right)\right], \tag{3.1b}$$

(b) $\theta_2 + \theta_2^* \sim 0$, $\theta_1 + \theta_1^* \sim +\infty$,

$$q_1 \to S_1^{2-} = \frac{\delta_3}{2\eta_1} e^{\frac{\theta_2-\theta_2^*-\ln\frac{\eta_5}{\eta_1}}{2}} e^{i(y-s)} \operatorname{sech}\left(\frac{\theta_2+\theta_2^*+\ln\frac{\eta_5}{\eta_1}}{2}\right), \tag{3.2a}$$

$$q_2 \to S_2^{2-} = \rho e^{ibs} e^{i(y-s)} \left[\frac{\xi_5+\xi_1}{2} + \frac{\xi_5-\xi_1}{2}\tanh\left(\frac{\theta_2+\theta_2^*+\ln\frac{\eta_5}{\eta_1}}{2}\right)\right]. \tag{3.2b}$$

B. After the interaction ($s \longrightarrow +\infty$),

(a) $\theta_1 + \theta_1^* \sim 0$, $\theta_2 + \theta_2^* \sim +\infty$

$$q_1 \to S_1^{1+} = \frac{\delta_4}{2\eta_4} e^{\frac{\theta_1-\theta_1^*-\ln\frac{\eta_5}{\eta_4}}{2}} e^{i(y-s)} \operatorname{sech}\left(\frac{\theta_1+\theta_1^*+\ln\frac{\eta_5}{\eta_4}}{2}\right), \tag{3.3a}$$

$$q_2 \to S_2^{1+} = \rho e^{ibs} e^{i(y-s)} \left[\frac{\xi_5+\xi_4}{2} + \frac{\xi_5-\xi_4}{2}\tanh\left(\frac{\theta_1+\theta_1^*+\ln\frac{\eta_5}{\eta_4}}{2}\right)\right], \tag{3.3b}$$

(b) $\theta_2 + \theta_2^* \sim 0$, $\theta_1 + \theta_1^* \sim -\infty$,

$$q_1 \to S_1^{2+} = \frac{\delta_2}{2} e^{\frac{\theta_2-\theta_2^*-\ln\eta_4}{2}} e^{i(y-s)} \operatorname{sech}\left(\frac{\theta_2+\theta_2^*+\ln\eta_4}{2}\right), \tag{3.4a}$$

$$q_2 \to S_2^{2+} = \rho\, e^{ibs}\, e^{i(y-s)} \left[\frac{\xi_4+1}{2} + \frac{\xi_4-1}{2}\tanh\left(\frac{\theta_2+\theta_2^*+\ln\eta_4}{2}\right)\right], \quad (3.4b)$$

where $S_1^{\tau\mp}$ ($\tau=1,2$) denote the status of the bright soliton S_1 before ($-$) and after ($+$) the interactions, while $S_2^{\tau\mp}$ ($\tau=1,2$) stand for the dark soliton S_2 before ($-$) and after ($+$) the interactions, respectively. The related parameters are shown in Appendix.

Through the parameters in Solutions (2.13), we can obtain that

$$\sqrt{\frac{\delta_1\delta_1^*}{4\eta_1}} = \sqrt{\frac{\delta_4\delta_4^*}{4\eta_4\eta_5}}, \quad |\rho|\sqrt{\frac{\xi_1-1}{2}\frac{\xi_1^*-1}{2}} = |\rho|\sqrt{\frac{\xi_5-\xi_4}{2}\frac{\xi_5^*-\xi_4^*}{2}},$$

$$\sqrt{\frac{\delta_3\delta_3^*}{4\eta_1\eta_5}} = \sqrt{\frac{\delta_2\delta_2^*}{4\eta_4}}, \quad |\rho|\sqrt{\frac{\xi_5-\xi_1}{2}\frac{\xi_5^*-\xi_1^*}{2}} = |\rho|\sqrt{\frac{\xi_4-1}{2}\frac{\xi_4^*-1}{2}},$$

which mean that the amplitudes/depths of solitons satisfy the relationships $|S_1^{1-}| = |S_1^{1+}|$, $|S_2^{1-}| = |S_2^{1+}|$, $|S_1^{2-}| = |S_1^{2+}|$ and $|S_2^{2-}| = |S_2^{2+}|$. From Table 1, we are easy to find that the velocities of both the bright and dark soltions keep unchanged before and after the interactions. Therefore, the interactions between two bright or two dark solitons are both elastic.

Table 1 Physical quantities of bright soliton S_1 and dark soliton S_2 before and after the interactions

Soliton	Amplitude/Depth	Velocity	Soliton	Amplitude/Depth	Velocity
S_1^{1-}	$\sqrt{\dfrac{\delta_1\delta_1^*}{4\eta_1}}$	$\dfrac{\rho\rho^*-8}{4(k_1k_1^*-ik_1+ik_1^*+1)}$	S_1^{1+}	$\sqrt{\dfrac{\delta_4\delta_4^*}{4\eta_4\eta_5}}$	$\dfrac{\rho\rho^*-8}{4(k_1k_1^*-ik_1+ik_1^*+1)}$
S_2^{1-}	$|\rho|\sqrt{\dfrac{\xi_1-1}{2}\dfrac{\xi_1^*-1}{2}}$	$\dfrac{\rho\rho^*-8}{4(k_1k_1^*-ik_1+ik_1^*+1)}$	S_2^{1+}	$|\rho|\sqrt{\dfrac{\xi_5-\xi_4}{2}\dfrac{\xi_5^*-\xi_4^*}{2}}$	$\dfrac{\rho\rho^*-8}{4(k_1k_1^*-ik_1+ik_1^*+1)}$
S_1^{2-}	$\sqrt{\dfrac{\delta_3\delta_3^*}{4\eta_1\eta_5}}$	$\dfrac{\rho\rho^*-8}{4(k_2k_2^*-ik_2+ik_2^*+1)}$	S_1^{2+}	$\sqrt{\dfrac{\delta_2\delta_2^*}{4\eta_4}}$	$\dfrac{\rho\rho^*-8}{4(k_2k_2^*-ik_2+ik_2^*+1)}$
S_2^{2-}	$|\rho|\sqrt{\dfrac{\xi_5-\xi_1}{2}\dfrac{\xi_5^*-\xi_1^*}{2}}$	$\dfrac{\rho\rho^*-8}{4(k_2k_2^*-ik_2+ik_2^*+1)}$	S_2^{2+}	$|\rho|\sqrt{\dfrac{\xi_4-1}{2}\dfrac{\xi_4^*-1}{2}}$	$\dfrac{\rho\rho^*-8}{4(k_2k_2^*-ik_2+ik_2^*+1)}$

4 Soliton interactions

Under the Scale Transformations (2.2), Solutions (2.13) will be converted to x-t plane. Due to the velocities of solitons and distance between the solitons, kinds of interactions will be analyzed: (i) when solitons own different velocities, oblique interactions will occur; (ii) when solitons share the same velocity and keep a adjacent distance, they will attract and repel each other periodically. Namely, the bound states of solitons will emerge [35, 36]; (iii) when solitons propagate with the same velocity and keep a large distance, parallel solitons will happen. These three kinds of interactions will be illustrated respectively.

Assuming $k_1 = k_{1R} + i\, k_{1I}$ and $k_2 = k_{2R} + i\, k_{2I}$, we rewrite the velocities of solitons as

$$V_1 = \frac{\rho\rho^* - 8}{4[k_{1R}^2 + (k_{1I} + 1)^2]}, \quad V_2 = \frac{\rho\rho^* - 8}{4[k_{2R}^2 + (k_{2I} + 1)^2]}, \qquad (4.1)$$

while the subscripts R and I represent the real and imaginary parts, with k_{1R}, k_{1I}, k_{2R} and k_{2I} being real constants. The distances between solitons can be influenced by $|\Delta\nu| = |\nu_1 - \nu_2|$.

Fig. 1 displays the oblique interactions between the two solitons which possess the velocities $V_1 = -\frac{7}{5}$ and $V_2 = -\frac{7}{20}$, respectively. After the interactions, the velocities and amplitudes of solitons keep invariant, while certain phase shifts occur. When we make the velocities as $V_1 = V_2 = -\frac{7}{8}$, the bound states of solitons and parallel solitons will happen, as seen as in Figs. 2-4. At the case of $|\Delta\nu| = 1$, two solitons are bounded together that we cannot make a distinction between them, which can be seen in Fig. 2. When the solitons have the distance of $|\Delta\nu| = \frac{11}{3}$ in Fig. 3, interaction strength between the left soliton with large amplitude/depth and the right soltion with small amplitude/depth becomes weaker than in Fig. 2. When the parameters are chosen as $|\Delta\nu| = 6$, the interaction strength between solitons tends to be neglected. As seen in Fig. 4, two parallel solitons propagate mutual independently.

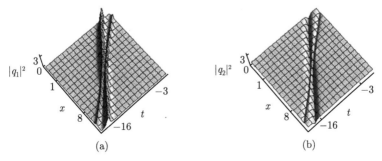

(a)   (b)

Fig. 1   Oblique interactions between the two solitons via Solutions (2.13) with $k_1 = \frac{1}{2}$, $k_2 = 1 + i$, $\delta_1 = 2$, $\delta_2 = 2$, $\nu_1 = -3$, $\nu_2 = 3$ and $\rho = 1$

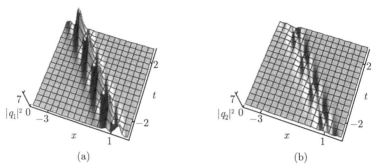

Fig. 2 Bound states of solitons via Solutions (2.13) with $k_1 = 1$, $k_2 = \dfrac{1}{2} - \dfrac{\sqrt{7}+2}{2}i$, $\delta_1 = 2$, $\delta_2 = 2$, $\nu_1 = -3$, $\nu_2 = -2$ and $\rho = 1$

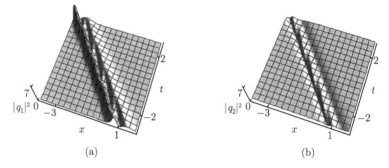

Fig. 3 Bound states of solitons via Solutions (2.13) with the same parameters as those in Fig. 2 except for $\nu_2 = \dfrac{2}{3}$

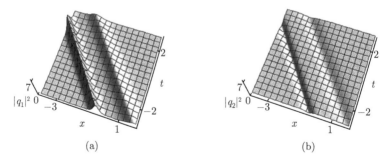

Fig. 4 Parallel solitons via Solutions (2.13) with the same parameters as those in Fig. 2 except for $\nu_2 = 3$

5 Conclusions

In this paper, with the help of Hirota method, we have investigated the coupled complex short pulse equations, i.e., Eq. (1.3), which describe the propagation of ultra-short optical pulses in cubic nonlinear media. The following results should be mentioned:

(i) The bright-dark One- and Two-Soliton Solutions (2.12) and (2.13) have been explicitly expressed based on Bilinear Forms (2.3) with the Scale Transformations (2.2).

(ii) Asymptotic Analysis (3.1a)-(3.4b) have been made on Solutions (2.13) to investigate the interactions between the two solitons. From Table 1, we conclude that interactions between two bright or two dark solitons are both elastic.

(iii) Oblique interactions [see Fig. 1], bound states of solitons [see Figs. 2 and 3] and parallel solitons [see Fig. 4] have been illustrated analytically and graphically. Types of interactions can be influenced by soliton velocities and distances between solitons.

We hope that our study will be helpful to the research and development of optical fiber communications.

Acknowledgments This work has been supported by the China Postdoctoral Science Foundation under Grant No. 2016M591121.

Appendix

The relevant coefficients in Solutions (2.12) and (2.13) are expressed as

$$\lambda = -\frac{1}{4}\rho\rho^*, \quad b = -1 + \frac{1}{4}\rho\rho^*, \quad \omega_1 = \frac{4 + 4i\,k_1 - \rho\rho^*}{4i + 4\,k_1}, \quad \omega_2 = \frac{4 + 4i\,k_2 - \rho\rho^*}{4i + 4\,k_2},$$

$$\xi_1 = \frac{k_1(-2i + k_1^*)}{k_1^*(2i + k_1)}, \quad \xi_2 = \frac{k_1(-2i + k_2^*)}{k_2^*(2i + k_1)}, \quad \xi_3 = \frac{k_2(-2i + k_1^*)}{k_1^*(2i + k_2)}, \quad \xi_4 = \frac{k_2(-2i + k_2^*)}{k_2^*(2i + k_2)},$$

$$\eta_1 = -\frac{\delta_1\delta_1^*}{8\lambda + (\xi_1 + \xi_1^*)\rho\rho^* - 4(\omega_1 + \omega_1^*)^2}, \quad \eta_2 = -\frac{\delta_1\delta_2^*}{8\lambda + (\xi_2 + \xi_3^*)\rho\rho^* - 4(\omega_1 + \omega_2^*)^2},$$

$$\eta_3 = -\frac{\delta_2\delta_1^*}{8\lambda + (\xi_3 + \xi_2^*)\rho\rho^* - 4(\omega_2 + \omega_1^*)^2}, \quad \eta_4 = -\frac{\delta_2\delta_2^*}{8\lambda + (\xi_4 + \xi_4^*)\rho\rho^* - 4(\omega_2 + \omega_2^*)^2},$$

$$\delta_3 = -\frac{(k_1 - k_2)[\delta_2\eta_1(k_1 + k_1^*)(2i + k_2 - k_1^*) - \delta_1\eta_3(k_2 + k_1^*)(2i + k_1 - k_1^*)]}{(k_1 + k_1^*)(k_2 + k_1^*)(2i + k_1 + k_2)},$$

$$\delta_4 = -\frac{(k_1 - k_2)[\delta_2\eta_2(k_1 + k_2^*)(2i + k_2 - k_2^*) - \delta_1\eta_4(k_2 + k_2^*)(2i + k_1 - k_2^*)]}{(k_1 + k_2^*)(k_2 + k_2^*)(2i + k_1 + k_2)},$$

$$\eta_5 = \frac{(k_1^* - k_2^*)[\delta_3\eta_2(k_2 + k_1^*)(2i + k_2 - k_2^*) - \delta_4\eta_1(k_2 + k_2^*)(2i + k_2 - k_1^*)]}{\delta_1(k_2 + k_1^*)(k_2 + k_2^*)(-2i + k_1^* + k_2^*)},$$

$$\xi_5 = \xi_1\xi_4.$$

References

[1] Hasegawa A, Tappert F. Transmission of stationary nonlinear optical physics in dispersive dielectric fibers I: anomalous dispersion [J]. Appl. Phys. Lett., 1973, 23: 142-144.

[2] Mollenauer L F, Stolen R H, Gordon J P. Experimental observation of picosecond pulse narrowing and solitons in optical fibers [J]. Phys. Rev. Lett., 1980, 45: 1095-1098.

[3] Agrawal G P. Nonlinear Fiber Optics. New York: Academic Press [M]. 1995.

[4] Kodama Y, Hasegawa A. Nonlinear pulse propagation in a monomode dielectric guide [J]. IEEE J. Quantum Elect., 1987, 23: 510-524.

[5] Mamyshev P V, Chernikov S V. Ultrashort-pulse propagation in optical fibers [J]. Opt. Lett., 1990, 15: 1076-1078.

[6] Li Z, Li L, Tian H, Zhou G. New types of solitary wave solutions for the higher order nonlinear Schrödinger equation [J]. Phys. Rev. Lett., 2000, 84: 4096-4099.

[7] Alterman D, Rauch J. Diffractive short pulse asymptotics for nonlinear wave equations [J]. Phys. Lett., A 2000, 264: 390-395.

[8] Brabec T, Krausz F. Nonlinear optical pulse propagation in the single-cycle regime [J]. Phys. Rev. Lett., 1997, 78: 3282-3285.

[9] Chernikov S V, Mamyshev P V. Femtosecond soliton propagation in fibers with slowly decreasing dispersion [J]. J. Opt. Soc. Am., B, 1991, 8: 1633-1641.

[10] Hile C V, Kath W L. Numerical solutions of Maxwell's equations for nonlinear optical pulse propagation [J]. J. Opt. Soc. Am., B, 1996, 13: 1135-1145.

[11] Kuetche V K, Youssoufa S, Kofane T C. Ultrashort optical waveguide excitations in uniaxial silica fibers: Elastic collision scenarios [J]. Phys. Rev., E, 2014, 90: 063203.

[12] Schäfer T, Wayne C E. Propagation of ultra-short optical pulses in cubic nonlinear media [J]. Physica, D, 2004, 196: 90-105.

[13] Sakovich A, Sakovich S. The short pulse equation is integrable [J]. J. Phys. Soc. Jpn., 2005, 74: 239-241.

[14] Brunelli J C. The bi-Hamiltonian structure of the short pulse equation [J]. Phys. Lett., A, 2006, 353: 475-478.

[15] Sakovich A, Sakovich S. Solitary wave solutions of the short pulse equation [J]. J. Phys., A, 2006, 39: L361-L367.

[16] Matsuno Y. Periodic solutions of the short pulse model equation [J]. J. Math. Phys., 2008, 49: 073508.

[17] Matsuno Y. Multiloop soliton and multibreather solutions of the short pulse model equation [J]. J. Phys. Soc. Jpn., 2007, 76: 084003.

[18] Feng B F, Maruno K, Ohta Y. Integrable discretization of the short pulse equation [J]. J. Phys., A, 2010, 43: 085203.

[19] Feng B F, Inoguchi J, Kajiwara K, Maruno K, Ohta Y. Discrete integrable systems and hodograph transformations arising from motions of discrete plane curves, J. Phys. A, 2011, 44: 395201.

[20] Yariv A, Yeh P. Optical Waves in Crystals: Propagation and Control of Laser Radiation [M], Wiley-Interscience, 1983.
[21] Feng B F. Complex short pulse and coupled complex short pulse equations, Physica D, 2015, 297: 62-75.
[22] Shen S F, Feng B F, Ohta Y. From the real and complex coupled dispersionless equations to the real and complex short pulse equations, Stud. Appl. Math. 2015, 136: 64-88.
[23] Christodoulides D N, Coskun T H, Mitchell M, Segev M. Theory of incoherent self-focusing in biased photorefractive media, Phys. Rev. Lett. 1997, 78: 646-649.
[24] Kumar V R, Radha R, Porsezian K. Intensity redistribution and shape changing collision in coupled femtosecond solitons, Eur. Phys. J. D, 2010, 57: 387-393.
[25] Wang Y F, Tian B, Wang M. Bell-polynomial approach and integrability for the coupled Gross-Pitaevskii equations in Bose-Einstein condensates, Stud. Appl. Math. 2013, 131: 119-134.
[26] Nakkeeran K. Exact dark soliton solutions for a family of N-coupled nonlinear Schrödinger equations in optical fiber media, Phys. Rev. E, 2001, 64: 046611.
[27] Manikandan N, Radhakrishnan R, Aravinthan K. Generalized dark-bright vector soliton solution to the mixed coupled nonlinear Schrödinger equations [J]. Phys. Rev., E, 2014, 90: 022902.
[28] Agalarov A, Zhulego V, Gadzhimuradov T. Bright, dark, and mixed vector soliton solutions of the general coupled nonlinear Schrödinger equations [J]. Phys. Rev., E, 2015, 91: 042909.
[29] Kivshar Y S, Quiroga-Teixeiro M L. Influence of cross-phase modulation on soliton switching in nonlinear optical fibers [J]. Opt. Lett., 1993, 18: 980-982.
[30] Ling L M, Feng B F, Zhu Z N. Multi-soliton, multi-breather and higher order rogue wave solutions to the complex short pulse equation [J]. Physica, D, 2016, 327: 13-29.
[31] Dai C Q, Xu Y J. Exact solutions for a Wick-type stochastic reaction Duffing equation [J]. Appl. Math. Model., 2015, 39: 7420-7426.
[32] Dai C Q, Wang Y Y. Controllable combined Peregrine soliton and Kuznetsov-Ma soliton in PT-symmetric nonlinear couplers with gain and loss [J]. Nonlinear Dyn., 2015, 80: 715-721.
[33] Dai C Q, Wang Y Y. Spatiotemporal localizations in (3+1)-dimensional PT-symmetric and strongly nonlocal nonlinear media [J]. Nonlinear Dyn., 2016, 83: 2453-2459.
[34] Hirota R. The Direct Method in Soliton Theory [M]. Cambridge Univ. Press, Cambridge, 2004.
[35] Sheppard A P, Kivshar Y S. Polarized dark solitons in isotropic Kerr media [J]. Phys. Rev., E, 1997, 55: 4773-4782.
[36] Stegeman G I, Segev M. Optical spatial solitons and their interactions: universality and diversity [J]. Science, 1999, 286: 1518-1523.

Global Existence of Smooth Solutions for the Magnetic Schrödinger Equation Arising from Hot Plasma*

Bian Dongfen (边东芬), Guo Boling (郭柏灵) and
Zhang Jingjun (张景军)

Abstract We consider a type of dispersive-dissipative system that arises in the infinite ion acoustic speed limit of the magnetic Zakharov system in a hot plasma. It is shown that this system admits a unique global smooth solution for suitably small initial data. Decay estimates for the solution are also obtained. In particular, the L^2 and L^∞ decay rates for the magnetic field are sharp.

Keywords global solution; decay estimate; magnetic effect

1 Introduction

In this paper, we are concerned with the equations which can be seen as a limit model of the magnetic Zakharov system

$$\begin{cases} \mathrm{i}E_t + \Delta E - nE + \mathrm{i}E \times B = 0, \\ c^{-2}n_{tt} - \Delta n = \Delta |E|^2, \\ \Delta B - \mathrm{i}\beta \nabla \times (\nabla \times (E \times \overline{E})) - \gamma \dfrac{\partial}{\partial t} \displaystyle\int_{\mathbb{R}^3} \dfrac{B(t,y)}{|x-y|^2}\, dy = 0. \end{cases} \quad (1.1)$$

Here, $E : \mathbf{R}^+ \times \mathbf{R}^3 \to \mathbf{C}^3$ is the slowly varying amplitude of the high-frequency electric field ($\overline{E}$ is the complex conjugate of E), $n : \mathbf{R}^+ \times \mathbf{R}^3 \to \mathbf{R}$ denotes the fluctuation of the ion-density from its equilibrium, and $B : \mathbf{R}^+ \times \mathbf{R}^3 \to \mathbf{R}^3$ is the self-generated magnetic field. The parameter c is proportional to the ion acoustic speed, and β, γ are positive constants. In the third equation of (1.1), $\nabla \times$ is the curl operator. For any $\mathbf{R}^3$ or $\mathbf{C}^3$ valued vectors f, g, the notation $f \times g$ means the cross product.

---

* J. Differential Equations, 2016, 261(9): 5202–5234. DOI:10.1016/j.jde.2016.07.024.

System (1.1) describes the spontaneous generation of a magnetic field in a hot plasma, and we refer to [16, 21] for its physical background. In the infinite ion acoustic speed limit (i.e., $c \to \infty$), the above system reduces to

$$\begin{cases} iE_t + \Delta E + |E|^2 E + iE \times B = 0, \\ \Delta B - i\beta \nabla \times (\nabla \times (E \times \overline{E})) - \gamma \dfrac{\partial}{\partial t} \displaystyle\int_{\mathbf{R}^3} \dfrac{B(t,y)}{|x-y|^2}\, dy = 0. \end{cases} \quad (1.2)$$

The limit $c \to \infty$ usually relates to the assumption that the plasma responds instantaneously to variations in the electric field so that the fluctuation of the density satisfies a stationary form $n \sim -|E|^2$.

The equation of B is different in other type of plasmas. Compared to system (1.1), the magnetic field B satisfies an elliptic equation in a cold plasma. More precisely, in cold plasmas, the equations describing the nonlinear interaction between plasma-wave and particles are

$$\begin{cases} iE_t + \Delta E - nE + iE \times B = 0, \\ c^{-2} n_{tt} - \Delta n = \Delta |E|^2, \\ \Delta B - i\eta \nabla \times (\nabla \times (E \times \overline{E})) + \kappa B = 0, \end{cases} \quad (1.3)$$

where $\eta > 0$ and $\kappa \leqslant 0$. Accordingly, its limit system as $c \to \infty$ reads

$$\begin{cases} iE_t + \Delta E + |E|^2 E + iE \times B = 0, \\ \Delta B - i\eta \nabla \times (\nabla \times (E \times \overline{E})) + \kappa B = 0. \end{cases} \quad (1.4)$$

Both (1.1) and (1.3) are generalized models of the classical Zakharov system [24], in which the effect of the self-generated magnetic field is not considered. Also, the limit systems (1.2) and (1.4) are generalized nonlinear Schrödinger equations with the magnetic effect. Omitting the effect of the magnetic field M, both (1.2) and (1.4) are reduced to the cubic Schrödinger equation

$$iE_t + \Delta E + |E|^2 E = 0. \quad (1.5)$$

Equation (1.5) has been studied by many researchers, see for example [4, 13, 18] and the references cited therein.

Let us recall the results that have been obtained for the magnetic Zakharov system (1.3). Laurey [22] established the local existence and uniqueness of solution $(E, n) \in C([0,T]; H^{s+1} \times H^s)$ with $s > \dfrac{d}{2}$ ($d = 2, 3$) for this system. Blow-up solutions of virial type for (1.3) were studied in [6]. See also [8] for the blow-up result in the $L_t^q L_x^p$ type norm and the discussion on the blow-up rate for (1.3). Gan-Guo-Huang [7] considered self-similar blow-up solutions and the nonlinear instability for a kind of periodic solutions to (1.3) in two dimensional case. Concerning on the limit system (1.4), we refer to Gan-Zhang [9] for the work on finite time blow-up, sharp threshold of global existence and instability of standing waves.

As for the system (1.1), Laurey [22] also obtained local well-posedness of the solution $(E, n, B) \in C([0, T]; H^{s+1} \times H^s \times H^s)$ with $s > \frac{3}{2}$. See also Kenig-Wang [19] for the existence of a local smooth solution in weighted space for the electric field. Han-Zhang-Gan-Guo [14] proved local well-posedness result for (1.1) with low regularity initial data. Nevertheless, as far as we know, there is no result of global existence for both (1.1) and (1.2). Hence, it would be interesting to study the global existence and large time behavior of solutions for these systems. In order to obtain a clear understanding on the effect of the magnetic field, we are mainly focused on the Cauchy problem of system (1.2) in this work. The initial data of (1.2) is given as

$$E(0,x) = E_0(x), \quad B(0,x) = B_0(x). \tag{1.6}$$

The main result of the paper is stated in the following theorem. For $x, y \geqslant 0$, the notation $x \lesssim y$ means that there exists a constant $C > 0$ such that $x \leqslant Cy$.

Theorem 1.1 *Let $N \geqslant 100$ and $0 < \delta \leqslant 10^{-3}$, then there exists a positive constant $\epsilon_0 \ll 1$ such that if the initial data satisfies*

$$\|E_0\|_{H^N} + \|xE_0\|_{L^2} + \||x|^2 E_0\|_{L^2} \leqslant \epsilon_0, \tag{1.7}$$

$$\|B_0\|_{H^{N-1}} + \|B_0\|_{L^1} \leqslant \epsilon_0, \tag{1.8}$$

then the Cauchy problem for (1.2) and (1.6) admits a unique global solution $(E, B) \in C(\mathbf{R}^+; H^N \times H^{N-1})$ satisfying for all $t \geqslant 0$,

$$\|E(t)\|_{H^N} \lesssim \epsilon_0(1+t)^\delta, \quad \|B(t)\|_{H^{N-1}} \lesssim \epsilon_0, \tag{1.9}$$

$$\|E(t)\|_{L^\infty} \lesssim \frac{\epsilon_0}{(1+t)^{7/6-\delta}}, \quad \|B(t)\|_{L^\infty} \lesssim \frac{\epsilon_0}{1+t}. \tag{1.10}$$

We remark that the L^∞ decay rate for B in (1.10) is optimal. Moreover, from our proof, we can also obtain the following sharp L^2 decay estimate

$$\|B\|_{\dot{H}^k} \lesssim \frac{\epsilon_0}{(1+t)^{1/2+k/3}}, \quad k = 0, 1, 2.$$

Theorem 1.1 is proved mainly on the framework of space-time resonance method developed by Germain-Masmoudi-Shatah[10]. The method is effective in proving global stability results for many well-known dispersive equations, such as nonlinear wave equations[23], water waves system[11], two-fluid system[12] and Zakharov system[15]. As usual, we introduce the profile of E in our analysis. However, as the equation for B in (1.2) has a dissipative structure, it is in general not workable to use the profile for B (and thus the weighted norms for the profile) as dispersive equations, so one should introduce new techniques to deal with the magnetic field. In fact, we will make use of the dissipative property, derivative nonlinear structure and the optimal decay rates to estimate the norms of B or the terms involving B. The arguments used here may be generalized to other dispersive-dissipative type system. In particular, combined with the work of [15], we can actually obtain the global existence and decay estimates of smooth solutions for the magnetic Zakharov system (1.1).

This paper is organized as follows. In the next section, we introduce some notations that will be used in the whole paper, then we write system (1.1) into an integral system by using the profile for the Schrödinger component, and introduce the work space and the linear decay estimates for the operators $e^{it\Delta}$ and $e^{-t|\nabla|^3}$. In Section 3, we present the energy estimates for the solution. Section 4 is devoted to dealing with the decay estimates for the magnetic field both in the L^∞ norm and L^2 norm. The weighted estimates for the electric field are given in Section 5.

2 Preliminaries

2.1 Notations

For $s \in \mathbf{R}$, H^s and $\dot{H}^s$ denote the inhomogeneous and homogeneous Sobolev spaces respectively, equipped with the norms

$$\|u\|_{H^s} := \|(1+|\xi|^2)^{s/2}\hat{u}\|_{L^2}, \quad \|u\|_{\dot{H}^s} := \||\xi|^s\hat{u}\|_{L^2}.$$

Here, $\hat{u} = \hat{u}(\xi)$ is the Fourier transform of u, namely,

$$\hat{u}(\xi) = \mathcal{F}u := \frac{1}{(2\pi)^{3/2}} \int_{\mathbf{R}^3} e^{-ix\cdot\xi} u(x)\,dx.$$

By this definition, we have (see e.g., [5]) $\widehat{uv} = (2\pi)^{-3/2}(\hat{u} * \hat{v})$.

Fix a radial, smooth function $\varphi(x)$ satisfying $0 \leqslant \varphi \leqslant 1$, $\varphi(x) = 1$ for $|x| \leqslant 1$ and $\mathrm{supp}\,\varphi \subset \{x \in \mathbf{R}^3; |x| \leqslant 2\}$. Then for $a > 0$, we denote by $P_{\leqslant a}$ and $P_{>a}$ the frequency projection operators defined by

$$\widehat{P_{\leqslant a}u} := \varphi(\xi/a)\hat{u}(\xi), \quad P_{>a} := 1 - P_{\leqslant a}.$$

For any radial, measurable function $\chi: \mathbf{R}^3 \to \mathbf{R}^+$, we denote the Fourier multiplier by
$$\chi(|\nabla|) := \mathcal{F}^{-1}\chi(\xi)\mathcal{F},$$
i.e., $\mathcal{F}(\chi(|\nabla|)u) := \chi(\xi)\widehat{u}(\xi)$. For example, $\mathcal{F}(|\nabla|^s u) := |\xi|^s \widehat{u}(\xi)$.

2.2 Reduction on system (1.2)

In order to prove Theorem 1.1, we should rewrite system (1.2) into an integral system. According to the Riesz potential, we see
$$\int_{\mathbf{R}^3} \frac{B(t,y)}{|x-y|^2}\, dy = c_0 |\nabla|^{-1} B, \quad c_0 > 0.$$

For simplicity, we set the constants β, γ and c_0 to be one. Hence, system (1.2) is now reduced to (note that $|\nabla|^2 = -\Delta$)
$$\begin{cases} \mathrm{i}E_t + \Delta E = -|E|^2 E - \mathrm{i}E \times B, \\ B_t + |\nabla|^3 B = -\mathrm{i}|\nabla|\nabla \times (\nabla \times (E \times \overline{E})). \end{cases} \tag{2.1}$$

For the Schrödinger equation, we define the profile
$$f = \mathrm{e}^{-\mathrm{i}t\Delta} E, \tag{2.2}$$
which implies $\|f\|_{H^s} = \|E\|_{H^s}$ for all $s \in \mathbf{R}$. Then by Duhamel's formula, we have
$$\begin{aligned} \widehat{f}(t,\xi) = &\widehat{f}(0,\xi) \\ &+ \frac{\mathrm{i}}{(2\pi)^3} \int_0^t \int_{\mathbf{R}^3 \times \mathbf{R}^3} \mathrm{e}^{\mathrm{i}s\phi(\xi,\eta,\sigma)} \widehat{f}(s,\xi-\eta)(\widehat{f}(s,\eta-\sigma) \cdot \overline{\widehat{f}}(s,\sigma))\, d\eta\, d\sigma\, ds \\ &- \frac{1}{(2\pi)^{3/2}} \int_0^t \int_{\mathbf{R}^3} \mathrm{e}^{\mathrm{i}s\psi_1(\xi,\eta)} \widehat{f}(s,\xi-\eta) \times \widehat{B}(s,\eta)\, d\eta\, ds, \end{aligned} \tag{2.3}$$
where the phases ϕ and ψ_1 are
$$\phi(\xi,\eta,\sigma) := |\xi|^2 - |\xi-\eta|^2 - |\eta-\sigma|^2 + |\sigma|^2 = 2\xi \cdot \eta + 2\eta \cdot \sigma - 2|\eta|^2, \tag{2.4}$$
$$\psi_1(\xi,\eta) := |\xi|^2 - |\xi-\eta|^2 = 2\xi \cdot \eta - |\eta|^2. \tag{2.5}$$

As the profile is not compatible with the dissipative operator in our analysis, so for the magnetic field B, we write
$$B(t) = \mathrm{e}^{-t|\nabla|^3} B_0 - \mathrm{i}\int_0^t \mathrm{e}^{-(t-s)|\nabla|^3} |\nabla|\nabla \times (\nabla \times (E \times \overline{E}))(s)\, ds. \tag{2.6}$$

Using (2.2) and taking Fourier transform, we also have

$$\widehat{B}(t,\xi) = e^{-t|\xi|^3}\widehat{B_0}$$
$$+ \frac{i}{(2\pi)^{3/2}} \int_0^t \int_{\mathbf{R}^3} e^{-(t-s)|\xi|^3} e^{is\psi_2(\xi,\eta)} |\xi|\xi \times (\xi \times (\widehat{f}(\xi-\eta) \times \overline{\widehat{f}}(\eta))) \, d\eta \, ds, \qquad (2.7)$$

where the phase ψ_2 is given by

$$\psi_2(\xi,\eta) := -|\xi-\eta|^2 + |\eta|^2 = 2\xi \cdot \eta - |\xi|^2. \qquad (2.8)$$

The integral identities (2.3) and (2.6) (or (2.7)) are the main equations that we will discuss later. Moreover, we will apply the following nonresonant conditions

$$\nabla_\xi \phi = \nabla_\sigma \phi, \qquad 2\xi = \nabla_\eta \psi_2, \qquad (2.9)$$

which are used to integrate by parts to get rid of the factor s. We remark that the nonresonant conditions can be identified as a natural generalization of the null forms introduced by Klainerman [20] (see the examples given by Pusateri-Shatah [23]).

2.3 Work space and linear decay estimates

Now we introduce our work space. Recall $N \geqslant 100$ and $0 < \delta \leqslant 10^{-3}$, then for any $T \in (0, +\infty]$, we define two norms associated to the work space

$$\|E\|_{X_T} := \sup_{t \in [0,T)} (1+t)^{-\delta} \|E(t)\|_{H^N}$$
$$+ \sup_{t \in [0,T)} \left((1+t)^{-\delta} \|xf(t)\|_{L^2} + (1+t)^{-2/3-\delta} \||x|^2 f(t)\|_{L^2} \right),$$
$$\|B\|_{Y_T} := \sup_{t \in [0,T)} \left(\|B(t)\|_{H^{N-1}} + (1+t)\|B(t)\|_{L^\infty} \right),$$

and set

$$A_T := \|E\|_{X_T} + \|B\|_{Y_T}. \qquad (2.10)$$

Decay estimates for the linear operators are important in our analysis. In order to present the linear decay estimates, we need the following lemma.

Lemma 2.1 *There hold that*

$$\|f\|_{L^1(\mathbf{R}^3)} \lesssim \|xf\|_{L^2(\mathbf{R}^3)}^{1/2} \||x|^2 f\|_{L^2(\mathbf{R}^3)}^{1/2}, \qquad (2.11)$$
$$\|f\|_{L^{4/3}(\mathbf{R}^3)} \lesssim \|f\|_{L^2(\mathbf{R}^3)}^{1/4} \|xf\|_{L^2(\mathbf{R}^3)}^{3/4}. \qquad (2.12)$$

Proof Let $a > 0$ be determined later. Using the fundamental estimates
$$\int_{|x|\leqslant a} \frac{1}{|x|^2}\,dx \lesssim a, \quad \int_{|x|\geqslant a} \frac{1}{|x|^4}\,dx \lesssim a^{-1},$$
we deduce by the Cauchy-Schwarz inequality that
$$\|f\|_{L^1(\mathbf{R}^3)} \leqslant \int_{|x|\leqslant a} |xf(x)| \cdot |x|^{-1}\,dx + \int_{|x|\geqslant a} |x|^2|f(x)| \cdot |x|^{-2}\,dx$$
$$\lesssim \|xf\|_{L^2} a^{1/2} + \||x|^2 f\|_{L^2} a^{-1/2}.$$
Then (2.11) follows immediately if we choose
$$a = \frac{\||x|^2 f\|_{L^2}}{\|xf\|_{L^2}}.$$
Here we may assume $\|xf\|_{L^2} \neq 0$, otherwise the estimate (2.11) holds obviously.

The proof for (2.12) is similar. In fact, using Hölder's inequality, we have
$$\|f\|_{L^{4/3}}^{4/3} \leqslant \int_{|x|\leqslant b} |f(x)|^{4/3} \cdot 1\,dx + \int_{|x|\geqslant b} |xf(x)|^{4/3} \cdot |x|^{-4/3}\,dx$$
$$\lesssim \|f\|_{L^2}^{4/3} b + \|xf\|_{L^2}^{4/3} b^{-1/3},$$
which gives the estimate (2.12) as desired provided that we set
$$b = \frac{\|xf\|_{L^2}}{\|f\|_{L^2}}.$$
This ends the proof of the lemma.

For the Schrödinger operator $e^{it\Delta}$, it is known that for $p \in [2, +\infty]$,
$$\|e^{it\Delta} f\|_{L^p(\mathbf{R}^3)} \lesssim \frac{1}{t^{3(1/2-1/p)}} \|f\|_{L^{p'}(\mathbf{R}^3)}, \quad \frac{1}{p} + \frac{1}{p'} = 1,$$
which can be found, for example, in [2]. Combining (2.2), (2.10) and Lemma 2.1, we obtain the following linear decay estimates
$$\|E\|_{L^\infty} = \|e^{it\Delta} f\|_{L^\infty} \lesssim \frac{1}{(1+t)^{7/6-\delta}} A_T, \tag{2.13}$$
$$\|E\|_{L^4} = \|e^{it\Delta} f\|_{L^4} \lesssim \frac{1}{(1+t)^{3/4-\delta}} A_T. \tag{2.14}$$

For the magnetic component, we have the following lemma.

Lemma 2.2 *For any $t > 0$ and $k = 0, 1, 2, \cdots$, there hold that*
$$\||\nabla|^k e^{-t|\nabla|^3} B_0\|_{L^\infty} \lesssim \frac{1}{t^{1+k/3}} \|B_0\|_{L^1}, \tag{2.15}$$

$$\|e^{-t|\nabla|^3} B_0\|_{\dot{H}^k} \lesssim \frac{1}{t^{1/2+k/3}} \|P_{\leqslant 1} B_0\|_{L^1} + e^{-t} \|P_{>1} B_0\|_{\dot{H}^k}$$

$$\lesssim \frac{1}{t^{1/2+k/3}} \|B_0\|_{L^1 \cap \dot{H}^k}, \qquad (2.16)$$

where the implicit constants depend only on k. Moreover, (2.15) also holds if $|\nabla|^k$ is replaced by the local operator ∇^k.

Proof We first prove (2.15). Define $K(x)$ by $\widehat{K}(\xi) := |\xi|^k e^{-|\xi|^3}$, then

$$\|K\|_{L^\infty} \lesssim \|\widehat{K}\|_{L^1} \lesssim 1,$$

and

$$|\xi|^k e^{-t|\xi|^3} = \frac{1}{t^{1+k/3}} \mathcal{F}[K(\frac{x}{t^{1/3}})].$$

Since

$$\mathcal{F}[|\nabla|^k e^{-t|\nabla|^3} B_0] = |\xi|^k e^{-t|\xi|^3} \widehat{B_0} = \frac{1}{t^{1+k/3}} \mathcal{F}[K(\frac{x}{t^{1/3}})] \cdot \mathcal{F} B_0,$$

by Hölder's inequality, we obtain

$$||\nabla|^k e^{-t|\nabla|^3} B_0| = \frac{1}{(2\pi)^{3/2} t^{1+k/3}} \left| \int_{\mathbf{R}^3} K(\frac{x-y}{t^{1/3}}) B_0(y) \, dy \right|$$

$$\lesssim t^{-1-k/3} \|K\|_{L^\infty} \|B_0\|_{L^1}$$

$$\lesssim t^{-1-k/3} \|B_0\|_{L^1},$$

which yields (2.15) as desired. Applying similar argument as above, we can also prove (2.15) with $|\nabla|^k$ replaced by ∇^k.

Now we show (2.16). Indeed,

$$\|e^{-t|\nabla|^3} B_0\|_{\dot{H}^k}^2 = \|e^{-t|\xi|^3} |\xi|^k \widehat{B_0}(\xi)\|_{L^2}^2$$

$$\lesssim \int_{|\xi| \leqslant 2} e^{-2t|\xi|^3} |\xi|^{2k} |\widehat{P_{\leqslant 1} B_0}(\xi)|^2 \, d\xi$$

$$+ \int_{|\xi| \geqslant 1} e^{-2t|\xi|^3} |\xi|^{2k} |\widehat{P_{>1} B_0}(\xi)|^2 \, d\xi$$

$$\lesssim \|\widehat{P_{\leqslant 1} B_0}\|_{L^\infty}^2 \cdot t^{-1-2k/3} \int_{\mathbf{R}^3} e^{-2|\eta|^3} |\eta|^{2k} \, d\eta$$

$$+ e^{-2t} \int_{\mathbf{R}^3} |\xi|^{2k} |\widehat{P_{>1} B_0}(\xi)|^2 \, d\xi$$

$$\lesssim t^{-1-2k/3} \|P_{\leqslant 1} B_0\|_{L^1}^2 + e^{-2t} \|P_{>1} B_0\|_{\dot{H}^k}^2.$$

Thus, the estimate (2.16) follows.

2.4 Strategy of the proof

In the following three sections, the main *a priori* estimate we want to prove is

$$A_T \lesssim \epsilon_0 + A_T^{3/2}, \tag{2.17}$$

where the implicit constant is independent of T, then a standard continuation argument implies the existence of global solution as stated in Theorem 1.1. Hence, according to the definition (2.10), one should estimate the energy norm for the solution, decay norm for B and weighted norm for f, respectively. Below we briefly illustrate the ideas in estimating these norms.

Since the regularities for E and B are different and the nonlinear term contains derivatives, it will lead to the loss of derivatives when estimating the high order energy norms for system (1.2) directly. To overcome this difficulty, we first estimate $\|B_t\|_{\dot{H}^{N-4}}$ by using the dissipative structure for the equation of B, then the $\dot{H}^{N-1}$ energy norm for B can be obtained via the bound of B_t. In the energy estimate for E, the problematic term comes from

$$\int_{\mathbf{R}^3} |\nabla|^N (E \times B) \cdot |\nabla|^N \overline{E} \, dx,$$

which requires $B \in H^N$ (note that we only have $B \in H^{N-1}$). This difficulty is conquered by introducing a cubic corrected quantity $M(t)$ as defined in (3.17). The key fact is that $\dfrac{d}{dt} M(t)$ will cancel the above problematic term, and also produce some error terms which can be well controlled. The energy estimate is presented in Section 3.

For the decay estimates of B, by Lemma 2.2, it reduces to show the key bound (4.6). Here, we essentially use the idea of the work [15] to prove (4.6). An important observation is that the derivative nonlinear structure implies a null resonance form for the phase ψ_2 (see (2.9)), which allows us to integrate by parts in η and produces a decay factor s^{-1}. Another important step is to introduce the decomposition for f in physical space as shown in (4.12). The advantage of such decomposition is to obtain a better decay rate by the support property (4.13). In particular, as we will see later, the term $\|e^{is\Delta} f_{\text{in}}\|_{L^\infty}$ has a decay rate $s^{-11/8+\delta}$, which is better than $\|e^{is\Delta} f\|_{L^\infty}$ that only decays as $s^{-7/6+\delta}$.

For weighted estimates of f, the cubic term in (2.3) is treated by using the non-resonant condition (see (2.9)), which allows us to integrate by parts in η and σ to eliminate the growth factor s (or s^2). But the quadratic term in (2.3) can not be estimated in this way as we have no information on the profile and the weighted norm of B. However, we observe that the sharp L^2 decay estimates of B can just cancel some of the growth factor of s and thus lead to the desired bounds for xf and $|x|^2 f$ (see Section 5 for these estimates).

3 Energy estimate

This section deals with the energy estimate for system (2.1). The main estimate is stated in Proposition 3.1 below. To prove this proposition, we need the following calculus inequalities [3, 17].

Lemma 3.1 *Assume $s > 0$ and $\|\cdot\|_{\dot{H}^{s,q}} := \||\nabla|^s \cdot \|_{L^q}$, then for any $\mathbf{R}^k$ or $\mathbf{C}^k$ valued functions f, g (or one of them is scalar valued), we have*

$$\|f \cdot g\|_{\dot{H}^{s,p}} \lesssim \|g\|_{L^{p_1}} \|f\|_{\dot{H}^{s,p_2}} + \|g\|_{\dot{H}^{s,p_3}} \|f\|_{L^{p_4}}, \quad (3.1)$$

$$\||\nabla|^s (f \cdot g) - f \cdot (|\nabla|^s g)\|_{L^p} \lesssim \|\nabla f\|_{L^{p_1}} \|g\|_{\dot{H}^{s-1,p_2}} + \|f\|_{\dot{H}^{s,p_3}} \|g\|_{L^{p_4}}, \quad (3.2)$$

with $p_2, p_3 \in (1, +\infty)$ such that

$$\frac{1}{p} = \frac{1}{p_1} + \frac{1}{p_2} = \frac{1}{p_3} + \frac{1}{p_4}.$$

Proposition 3.1 *Assume that $(E, B) \in C([0,T]; H^N \times H^{N-1})$ is a smooth solution of system (2.1) satisfying $A_T \ll 1$, where A_T is defined by (2.10). Then we have*

$$\sup_{t \in [0,T]} ((1+t)^{-\delta} \|E(t)\|_{H^N} + \|B(t)\|_{H^{N-1}}) \lesssim \|E_0\|_{H^N} + \|B_0\|_{H^{N-1}} + A_T^{3/2}, \quad (3.3)$$

where the implicit constant C is independent of T.

Proof We first estimate the H^{N-1} norm for B. By the Cauchy-Schwarz inequality and (3.1), we deduce from the second equation of (2.1) that

$$\frac{1}{2} \frac{d}{dt} \|B\|_{L^2}^2 + \||\nabla|^{3/2} B\|_{L^2}^2 = -i \int_{\mathbf{R}^3} |\nabla|\nabla \times (\nabla \times (E \times \overline{E})) \cdot B \, dx$$
$$\lesssim \|E\|_{H^3} \|E\|_{L^\infty} \|B\|_{L^2},$$

which, by (2.10) and (2.13), yields

$$\frac{d}{dt} \|B\|_{L^2} \lesssim \|E\|_{H^3} \|E\|_{L^\infty} \lesssim (1+t)^{-(7/6 - 2\delta)} A_T^2.$$

So we integrate this inequality to obtain

$$\|B(t)\|_{L^2} \lesssim \|B_0\|_{L^2} + A_T^2. \tag{3.4}$$

Now we estimate $\|B\|_{\dot{H}^{N-1}}$. To this end, we differentiate the equation of B with respect to t and get

$$B_{tt} + |\nabla|^3 B_t = -\mathrm{i}|\nabla|\nabla \times (\nabla \times (E \times \overline{E})_t).$$

Taking the energy estimate at $\dot{H}^{N-4}$ level for B_t and using the Cauchy-Schwarz inequality, we have

$$\frac{1}{2}\frac{d}{dt}\||\nabla|^{N-4}B_t\|_{L^2}^2 + \||\nabla|^{N-5/2}B_t\|_{L^2}^2$$
$$= -\mathrm{i}\int_{\mathbf{R}^3} |\nabla|^{N-3}\nabla \times (\nabla \times (E \times \overline{E})_t) \cdot |\nabla|^{N-4}B_t\, dx$$
$$= -\mathrm{i}\int_{\mathbf{R}^3} |\nabla|^{N-9/2}\nabla \times (\nabla \times (E \times \overline{E})_t) \cdot |\nabla|^{N-5/2}B_t\, dx$$
$$\leqslant \frac{1}{2}\||\nabla|^{N-5/2}B_t\|_{L^2}^2 + \frac{1}{2}\||\nabla|^{N-5/2}(E \times \overline{E})_t\|_{L^2}^2,$$

from which we obtain (using (3.1))

$$\frac{d}{dt}\||\nabla|^{N-4}B_t\|_{L^2}^2 \leqslant \||\nabla|^{N-5/2}(E \times \overline{E})_t\|_{L^2}^2$$
$$\lesssim \|E_t\|_{\dot{H}^{N-5/2}}^2 \|E\|_{L^\infty}^2 + \|E_t\|_{L^\infty}^2 \|E\|_{\dot{H}^{N-5/2}}^2. \tag{3.5}$$

In view of the equation of E, we have

$$E_t = \mathrm{i}\Delta E + \mathrm{i}|E|^2 E - E \times B, \tag{3.6}$$

so there holds

$$\|E_t\|_{H^{N-2}} \lesssim \|E\|_{H^N} + \|E\|_{H^{N-2}}\|E\|_{L^\infty}^2 + \|E\|_{H^{N-2}}\|B\|_{H^{N-2}}$$
$$\leqslant \|E\|_{H^N}(1 + \|E\|_{L^\infty}^2 + \|B\|_{H^{N-2}})$$
$$\lesssim \|E\|_{H^N}$$
$$\lesssim (1+t)^\delta A_T, \tag{3.7}$$

due to the assumption $A_T \ll 1$. One also sees from (3.6) that

$$\|E_t\|_{L^\infty} \leqslant \|\Delta E\|_{L^\infty} + \|E\|_{L^\infty}^3 + \|E\|_{L^\infty}\|B\|_{L^\infty}$$
$$\lesssim \|\Delta E\|_{L^\infty} + (1+t)^{-(7/2-3\delta)}A_T^3 + (1+t)^{-(13/6-\delta)}A_T^2$$

$$\lesssim (1+t)^{-(25/24-\delta)} A_T, \qquad (3.8)$$

where in the last step we used Bernstein's inequality (see e.g., [1, Lemma 2.1]) to obtain

$$\|\Delta E\|_{L^\infty} \leqslant \|P_{\leqslant (1+t)^{1/16}} \Delta E\|_{L^\infty} + \|P_{> (1+t)^{1/16}} \Delta E\|_{L^\infty}$$
$$\lesssim (1+t)^{1/8} \|P_{\leqslant (1+t)^{1/16}} E\|_{L^\infty} + (1+t)^{-(N-4)/16} \|P_{>(1+t)^{1/16}} |\nabla|^{N-2} E\|_{L^\infty}$$
$$\lesssim (1+t)^{-(25/24-\delta)} A_T + (1+t)^{-(N-4)/16+\delta} A_T$$
$$\lesssim (1+t)^{-(25/24-\delta)} A_T. \qquad (3.9)$$

Inserting (3.7) and (3.8) into (3.5), we get

$$\frac{d}{dt} \||\nabla|^{N-4} B_t\|_{L^2}^2 \lesssim (1+t)^{-(25/12-4\delta)} A_T^4. \qquad (3.10)$$

Note that $(E, B) \in C([0,T]; H^N \times H^{N-1})$ is a smooth solution of system (2.1). Here, we remark that local existence of smooth solution for system (2.1) in such space can be established by applying the arguments in [22]. Hence, the equation of B implies $B_t \in C([0,T]; H^{N-4})$, and taking $t \to 0^+$ gives

$$\|B_t(0)\|_{\dot{H}^{N-4}} \lesssim \||\nabla|^3 B(0)\|_{\dot{H}^{N-4}} + \||\nabla|\nabla \times \nabla \times (E(0) \times \overline{E(0)})\|_{\dot{H}^{N-4}}$$
$$\lesssim \|B_0\|_{H^{N-1}} + \|E_0\|_{H^{N-1}} \|E_0\|_{L^\infty}$$
$$\lesssim \|B_0\|_{H^{N-1}} + \|E_0\|_{H^{N-1}}.$$

So integrating (3.10) yields

$$\||\nabla|^{N-4} B_t\|_{L^2} \lesssim \|B_0\|_{H^{N-1}} + \|E_0\|_{H^{N-1}} + A_T^2. \qquad (3.11)$$

Moreover, since the equation for B can be rewritten as

$$B = -|\nabla|^{-3} B_t - i|\nabla|^{-2} \nabla \times \nabla(E \times \overline{E}),$$

then by (3.1), (2.10) and (2.13),

$$\|B\|_{\dot{H}^{N-1}} \lesssim \|B_t\|_{\dot{H}^{N-4}} + \|E \times \overline{E}\|_{\dot{H}^{N-1}}$$
$$\lesssim \|B_t\|_{\dot{H}^{N-4}} + \|E\|_{H^{N-1}} \|E\|_{L^\infty}$$
$$\lesssim \|B_t\|_{\dot{H}^{N-4}} + A_T^2.$$

This bound together with (3.11) give

$$\|B\|_{\dot{H}^{N-1}} \lesssim \|B_0\|_{H^{N-1}} + \|E_0\|_{H^{N-1}} + A_T^2. \qquad (3.12)$$

Hence, the desired bound for $\|B\|_{H^{N-1}}$ in (3.3) clearly follows from (3.4) and (3.12).

Now we turn to estimate $\|E\|_{H^N}$. Multiplying the first equation of (2.1) with $\overline{E}$ and integrating the imaginary part over $\mathbf{R}^3$, we see

$$\frac{1}{2}\frac{d}{dt}\|E\|_{L^2}^2 = -\mathrm{Im}\int_{\mathbf{R}^3} \mathrm{i}(E\times B)\cdot \overline{E}\,dx = \mathrm{Im}\int_{\mathbf{R}^3} \mathrm{i}(E\times \overline{E})\cdot B\,dx = 0,$$

since $E\times \overline{E}$ is purely imaginary. This shows the L^2 norm of E is conserved, i.e.,

$$\|E(t)\|_{L^2} = \|E(0)\|_{L^2}. \tag{3.13}$$

It remains to estimate the $\dot{H}^N$ norm of E. Again, from the equation of E, we can obtain

$$\frac{1}{2}\frac{d}{dt}\||\nabla|^N E\|_{L^2}^2 = -\mathrm{Im}\int_{\mathbf{R}^3} |\nabla|^N(|E|^2 E)\cdot |\nabla|^N \overline{E}\,dx$$

$$-\mathrm{Im}\int_{\mathbf{R}^3} \mathrm{i}|\nabla|^N(E\times B)\cdot |\nabla|^N \overline{E}\,dx$$

$$= -\mathrm{Im}\int_{\mathbf{R}^3} |\nabla|^N(|E|^2 E)\cdot |\nabla|^N \overline{E}\,dx \tag{3.14}$$

$$-\mathrm{Im}\int_{\mathbf{R}^3} \mathrm{i}[|\nabla|^N(E\times B) - (E\times |\nabla|^N B)]\cdot |\nabla|^N \overline{E}\,dx \tag{3.15}$$

$$-\mathrm{Im}\int_{\mathbf{R}^3} \mathrm{i}(E\times |\nabla|^N B)\cdot |\nabla|^N \overline{E}\,dx.$$

By the Cauchy-Schwarz inequality and (3.1), the term (3.14) can be estimated as

$$|(3.14)| \lesssim \|E\|_{\dot{H}^N}^2 \|E\|_{L^\infty}^2 \lesssim (1+t)^{-(7/3-4\delta)} A_T^4.$$

For the term (3.15), we use commutator estimate (3.2) to obtain

$$|(3.15)| \lesssim (\|\nabla E\|_{L^\infty}\|B\|_{\dot{H}^{N-1}} + \|E\|_{\dot{H}^N}\|B\|_{L^\infty})\|E\|_{\dot{H}^N}.$$

By (2.13) and (3.9), we see

$$\|\nabla E\|_{L^\infty} \lesssim \|E\|_{L^\infty} + \|\Delta E\|_{L^\infty} \lesssim (1+t)^{-(25/24-\delta)} A_T,$$

then

$$|(3.15)| \lesssim (1+t)^{-(25/24-\delta)} A_T^3 + (1+t)^{-1+2\delta} A_T^3 \lesssim (1+t)^{-1+2\delta} A_T^3.$$

Hence, we obtain

$$\frac{d}{dt}\||\nabla|^N E\|_{L^2}^2 \leqslant C(1+t)^{-1+2\delta} A_T^3 - 2\mathrm{Im}\int_{\mathbf{R}^3} \mathrm{i}(E\times |\nabla|^N B)\cdot |\nabla|^N \overline{E}\,dx. \tag{3.16}$$

To eliminate the last integral term in (3.16), we introduce the quantity

$$M(t) := -2\text{Im} \int_{\mathbf{R}^3} i(E \times |\nabla|^{N-3}B) \cdot |\nabla|^N \overline{E} \, dx, \tag{3.17}$$

then using the equation of B, we can get

$$\frac{d}{dt}M(t) = 2\text{Im} \int_{\mathbf{R}^3} i(E \times |\nabla|^N B) \cdot |\nabla|^N \overline{E} \, dx$$

$$- 2\text{Im} \int_{\mathbf{R}^3} (E \times |\nabla|^{N-2} \nabla \times (\nabla \times (E \times \overline{E}))) \cdot |\nabla|^N \overline{E} \, dx \tag{3.18}$$

$$- 2\text{Im} \int_{\mathbf{R}^3} i(E_t \times |\nabla|^{N-3} B) \cdot |\nabla|^N \overline{E} \, dx \tag{3.19}$$

$$- 2\text{Im} \int_{\mathbf{R}^3} i(E \times |\nabla|^{N-3} B) \cdot |\nabla|^N \overline{E_t} \, dx. \tag{3.20}$$

For the terms (3.18) and (3.19), it is easy to see

$$|(3.18)| \lesssim \|E\|_{\dot{H}^N}^2 \|E\|_{L^\infty}^2 \lesssim (1+t)^{-(7/3-4\delta)} A_T^4,$$

$$|(3.19)| \lesssim \|E_t\|_{L^\infty} \|B\|_{\dot{H}^{N-3}} \|E\|_{\dot{H}^N} \lesssim (1+t)^{-(25/24-2\delta)} A_T^3,$$

where we have used the bound (3.8) in the estimate for (3.19). For the term (3.20), we use the Cauchy-Schwarz inequality, (3.1), (3.7) and (3.9) to obtain

$$|(3.20)| = |2\text{Im} \int_{\mathbf{R}^3} i|\nabla|^2 (E \times |\nabla|^{N-3} B) \cdot |\nabla|^{N-2} \overline{E_t} \, dx|$$

$$\lesssim (\|\Delta E\|_{L^\infty} \|B\|_{\dot{H}^{N-3}} + \|E\|_{L^\infty} \|B\|_{\dot{H}^{N-1}}) \|E_t\|_{\dot{H}^{N-2}}$$

$$\lesssim (1+t)^{-(25/24-2\delta)} A_T^3.$$

In view of these bounds, we have

$$\frac{d}{dt}M(t) \leqslant 2\text{Im} \int_{\mathbf{R}^3} i(E \times |\nabla|^N B) \cdot |\nabla|^N \overline{E} \, dx + C(1+t)^{-(25/24-2\delta)} A_T^3. \tag{3.21}$$

Now it follows from (3.16) and (3.21) that

$$\frac{d}{dt}(M(t) + \|E\|_{\dot{H}^N}^2) \lesssim (1+t)^{-1+2\delta} A_T^3,$$

which implies

$$M(t) + \|E(t)\|_{\dot{H}^N}^2 \lesssim M(0) + \|E(0)\|_{\dot{H}^N}^2 + (1+t)^{2\delta} A_T^3. \tag{3.22}$$

By (3.17) and Hölder's inequality, we notice that for all $t \in [0, T)$,

$$|M(t)| \lesssim \|E\|_{L^\infty} \|B\|_{\dot{H}^{N-3}} \|E\|_{\dot{H}^N} \lesssim \|E\|_{L^\infty} \|B\|_{H^{N-1}} \|E\|_{H^N} \lesssim A_T^3.$$

Hence, we obtain from (3.22)

$$\|E(t)\|_{\dot{H}^N} \lesssim \|E_0\|_{\dot{H}^N} + (1+t)^\delta A_T^{3/2},$$

which, by combining (3.13), yields the desired bound for $\|E\|_{H^N}$.

4 Decay estimates for the magnetic field

The main aim of this section is to prove the following proposition.

Proposition 4.1 *Assume $(E, B) \in C([0, T); H^N \times H^{N-1})$ satisfies system (2.1) with $T > 0$. If $A_T \ll 1$, then for all $t \in [0, T)$,*

$$(1+t)\|B(t)\|_{L^\infty} \lesssim \|B_0\|_{L^1 \cap H^2} + A_T^2, \tag{4.1}$$

where the implicit constant is independent of T.

Proof When $t \leqslant 4$, we use (2.6) and the Sobolev embedding $H^2(\mathbf{R}^3) \hookrightarrow L^\infty(\mathbf{R}^3)$ to obtain

$$\begin{aligned}\|B(t)\|_{L^\infty} &\lesssim \|e^{-t|\nabla|^3} B_0\|_{H^2} + \int_0^t \|e^{-(t-s)|\nabla|^3}|\nabla|\nabla \times \nabla \times (E \times \overline{E})\|_{H^2}\, ds \\ &\lesssim \|B_0\|_{H^2} + \int_0^4 \|E \times \overline{E}\|_{H^5}\, ds \\ &\lesssim \|B_0\|_{H^2} + A_T^2.\end{aligned} \tag{4.2}$$

Then the bound (4.1) clearly suffices in this case since $1 + t \sim 1$, if $0 \leqslant t \leqslant 4$.

Now let $t > 4$. We rewrite the expression (2.6) as

$$\begin{aligned}B(t) = {}& e^{-t|\nabla|^3} B_0 - i \int_0^1 e^{-(t-s)|\nabla|^3}|\nabla|\nabla \times (\nabla \times (E \times \overline{E}))(s)\, ds \\ &- i \int_1^{t-1} e^{-(t-s)|\nabla|^3}|\nabla|\nabla \times (\nabla \times (E \times \overline{E}))(s)\, ds \\ &- i \int_{t-1}^t e^{-(t-s)|\nabla|^3}|\nabla|\nabla \times (\nabla \times (E \times \overline{E}))(s)\, ds.\end{aligned} \tag{4.3}$$

In view of (2.15), it is easy to see

$$\|e^{-t|\nabla|^3} B_0\|_{L^\infty} \lesssim t^{-1}\|B_0\|_{L^1}, \tag{4.4}$$

and

$$\left\|\int_0^1 e^{-(t-s)|\nabla|^3}|\nabla|\nabla \times (\nabla \times (E \times \overline{E}))(s)\, ds\right\|_{L^\infty}$$

$$\lesssim \int_0^1 \frac{1}{t-s} \||\nabla|\nabla \times (\nabla \times (E \times \overline{E}))(s)\|_{L^1}\, ds$$

$$\lesssim \int_0^1 \frac{1}{t-s} \|E(s)\|_{H^3}^2\, ds$$

$$\lesssim \frac{1}{t-1} A_T^2$$

$$\sim \frac{1}{t} A_T^2. \tag{4.5}$$

To estimate the second integral term in (4.3), we use the claim

$$\||\nabla|\nabla \times (\nabla \times (E \times \overline{E}))(s)\|_{L^1} \lesssim s^{-7/6} A_T^2, \quad 1 \leqslant s < T. \tag{4.6}$$

Assuming now (4.6) holds, and also using (2.15), we have

$$\left\| \int_1^{t-1} e^{-(t-s)|\nabla|^3} |\nabla|\nabla \times (\nabla \times (E \times \overline{E}))(s)\, ds \right\|_{L^\infty}$$

$$\lesssim A_T^2 \int_1^{t-1} \frac{1}{t-s} \cdot \frac{1}{s^{7/6}}\, ds$$

$$\lesssim A_T^2 \int_1^{t/2} \frac{1}{t-s} \cdot \frac{1}{s^{7/6}}\, ds + A_T^2 \int_{t/2}^{t-1} \frac{1}{t-s} \cdot \frac{1}{s^{7/6}}\, ds$$

$$\lesssim \frac{1}{t} A_T^2. \tag{4.7}$$

Similar to (4.2), the last term in (4.3) is estimated by Sobolev embedding

$$\left\| \int_{t-1}^{t} e^{-(t-s)|\nabla|^3} |\nabla|\nabla \times (\nabla \times (E \times \overline{E}))(s)\, ds \right\|_{L^\infty}$$

$$\lesssim \int_{t-1}^{t} \|(E \times \overline{E})(s)\|_{H^5}\, ds$$

$$\lesssim \int_{t-1}^{t} \|E(s)\|_{H^5} \|E(s)\|_{L^\infty}\, ds$$

$$\lesssim A_T^2 \int_{t-1}^{t} \frac{1}{(1+s)^{7/6-2\delta}}\, ds$$

$$\lesssim \frac{1}{t^{7/6-2\delta}} A_T^2. \tag{4.8}$$

Inserting the estimates (4.4), (4.5), (4.7) and (4.8) into (4.3), we thus get for $t > 4$,

$$\|B(t)\|_{L^\infty} \lesssim t^{-1} \|B_0\|_{L^1} + t^{-1} A_T^2. \tag{4.9}$$

Then the desired bound (4.1) follows from (4.2) and (4.9).

Now, in order to finish the proof of Proposition 4.1, it remains to show the estimate (4.6). By Bernstein's inequality, we see that

$$\|P_{>s^{1/80}}|\nabla|\nabla \times (\nabla \times (E \times \overline{E}))(s)\|_{L^1}$$
$$\lesssim s^{-(N-3)/80}\||\nabla|^{N-3}|\nabla|\nabla \times (\nabla \times (E \times \overline{E}))(s)\|_{L^1}$$
$$\lesssim s^{-(N-3)/80}\|E\|_{H^N}^2$$
$$\lesssim s^{-(N-3)/80+2\delta}A_T^2,$$

which clearly suffices the desired bound (4.6) since $\frac{N-3}{80} - 2\delta > \frac{7}{6}$ for $N \geqslant 100$ and $0 < \delta \leqslant 10^{-3}$. Hence, to prove (4.6), it reduces to show

$$\|P_{\leqslant s^{1/80}}|\nabla|\nabla \times (\nabla \times (E \times \overline{E}))(s)\|_{L^1} \lesssim s^{-7/6}A_T^2, \quad 1 \leqslant s < T. \tag{4.10}$$

Using the profile f (see (2.2)), we have

$$\mathcal{F}[P_{\leqslant s^{1/80}}|\nabla|\nabla \times (\nabla \times (E \times \overline{E}))(s)](\xi)$$
$$= -\frac{1}{(2\pi)^{3/2}}\int_{\mathbf{R}^3} e^{is\psi_2(\xi,\eta)}\varphi(\xi/s^{1/80})|\xi|\xi \times (\xi \times (\widehat{f}(\xi-\eta) \times \overline{\widehat{f}(\eta)}))\,d\eta,$$

where the phase ψ_2 is given by (2.8) and φ is the smooth cut-off function defined in Section 2.1. Let $\xi = (\xi_1, \xi_2, \xi_3)$ and $f = (f_1, f_2, f_3)$, then by expanding the term

$$\xi \times (\xi \times (\widehat{f}(\xi-\eta) \times \overline{\widehat{f}(\eta)})),$$

we see clearly that in order to show (4.10), it is sufficient to prove for $1 \leqslant s < T$ and for all $i, i', j, j' = 1, 2, 3$,

$$\|\mathcal{F}^{-1}\int_{\mathbf{R}^3} e^{is\psi_2(\xi,\eta)}\varphi(\xi/s^{1/80})|\xi|\xi_i\xi_{i'}\widehat{f_j}(\xi-\eta)\overline{\widehat{f_{j'}}(\eta)}))\,d\eta\|_{L^1} \lesssim s^{-7/6}A_T^2. \tag{4.11}$$

To this end, we should introduce the decomposition for f in physical space. More precisely, we define

$$f_{\text{in}}(s,x) := \varphi(|x|/s^{1/4})f(s,x),$$
$$f_{\text{out}}(s,x) := f(s,x) - f_{\text{in}}(s,x), \tag{4.12}$$

so there holds

$$\operatorname{supp} f_{\text{in}} \subset \{|x|; |x| \leqslant 2s^{1/4}\}, \quad \operatorname{supp} f_{\text{out}} \subset \{|x|; |x| \geqslant s^{1/4}\}. \tag{4.13}$$

Using (4.12), we decompose the nonlinear term $f \times \bar{f}$ into

$$f \times \bar{f} = f_{\text{out}} \times \overline{f_{\text{out}}} + f_{\text{in}} \times \overline{f_{\text{out}}} + f_{\text{out}} \times \overline{f_{\text{in}}} + f_{\text{in}} \times \overline{f_{\text{in}}}.$$

Applying such decomposition to (4.11), we have

$$\text{LHS of (4.11)} \lesssim \|I_1\|_{L^1} + \|I_2\|_{L^1} + \|I_3\|_{L^1} + \|I_4\|_{L^1} \tag{4.14}$$

with

$$\begin{aligned}
\widehat{I_1} &:= \int_{\mathbb{R}^3} e^{is\psi_2(\xi,\eta)} \varphi(\xi/s^{1/80}) |\xi| \xi_i \xi_{i'} \widehat{f_{j,\text{out}}}(\xi-\eta) \overline{\widehat{f_{j',\text{out}}}}(\eta) \, d\eta, \\
\widehat{I_2} &:= \int_{\mathbb{R}^3} e^{is\psi_2(\xi,\eta)} \varphi(\xi/s^{1/80}) |\xi| \xi_i \xi_{i'} \widehat{f_{j,\text{in}}}(\xi-\eta) \overline{\widehat{f_{j',\text{out}}}}(\eta) \, d\eta, \\
\widehat{I_3} &:= \int_{\mathbb{R}^3} e^{is\psi_2(\xi,\eta)} \varphi(\xi/s^{1/80}) |\xi| \xi_i \xi_{i'} \widehat{f_{j,\text{out}}}(\xi-\eta) \overline{\widehat{f_{j',\text{in}}}}(\eta) \, d\eta, \\
\widehat{I_4} &:= \int_{\mathbb{R}^3} e^{is\psi_2(\xi,\eta)} \varphi(\xi/s^{1/80}) |\xi| \xi_i \xi_{i'} \widehat{f_{j,\text{in}}}(\xi-\eta) \overline{\widehat{f_{j',\text{in}}}}(\eta) \, d\eta,
\end{aligned} \tag{4.15}$$

where $f_{j,\text{in}}$ (or $f_{j,\text{out}}$) denotes the jth component of f_{in} (or f_{out}). In estimating these four terms, one of the key observations is the nonresonant structure for the phase ψ_2, that is

$$\xi_i = \frac{1}{2} \partial_{\eta_i} \psi_2, \quad i = 1, 2, 3, \tag{4.16}$$

or

$$|\xi| = \frac{1}{2|\xi|} \nabla_\eta \psi_2 \cdot \xi. \tag{4.17}$$

We first deal with the term I_1. Using the relation (4.16) to integrate by parts in η_i, we have

$$\begin{aligned}
\widehat{I_1} &= \frac{1}{2} \int_{\mathbb{R}^3} e^{is\psi_2(\xi,\eta)} \varphi(\xi/s^{1/80}) \partial_{\eta_i} \psi_2 |\xi| \xi_{i'} \widehat{f_{j,\text{out}}}(\xi-\eta) \overline{\widehat{f_{j',\text{out}}}}(\eta) \, d\eta \\
&= \int_{\mathbb{R}^3} \frac{\partial_{\eta_i} e^{is\psi_2(\xi,\eta)}}{2is} \varphi(\xi/s^{1/80}) |\xi| \xi_{i'} \widehat{f_{j,\text{out}}}(\xi-\eta) \overline{\widehat{f_{j',\text{out}}}}(\eta) \, d\eta \\
&= -\frac{1}{2is} \int_{\mathbb{R}^3} e^{is\psi_2(\xi,\eta)} \varphi(\xi/s^{1/80}) |\xi| \xi_{i'} \partial_{\eta_i} \widehat{f_{j,\text{out}}}(\xi-\eta) \overline{\widehat{f_{j',\text{out}}}}(\eta) \, d\eta \\
&\quad - \frac{1}{2is} \int_{\mathbb{R}^3} e^{is\psi_2(\xi,\eta)} \varphi(\xi/s^{1/80}) |\xi| \xi_{i'} \widehat{f_{j,\text{out}}}(\xi-\eta) \partial_{\eta_i} \overline{\widehat{f_{j',\text{out}}}}(\eta) \, d\eta,
\end{aligned}$$

which implies

$$I_1 = (2\pi)^{3/2}(2is)^{-1} P_{\leqslant s^{1/80}} |\nabla| \partial_{x_{i'}} [e^{is\Delta}(x_i f_{j,\text{out}}) \cdot e^{-is\Delta} \overline{f_{j',\text{out}}}]$$

$$+ (2\pi)^{3/2}(2\mathrm{i}s)^{-1}P_{\leqslant s^{1/80}}|\nabla|\partial_{x_{i'}}[\mathrm{e}^{\mathrm{i}s\Delta}f_{j,\mathrm{out}}\cdot\mathrm{e}^{-\mathrm{i}s\Delta}(x_i\overline{f_{j',\mathrm{out}}})].$$

It follows from the support property of f_{out} (see (4.13)) that

$$\|f_{\mathrm{out}}\|_{L^2} \lesssim s^{-1/4}\|xf\|_{L^2} \lesssim s^{-1/4+\delta}A_T,$$

hence, combining with Bernstein's inequality and the Cauchy-Schwarz inequality, we can obtain

$$\begin{aligned}\|I_1\|_{L^1} &\lesssim s^{-1}s^{2/80}\|\mathrm{e}^{\mathrm{i}s\Delta}(x_if_{j,\mathrm{out}})\|_{L^2}\|\mathrm{e}^{-\mathrm{i}s\Delta}\overline{f_{j',\mathrm{out}}}\|_{L^2}\\ &\quad + s^{-1}s^{2/80}\|\mathrm{e}^{\mathrm{i}s\Delta}f_{j,\mathrm{out}}\|_{L^2}\|\mathrm{e}^{-\mathrm{i}s\Delta}\overline{x_if_{j',\mathrm{out}}}\|_{L^2}\\ &\lesssim s^{-1}s^{1/40}\|f_{\mathrm{out}}\|_{L^2}\|xf_{\mathrm{out}}\|_{L^2}\\ &\lesssim s^{-1}s^{1/40}s^{-1/4+\delta}s^{\delta}A_T^2\\ &\lesssim s^{-7/6}A_T^2.\end{aligned} \quad (4.18)$$

To estimate I_2, note that
$$\nabla_\xi \psi_2 = 2(\eta - \xi),$$
so we further make the high frequency cutoff with respect to $\xi-\eta$ in $\widehat{I_2}$ and obtain

$$I_2 = I_{2,\mathrm{low}} + I_{2,\mathrm{high}}$$

with

$$\widehat{I_{2,\mathrm{low}}} := \int_{\mathbf{R}^3} \mathrm{e}^{\mathrm{i}s\psi_2(\xi,\eta)}\varphi(\xi/s^{1/80})\varphi((\xi-\eta)/s^{1/80})|\xi|\xi_i\xi_{i'}\widehat{f_{j,\mathrm{in}}}(\xi-\eta)\overline{\widehat{f_{j',\mathrm{out}}}}(\eta)d\eta,$$

$$\widehat{I_{2,\mathrm{high}}} := \int_{\mathbf{R}^3} \mathrm{e}^{\mathrm{i}s\psi_2(\xi,\eta)}\varphi(\xi/s^{1/80})(1-\varphi((\xi-\eta)/s^{1/80}))|\xi|\xi_i\xi_{i'}\widehat{f_{j,\mathrm{in}}}(\xi-\eta)\overline{\widehat{f_{j',\mathrm{out}}}}(\eta)d\eta.$$

For the high frequency part, it is easy to see

$$\begin{aligned}\|I_{2,\mathrm{high}}\|_{L^1} &\lesssim s^{3/80}\|P_{>s^{1/80}}\mathrm{e}^{\mathrm{i}s\Delta}f_{j,\mathrm{in}}\|_{L^2}\|\mathrm{e}^{-\mathrm{i}s\Delta}\overline{f_{j',\mathrm{out}}}\|_{L^2}\\ &\lesssim s^{-(N-3)/80}\|f_{\mathrm{in}}\|_{H^N}\|f\|_{L^2}\\ &\lesssim s^{-(N-3)/80+\delta}A_T^2\\ &\lesssim s^{-7/6}A_T^2.\end{aligned} \quad (4.19)$$

While for the low frequency part, we use (2.11) and Plancherel's identity to obtain

$$\|I_{2,\mathrm{low}}\|_{L^1} \lesssim \|xI_{2,\mathrm{low}}\|_{L^2}^{1/2}\||x|^2 I_{2,\mathrm{low}}\|_{L^2}^{1/2} \sim \|\nabla_\xi \widehat{I_{2,\mathrm{low}}}\|_{L^2}^{1/2}\|\Delta_\xi \widehat{I_{2,\mathrm{low}}}\|_{L^2}^{1/2}. \quad (4.20)$$

Applying ∂_{ξ_k} ($k \in \{1,2,3\}$) to $\widehat{I_{2,\text{low}}}$ gives

$$\partial_{\xi_k}\widehat{I_{2,\text{low}}}$$
$$= \int_{\mathbf{R}^3} e^{is\psi_2}\varphi(\xi/s^{1/80})\varphi((\xi-\eta)/s^{1/80})|\xi|\xi_i\xi_{i'}\partial_{\xi_k}\widehat{f_{j,\text{in}}}(\xi-\eta)\overline{\widehat{f_{j',\text{out}}}(\eta)}\,d\eta \quad (4.21)$$
$$+ \int_{\mathbf{R}^3} e^{is\psi_2}\partial_{\xi_k}[\varphi(\xi/s^{1/80})\varphi((\xi-\eta)/s^{1/80})|\xi|\xi_i\xi_{i'}]\widehat{f_{j,\text{in}}}(\xi-\eta)\overline{\widehat{f_{j',\text{out}}}(\eta)}\,d\eta \quad (4.22)$$
$$+ \int_{\mathbf{R}^3} is\partial_{\xi_k}\psi_2 e^{is\psi_2}\varphi(\xi/s^{1/80})\varphi((\xi-\eta)/s^{1/80})|\xi|\xi_i\xi_{i'}\widehat{f_{j,\text{in}}}(\xi-\eta)\overline{\widehat{f_{j',\text{out}}}(\eta)}\,d\eta, \quad (4.23)$$

where $\partial_{\xi_k}\psi_2 = 2(\eta_k - \xi_k)$.

Estimate for (4.21) By (4.16), we integrate by parts in η_i and obtain the following contributions for the term (4.21)

$$-\frac{1}{2is}\int_{\mathbf{R}^3} e^{is\psi_2}\varphi(\xi/s^{1/80})\partial_{\eta_i}[\varphi((\xi-\eta)/s^{1/80})]|\xi|\xi_{i'}\partial_{\xi_k}\widehat{f_{j,\text{in}}}(\xi-\eta)\overline{\widehat{f_{j',\text{out}}}(\eta)}\,d\eta \quad (4.24)$$
$$-\frac{1}{2is}\int_{\mathbf{R}^3} e^{is\psi_2}\varphi(\xi/s^{1/80})\varphi((\xi-\eta)/s^{1/80})|\xi|\xi_{i'}\partial_{\eta_i\xi_k}\widehat{f_{j,\text{in}}}(\xi-\eta)\overline{\widehat{f_{j',\text{out}}}(\eta)}\,d\eta \quad (4.25)$$
$$-\frac{1}{2is}\int_{\mathbf{R}^3} e^{is\psi_2}\varphi(\xi/s^{1/80})\varphi((\xi-\eta)/s^{1/80})|\xi|\xi_{i'}\partial_{\xi_k}\widehat{f_{j,\text{in}}}(\xi-\eta)\partial_{\eta_i}\overline{\widehat{f_{j',\text{out}}}(\eta)}\,d\eta. \quad (4.26)$$

For the term (4.24), note that

$$\partial_{\eta_i}[\varphi((\xi-\eta)/s^{1/80})] = -s^{-1/80}\partial_{\zeta_i}\varphi(\zeta)|_{\zeta=(\xi-\eta)/s^{1/80}}$$
$$=: -s^{-1/80}\chi(\zeta)|_{\zeta=(\xi-\eta)/s^{1/80}},$$

where $\chi(\zeta)$ is supported in the annulus $\{\zeta; 1 \leqslant |\zeta| \leqslant 2\}$, so when returning to the physical space, one has

$$\mathcal{F}^{-1}(4.24) \sim s^{-1}s^{-1/80}P_{\leqslant s^{1/80}}|\nabla|\partial_{x_{i'}}[\chi(|\nabla|/s^{1/80})e^{is\Delta}(x_k f_{j,\text{in}})\cdot e^{-is\Delta}\overline{f_{j',\text{out}}}].$$

We then use Plancherel's identity, Bernsetin's inequality and Hölder's inequality to obtain

$$\|(4.24)\|_{L^2} \lesssim s^{-1}s^{-1/80}s^{2/80}\|e^{is\Delta}(x_k f_{j,\text{in}})\|_{L^2}\|e^{is\Delta}f_{j',\text{out}}\|_{L^\infty}$$
$$\lesssim s^{-1}s^{1/80}s^\delta s^{-7/6+\delta}A_T^2.$$

Similarly, we can estimate the terms (4.25) and (4.26) as

$$\|(4.25)\|_{L^2} \sim s^{-1}\|P_{\leqslant s^{1/80}}|\nabla|\partial_{x_{i'}}[P_{\leqslant s^{1/80}}e^{is\Delta}(x_k x_i f_{j,\text{in}})\cdot e^{-is\Delta}\overline{f_{j',\text{out}}}]\|_{L^2}$$

$$\lesssim s^{-1}s^{2/80}\|x_kx_if_{j,\mathrm{in}}\|_{L^2}\|e^{is\Delta}f_{j',\mathrm{out}}\|_{L^\infty}$$
$$\lesssim s^{-1}s^{1/40}s^{1/4+\delta}s^{-7/6+\delta}A_T^2,$$
$$\|(4.26)\|_{L^2} \sim s^{-1}\|P_{\leqslant s^{1/80}}|\nabla|\partial_{x_{i'}}[P_{\leqslant s^{1/80}}e^{is\Delta}(x_kf_{j,\mathrm{in}})\cdot e^{-is\Delta}\overline{(x_if_{j',\mathrm{out}})}]\|_{L^2}$$
$$\lesssim s^{-1}s^{2/80}\|e^{is\Delta}(x_kf_{j,\mathrm{in}})\|_{L^\infty}\|e^{is\Delta}(x_if_{j',\mathrm{out}})\|_{L^2}$$
$$\lesssim s^{-1}s^{1/40}s^{-9/8+\delta}s^\delta A_T^2,$$

where in the above estimates, we have used

$$\|x_kx_if_{j,\mathrm{in}}\|_{L^2} \lesssim s^{1/4}\|xf\|_{L^2} \lesssim s^{1/4+\delta}A_T,$$
$$\|e^{is\Delta}(x_kf_{j,\mathrm{in}})\|_{L^\infty} \lesssim s^{-3/2}\|x_kf_{j,\mathrm{in}}\|_{L^1} \lesssim s^{-3/2}s^{3/8}\|xf\|_{L^2} \lesssim s^{-9/8+\delta}A_T.$$

Estimate for (4.22) A direct computation gives

$$(4.22) = \int_{\mathbf{R}^3} e^{is\psi_2}(m_1(\xi,\eta)+\cdots+m_5(\xi,\eta))\widehat{f_{j,\mathrm{in}}}(\xi-\eta)\overline{\widehat{f_{j',\mathrm{out}}}(\eta)}\,d\eta,$$

where

$$m_1(\xi,\eta) := s^{-1/80}\partial_{\zeta_k}\varphi(\zeta)|_{\zeta=\xi/s^{1/80}}\cdot\varphi((\xi-\eta)/s^{1/80})|\xi|\xi_i\xi_{i'},$$
$$m_2(\xi,\eta) := \varphi(\xi/s^{1/80})\cdot s^{-1/80}\partial_{\zeta_k}\varphi(\zeta)|_{\zeta=(\xi-\eta)/s^{1/80}}\cdot|\xi|\xi_i\xi_{i'},$$
$$m_3(\xi,\eta) := \varphi(\xi/s^{1/80})\varphi((\xi-\eta)/s^{1/80})\frac{\xi_k}{|\xi|}\xi_i\xi_{i'},$$
$$m_4(\xi,\eta) := \varphi(\xi/s^{1/80})\varphi((\xi-\eta)/s^{1/80})|\xi|\delta_{ik}\xi_{i'},$$
$$m_5(\xi,\eta) := \varphi(\xi/s^{1/80})\varphi((\xi-\eta)/s^{1/80})|\xi|\xi_i\delta_{i'k}.$$

As the arguments for these five terms are similar, here we only take m_3 for an example. Again using (4.16), after integration by parts in η_i, there holds

$$\int_{\mathbf{R}^3} e^{is\psi_2}m_3(\xi,\eta)\widehat{f_{j,\mathrm{in}}}(\xi-\eta)\overline{\widehat{f_{j',\mathrm{out}}}(\eta)})\,d\eta$$
$$= -\frac{1}{2is}\int_{\mathbf{R}^3} e^{is\psi_2}\varphi(\xi/s^{1/80})\varphi((\xi-\eta)/s^{1/80})\frac{\xi_k}{|\xi|}\xi_{i'}\partial_{\eta_i}\widehat{f_{j,\mathrm{in}}}(\xi-\eta)\overline{\widehat{f_{j',\mathrm{out}}}(\eta)}\,d\eta \quad (4.27)$$
$$-\frac{1}{2is}\int_{\mathbf{R}^3} e^{is\psi_2}\varphi(\xi/s^{1/80})\varphi((\xi-\eta)/s^{1/80})\frac{\xi_k}{|\xi|}\xi_{i'}\widehat{f_{j,\mathrm{in}}}(\xi-\eta)\partial_{\eta_i}\overline{\widehat{f_{j',\mathrm{out}}}(\eta)}\,d\eta \quad (4.28)$$
$$-\frac{1}{2is}\int_{\mathbf{R}^3} e^{is\psi_2}\varphi(\xi/s^{1/80})\partial_{\eta_i}[\varphi((\xi-\eta)/s^{1/80})]\frac{\xi_k}{|\xi|}\xi_{i'}\widehat{f_{j,\mathrm{in}}}(\xi-\eta)\overline{\widehat{f_{j',\mathrm{out}}}(\eta)}\,d\eta, \tag{4.29}$$

from which we can see

$$\mathcal{F}^{-1}(4.27) \sim s^{-1}P_{\leqslant s^{1/80}}R_k\partial_{x_{i'}}[P_{\leqslant s^{1/80}}e^{is\Delta}(x_if_{j,\mathrm{in}})\cdot e^{-is\Delta}\overline{f_{j',\mathrm{out}}}],$$

$$\mathcal{F}^{-1}(4.28) \sim s^{-1} P_{\leqslant s^{1/80}} R_k \partial_{x_{i'}} [P_{\leqslant s^{1/80}} e^{is\Delta} f_{j,\text{in}} \cdot e^{-is\Delta} (x_i \overline{f_{j',\text{out}}})],$$

$$\mathcal{F}^{-1}(4.29) \sim s^{-1} s^{-1/80} P_{\leqslant s^{1/80}} R_k \partial_{x_{i'}} [\chi(|\nabla|/s^{1/80}) e^{is\Delta} f_{j,\text{in}} \cdot e^{-is\Delta} \overline{f_{j',\text{out}}}],$$

where $\chi(\xi) := \partial_{\xi_i} \varphi(\xi)$, and $R_k = \partial x_k / |\nabla|$ denotes the Riesz transform. Then by the L^2 boundedness of the Riesz transform, Bernsetin's inequality and Hölder's inequality, we have

$$\|(4.27)\|_{L^2} \lesssim s^{-1} s^{1/80} \|e^{is\Delta}(x_i f_{j,\text{in}})\|_{L^2} \|e^{is\Delta} f_{j',\text{out}}\|_{L^\infty} \lesssim s^{-1} s^{1/80} s^\delta s^{-7/6+\delta} A_T^2,$$

$$\|(4.28)\|_{L^2} \lesssim s^{-1} s^{1/80} \|e^{is\Delta} f_{j,\text{in}}\|_{L^\infty} \|e^{is\Delta}(x_i f_{j',\text{out}})\|_{L^2} \lesssim s^{-1} s^{1/80} s^{-7/6+\delta} s^\delta A_T^2,$$

$$\|(4.29)\|_{L^2} \lesssim s^{-1} s^{-1/80} s^{1/80} \|e^{is\Delta} f_{j,\text{in}}\|_{L^2} \|e^{is\Delta} f_{j',\text{out}}\|_{L^\infty} \lesssim s^{-1} s^{-7/6+\delta} A_T^2.$$

Estimate for (4.23) Note that the term (4.23) contains a factor s, we then use (4.16) to integrate by parts in η_i and $\eta_{i'}$ and obtain

$$(4.23) = \frac{1}{2is} \int_{\mathbf{R}^3} e^{is\psi_2(\xi,\eta)} m_6(\xi,\eta) \widehat{f_{j,\text{in}}}(\xi-\eta) \widehat{\overline{f_{j',\text{out}}}}(\eta) \, d\eta \tag{4.30}$$

$$+ \frac{1}{2is} \int_{\mathbf{R}^3} e^{is\psi_2(\xi,\eta)} m_7(\xi,\eta) \partial_{\eta_{i'}} \widehat{f_{j,\text{in}}}(\xi-\eta) \widehat{\overline{f_{j',\text{out}}}}(\eta) \, d\eta \tag{4.31}$$

$$+ \frac{1}{2is} \int_{\mathbf{R}^3} e^{is\psi_2(\xi,\eta)} m_7(\xi,\eta) \widehat{f_{j,\text{in}}}(\xi-\eta) \partial_{\eta_{i'}} \widehat{\overline{f_{j',\text{out}}}}(\eta) \, d\eta \tag{4.32}$$

$$+ \frac{1}{2is} \int_{\mathbf{R}^3} e^{is\psi_2(\xi,\eta)} m_8(\xi,\eta) \partial_{\eta_i} \widehat{f_{j,\text{in}}}(\xi-\eta) \widehat{\overline{f_{j',\text{out}}}}(\eta) \, d\eta \tag{4.33}$$

$$+ \frac{1}{2is} \int_{\mathbf{R}^3} e^{is\psi_2(\xi,\eta)} m_8(\xi,\eta) \widehat{f_{j,\text{in}}}(\xi-\eta) \partial_{\eta_i} \widehat{\overline{f_{j',\text{out}}}}(\eta) \, d\eta \tag{4.34}$$

$$+ \frac{1}{2is} \int_{\mathbf{R}^3} e^{is\psi_2(\xi,\eta)} m_9(\xi,\eta) \partial_{\eta_i} \widehat{f_{j,\text{in}}}(\xi-\eta) \partial_{\eta_{i'}} \widehat{\overline{f_{j',\text{out}}}}(\eta) \, d\eta \tag{4.35}$$

$$+ \frac{1}{2is} \int_{\mathbf{R}^3} e^{is\psi_2(\xi,\eta)} m_9(\xi,\eta) \partial_{\eta_{i'}} \widehat{f_{j,\text{in}}}(\xi-\eta) \partial_{\eta_i} \widehat{\overline{f_{j',\text{out}}}}(\eta) \, d\eta \tag{4.36}$$

$$+ \frac{1}{2is} \int_{\mathbf{R}^3} e^{is\psi_2(\xi,\eta)} m_9(\xi,\eta) \partial_{\eta_i \eta_{i'}} \widehat{f_{j,\text{in}}}(\xi-\eta) \widehat{\overline{f_{j',\text{out}}}}(\eta) \, d\eta \tag{4.37}$$

$$+ \frac{1}{2is} \int_{\mathbf{R}^3} e^{is\psi_2(\xi,\eta)} m_9(\xi,\eta) \widehat{f_{j,\text{in}}}(\xi-\eta) \partial_{\eta_i \eta_{i'}} \widehat{\overline{f_{j',\text{out}}}}(\eta) \, d\eta, \tag{4.38}$$

where

$$m_6(\xi,\eta) := |\xi| \varphi(\xi/s^{1/80}) \partial_{\eta_i \eta_{i'}} [\varphi((\xi-\eta)/s^{1/80})(\eta_k - \xi_k)],$$

$$m_7(\xi,\eta) := |\xi| \varphi(\xi/s^{1/80}) \partial_{\eta_i} [\varphi((\xi-\eta)/s^{1/80})(\eta_k - \xi_k)],$$

$$m_8(\xi,\eta) := |\xi| \varphi(\xi/s^{1/80}) \partial_{\eta_{i'}} [\varphi((\xi-\eta)/s^{1/80})(\eta_k - \xi_k)],$$

$$m_9(\xi,\eta) := |\xi| \varphi(\xi/s^{1/80}) \varphi((\xi-\eta)/s^{1/80})(\eta_k - \xi_k).$$

The estimates for (4.30), (4.31)–(4.34), (4.35)–(4.36) and (4.37) can be treated in a similar way as (4.29), (4.27)–(4.28), (4.26) and (4.25), respectively, so we omit the details. For the last term, we use Hölder's inequality to get

$$\|(4.38)\|_{L^2} \sim s^{-1}\|P_{\leqslant s^{1/80}}|\nabla|[P_{\leqslant s^{1/80}}\partial_{x_k} e^{is\Delta} f_{j,\text{in}} \cdot e^{-is\Delta}(x_i x_{i'} \overline{f_{j',\text{out}}})]\|_{L^2}$$
$$\lesssim s^{-1} s^{2/80} \|e^{is\Delta} f_{\text{in}}\|_{L^\infty} \|e^{is\Delta}|x|^2 f_{\text{out}}\|_{L^2}$$
$$\lesssim s^{-1} s^{1/40} s^{-3/2} \|f_{\text{in}}\|_{L^1} s^{2/3+\delta} A_T$$
$$\lesssim s^{-1} s^{1/40} s^{-3/2} s^{1/8+\delta} s^{2/3+\delta} A_T^2$$
$$= s^{-41/24} s^{1/40+2\delta} A_T^2,$$

where we have used

$$\|f_{\text{in}}\|_{L^1} \lesssim \|xf\|_{L^2} \||x|^{-1}\|_{L^2(|x|\lesssim s^{1/4})} \lesssim s^{1/8+\delta} A_T.$$

Combining the estimates for (4.24)–(4.38) gives (note that the worst bound comes from the term (4.38))

$$\|xI_{2,\text{low}}\|_{L^2} = \|\nabla_\xi \widehat{I_{2,\text{low}}}\|_{L^2} \lesssim s^{-41/24} s^{1/40+2\delta} A_T^2. \tag{4.39}$$

To estimate the L^2 norm of $|x|^2 I_{2,\text{low}}$, we further apply ∂_{ξ_k} to the terms (4.24)–(4.38). Observe that when ∂_{ξ_k} hits the function $\widehat{f_{\text{in}}}$, it produces a growth factor $s^{1/4}$, while ∂_{ξ_k} hits the phase ψ_2, the extra growth factor is $s^{1+1/80}$, and the bounds for (4.24)–(4.38) still hold when ∂_{ξ_k} hits the Fourier symbols such as $m_1, m_2, \cdots$. Hence, we conclude that

$$\||x|^2 I_{2,\text{low}}\|_{L^2} \lesssim s^{-17/24} s^{3/80+2\delta} A_T^2. \tag{4.40}$$

It follows from (4.20), (4.39) and (4.40) that

$$\|I_{2,\text{low}}\|_{L^1} \lesssim \|xI_{2,\text{low}}\|_{L^2}^{1/2} \||x|^2 I_{2,\text{low}}\|_{L^2}^{1/2} \lesssim s^{-29/24} s^{5/160+2\delta} A_T^2,$$

which, combined with (4.19), yields

$$\|I_2\|_{L^1} \lesssim (s^{-7/6} + s^{-29/24} s^{5/160+2\delta}) A_T^2 \lesssim s^{-7/6} A_T^2. \tag{4.41}$$

It is worth mentioning that the support property of the function f_{in} plays an important role in the above estimate for I_2. So for the term I_3 (see (4.15)), we make the change of variables $\xi - \eta \to \eta$ and $\eta \to \xi - \eta$, then

$$\widehat{I_3} = \int_{\mathbf{R}^3} e^{-is\psi_2(\xi,\eta)} \varphi(\xi/s^{1/80}) |\xi| \xi_i \xi_{i'} \widehat{f_{j,\text{out}}}(\eta) \widehat{f_{j',\text{in}}}(\xi - \eta) \, d\eta.$$

That is, we put the input frequency $\xi - \eta$ into the function $\overline{f_{\text{in}}}$. Hence, by symmetry, we see that the term I_3 can be estimated similarly as I_2. For the term I_4, as both f and $\overline{f}$ have localized support, so we can also apply analogous treatment as I_2 to estimate I_4. Since the arguments for I_3 and I_4 are the same as above, we omit further details.

Therefore, the bound (4.10) follows from (4.14), (4.18) and (4.41). This ends the proof of Proposition 4.1.

In the following proposition, we deal with the decay estimates for $\|\nabla B\|_{L^2}$ and $\|\Delta B\|_{L^2}$, which will be used in the next section.

Proposition 4.2 *Under the same assumptions as Proposition 4.1, there hold that*

$$(1+t)^{5/6}\|\nabla B(t)\|_{L^2} \lesssim \|B_0\|_{L^1 \cap \dot{H}^1} + A_T^2, \qquad (4.42)$$

$$(1+t)^{7/6}\|\Delta B(t)\|_{L^2} \lesssim \|B_0\|_{L^1 \cap \dot{H}^2} + A_T^2 \qquad (4.43)$$

for all $t \in [0, T)$, where the implicit constant is independent of T.

Proof We only prove (4.42), since the proof for (4.43) is similar. From (2.6), we have

$$\nabla B(t) = e^{-t|\nabla|^3}\nabla B_0 - i \int_0^t e^{-(t-s)|\nabla|^3}\nabla|\nabla|\nabla \times (\nabla \times (E \times \overline{E}))(s)\,ds.$$

As $1 + t \sim 1$ for $t \lesssim 1$, the bound (4.42) clearly holds if $0 \leq t \leq 4$. Assume now $t > 4$, then we split ∇B into

$$\nabla B(t) = e^{-t|\nabla|^3}\nabla B_0 + J_1(t) + J_2(t) + J_3(t) \qquad (4.44)$$

with

$$J_1(t) := -i \int_0^1 e^{-(t-s)|\nabla|^3}\nabla|\nabla|\nabla \times (\nabla \times (E \times \overline{E}))(s)\,ds,$$

$$J_2(t) := -i \int_1^{t-1} e^{-(t-s)|\nabla|^3}\nabla|\nabla|\nabla \times (\nabla \times (E \times \overline{E}))(s)\,ds,$$

$$J_3(t) := -i \int_{t-1}^t e^{-(t-s)|\nabla|^3}\nabla|\nabla|\nabla \times (\nabla \times (E \times \overline{E}))(s)\,ds.$$

According to (2.16), we see

$$\|e^{-t|\nabla|^3}\nabla B_0\|_{L^2} \lesssim \frac{1}{t^{5/6}}\|B_0\|_{L^1 \cap \dot{H}^1}, \qquad (4.45)$$

$$\|J_1(t)\|_{L^2} \lesssim \int_0^1 \frac{1}{(t-s)^{5/6}}\||\nabla|\nabla \times (\nabla \times (E \times \overline{E}))(s)\|_{L^1 \cap \dot{H}^1}\,ds$$

$$\lesssim \int_0^1 \frac{1}{(t-s)^{5/6}} \|E(s)\|_{H^4}^2 \, ds$$

$$\lesssim \frac{1}{(t-1)^{5/6}} A_T^2$$

$$\sim \frac{1}{t^{5/6}} A_T^2, \tag{4.46}$$

and

$$\|J_2(t)\|_{L^2} \lesssim \int_1^{t-1} \frac{1}{(t-s)^{5/6}} \||\nabla|\nabla \times (\nabla \times (E \times \overline{E}))(s)\|_{L^1} \, ds \tag{4.47}$$

$$+ \int_1^{t-1} \frac{1}{(t-s)^{5/6}} \|P_{>s^{1/80}}|\nabla|\nabla \times (\nabla \times (E \times \overline{E}))(s)\|_{\dot{H}^1} \, ds \tag{4.48}$$

$$+ \int_1^{t-1} \frac{1}{(t-s)^{5/6}} \|P_{\leqslant s^{1/80}}|\nabla|\nabla \times (\nabla \times (E \times \overline{E}))(s)\|_{\dot{H}^1} \, ds. \tag{4.49}$$

By (4.6), we estimate the term (4.47) as

$$|(4.47)| \lesssim A_T^2 \int_1^{t-1} \frac{1}{(t-s)^{5/6}} \frac{1}{s^{7/6}} \, ds \lesssim \frac{1}{t^{5/6}} A_T^2.$$

The term (4.48) is estimated by the Bernstein inequality and (3.1),

$$|(4.48)| \lesssim \int_1^{t-1} \frac{1}{(t-s)^{5/6}} \frac{1}{s^{(N-4)/80}} \|E \times \overline{E}\|_{\dot{H}^N} \, ds$$

$$\lesssim \int_1^{t-1} \frac{1}{(t-s)^{5/6}} \frac{1}{s^{(N-4)/80}} \|E\|_{H^N} \|E\|_{L^\infty} \, ds$$

$$\lesssim A_T^2 \int_1^{t-1} \frac{1}{(t-s)^{5/6}} \frac{1}{s^{(N-4)/80}} s^\delta \frac{1}{s^{7/6-\delta}} \, ds$$

$$\lesssim \frac{1}{t^{5/6}} A_T^2.$$

For (4.49), we use Hölder's inequality and (2.14) to obtain

$$|(4.49)| \lesssim A_T^2 \int_1^{t-1} \frac{1}{(t-s)^{5/6}} s^{4/80} \|E\|_{L^4}^2 \, ds$$

$$\lesssim A_T^2 \int_1^{t-1} \frac{1}{(t-s)^{5/6}} s^{1/20} \frac{1}{s^{3/2-2\delta}} \, ds$$

$$\lesssim \frac{1}{t^{5/6}} A_T^2.$$

Hence, it follows from the above three estimates that

$$\|J_2(t)\|_{L^2} \lesssim t^{-5/6} A_T^2. \tag{4.50}$$

It remains to estimate the term J_3. Indeed, by Bernsten's inequality and (2.13)–(2.14), we have

$$\|J_3\|_{L^2} \lesssim \|P_{>1/80} J_3\|_{L^2} + \|P_{\leqslant 1/80} J_3\|_{L^2}$$
$$\lesssim \int_{t-1}^{t} s^{-(N-4)/80} \|E \times \overline{E}\|_{H^N} \, ds + \int_{t-1}^{t} s^{4/80} \|E \times \overline{E}\|_{L^2} \, ds$$
$$\lesssim \int_{t-1}^{t} s^{-(N-4)/80} \|E\|_{H^N} \|E\|_{L^\infty} \, ds + \int_{t-1}^{t} s^{4/80} \|E\|_{L^4}^2 \, ds$$
$$\lesssim A_T^2 \int_{t-1}^{t} \frac{1}{s^{(N-4)/80}} s^\delta \frac{1}{s^{7/6-\delta}} \, ds + A_T^2 \int_{t-1}^{t} s^{1/20} \frac{1}{s^{3/2-2\delta}} \, ds$$
$$\lesssim t^{-5/6} A_T^2. \tag{4.51}$$

Therefore, combing (4.44), (4.45), (4.46), (4.50) and (4.51) gives (4.42) as desired.

Remark In view of the linear decay estimates (see Lemma 2.2), we see the decay rates obtained in (4.1), (4.42) and (4.43) are optimal. As shown in the next section, these sharp decay estimates are crucial in proving the weighted bounds for f. Due to the restriction of the nonlinear term $|\nabla|\nabla \times \nabla \times (E \times \overline{E})$, we can not obtain the optimal decay estimate for $\|B\|_{\dot{H}^k}$ with $k \geqslant 3$. However, in our analysis, Propositions 4.1 and 4.2 are sufficient to close the argument.

5 Weighted estimates for the electric field

In this section, we prove the weighted estimates for the profile of E. For simplicity, we rewrite (2.3) as

$$f(t) = f(0) + \frac{i}{(2\pi)^3} F_1(t) - \frac{1}{(2\pi)^{3/2}} F_2(t), \tag{5.1}$$

where

$$\widehat{F_1}(t,\xi) := \int_0^t \int_{\mathbf{R}^3 \times \mathbf{R}^3} e^{is\phi(\xi,\eta,\sigma)} \widehat{f}(\xi-\eta)(\widehat{f}(\eta-\sigma) \cdot \overline{\widehat{f}}(\sigma)) \, d\eta \, d\sigma \, ds,$$
$$\widehat{F_2}(t,\xi) := \int_0^t \int_{\mathbf{R}^3} e^{is\psi_1(\xi,\eta)} \widehat{f}(\xi-\eta) \times \widehat{B}(\eta) \, d\eta \, ds. \tag{5.2}$$

Proposition 5.1 *Let (E,B) be the solution of system (2.1) on $[0,T) \times \mathbf{R}^3$ with $T > 0$. If $A_T \ll 1$ and $\|B_0\|_{L^1} \ll 1$, then*

$$\sup_{t \in [0,T)} ((1+t)^{-\delta} \|xf(t)\|_{L^2} + (1+t)^{-2/3-\delta} \||x|^2 f(t)\|_{L^2})$$
$$\lesssim \|xE_0\|_{L^2} + \||x|^2 E_0\|_{L^2} + \|B_0\|_{L^1 \cap H^2} + A_T^2. \tag{5.3}$$

Proof Note that $f(0) = E_0$, so

$$\|xf(0)\|_{L^2} = \|xE_0\|_{L^2}, \quad \||x|^2 f(0)\|_{L^2} = \||x|^2 E_0\|_{L^2}. \tag{5.4}$$

According to (5.1), it remains to estimate the weighted L^2 norm of the quantities F_1, and F_2.

We first estimate the cubic term F_1. Indeed, for this term we have stronger bound

$$\sup_{t\in[0,T]} \left(\|xF_1(t)\|_{L^2} + \||x|^2 F_1(t)\|_{L^2} \right) \lesssim A_T^3. \tag{5.5}$$

Note that the case $0 \leqslant t \leqslant 1$ or the contribution of the time integral from 0 to 1 can be easily treated. So in order to prove (5.5), we assume $t > 1$ and only consider the contribution of the time integral from 1 to t. Moreover, by taking the components into account, it is sufficient to consider the quantity

$$\widehat{G_{ij}}(t,\xi) := \int_1^t \int_{\mathbf{R}^3 \times \mathbf{R}^3} e^{is\phi(\xi,\eta,\sigma)} \widehat{f_i}(\xi - \eta) \widehat{f_j}(\eta - \sigma) \overline{\widehat{f_j}}(\sigma) \, d\eta \, d\sigma \, ds$$

for any fixed $i, j = 1, 2, 3$. Recall first the phase ϕ (see (2.4)) satisfies

$$\nabla_\xi \phi = 2\eta = \nabla_\sigma \phi. \tag{5.6}$$

Taking ∇_ξ to $\widehat{G_{ij}}$ and using the above relation, we obtain

$$\nabla_\xi \widehat{G_{ij}} = \int_1^t \int_{\mathbf{R}^3 \times \mathbf{R}^3} e^{is\phi} \nabla_\xi \widehat{f_i}(s, \xi - \eta) \widehat{f_j}(s, \eta - \sigma) \overline{\widehat{f_j}}(s, \sigma) \, d\eta \, d\sigma \, ds$$

$$+ \int_1^t \int_{\mathbf{R}^3 \times \mathbf{R}^3} is \nabla_\xi \phi e^{is\phi} \widehat{f_i}(s, \xi - \eta) \widehat{f_j}(s, \eta - \sigma) \overline{\widehat{f_j}}(s, \sigma) \, d\eta \, d\sigma \, ds$$

$$= \int_1^t \int_{\mathbf{R}^3 \times \mathbf{R}^3} e^{is\phi} \nabla_\xi \widehat{f_i}(s, \xi - \eta) \widehat{f_j}(s, \eta - \sigma) \overline{\widehat{f_j}}(s, \sigma) \, d\eta \, d\sigma \, ds \tag{5.7}$$

$$- \int_1^t \int_{\mathbf{R}^3 \times \mathbf{R}^3} e^{is\phi} \widehat{f_i}(s, \xi - \eta) \nabla_\sigma \widehat{f_j}(s, \eta - \sigma) \overline{\widehat{f_j}}(s, \sigma) \, d\eta \, d\sigma \, ds \tag{5.8}$$

$$- \int_1^t \int_{\mathbf{R}^3 \times \mathbf{R}^3} e^{is\phi} \widehat{f_i}(s, \xi - \eta) \widehat{f_j}(s, \eta - \sigma) \nabla_\sigma \overline{\widehat{f_j}}(s, \sigma)) \, d\eta \, d\sigma \, ds. \tag{5.9}$$

Hence, using Plancherel's identity and Hölder's inequality, there holds

$$\|\nabla_\xi \widehat{G_{ij}}\|_{L^2} \lesssim \|(5.7)\|_{L^2} + \|(5.8)\|_{L^2} + \|(5.9)\|_{L^2}$$

$$\lesssim \int_1^t \|xf\|_{L^2} \|E\|_{L^\infty}^2 \, ds$$

$$\lesssim A_T^3 \int_1^t (1+s)^\delta \cdot (1+s)^{-7/3+2\delta}\, ds$$
$$\lesssim A_T^3. \tag{5.10}$$

Using again (5.6), we can obtain after a direct computation,

$$\Delta_\xi \widehat{G_{ij}} = \int_1^t \int_{\mathbf{R}^3 \times \mathbf{R}^3} e^{is\phi} \Delta_\xi \widehat{f_i}(s, \xi - \eta) \widehat{f_j}(s, \eta - \sigma) \overline{\widehat{f_j}}(s, \sigma)\, d\eta\, d\sigma\, ds \tag{5.11}$$

$$+ \int_1^t \int_{\mathbf{R}^3 \times \mathbf{R}^3} e^{is\phi} \widehat{f_i}(s, \xi - \eta) \Delta_\sigma \widehat{f_j}(s, \eta - \sigma) \overline{\widehat{f_j}}(s, \sigma)\, d\eta\, d\sigma\, ds \tag{5.12}$$

$$+ \int_1^t \int_{\mathbf{R}^3 \times \mathbf{R}^3} e^{is\phi} \widehat{f_i}(s, \xi - \eta) \widehat{f_j}(s, \eta - \sigma) \Delta_\sigma \overline{\widehat{f_j}}(s, \sigma)\, d\eta\, d\sigma\, ds \tag{5.13}$$

$$+ 2 \int_1^t \int_{\mathbf{R}^3 \times \mathbf{R}^3} e^{is\phi} \widehat{f_i}(s, \xi - \eta) \nabla_\sigma \widehat{f_j}(s, \eta - \sigma) \cdot \nabla_\sigma \overline{\widehat{f_j}}(s, \sigma)\, d\eta\, d\sigma\, ds \tag{5.14}$$

$$- 2 \int_1^t \int_{\mathbf{R}^3 \times \mathbf{R}^3} e^{is\phi} \nabla_\xi \widehat{f_i}(s, \xi - \eta) \cdot \nabla_\sigma \widehat{f_j}(s, \eta - \sigma) \overline{\widehat{f_j}}(s, \sigma)\, d\eta\, d\sigma\, ds \tag{5.15}$$

$$- 2 \int_1^t \int_{\mathbf{R}^3 \times \mathbf{R}^3} e^{is\phi} \nabla_\xi \widehat{f_i}(s, \xi - \eta) \widehat{f_j}(s, \eta - \sigma) \cdot \nabla_\sigma \overline{\widehat{f_j}}(s, \sigma)\, d\eta\, d\sigma\, ds. \tag{5.16}$$

For the first three terms, by Hölder's inequality, we have

$$\|(5.11)\|_{L^2} + \|(5.12)\|_{L^2} + \|(5.13)\|_{L^2} \lesssim \int_1^t \||x|^2 f\|_{L^2} \|E\|_{L^\infty}^2\, ds$$
$$\lesssim A_T^3 \int_1^t (1+s)^{2/3+\delta}(1+s)^{-7/3+2\delta}\, ds$$
$$\lesssim A_T^3. \tag{5.17}$$

For the last three terms, we use the estimate

$$\|e^{is\Delta}(xf)\|_{L^4} \lesssim (1+s)^{-3/4} \|xf\|_{L^2}^{1/4} \||x|^2 f\|_{L^2}^{3/4} \lesssim (1+s)^{-1/4+\delta} A_T \tag{5.18}$$

to obtain

$$\|(5.14)\|_{L^2} + \|(5.15)\|_{L^2} + \|(5.16)\|_{L^2} \lesssim \int_1^t \|E\|_{L^\infty} \|e^{is\Delta}(xf)\|_{L^4}^2\, ds$$
$$\lesssim A_T^3 \int_1^t (1+s)^{-7/6+\delta}(1+s)^{-1/2+2\delta}\, ds$$
$$\lesssim A_T^3. \tag{5.19}$$

Now, combining (5.10), (5.17) and (5.19), we thus get (5.5).

Next, we estimate the quadratic term F_2 defined in (5.2). Again, it is sufficient to consider the contribution of the time integral from 1 to t as the arguments for the other cases are easy. For any fixed $i, j = 1, 2, 3$, we set

$$\widehat{H_{ij}}(t, \xi) := \int_1^t \int_{\mathbf{R}^3} e^{is\psi_1(\xi,\eta)} \widehat{f_i}(\xi - \eta) \widehat{B_j}(\eta) \, d\eta \, ds.$$

Note that the phase given by (2.5) satisfies

$$\nabla_\xi \psi_1 = 2\eta, \quad \Delta_\xi \psi_1 = 0. \tag{5.20}$$

Hence, we compute

$$\nabla_\xi \widehat{H_{ij}} = \int_1^t \int_{\mathbf{R}^3} e^{is\psi_1} \nabla_\xi \widehat{f_i}(\xi - \eta) \widehat{B_j}(\eta) \, d\eta \, ds$$
$$+ 2 \int_1^t \int_{\mathbf{R}^3} s e^{is\psi_1} \widehat{f_i}(\xi - \eta) \widehat{\nabla B_j}(\eta) \, d\eta \, ds,$$

which, by Hölder's inequality, (2.13) and (4.42), gives

$$\|\nabla_\xi \widehat{H_{ij}}\|_{L^2} \lesssim \int_1^t \|e^{is\Delta}(xf_i)\|_{L^2} \|B_j\|_{L^\infty} \, ds + \int_1^t s \|e^{is\Delta} f_i\|_{L^\infty} \|\nabla B_j\|_{L^2} \, ds$$
$$\lesssim A_T^2 \int_1^t \frac{(1+s)^\delta}{1+s} \, ds + \int_1^t s \frac{A_T}{(1+s)^{7/6-\delta}} \cdot \frac{\|B_0\|_{L^1 \cap \dot{H}^1} + A_T^2}{(1+s)^{5/6}} \, ds$$
$$\lesssim (1+t)^\delta A_T^2 + (1+t)^\delta (A_T \|B_0\|_{L^1 \cap \dot{H}^1} + A_T^3)$$
$$\lesssim (1+t)^\delta (A_T^2 + \|B_0\|_{L^1 \cap H^1}), \tag{5.21}$$

where in the last step we have used the small assumptions on A_T and B_0. Moreover, by (5.20), one can get after a straightforward calculation

$$\Delta_\xi \widehat{H_{ij}} = \int_1^t \int_{\mathbf{R}^3} e^{is\psi_1} \Delta_\xi \widehat{f_i}(\xi - \eta) \widehat{B_j}(\eta) \, d\eta \, ds \tag{5.22}$$

$$+ 4 \int_1^t \int_{\mathbf{R}^3} s^2 e^{is\psi_1} \widehat{f_i}(\xi - \eta) \widehat{\Delta B_j}(\eta) \, d\eta \, ds \tag{5.23}$$

$$+ 4 \int_1^t \int_{\mathbf{R}^3} s e^{is\psi_1} \nabla_\xi \widehat{f_i}(\xi - \eta) \cdot \widehat{\nabla B_j}(\eta) \, d\eta \, ds. \tag{5.24}$$

The term (5.22) is estimated by a direct $L^2 \times L^\infty$ estimate

$$\|(5.22)\|_{L^2} \sim \|\int_1^t e^{-is\Delta} [e^{is\Delta}(|x|^2 f_i) B_j] \, ds\|_{L^2}$$
$$\lesssim \int_1^t \|e^{is\Delta}(|x|^2 f_i)\|_{L^2} \|B_j\|_{L^\infty} \, ds$$

$$\lesssim A_T^2 \int_1^t (1+s)^{2/3+\delta} \frac{1}{1+s} \, ds$$
$$\lesssim (1+t)^{2/3+\delta} A_T^2. \tag{5.25}$$

For (5.23), we use (2.13) and (4.43) to obtain

$$\|(5.23)\|_{L^2} \sim \left\| \int_1^t s^2 e^{-is\Delta}[(e^{is\Delta} f_i)(\Delta B_j)] \, ds \right\|_{L^2}$$
$$\lesssim \int_1^t s^2 \|e^{is\Delta} f\|_{L^\infty} \|\Delta B\|_{L^2} \, ds$$
$$\lesssim \int_1^t s^2 \frac{A_T}{(1+s)^{7/6-\delta}} \frac{1}{(1+s)^{7/6}} (\|B_0\|_{L^1 \cap \dot{H}^2} + A_T^2) \, ds$$
$$\lesssim (1+t)^{2/3+\delta}(A_T^2 + \|B_0\|_{L^1 \cap H^2}). \tag{5.26}$$

By (4.42), (4.43) and the interpolation inequality,

$$\|\nabla B(s)\|_{L^4} \lesssim \|\nabla B(s)\|_{L^2}^{1/4} \|\Delta B(s)\|_{L^2}^{3/4}$$
$$\lesssim (1+s)^{-13/12}(\|B_0\|_{L^1 \cap H^2} + A_T^2),$$

we can estimate the term (5.24) as

$$\|(5.24)\|_{L^2} \sim \left\| \int_1^t s e^{-is\Delta}[(e^{is\Delta}(xf_i))(\nabla B_j)] \, ds \right\|_{L^2}$$
$$\lesssim \int_1^t s \|e^{is\Delta}(xf)\|_{L^4} \|\nabla B\|_{L^4} \, ds$$
$$\lesssim A_T^2 \int_1^t s \frac{1}{(1+s)^{1/4-\delta}} A_T \cdot \frac{1}{(1+s)^{13/12}} (\|B_0\|_{L^1 \cap H^2} + A_T^2) \, ds$$
$$\lesssim (1+t)^{2/3+\delta}(A_T^2 + \|B_0\|_{L^1 \cap H^2}), \tag{5.27}$$

where in the second inequality, we have also used the estimate (5.18). Combining the estimates (5.21) and (5.25)–(5.27) gives

$$\sup_{t \in [0,T]} ((1+t)^{-\delta} \|xF_2(t)\|_{L^2} + (1+t)^{-2/3-\delta} \||x|^2 F_2(t)\|_{L^2}) \lesssim A_T^2 + \|B_0\|_{L^1 \cap H^2}. \tag{5.28}$$

Therefore, the desired bound (5.3) follows from (5.4), (5.5) and (5.28).

Finally, we can finish the proof of Theorem 1.1 with the help of the *a priori* estimates obtained in Sections 3–5. In fact, we first note that the local existence of smooth solution for system (1.2) can be established similarly as [22]. Then combining

Proposition 3.1, Proposition 4.1 and Proposition 5.1, and using the conditions (1.7)–(1.8), we thus obtain the bound (2.17) as desired. Hence, by a standard continuation argument, we can get $A_\infty \lesssim \epsilon_0$ provided that the initial data is sufficiently small, which yields the results of Theorem 1.1. For the sake of simplicity, we omit further details.

Acknowledgments The authors would like to thank the referee's valuable comments and suggestions which improved the presentation considerably. Bian D. is supported by NSFC under Grant Nos.11501028 and 11471323, and China Postdoctoral Science Foundation (Grant No.2015M570938). Zhang J. is supported by the NSFC under Grant Nos.11201185 and 11471057.

References

[1] Bahouri H, Chemin J, Danchin R. Fourier analysis and nonlinear partial differential equations [M]. Grundlehren der Mathematischen Wissenschaften, vol. 343, Heidelberg: Springer-Verlag, 2011.

[2] Cazenave T. Semilinear Schrödinger equations [M]. Providence: American Mathematical Society, 2003.

[3] Coifman R, Meyer Y. Nonlinear harmonic analysis operator theory and P.D.E. [M]. In Beijing Lectures in Harmonic Analysis. Princeton University Press, 1986.

[4] Cazenave T, Weissler F. The Cauchy problem for the critical nonlinear Schrödinger equation in H^s [J]. Nonlinear Anal., 1990, 14: 807-836.

[5] Evans L C. Partial differential equations [M]. Graduate Studies in Mathematics, Vol. 19. Providence: American Mathematical Society, 1998.

[6] Gan Z, Guo B, Han L, Zhang J. Virial type blow-up solutions for the Zakharov system with magnetic field in a cold plasma [J]. J. Funct. Anal., 2011, 261: 2508-2528.

[7] Gan Z, Guo B, Huang D. Blow-up and nonlinear instability for the magnetic Zakharov system [J]. J. Funct. Anal., 2013, 265: 953-982.

[8] Gan Z, Ma Y, Zhong T. Some remarks on the blow-up rate for the 3D magnetic Zakharov system [J]. J. Funct. Anal., 2015, 269: 2505-2529.

[9] Gan Z, Zhang J. Nonlocal nonlinear Schrödinger equation in $\mathbf{R}^3$ [J]. Arch. Rational Mech. Anal., 2013, 209: 1-39.

[10] Germain P, Masmoudi N, Shatah J. Global solutions for 3D quadratic Schrödinger equations [J]. Int. Math. Res. Not., 2009: 414-432.

[11] Germain P, Masmoudi N, Shatah J. Global solutions for the gravity water waves equation in dimension 3 [J]. Ann. Math., 2012, 175: 691-754.

[12] Guo Y, Ionescu A D, Pausader. Global solutions of the Euler-Maxwell two-fluid system in 3D [J]. Ann. Math., 2016, 183: 377-498.

[13] Hayashi N, Naumkin P I. Asymptotics for large time of solutions to the nonlinear Schrödinger and Hartree equations [J]. Amer. J. Math., 1998, 120(2): 369-389.

[14] Han L, Zhang J, Gan Z, Guo B. Cauchy problem for the Zakharov system arising from hot plasma with low regularity data [J]. Commun. Math. Sci., 2013, 11: 403-420.

[15] Hani Z, Pusateri F, Shatah J. Scattering for the Zakharov system in 3 dimensions [J]. Commun. Math. Phys., 2013, 322: 731-753.

[16] He X. The pondermotive force and magnetic field generation effects resulting from the non-linear interaction between plasma-wave and particles (in Chinese) [J]. Acta Phys. Sinica, 1983, 32: 325-337.

[17] Kato T. Liapunov functions and monotonicity in the Navier-Stokes equations [M]. Lecture Notes in Mathematics, vol. 1450. Berlin: Springer-Verlag, 1990.

[18] Kato J, Pusateri F. A new proof of long range scattering for critical nonlinear Schrödinger equations [J]. Diff. Integral Equa., 2011, 24: 923-940.

[19] Kenig C, Wang W. Existence of local smooth solution for a generalized Zakharov system [J]. J. Fourier Anal. Appl., 1998, 4: 469-490.

[20] Klainerman S. The null condition and global existence to nonlinear wave equations [J]. Lect. Appl. Math., 1986, 23: 293-326.

[21] Kono M, Skoric M M, Haar D T. Spontaneous excitation of magnetic fields and collapse dynamics in a Langmuir plasma [J]. J. Plasma Phys., 1981, 26: 123-146.

[22] Laurey C. The Cauchy problem for a generalized Zakharov system [J]. Diff. Integral Equ., 1995, 8(1): 105-130.

[23] Pusateri F, Shatah J. Space-time resonances and the null condition for first order systems of wave equations [J]. Comm. Pure Appl. Math., 2013, 66: 1495-1540.

[24] Zakharov V E. Collapse of Langmuir waves [J]. Sov. Phys. JETP., 1972, 35: 908-914.

Approximate Solution of Bagley-Torvik Equations with Variable Coefficients and Three-point Boundary-value Conditions*

Huang Qiongao (黄琼敖), Zhong Xianci (钟献词) and
Guo Boling (郭柏灵)

Abstract The fractional Bagley-Torvik equation with variable coefficients is investigated under three-point boundary-value conditions. By using the integration method, the considered problems are transformed into Fredholm integral equations of the second kind. It is found that when the fractional order is $1 < \alpha < 2$, the obtained Fredholm integral equation is with a weakly singular kernel. When the fractional order is $0 < \alpha < 1$, the given Fredholm integral equation is with a continuous kernel or a weakly singular kernel depending on the applied boundary-value conditions. The uniqueness of solution for the obtained Fredholm integral equation of the second kind with weakly singular kernel is addressed in continuous function spaces. A new numerical method is further proposed to solve Fredholm integral equations of the second kind with weakly singular kernels. The approximate solution is made and its convergence and error estimate are analyzed. Several numerical examples are computed to show the effectiveness of the solution procedures.

Keywords Bagley-Torvik equation; three-point boundary-value problem; Fredholm integral equation; weakly singular kernels; approximate solution

1 Introduction

Fractional calculus has profound physics background and rich connotation for mathematical theory and applications [1–3]. In simulating the motion of a rigid plate

---

* Int. J. Appl. Comput. Math., 2016, 2:327–347. DOI:10.1007/s40819-015-0063-5.

immersed in a Newtonian fluid, Torvik and Bagley [4] gave a fractional differential equation as

$$Ay''(t) + BD^\alpha y(t) + Cy(t) = F(t), \quad \alpha = \frac{3}{2}, \quad (1.1)$$

where A, B and C are constants depending on mass and area of the plate, stiffness of spring, fluid density and viscosity, $F(t)$ is a known function denoting the external force, $y(t)$ stands for the displacement of the plate and it should be solved. The Bagley-Torvik equation (1.1) has been generalized as $\alpha \in (0, 2)$ and many methods have been proposed for its general solution and the special solutions with initial value conditions [5-15].

On the other hand, except for initial value problems, one may measure the displacements of the plate at some points and give multipoint boundary value problems of the Bagley-Torvik equation. It is noted that for the generalized nonlinear Bagley-Torvik equation, Stanek [16] has studied the two-point boundary value problem. Moreover, the coefficients A, B and C may also change with the changes of fluid density and viscosity. That is, A, B and C may be functions with respect to t. Consequently, here we generally consider the following three-point boundary value problems of the Bagley-Torvik equation with variable coefficients:

$$\varphi''(x) + p(x)D^\alpha \varphi(x) + q(x)\varphi(x) = g(x), \quad 0 < \alpha < 2, \quad x \in [a, b], \quad (1.2)$$

with

$$\varphi(a) = \alpha_1, \quad \varphi(b) + \lambda\varphi(\xi) = \beta_1, \quad \xi \in (a, b), \quad (1.3)$$

or

$$\varphi(a) + \mu\varphi(\xi) = \alpha_1, \quad \varphi(b) = \beta_1, \quad \xi \in (a, b), \quad (1.4)$$

where $p(x)$, $q(x)$ and $g(x)$ are known functions, α_1, β_1, λ, μ and ξ are known constants. For convenience, hereafter the fractional derivative $D^\alpha \varphi(x)$ is rewritten as $_aD_x^\alpha \varphi(x)$ and we adopt the definition of the Riemann-Liouville fractional derivative, namely

$$_aD_x^\alpha \varphi(x) = \frac{1}{\Gamma(n-\alpha)} \frac{d^n}{dx^n} \int_a^x \frac{\varphi(s)}{(x-s)^{\alpha-n+1}} ds, \quad n-1 < \alpha < n,$$

where Γ is the Euler γ function.

Furthermore, it is seen that many numerical methods have been given to evaluate fractional order integrals and the solution of fractional order differential equations [17-25]. As shown in the above mentioned works, the principal difficulty in evaluating fractional order integrals is how to deal with the weakly singular kernel. In the typical approaches, the fractional order integrals are approximated by the sum of a discrete convolution and a few correction terms, or the convolution kernel $t^{\alpha-1}(0<\alpha<1)$ is represented as

$$t^{\alpha-1} = \frac{1}{\Gamma(1-\alpha)} \int_0^\infty e^{-\xi t}\xi^{-\alpha}\,d\xi.$$

Furthermore, in numerically solving fractional order differential equations, some methods involving of the difference method, the Adomian decomposition method, the homotopy function method and the iterative method are proposed. In the present paper, we will transform the three-point boundary value problems for the fractional Bagley-Torvik equation (1.2) with the boundary-value conditions (1.3) or (1.4) into Fredholm integral equations of the second kind. Uniqueness and approximation of the solutions for the obtained Fredholm integral equations will be investigated in detail. Numerical results will carry out to verify the proposed solution procedures.

The paper is organized as follows. In Section 2, we transform the boundary value problems (1.2) with (1.3) or (1.4) into a Fredholm integral equation of the second kind by using the integration method. The uniqueness of the solution for the given Fredholm integral equation of the second kind with weakly singular kernel is shown in $\mathbf{C}[a,b]$ by using the contraction operator theorem. Section 3 shows a new numerical method for solving the obtained Fredholm integral equations of the second kind. The approximate solution is made and its convergence and error estimate are considered. Some numerical examples are carried out in Section 4 to show the effectiveness of the proposed methods. Section 5 contains some main conclusions.

2   Fredholm integral equations with weakly singular kernels

In analyzing multi-point boundary value problems of fractional differential equations, a typical method is to express the solutions as integrals [26–28]. Now let us transform the three-point boundary-value problems (1.2) with (1.3) and (1.4) respectively to Fredholm integral equations of the second kind. It is convenient to define the auxiliary functions as

$$V(x,t) = \begin{cases} V_1(x,t), & 0<\alpha<1, \\ V_2(x,t), & 1\leqslant \alpha<2, \end{cases}$$

$$\Lambda(x,t) = \begin{cases} \Lambda_1(x,t), & 0<\alpha<1, \\ \Lambda_2(x,t), & 1\leqslant \alpha<2, \end{cases}$$

where

$$V_1(x,t) = \frac{1}{\Gamma(1-\alpha)}\left[\Gamma(1-\alpha)(x-t)q(t) + \int_t^x \frac{p(s)-(x-s)p'(s)}{(s-t)^\alpha}ds\right],$$

$$\Lambda_1(x,t) = \frac{1}{\Gamma(1-\alpha)}\left[\frac{(\xi-x)p(\xi)}{(\xi-t)^\alpha} + \int_\xi^x \frac{p(s)-(x-s)p'(s)}{(s-t)^\alpha}ds\right],$$

and

$$V_2(x,t) = \frac{1}{\Gamma(2-\alpha)}\left[\Gamma(2-\alpha)(x-t)q(t) + \frac{p(x)}{(x-t)^{\alpha-1}}\right.$$
$$\left. + \int_t^x \frac{(x-s)p''(s)-2p'(s)}{(s-t)^{\alpha-1}}ds\right],$$

$$\Lambda_2(x,t) = \frac{1}{\Gamma(2-\alpha)}\left[\frac{(x-\xi)p'(\xi)-p(\xi)}{(\xi-t)^{\alpha-1}} + \frac{p(x)}{(x-t)^{\alpha-1}}\right.$$
$$\left. + \int_\xi^x \frac{(x-s)p''(s)-2p'(s)}{(s-t)^{\alpha-1}}ds\right].$$

2.1 The transformation method

Applying the integration method, we give the following two theorems.

Theorem 2.1 *If $(b-a)+\lambda(\xi-a) \neq 0$ and $0 < \alpha < 2$, the boundary value problem*

$$\begin{cases} \varphi''(x) + p(x)D^\alpha \varphi(x) + q(x)\varphi(x) = g(x), & x \in [a,b], \\ \varphi(a) = \alpha_1, \quad \varphi(b) + \lambda\varphi(\xi) = \beta_1, & \xi \in (a,b) \end{cases} \quad (2.1)$$

is equivalent to the following Fredholm integral equation of the second kind

$$\varphi(x) + \int_a^b K_1(x,t)\varphi(t)\,dt = f_1(x), \quad (2.2)$$

where

$$K_1(x,t) = \begin{cases} \dfrac{[(b-a)+\lambda(\xi-a)]V(x,t)+(a-x)[V(b,t)+\lambda V(\xi,t)]}{(b-a)+\lambda(\xi-a)}, \\ \qquad for \qquad a \leqslant t \leqslant \min\{x,\xi\} \leqslant b, \\[4pt] \dfrac{(a-x)V(b,t)}{(b-a)+\lambda(\xi-a)}, \qquad for\ a \leqslant \max\{x,\xi\} \leqslant t \leqslant b, \\[4pt] \dfrac{[(b-a)+\lambda(\xi-a)]V(x,t)+(a-x)V(b,t)}{(b-a)+\lambda(\xi-a)}, \\ \qquad for \qquad a \leqslant \xi \leqslant t \leqslant x \leqslant b, \\[4pt] \dfrac{(a-x)[V(b,t)+\lambda V(\xi,t)]}{(b-a)+\lambda(\xi-a)}, \qquad for\ a \leqslant x \leqslant t \leqslant \xi \leqslant b, \end{cases}$$

$$f_1(x) = \frac{(b-x)+\lambda(\xi-x)}{(b-a)+\lambda(\xi-a)}\alpha_1 + \frac{(x-a)}{(b-a)+\lambda(\xi-a)}\beta_1 + \int_a^b G_1(x,t)g(t)\,dt,$$

with

$$G_1(x,t) = \begin{cases} \dfrac{(t-a)[(x-b)+\lambda(x-\xi)]}{(b-a)+\lambda(\xi-a)}, & a \leqslant t \leqslant \min\{x,\xi\} \leqslant b, \\[2mm] \dfrac{(a-x)(b-t)}{(b-a)+\lambda(\xi-a)}, & a \leqslant \max\{x,\xi\} \leqslant t \leqslant b, \\[2mm] \dfrac{(b-x)(a-t)+\lambda(\xi-a)(x-t)}{(b-a)+\lambda(\xi-a)}, & a \leqslant \xi \leqslant t \leqslant x \leqslant b, \\[2mm] \dfrac{(a-x)[(b-t)+\lambda(\xi-t)]}{(b-a)+\lambda(\xi-a)}, & a \leqslant x \leqslant t \leqslant \xi \leqslant b. \end{cases}$$

Proof We first consider the case of $1 \leqslant \alpha < 2$ and integrate both sides of the differential equation in (2.1) with respect to x from a to x twice. It follows that

$$\varphi(x) + \int_a^x \left\{ (x-t)q(t) + \frac{1}{\Gamma(2-\alpha)} \left[\frac{p(x)}{(x-t)^{\alpha-1}} \right. \right.$$
$$\left. \left. + \int_t^x \frac{(x-s)p''(s) - 2p'(s)}{(s-t)^{\alpha-1}} ds \right] \right\} \varphi(t)\,dt$$
$$= \int_a^x (x-t)g(t)\,dt + \varphi(a)$$
$$+ \left[\frac{p(a)}{\Gamma(2-\alpha)} \frac{d}{dt} \int_a^t \frac{\varphi(s)}{(t-s)^{\alpha-1}} ds \bigg|_{t=a} + \varphi'(a) \right](x-a). \qquad (2.3)$$

Let $x = b$ in (2.3) and one has

$$\frac{p(a)}{\Gamma(2-\alpha)} \frac{d}{dt} \int_a^t \frac{\varphi(s)}{(t-s)^{\alpha-1}} ds \bigg|_{t=a} + \varphi'(a)$$
$$= \frac{1}{(b-a)} \left[\int_a^b V_2(b,t)\varphi(t)\,dt + \varphi(b) - \varphi(a) - \int_a^b (b-t)g(t)\,dt \right]. \qquad (2.4)$$

From (2.4) and (2.3), it is seen that

$$\varphi(x) + \int_a^x V_2(x,t)\varphi(t)\,dt - \int_a^b \frac{x-a}{b-a} V_2(b,t)\varphi(t)\,dt$$
$$= \int_a^x (x-t)g(t)\,dt - \int_a^b \frac{x-a}{b-a}(b-t)g(t)\,dt + \frac{b-x}{b-a}\varphi(a) + \frac{x-a}{b-a}\varphi(b). \qquad (2.5)$$

Then we assume $x = \xi$ in (2.3) and get

$$\varphi(x) + \int_a^x V_2(x,t)\varphi(t)\,dt - \int_a^\xi \frac{x-a}{\xi-a} V_2(\xi,t)\varphi(t)\,dt$$

$$= \int_a^x (x-t)g(t)\,dt - \int_a^\xi \frac{x-a}{\xi-a}(\xi-t)g(t)\,dt + \frac{\xi-x}{\xi-a}\varphi(a) + \frac{x-a}{\xi-a}\varphi(\xi). \tag{2.6}$$

With the knowledge of (2.5), (2.6) and the boundary conditions in (2.1), one gives

$$\varphi(x) + \int_a^x V_2(x,t)\varphi(t)\,dt + \int_a^b \frac{(a-x)}{(b-a)+\lambda(\xi-a)} V_2(b,t)\varphi(t)\,dt$$
$$+ \int_a^\xi \frac{\lambda(a-x)}{(b-a)+\lambda(\xi-a)} V_2(\xi,t)\varphi(t)\,dt = f_1(x). \tag{2.7}$$

In the end, for various x, ξ and t, we can give a series of results. That is,

(1) for $a \leqslant t \leqslant \min\{x,\xi\} \leqslant b$,

$$\varphi(x) + \int_a^{\min\{x,\xi\}} V_2(x,t)\varphi(t)\,dt + \int_a^{\min\{x,\xi\}} \frac{[(a-x)[V_2(b,t)+\lambda V_2(\xi,t)]]}{(b-a)+\lambda(\xi-a)}\varphi(t)\,dt$$
$$= \frac{(b-x)+\lambda(\xi-x)}{(b-a)+\lambda(\xi-a)}\alpha_1 + \frac{(x-a)}{(b-a)+\lambda(\xi-a)}\beta_1$$
$$+ \int_a^{\min\{x,\xi\}} \frac{(t-a)[(x-b)+\lambda(x-\xi)]}{(b-a)+\lambda(\xi-a)} g(t)\,dt, \tag{2.8}$$

(2) for $a \leqslant \max\{x,\xi\} \leqslant t \leqslant b$,

$$\varphi(x) + \int_{\max\{x,\xi\}}^b \frac{(a-x)V_2(b,t)}{(b-a)+\lambda(\xi-a)}\varphi(t)\,dt$$
$$= \frac{(b-x)+\lambda(\xi-x)}{(b-a)+\lambda(\xi-a)}\alpha_1 + \frac{(x-a)}{(b-a)+\lambda(\xi-a)}\beta_1$$
$$+ \int_{\max\{x,\xi\}}^b \frac{(a-x)(b-t)}{(b-a)+\lambda(\xi-a)} g(t)\,dt, \tag{2.9}$$

(3) for $a \leqslant \xi \leqslant t \leqslant x \leqslant b$,

$$\varphi(x) + \int_\xi^x \frac{[(b-a)+\lambda(\xi-a)]V_2(x,t) + (a-x)V_2(b,t)}{(b-a)+\lambda(\xi-a)}\varphi(t)\,dt$$
$$= \frac{(b-x)+\lambda(\xi-x)}{(b-a)+\lambda(\xi-a)}\alpha_1 + \frac{(x-a)}{(b-a)+\lambda(\xi-a)}\beta_1$$
$$+ \int_\xi^x \frac{(b-x)(a-t)+\lambda(\xi-a)(x-t)}{(b-a)+\lambda(\xi-a)} g(t)\,dt, \tag{2.10}$$

(4) for $a \leqslant x < t < \xi \leqslant b$,

$$\varphi(x) + \int_x^\xi \frac{(a-x)[V_2(b,t)+\lambda V_2(\xi,t)]}{(b-a)+\lambda(\xi-a)}\varphi(t)\,dt$$
$$= \frac{(b-x)+\lambda(\xi-x)}{(b-a)+\lambda(\xi-a)}\alpha_1 + \frac{(x-a)}{(b-a)+\lambda(\xi-a)}\beta_1$$

$$+ \int_x^\xi \frac{(a-x)[(b-t)+\lambda(\xi-t)]}{(b-a)+\lambda(\xi-a)} g(t)\, dt. \tag{2.11}$$

So the theorem is proved after rewriting (2.8)-(2.11) as (2.2). Similarly, for the case of $0 < \alpha < 1$, we can transform the three-point boundary value problem (2.1) into Fredholm integral equation of the second kind. The detail procedures are omitted here for saving spaces.

Theorem 2.2 *If $(b-a)+\mu(b-\xi) \neq 0$ and $0 < \alpha < 2$, the boundary value problem*

$$\begin{cases} \varphi''(x) + p(x)D^\alpha \varphi(x) + q(x)\varphi(x) = g(x), & x \in [a,b], \\ \varphi(a) + \mu\varphi(\xi) = \alpha_1, \quad \varphi(b) = \beta_1, & \xi \in (a,b), \end{cases} \tag{2.12}$$

is equivalent to the following Fredholm integral equation of the second kind

$$\varphi(x) + \int_a^b K_2(x,t)\varphi(t)\, dt = f_2(x), \tag{2.13}$$

where

$$K_2(x,t) = \begin{cases} \dfrac{[(b-a)V(x,t)+(a-x)V(b,t)]+\mu[(b-\xi)\Lambda(x,t)+(\xi-x)\Lambda(b,t)]}{(b-a)+\mu(b-\xi)}, \\ \qquad for \qquad a \leqslant t \leqslant \min\{x,\xi\} \leqslant b, \\ \dfrac{(a-x)+\mu(\xi-x)}{(b-a)+\mu(b-\xi)} V(b,t), \qquad for \qquad a \leqslant \max\{x,\xi\} \leqslant t \leqslant b, \\ \dfrac{[(b-a)V(x,t)+(a-x)V(b,t)]+\mu[(b-\xi)V(x,t)+(\xi-x)V(b,t)]}{(b-a)+\mu(b-\xi)}, \\ \qquad for \qquad a \leqslant \xi \leqslant t \leqslant x \leqslant b, \\ \dfrac{[(a-x)V(b,t)+\mu(\xi-x)\Lambda(b,t)]+\mu(\xi-b)[V(x,t)-\Lambda(x,t)]}{(b-a)+\mu(b-\xi)}, \\ \qquad for \qquad a \leqslant x \leqslant t \leqslant \xi \leqslant b, \end{cases}$$

$$f_2(x) = \frac{(b-x)}{(b-a)+\mu(b-\xi)} \alpha_1 + \frac{(x-a)+\mu(x-\xi)}{(b-a)+\mu(b-\xi)} \beta_1 + \int_a^b G_2(x,t)g(t)\, dt,$$

with

$$G_2(x,t) = \begin{cases} \dfrac{(b-x)(a-t)}{(b-a)+\mu(b-\xi)}, & a \leqslant t \leqslant \min\{x,\xi\} \leqslant b, \\ \dfrac{(a-x)+\mu(\xi-x)}{(b-a)+\mu(b-\xi)}(b-t), & a \leqslant \max\{x,\xi\} \leqslant t \leqslant b, \\ \dfrac{(b-x)[(a-t)+\mu(\xi-t)]}{(b-a)+\mu(b-\xi)}, & a \leqslant \xi \leqslant t \leqslant x \leqslant b, \\ \dfrac{(a-x)(b-t)+\mu(\xi-b)(x-t)}{(b-a)+\mu(b-\xi)}, & a \leqslant x \leqslant t \leqslant \xi \leqslant b. \end{cases}$$

Proof For $1 \leqslant \alpha < 2$, we integrate both sides of the differential equation in (2.12) with respect to x from ξ to x twice, and arrive at

$$\varphi(x) + \int_a^x \left[\frac{1}{\Gamma(2-\alpha)} \frac{p(x)}{(x-t)^{\alpha-1}}\right] \varphi(t) \, dt$$

$$+ \int_\xi^x \left[(x-t)q(t) + \frac{1}{\Gamma(2-\alpha)} \int_t^x \frac{(x-s)p''(s) - 2p'(s)}{(s-t)^{\alpha-1}} \, ds\right] \varphi(t) \, dt$$

$$+ \int_a^\xi \frac{1}{\Gamma(2-\alpha)} \left[\frac{(x-\xi)p'(\xi) - p(\xi)}{(\xi-t)^{\alpha-1}} + \int_\xi^x \frac{(x-s)p''(s) - 2p'(s)}{(s-t)^{\alpha-1}} \, ds\right] \varphi(t) \, dt$$

$$= \int_\xi^x (x-t)g(t) \, dt + \left[\frac{p(\xi)}{\Gamma(2-\alpha)} \frac{d}{dt} \int_a^t \frac{\varphi(s)}{(t-s)^{\alpha-1}} \, ds \bigg|_{t=\xi} + \varphi'(\xi)\right](x-\xi) + \varphi(\xi),$$

(2.14)

or

$$\varphi(x) + \int_\xi^x V_2(x,t)\varphi(t) \, dt + \int_a^\xi \Lambda_2(x,t)\varphi(t) \, dt$$

$$= \int_\xi^x (x-t)g(t) \, dt + \left[\frac{p(\xi)}{\Gamma(2-\alpha)} \frac{d}{dt} \int_a^t \frac{\varphi(s)}{(t-s)^{\alpha-1}} \, ds \bigg|_{t=\xi} + \varphi'(\xi)\right](x-\xi) + \varphi(\xi).$$

(2.15)

Assume $x = b$ in (2.15) and one has

$$\frac{p(\xi)}{\Gamma(2-\alpha)} \frac{d}{dt} \int_a^t \frac{\varphi(s)}{(t-s)^{\alpha-1}} \, ds \bigg|_{t=\xi} + \varphi'(\xi)$$

$$= \frac{1}{(b-\xi)} \left[\varphi(b) - \varphi(\xi) - \int_\xi^b (b-t)g(t) \, dt + \int_a^\xi \Lambda_2(b,t)\varphi(t) \, dt + \int_\xi^b V_2(b,t)\varphi(t) \, dt\right].$$

(2.16)

According to (2.15) and (2.16), it is found that

$$\varphi(x) + \int_\xi^x V_2(x,t)\varphi(t) \, dt - \frac{x-\xi}{b-\xi} \int_\xi^b V_2(b,t)\varphi(t) \, dt$$

$$+ \int_a^\xi \Lambda_2(x,t)\varphi(t) \, dt - \frac{x-\xi}{b-\xi} \int_a^\xi \Lambda_2(b,t)\varphi(t) \, dt$$

$$= \int_\xi^x (x-t)g(t) \, dt - \frac{x-\xi}{b-\xi} \int_\xi^b (b-t)g(t) \, dt + \frac{b-x}{b-\xi}\varphi(\xi) + \frac{x-\xi}{b-\xi}\varphi(b).$$

(2.17)

Making use of the boundary conditions in (2.12), one can obtain

$$\varphi(x) + \int_a^x \frac{(b-a)V_2(x,t)}{(b-a) + \mu(b-\xi)}\varphi(t) \, dt + \int_a^b \frac{(a-x)V_2(b,t)}{(b-a) + \mu(b-\xi)}\varphi(t) \, dt$$

$$+ \int_{\xi}^{x} \frac{\mu(b-\xi)V_2(x,t)}{(b-a)+\mu(b-\xi)}\varphi(t)\,dt + \int_{\xi}^{b} \frac{\mu(\xi-x)V_2(b,t)}{(b-a)+\mu(b-\xi)}\varphi(t)\,dt$$
$$+ \int_{a}^{\xi} \frac{\mu[(b-\xi)\Lambda_2(x,t)+(\xi-x)\Lambda_2(b,t)]}{(b-a)+\mu(b-\xi)}\varphi(t)\,dt = f_2(x). \qquad (2.18)$$

Similar to Theorem 2.1, we can rewrite (2.18) as (2.13) by considering the various values of x, ξ and t. Furthermore, the results for $0 < \alpha < 1$ can be obtained and the theorem is proved.

It is seen from Theorems 2.1 and 2.2 that one has transformed the three-point boundary value problems for the generalized Bagley-Torvik equation into Fredholm integral equations of the second kind. It is interesting to find that the kernel $K_1(x,t)$ is continuous for $\alpha \in (0,1)$ and weakly singular for $\alpha \in (1,2)$, and the kernel $K_2(x,t)$ is weakly singular for $\alpha \in (0,1) \cup (1,2)$. In addition, when $\alpha = 1$, corresponding to three-point boundary value problems of second-order ordinary differential equations, the obtained results in Theorems 2.1 and 2.2 can be degenerated to those in [29].

2.2 Uniqueness of the solution

In the subsection, let us focus on the uniqueness of the solutions of the obtained Fredholm integral equations (2.2) and (2.13) in the space $\mathbf{C}[a,b]$. When the derived Fredholm integral equation is with the continuous kernel, the uniqueness theorem of its solution can be given similar to those in [29] and the detail procedure is omitted here. When the derived Fredholm integral equation is with the weakly singular kernel, the uniqueness of its solution will be investigated in detail. One can see that the uniqueness of the solution for Fredholm integral equations with weakly singular kernels has been investigated comprehensively [30, 31]. However, the kernels in (2.2) and (2.13) are different from the typical weakly singular kernels, and it is significant to address the uniqueness of the solution for the obtained Fredholm integral equations with weakly singular kernels. In what follows, according to the contraction operator theorem in continuous function spaces, the theorem about the uniqueness of the solution will be given.

Theorem 2.3 *It is assumed that $1 < \alpha < 2$ and $p(x) \in \mathbf{C}^2[a,b]$, $q(x) \in \mathbf{L}^1[a,b]$, $g(x) \in \mathbf{L}^1[a,b]$ for finite a and b. When*

$$\max_{a \leqslant x \leqslant b} \int_a^b |K_i(x,t)|\,dt < 1, \quad i = 1, 2,$$

the Fredholm integral equations of the second kind (2.2) and (2.13) have unique solutions in $\mathbf{C}[a,b]$.

Proof For convenience, Eq. (2.2) is written as

$$\varphi = f_1 - \mathbf{K}\varphi = \mathbf{T}\varphi$$

with

$$\mathbf{K}\varphi = \int_a^b K_1(x,t)\varphi(t)\,dt.$$

It is seen that the operator $\mathbf{K} : \mathbf{C}[a,b] \to \mathbf{C}[a,b]$ is linear and bounded. In fact, since $p(x) \in \mathbf{C}^2[a,b]$, $q(x) \in \mathbf{L}^1[a,b]$, $(x-t) \in \mathbf{C}[a,b]$, $1/(x-t)^{\alpha-1} \in \mathbf{L}^1[a,b]$ and a and b are finite, it is obtained that $V_2(x,t), \Lambda_2(x,t) \in \mathbf{L}^1[a,b;a,b]$. Hence one gets $K_1(x,t) \in \mathbf{L}^1[a,b;a,b]$. Furthermore, let

$$\max_{a \leqslant x \leqslant b} \int_a^b |K_1(x,t)|\,dt = c, \quad c \geqslant 0.$$

For $\varphi(x) \in \mathbf{C}[a,b]$, it gives

$$\|\mathbf{K}\varphi\|_\infty = \left\| \int_a^b K_1(x,t)\varphi(t)\,dt \right\|_\infty \leqslant \max_{a \leqslant x \leqslant b} \int_a^b |K_1(x,t)|\,dt \cdot \|\varphi\|_\infty = c\|\varphi\|_\infty.$$

So the operator $\mathbf{K}$ is linear and bounded, and

$$\|\mathbf{K}\|_\infty \leqslant \max_{a \leqslant x \leqslant b} \int_a^b |K_1(x,t)|\,dt = c.$$

For all $\varphi_1, \varphi_2 \in \mathbf{C}$, one has

$$\|\mathbf{T}\varphi_1 - \mathbf{T}\varphi_2\|_\infty = \|\mathbf{K}(\varphi_1 - \varphi_2)\|_\infty \leqslant \|\mathbf{K}\|_\infty \cdot \|\varphi_1 - \varphi_2\|_\infty \leqslant c\|\varphi_1 - \varphi_2\|_\infty.$$

When $c < 1$, the Fredholm integral equation of the second kind (2.2) has a unique solution in $\mathbf{C}[a,b]$ by using Banach fixed point theorem. Similarly, for Eq. (2.13), it is easy to prove that there is a unique solution in $\mathbf{C}[a,b]$. This completes the proof.

Moreover, we have the following theorem.

Theorem 2.4 Under the assumption of $0 < \alpha < 1$ and $p(x) \in \mathbf{C}[a,b]$, $q(x) \in \mathbf{L}^1[a,b]$, $g(x) \in \mathbf{L}^1[a,b]$ for finite a and b, the Fredholm integral equations of the second kind (2.2) and (2.13) have unique solutions in $\mathbf{C}[a,b]$ when

$$\max_{a \leqslant x \leqslant b} \int_a^b |K_i(x,t)|\,dt < 1, \quad i = 1,2.$$

The proof of Theorem 2.4 is similar to that of Theorem 2.3 and it has been omitted here. From Theorems 2.3 and 2.4, it is seen that one only gives the sufficient conditions for the uniqueness of solution in $\mathbf{C}[a,b]$. The applied method is different from those in [30, 31], where the iterative method is applied. The reason is based on the fact that the kernels in Eqs. (2.2) and (2.13) are different from the typical weakly singular kernel $1/|x-t|$. They have the following style

$$K(x,t) = \frac{A_0(x,t)}{(x-t)^{\alpha_0}} + \frac{B_0(x,t)}{(b-t)^{\alpha_0}} + \frac{C_0(x,t)}{(\xi-t)^{\alpha_0}},$$

with $0 < \alpha_0 < 1$ and the piecewise continuous functions $A_0(x,t)$, $B_0(x,t)$ and $C_0(x,t)$.

3 A new numerical method

In what follows, we turn our attention to the numerical solutions of Eqs. (2.12) and (2.13). Recently, the piecewise Taylor-series expansion method is proposed to solve Fredholm integral equations of the second kind [29]. Unfortunately, the kernels in Eqs. (2.12) and (2.13) may be weakly singular, then the proposed method in [29] cannot be used directly. Consequently, we will modify the numerical method in [29] so that it can be used to solve Fredholm integral equations of the second kind with weakly singular kernels.

3.1 Construction of the approximate solution

Generally Fredholm integral equations of the second kind can be written as:

$$\varphi(x) + \int_a^b K(x,t)\varphi(t)\,dt = f(x), \quad x \in [a,b], \tag{3.1}$$

where $f(x)$ is a known function on the finite interval $[a,b]$, $\varphi(x)$ is unknown to be solved, and the kernel $K(x,t)$ is weakly singular. Different from the differential method in [29], we integrate both sides of Eq.(3.1) $i(i=1,2,\cdots,n)$ times and get

$$\int_a^x \frac{(x-t)^{i-1}}{(i-1)!}\varphi(t)\,dt + \int_a^x \frac{(x-s)^{i-1}}{(i-1)!}\left[\int_a^b K(s,t)\varphi(t)\,dt\right]ds$$
$$= \int_a^x \frac{(x-t)^{i-1}}{(i-1)!}f(t)\,dt. \tag{3.2}$$

Exchanging the integral order in (3.2), one has

$$\int_a^x \frac{(x-t)^{i-1}}{(i-1)!}\varphi(t)\,dt + \int_a^b \left[\int_a^x \frac{(x-s)^{i-1}}{(i-1)!}K(s,t)\,ds\right]\varphi(t)\,dt$$

$$= \int_a^x \frac{(x-t)^{i-1}}{(i-1)!} f(t)\, dt. \qquad (3.3)$$

Similar to those in [29], we choose a series of quadrature points such as

$$a = x_0 < x_1 < \cdots < x_m = b \ (m \geqslant 1),$$

where $x_q = a + qh$ $(q = 0, 1, \cdots, m)$ with $h = (b-a)/m$. Letting $x = x_k (k = 1, 2, \cdots, m)$, Eq. (3.3) can be further written as

$$\sum_{q=0}^{k-1} \int_{x_q}^{x_{q+1}} \frac{(x_k - t)^{i-1}}{(i-1)!} \varphi(t)\, dt$$

$$+ \sum_{q=0}^{m-1} \sum_{p=0}^{k-1} \int_{x_q}^{x_{q+1}} \left[\int_{x_p}^{x_{p+1}} \frac{(x_k - s)^{i-1}}{(i-1)!} K(s,t)\, ds \right] \varphi(t)\, dt$$

$$= \sum_{q=0}^{k-1} \int_{x_q}^{x_{q+1}} \frac{(x_k - t)^{i-1}}{(i-1)!} f(t)\, dt. \qquad (3.4)$$

Furthermore we let $t = x_q + h\eta$ and $s = x_p + h\zeta$, Eq. (3.4) can be written as

$$\sum_{q=0}^{k-1} \frac{h^i}{(i-1)!} \int_0^1 (k - q - \eta)^{i-1} \varphi(x_q + h\eta)\, d\eta$$

$$+ \sum_{q=0}^{m-1} \sum_{p=0}^{k-1} \frac{h^{i+1}}{(i-1)!} \int_0^1 \int_0^1 (k - p - \zeta)^{i-1} K(x_p + h\zeta, x_q + h\eta) \varphi(x_q + h\eta)\, d\zeta\, d\eta$$

$$= \sum_{q=0}^{k-1} \frac{h^i}{(i-1)!} \int_0^1 (k - q - \eta)^{i-1} f(x_q + h\eta)\, d\eta. \qquad (3.5)$$

It is assumed that $\varphi(x_q + h\eta)$ can be expressed as a Taylor-series expansion with Lagrange remainder. That is, we have

$$\varphi(x_q + h\eta) = \sum_{j=0}^n \frac{\varphi^{(j)}(x_q)}{j!} (h\eta)^j + \frac{\varphi^{(n+1)}(\theta_q)}{(n+1)!} (h\eta)^{n+1}, \quad x_q \leqslant \theta_q \leqslant x_q + h\eta. \qquad (3.6)$$

Insertion of Eq. (3.6) into Eq. (3.5) yields

$$\sum_{q=0}^{k-1} \frac{h^i}{(i-1)!} \int_0^1 (k - q - \eta)^{i-1} \left[\sum_{j=0}^n \frac{\varphi^{(j)}(x_q)}{j!} (h\eta)^j + \frac{\varphi^{(n+1)}(\theta_q)}{(n+1)!} (h\eta)^{n+1} \right] d\eta$$

$$+ \sum_{q=0}^{m-1} \sum_{p=0}^{k-1} \frac{h^{i+1}}{(i-1)!} \int_0^1 \int_0^1 (k - p - \zeta)^{i-1} K(x_p + h\zeta, x_q + h\eta)$$

$$\times \left[\sum_{j=0}^{n} \frac{\varphi^{(j)}(x_q)}{j!} (h\eta)^j + \frac{\varphi^{(n+1)}(\theta_q)}{(n+1)!} (h\eta)^{n+1} \right] d\zeta\, d\eta$$

$$= \sum_{q=0}^{k-1} \frac{h^i}{(i-1)!} \int_0^1 (k-q-\eta)^{i-1} f(x_q + h\eta)\, d\eta. \tag{3.7}$$

In Eq.(3.7), one replaces $\varphi^{(j)}(x_q)$ by the numerical solution $\varphi_q^{(j)}$ and obtains a discrete format as

$$\sum_{q=0}^{k-1} \sum_{j=0}^{n} \frac{\varphi_q^{(j)} h^{i+j}}{(i-1)!\,j!} \int_0^1 (k-q-\eta)^{i-1} \eta^j\, d\eta$$

$$+ \sum_{q=0}^{m-1} \sum_{p=0}^{k-1} \sum_{j=0}^{n} \frac{\varphi_q^{(j)} h^{i+j+1}}{(i-1)!\,j!} \int_0^1 \int_0^1 (k-p-\zeta)^{i-1} K(x_p + h\zeta, x_q + h\eta) \eta^j\, d\zeta\, d\eta$$

$$= \sum_{q=0}^{k-1} \frac{h^i}{(i-1)!} \int_0^1 (k-q-\eta)^{i-1} f(x_q + h\eta)\, d\eta, \tag{3.8}$$

where $i = 1, 2, \cdots, n$ and $k = 1, 2, \cdots, m$. Moreover, applying Eq. (3.6) to (3.1), Eq. (3.1) can be approximated by the following discrete formation

$$\varphi_{k-1}^{(0)} + \sum_{q=0}^{m-1} \sum_{j=0}^{n} \frac{\varphi_q^{(j)} h^{j+1}}{j!} \int_0^1 K(x_{k-1}, x_q + h\eta) \eta^j\, d\eta = f(x_{k-1}), \tag{3.9}$$

for $q = 0, 1, \cdots, (m-1)$. Once the linear system of (3.8) and (3.9) with $m(n+1)$ equations is solved, the approximate solution of $\varphi(x)$ can be further given as

$$\varphi_{m,n}(x) = f(x) - \sum_{q=0}^{m-1} \sum_{j=0}^{n} \frac{\varphi_q^{(j)} h^{j+1}}{j!} \int_0^1 K(x, x_q + h\eta) \eta^j\, d\eta, \quad a \leqslant x \leqslant b. \tag{3.10}$$

3.2 Convergence and error estimate

Now we focus on the convergence and error estimate of the approximate solution $\varphi_{m,n}(x)$. First, we rewrite Eqs. (3.8) and (3.9) as

$$(A+B)\tilde{\Phi} = F, \tag{3.11}$$

where

$$A = [a_{st}]_{m(n+1) \times m(n+1)},$$
$$B = [b_{st}]_{m(n+1) \times m(n+1)},$$

$$F = F_{m(n+1)\times 1} = \left[f(x_0), \cdots, \sum_{q=0}^{m-1} \frac{h^n}{(n-1)!} \int_0^1 (m-q-\eta)^{n-1} f(x_q + h\eta) \, d\eta \right]^{\mathrm{T}},$$

$$\tilde{\Phi} = [\varphi_q^{(j)}]_{m(n+1)\times 1} = \left[\varphi_0^{(0)}, \varphi_1^{(0)}, \cdots, \varphi_{m-1}^{(n)} \right]^{\mathrm{T}},$$

with

$$a_{11} = 1,$$

$$a_{12} = 0,$$

$$\cdots\cdots$$

$$a_{m(n+1),m(n+1)} = \frac{h^{2n}}{(n-1)!n!} \int_0^1 (1-\eta)^{n-1} \eta^n \, d\eta,$$

$$b_{11} = h \int_0^1 K(x_0, x_0 + h\eta) \, d\eta,$$

$$b_{12} = h \int_0^1 K(x_0, x_1 + h\eta) \, d\eta,$$

$$\cdots\cdots$$

$$b_{m(n+1),m(n+1)} = \sum_{p=0}^{m-1} \frac{h^{2n+1}}{(n-1)!n!} \int_0^1 \int_0^1 (m-p-\zeta)^{n-1} K(x_p + h\zeta, x_{m-1} + h\eta) \eta^n \, d\zeta d\eta.$$

Moreover, by considering the Lagrange remainder, it follows that

$$(A + B)\Phi = F - R, \tag{3.12}$$

where

$$\Phi = [\varphi^{(j)}(x_q)]_{m(n+1)\times 1} = [\varphi^{(0)}(x_0), \varphi^{(0)}(x_1), \cdots, \varphi^{(n)}(x_{m-1})]^{\mathrm{T}},$$

$$R = [r_s]_{m(n+1)\times 1},$$

with

$$r_1 = \sum_{q=0}^{m-1} \frac{h^{n+2}}{(n+1)!} \int_0^1 K(x_0, x_q + h\eta) \varphi^{(n+1)}(\theta_q) \eta^{n+1} \, d\eta,$$

$$r_2 = \sum_{q=0}^{m-1} \frac{h^{n+2}}{(n+1)!} \int_0^1 K(x_1, x_q + h\eta) \varphi^{(n+1)}(\theta_q) \eta^{n+1} \, d\eta,$$

$$\cdots\cdots$$

$$r_{m(n+1)} = \sum_{q=0}^{m-1} \frac{h^{2n+1}}{(n-1)!(n+1)!} \int_0^1 (m-q-\eta)^{n-1} \varphi^{(n+1)}(\theta_q) \eta^{n+1} \, d\eta$$

$$+ \sum_{q=0}^{m-1} \sum_{p=0}^{m-1} \frac{h^{2n+2}}{(n-1)!(n+1)!} \int_0^1 \int_0^1 (m-p-\zeta)^{n-1}$$

$$\cdot K(x_p + h\zeta, x_q + h\eta) \varphi^{(n+1)}(\theta_q) \eta^{n+1} \, d\zeta d\eta.$$

It is seen that the matrix A is nonsingular, and one lets

$$\mathcal{A} : \mathbf{R}^{m(n+1)} \to \mathbf{R}^{m(n+1)}$$

be a mapping. Given $x \in \mathbf{R}^{m(n+1)}$ we definite

$$\mathcal{A}x = Ax.$$

It is easy to prove that $\mathcal{A}$ is a linear operator, and $\mathcal{A}$ is bijective. In addition, one can calculate that

$$\|A\|_\infty = \max\left\{1, \max_{1\leqslant k\leqslant m, 1\leqslant i\leqslant n} \sum_{j=0}^{n} \sum_{q=0}^{k-1} \frac{h^{i+j}}{(i-1)!j!} \left| \int_0^1 (k-q-\eta)^{i-1} \eta^j \, d\eta \right| \right\}$$

$$\leqslant \max\left\{1, \max_{1\leqslant k\leqslant m, 1\leqslant i\leqslant n} \sum_{j=0}^{n} \sum_{q=0}^{k-1} \frac{h^{i+j}}{(i-1)!j!} \int_0^1 (k-q-\eta)^{i-1} \, d\eta \right\}$$

$$= \max\left\{1, \sum_{j=0}^{n} \frac{h^j}{j!} \cdot \max_{1\leqslant k\leqslant m, 1\leqslant i\leqslant n} \frac{h^i}{i!} \sum_{q=0}^{k-1} \left[(k-q)^i - (k-q-1)^i\right] \right\}$$

$$\leqslant \max\left\{1, \sum_{j=0}^{+\infty} \frac{h^j}{j!} \cdot \max_{1\leqslant k\leqslant m, 1\leqslant i\leqslant n} \frac{(hk)^i}{i!} \right\} \leqslant \max\left\{1, e^h \cdot \sum_{i=1}^{n} \frac{(b-a)^i}{i!} \right\}$$

$$\leqslant \max\{1, e^{h+b-a}\} \leqslant e^{h+b-a} \leqslant e^{2(b-a)} < +\infty. \tag{3.13}$$

So $\mathcal{A}$ is a bounded operator in $\mathbf{R}^{m(n+1)}$. According to the inverse operator theorem in Banach spaces [32], the inverse operator $\mathcal{A}^{-1}$ is bounded. Now we give the following theorem:

Theorem 3.1 *We let $\|A^{-1}\|_\infty \leqslant L$ and have the following conditions*

$$\max_{q=0,1,\cdots,m-1} \max_{a\leqslant x\leqslant b} \int_0^1 |K(x_{k-1}, x_q + h\eta)| \, d\eta = M < \frac{1}{L(b-a)e^{h+(b-a)}} < +\infty,$$

$$\|\varphi^{(n+1)}(x)\|_\infty = \max_{a\leqslant x\leqslant b} |\varphi^{(n+1)}(x)| = N < +\infty.$$

The approximate solution $\varphi_{m,n}(x)$ in (3.10) is convergent to the exact solution $\varphi(x)$. That is, we get
$$\lim_{n\to+\infty}\|\varphi_{m,n}(x)-\varphi(x)\|_\infty=0,$$
and
$$\lim_{m\to+\infty}\|\varphi_{m,n}(x)-\varphi(x)\|_\infty=0.$$
Moreover, the following error estimate can be obtained
$$\|\varphi_{m,n}(x)-\varphi(x)\|_\infty\leqslant\frac{(b-a)MN[1+Lhe^{h+(b-a)}]}{[1-(b-a)MLe^{h+(b-a)}]}\cdot\frac{h^{n+1}}{(n+1)!}.$$

Proof One can calculate that
$$\|B\|_\infty=\max\left\{\max_{1\leqslant k\leqslant m}\sum_{j=0}^{n}\sum_{q=0}^{m-1}\frac{h^{j+1}}{j!}\left|\int_0^1 K(x_{k-1},x_q+h\eta)\eta^j\,d\eta\right|,\right.$$
$$\max_{1\leqslant k\leqslant m,1\leqslant i\leqslant n}\sum_{j=0}^{n}\sum_{q=0}^{m-1}\sum_{p=0}^{k-1}\frac{h^{i+j+1}}{(i-1)!j!}$$
$$\left.\times\left|\int_0^1\int_0^1(k-p-\zeta)^{i-1}K(x_p+h\zeta,x_q+h\eta)\eta^j\,d\zeta\,d\eta\right|\right\}$$
$$\leqslant\max\left\{hmMe^h,hmM\sum_{j=0}^{n}\frac{h^j}{j!}\right.$$
$$\left.\times\max_{1\leqslant k\leqslant m,1\leqslant i\leqslant n}\frac{h^i}{(i-1)!}\sum_{p=0}^{k-1}\int_0^1(k-p-\zeta)^{i-1}\,d\zeta\right\}$$
$$\leqslant\max\left\{hmMe^h,hmMe^h\cdot\sum_{i=1}^{n}\frac{(b-a)^i}{i!}\right\}\leqslant(b-a)Me^{h+(b-a)},$$
and
$$\|R\|_\infty=\max\left\{\max_{1\leqslant k\leqslant m}\left|\sum_{q=0}^{m-1}\frac{h^{n+2}}{(n+1)!}\int_0^1 K(x_{k-1},x_q+h\eta)\varphi^{(n+1)}(\theta_q)\eta^{n+1}\,d\eta\right|,\right.$$
$$\max_{1\leqslant k\leqslant m,1\leqslant i\leqslant n}\left|\sum_{q=0}^{k-1}\frac{h^{i+n+2}}{(i-1)!(n+1)!}\int_0^1(k-q-\eta)^{i-1}\varphi^{(n+1)}(\theta_q)\eta^{n+1}\,d\eta\right.$$
$$+\sum_{q=0}^{m-1}\sum_{p=0}^{k-1}\frac{h^{i+n+2}}{(i-1)!(n+1)!}\int_0^1\int_0^1(k-p-\zeta)^{i-1}K(x_p+h\zeta,x_q+h\eta)$$
$$\left.\left.\varphi^{(n+1)}(\theta_q)\eta^{n+1}\,d\zeta\,d\eta\right|\right\}$$

$$\leqslant \max\left\{\frac{(b-a)MNh^{n+1}}{(n+1)!}, \frac{Nh^{n+2}}{(n+1)!}\mathrm{e}^{b-a}+\frac{(b-a)MNh^{n+1}}{(n+1)!}\mathrm{e}^{b-a}\right\}$$
$$\leqslant \frac{Nh^{n+1}\mathrm{e}^{b-a}[h+(b-a)M]}{(n+1)!}.$$

Then one has $\|BA^{-1}\|_\infty \leqslant \|B\|_\infty\|A^{-1}\|_\infty < 1$. It is easy to prove that $I+BA^{-1}$ is a strictly diagonally dominant matrix, where I is a unity matrix. Hence, it is found that $I+BA^{-1}$ is a nonsingular matrix. Furthermore, one can calculate that

$$\begin{aligned}\|\Phi-\tilde{\Phi}\|_\infty &= \|[(I+BA^{-1})A]^{-1}R\|_\infty \\ &\leqslant \|A^{-1}\|_\infty \|(I+BA^{-1})^{-1}\|_\infty \|R\|_\infty \\ &\leqslant \frac{\|A^{-1}\|_\infty \|R\|_\infty}{1-\|BA^{-1}\|_\infty} \leqslant \frac{\|A^{-1}\|_\infty \|R\|_\infty}{1-\|B\|_\infty\|A^{-1}\|_\infty} \\ &\leqslant \frac{NLh^{n+1}\mathrm{e}^{b-a}[h+(b-a)M]}{[1-(b-a)ML\mathrm{e}^{h+(b-a)}](n+1)!}.\end{aligned}$$

It is easy to obtain

$$\begin{aligned}&\|\varphi(x)-\varphi_{m,n}(x)\|_\infty \\ &= \left\|\sum_{q=0}^{m-1} h\int_0^1 K(x,x_q+h\eta)\left[\varphi(x_q+h\eta)-\sum_{j=0}^n \frac{\varphi_q^{(j)}h^j\eta^j}{j!}\right]d\eta\right\|_\infty \\ &\leqslant \left\|\sum_{q=0}^{m-1} h\int_0^1 K(x,x_q+h\eta)\left[\varphi(x_q+h\eta)-\sum_{j=0}^n \frac{\varphi^{(j)}(x_q)h^j\eta^j}{j!}\right.\right.\\ &\qquad\left.\left.+\sum_{j=0}^n \frac{\varphi^{(j)}(x_q)h^j\eta^j}{j!}-\sum_{j=0}^n \frac{\varphi_q^{(j)}h^j\eta^j}{j!}\right]d\eta\right\|_\infty \\ &\leqslant \left\|\sum_{q=0}^{m-1}\frac{h^{n+2}}{(n+1)!}\int_0^1 K(x,x_q+h\eta)\varphi^{(n+1)}(\theta_q)\eta^{n+1}\,d\eta\right\|_\infty \\ &\quad+\left\|\sum_{q=0}^{m-1}\sum_{j=0}^n \frac{h^{j+1}}{j!}\left[\varphi^{(j)}(x_q)-\varphi_q^{(j)}\right]\int_0^1 K(x,x_q+h\eta)\eta^j\,d\eta\right\|_\infty \\ &\leqslant \frac{(b-a)MNh^{n+1}}{(n+1)!}+hM\sum_{q=0}^{m-1}\sum_{j=0}^n \frac{h^j}{j!}\cdot\|\varphi^{(j)}(x_q)-\varphi_q^{(j)}\|_\infty \\ &\leqslant \frac{(b-a)MNh^{n+1}}{(n+1)!}+(b-a)M\mathrm{e}^h\cdot\frac{NLh^{n+1}\mathrm{e}^{b-a}[h+(b-a)M]}{[1-(b-a)ML\mathrm{e}^{h+(b-a)}](n+1)!} \\ &\leqslant \frac{(b-a)MN[1+Lh\mathrm{e}^{h+(b-a)}]}{[1-(b-a)ML\mathrm{e}^{h+(b-a)}]}\cdot\frac{h^{n+1}}{(n+1)!}.\end{aligned}$$

In the end, we have
$$\lim_{n\to+\infty} \|\varphi_{m,n}(x) - \varphi(x)\|_\infty = 0,$$
and
$$\lim_{h\to 0} \|\varphi_{m,n}(x) - \varphi(x)\|_\infty = 0.$$

The proof is completed.

Similar to those in [29], a pair of feasible values for m (i.e., h) and n can be chosen to obtain a good approximation of the exact solution. The above observations will be further verified by using the numerical examples in the next section.

4 Numerical results

In order to show the effectiveness of the proposed methods, we give several numerical examples. First, we compare the proposed method in the present paper with that in [29] and give Example 4.1.

Example 4.1 [29] A three-point boundary value problem for a second-order ordinary differential equation with variable coefficients is given as
$$\begin{cases} \varphi''(x) - (1+\sin x)\varphi(x) = -\mathrm{e}^x \sin x, \\ \varphi(0) = 1, \quad \varphi(1) + \varphi(1/2) = \mathrm{e} + \mathrm{e}^{1/2}, \end{cases}$$
with $x \in [0,1]$ and the exact solution $\varphi(x) = \mathrm{e}^x$.

As shown in [29], the solution is unique and the approximate solution has been calculated. Here one computes the example again by using the present method and the results are given in Table 1. The parameters m and n are still chosen as $(m,n) = (2,2)$, $(m,n) = (2,3)$, $(m,n) = (2,4)$, $(m,n) = (4,2)$, $(m,n) = (4,3)$ and $(m,n) = (4,4)$ respectively. As compared to the observations in Table 1 and those in [29], the results in Table 1 have more accuracy than those in [29] for the same pair parameters of (m,n). It is further seen that the proposed method in the present study can be used not only to Fredholm integral equations of the second kind with continuous kernels, but also to those with weakly singular kernels. That is to say, the present method is more adaptable than that in [29]. To further show the effectiveness of the present method, the exact and the approximate solutions are plotted in Fig. 1 for $(m,n) = (1,0)$ with $y = \varphi(x)$ or $\varphi_{m,n}(x)$.

Second, we consider the generalized Bagley-Torvik equation with three-point boundary value conditions for $\alpha = 3/2$ and give Example 4.2.

Table 1 The absolute errors of approximate and exact solutions
for Example 4.1

x	$|\varphi(x) - \varphi_{m,n}(x)|$ for various pairs of (m,n)					
	(2, 2)	(2, 3)	(2, 4)	(4, 2)	(4, 3)	(4, 4)
0.10	2.4409e-6	5.3333e-8	9.7388e-10	2.3944e-7	1.5073e-09	2.9159e-12
0.20	9.4559e-6	1.0982e-7	5.2815e-10	2.0862e-7	2.2121e-09	1.0288e-11
0.30	1.4172e-5	6.0804e-8	2.4506e-09	7.9467e-8	1.1769e-09	1.0706e-11
0.40	8.1844e-6	1.8787e-7	1.7840e-09	5.4620e-7	8.9138e-10	2.3890e-11
0.50	6.4122e-7	2.2406e-9	2.4513e-12	1.9920e-8	1.8241e-11	4.5606e-15
0.60	4.5172e-6	1.2861e-7	2.3540e-09	5.3349e-7	3.6835e-09	6.1419e-12
0.70	2.0468e-5	2.6723e-7	1.0004e-09	4.7660e-7	4.9360e-09	2.2609e-11
0.80	3.1420e-5	1.0660e-7	5.4493e-09	1.7008e-7	2.5689e-09	2.3686e-11
0.90	1.9112e-5	3.9152e-7	3.5898e-09	1.1865e-6	1.6643e-09	5.0942e-11
1.00	6.4122e-7	2.2406e-9	2.4513e-12	1.9920e-8	1.8241e-11	4.5607e-15

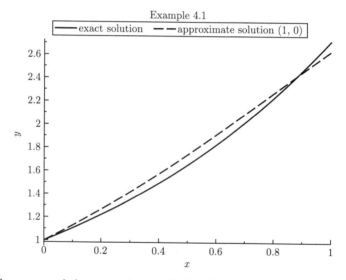

Fig. 1 The exact and the approximate solutions for $(m,n) = (1,0)$ with $y = \varphi(x)$ or $\varphi_{m,n}(x)$

Example 4.2 Assume that a three-point boundary value problem for the generalized Bagley-Torvik equation is expressed as

$$\begin{cases} \varphi''(x) + \sqrt{\pi} x^2 D^{\frac{3}{2}} \varphi(x) + (1 - 4x^{\frac{1}{2}})\varphi(x) = x^2 + 2, \\ \varphi(0) = 0, \quad \varphi(1/5) + \varphi(1/10) = 1/20, \end{cases}$$

with $x \in \left[0, \dfrac{1}{5}\right]$ and the exact solution $\varphi(x) = x^2$.

Now we deal with the uniqueness of the solution of Example 4.2. It is calculated that
$$|V(x,t)| = \left|(x-t)(1-4\sqrt{t}) - 8t\sqrt{x-t} + \frac{x^2}{\sqrt{x-t}}\right|.$$
According to Theorem 2.1 one has
$$\max_{0 \leqslant x \leqslant \frac{1}{5}} \int_0^{\frac{1}{5}} |K_1(x,t)|\, dt$$
$$= \max_{0 \leqslant x \leqslant \frac{1}{5}} \int_0^x |V(x,t)|\, dt + \frac{10x}{3} \int_0^{\frac{1}{5}} \left|V(\frac{1}{5},t)\right|\, dt + \frac{10x}{3} \int_0^{\frac{1}{10}} \left|V(\frac{1}{10},t)\right|\, dt$$
$$\leqslant \max_{0 \leqslant x \leqslant \frac{1}{5}} \left[\frac{26}{5}x^{\frac{5}{2}} + \frac{1}{2}x^2 + \left(\frac{1}{12} + \frac{52\sqrt{5}}{375} + \frac{13\sqrt{10}}{750}\right)x\right] \approx 0.2027 < 1.$$

It is seen that the boundary value problem has a unique solution by using Theorem 2.3. Moreover, the absolute errors of the approximate and exact solutions are shown in Table 2. One can find from Table 2 that a good approximation has been obtained by using the proposed method. As shown in Fig. 2, the absolute errors between the approximate and exact solutions are very small even for $(m,n) = (1,0)$.

Table 2 The absolute errors of approximate and exact solutions for Example 4.2

x	$|\varphi(x) - \varphi_{m,n}(x)|$ for various pairs of (m,n)					
	(2,0)	(2,1)	(4,0)	(4,1)	(8,0)	(8,1)
0.02	9.8220e-6	1.7560e-6	4.8489e-6	3.3231e-7	2.0006e-6	6.0631e-8
0.04	1.9457e-5	3.6174e-6	9.5112e-6	6.2413e-7	3.8910e-6	1.3845e-7
0.06	2.8426e-5	5.2283e-6	1.4591e-5	1.1829e-6	5.8598e-6	2.2686e-7
0.08	3.5928e-5	5.7881e-6	1.7301e-5	1.6487e-6	7.9866e-6	2.7619e-7
0.10	4.0843e-5	4.0770e-6	1.6630e-5	7.4029e-7	6.2251e-6	1.3286e-7
0.12	6.5460e-5	1.4499e-5	2.9073e-5	2.9104e-6	7.9716e-6	3.7253e-7
0.14	6.2870e-5	1.9870e-5	1.8837e-5	2.2855e-6	1.1839e-5	6.1163e-7
0.16	4.5301e-5	2.0809e-5	4.1940e-5	3.2812e-6	1.7801e-5	7.3445e-7
0.18	7.1547e-5	1.4117e-5	3.5404e-5	4.6672e-6	2.2642e-5	6.3016e-7
0.20	4.0843e-5	4.0770e-6	1.6630e-5	7.4029e-7	6.2251e-6	1.3286e-7

Third, we consider $\alpha = 1/2$ and give Example 4.3.

Example 4.3 A three-point boundary value problem for the fractional Bagley-Torvik equation with variable coefficients is written as
$$\begin{cases} \varphi''(x) - 5\sqrt{\pi}xD^{\frac{1}{2}}\varphi(x) + 16x^{\frac{1}{2}}\varphi(x) = 6x, \\ \varphi(0) + 2\varphi(1/10) = 1/500, \quad \varphi(1/5) = 1/125, \end{cases}$$

with $x \in \left[0, \dfrac{1}{5}\right]$ and the exact solution $\varphi(x) = x^3$.

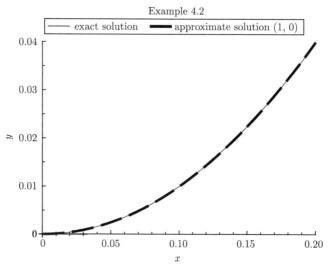

Fig. 2 The exact and the approximate solutions for $(m, n) = (1, 0)$ with $y = \varphi(x)$ or $\varphi_{m,n}(x)$

Now we deal with the uniqueness of the solution of Example 4.3. It is calculated that
$$|V(x,t)| = \left|16\sqrt{t}(x-t) + \frac{10}{3}(x-4t)\sqrt{x-t}\right|,$$
$$|\Lambda(x,t)| = \left|\frac{10}{3}(x-4t)\sqrt{x-t} + \frac{2}{3}(20t - 15x + 1)\sqrt{\frac{1}{10} - t} + \frac{1}{20}\frac{10x-1}{\sqrt{\frac{1}{10} - t}}\right|.$$

According to Theorem 2.2 one has
$$\max_{0 \leqslant x \leqslant \frac{1}{5}} \int_0^{\frac{1}{5}} |K_2(x,t)|\, dt$$
$$\leqslant \max_{0 \leqslant x \leqslant \frac{1}{5}} \left\{ \frac{1}{2} \int_0^x |V(x,t)|\, dt + \frac{5x}{2} \int_0^{\frac{1}{5}} \left|V(\frac{1}{5}, t)\right| dt + \frac{1}{2} \int_{\frac{1}{10}}^x |V(x,t)|\, dt \right.$$
$$+ 5\left(x + \frac{1}{10}\right) \int_{\frac{1}{10}}^{\frac{1}{5}} \left|V(\frac{1}{5}, t)\right| dt + \frac{1}{2} \int_0^{\frac{1}{10}} |\Lambda(x,t)|\, dt$$
$$\left. + 5\left(x + \frac{1}{10}\right) \int_0^{\frac{1}{10}} \left|\Lambda(\frac{1}{5}, t)\right| dt \right\} \approx 0.7693 < 1.$$

So the boundary value problem also has a unique solution by using Theorem 2.4. Furthermore, the numerical results are carried out and shown in Table 3 and Fig. 3

by using the present method. As shown in Tables 1–3, one can see that when m or n is increasing, the absolute errors $|\varphi(x) - \varphi_{m,n}(x)|$ is decreasing. The obtained results are in good agreement with the theoretical analysis in Theorem 3.1.

Table 3 **The absolute errors of approximate and exact solutions for Example 4.3**

x	$|\varphi(x) - \varphi_{m,n}(x)|$ for various pairs of (m, n)					
	$(2, 0)$	$(2, 1)$	$(2, 2)$	$(4, 0)$	$(4, 1)$	$(4, 2)$
0.00	1.8750e-6	1.6321e-6	3.3402e-9	1.3150e-6	1.9344e-7	8.1180e-11
0.02	1.3157e-6	9.3094e-7	2.0659e-9	9.3937e-7	1.1255e-7	2.5205e-10
0.04	7.5519e-7	2.6518e-7	1.1401e-8	5.6299e-7	4.0605e-8	1.2628e-09
0.06	3.9308e-7	1.0391e-7	3.9217e-8	2.0555e-7	2.7457e-8	1.1225e-09
0.08	6.7104e-7	3.9608e-7	5.6790e-8	1.7126e-7	1.1906e-7	3.7779e-10
0.10	1.3749e-6	8.1603e-7	1.6701e-9	6.5748e-7	9.6719e-8	4.0590e-11
0.12	2.4253e-6	1.4781e-6	5.1651e-8	4.8432e-7	1.8599e-7	3.4848e-09
0.14	3.5371e-6	2.5651e-6	7.0735e-8	8.0560e-7	2.5485e-7	1.4666e-09
0.16	3.6703e-6	3.3204e-6	1.8646e-8	5.0344e-7	7.7106e-8	3.3408e-09
0.18	3.5553e-6	2.8250e-6	6.3534e-8	1.1779e-6	3.0832e-7	3.5001e-09

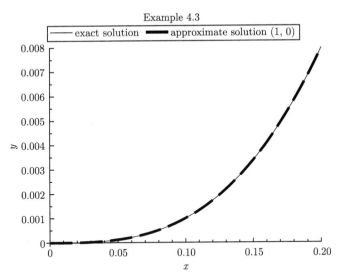

Fig. 3 The exact and the approximate solutions for $(m, n) = (1, 0)$ with $y = \varphi(x)$ or $\varphi_{m,n}(x)$

5 Conclusions

To simulate a real case, the Bagley-Torvik equation is generalized by equipping variable coefficients and three-point boundary conditions. The integration method is proposed to transform the three point boundary value problems into Fredholm integral equations of the second kind. Different from those for ordinary differential equations, the obtained Fredholm integral equations of the second kind may be with weakly singular kernels depending on the fractional order and boundary value conditions. A new numerical method for Fredholm integral equations of the second kind with weakly singular kernels has been further proposed. The approximate solution has been constructed, and its convergence and error estimate have been made. Numerical results have been shown that the proposed solution procedures are effective.

Acknowledgements   The authors would like to thank the reviewer for the valuable suggestions and comments of improving the paper. The work was supported by the National Natural Science Foundation of China (No. 11362002), the project of outstanding young teachers' training in higher education institutions of Guangxi, and the project of Guangxi Colleges and Universities Key Laboratory of Mathematics and Its Applications.

References

[1] Ortigueira M D. Fractional Calculus for Scientists and Engineers [M]. Berlin: Springer-Verlag, 2011.

[2] Teodor M A, Stevan P, Bogoljub S, Dusan Z. Fractional Calculus with Applications in Mechanics: Vibrations and Diffusion Processes [M]. New Jersey: Wiley-ISTE, 2014.

[3] Uchaikin V V. Fractional Derivatives for Physicists and Engineers [M]. Beijing: Higher Education Press, 2013.

[4] Torvik P J, Bagley R L. On the appearance of the fractional derivative in the behavior of real materials [J]. ASME Trans. J. Appl. Mech., 1984, 51: 294-298.

[5] Diethelm K, Ford J. Numerical solution of the Bagley-Torvik equation [J]. BIT Numer. Math., 2002, 42(3): 490-507.

[6] Ray S S, Bera R K. Analytical solution of the Bagley-Torvik equation by Adomian decomposition method [J]. Appl. Math. Comput., 2005, 168: 398-410.

[7] Zolfaghari M, Ghaderi R, SheikholEslami A, Ranjbar A, Hosseinnia S H, Momani S, Sadati J. Application of the enhanced homotopy perturbation method to solve the fractional-order Bagley-Torvik differential equation[J]. Phys. Scr., 2009, T136:014032.

[8] Wang Z H, Wang X. General solution of the Bagley-Torvik equation with fractional-order derivative [J]. Commun. Nonlinear Sci. Numer. Simulat., 2010, 15: 1279-1285.

[9] Cenesiz Y, Keskin Y, Kurnaz A. The solution of the Bagley-Torvik equation with the generalized Taylor collocation method [J]. J. Franklin Institute, 2010, 347: 452-466.

[10] Raja M A Z, Khan J A, Qureshi I M. Solution of fractional order system of Bagley-Torvik equation using evolutionary computational intelligence [J]. Math. Probl. Eng., 2011, 2011: 675075.

[11] Ray S S. On Haar wavelet operational matrix of general order and its application for the numerical solution of fractional Bagley-Torvik equation [J]. Appl. Math. Comput., 2012, 218: 5239-5248.

[12] Mekkaoui T, Hammouch Z. Approximate analytical solutions to the Bagley-Torvik equation by the fractional iteration method [J]. An. Univ. Craiova Ser. Mat. Inform., 2012, 39(2): 251-256.

[13] Atanackovic T M, Zorica D. On the Bagley-Torvik equation [J]. ASME J. Appl. Mech., 2013, 80:041013.

[14] Setia A, Liu Y C, Vatsala A S. The solution of the Bagley-Torvik equation by using second kind Chebyshev wavelet [C]. IEEE 11th International Conference on Information Technology: New Generations, 2014: 443-446.

[15] Cermak J, Kisel T. Exact and discretized stability of the Bagley-Torvik equation [J]. J. Comput. Appl. Math., 2014, 269: 53-67.

[16] Stanek S. Two-point boundary value problems for the generalized Bagley-Torvik fractional differential equation [J]. Cent. Eur. J. Math., 2013, 11(3): 574-593.

[17] Keshavarz E, Ordokhani Y, Razzaghi M. Bernoulli wavelet operational matrix of fractional order integration and its applications in solving the fractional order differential equations [J]. Appl. Math. Model., 2014, 38: 6038-6051.

[18] Lubich C. Discretized fractional calculus [J]. SIAM J. Math. Anal., 1986, 17: 704-719.

[19] Podlubny I. Fractional Differential Equations [M]. New York: Academic Press, 1993.

[20] Kazem S, Abbasbandy S, Kumar S. Fractional-order Legendre functions for solving fractional-order differential equations [J]. Appl. Math. Model., 2013, 37: 5498-5510.

[21] Diethelm K. Generalized compound quadrature formulae for finite-part integrals [J]. IMA J. Numer. Anal., 1997, 17: 479-493.

[22] Diethelm K, Ford N J, Freed A D, Luchko Y. Algorithms for the fractional calculus: A selection of numerical methods [J]. Comput. Methods Appl. Mech. Engrg., 2005, 194: 743-773.

[23] Lopez-Fernandez M, Lubich C, Schadle A. Adaptive, fast, and oblivious convolution in evolution equations with memory [J]. SIAM J. Sci. Comput., 2008, 30: 1015-1037.

[24] Ma X H, Huang C M. Spectral collocation method for linear fractional integro-differential equations [J]. Appl. Math. Model., 2014, 38: 1434-1448.

[25] Li J R. A fast time stepping method for evaluating fractional integrals [J]. SIAM J. Sci. Comput., 2010, 31: 4696-4714.

[26] Bai Z B. On positive solutions of a nonlocal fractional boundary value problem [J]. Nonlinear Anal., 2010, 72: 916-924.

[27] Jia Y L, Zhang X Q. Positive solutions for a class of fractional differential equation multi-point boundary value problems with changing sign nonlinearity [J]. J. Appl. Math. Comput., 2015, 47: 15-31.

[28] Ahmad B, Ntouyas S K. On Hadamard fractional integro-differential boundary value problems [J]. J. Appl. Math. Comput., 2015, 47: 119-131.

[29] Zhong X C, Huang Q A. Approximate solution of three-point boundary value problems for second-order ordinary differential equations with variable coefficients [J]. Appl. Math. Comput., 2014, 247: 18-29.

[30] Richter G R. On weakly singular Fredholm integral equations with displacement kernels [J]. J. Math. Anal. Appl., 1976, 55: 32-42.

[31] Goursat E. A Course in Mathematical Analysis, Vol. III Part 2: Integral Equations, Calculus of Variations [M]. New York: Dover, 1964.

[32] Lang S. Real and Functional analysis [M]. Berlin: Springer-Verlag, 1993.

Blow-up of Smooth Solution to Quantum Hydrodynamic Models in $\mathbf{R}^d$[*]

Guo Boling (郭柏灵) and Wang Guangwu (王光武)

Abstract In this paper we firstly investigate the local-in-time existence of smooth solution for the quantum hydrodynamic models (QHD) in $\mathbf{R}^d$. Then we prove that any smooth solution of QHD model which satisfies suitable conditions will blow up in finite time. The model can be derived from nonlinear Schrödinger equation by a Madelung transformation. The main idea is based on the construction of approximate solution and energy inequality.

Keywords quantum hydrodynamic model; local-in-time existence; blow-up of the smooth solution

1 Introduction

In 1927, Madelung [20] gave a fluid-dynamical description of quantum systems governed by the Schrödinger equation for the wave function ψ:

$$i\varepsilon\partial_t\psi = -\frac{\varepsilon^2}{2}\Delta\psi + h(|\psi|^2)\psi, \quad x \in \mathbf{R}^d, \tag{1.1}$$

$$\psi(x,0) = \psi_0, \quad \text{in } \mathbf{R}^d, \tag{1.2}$$

where $d \geqslant 1$, $\varepsilon > 0$ is the scaled Planck constant, and h is an integrable function such that

$$h'(\rho) = \frac{P'(\rho)}{\rho}. \tag{1.3}$$

Equation (1.1) prescribes multi-particle approximations in the mean field theory of quantum mechanics, when one considers a large number of quantum particles acting

---

[*] J. Diff. Eqns., 2016, 261(7): 3815–3842.

in unison and takes into account only a finite number of particle-particle interactions. In fact the complex-valued wave function ψ can be rewritten as

$$\psi = \sqrt{\rho}\exp\left(\frac{iS}{\varepsilon}\right), \tag{1.4}$$

where $\rho = |\psi|^2$ is particle density, and S is some phase function.

From (1.4) we have

$$\psi_t = (\sqrt{\rho})_t \exp\left(\frac{iS}{\varepsilon}\right) + \sqrt{\rho}\exp\left(\frac{iS}{\varepsilon}\right)\frac{iS_t}{\varepsilon}, \tag{1.5}$$

$$\Delta\psi = \Delta(\sqrt{\rho})\exp\left(\frac{iS}{\varepsilon}\right) + 2\nabla\sqrt{\rho}\cdot\nabla S\frac{i}{\varepsilon}\exp\left(\frac{iS}{\varepsilon}\right)$$

$$+ \sqrt{\rho}\exp\left(\frac{iS}{\varepsilon}\right)\frac{-1}{\varepsilon^2}(\nabla S \otimes \nabla S) + \sqrt{\rho}\exp\left(\frac{iS}{\varepsilon}\right)\frac{i\nabla(\nabla S)}{\varepsilon}. \tag{1.6}$$

Then by plugging (1.5) and (1.6) into (1.1), the imaginary and real parts of the resulting equation are

$$\rho_t = -\nabla\rho\cdot\nabla S - \rho\nabla\cdot(\nabla S), \tag{1.7}$$

$$S_t = \frac{\varepsilon^2}{2}\frac{\Delta\sqrt{\rho}}{\sqrt{\rho}} - \frac{1}{2}(\nabla S \otimes \nabla S) - h(\rho). \tag{1.8}$$

Then we set $u = \nabla S$, we can formally get the equations about ρ, u:

$$\rho_t + \mathrm{div}(\rho u) = 0, \tag{1.9}$$

$$(\rho u)_t + \mathrm{div}(\rho u \otimes u) + \nabla P(\rho) = \frac{\varepsilon^2}{2}\rho\nabla\left(\frac{\Delta\sqrt{\rho}}{\sqrt{\rho}}\right). \tag{1.10}$$

Here $t > 0$ and $x \in \mathbf{R}^d$ and $(u\otimes u)_{ij} = u_i u_j$. The pressure function $P(\rho) = A\rho^\gamma$ ($A > 0$ and $\gamma > 1$ are constants).

Initial data is assigned which is constants outside the bounded set $\{|x| \geqslant R\}$:

$$\rho(x,0) = \rho_0(x) > 0, \quad \rho_0(x) = \bar{\rho}, \quad |x| \geqslant R, \tag{1.11}$$

$$u(x,0) = u_0(x), \quad u_0(x) = \bar{u}(=0), \quad |x| \geqslant R. \tag{1.12}$$

The formal equivalence between (1.1)-(1.2) and (1.9)-(1.12) is explained more in details in the following lemma:

Lemma 1.1 *Let ψ be a solution to (1.1)-(1.2) with initial datum $\psi_0 := \sqrt{\rho_0}\exp\cdot\left(\frac{iS_0}{\varepsilon}\right)$; the functions $\rho := |\psi|^2$ and $u := \varepsilon\frac{Im(\bar{\psi}\nabla\psi)}{|\psi|^2}$ solve (1.9)-(1.12) with initial* data $\rho_0 = |\psi_0|^2$ and $u_0 = \nabla S_0$ as long as $\rho > 0$.

Vice versa let (ρ, u) be a solution to (1.9)-(1.12), where $u = \nabla S$ and $\rho > 0$ in $\mathbf{R}^d$, $t > 0$; the function $\psi := \sqrt{\rho}\exp\left(\dfrac{iS}{\varepsilon}\right)$ solves (1.1)-(1.2).

The equations (1.9)-(1.10) can be interpreted as the pressureless Euler equations including the quantum Bohm potential $\dfrac{\varepsilon^2}{2}\dfrac{\Delta\sqrt{\rho}}{\sqrt{\rho}}$. It has been used for the modeling of superfluid phenomena like Helium II [16, 17, 19]. They also can be described by a dispersive perturbation associated to the Hamilton-Jacobi system for compressible fluid dynamics, sometimes referred as a dispersive perturbation of the Eikonal equation for the evolution of amplitude and phase velocity of quantum wave guides. This system (1.9)-(1.10) can be also used to describe the Bose-Einstein condensation in trapped gases [3].

Gardner [7] and Gasser et al. [8, 9] have derived a Madelung-type equations which include an exterior electric field and a momentum relaxation term taking into account interactions of the electrons in the semiconductor crystal, and are self-consistently coupled to the Poisson equation for the electrostatic potential ϕ:

$$\rho_t + \text{div}(\rho u) = 0, \tag{1.13}$$

$$(\rho u)_t + \text{div}(\rho u \otimes u) + \nabla P(\rho) = \dfrac{\varepsilon^2}{2}\rho\nabla\left(\dfrac{\Delta\sqrt{\rho}}{\sqrt{\rho}}\right) + \rho\nabla\phi - \dfrac{\rho u}{\tau}, \tag{1.14}$$

$$\lambda^2 \Delta\phi = \rho - \mathcal{C}(x). \tag{1.15}$$

Remark 1.1 *The equations (1.13)-(1.15) are referred as the quantum Euler-Poisson system or as the quantum hydrodynamic model. It is not difficult to find that (1.10) and (1.14) are different from the last two term: the exterior electric field $\rho\nabla\phi$ and the relaxation term $\dfrac{\rho u}{\tau}$.*

In the recent years, equations (1.13)-(1.15) has attracted a lot of interest and many papers have been published. In [13] Jüngel and Li proved local-in-time existence of smooth solutions to (1.13)-(1.15) for the one-dimensional case with Dirichlet and Neumann boundary conditions for the particle density ρ. Moreover, if the initial data are close enough to the steady-state solution, the local-in-time solutions will exist globally in time(so-called small solutions). In [18], Li and Marcati proved the local-in-time existence of the solutions for equations (1.13)-(1.15) in the multi-dimensional torus for the irrotational velocity, and they also proved that local-in-time solutions exit globally and converge toward the corresponding steady-state as $t \to \infty$ under a "subsonic" type stability condition. In [12], Huang, Li, Matsumura and Odanaka proved the local-in-time and global-in-time solutions of equations (1.13)-(1.15) in $\mathbf{R}^n$ for rotational fluid. Here we will emphasize that because of the existence of the relaxation term we can show the local-in-time solution is globally existing near the equilibrium state and exponentially decaying to the equilibrium state.

In [1], Antonelli and Marcati have proved the existence of the global weak solutions of (1.13)-(1.15) with the finite energy in three dimension. In [2], Antonelli and Marcati have given the finite energy weak solution of the equations (1.9)-(1.10) with Poisson equation and (1.13)-(1.15) in two dimension. For the multi-dimensional case, Gasser and Markowich [8] have proved the existence of the global finite energy weak solution to the system (1.9)-(1.10) with a electric potential term. We should emphasize that there is no contradiction between the existence of the global weak solution to the blow-up of the smooth solution, because we cannot derive the existence of the global smooth solution from the existence of the global weak solution. Indeed, From [1, 2, 8], we know that for the weak solution we only have that $\sqrt{\rho} \in L^2_{loc}((0,T); H^1_{loc}(\mathbf{R}^d))$, $\sqrt{\rho} u \in L^2_{loc}([0,T); L^2_{loc}(\mathbf{R}^d))$, and the (1.9)-(1.10) holds in the distribution sense. But from the proof of the blow-up in this paper, we need that $\rho \in C([0,T); C^3(\mathbf{R}^d))$ and $u \in C([0,T); C^2(\mathbf{R}^d))$.

In [5, 6], Gamba and Jüngel studied the positive steady-state solutions to the potential flow of (1.9)-(1.10). The existence of the solutions to one-dimensional stationary viscous quantum hydrodynamic model has been shown in [11, 14].

In [4], Gamba, Gualdani and Zhang proved that the smooth solution of (1.9)-(1.10) will blow-up in finite time for some boundary condition in a bounded domain. For the focusing nonlinear Schrödinger equation (1.1), Glassey in [10] showed that the smooth solution will blow up in finite time. The main ideal of the proof in [4] or [10] is that the non-negative quantity $I(t)$ will be negative if the smooth solution exists long enough. This idea also have been used in the study of the finite time blowing up of smooth solutions to the compressible Navier-Stokes equations by Xin [27].

In [23], Sideris showed that a full Euler fluid will develop singularities if, on average, it is slightly compressed and out-going near the wave front or supersonic in some region.

In the former part of this paper, we will show the local-in-time existence of (1.9)-(1.10), The main method about local-in-time existence is similar to [12]. It is convenient to make use of the variable transformation $\rho = \zeta^2$ in (1.9)-(1.12). Then we derive the corresponding IVP for (ζ, u):

$$2\zeta \cdot \partial_t \zeta + \nabla \cdot (\zeta^2 u) = 0, \tag{1.16}$$

$$\partial_t u + (u \cdot \nabla) u + \nabla h(\zeta^2) = \frac{\varepsilon^2}{2} \nabla \left(\frac{\Delta \zeta}{\zeta} \right), \tag{1.17}$$

$$\zeta(x, 0) = \zeta_0(x) = \sqrt{\rho_0}, \quad u(x, 0) = u_0(x). \tag{1.18}$$

Then the first result is the following local-in-time existence theorem.

Theorem 1.1 (local existence) *Suppose* $P(\rho) = A\rho^\gamma (A > 0, \gamma > 1)$. *And* $\zeta_* = \inf_{x \in \mathbf{R}^d} \zeta_0(x) > 0$.

(1) *For* $d = 1, 2, 3$, *assume* $(\rho_0, u_0) \in H^6(\mathbf{R}^d) \times \mathcal{H}^5(\mathbf{R}^d)$. *Then, there exists* $T_{**} > 0$, *such that there exists a unique solution* (ζ, u) *to the IVP* (1.16)-(1.18), *with* $\zeta > 0$, *which satisfies*

$$\zeta - \sqrt{\bar{\rho}} \in C^i([0, T_{**}]; H^{6-2i}(\mathbf{R}^d)) \cap C^3([0, T_{**}]; L^2(\mathbf{R}^d)), \quad i = 0, 1, 2; \tag{1.19}$$

$$u \in C^i([0, T_{**}]; \mathcal{H}^{5-2i}(\mathbf{R}^d)), \quad i = 0, 1, 2. \tag{1.20}$$

(2) *For* $d > 3$, *assume* $(\rho_0, u_0) \in H^{2d}(\mathbf{R}^d) \times \mathcal{H}^{2d-1}(\mathbf{R}^d)$. *Then, there exists* $T_{**} > 0$, *such that there exists a unique solution* (ζ, u) *to the IVP* (1.16)-(1.18), *with* $\zeta > 0$, *which satisfies*

$$\zeta - \sqrt{\bar{\rho}} \in C^i([0, T_{**}]; H^{2d-2i}(\mathbf{R}^d)) \cap C^d([0, T_{**}]; L^2(\mathbf{R}^d)), \tag{1.21}$$

$$u \in C^i([0, T_{**}]; \mathcal{H}^{2d-1-2i}(\mathbf{R}^d)), \tag{1.22}$$

where $i = 0, 1, \cdots, d - 1$. Hereafter

$$\mathcal{H}^k(\mathbf{R}^d) = \begin{cases} \{f \in L^q(\mathbf{R}^d), Df \in H^{k-1}(\mathbf{R}^d)\}, & q = \dfrac{2d}{d-2}, \quad d > 3, \\ \{f \in L^q(\mathbf{R}^d), Df \in H^{k-1}(\mathbf{R}^d)\}, & q = 6, \quad d = 1, 2, 3, \end{cases} \quad k \geqslant 1.$$

In the final installment of this article, we will give some results about the blow-up of the smooth solution to the equations (1.9)-(1.12) in finite time. The main method is similar to [23].

From the initial condition (1.11)-(1.12) we know that the maximum speed of propagation of the front of a smooth disturbance is governed by the sound speed

$$\sigma = [P(\bar{\rho})]^{\frac{1}{2}} = [A\bar{\rho}^\gamma]^{\frac{1}{2}}, \tag{1.23}$$

since $\bar{u} = 0$. More precisely, letting

$$D(t) = \{x \in \mathbf{R}^d : |x| \geqslant R + \sigma t\}. \tag{1.24}$$

We have the following proposition:

Proposition 1.1 *If (ρ, u) is a smooth solution of (1.9)-(1.12) on $D(t)$, for $0 \leqslant t \leqslant T$, then $(\rho, u) \equiv (\bar{\rho}, 0)$ on $D(t)$, $0 \leqslant t < T$.*

This is a consequence of local energy estimates; see the proposition in [24].

Our first result on formation of singularities, arising form "large" initial disturbances, relies on the relationship among the quantities

$$m(t) = \int_{\mathbf{R}^d} (\rho(x,t) - \bar{\rho}) dx, \tag{1.25}$$

$$F(t) = \int_{\mathbf{R}^d} x \cdot \rho u(x) dx, \tag{1.26}$$

which represents mass and the radial component of momentum respectively.

Theorem 1.2 *Suppose (ρ, u) is a smooth solution of system (1.9)-(1.12) for $0 \leqslant t \leqslant T$. If*

$$m(0) \geqslant 0, \tag{1.27}$$

$$F(0) \geqslant (d+1)\beta_d \sigma R^{d+1} \max \rho_0(x), \quad \beta_d \text{ is the volume of unit ball in } \mathbf{R}^d, \tag{1.28}$$

then the life span T is finite. Furthermore, T satisfies

$$T \leqslant \frac{1}{\sigma} \left(\frac{F(0) R^{d+1}}{F(0) - \beta_d (d+1) \sigma \max \rho^0 R^{d+1}} \right)^{\frac{1}{d+1}} - \frac{1}{\sigma} R.$$

To illustrate one way in which (1.27)-(1.28) can be satisfied, take as initial conditions

$$\rho^0 = \bar{\rho}.$$

Then $m(0) = 0$ and (1.28) holds if

$$\int_{\mathbf{R}^d} x \cdot u_0(x) dx \geqslant (d+1)\beta_d R^{d+1}.$$

Comparing both sides, we find that the initial flow velocity must be supersonic in some region.

From (1.9)-(1.10) we can define the energy

$$E(x,t) = \int_{\mathbf{R}^d} \left(\frac{1}{2}\rho|u|^2 + \frac{A}{\gamma-1}\rho^\gamma + \frac{\varepsilon^2}{2}|\nabla\sqrt{\rho}|^2 \right) dx. \tag{1.29}$$

Moreover we can define

$$e(t) = \int_{\mathbf{R}^d} (E(x,t) - \bar{E}) dx. \tag{1.30}$$

We will also prove a slight different "large data" result:

Theorem 1.3 *Suppose (ρ, u) is a smooth solution of (1.9)-(1.12) for $0 \leqslant t \leqslant T$. If*

$$m(0) \geqslant 0, \tag{1.31}$$

$$e(0) \geqslant 0, \quad 1 < \gamma < \frac{d+2}{2}, \tag{1.32}$$

$$F(0) \geqslant \frac{2\beta_d}{2 - d(\gamma-1)} R^{d+1} \max \rho^0(x), \quad \beta_d \text{ is the volume of unit ball in } \mathbf{R}^d, \tag{1.33}$$

then the life span T is finite. Moreover, the life span T satisfies

$$T \leqslant \frac{1}{\sigma} \left(\frac{(2 - d(\gamma-1))F(0)R^{d+1}}{(2 - d(\gamma-1))F(0) - 2\beta_d(d+1)\sigma \max \rho^0 R^{d+1}} \right)^{\frac{1}{d+1}} - \frac{1}{\sigma} R.$$

This paper is arranged as follows. In section 2, we present necessary results on divergence equation, vorticity equation, and a semi-linear fourth-order wave equation on $\mathbf{R}^d$; then list some known calculus inequalities. Section 3 is devoted to the proof of local-in-time existence. We first construct an approximate solution series, then prove the convergence. In section 4, we proof the blow-up of the smooth solutions.

Notation: C always denotes generic positive constant. $L^s(\mathbf{R}^d)$, $1 \leqslant s \leqslant \infty$, is the space of s-powers integral functions on $\mathbf{R}^d$ with norm $\|\cdot\|_{L^s}$. Particularly, the norm of the space of square integral functions on $\mathbf{R}^d$ is denoted by $\|\cdot\|$. $H^k(\mathbf{R}^d)$ with integer $k \geqslant 1$ denotes the usual Sobolev space of function f, satisfying $\partial_x^k f \in L^2(0 \leqslant i \leqslant k)$, with norm

$$\|f\|_k = \sqrt{\sum_{0 \leqslant |l| \leqslant m} \|D^l f\|^2},$$

here and after $D^\alpha = \partial_1^{\alpha_1} \partial_2^{\alpha_2} \cdots \partial_d^{\alpha_d}$ for $|\alpha| = \alpha_1 + \alpha_2 + \cdots + \alpha_d$ and $\partial_j = \partial_{x_j}$, $j = 1, 2, \cdots d$, for abbreviation. Especially, $\|\cdot\|_0 = \|\cdot\|$. Moreover, $W^{k,p}$, with $k \geqslant 1$, $p \geqslant 1$, denotes the space of functions with $D^l f \in L^p$, $0 \leqslant |l| \leqslant k$,

$$\mathcal{H}^k(\mathbf{R}^d) = \begin{cases} \{f \in L^q(\mathbf{R}^d), Df \in H^{k-1}(\mathbf{R}^d)\}, & q = \frac{2d}{d-2}, \quad d > 3, \\ \{f \in L^q(\mathbf{R}^d), Df \in H^{k-1}(\mathbf{R}^d)\}, & q = 6, \quad d = 1, 2, 3. \end{cases} \quad k \geqslant 1.$$

Let $T > 0$ and $\mathcal{B}$ be a Banach space. $C^k(0, T; \mathcal{B})(C^k([0, T]; \mathcal{B})$ resp.) denotes the space of $\mathcal{B}$-valued k-times continuously differentiable functions on $(0, T)$ (or $[0, T]$, resp.), $L^2([0, T]; \mathcal{B})$ the space of $\mathcal{B}$-valued L^2-functions on $[0, T]$, and $H^k([0, T]; \mathcal{B})$ the spaces of $f(x, t)$ with $\partial_t^i f \in L^2([0, T]; \mathcal{B})$, $1 \leqslant i \leqslant k$, $1 \leqslant p \leqslant \infty$.

2 Preliminaries

In this section, we list the existence and uniqueness of solution of divergence equation and vorticity equation in $\mathbf{R}^d$ without proof, mention the orthogonal decomposition of velocity vector field, and then turn to prove the well-posedness for an abstract second-order semi-linear wave equation. Finally, some useful calculus inequalities are listed.

First, we have the theorem on the divergence operator and vorticity operator in $\mathbf{R}^d$:

Theorem 2.1 *Let $f \in H^s(\mathbf{R}^d)$, $s \geqslant d/2$. There is a unique solution u of the divergence equation*

$$\nabla \cdot u = f, \quad \nabla \times u = 0, \quad u(x) \to 0, \quad |x| \to \infty, \tag{2.1}$$

satisfying

$$\begin{cases} \|u\|_{L^q} \leqslant C\|Du\|, \quad q = \dfrac{2d}{d-2}, \quad d > 3, \\ \|u\|_{L^q} \leqslant C\|Du\|, \quad q = 6, \quad d = 1, 2, 3, \end{cases} \quad \|Du\|_{H^s(\mathbf{R}^d)} \leqslant C\|f\|_{H^s(\mathbf{R}^d)}. \tag{2.2}$$

Theorem 2.2 *Let $f \in H^s(\mathbf{R}^d)$, $s \geqslant d/2$. There is a unique solution u of the vorticity equation*

$$\nabla \times u = f, \quad \nabla \cdot u = 0, \quad u(x) \to 0, \quad |x| \to \infty, \tag{2.3}$$

satisfying

$$\begin{cases} \|u\|_{L^q} \leqslant C\|Du\|, \quad q = \dfrac{2d}{d-2}, \quad d > 3, \\ \|u\|_{L^q} \leqslant C\|Du\|, \quad q = 6, \quad d = 1, 2, 3, \end{cases} \quad \|Du\|_{H^s(\mathbf{R}^d)} \leqslant C\|f\|_{H^s(\mathbf{R}^d)}. \tag{2.4}$$

Theorem 2.3 *For $u \in \mathcal{H}^s(\mathbf{R}^d)$, $s \geqslant d/2$, it has a unique decomposition consisting of the gradient vector field $v \in \mathcal{H}^s(\mathbf{R}^d)$ and the divergence free vector field $z \in \mathcal{H}^s(\mathbf{R}^d)$, i.e.,*

$$u = v + z = \mathcal{Q}u + \mathcal{P}u, \quad \mathcal{Q} = \mathcal{I} - \mathcal{P}, \quad \nabla \cdot \mathcal{P}u = 0.$$

Proof The proof of the above theorems can be made by using the standard arguments and the Riesz's potential theory. The reader can refer to [22, 26], we omit the details here. □

Finally, let us turn to consider an abstract initial value problem in Hilbert space $L^2(\mathbf{R}^d)$

$$u'' + Au + \mathcal{L}u' = F(t), \tag{2.5}$$

$$u(0) = u_0, \quad u'(0) = u_1, \tag{2.6}$$

where u' denotes $\dfrac{d}{du}$, and the operator A is given by

$$Au = \nu_0 \Delta^2 u + \nu_1 u, \tag{2.7}$$

where Δ the Laplace operator on $\mathbf{R}^d$, and $\nu_0, \nu_1 > 0$ constants. The domain of linear operator A is $D(A) = H^4(\mathbf{R}^d)$. Related to the operator A, we define a continuous and symmetric bilinear form $a(u,v)$ on $H^2(\mathbf{R}^d)$

$$a(u,v) = \int_{\mathbf{R}^d} (\nu_0 \Delta u \Delta v + \nu_1 uv) dx, \quad \forall\, u, v \in H^2(\mathbf{R}^d), \tag{2.8}$$

which is coercive, i.e.,

$$\exists \nu > 0, \quad a(u,b) \geqslant \nu \|u\|_2, \quad \forall\, u \in H^2(\mathbf{R}^d). \tag{2.9}$$

Related to $\mathcal{L}u$ and $F(t)$, we have

$$<\mathcal{L}u, u> = \int_{\mathbf{R}^d} (b(x,t) \cdot \nabla u) v dx, \quad u, v \in H^2(\mathbf{R}^d), \tag{2.10}$$

$$<F(t), v> = \int_{\mathbf{R}^d} f(x,t) v dx, \quad v \in H^2(\mathbf{R}^d), \tag{2.11}$$

where $b : R \times [0,T] \to \mathbf{R}^d$ and $f : R \times [0,T] \to R$ are measurable functions.

Note that the space $H^4(R^3)$ is separable and has a complete basis $\{r_j\}_{j \geqslant 1}$. Applying the Faedo-Galerkin method [18, 28], we can obtain the existence of solutions of (2.5)-(2.6).

Theorem 2.4 *Let $T > 0$ and assume that*

$$F \in C^1([0,T]; L^2(\mathbf{R}^d)), \quad b \in C^1([0,T]; \mathcal{H}^d(\mathbf{R}^d)). \tag{2.12}$$

Then, if $u_0 \in H^4(\mathbf{R}^d)$ and $u_1 \in H^2(\mathbf{R}^d)$, the solution of (2.5)-(2.6) exists and satisfies

$$u \in C^i([0,T]; H^{4-2j}(\mathbf{R}^d)) \cap C^2([0,T]; L^2(\mathbf{R}^d)), \quad j = 0, 1. \tag{2.13}$$

Moreover, assume

(1) *for* $d = 1, 2, 3$,
$$F \in C^1([0,T]; H^2(\boldsymbol{R}^d)). \tag{2.14}$$

Then, if $u_0 \in H^6(\boldsymbol{R}^d)$ *and* $u_1 \in H^4(\boldsymbol{R}^d)$, *it holds*
$$u \in C^i([0,T]; H^{6-2j}(\boldsymbol{R}^d)) \cap C^3([0,T]; L^2(\boldsymbol{R}^d)), \quad j = 0, 1, \cdots, 2. \tag{2.15}$$

(2) *for* $d > 3$,
$$F \in C^1([0,T]; H^{2d-4}(\boldsymbol{R}^d)). \tag{2.16}$$

Then, if $u_0 \in H^{2d}(\boldsymbol{R}^d)$ *and* $u_1 \in H^{2d-2}(\boldsymbol{R}^d)$, *it holds*
$$u \in C^i([0,T]; H^{2d-2j}(\boldsymbol{R}^d)) \cap C^d([0,T]; L^2(\boldsymbol{R}^d)), \quad j = 0, 1, \cdots, d-1. \tag{2.17}$$

Proof The (2.15) follows from (2.13) with some modification when we consider the similar problem for new variable $v \in D^\alpha$ (for $d = 1, 2, 3, \alpha = 2$, for $d > 3$, $\alpha = d-1$). The (2.13) can be proved by applying the Faedo-Galerkin method. We omit the details here. □

Finally, we list below the Moser-type calculus inequalities [15, 21].

Lemma 2.1 (1) *Let* $f, g \in L^\infty \cap H^s(\boldsymbol{R}^d)$, $s \geqslant d/2$. *Then, it holds*
$$\|D^\alpha(fg)\| \leqslant C\|g\|_{L^\infty}\|D^\alpha f\| + C\|f\|_{L^\infty}\|D^\alpha g\|, \tag{2.18}$$
$$\|D^\alpha(fg) - fD^\alpha g\| \leqslant C\|g\|_{L^\infty}\|D^\alpha f\| + C\|f\|_{L^\infty}\|D^{\alpha-1}f\|, \tag{2.19}$$

for $1 \leqslant |\alpha| \leqslant s$.

(2) *Let* $u \in \mathcal{H}^1(\boldsymbol{R}^d)$, *then it holds*
$$\begin{cases} \|u\|_{L^q} \leqslant C\|Du\|, & q = \dfrac{2d}{d-2}, \quad d > 3, \\ \|u\|_{L^q} \leqslant C\|Du\|, & q = 6, \quad d = 1, 2, 3. \end{cases} \tag{2.20}$$

3 Local-in-time existence

This section is concerned with the proof of Theorem 1.1. Instead, we shall prove the well-posedness for a new extended problem, derived based on (1.16)-(1.18), for $U = (v, z, \varphi, \zeta, u)$,

$$\nabla \cdot v = r(t), \quad \nabla \times v = 0, \quad \text{and} \quad v(x,t) = 0, \quad |x| \geqslant R, \tag{3.1}$$

$$\begin{cases} z = B_0 \int_{\mathbf{R}^d} |x-y|^{-d}(x-y) \times \omega(y,t) dy, \\ \omega' + (v+z) \cdot \nabla \omega + \omega \nabla \cdot v - (\omega \cdot \nabla)[v+z] = 0, \\ \omega(x,0) = \omega_0(x) =: \nabla \times u_0, \end{cases} \quad (3.2)$$

$$\begin{cases} \varphi' + \dfrac{1}{2}(\nabla \cdot v)\varphi + u \cdot \nabla \zeta = 0, \\ \varphi(x,0) = \zeta_0(x), \end{cases} \quad (3.3)$$

$$\begin{cases} \zeta'' + \nu \Delta^2 \zeta + \nu \zeta + k(t) \cdot \nabla \zeta' = h(t), \\ \zeta(x,0) = \zeta_0(x), \quad \zeta'(x,0) = \zeta_1(x) =: -\dfrac{1}{2}\zeta_0 \nabla \cdot u_0 - u_0 \nabla \zeta_0, \end{cases} \quad (3.4)$$

$$\begin{cases} u' = g(t), \\ u(x,0) = u_0(x), \end{cases} \quad (3.5)$$

where $\nu = \dfrac{1}{4}\varepsilon^2$, and

$$r(t) = r(x,t) = -\frac{2(\zeta' + u \cdot \nabla \zeta)}{\varphi}, \quad (3.6)$$

$$k(t) = k(x,t) = u(x,t) + v(x,t) + z(x,t), \quad (3.7)$$

$$\begin{aligned} h(t) = h(x,t) =& \frac{1}{\varphi}\zeta'(\zeta' + u \cdot \nabla \zeta) + \frac{\varepsilon^2}{4}\frac{|\Delta \zeta|^2}{\varphi} + \frac{1}{2}\frac{\Delta P(\zeta^2)}{\varphi} + \nu \zeta \\ &+ \frac{1}{2}\nabla \zeta \cdot \nabla(|v+z|^2) - \frac{1}{2}\zeta|\omega|^2 - [v+z] \cdot \nabla(u \cdot \nabla \zeta) \\ &+ \frac{1}{\varphi}(\zeta' + u \cdot \nabla \zeta)(v \cdot \nabla \zeta) - \nabla \zeta \cdot ([v+z] \times \omega) \\ &+ \frac{1}{2}\zeta \nabla(v+z) : \nabla(v+z), \end{aligned} \quad (3.8)$$

$$g(t) = g(x,t) = -\frac{1}{2}\nabla(|v+z|^2) + [v+z] \times \omega - \nabla h(\zeta^2) + \frac{1}{2}\varepsilon^2\left(\frac{\nabla \Delta \zeta}{\varphi} - \frac{\Delta \zeta}{\varphi^2}\nabla \zeta\right). \quad (3.9)$$

The most important fact which will be belown in Section 3.2 is to note that the above extended system for $U = (v, z, \varphi, \zeta, u)$ is equivalent to the equations (1.16)-(1.18) of (ζ, u) for classical solutions when $u = v + z$ and $\zeta = \varphi > 0$.

The main result in this section is

Theorem 3.1 (1) *For $d = 1, 2, 3$, assume that $(\zeta_0 - \sqrt{\bar{\rho}}, u_0) \in H^6(\mathbf{R}^d) \times \mathcal{H}^5(\mathbf{R}^d)$ satisfying*

$$\zeta^* = \sup_{x \in \mathbf{R}^d} \zeta_0(x), \quad \zeta_* = \inf_{x \in \mathbf{R}^d} \zeta_0(x) > 0. \quad (3.10)$$

Then, there is a uniform time T_{**} such that there exists a solution series $U = (v, z, \varphi, \zeta, u)$ which solves uniformly the systems (3.1)-(3.5) for $t \in [0, T_{**}]$ and satisfies

$$\begin{cases} v \in C^j([0, T_{**}]; \mathcal{H}^{4-j}(\mathbf{R}^d)) \cap C^2([0, T_*]; \mathcal{H}^1(\mathbf{R}^d)), \\ z \in C^l([0, T_{**}]; \mathcal{H}^{4-l}(\mathbf{R}^d)), \quad \omega \in C^l([0, T_1]; H^{3-l}(\mathbf{R}^d)), \\ u \in C^1([0, T_{**}]; \mathcal{H}^3(\mathbf{R}^d)) \cap C^2([0, T_{**}]; \mathcal{H}^1(\mathbf{R}^d)), \\ \varphi - \sqrt{\bar{\rho}} \in C^1([0, T_*]; H^3(\mathbf{R}^d)) \cap C^2([0, T_{**}]; H^2(\mathbf{R}^d)) \cap C^3([0, T_{**}]; L^2(\mathbf{R}^d)), \\ \zeta - \sqrt{\bar{\rho}} \in C^k([0, T_*]; H^{6-2k}(\mathbf{R}^d)) \cap C^3([0, T_{**}]; L^2(\mathbf{R}^d)), \end{cases} \quad (3.11)$$

where $j = 0, 1$, $l = 0, 1, 2$, $k = 0, 1, 2$.

(2) For $d > 3$, assume that $(\zeta_0 - \sqrt{\bar{\rho}}, u_0) \in H^{2d}(\mathbf{R}^d) \times \mathcal{H}^{2d-1}(\mathbf{R}^d)$ satisfying

$$\zeta^* = \sup_{x \in \mathbf{R}^d} \zeta_0(x), \quad \zeta_* = \inf_{x \in \mathbf{R}^d} \zeta_0(x) > 0. \quad (3.12)$$

Then, there is a uniform time T_{**} such that there exists a solution series $U = (v, z, \varphi, \zeta, u)$ which solves uniformly the systems (3.1)-(3.5) for $t \in [0, T_{**}]$ and satisfies

$$\begin{cases} v \in C^l([0, T_{**}]; \mathcal{H}^{2d-2-j}(\mathbf{R}^d)), \\ z \in C^l([0, T_{**}]; \mathcal{H}^{2d-2-l}(\mathbf{R}^d)), \quad \omega \in C^l([0, T_1]; H^{d-l}(\mathbf{R}^d)), \\ u \in C^l([0, T_{**}]; \mathcal{H}^{2d-3-l}(\mathbf{R}^d)), \\ \varphi - \sqrt{\bar{\rho}} \in C^l([0, T_{**}]; H^{2d-3-l}(\mathbf{R}^d)), \\ \zeta - \sqrt{\bar{\rho}} \in C^l([0, T_*]; H^{2d-2l}(\mathbf{R}^d)), \end{cases} \quad (3.13)$$

where $l = 0, 1, \cdots, d - 1$.

We will show the construction of the extended systems (3.1)-(3.9) based on (1.16)-(1.18) in Section 3.1. Then we define an iterative scheme of approximate solution sequence of the extended system and obtain the uniform estimates, and then prove Theorem 1.1 in Section 3.2.

3.1 Construction of new problems

We construct the extended systems (3.1)-(3.9) based on (1.16)-(1.18) by modifying the main idea in [13, 18]. For general smooth fluid dynamics, the velocity vector field can be (uniquely) decomposed into the gradient field and the divergence free vector field:

$$u = v + z = \mathcal{Q}u + \mathcal{P}u = \nabla S + z, \quad \nabla \cdot z = 0. \quad (3.14)$$

Equation (1.17) for the velocity vector field u can be written as

$$\partial_t u + \frac{1}{2}\nabla(|u|^2) - u \times (\nabla \times u) + \nabla h(\zeta^2) = \frac{\varepsilon^2}{2}\nabla\left(\frac{\Delta \zeta}{\zeta}\right), \tag{3.15}$$

where we use the relation of the convection term

$$(u \cdot \nabla)u = \frac{1}{2}\nabla(|u|^2) - u \times (\nabla \times u). \tag{3.16}$$

Taking *curl* on (3.15) and letting $w = \nabla \times u$, we have

$$\partial_t w + u \cdot \nabla w + w \nabla \cdot u - (w \cdot \nabla)u = 0. \tag{3.17}$$

For smooth $\zeta > 0$ (1.16) is equivalent to

$$2\partial_t \zeta + 2u \cdot \nabla \zeta + \eta \nabla \cdot u = 0. \tag{3.18}$$

Based on (3.17)-(3.18), we show the ideas on how to construct the extended systems (3.1)-(3.9) for $U = (v, z, \varphi, \zeta, u)$ to dealt with on the basis of Section 3.2. Given u and ζ, we introduce new equations for "density" $\varphi > 0$ and gradient velocity vector field v in terms of divergence free vector field z as

$$\partial_t \varphi + \frac{1}{2}\varphi \nabla \cdot v + u \cdot \nabla \zeta = 0, \quad \varphi(x,0) = \zeta_0(x) > 0. \tag{3.19}$$

$$\nabla \cdot v = -\frac{2(\partial_t \zeta + u \cdot \nabla \zeta)}{\varphi}, \quad \nabla \times v = 0, \text{ and } v(x,t) = 0, \ |x| \geq R. \tag{3.20}$$

The new divergence free vector field z is represented by its vorticity (which will still be denoted by w) $w = \nabla \times z$ as

$$z(x,t) = B_0 \int_{\mathbf{R}^d} |x-y|^{-d}(x-y) \times w(y,t)dy, \tag{3.21}$$

where B_0 is a constant matrix, and the vorticity vector field $w = \nabla \times z$ solves the following equation

$$\partial_t w + (v+z) \cdot \nabla w + w \nabla \cdot v - (w \cdot \nabla)[v+z] = 0, \tag{3.22}$$

$$w(x,0) = \nabla \times u_1(x), \tag{3.23}$$

which is obtained by taking *curl* on (3.17) after replacing $z + v$ for u.

We need to pay attention to, however, that we should be able to determine u and ζ again so long as we can solve the above equations for v, z and φ. Namely, we will propose the corresponding two equations for u and ζ based on (3.19),(3.20) and (3.21) as follows. In fact, we can construct the expended equation for the velocity u as

$$\partial_t u + \frac{1}{2}\nabla(|v+z|^2) - (v+z) \times \omega + \nabla h(\zeta^2) = \frac{\varepsilon^2}{2}\left(\frac{\nabla\Delta\zeta}{\varphi} - \frac{\Delta\zeta\nabla\zeta}{\varphi^2}\right), \tag{3.24}$$

$$u(x,0) = u_0(x), \tag{3.25}$$

which is derived from (3.15) by substituting $v+z$ for u into the convection term through the relation (3.16), and by using the equality

$$\nabla\left(\frac{\Delta\zeta}{\zeta}\right) = \left(\frac{\nabla\Delta\zeta}{\zeta} - \frac{\Delta\zeta\nabla\zeta}{\zeta^2}\right), \tag{3.26}$$

and replacing $\frac{1}{\zeta}$ by $\frac{1}{\varphi}$ on the right hand side of (3.26). And we construct the equation for the density ζ as

$$\begin{aligned}&\zeta_{tt} + \frac{1}{4}\varepsilon^2\Delta^2\zeta + \frac{1}{4}\varepsilon^2\frac{|\Delta\zeta|^2}{\varphi} - \frac{1}{2\varphi}\Delta P(\zeta^2) + (u+v+z)\cdot\nabla\zeta_t \\ &- \frac{\zeta_t}{\varphi}(\zeta_t + u\cdot\nabla\zeta) - \frac{1}{2}\nabla\zeta\cdot\nabla(|v+z|^2) + \nabla\zeta\cdot([v+z]\times\omega) \\ &- \frac{1}{2}\zeta\nabla(v+z):\nabla(v+z) + \frac{1}{2}\zeta|\omega|^2 + (v+z)\cdot\nabla(u\cdot\nabla\zeta) \\ &- \frac{1}{\varphi}(\zeta_t + u\cdot\nabla\zeta)([v+z]\cdot\nabla\zeta) = 0,\end{aligned} \tag{3.27}$$

with initial data

$$\zeta(x,0) = \zeta_0, \quad \zeta_t(x,t) = \zeta_1 =: -\frac{1}{2}\zeta_0\nabla u_0 - u_0\cdot\nabla\zeta_0, \tag{3.28}$$

where for $v = (v^1,\cdots,v^d)$,

$$\nabla v : \nabla v = \sum_{ij}|\partial_i v^j|^2.$$

In fact, by differentiating (3.18) with respect to time, replacing u_t in terms of (3.15) where the unknown u of the convection term is substituted by $(v+z)$, using (3.16) and replacing the term $\frac{1}{2}\zeta\left[\frac{1}{2}\Delta(|v+z|^2) - \nabla\cdot([v+z]\times\omega)\right]$ by

$$\frac{1}{2}\zeta\nabla(v+z):\nabla(v+z) - \frac{1}{2}\zeta|\omega|^2 - (v+z)\cdot\nabla\zeta_t$$
$$- (v+z)\cdot\nabla(u\cdot\nabla\zeta) + \frac{1}{\zeta}(\zeta_t + u\cdot\nabla\zeta)([v+z]\cdot\nabla\zeta),$$

and by using the relation

$$\nabla \cdot \left[\zeta^2 \nabla \left(\frac{\Delta \zeta}{\zeta}\right)\right] = \zeta \left[\Delta^2 \zeta - \frac{|\Delta \zeta|^2}{\zeta}\right]$$

and by replacing all $\frac{1}{\zeta}$ in the resulting equation by $\frac{1}{\varphi}$, we can obtain (3.27).

So far, we have constructed the extended and closed systems (3.1)-(3.9) for $U = (v, z, \varphi, \zeta, u)$ which consists of two ODEs (3.19) for φ and (3.24) for u, a wave type equation (3.27) for ζ, a divergence equation (3.20) for v, and a formula (3.21) for z in terms of ω which solves a hyperbolic equation (3.22).

3.2 Iteration scheme and local existence

We define an iterative scheme of approximate solution sequence of the extended system an obtain the uniform estimates, and then prove Theorem 3.1 and Theorem 1.1. We consider the corresponding problem for an approximate solution sequence $\{U^i\}_{i=1}^{\infty}$ with

$$U^p = (v^p, z^p, \varphi^p, \zeta^p, u^p)$$

based on the extended system (3.1)-(3.9) constructed in Section 3.1.

We construct an iterative scheme for the solution

$$U^{p+1} = (v^{p+1}, z^{p+1}, \varphi^{p+1}, \zeta^{p+1}, u^{p+1}), \quad p \geq 1,$$

on $\mathbf{R}^d$ by solving the following problems

$$\nabla \cdot v^{p+1} = r^p(t), \quad \nabla \times v^{p+1} = 0, \quad \text{and} \quad v^{p+1}(x, t) = 0, \quad |x| \geq R, \tag{3.29}$$

$$\begin{cases} z^{p+1} = B_0 \int_{\mathbf{R}^d} |x-y|^{-d}(x-y) \times \omega^{p+1}(y, t) dy, \\ \omega_t^{p+1} + (v^p + z^p) \cdot \nabla \omega^{p+1} + \omega^{p+1} \nabla \cdot v^p - (\omega^{p+1} \cdot \nabla)[v^p + z^p] = 0, \\ \omega^{p+1}(x, 0) = \omega_0(x) =: \nabla \times u_0(x), \end{cases} \tag{3.30}$$

$$\begin{cases} \varphi_t^{p+1} + \frac{1}{2}(\nabla \cdot v^p)\varphi_{p+1} + u^p \cdot \nabla \zeta^p = 0, \\ \varphi^{p+1}(x, 0) = \zeta_0(x), \end{cases} \tag{3.31}$$

$$\begin{cases} \zeta_{tt}^{p+1} + \nu \Delta^2 \zeta^{p+1} + \nu \zeta^{p+1} + k^p(t) \cdot \nabla \zeta_t^{p+1} = h^p(t), \\ \zeta^{p+1}(x, 0) = \zeta_0(x), \\ \zeta_t^{p+1}(x, 0) = \zeta_1(x) =: -\frac{1}{2}\zeta_0 \nabla \cdot u_0 - u_0 \cdot \nabla \zeta_0, \end{cases} \tag{3.32}$$

$$\begin{cases} u_t^{p+1} = g^p(t), \\ u^{p+1}(x,0) = u_0(x), \end{cases} \quad (3.33)$$

where $v = \dfrac{1}{4}\varepsilon^2$ and

$$r^p(t) = r^p(x,t) = -\frac{2(\zeta_t^p + u^p \cdot \nabla \zeta^p)}{\varphi^p}, \quad (3.34)$$

$$k^p(t) = k^p(x,t) = u^p(x,t) + v^p(x,t) + z^p(x,t), \quad (3.35)$$

$$\begin{aligned} h^p(t) = h^p(x,t) =& \frac{1}{\varphi^p}\zeta_t^p(\zeta_t^p + u^p \cdot \nabla\zeta^p) + \frac{\varepsilon^2}{4}\frac{|\Delta\zeta^p|^2}{\varphi^p} - \frac{1}{2}\frac{\Delta P((\zeta^p)^2)}{\varphi^p} + \nu\zeta^p \\ &+ \frac{1}{2}\nabla\zeta^p \cdot \nabla(|v^p + z^p|^2) - \frac{1}{2}\zeta^p|\omega^p|^2 - [v^p + z^p]\cdot\nabla(u^p\cdot\nabla\zeta^p) + \\ &\frac{1}{\varphi^p}(\zeta_t^p + u^p\cdot\nabla\zeta^p)(v^p\cdot\nabla\zeta^p) - \nabla\zeta^p\cdot([v^p + z^p]\times\omega^p) \\ &+ \frac{1}{2}\zeta^p\nabla(v^p + z^p):\nabla(v^p + z^p), \end{aligned} \quad (3.36)$$

$$\begin{aligned} g^p(t) = g^p(x,t) =& -\frac{1}{2}\nabla(|v^p + z^p|^2) + [v^p + z^p]\times\omega^p - \nabla h((\zeta^p)^2) \\ &+ \frac{1}{2}\varepsilon^2\left(\frac{\nabla\Delta\zeta^p}{\varphi^p} - \frac{\Delta\zeta^p}{(\varphi^p)^2}\nabla\zeta^p\right). \end{aligned} \quad (3.37)$$

Here, we also note that functions $r^p(0), k^p(0), h^p(0), g^p(0), q^p(0)$ only depend on initial data (ζ_0, u_0).

The main result in this subsection is

Lemma 3.1 *Let the assumption of Theorem 3.1 holds. Then there is a uniform time T_* such that there exists a solution series $\{U^p\}_{p=1}^\infty$ which solves uniformly the systems (3.29)-(3.34) for $t \in [0, T_*]$ and satisfies*

(1) *for $d = 1, 2, 3$,*

$$\begin{cases} v^p \in C^j([0,T_*]; \mathcal{H}^{4-j}(\mathbf{R}^d)) \cap C^2([0,T_*]; \mathcal{H}^1(\mathbf{R}^d)), \\ z^p \in C^k([0,T_*]; \mathcal{H}^{4-j}(\mathbf{R}^d)), \quad \omega^p \in C^k([0,T_*]; H^{3-k}(\mathbf{R}^d)), \\ u^p \in C^1([0,T_*]; \mathcal{H}^3(\mathbf{R}^d)) \cap C^2([0,T_*]; \mathcal{H}^1(\mathbf{R}^d)), \\ \varphi^p - \sqrt{\bar\rho} \in C^1([0,T_*]; H^3(\mathbf{R}^d)) \cap C^2([0,T_*]; H^2(\mathbf{R}^d)) \cap C^3([0,T_*]; L^2(\mathbf{R}^d)), \\ \zeta^p - \sqrt{\bar\rho} \in C^l([0,T_*]; H^{6-2l}(\mathbf{R}^d)) \cap C^3([0,T_*]; L^2(\mathbf{R}^d)), \end{cases} \quad (3.38)$$

where $j = 0, 1$, $k = 0, 1, 2$, $l = 0, 1, 2$. Furthermore, the solution series $\{U^p\}_{p=1}^\infty$ is bounded uniformly for each $p \geqslant 1$ by

$$\begin{cases} \|D(v_t^p, v^p, u_t^p)(t)\|_{H^2}^2 + \|Dv^p(t)\|_{H^3}^2 + \|D(v_{tt}^p, u_{tt}^p)(t)\|^2 \leqslant M_*, \\ \|(\omega^p, \omega_t^p, \omega_{tt}^p)(t)\|_{H^3 \times H^2 \times H^1}^2 + \|D(z^p, z_t^p, z_{tt}^p)(t)\|_{H^3 \times H^2 \times H^1}^2 \leqslant M_*, \\ \|(\zeta^p - \sqrt{\bar{\rho}}, \zeta_t^p, \zeta_{tt}^p, \zeta_{ttt}^p)(t)\|_{H^6 \times H^4 \times H^2 \times L^2}^2 \leqslant M_*, \\ \|(\varphi^p - \sqrt{\bar{\rho}}, \varphi_t^p, \varphi_{tt}^p, \varphi_{ttt}^p)(t)\|_{H^6 \times H^4 \times H^2 \times L^2}^2 \leqslant M_*, \end{cases} \quad (3.39)$$

with M_* a positive constant independent of U^p, $p \geqslant 1$.

(2) For $d > 3$,

$$\begin{cases} v^p \in C^j([0, T_*]; \mathcal{H}^{2d-2-j}(\mathbf{R}^d)), \\ z^p \in C^k([0, T_*]; \mathcal{H}^{2d-2-j}(\mathbf{R}^d)), \quad \omega^p \in C^k([0, T_*]; H^{d-k}(\mathbf{R}^d)), \\ u^p \in C^j([0, T_*]; \mathcal{H}^{2d-3-j}(\mathbf{R}^d)), \\ \varphi^p - \sqrt{\bar{\rho}} \in C^j([0, T_*]; H^{2d-3-j}(\mathbf{R}^d)), \\ \zeta^p - \sqrt{\bar{\rho}} \in C^l([0, T_*]; H^{2d-2l}(\mathbf{R}^d)), \end{cases} \quad (3.40)$$

where $j = 0, 1, \cdots, d-2$, $k = 0, 1, \cdots, d-1$, $l = 0, 1, \cdots, d$. Furthermore, the solution series $\{U^p\}_{p=1}^{\infty}$ is bounded uniformly for each $p \geqslant 1$ by

$$\begin{cases} \|D(v_t^p, v^p, u_t^p)(t)\|_{H^{d-2}}^2 + \|Dv^p(t)\|_{H^d}^2 + \|D(v_{tt}^p, u_{tt}^p, \cdots, u_{t\cdots t}^p)(t)\|^2 \leqslant M_*, \\ \|(\omega^p, \cdots, \omega_{t\cdots t}^p)(t)\|_{H^d \times \cdots \times H^1}^2 + \|D(z^p, \cdots, z_{tt}^p)(t)\|_{H^d \times \cdots \times H^1}^2 \leqslant M_*, \\ \|(\zeta^p - \sqrt{\bar{\rho}}, \zeta_t^p, \cdots, \zeta_{t\cdots t}^p)(t)\|_{H^{2d} \times H^{2d-2} \times \cdots \times L^2}^2 \leqslant M_*, \\ \|(\varphi^p - \sqrt{\bar{\rho}}, \varphi_t^p, \cdots, \varphi_{t\cdots t}^p)(t)\|_{H^{2d} \times H^{2d-2} \times \cdots \times L^2}^2 \leqslant M_*. \end{cases} \quad (3.41)$$

Remark 3.1 Here, the highest order time derivative (3.41) is $d - 1$. Note that by (2.20) we can obtain automatically the L^q-norm $\left(d = 1, 2, 3, q = 6; d > 3, q = \dfrac{2d}{d-2}\right)$ of the unknown $v^{p+1}, z^{p+1}, u^{p+1}$ and there time derivatives so long as (3.39) ($d = 1, 2, 3$) or (3.41) holds, i.e.,

(1) for $d = 1, 2, 3$,

$$\|(v^{p+1}, v_t^{p+1}, v_{tt}^{p+1})\|_{L^6} \leqslant C\|D(v^{p+1}, v_t^{p+1}, v_{tt}^{p+1})\|_{L^2}, \quad (3.42)$$

$$\|(z^{p+1}, z_t^{p+1}, z_{tt}^{p+1})\|_{L^6} \leqslant C\|D(z^{p+1}, z_t^{p+1}, z_{tt}^{p+1})\|_{L^2}, \quad (3.43)$$

$$\|(u^{p+1}, u_t^{p+1}, u_{tt}^{p+1})\|_{L^6} \leqslant C\|D(u^{p+1}, u_t^{p+1}, u_{tt}^{p+1})\|_{L^2}. \quad (3.44)$$

(2) For $d > 3$, $q = \dfrac{2d}{d-2}$,

$$\|(v^{p+1}, v_t^{p+1}, \cdots, v_{t\cdots t}^{p+1})\|_{L^q} \leqslant C\|D(v^{p+1}, v_t^{p+1}, v_{t\cdots t}^{p+1})\|_{L^2}, \quad (3.45)$$

$$\|(z^{p+1}, z_t^{p+1}, \cdots, z_{t\cdots t}^{p+1})\|_{L^q} \leqslant C\|D(z^{p+1}, z_t^{p+1}, z_{t\cdots t}^{p+1})\|_{L^2}, \tag{3.46}$$

$$\|(u^{p+1}, u_t^{p+1}, \cdots, u_{t\cdots t}^{p+1})\|_{L^q} \leqslant C\|D(u^{p+1}, u_t^{p+1}, u_{t\cdots t}^{p+1})\|_{L^2}. \tag{3.47}$$

Proof We prove Lemma 3.1 and verify the a-priori estimates (3.39) in terms of energy method and induction arguments as follows:

First of all, we consider the case $p = 0$. We choose

$$U^0 = (v^0, z^0, \varphi^0, \zeta^0, u^0) = (\mathcal{Q}u_0, \mathcal{P}u_0, \zeta_0, \zeta_0, u_0),$$

which obviously satisfies (3.38)-(3.39)(or (3.40)-(3.41)) for $t \in [0, T_0]$ with M_* replaced by a constant $B_1 > 0$. Starting with $U^0 = (\mathcal{Q}u_0, \mathcal{P}u_0, \zeta_0, \zeta_0, u_0)$ and solving the problems (3.29)-(3.33) for $p = 0$, we shall prove the (local in time) existence of solution $U^1 = (v^1, z^1, \zeta^1, \varphi^1, u^1)$ which also satisfies (3.38)-(3.39)(or (3.40)-(3.41)) for $t \in [0, T_0]$ with $T_0 > 0$ and with M_* replaced by another constant $\tilde{M}_1 > 0$. In fact, for $U^0 = (\mathcal{Q}u_0, \mathcal{P}u_0, \zeta_0, \zeta_0, u_0)$ the function r^0, k^0, h^0, g^0, q^0 depend only on the initial data (ζ^0, u^0), i.e.,

$$r^0(x,t) = \tilde{r}^0(x), \quad k^0(x,t) = \tilde{k}^0(x), \quad h^0(x,t) = \tilde{h}^0(x),$$
$$g^0(x,t) = \tilde{g}^0(x), \quad q^0(x,t) = \tilde{q}^0(x),$$

satisfying

$$\begin{cases} \|\tilde{r}^0\|_2^2 + \|\tilde{k}^0\|_{L^6}^6 + \|D\tilde{k}_1\|_3^2 + \|\tilde{h}^0\|_3^2 + \|D\tilde{g}^0\|_3^2 + \|\tilde{g}^0\|_{L^6}^6 + \|\tilde{q}^0\|_2^2 \leqslant NI_0^4, \\ \quad d = 0, 1, 2, 3; \\ \|\tilde{r}^0\|_{2d-4}^2 + \|\tilde{k}^0\|_{L^q}^q + \|D\tilde{k}_1\|_{2d-3}^2 + \|\tilde{h}^0\|_{2d-3}^2 + \|D\tilde{g}^0\|_{2d-3}^2 + \|\tilde{g}^0\|_{L^q}^q + \|\tilde{q}^0\|_{2d-4}^2 \\ \quad \leqslant NI_0^4, \quad d \geqslant 3, \end{cases} \tag{3.48}$$

here and after $N > 0$ denotes a generic constant independent of $U^p (p \geqslant 1)$,

$$I_0 = \begin{cases} \|(\zeta_0 - \sqrt{\bar{\rho}})\|_6^2 + \|\nabla\zeta_0\|_5^2 + \|u_0\|_5^2, & d = 1, 2, 3, \\ \|(\zeta_0 - \sqrt{\bar{\rho}})\|_{2d}^2 + \|\nabla\zeta_0\|_{2d-1}^2 + \|u_0\|_{2d-1}^2, & d \geqslant 3, \end{cases} \tag{3.49}$$

The system (3.29)-(3.33) with $p = 0$ are linear with $U^1 = (v^1, z^1, \zeta^1, \varphi^1, u^1)$. It can be solved with the help of the estimates (3.48) about the right hand side terms as follows. Applying Theorem 2.1 to the divergence equation (3.29) with $r^0(x,t)$ replaced by $\tilde{r}^0(x)$ and (3.33) with $q^0(x,t)$ replaced by $\tilde{q}^0(x)$, we obtain the existence of solution

$$v^1 \in \begin{cases} C^j([0,T_0]; \mathcal{H}^{4-j}(\mathbf{R}^d)) \cap C^2([0,T_0]; \mathcal{H}^1(\mathbf{R}^d)), & j = 0, 1, \quad d = 1, 2, 3, \\ C^j([0,T_0]; \mathcal{H}^{2d-2-j}), & j = 0, 1, \cdots, d-1, \quad d \geqslant 3. \end{cases}$$

Making use of the theory of linear hyperbolic system, we show the existence of

$$\omega^1 \in \begin{cases} C^j([0,T_0]; H^{3-j}(\mathbf{R}^d)), & j=0,1,2, \ d=1,2,3, \\ C^j([0,T_0]; H^{2d-3-j}(\mathbf{R}^d)), & j=0,1,\cdots,d-1, \ d \geqslant 3 \end{cases}$$

of $(3.30)_{2,3}$, which together with $(3.30)_1$ and (2.4) implies the existence of

$$z^1 \in \begin{cases} C^j([0,T_0]; \mathcal{H}^{4-j}(\mathbf{R}^d)), & j=0,1,2, \ d=1,2,3, \\ C^j([0,T_0]; \mathcal{H}^{2d-2-j}(\mathbf{R}^d)), & j=0,1,\cdots,d-1, \ d \geqslant 3. \end{cases}$$

By the theory of linear ODE system, we prove the existence of solution

$$u^1 \in \begin{cases} C^1([0,T_0]; \mathcal{H}^3(\mathbf{R}^d)) \cap C^2([0,T_0]; \mathcal{H}^1(\mathbf{R}^d)), & d=1,2,3, \\ C^j([0,T_0]; \mathcal{H}^{2d-2-j}(\mathbf{R}^d)), & j=0,1,\cdots,d-1, \ d \geqslant 3 \end{cases}$$

of (3.33) for $g^0(x,t) = \tilde{g}^0(x)$, and solution of (3.31) satisfying

$$\varphi^1 - \sqrt{\bar{\rho}} \in \begin{cases} C^1([0,T_0]; H^3(\mathbf{R}^d)) \cap C^2([0,T_0]; H^2(\mathbf{R}^d)) \cap C^3([0,T_0]; L^2(\mathbf{R}^d)), \\ \qquad\qquad\qquad\qquad d=0,1,2,3; \\ C^j([0,T_0]; H^{2d-3-j}(\mathbf{R}^d)), \quad j=1,\cdots,d-1, \ d \geqslant 3. \end{cases}$$

Finally, applying Theorem 2.4 with $b(x,t) = 2u^0(x)$ in (2.10) and $f(x,t) = \tilde{h}^0(x)$ in (2.11) to semi-linear wave equation (3.32), we conclude the existence of solution

$$\zeta^1 - \sqrt{\bar{\rho}} \in \begin{cases} C^j([0,T_0]; H^{6-2j}(\mathbf{R}^d)), & j=0,1,2, \ d=1,2,3, \\ C^j([0,T_0]; H^{2d-2j}(\mathbf{R}^d)), & j=0,1,\cdots,d-1, \ d \geqslant 3. \end{cases}$$

Moreover, based on the estimates (3.48), we conclude that there is a constant $\tilde{M}_1$ such that U^1 satisfies (3.39) where $p=1$, $M_* = \tilde{M}_1$ and $T_* = T_0$.

Next, let us prove the estimates for $p \geqslant 1$. Assume that $\{U^i\}_{i=1}^p (p \geqslant 1)$ exists uniformly for $t \in [0, T_0]$, solves the system (3.29)-(3.33), and satisfies (3.38)-(3.39)(or (3.40)-(3.41)) with(the constant M_* replaced by) the upper bound $\tilde{M}_p (\geqslant \max_{1 \leqslant j \leqslant p-1}\{\tilde{M}_j\})$. We shall prove that it still holds for U^{p+1} for $t \in [0, T_1]$. In fact, the system (3.33) are linear with $U^{p+1} = (v^{p+1}, z^{p+1}, \varphi^{p+1}, \zeta^{p+1}, u^{p+1})$ for given U^p. In analogy, the application of Theorem 2.1 to divergence equation (3.29) for v^{p+1}, theory of linear ODE system to (3.31) for φ^{p+1} and (3.33) for u^{p+1}, and Theorem 2.4 to wave equation (3.32) for ζ^{p+1} with $f(x,t) = h^p(t)$ and $b(x,t) = k^p(t)$, shows that $U^{p+1} = (v^{p+1}, z^{p+1}, \zeta^{p+1}, \varphi^{p+1}, u^{p+1})$ exists for $t \in [0, T_0]$ and satisfies

(1) for $d = 1, 2, 3$,

$$\begin{cases} v^{p+1} \in C^j([0, T_0]; \mathcal{H}^{4-j}(\mathbf{R}^d)) \cap C^2([0, T_0]; \mathcal{H}^1(\mathbf{R}^d)), \quad j = 0, 1, \\ z^{p+1} \in C^j([0, T_0]; \mathcal{H}^{4-j}(\mathbf{R}^d)), \quad \omega^{p+1} \in C^j([0, T_0]; H^{3-j}(\mathbf{R}^d)), \quad j = 0, 1, 2, \\ u^{p+1} \in C^1([0, T_0]; \mathcal{H}^3(\mathbf{R}^d)) \cap C^2([0, T_0]; \mathcal{H}^d(\mathbf{R}^d)), \\ \varphi^{p+1} - \sqrt{\bar{\rho}} \in C^1([0, T_0]; H^3(\mathbf{R}^d)) \cap C^2([0, T_0]; H^2(\mathbf{R}^d)) \cap C^3([0, T_0]; L^2(\mathbf{R}^d)), \\ \zeta^{p+1} - \sqrt{\bar{\rho}} \in C^l([0, T_0]; H^{6-2l}(\mathbf{R}^d)) \cap C^3([0, T_0]; L^2(\mathbf{R}^d)), \quad l = 0, 1, 2, \end{cases}$$

(2) for $d \geqslant 3$,

$$\begin{cases} v^{p+1} \in C^j([0, T_0]; \mathcal{H}^{2d-2-j}(\mathbf{R}^d)), \\ z^{p+1} \in C^j([0, T_0]; \mathcal{H}^{2d-2-j}(\mathbf{R}^d)), \quad \omega^{p+1} \in C^j([0, T_0]; H^{2d-3-j}(\mathbf{R}^d)), \\ u^{p+1} \in C^l([0, T_0]; \mathcal{H}^{2d-3-l}(\mathbf{R}^d)), \\ \varphi^{p+1} - \sqrt{\bar{\rho}} \in C^l([0, T_0]; H^{2d-3-l}(\mathbf{R}^d)), \\ \zeta^{p+1} - \sqrt{\bar{\rho}} \in C^l([0, T_0]; H^{2d-2l}(\mathbf{R}^d)), \end{cases}$$

where $j = 0, 1, \cdots, d-2$, and $l = 0, 1, \cdots, d-1$.

What we do next is to obtain a uniform upper bound of U^{j+1}, $1 \leqslant j \leqslant p$, for a fixed time period.

Let us verify the L^2 norm of the initial value of $\zeta^{p+1}, \zeta_t^{p+1}, \zeta_{tt}^{p+1}$ first, where the initial value $\hat{\zeta}$ of ζ_{tt}^{p+1} is obtain through $(3.32)_1$ at $t = 0$

$$\hat{\zeta} = -\nu \Delta^2 \zeta_0 - \nu \zeta_0 - 2 u_0 \cdot \nabla \zeta_1 + \tilde{h}(0), \tag{3.50}$$

with $\tilde{h}(0) = h^p(0)$ only depending on (ζ_0, u_0). Hence, the initial values of $\zeta^{p+1}, \zeta_t^{p+1}, \zeta_{tt}^{p+1}$ only depend on (ζ_0, u_0). Obviously, there is a constant $M_2 > 0$ such that the initial values of $\zeta^{p+1}, \zeta_t^{p+1}, \zeta_{tt}^{p+1}$ for $p \geqslant 1$ are bounded by

$$\begin{cases} \max\left\{\|\zeta_0 - \sqrt{\bar{\rho}}\|_2^2, \|\zeta_1\|_2^2, \|\hat{\zeta}\|^2, \|Du_0\|_4^2\right\} \leqslant M_2 I_0, & d = 1, 2, 3, \\ \max\left\{\|\zeta_0 - \sqrt{\bar{\rho}}\|_{2d-4}^2, \|\zeta_1\|_{2d-4}^2, \|\hat{\zeta}\|_{2d-6}^2, \|Du_0\|_{2d-2}^2\right\} \leqslant M_2 I_0, & d \geqslant 3 \end{cases} \tag{3.51}$$

Here, we recall that I_0 is defined by (3.49).

Set

$$\left.\begin{aligned} M_0 &= 16 N I_0 \cdot \max\{1, \nu^{-1}\}, \\ M_1 &= 2N(I_0 + 1 + M_0)^7 \cdot \max\{1, \nu^{-2}\}, \end{aligned}\right\} \tag{3.52}$$

$$\zeta^* =: \sup_{x \in \mathbf{R}^d} \zeta_0(x) > 0, \quad \zeta_* =: \inf_{x \in \mathbf{R}^d} \zeta_0(x) > 0, \tag{3.53}$$

and choose
$$T_* = \min\left\{1, \frac{\zeta_*}{4M_0}, \frac{2I_0}{M_3}, \frac{\ln 2}{NM_4}\right\}, \qquad (3.54)$$
where
$$M_3 = (I_0 + 1 + M_0 + M_1)^{14}, \quad M_4 = 2(I_0 + 1 + M_0 + M_1)^8, \qquad (3.55)$$
and we choose the generic constant $N > M_2$ independent of U^p, $p \geqslant 1$.

The main hard work in this part is to apply energy method to the coupled system (3.29)-(3.33) for $U^{p+1} = (v^{p+1}, z^{p+1}, \varphi^{p+1}, \zeta^{p+1}, u^{p+1})$ in a similar argument to the above step for U^2, and obtain after a tedious but straightforward computation (We omit the details here, one can also ref to [18] for the main steps for irrotational case) the following statement:

If it hold for $\{U^j\}_{j=1}^p (p \geqslant 2)$, solving problems (3.29)-(3.33), that
(1) $d = 1, 2, 3$,
$$\begin{cases} \|(Dz^j, \omega^j)(t)\|_3^2 + \|Du^j(t)\|_2^2 \leqslant M_0, \\ \|(\zeta^j - \sqrt{\bar\rho}, \zeta_t^j)(t)\|_4^2 + \|\zeta_{tt}^j\|_2^2 \leqslant M_0, \\ \|Dv^j(t)\|_3^2 + \|D\Delta\zeta^j(t)\|_1^2 \leqslant M_1, \end{cases} \qquad (3.56)$$

(2) $d \geqslant 3$,
$$\begin{cases} \|(Dz^j, \omega^j)(t)\|_{2d-3}^2 + \|Du^j(t)\|_{2d-4}^2 \leqslant M_0, \\ \|(\zeta^j - \sqrt{\bar\rho}, \zeta_t^j)(t)\|_{2d-2}^2 + \|\zeta_{tt}^j\|_{2d-4}^2 \leqslant M_0, \\ \|Dv^j(t)\|_{2d-3}^2 + \|D\Delta\zeta^j(t)\|_{2d-5}^2 \leqslant M_1, \end{cases}$$

for $1 \leqslant j \leqslant p$ and $t \in [0, T_*]$, then it is also holds for U^{p+1} that
(1) $d = 1, 2, 3$,
$$\begin{cases} \|(Dz^{p+1}, \omega^{p+1})(t)\|_3^2 + \|Du^{p+1}\|_2^2 \leqslant M_0, \\ \|(\zeta^{p+1} - \sqrt{\bar\rho}, \zeta_t^{p+1})(t)\|_4^2 + \|\zeta_{tt}^{p+1}(t)\|_2^2 \leqslant M_0, \\ \|Dv^{p+1}(t)\|_3^2 + \|D\Delta\zeta^{p+1}(t)\|_1^2 \leqslant M_1, \end{cases} \qquad (3.57)$$

(2) $d \geqslant 3$,
$$\begin{cases} \|(Dz^{p+1}, \omega^{p+1})(t)\|_{2d-3}^2 + \|Du^{p+1}\|_{2d-4}^2 \leqslant M_0, \\ \|(\zeta^{p+1} - \sqrt{\bar\rho}, \zeta_t^{p+1})(t)\|_{2d-2}^2 + \|\zeta_{tt}^{p+1}(t)\|_{2d-5}^2 \leqslant M_0, \\ \|Dv^{p+1}(t)\|_{2d-3}^2 + \|D\Delta\zeta^{p+1}(t)\|_{2d-5}^2 \leqslant M_1, \end{cases} \qquad (3.58)$$

for $t \in [0, T_*]$. Here M_0 and M_1 are given by (3.52) and (3.53).

Furthermore, with the help of the above statement, we can conclude, by a direct complicated computation, that the approximate solution sequence $U^{p+1} = (v^{p+1},$

$z^{p+1}, \varphi^{p+1}, \zeta^{p+1}, u^{p+1})$, $p \geqslant 1$, which solves (3.29)-(3.33) uniformly for $t \in [0, T_*]$ with T_* defined by (3.54), is uniformly bounded for $t \in [0, T_*]$, and satisfies (3.33)-(3.39) with the constant $M_* > 0$ chosen by

$$M_* = N(I_0 + 1 + M_0 + M_1)^{15}. \tag{3.59}$$

In addition, for given $(x,t) \in \mathbf{R}^d \times [0, T_*]$ using the standard argument of ODE, we obtain φ^{j+1} from (3.31), that is,

$$\begin{cases} \varphi^{j+1}(x,t) = (\zeta_0(x) - \int_0^t e^{\frac{1}{2}\int_0^s \nabla \cdot v^j(x,\xi)d\xi} u^j \cdot \nabla \zeta^j(x,s)ds)e^{-\frac{1}{2}\int_0^t \nabla \cdot v^j(x,s)ds}, \\ \varphi^{j+1} - \sqrt{\bar{\rho}} \in \begin{cases} C^1([0,T_1]; H^3(\mathbf{R}^d)) \cap C^2([0,T_1]; H^2(\mathbf{R}^d)) \cap C^3([0,T_1]; L^2(\mathbf{R}^d)), \\ d = 1,2,3; \\ C^j([0,T_1]; H^{2d-3-j}(\mathbf{R}^d)), \quad j = 0, 1, \cdots, d-1, \quad d \geqslant 3, \end{cases} \end{cases} \tag{3.60}$$

which together with (3.56) gives rise to the uniform positivity for $(x,t) \in \mathbf{R}^d \times [0, T_*]$

$$\frac{1}{4}\zeta_* \leqslant \varphi^{j+1}(x,t) \leqslant 2(\zeta^* + \zeta_*). \tag{3.61}$$

Recall here that M_0, M_1, ζ^* and ζ_* are defined by (3.52)-(3.53) respectively and $N > 0$ is a generic constant, which are independent of U^{p+1}, $p \geqslant 1$. Thus, the proof of Lemma 3.1 is completed. □

Proof of Theorem 3.1 By Lemma 3.1, we obtain an approximate solution sequence of $\{U^i\}_{i=1}^\infty$ satisfying (3.38)-(3.39)(or (3.40)-(3.41)). To prove the uniform convergence of the whole sequence, we need to estimate the difference of the approximate solution sequence

$$Y^{p+1} = (\bar{v}^{p+1}, \bar{\varphi}^{p+1}, \bar{\zeta}^{p+1}, \bar{u}^{p+1}) := U^{p+1} - U^p, \quad p \geqslant 1.$$

based on Lemma 3.1. In fact, let

$$\bar{v}^{p+1} = v^{p+1} - v^p, \quad \bar{z}^{p+1} = z^{p+1} - z^p, \quad \bar{\varphi}^{p+1} = \varphi^{p+1} - \varphi^p$$
$$\bar{\zeta}^{p+1} = \zeta^{p+1} - \zeta^p, \bar{u}^{p+1} = u^{p+1} - u^p,$$

then by using Lemma 3.1 and repeating the similar arguments as above, we can show after a tedious computation that there is a time $0 < T_{**} \leqslant T_*$ such that the difference Y^{p+1}, $p \geqslant 1$, of the approximate solution sequence satisfies the following estimates:

(1) $d = 1, 2, 3$,

$$\begin{cases} \sum_{p=1}^{\infty} \left(\|\bar{u}^{p+1}(t)\|^2_{\mathcal{H}^3(\mathbf{R}^d)} + \|\bar{u}_t^{p+1}\|^2_{\mathcal{H}^2(\mathbf{R}^d)} \right) \leqslant C_*, & (3.62) \\ \sum_{p=1}^{\infty} \left(\|(\bar{v}^{p+1}, \bar{z}^{p+1})\|^2_{\mathcal{H}^4(\mathbf{R}^d)} + \|(\bar{v}_t^{p+1}, \bar{z}_t^{p+1})\|^2_{\mathcal{H}^3(\mathbf{R}^d)} \right) \leqslant C_*, & (3.63) \\ \sum_{p=1}^{\infty} \left(\|\bar{\zeta}^{p+1}\|^2_{C^i([0,T_{**}];H^{6-2i})} + \|\bar{\varphi}^{p+1}\|^2_{C^1([0,T_{**}];H^4)} + \|\bar{\varphi}_{tt}^{p+1}\|^2_{C([0,T_{**}];H^2)} \right) \leqslant C_*, & \\ & (3.64) \end{cases}$$

where $i = 0, 1, 2$;

(2) $d \geqslant 3$,

$$\begin{cases} \sum_{p=1}^{\infty} \left(\|\bar{u}^{p+1}(t)\|^2_{\mathcal{H}^{2d-3}(\mathbf{R}^d)} + \|\bar{u}_t^{p+1}\|^2_{\mathcal{H}^{2d-4}(\mathbf{R}^d)} \right) \leqslant C_*, & (3.65) \\ \sum_{p=1}^{\infty} \left(\|(\bar{v}^{p+1}, \bar{z}^{p+1})\|^2_{\mathcal{H}^{2d-2}(\mathbf{R}^d)} + \|(\bar{v}_t^{p+1}, \bar{z}_t^{p+1})\|^2_{\mathcal{H}^d(\mathbf{R}^d)} \right) \leqslant C_*, & (3.66) \\ \sum_{p=1}^{\infty} \left(\|\bar{\zeta}^{p+1}\|^2_{C^i([0,T_{**}];H^{2d-2i})} + \|\bar{\varphi}^{p+1}\|^2_{C^1([0,T_{**}];H^{2d-2})} + \|\bar{\varphi}_{tt}^{p+1}\|^2_{C([0,T_{**}];H^{2d-4})} \right) \leqslant C_*, & \\ & (3.67) \end{cases}$$

where $i = 0, 1, \cdots, d-1$, and $C_* = C_*(N, M_*)$ denotes a positive constant depending on N and M_*.

By applying Ascoli-Arzela Theorem (to time variable) and Rellich-Kondrachov theorem (to spatial variable) [25], we can prove by the standard argument [23] that there is exists a (unique) $U = (v, z, \varphi, \zeta, u)$ such that as $p \to \infty$.

(1) $d = 1, 2, 3$,

$$\begin{cases} v^p \to v & in \ C^i([0, T_{**}]; \mathcal{H}^{4-i-\sigma}(\mathbf{R}^d)), \\ z^p \to z & in \ C^i([0, T_{**}]; \mathcal{H}^{4-i-\sigma}(\mathbf{R}^d)), \\ u^p \to u & in \ C^i([0, T_{**}]; \mathcal{H}^{3-\sigma}(\mathbf{R}^d)), \\ \varphi^p \to \varphi & in \ C^1([0, T_{**}]; H^{3-\sigma}(\mathbf{R}^d)) \cap C^2([0, T_{**}]; H^{2-\sigma}(\mathbf{R}^d)), \\ \zeta^p \to \zeta^p & in \ C^j([0, T_{**}]; H^{6-2j-\sigma}(\mathbf{R}^d)), \end{cases}$$

where $i = 0, 1$ and $j - 0, 1, 2$;

(2) $d \geqslant 3$,
$$\begin{cases} v^p \to v & in \ C^i([0, T_{**}]; \mathcal{H}^{2d-2-i-\sigma}(\mathbf{R}^d)), \\ z^p \to z & in \ C^i([0, T_{**}]; \mathcal{H}^{2d-2-i-\sigma}(\mathbf{R}^d)), \\ u^p \to u & in \ C^i([0, T_{**}]; \mathcal{H}^{2d-3-\sigma}(\mathbf{R}^d)), \\ \varphi^p \to \varphi & in \ C^j([0, T_{**}]; H^{2d-3-j-\sigma}(\mathbf{R}^d)), \\ \zeta^p \to \zeta^p & in \ C^j([0, T_{**}]; H^{2d-2j-\sigma}(\mathbf{R}^d)), \end{cases}$$

where $i = 0, 1, \cdots, d-2$, $j = 0, 1, \cdots, d-1$, and $0 < \sigma < \frac{1}{2}$. Moreover, by (3.61) it holds
$$\varphi(x, t) \geqslant \frac{1}{4}\zeta_* > 0, \quad (x, t) \in \mathbf{R}^d \times [0, T_{**}]. \tag{3.68}$$

Passing into limit $p \to \infty$ in (3.29)-(3.33), we obtain the (short time) existence and uniqueness of classical solution of the extended system (3.1)-(3.5) constructed in Section 3.1. The proof of Theorem 3.1 is completed. □

Proof of Theorem 1.1 By Theorem 3.1, we have the existence and uniqueness of short time strong solution $(v, z, \zeta, \varphi, u)$ of the extended system (3.1)-(3.5) with initial data
$$(v, z, \zeta, \varphi, u)(x, 0) = (\mathcal{Q}u_0, \mathcal{P}u_0, \zeta_0, \zeta_0, u_0)(x).$$

The most important fact we show below is that the extended systems (3.1)-(3.5) for $U = (v, z, \zeta, \varphi, u)$ are equivalent to (1.16)-(1.18) for (ζ, u) for classical solutions. That is, it holds
$$\zeta = \varphi, \quad u = v + z, \quad t \geqslant 0, \quad \text{as long as} \ [\varphi - \zeta](0) = 0, \ [u - v - z](0) = 0. \tag{3.69}$$

In fact, erasing the common term $u \cdot \nabla \zeta$ in the ODE equation (3.3) for φ and the divergence equation for v, i.e.,
$$\varphi_t + u \cdot \nabla \zeta + \frac{1}{2}\varphi \nabla \cdot v = 0, \quad \nabla \cdot v = -\frac{2(\zeta_t + u \cdot \nabla \zeta)}{\varphi}, \tag{3.70}$$

we obtain
$$(\varphi - \zeta)_t(x, t) = 0, \quad t \in [0, T_{**}]. \tag{3.71}$$

By (3.71), (3.68) and the fact
$$\varphi(x, 0) = \zeta(x, 0) = \zeta_0(x) \Rightarrow (\varphi - \zeta)(x, 0) = 0,$$

We obtain
$$\zeta(x, t) = \varphi(x, t) \geqslant \frac{1}{4}\zeta_* > 0, \quad (x, t) \in \mathbf{R}^d \times [0, T_{**}], \tag{3.72}$$

$$\zeta_t + u \cdot \nabla \zeta + \frac{1}{2}\zeta \nabla \cdot v = 0, \quad (x,t) \in \mathbf{R}^d \times [0, T_{**}]. \tag{3.73}$$

With the help of $(3.2)_1$ which gives $w = \nabla \times z$, (3.72) and (3.73), we obtain from (3.2) for u the following equation

$$\partial_t u + \frac{1}{2}\nabla(|v+z|^2) - (v+z) \times (\nabla \times z) + \nabla h(\zeta^2) + u = \frac{\varepsilon^2}{2}\nabla\left(\frac{\Delta \zeta}{\zeta}\right). \tag{3.74}$$

Taking *curl* to (3.74) and making the difference between the resulted equation and $(3.2)_2$ for $w = \nabla \times z$, we have

$$\nabla \times (u - z)_t + \nabla \times (u - z) = 0, \quad t \geqslant 0. \tag{3.75}$$

Similarly, recombing the various term in (3.27) for ζ we can obtain, with the help (3.72) and (3.26), that

$$\zeta_t - \frac{1}{4\zeta}\nabla \cdot \left(\zeta^2 \frac{1}{2}\nabla(|v+z|^2) - \zeta^2(v+z) \times (\nabla \times z)\right) - \frac{1}{2\zeta}\Delta P(\zeta^2)$$
$$+ u \cdot \nabla \zeta_t + \frac{1}{2}\zeta_t(\nabla \cdot v) + \frac{1}{4\zeta}\varepsilon^2 \nabla \cdot \left(\zeta^2 \nabla\left(\frac{\Delta \zeta}{\zeta}\right)\right) = 0. \tag{3.76}$$

From (3.73) we have $\zeta_t = -u \cdot \nabla \zeta - \frac{1}{2}\zeta \nabla \cdot v$. Substituting it into (3.76) and representing u_t by (3.74), we obtain after a computation that

$$\nabla \cdot (u-v)_t + \nabla \cdot (u-v) = 0. \tag{3.77}$$

Since

$$\nabla \cdot (u-v)(x,0) = 0, \quad \nabla \times (u-z)(x,0) = 0,$$

it follows from (3.75),(3.76) that

$$\nabla \cdot (u-v)(x,t) = 0, \quad \nabla \times (u-z)(x,t) = 0, \quad (x,t) \in \mathbf{R}^d \times [0, T_{**}]. \tag{3.78}$$

After decomposing the velocity u into $u_g + u_r =: \mathcal{Q}u + \mathcal{P}u$ and using Theorem 2.1-Theorem 2.3, we conclude from (3.78) that

$$u = v + z, \quad \mathcal{Q}u = v, \quad \mathcal{P}u = z, \quad (x,t) \in \mathbf{R}^d \times [0, T_{**}]. \tag{3.79}$$

Thus, by (3.79), (3.74), and (3.16), we recover the equation for u which is exactly (3.15). Multiplying (3.73) with ζ and using (3.78) we recover the equation for ζ (which is exactly (1.16))

$$\partial_t(\zeta^2) + \nabla \cdot (\zeta^2 u) = 0. \tag{3.80}$$

Therefore (ζ, u) with $\zeta \geq \frac{1}{2}\zeta_*$ is the unique local (in time) solution of IVP (1.16)-(1.18). Again by a straightforward computation once more, we can find

(1) $d = 1, 2, 3$,

$$\begin{cases} \zeta \in C^i([0, T_{**}]; H^{6-2i}(\mathbf{R}^d)) \cap C^3([0, T_{**}]; L^2(\mathbf{R}^d)), & i = 0, 1, 2; \\ u \in C^i([0, T_{**}]; \mathcal{H}^{5-2i}(\mathbf{R}^d)), & i = 0, 1, 2, \end{cases}$$

(2) $d \geq 3$,

$$\begin{cases} \zeta \in C^i([0, T_{**}]; H^{2d-2i}(\mathbf{R}^d)), & i = 0, 1, \cdots, d-1; \\ u \in C^i([0, T_{**}]; \mathcal{H}^{2d-1-2i}(\mathbf{R}^d)), & i = 0, 1, \cdots, d-1. \end{cases}$$

The proof of Theorem 1.1 is completed. □

4 Blow-up of the smooth solution

Let (ρ, u) be a smooth solution of (1.9)-(1.12) for $0 \leq t < T$.

First we will show that for any $0 \leq t \leq T$,

$$m(t) = m(0), \tag{4.1}$$

$$e(t) = e(0). \tag{4.2}$$

In fact, by integrating by parts, we have

$$\frac{dm(t)}{dt} = \int_{\mathbf{R}^d} (\rho - \bar{\rho})_t dx = \int_{\mathbf{R}^d} \mathrm{div}((\rho - \bar{\rho})u) dx = 0. \tag{4.3}$$

On the other hand, multiplying (1.10) by u and integrating by parts, we can obtain

$$\frac{d}{dt} \int_{\mathbf{R}^d} \{\frac{1}{2}\rho|u|^2 + \frac{A}{\gamma-1}(\rho^\gamma - \bar{\rho}^\gamma) + \frac{\varepsilon^2}{2}|\nabla\sqrt{\rho}|^2\} dx = 0,$$

i.e.,

$$\frac{de(t)}{dt} = \int_{\mathbf{R}^d} (E - \bar{E})_t dx = 0. \tag{4.4}$$

Similarly, we can derive

$$F'(t)$$
$$= \int_{\mathbf{R}^d} x \cdot (\rho u)_t dx$$
$$= -\int_{\mathbf{R}^d} x \cdot \mathrm{div}(\rho u \otimes u) dx - \int_{\mathbf{R}^d} x \cdot \nabla P(\rho) dx + \frac{\varepsilon^2}{2} \int_{\mathbf{R}^d} x \cdot \rho \nabla \left(\frac{\Delta\sqrt{\rho}}{\sqrt{\rho}}\right) dx$$

$$= \int_{\mathbf{R}^d} \rho|u|^2 dx + d\int_{\mathbf{R}^d}(P(\rho) - P(\bar{\rho}))dx - \frac{\varepsilon^2}{2}\int_{\mathbf{R}^d} x\cdot\operatorname{div}(x\rho)\left(\frac{\Delta\rho}{\rho}\right)dx$$

$$= \int_{\mathbf{R}^d} \rho|u|^2 dx + d\int_{\mathbf{R}^d}(P(\rho) - P(\bar{\rho}))dx$$

$$- \frac{\varepsilon^2}{2}\int_{\mathbf{R}^d}(d\sqrt{\rho}\Delta\sqrt{\rho}dx + x\cdot 2\nabla\sqrt{\rho}\Delta\sqrt{\rho})dx$$

$$= \int_{\mathbf{R}^d} \rho|u|^2 dx + d\int_{\mathbf{R}^d}(P(\rho) - P(\bar{\rho}))dx$$

$$+ \frac{\varepsilon^2}{2}\int_{\mathbf{R}^d}(d|\nabla\sqrt{\rho}|^2 dx + 2|\nabla\sqrt{\rho}|^2 - x\cdot\nabla|\nabla\sqrt{\rho}|^2)dx$$

$$= \int_{\mathbf{R}^d} \rho|u|^2 dx + d\int_{\mathbf{R}^d}(P(\rho) - P(\bar{\rho}))dx + \varepsilon^2\int_{\mathbf{R}^d}|\nabla\sqrt{\rho}|^2 dx. \tag{4.5}$$

The proof of Theorem 1.2 The goal is to obtain a differential inequality for $F(t)$.

Define $B(t) = \{x : |x| \leqslant R + \sigma t\} = D(t)^c$. The Proposition 1 says that $(\rho, u) = (\bar{\rho}, 0)$ outside $B(t)$. By Jensen's inequality, and (4.1) we have

$$\int_{B(t)} pdx = A\int_{B(t)} \rho^\gamma dx \geqslant A(\operatorname{vol} B(t))^{1-\gamma}\left(\int_{B(t)} \rho dx\right)^\gamma$$

$$= A(\operatorname{vol} B(t))^{1-\gamma}(m(0) + \operatorname{vol} B(t)\bar{\rho})^\gamma \geqslant \bar{p}\operatorname{vol} B(t) = \int_{B(t)} \bar{p}dx. \tag{4.6}$$

From (4.5) we obtain

$$F'(t) \geqslant \int_{\mathbf{R}^d} \rho|u|^2 dx. \tag{4.7}$$

Next, we apply Cauchy-Schwartz inequality to obtain

$$F(t)^2 = \left(\int_{B(t)} x\cdot\rho u dx\right)^2 \leqslant \left(\int_{B(t)} |x|^2 \rho dx\right)\left(\int_{B(t)} \rho|u|^2 dx\right). \tag{4.8}$$

With the aid of (4.1) we have

$$\int_{B(t)} |x|^2 \rho dx$$

$$\leqslant (R + \sigma t)^2 \int_{B(t)} \rho dx = (R + \sigma t)^2 \left(m(0) + \int_{B(t)} \bar{\rho}dx\right)$$

$$= (R + \sigma t)^2 \left(\int_{B(t)}(\rho_0 - \bar{\rho})dx + \int_{B(t)} \bar{\rho}dx\right) \leqslant \beta_d(R + \sigma t)^{d+2}\max\rho_0(x), \tag{4.9}$$

where β_d is the volume of unit ball in $\mathbf{R}^d$,

$$\beta_n = \begin{cases} \dfrac{(2\pi)^m}{(2m)!!}, & d = 2m, \\ \dfrac{2(2\pi)^m}{(2m+1)!!}, & d = 2m+1. \end{cases}$$

Combining (4.7)-(4.9), we give the inequality

$$F'(t) \geqslant \left[\beta_d (R+\sigma t)^{d+2} \max \rho_0(x)\right]^{-1} F(t)^2, \tag{4.10}$$

$F(0) > 0$, so (4.10) implies that $F(t)$ remains positive on $0 \leqslant t < T$. Dividing by $F(t)^2$ and integrating from 0 to T in (4.10) results in

$$\frac{1}{F(0)} \geqslant \frac{1}{F(0)} - \frac{1}{F(t)} \geqslant \frac{1}{(\beta_d(d+1)\sigma \max \rho^0)} \left(\frac{1}{R^{d+1}} - \frac{1}{(R+\sigma T)^{d+1}} \right). \tag{4.11}$$

From (4.11) we can get

$$\frac{1}{(R+\sigma T)^{d+1}} \geqslant \frac{1}{R^{d+1}} - \frac{\beta_d(d+1)\sigma \max \rho^0}{F(0)} = \frac{F(0) - \beta_d(d+1)\sigma \max \rho^0 R^{d+1}}{F(0) R^{d+1}}.$$

If (1.28) holds, i.e., $F(0) \geqslant \beta_d(d+1)\sigma \max \rho^0 R^{d+1}$, we have the finite life span time

$$T \leqslant \frac{1}{\sigma} \left(\frac{F(0) R^{d+1}}{F(0) - \beta_d(d+1)\sigma \max \rho^0 R^{d+1}} \right)^{\frac{1}{d+1}} - \frac{1}{\sigma} R. \tag{4.12}$$

It is now clear that T cannot become arbitrarily large without contradicting the assumption (1.28). This proves Theorem 1.2. $\square$

The proof of Theorem 1.3 Suppose (ρ, u) is a smooth solution of (1.9)-(1.12). Then (4.10) and (4.11) still holds. Moreover, we obtain

$$\begin{aligned} F'(t) &= \int_{\mathbf{R}^d} \rho |u|^2 dx + d \int_{\mathbf{R}^d} (P(\rho) - P(\bar\rho)) dx + \varepsilon^2 \int_{\mathbf{R}^d} |\nabla \sqrt{\rho}|^2 dx \\ &= \int_{\mathbf{R}^d} \left(\frac{2 - d(\gamma-1)}{2} \rho |u|^2 + d(\gamma-1) e(t) + \varepsilon^2 \frac{2 - d(\gamma-1)}{2} |\nabla \sqrt{\rho}|^2 \right) dx \\ &= \int_{\mathbf{R}^d} \left(\frac{2 - d(\gamma-1)}{2} \rho |u|^2 + d(\gamma-1) e(0) + \varepsilon^2 \frac{2 - d(\gamma-1)}{2} |\nabla \sqrt{\rho}|^2 \right) dx. \end{aligned} \tag{4.13}$$

By the assumption (1.32), there holds

$$F'(t) \geqslant \frac{2 - d(\gamma-1)}{2} \int_{\mathbf{R}^d} \rho |u|^2 dx. \tag{4.14}$$

Using (4.8) and (4.9), we have

$$F'(t) \geqslant \frac{2 - d(\gamma - 1)}{2\beta_d (R + \sigma t)^{d+2} \max \rho_0(x)} F(t)^2, \qquad (4.15)$$

Dividing by $F(t)^2$ and integrating from 0 to T in (4.15) results in

$$\frac{1}{F(0)} \geqslant \frac{1}{F(0)} - \frac{1}{F(t)} \geqslant \frac{2 - d(\gamma - 1)}{(2\beta_d(d+1)\sigma \max \rho^0)} \left(\frac{1}{R^{d+1}} - \frac{1}{(R+\sigma T)^{d+1}} \right). \qquad (4.16)$$

Furthermore we have

$$\frac{1}{(R+\sigma T)^{d+1}} \geqslant \frac{1}{R^{d+1}} - \frac{2\beta_d(d+1)\sigma \max \rho^0}{(2 - d(\gamma - 1))F(0)}$$

$$= \frac{(2 - d(\gamma - 1))F(0) - 2\beta_d(d+1)\sigma \max \rho^0 R^{d+1}}{(2 - d(\gamma - 1))F(0)R^{d+1}}.$$

If (1.33) holds, we can obtain the life span T satisfies

$$T \leqslant \frac{1}{\sigma} \left(\frac{(2 - d(\gamma - 1))F(0)R^{d+1}}{(2 - d(\gamma - 1))F(0) - 2\beta_d(d+1)\sigma \max \rho^0 R^{d+1}} \right)^{\frac{1}{d+1}} - \frac{1}{\sigma} R. \qquad (4.17)$$

It is now clear that T cannot become arbitrarily large without contradicting the assumption (1.32). This proves Theorem 1.3. $\square$

Acknowledgement The authors would like to express their sincere thanks to Professor Yan Guo in Brown University, for his instructive advice and useful suggestions.

References

[1] P. Antonelli, P. Marcati. On the finite energy weak solutions to a system in quantum fluid dynamics. Comm. Math. Phys., 287(2) (2009), 657-686.

[2] P. Antonelli, P. Marcati. The quantum hydrodynamics system in two space in two space dimensions. Archive for Rational Mechanics and Analysis, 203 (2012), 499-527.

[3] R. Feynman, S. Giorgini, L. Pitaevskii and S. Stringari. Theory of Bose-Einstein condensation in trapped gases. Rev. Mode. Phys., 71 (1999), 463-512.

[4] I. M. Gamba, M. P. Gualdani and P. Zhang. On the blowing up of solutions to quantum hydrodynamic models on bounded domains. Monatsh Math, 157 (2009), 37-54.

[5] I.M. Gamba and A. Jüngel. Asymptotic limits in quantum trajectory models. Comm. P.D.E., 27 (2002), 669-691.

[6] I. M. Gamba and A.Jüngel. Positive solutions to singular second and third order differential equations for quantum fluids. Arch. Ration. Mech. Anal., 156 (2001), 183-203.

[7] C. Gardner. The quantum hydrodynamic model for semiconductors devices. SIAM J. Appl. Math., 54 (1994), 409-427.

[8] I. Gasser and P. Markowich. Quantum hydrodynamics, Wigner transforms and the classical limit. Asymptotic Anal., 14 (1997), 97-116.

[9] I. Gasser, P. A. Markowich, and C. Ringhofer. Closure conditions for classical and quantum moment hierarchies in the small temperature limit. Transp. Theory Stat. Phys., 25 (1996), 409-423.

[10] R. T. Glassey. On the blowing up of solutions to the Cauchy problem for nonlinear Schrödinger equations. J. Math. Phys, 18 (1977), 1794-1797.

[11] M. P. Gualdani and A. Jüngel. Analysis of the viscous quantum hydrodynamic equations for semiconductors. Eur. J. Appl. Math., 15 (2004), 577-595.

[12] F. M. Huang, H. L. Li, A. Matsumura and S. Odanaka. Well-posedness and stability of multi-dimensional quantum hydrodynamics for semiconductors in R^3. Series in Contemporary Applied Mathematics CAM 15, High Education Press, Beijing, 2010.

[13] A. Jüngel and H. L. Li. Quantum Euler-Poisson systems: global existence and exponential decay. Quart. Appl. Math., 62(3) (2004), 569-600.

[14] A. Jüngel and J. P. Milisic. Physical and numerical viscosity for quantum hydrodynamics. Comm. Math. Sci., 5(2) (2007), 447-471.

[15] T. Kato. Qusi-linear equations of evolution, with applications to partial differential equations. Lecture Notes in Math., 448 (1975), 25-70.

[16] L. D. Landau. Theory of the superfluidity of Helium II. Phys. Rev. 60 (1941), 356.

[17] L. D. Landau and E. M. Lifshitz. Quantum mechanics: non-relativistic theory. New York, Pergamon Press, 1977.

[18] H. L. Li and P. Marcati. Existence and asymptotic behavior of multi-dimensional quantum hydrodynamic model for semiconductors. Comm. Math. Phys., 245 (2004), 215-247.

[19] M. Loffredo and L. Morato. On the creation of quantum vortex lines in rotating HeII. Il nouvo cimento, 108(B) (1993), 205-215.

[20] E. Madelung. Quantuentheorie in hydrodynamischer form. Z. Physik, 40 (1927), 322.

[21] A. Majda. Compressible fluid flow and systems of conversation laws in several space variables. Springer-Verlag, 1984.

[22] A. Majda and A. Bertozzi. Vorticity and incompressible flow. Cambridge University Press, Cambriadge, 2002.

[23] T. C. Sideris. Formation of singularities in three-dimensional compressible fluids. Comm. Math. Phys, 101 (1985), 475-485.

[24] T. C. Sideris. Formation of singularities of solutions to nonlinear hyperbolic equations. Arch. Ration. Mech. Anal., 86 (1984), 369-381.

[25] J. Simon. Compact sets in space $L^p(0,T;B)$. Ann. Math. Pura. Appl., 146(1987), 65-96.

[26] E. M. Stein. Singular integrals and differentiability properties of functions. Princeton University Press, New Jersey, 1970.

[27] Z. P. Xin. Blow up of smooth solutions to the compressible Navier-Stokes equation with compact density. Comm. Pure Appl. Math., 51 (1998), 229-240.

[28] E. Zeidler. Nonlinear functional analysis and its applications,II: Nonlinear monotone operators. Springer-Verlag, 1990.

Existence and Uniqueness of Global Solutions of the Modified KS-CG Equations for Flames Governed by A Sequential Reaction*

Guo Boling (郭柏灵) and Xie Binqiang (解斌强)

Abstract In this paper, we are concerned with the existence and uniqueness of global solutions of the modified KS-CGL equations for flames governed by a sequential reaction, where the term $|P|^2P$ is replaced with the generalized form $|P|^{2\sigma}P$, see [18]. The main novelty compared with [18] in this paper is to control the norms of the first order of the solutions and extend the global well-posedness to three dimensional space.

Keywords existence and uniqueness; modified KS-CGL; global solutions.

1 Introduction

This paper is devoted to the existence and uniqueness of global solutions for the following coupled modified Kuramoto-Sivashinsky-complex Ginzburg-Landau(GKS-CGL)equations for flames

$$\partial_t P = \xi P + (1+i\mu)\Delta P - (1+i\nu)|P|^{2\sigma}P - \nabla P \nabla Q - r_1 P \Delta Q - gr_2 P \Delta^2 Q, \quad (1.1)$$

$$\partial_t Q = -\Delta Q - g\Delta^2 Q + \delta\Delta^3 Q - \frac{1}{2}|\nabla Q|^2 - \eta|P|^2, \quad (1.2)$$

with the periodic initial conditions

$$P(x+Le_i, t) = P(x,t), \quad Q(x+Le_i, t) = Q(x,t), \quad x \in \Omega, t \geqslant 0, \quad (1.3)$$

$$P(x,0) = P_0(x), \quad Q(x,0) = Q_0(x), \quad x \in \Omega, \quad (1.4)$$

---

*Ann. of Appl. Math, 2016, 32(1): 1–19.

where Ω is a box with length L denoted $T^n(n = 1, 2, 3)$. The complex function $P(x, t)$ denotes the rescaled amplitude of the flame oscillations, and the real function $Q(x, t)$ is the deformation of the first front. The Landau coefficients μ, ν and the coupling coefficient $\eta > 0$ are real, while r_1 and r_2 are complex parameters of the form $r_1 = r_{1r} + ir_{1i}, r_2 = r_{2r} + ir_{2i}$, respectively. The coefficient $g > 0$ is proportional to the supercriticality of the oscillatory mode, $\delta > 0$ is a contant, $L > 0$ is the period and e_i is the standard coordinate vector, and the coefficient $\xi = \pm 1$. The parameter σ, μ, ν satisfy

$$(A1): 1 < \sigma < \frac{1}{\sqrt{1 + (1 + \frac{\mu - \nu}{2}^2) - 1}},$$

$$(A2): \sigma \leqslant \frac{\sqrt{1 + \mu^2}}{4\sqrt{1 + \mu^2} - 4}.$$

For $\delta = 0$, the coupled GKS-CGL equations (1.1) (1.2) are reduced to the classical KS-CGL equations [1], which decribe the nonlinear interaction between the monotonic and oscillatory modes of instability of the two uniformly propagating flame fronts in a sequential reaction. Specifically, they describe both the long-wave evolution of the oscillatory mode near the oscillatory instability threshold and the evolution of the monotonic mode. For the background of the uniformly propagating premixed flame fronts and the derivation of the KS-CGL model, one refer to [1,2,3,4] for details. If there exist no coupling with the monotonic model, then Eq.(1.1) is the well-known CGL equation that describe the weakly nonlinear evolution of a long-scaled instability[5]. For $\delta = 0$ and the coupled coefficient $\eta = 0$ in Eq.(1.2), the Eq.(1.2) reduces to the well known KS equation[6], which governs the flame front's spatio-temporal evolution and produces monotonic instability. It's seen that the coupled GKS-CGL equations (1.1)-(1.2) can better describe the dynamical behavior for flames governed by a sequential reaction, since they generalize the KS equation, the CGL equation, and the KS-CGL equations.

So far, the mathematical analysis and physical study about the CGL equation and KS equation have been done by many researchers. For example, the existence of global solutions and the attractor for the CGL equation are studied in [7-11]. For some other results, see [12-15] and reference therein. However, little progress has been made for the coupled KS-CGL equations which are derived to describe the nonlinear evolution for flames by A.A.Golovin, eta[1], who have studied the traveling waves of the coupled equations numerically and the spiral waves in [16], where new types of instabilities are exhibited. Meanwhile, there is few work to consider mathematical analytical properties of the KS-CGL equations and the generalized KS-CGL equations, even existence and uniqueness of the solutions.

In [17], the Littlewood-Paley theory is used to obtain the local solution and global solution with small initial conditions for the coupled KS-Burgers which are derived by [1], while in [18] an additional sixth order term is added to control the nonlinear term estimate through which one can get the global smooth solution of the generalized GKS-CGL equation. In this paper, via delicate a prior estimates and the Galerkin method, we consider the global smooth solution of GKS-CGL equations where the third term is replaced by $(1+i\nu)|P|^{2\sigma}P$, that is, we study the system (1.1)-(1.4) and extend the global well-posedness to the three dimensional space.

The rest paper is organized as follows. In Section 2, we briefly give some notations and preliminaries. In Section 3, local solutions are constructed by the contraction mapping theorem. A prior estimates for the solutions of the periodic initial value problem (1.1)-(1.4) are obtained in Section 4. In Section 5, we deduce from the so-called continuity method that the existence and uniqueness of the global solutions of the periodic initial value problem (1.1)-(1.4).

2 Notaions and preliminaries

For convenience of the reader we will recall some notations and preliminaries which will be used in the sequel.

Let L_{per}^k and H_{per}^k $(k=1,2,\cdots)$ denote the Sobolev spaces of L-periodic, complex-valued functions endowed with norms

$$\|u\|_p = \left(\int_\Omega |u|^p dx\right)^{1/p}, \|u\|_{H^k} = \left(\sum_{|\alpha|\leqslant k}\|D^\alpha u(x)\|\right)^2,$$

here we write $\|u\| = \|u\|_{L^2} = \sqrt{(u,u)}$, where the inner product $(\cdot,\cdot)$ is defined by $(u,v) = \int_\Omega u(x)\overline{v(x)}dx$ and $\overline{v}$ denotes the complex conjugate of v.

Now, we give some useful inequalities.

Lemma 2.1 (Young's inequality with ε)[19]   Let $a>0, b>0, 1<p,q<\infty, \dfrac{1}{p}+\dfrac{1}{q}=1$. Then

$$ab \leqslant \varepsilon a^p + C(\varepsilon)b^q,$$

for $C(\varepsilon) = (\varepsilon p)^{-q/p} q^{-1}$.

Lemma 2.2 (Gagliardo-Nirenberg inequality)[20] Let Ω be a bounded domain with $\partial \Omega$ in C^m, and let u be any function in $W^{m,r} \cap L^q(\Omega)$, $1 \leqslant q, r \leqslant \infty$. For any integer j, $0 \leqslant j < m$ and for any number a in the interval $j/m \leqslant a \leqslant 1$, set

$$\frac{1}{p} = \frac{j}{n} + a\left(\frac{1}{r} - \frac{m}{n}\right) + (1-a)\frac{1}{q}.$$

If $m - j - n/r$ is not nonnegative integer, then

$$\|D^j u\|_p \leqslant C \|u\|_{W^{m,r}}^a \|u\|_{L^q}^{1-a}. \tag{2.1}$$

If $m - j - n/r$ is a nonnegative integer, then (2.1) holds for $a = j/m$. The constant C depends only on Ω, r, q, j and a.

In the sequel, we will use the following inequalities which are the specific cases of the Gagliardo-Nirenberg inequality:

$$\|D^j u\|_\infty \leqslant C \|u\|_{H^m}^a \|u\|^{1-a}, \quad ma = j + n/2, \tag{2.2}$$

$$\|D^j u\|_2 \leqslant C \|u\|_{H^m}^a \|u\|^{1-a}, \quad ma = j, \tag{2.3}$$

$$\|D^j u\|_4 \leqslant C \|u\|_{H^m}^a \|u\|^{1-a}, \quad ma = j + n/4. \tag{2.4}$$

3 Local solution

In this section, we will use the contraction mapping principle to prove the local solution for the system (1.1)-(1.4).

For convenience, we rewrite the system as an abstract form

$$\frac{dP}{dt} + AP = M(p, Q), \tag{3.1}$$

$$\frac{dQ}{dt} + BQ = N(p, Q), \tag{3.2}$$

where $A = -(1+i\mu)\Delta$ with domain $D(A) = H^1(\Omega)$, $M(p,Q) = \xi P - (1+i\nu)|P|^{2\sigma} P - \nabla P \nabla Q - r_1 P \Delta Q - g r_2 P \Delta^2 Q$, $B = -\delta \Delta^3$ with domain $D(A) = H^5(\Omega)$, $N(p,Q) = -\Delta Q - g \Delta^2 Q - \frac{1}{2}|\nabla Q|^2 - \eta |P|^2$. Then operator $-A$ and $-B$ generate uniformly bounded analytic semigroup $S_1(t) = e^{-At}$ and $S_2(t) = e^{-Bt}$ for $t \geqslant 0$, respectively. Therefore, we deduce from (3.1), (3.2) that

$$P = S_1(t) P_0 + \int_0^t S_1(t-s) M(P, Q) ds, \tag{3.3}$$

$$Q = S_2(t)Q_0 + \int_0^t S_2(t-s)N(P,Q)ds. \tag{3.4}$$

Then, define a mapping

$$\Re : (P, Q) \to (\overline{P}, \overline{Q}), \tag{3.5}$$

where

$$\overline{P} = S_1(t)P_0 + \int_0^t S_1(t-s)M(P,Q)ds, \tag{3.6}$$

$$\overline{Q} = S_2(t)Q_0 + \int_0^t S_2(t-s)N(P,Q)ds, \tag{3.7}$$

and define a normed linear space

$$\Im = \{(P, Q) \in C([0,T]; H^1(\Omega)) \times C([0,T]; H^5(\Omega))|,$$
$$\sup_{[0,T]} \|P\|_{H^1} \leqslant C\|P_0\|_{H^1} + 1 \equiv R, \sup_{[0,T]} \|Q\|_{H^5} \leqslant C\|Q_0\|_{H^1} + 1 \equiv R\},$$

therefore,

$$\|\overline{P}\|_{H^1}$$
$$\leqslant C\|P_0\|_{H^1} + \int_0^t \|A^{1/2}e^{-A(t-\tau)}M(P,Q)\|_{L^2}d\tau$$
$$\leqslant C\|P_0\|_{H^1} + T^{1/2}\|M(P,Q)\|_{L^2},$$

and

$$\|M(P,Q)\|_{L^2}$$
$$= \|\xi P - (1+i\nu)|P|^{2\sigma}P - \nabla P \nabla Q - r_1 P\Delta Q - gr_2 P\Delta^2 Q\|_{L^2}$$
$$\leqslant C\|P\|_{L^2} + R^{2\sigma}\|P\|_{H^1} + C\|\nabla P\|_{L^2}\|Q\|_{H^4}$$
$$\quad + C\|P\|_{L^2}\|Q\|_{H^4} + \|P\|_{L^4}\|\Delta^2 Q\|_{L^4}$$
$$\leqslant C\|P\|_{L^2} + R^{2\sigma}\|P\|_{H^1} + C\|\nabla P\|_{L^2}\|Q\|_{H^4}$$
$$\quad + C\|P\|_{L^2}\|Q\|_{H^4} + \|P\|_{H^1}\|Q\|_{H^5}.$$

Here the Sobolev imbedding theorem $H^1 \hookrightarrow L^{2(2\sigma+1)}$ is used, then we obtain

$$\|\overline{P}\|_{H^1} \leqslant C\|P_0\|_{H^1} + CT^{1/2}(1 + R^{2\sigma} + R)R. \tag{3.8}$$

Therefore, when T is small, $\Re$ maps $\Im$ into itself. Similarly, we get

$$\|\overline{Q}\|_{H^5} \leqslant C\|Q_0\|_{H^5} + \int_0^t \|A^{5/6}e^{-B(t-\tau)}N(P,Q)\|_{L^2}d\tau$$

$$\leqslant C\|Q_0\|_{H^5} + T^{1/6}\|N(P,Q)\|_{L^2},$$

$$\|N(P,Q)\|_{L^2}$$
$$= \|\Delta Q + g\Delta^2 Q - \frac{1}{2}|\nabla Q|^2 - \eta|P|^2\|_{L^2}$$
$$\leqslant C\|\Delta Q\|_{L^2} + C\|\Delta^2 Q\|_{L^2}$$
$$+ C\|\nabla Q\|_{L^\infty}\|\nabla Q\|_{L^2} + C\|P\|_{L^4}^2,$$

then we have

$$\|\overline{Q}\|_{H^5} \leqslant C\|Q_0\|_{H^5} + CT^{1/6}(1+R)R.$$

Therefore, when T is small, $\Re$ maps $\Im$ into itself.

Now, we prove $\Re$ is a contractive mapping on $\Im$, since

$$\|\Re(P_1) - \Re(P_2)\|_{H^1}$$
$$\leqslant \int_0^t \|A^{1/2}e^{-A(t-\tau)}M(P_1-P_2, Q_1-Q_2)\|_{L^2}d\tau,$$

$$\|M(P_1-P_2, Q_1-Q_2)\|_{L^2}$$
$$\leqslant \|\xi(P_1-P_2) - (1+i\nu)(|P_1|^{2\sigma}P_1 - |P_2|^{2\sigma}P_2)$$
$$- (\nabla P_1 \nabla Q_1 - \nabla P_2 \nabla Q_2) - r_1(P_1\Delta Q_1 - P_2\Delta Q_2)$$
$$- gr_2(P_1\Delta^2 Q_1 - P_2\Delta^2 Q_2)\|_{L^2},$$

where

$$\|(1+i\nu)(|P_1|^{2\sigma}P_1 - |P_2|^{2\sigma}P_2)\|_{L^2}$$
$$\leqslant CR^{2\sigma}\|P_1 - P_2\|_{L^2}$$
$$\leqslant CR^{2\sigma}\|P_1 - P_2\|_{H^1},$$

$$\|\nabla P_1 \nabla Q_1 - \nabla P_2 \nabla Q_2\|_{L^2}$$
$$= \|\nabla P_1 \nabla Q_1 - \nabla P_1 \nabla Q_2 + \nabla P_1 \nabla Q_2 - \nabla P_2 \nabla Q_2\|_{L^2}$$
$$\leqslant \|\nabla Q_1 - \nabla Q_2\|_{L^\infty}\|\nabla P_1\|_{L^2}$$
$$+ \|\nabla P_1 - \nabla P_2\|_{L^2}\|\nabla Q_2\|_{L^\infty}$$
$$\leqslant CR(\|P_1 - P_2\|_{H^1} + \|Q_1 - Q_2\|_{H^5}),$$

$$\|r_1(P_1\Delta Q_1 - P_2\Delta Q_2)\|_{L^2}$$
$$= \|r_1(P_1\Delta Q_1 - P_2\Delta Q_1 + P_2\Delta Q_1 - P_2\Delta Q_2)\|_{L^2}$$
$$\leqslant C\|P_1 - P_2\|_{L^2}\|\Delta Q_1\|_{L^\infty} + C\|\Delta Q_1 - \Delta Q_2\|_{L^\infty}\|\nabla P_2\|_{L^2}$$
$$\leqslant CR(\|P_1 - P_2\|_{H^1} + \|Q_1 - Q_2\|_{H^5}),$$

$$\|gr_2(P_1\Delta^2 Q_1 - P_2\Delta^2 Q_2)\|_{L^2}$$
$$= \|gr_2(P_1\Delta^2 Q_1 - P_2\Delta^2 Q_1 + P_2\Delta^2 Q_1 - P_2\Delta^2 Q_2)\|_{L^2}$$
$$\leqslant C\|P_1 - P_2\|_{L^4}\|\Delta Q_1\|_{L^4} + C\|\Delta^2 Q_1 - \Delta^2 Q_2\|_{L^4}\|P_2\|_{L^4}$$
$$\leqslant C\|P_1 - P_2\|_{H^1}\|Q_1\|_{H^5} + C\|Q_1 - Q_2\|_{H^5}\|P_2\|_{H^1}$$
$$\leqslant CR(\|P_1 - P_2\|_{H^1} + \|Q_1 - Q_2\|_{H^5}),$$

then we have

$$\|\Re(P_1) - \Re(P_2)\|_{H^1} \leqslant CT^{1/2}[(R + R^{2\sigma})\|P_1 - P_2\|_{H^1} + R\|Q_1 - Q_2\|_{H^5}]. \tag{3.9}$$

Similarly, we get

$$\|\Re(Q_1) - \Re(Q_2)\|_{H^5} \leqslant CT^{1/2}R(\|P_1 - P_2\|_{H^1} + \|Q_1 - Q_2\|_{H^5}). \tag{3.10}$$

Adding the above two inequalities, we finally deduce

$$\|\Re(P_1) - \Re(P_2)\|_{H^1} + \|\Re(Q_1) - \Re(Q_2)\|_{H^5}$$
$$\leqslant CT^{1/2}[(R + R^{2\sigma})\|P_1 - P_2\|_{H^1} + R\|Q_1 - Q_2\|_{H^5}]. \tag{3.11}$$

By taking T so small that $CT^{1/2}(R + R^{2\sigma}) \leqslant \dfrac{1}{2}$, then $\Re$ is a contraction on $\Im$. We deduce from the contraction mapping principle that there exists a fixed point of $\Re$ on $\Im$, that is, there exists a unique local solution to the system (1.1)-(1.4) such that

$$(P, Q) \in C([0, T]; H^1(\Omega)) \times C([0, T]; H^5(\Omega)), \tag{3.12}$$

where T depends on $\|P_0\|_{H^1(\Omega)}, \|Q_0\|_{H^5(\Omega)}$.

4 A priori estimates

In this section, we will derive the a priori estimates for the solutions of the problem(1.1)-(1.4). Firstly, we have

Lemma 4.1 Let $P_0(x) \in L^2_{per}(\Omega)$, $Q_0(x) \in H^1_{per}(\Omega)$ and suppose that $\sigma > 1$ and Ω is a bounded domain with $\partial \Omega$ in C^m. Then

$$\|P\|^2 \leqslant e^{Ct}(\|P_0\|^2 + \|\nabla Q_0\|^2 + Ct),$$
$$\|\nabla Q\|^2 \leqslant e^{Ct}(\|P_0\|^2 + \|\nabla Q_0\|^2 + Ct), \qquad (4.1)$$

for C is a positive constant.

Proof Firstly we differentiate Eq.(1.2) with respect to x once and set

$$W = \nabla Q, \qquad (4.2)$$

then Eqs.(1.1) and (1.2) can be rewritten as

$$\partial_t P = \xi P + (1+i\mu)\Delta P - (1+i\nu)|P|^{2\sigma}P - \nabla PW - r_1 P \nabla W - gr_2 P \nabla \Delta W, \qquad (4.3)$$
$$\partial_t W = -\Delta W - g\Delta^2 W + \delta \Delta^3 W - W \nabla W - \eta \nabla(|P|^2). \qquad (4.4)$$

Multiplying (4.3) by $\overline{P}$, integrating with respect to x over Ω and taking the real part, we obtain

$$\frac{1}{2}\frac{d}{dt}\|P\|^2 = \operatorname{Re} \int_\Omega P_t \overline{P} dx$$
$$= \xi \|P\|^2 - \|\nabla P\|^2 - \int_\Omega |P|^{2\sigma+2} dx - \operatorname{Re} \int_\Omega \nabla P \overline{P} W dx$$
$$- r_{1r} \int_\Omega |P|^2 \nabla W dx - gr_{2r} \int_\Omega |P|^2 \nabla \Delta W dx, \qquad (4.5)$$

where

$$-\operatorname{Re} \int_\Omega \nabla P \overline{P} W dx = \frac{1}{2} \int_\Omega |P|^2 \nabla W dx. \qquad (4.6)$$

On the other hand, multiplying (4.4) by W and integrating over Ω, we have

$$\frac{1}{2}\frac{d}{dt}\|W\|^2 = \int_\Omega W_t W dx$$
$$= \|\nabla W\|^2 - g\|\Delta W\|^2 - \delta \|\nabla \Delta W\|^2$$
$$- \int_\Omega W^2 \nabla W dx - \eta \int_\Omega \nabla(|P|^2) W dx, \qquad (4.7)$$

where

$$\int_\Omega W^2 \nabla W dx = \frac{1}{3}\int_\Omega \nabla W^3 dx = 0, \qquad (4.8)$$

and

$$-\eta \int_\Omega \nabla(|P|^2) W dx = \eta \int_\Omega |P|^2 \nabla(W) dx. \qquad (4.9)$$

Adding (4.5) and (4.7) together, and noticing (4.6), (4.8) and (4.9), we deduce

$$\frac{d}{dt}(\|P\|^2 + \|W\|^2)$$
$$= 2\xi\|P\|^2 - 2\|\nabla P\|^2 - 2\int_\Omega |P|^{2\sigma+2}dx$$
$$+ 2\|\nabla W\|^2 - 2g\|\Delta W\|^2 - 2\delta\|\nabla\Delta W\|^2$$
$$+ (1 + 2\eta - 2r_{1,r})\int_\Omega |P|^2 \nabla W dx - 2gr_{2r}\int_\Omega |P|^2 \nabla\Delta W dx. \tag{4.10}$$

According to the Young inequality (2.2) and the Holder inequality, we have

$$\left|(1 + 2\eta - 2r_{1,r})\int_\Omega |P|^2 \nabla W dx\right| \tag{4.11}$$

$$\leqslant |1 + 2\eta - 2r_{1,r}| \left(\left|\int_\Omega |P|^4 dx\right|\right)^{\frac{1}{2}} \|\nabla W\| \tag{4.12}$$

$$\leqslant \frac{1}{2}\left|\int_\Omega |P|^4 dx + \frac{|1 + 2\eta - 2r_{1,r}|^2}{2}\|\nabla W\|^2 \tag{4.13}$$

$$\leqslant \frac{1}{2}\left|\int_\Omega |P|^{2\sigma+2} dx + C + \frac{|1 + 2\eta - 2r_{1,r}|^2}{2}\|\nabla W\|^2, \tag{4.14}$$

and

$$\left|- 2gr_{2r}\int_\Omega |P|^2 \nabla\Delta W dx\right| \tag{4.15}$$

$$\leqslant |2gr_{2r}| \left(\left|\int_\Omega |P|^4 dx\right|\right)^{\frac{1}{2}} \|\nabla\Delta W\| \tag{4.16}$$

$$\leqslant \frac{\delta}{2}\|\nabla\Delta W\|^2 + \frac{|2gr_{2r}|^2}{2\delta}\int_\Omega |P|^4 dx \tag{4.17}$$

$$\leqslant \frac{\delta}{2}\|\nabla\Delta W\|^2 + \frac{1}{2}\int_\Omega |P|^{2\sigma+2} dx + C. \tag{4.18}$$

Combining (4.10)-(4.12) and $|\xi| = 1$, we have

$$\frac{d}{dt}\left(\|P\|^2 + \|W\|^2\right) \leqslant 2\|P\|^2 + \left(1 + \frac{|1 + 2\eta - 2r_{1,r}|^2}{2}\right)\|\nabla W\|^2$$
$$- 2\|\nabla P\|^2 - \int_\Omega |P|^{2\sigma+2} dx$$
$$- 2g\|\Delta W\|^2 - \frac{3}{2}\delta\|\nabla\Delta W\|^2 + C, \tag{4.19}$$

using the Gagliardo-Nirenberg inequality (2.3), we deduce

$$\left(1 + \frac{|1 + 2\eta - 2r_{1,r}|^2}{2}\right)\|\nabla W\|^2 \leqslant \left(1 + \frac{|1 + 2\eta - 2r_{1,r}|^2}{2}\right)\|\nabla\Delta W\|^{\frac{2}{3}}\|W\|^{\frac{4}{3}}$$

$$\leqslant \frac{\delta}{2}\|\nabla\Delta W\|^2 + C\|W\|^2. \tag{4.20}$$

By (4.13) and (4.14), we obtain

$$\frac{d}{dt}(\|P\|^2 + \|W\|^2) + \int_\Omega |P|^{2\sigma+2}dx + 2g\|\Delta W\|^2 + \delta\|\nabla\Delta W\|^2$$
$$\leqslant C(\|P\|^2 + \|W\|^2) + C, \tag{4.21}$$

we deduce from the Gronwall's inequality that

$$\|P\|^2 + \|W\|^2 \leqslant e^{Ct}(\|P_0\|^2 + \|W_0\|^2 + Ct). \tag{4.22}$$

Lemma 4.2 *Under the assumptions of Lemma 3.1, we have*

$$\frac{1}{2\sigma+2}\frac{d}{dt}\int_\Omega |P|^{2\sigma+2}dx + \frac{1}{8}\int_\Omega |P|^{4\sigma+2}$$
$$\leqslant -\frac{1}{4}\int_\Omega |P|^{2\sigma-2}[(1+2\sigma)|\nabla|P|^2|^2$$
$$- 2\mu\sigma\nabla|P|^2 \cdot i(P\nabla\overline{P} - \overline{P}\nabla P)$$
$$+ |P\nabla\overline{P} - \overline{P}\nabla P|^2]$$
$$+ \frac{1}{8}\|\Delta P\|^2 + \frac{1}{3}\delta\|\nabla\Delta^2 W\|^2 + C, \tag{4.23}$$

where C is a positive constant.

Proof Since

$$\frac{1}{2\sigma+2}\frac{d}{dt}\int_\Omega |P|^{2\sigma+2}dx = \xi\|P\|^{2\sigma+2} + \mathrm{Re}\int_\Omega (1+i\mu)|P|^{2\sigma}\overline{P}\Delta P dx$$
$$- \int_\Omega |P|^{4\sigma+2}dx - \mathrm{Re}\int_\Omega |P|^{2\sigma}\overline{P}\nabla PW dx$$
$$+ r_{1,r}\int_\Omega |P|^{2\sigma+2}W dx + gr_{2,r}\int_\Omega |P|^{2\sigma+2}\nabla\Delta W dx, \tag{4.24}$$

for the second term of RSH of the (4.18), we have

$$\mathrm{Re}\int_\Omega (1+i\mu)|P|^{2\sigma}\overline{P}\Delta P dx$$
$$= -\frac{1}{4}\int_\Omega |P|^{2\sigma-2}[(1+2\sigma)|\nabla|P|^2|^2$$
$$- 2\mu\sigma\nabla|P|^2 \cdot i(P\nabla\overline{P} - \overline{P}\nabla P)$$
$$+ |P\nabla\overline{P} - \overline{P}\nabla P|^2],$$

for the remaining four terms, using the Young and Gagliargo-Nirenberg inequality, we have

$$|Re\int_\Omega |P|^{2\sigma}\overline{P}\nabla PW dx|$$
$$\leqslant \frac{1}{4}\int_\Omega |P|^{4\sigma+2} + C\int_\Omega |\nabla P|^2|W|^2 dx$$
$$\leqslant \frac{1}{4}\int_\Omega |P|^{4\sigma+2} + C\|\nabla P\|^2\|W\|_\infty^2$$
$$\leqslant \frac{1}{4}\int_\Omega |P|^{4\sigma+2} + C\|\Delta P\|\|P\|\|\nabla\Delta^2 W\|^{\frac{n}{5}}\|W\|^{\frac{10-n}{5}}$$
$$\leqslant \frac{1}{4}\int_\Omega |P|^{4\sigma+2} + \frac{1}{8}\|\Delta P\|^2 + C\|\nabla\Delta^2 W\|^{\frac{2n}{5}}$$
$$\leqslant \frac{1}{4}\int_\Omega |P|^{4\sigma+2} + \frac{1}{8}\|\Delta P\|^2 + \frac{1}{9}\delta\|\nabla\Delta^2 W\|^2 + C, \qquad (4.25)$$

$$|r_{1,r}\int_\Omega |P|^{2\sigma+2}W dx|$$
$$\leqslant \frac{1}{4}\int_\Omega |P|^{4\sigma+2} + C\int_\Omega |P|^2|W|^2 dx$$
$$\leqslant \frac{1}{4}\int_\Omega |P|^{4\sigma+2} + C\|P\|^2\|W\|_\infty^2$$
$$\leqslant \frac{1}{4}\int_\Omega |P|^{4\sigma+2} + C\|\nabla\Delta^2 W\|^{\frac{n}{5}}\|W\|^{\frac{10-n}{5}}$$
$$\leqslant \frac{1}{4}\int_\Omega |P|^{4\sigma+2} + \frac{1}{9}\delta\|\nabla\Delta^2 W\|^2 + C, \qquad (4.26)$$

and

$$|gr_{2,r}\int_\Omega |P|^{2\sigma+2}\nabla\Delta W dx|$$
$$\leqslant \frac{1}{4}\int_\Omega |P|^{4\sigma+2} + C\int_\Omega |P|^2|\nabla\Delta W|^2 dx$$
$$\leqslant \frac{1}{4}\int_\Omega |P|^{4\sigma+2} + C\|P\|^2\|\nabla\Delta W\|_\infty^2$$
$$\leqslant \frac{1}{4}\int_\Omega |P|^{4\sigma+2} + C\|\nabla\Delta W\|_\infty^2$$
$$\leqslant \frac{1}{4}\int_\Omega |P|^{4\sigma+2} + C\|\nabla\Delta^2 W\|^{\frac{n+6}{5}}\|W\|^{\frac{4-n}{5}}$$
$$(n\leqslant 4)$$
$$\leqslant \frac{1}{4}\int_\Omega |P|^{4\sigma+2} + \frac{1}{9}\delta\|\nabla\Delta^2 W\|^2 + C, \qquad (4.27)$$

and
$$|\xi\|P\|^{2\sigma+2}| \leqslant \frac{1}{8}\int_\Omega |P|^{4\sigma+2} + C. \tag{4.28}$$

Combing the above estimates, we complete the proof.

Lemma 4.3 *Let $P_0(x) \in H^1_{per}(\Omega) \cap L^{2\sigma+2}_{per}(\Omega)$, $Q_0(x) \in H^3_{per}(\Omega)$ and suppose that $\sigma > 1$ and Ω is a bounded domain with $\partial\Omega$ in C^m. Then*

$$\|\nabla P\|_2 \leqslant e^{Ct}(\|\nabla P_0\|_2 + \|\Delta Q_0\|_2 + \|\nabla\Delta Q_0\|_2 + \|P_0\|^{2\sigma+2}_{2\sigma+2} + Ct), \tag{4.29}$$

$$\|\Delta Q\|_2 + \|\nabla\Delta Q\|_2 \leqslant e^{Ct}(\|\nabla P_0\|_2 + \|\Delta Q_0\|_2 + \|\nabla\Delta Q_0\|_2 + \|P_0\|^{2\sigma+2}_{2\sigma+2} + Ct), \tag{4.30}$$

$$\|P\|^{2\sigma+2}_{2\sigma+2} \leqslant e^{Ct}(\|\nabla P_0\|_2 + \|\Delta Q_0\|_2 + \|\nabla\Delta Q_0\|_2 + \|P_0\|^{2\sigma+2}_{2\sigma+2} + Ct), \tag{4.31}$$

where C is a positive constant.

Proof Multiplying (4.3) by $-\Delta\overline{P}$, integrating with respect to x over Ω and taking the real part, we obtain

$$\frac{1}{2}\frac{d}{dt}\|\nabla P\|^2 = \xi\|\nabla P\|^2 - \|\Delta P\|^2 - \mathrm{Re}\int_\Omega (1+i\nu)|P|^{2\sigma}P\Delta\overline{P}dx$$
$$+ \mathrm{Re}\int_\Omega \nabla P\Delta\overline{P}Wdx + \mathrm{Re}\int_\Omega r_1 P\Delta\overline{P}\nabla W dx$$
$$+ \mathrm{Re}\int_\Omega gr_2 P\Delta\overline{P}\nabla\Delta W dx. \tag{4.32}$$

Multiplying (4.4) by $-\Delta W$ and $-\Delta^2 W$, and integrating over Ω respectively, we deduce

$$\frac{1}{2}\frac{d}{dt}\|\nabla W\|^2 = \|\Delta W\|^2 - g\|\nabla\Delta W\|^2 - \delta\|\Delta^2 W\|^2$$
$$+ \int_\Omega W\nabla W\Delta W dx + \eta\int_\Omega \nabla(|P|^2)\Delta W dx, \tag{4.33}$$

$$\frac{1}{2}\frac{d}{dt}\|\Delta W\|^2 = \|\nabla\Delta W\|^2 - g\|\Delta^2 W\|^2 - \delta\|\nabla\Delta^2 W\|^2$$
$$- \int_\Omega W\nabla W\Delta^2 W dx - \eta\int_\Omega \nabla(|P|^2)\Delta^2 W dx. \tag{4.34}$$

Adding (4.27), (4.28) and (4.29), we get

$$\frac{d}{dt}\left(\frac{1}{2}\|\nabla P\|^2 + \frac{1}{2}\|\nabla W\|^2 + \frac{1}{2}\|\Delta W\|^2\right)$$
$$= \xi\|\nabla P\|^2 - \|\Delta P\|^2 + \|\Delta W\|^2$$
$$- (g-1)\|\nabla\Delta W\|^2 - (\delta+g)\|\Delta^2 W\|^2$$
$$- \delta\|\nabla\Delta^2 W\|^2 + Re\int_\Omega (1+i\nu)|P|^{2\sigma}P\Delta\overline{P}dx$$
$$+ Re\int_\Omega \nabla P\Delta\overline{P}Wdx + Re\int_\Omega r_1 P\Delta\overline{P}\nabla Wdx$$
$$+ Re\int_\Omega gr_2 P\Delta\overline{P}\nabla\Delta Wdx + \int_\Omega W\nabla W\Delta Wdx$$
$$+ \eta\int_\Omega \nabla(|P|^2)\Delta Wdx - \int_\Omega W\nabla W\Delta^2 Wdx$$
$$- \eta\int_\Omega \nabla(|P|^2)\Delta^2 Wdx. \tag{4.35}$$

Now we need to control the right hand side of (4.30). Firstly, for the seventh term, we obtain

$$Re\int_\Omega (1+i\nu)|P|^{2\sigma}P\Delta\overline{P}dx$$
$$= -\frac{1}{4}\int_\Omega |P|^{2\sigma-2}[(1+2\sigma)|\nabla|P|^2|^2$$
$$- 2\nu\sigma\nabla|P|^2 \cdot i(\overline{P}\nabla P - P\nabla\overline{P})$$
$$+ |\overline{P}\nabla P - P\nabla\overline{P}|^2]. \tag{4.36}$$

Meanwhile, according to the Young' inequality and Gagliardo-Nirenberg inequality, we obtain the following estimates

$$2\|\Delta W\|^2 - 2(g-1)\|\nabla\Delta W\|^2$$
$$+ 2Re\int_\Omega \nabla P\Delta\overline{P}Wdx + 2Re\int_\Omega r_1 P\Delta\overline{P}\nabla Wdx$$
$$+ Re\int_\Omega gr_2 P\Delta\overline{P}\nabla\Delta Wdx + 2\int_\Omega W\nabla W\Delta Wdx$$
$$\leqslant 2\|\Delta W\|^2 + 2|g-1|\|\nabla^4 W\|^{\frac{3}{2}}\|W\|^{\frac{1}{2}} + 2\|W\|_\infty\|\nabla P\|\|\Delta P\|$$
$$+ 2|r_1|\|\nabla W\|_\infty\|P\|\|\Delta P\| + 2|r_2|\|\nabla\Delta W\|_\infty\|P\|\|\Delta P\|$$
$$+ 2\|\nabla W\|_\infty\|W\|\|\Delta W\|$$
$$\leqslant 2\|\Delta W\|^2 + \frac{\delta+g}{4}\|\nabla^4 W\|^2$$

$$+ 2C\|\nabla^4 W\|^{\frac{n}{8}}\|W\|^{\frac{8-n}{8}}\|\nabla^2 P\|^{\frac{1}{2}}\|P\|^{\frac{1}{2}}\|\Delta P\|$$
$$+ 2|r_1|\|P\|\|\Delta P\|\|\nabla^5 W\|^{\frac{1+n/2}{5}}\|W\|^{\frac{4-n/2}{5}}$$
$$+ 2|r_2|\|P\|\|\Delta P\|\|\nabla^5 W\|^{\frac{3+n/2}{5}}\|W\|^{\frac{2-n/2}{5}}$$
$$+ 2\|W\|\|\Delta W\|\|\nabla^5 W\|^{\frac{1+n/2}{5}}\|W\|^{\frac{4-n/2}{5}}$$
$$\leqslant 2\|\Delta W\|^2 + \frac{\delta+g}{4}\|\Delta^2 W\|^2 + C\|\Delta P\|^{\frac{3}{2}\times\frac{16}{16-n}}$$
$$\frac{1}{8}\|\Delta P\|^2 + C\|\nabla^5 W\|^{\frac{2+n}{5}}$$
$$+ \frac{1}{8}\|\Delta P\|^2 + C\|\nabla^5 W\|^{\frac{6+n}{5}} + \frac{1}{4}\|\Delta W\|^2 + C\|\nabla^5 W\|^{\frac{2+n}{5}}$$
$$\leqslant \frac{3}{8}\|\Delta P\|^2 + \frac{\delta+g}{4}\|\Delta^2 W\|^2 + \frac{1}{3}\delta\|\nabla\Delta^2 W\|^2 + \frac{9}{4}\|\Delta W\|^2 + C \quad (n\leqslant 4). \tag{4.37}$$

Next, for the remaining two terms, we obtain

$$\left|2\eta\int_\Omega \nabla(|P|^2)\Delta W\,dx\right|$$
$$= |2\eta\int_\Omega |P|^2\nabla\Delta W\,dx|$$
$$\leqslant 2\eta\|\nabla\Delta W\|\|P\|_4^2$$
$$\leqslant C\|\nabla^5 W\|^{\frac{3}{5}}\|W\|^{\frac{2}{5}}\|\nabla^2 P\|^{\frac{n}{4}}\|P\|^{\frac{8-n}{4}}$$
$$\leqslant \frac{1}{9}\delta\|\nabla\Delta^2 W\|^2 + C\|\nabla^2 P\|^{\frac{n}{4}\times\frac{10}{7}}$$
$$(n\leqslant 28/5)$$
$$\leqslant \frac{1}{9}\delta\|\nabla\Delta^2 W\|^2 + \frac{1}{8}\|\Delta P\|^2 + C, \tag{4.38}$$

and

$$\left|-2\int_\Omega W\nabla W\Delta^2 W\,dx - 2\eta\int_\Omega \nabla(|P|^2)\Delta^2 W\,dx\right|$$
$$\leqslant 2|\int_\Omega W^2\nabla\Delta^2 W\,dx| + 2\eta|\int_\Omega |P|^2\nabla\Delta^2 W\,dx|$$
$$\leqslant \frac{1}{9}\delta\|\Delta\|\|\nabla\Delta^2 W\|^2 + C(\|W\|_4^4 + \|P\|_4^4)$$
$$\leqslant \frac{1}{9}\delta\|\Delta\|\|\nabla\Delta^2 W\|^2 + C\left(\|\nabla^4 W\|^{\frac{n}{4}}\|W\|^{\frac{16-n}{4}} + \|\nabla^2 P\|^{\frac{n}{2}}\|P\|^{\frac{8-n}{2}}\right)$$
$$(n\leqslant 4)$$
$$\leqslant \frac{1}{9}\delta\|\nabla\Delta^2 W\|^2 + \frac{\delta+g}{4}\|\Delta^2 W\|^2 + \frac{1}{8}\|\Delta P\|^2 + C. \tag{4.39}$$

Combing the above estimates, we have

$$\frac{d}{dt}(\|\nabla P\|^2 + \|\nabla W\|^2 + \|\Delta W\|^2)$$

$$+ \frac{1}{4}\|\Delta P\|^2 + \frac{\delta+g}{2}\|\Delta^2 W\|^2 + \frac{1}{9}\delta\|\nabla\Delta^2 W\|^2$$
$$\leqslant -\frac{1}{4}\int_\Omega |P|^{2\sigma-2}[(1+2\sigma)|\nabla|P|^2|^2 - 2\nu\sigma\nabla|P|^2 \cdot i(\overline{P}\nabla P - P\nabla\overline{P})$$
$$+ |\overline{P}\nabla P - P\nabla\overline{P}|^2]$$
$$+ C(\|\nabla P\|^2 + \|\nabla W\|^2 + \|\Delta W\|^2) + C. \tag{4.40}$$

Adding (4.17) and (4.35), we have

$$\frac{d}{dt}\left(\frac{1}{2}\|\nabla P\|^2 + \frac{1}{2}\|\nabla W\|^2 + \frac{1}{2}\|\Delta W\|^2 + \frac{1}{2\sigma+2}\frac{d}{dt}\int_\Omega |P|^{2\sigma+2}dx\right)$$
$$+ \frac{1}{4}\|\Delta P\|^2 + \frac{\delta+g}{2}\|\Delta^2 W\|^2 + \frac{1}{9}\delta\|\nabla\Delta^2 W\|^2 + \frac{1}{8}\int_\Omega |P|^{4\sigma+2}$$
$$\leqslant -\frac{1}{4}\int_\Omega |P|^{2\sigma-2}[2(1+2\sigma)|\nabla|P|^2|^2 - 2(\nu-\mu)\sigma\nabla|P|^2 \cdot i(\overline{P}\nabla P - P\nabla\overline{P})$$
$$+ 2|\overline{P}\nabla P - P\nabla\overline{P}|^2]dx + C(\|\nabla P\|^2 + \|\nabla W\|^2 + \|\Delta W\|^2) + C, \tag{4.41}$$

note that if the assumption (A1) holds, the term $|P|^{2\sigma-2}[2(1+2\sigma)|\nabla|P|^2|^2 - 2(\nu-\mu)\sigma\nabla|P|^2 \cdot i(\overline{P}\nabla P - P\nabla\overline{P}) + 2|\overline{P}\nabla P - P\nabla\overline{P}|^2]$ is positive, therefore we can omit this term and then integrating with respect to t to get the final estimate

$$\|\nabla P\|^2 + \|\nabla W\|^2 + \|\Delta W\|^2 + \|P\|_{2\sigma+2}^{2\sigma+2}$$
$$\leqslant e^{Ct}(\|\nabla P_0\|^2 + \|\nabla W_0\|^2 + \|\Delta W_0\|^2 + \|P_0\|_{2\sigma+2}^{2\sigma+2} + Ct). \tag{4.42}$$

The following Lemma has been proved in [18], we omit the proof here.

Lemma 4.4 *Under the assumptions of Lemma 3.3, we have*

$$\|P\|_{H^1_{per}} \leqslant C, \quad \|\nabla Q\|_\infty \leqslant C, \tag{4.43}$$

where C is a positive constant.

Lemma 4.5 *Let* $s \leqslant \dfrac{2\sqrt{1+\mu^2}}{\sqrt{1+\mu^2}-1}$, *then*

$$\|P\|_s^s \leqslant C, \tag{4.44}$$

where C is a positive constant.

Proof By replacing the exponent $\sigma+2$ in Lemma 4.2 by s, we have

$$\frac{1}{s}\frac{d}{dt}\int_\Omega |P|^s dx + \frac{1}{8}\int_\Omega |P|^{s+2\sigma}$$

$$\leqslant -\frac{1}{4}\int_{\Omega}|P|^{s-4}[(s-1)|\nabla|P|^2|^2 - 2\mu(s-2)\nabla|P|^2 \cdot i(P\nabla\overline{P}$$
$$- \overline{P}\nabla P) + |P\nabla\overline{P} - \overline{P}\nabla P|^2]dx + \frac{1}{8}\|\Delta P\|^2 + \frac{1}{3}\delta\|\nabla\Delta^2 W\|^2 + C, \quad (4.45)$$

note that if the assumption $s \leqslant \dfrac{2\sqrt{1+\mu^2}}{\sqrt{1+\mu^2}-1}$ holds, the term $|P|^{s-4}[(s-1)|\nabla|P|^2|^2 - 2\mu(s-2)\nabla|P|^2 \cdot i(P\nabla\overline{P} - \overline{P}\nabla P) + |P\nabla\overline{P} - \overline{P}\nabla P|^2]$ is positive, then we can omit this term. By (4.40), (4.36) and the Gronwall's inequality, we complete the proof.

Lemma 4.6 Let $P_0(x) \in H^2_{per}(\Omega)$, $Q_0(x) \in H^4_{per}(\Omega)$ and suppose that $\sigma > 1$ and Ω is a bounded domain with $\partial\Omega$ in C^m. Then

$$\|\Delta P\|^2 + \|\Delta^2 Q\|^2 \leqslant e^{Ct}(\|\Delta P_0\|^2 + \|\Delta^2 Q_2\|^2 + ct), \quad (4.46)$$

where C is a positive constant.

Proof Multiplying (4.3) by $\Delta^2 \overline{P}$, integrating over Ω and taking the real part, we obtain

$$\frac{1}{2}\frac{d}{dt}\|\Delta P\|^2 = \xi\|\Delta P\|^2 - \|\nabla\Delta P\|^2 - \mathrm{Re}\int_{\Omega}(1+i\nu)|P|^{2\sigma}P\Delta^2\overline{P}dx$$
$$- \mathrm{Re}\int_{\Omega}\nabla P\Delta^2\overline{P}W dx - \mathrm{Re}\int_{\Omega}r_1 P\Delta^2\overline{P}\nabla W dx$$
$$- \mathrm{Re}\int_{\Omega}gr_2 P\Delta^2\overline{P}\nabla\Delta W dx. \quad (4.47)$$

Multiplying (4.4) by $-\Delta^3 W$ and integrating over Ω, we deduce

$$\frac{1}{2}\frac{d}{dt}\|\nabla\Delta W\|^2 = \|\Delta^2 W\|^2 - g\|\nabla\Delta^2 W\|^2 - \delta\|\Delta^3 W\|^2$$
$$+ \int_{\Omega}W\nabla W\Delta^3 W dx + \eta\int_{\Omega}\nabla(|P|^2)\Delta^3 W dx. \quad (4.48)$$

Adding the above two equalities yields

$$\frac{d}{dt}(\|\Delta P\|^2 + \|\nabla\Delta W\|^2)$$
$$= 2\xi\|\Delta P\|^2 - 2\|\nabla\Delta P\|^2 + 2\|\Delta^2 W\|^2 - 2g\|\nabla\Delta^2 W\|^2 - 2\delta\|\Delta^3 W\|^2$$
$$- 2\mathrm{Re}\int_{\Omega}(1+i\nu)|P|^{2\sigma}P\Delta^2\overline{P}dx - 2\mathrm{Re}\int_{\Omega}\nabla P\Delta^2\overline{P}W dx$$
$$- 2\mathrm{Re}\int_{\Omega}r_1 P\Delta^2\overline{P}\nabla W dx - 2\mathrm{Re}\int_{\Omega}gr_2 P\Delta^2\overline{P}\nabla\Delta W dx$$
$$+ 2\int_{\Omega}W\nabla W\Delta^3 W dx + 2\eta\int_{\Omega}\nabla(|P|^2)\Delta^3 W dx, \quad (4.49)$$

according to the Gagliardo-Nirenberg inequality, Lemma 4.1, Lemma 4.3, Lemma 4.5 and the assumption A2, we have

$$\left|-2Re\int_\Omega (1+i\nu)|P|^{2\sigma}P\Delta^2\overline{P}dx\right|$$
$$\leqslant 6|1+i\nu|\int_\Omega |P|^{2\sigma}|\nabla P||\nabla\Delta P|dx$$
$$\leqslant \frac{1}{3}\|\nabla\Delta P\|^2 + C\||P|^{2\sigma}\|_4^2\|\nabla P\|_4^2$$
$$\leqslant \frac{1}{3}\|\nabla\Delta P\|^2 + C\|\nabla\Delta P\|^{\frac{n}{4}}\|\nabla P\|^{\frac{8-n}{4}}$$
$$\leqslant \frac{2}{3}\|\nabla\Delta P\|^2 + C, \tag{4.50}$$

and

$$\left|-2Re\int_\Omega \nabla P\Delta^2\overline{P}Wdx\right|$$
$$= |2Re\int_\Omega \nabla P\nabla\Delta\overline{P}\nabla Wdx + 2Re\int_\Omega \Delta P\nabla\Delta\overline{P}Wdx|$$
$$\leqslant 2\|\nabla W\|_\infty\|\nabla P\|\|\nabla\Delta P\| + 2\|W\|_\infty\|\Delta P\|\|\nabla\Delta P\|$$
$$\leqslant \frac{1}{3}\|\nabla\Delta P\|^2 + C\|\nabla\Delta W\|^{\frac{n}{2}}\|\nabla W\|^{\frac{4-n}{2}} + C\|\Delta P\|^2$$
$$(n\leqslant 4)$$
$$\leqslant \frac{1}{3}\|\nabla\Delta P\|^2 + C\|\nabla\Delta W\|^2 + C\|\Delta P\|^2 + C. \tag{4.51}$$

For the remaining terms of the RHS of (4.44), we have

$$|2\|\Delta^2 W\|^2 - 2Re\int_\Omega r_1 P\Delta^2\overline{P}\nabla Wdx - 2Re\int_\Omega gr_2 P\Delta^2\overline{P}\nabla\Delta Wdx|$$
$$\leqslant 2\|\Delta^2 W\|^2 + |2Re\int_\Omega r_1(P\nabla\Delta\overline{P}\Delta W + \nabla P\nabla\Delta\overline{P}\nabla W)dx$$
$$+ |2Re\int_\Omega gr_2(P\nabla\Delta\overline{P}\Delta^2 W + \nabla P\nabla\Delta\overline{P}\Delta W)dx$$
$$\leqslant 2\|\Delta^2 W\|^2 + 2|r_1|\|P\|\|\nabla\Delta P\|\|\Delta W\|_\infty$$
$$+ 2|r_1|\|\nabla P\|\|\nabla\Delta P\|\|\nabla W\|_\infty + 2g|r_2|\|P\|\|\nabla\Delta P\|\|\Delta_2 W\|_\infty$$
$$+ 2g|r_2|\|\nabla P\|\|\nabla\Delta P\|\|\nabla\Delta W\|_\infty$$
$$\leqslant 2\|\Delta^2 W\|^2 + \frac{1}{3}\|\nabla\Delta P\|^2 + C\|\nabla W\|_{H^3}^{\frac{2+n}{3}}\|\nabla W\|^{\frac{4-n}{3}}$$
$$+ C\|\nabla W\|_{H^2}^{\frac{n}{2}}\|\nabla W\|^{\frac{4n}{2}} + C\|\nabla W\|_{H^5}^{\frac{6+n}{5}}\|\nabla W\|^{\frac{4n}{5}}$$
$$+ C\|\nabla W\|_{H^4}^{\frac{4+n}{4}}\|\nabla W\|^{\frac{4n}{4}}$$

$$\leqslant \frac{1}{3}\|\nabla\Delta P\|^2 + g\|\nabla\Delta^2 W\|^2 + \frac{\delta}{4}\|\Delta^3 W\|^2 + C\|\nabla\Delta W\|^2 + C, \tag{4.52}$$

and

$$\left|2\int_\Omega W\nabla W\Delta^3 W dx + 2\eta\int_\Omega \nabla(|P|^2)\Delta^3 W dx\right|$$
$$\leqslant 2\|W\|_\infty\|\nabla W\|\|\Delta^3 W\| + 4\eta\|P\|_\infty\|\nabla P\|\|\Delta^3 W\|$$
$$\leqslant \frac{\delta}{2}\|\Delta^3 W\| + C, \tag{4.53}$$

Then combining (4.44)-(4.48) and noticing $|\xi| = 1$, we have

$$\frac{d}{dt}(\|\Delta P\|^2 + \|\nabla\Delta W\|^2) \leqslant C(\|\Delta P\|^2 + \|\nabla\Delta W\|^2) + C. \tag{4.54}$$

Lemma 4.7   Let $P_0(x) \in H^2_{per}(\Omega)$, $Q_0(x) \in H^4_{per}(\Omega)$ and suppose that $\sigma > 1$ and Ω is a bounded domain with $\partial\Omega$ in C^m. Then

$$\|P\|_\infty \leqslant C, \quad \|\Delta Q\|_\infty \leqslant C, \tag{4.55}$$

where C is a positive constant.

Lemma 4.8   Let $P_0(x) \in H^2_{per}(\Omega)$, $Q_0(x) \in H^4_{per}(\Omega)$ and suppose that $\sigma > 1$ be a bounded domain with $\partial\Omega$ in C^m. Then

$$\|P\|_{H^2_{per}} \leqslant C, \quad \|Q\|_{H^4_{per}} \leqslant C, \tag{4.56}$$

for C is a positive constant.

For the proof of Lemmas 4.7 and 4.8 in detail, one can refer to [18].

Lemma 4.9   Let $P_0(x) \in H^3_{per}(\Omega)$, $Q_0(x) \in H^5_{per}(\Omega)$ and suppose that $\sigma > 1$ and Ω is a bounded domain with $\partial\Omega$ in C^m. Then

$$\|\nabla\Delta P\| + \|\nabla\Delta^2 Q\| \leqslant e^{Ct}(\|\nabla\Delta P_0\| + \|\nabla\Delta^2 Q_0\| + Ct), \tag{4.57}$$

where C is a positive constant.

Proof   Multiplying (4.3) by $-\Delta^3\overline{P}$, integrating over Ω and taking the real part, we obtain

$$\frac{1}{2}\frac{d}{dt}\|\nabla\Delta P\|^2$$
$$= \xi\|\nabla\Delta P\|^2 - \|\Delta^2 P\|^2 - \mathrm{Re}\int_\Omega (1+i\nu)|P|^{2\sigma}P\Delta^3\overline{P}dx$$

$$- Re \int_\Omega \nabla P \Delta^3 \overline{P} W dx - Re \int_\Omega r_1 P \Delta^3 \overline{P} \nabla W dx$$
$$- Re \int_\Omega gr_2 P \Delta^3 \overline{P} \nabla \Delta W dx. \tag{4.58}$$

Multiplying (4.4) by $\Delta^4 W$ and integrating over Ω, we deduce

$$\frac{1}{2}\frac{d}{dt}\|\Delta^2 W\|^2$$
$$= \|\nabla \Delta^2 W\|^2 - g\|\Delta^3 W\|^2 - \delta\|\Delta^3 W\|^2$$
$$- \int_\Omega W \nabla W \Delta^4 W dx - \eta \int_\Omega \nabla(|P|^2) \Delta^4 W dx. \tag{4.59}$$

Adding the above two equalities arrives at

$$\frac{d}{dt}(\|\nabla \Delta P\|^2 + \|\Delta^2 W\|^2)$$
$$= 2\xi\|\nabla \Delta P\|^2 - 2\|\Delta^2 P\|^2 + 2\|\nabla \Delta^2 W\|^2$$
$$- 2g\|\Delta^3 W\|^2 - 2\delta\|\Delta^3 W\|^2 - 2Re \int_\Omega (1+i\nu)|P|^{2\sigma} P \Delta^3 \overline{P} dx$$
$$- 2Re \int_\Omega \nabla P \Delta^3 \overline{P} W dx - 2Re \int_\Omega r_1 P \Delta^3 \overline{P} \nabla W dx$$
$$- 2Re \int_\Omega gr_2 P \Delta^3 \overline{P} \nabla \Delta W dx - 2 \int_\Omega W \nabla W \Delta^4 W dx$$
$$- 2\eta \int_\Omega \nabla(|P|^2) \Delta^4 W dx. \tag{4.60}$$

In order to control the RHS of (4.44), using the previous lemmas, we have

$$\left| 2Re \int_\Omega (1+i\nu)|P|^{2\sigma} P \Delta^3 \overline{P} dx \right|$$
$$\leqslant 2|1+i\nu| \int_\Omega (6|\nabla P|^2 |P| + 3|P|^2|\Delta P|)|P|^{2\sigma-2}|\Delta^2 P| dx$$
$$\leqslant \frac{1}{8}\|\Delta^2 P\|^2 + C(\|\nabla P|^4\|\|P\|_\infty^2 + \||\Delta P|^2\|\|P\|_\infty^4)\||P|^{4\sigma-4}\|$$
$$\leqslant \frac{1}{8}\|\Delta^2 P\|^2 + C(\|\nabla P\|_8^4 + \|\Delta P\|_4^2)$$
$$\leqslant \frac{1}{8}\|\Delta^2 P\|^2 + C(\|\Delta^2 \nabla P\|^{\frac{n}{2}}\|\nabla P\|^{\frac{8-n}{2}} + \|\Delta^2 P\|^{\frac{n}{4}}\|\Delta P\|^{\frac{8-n}{4}}$$
$$\leqslant \frac{3}{8}\|\Delta^2 P\|^2 + C, \tag{4.61}$$

and

$$\left| 2Re \int_\Omega \nabla P \Delta^3 \overline{P} W dx \right|$$

$$= \left|2Re\int_\Omega (\Delta W\nabla P + 2\Delta P\nabla W + W\nabla\Delta P)\Delta^2\overline{P}dx\right|$$

$$\leqslant 2(\|\Delta W\|\|\nabla P\|_\infty + 2\|\Delta P\|\|\nabla W\|_\infty + \|\nabla\Delta P\|\|W\|_\infty)\|\Delta^2 P\|$$

$$\leqslant \frac{1}{8}\|\Delta^2 P\|^2 + C\|\nabla\Delta P\|^{\frac{n}{2}}\|P\|^{\frac{4-n}{2}} + C\|\nabla\Delta P\|^2 + C$$

$$\leqslant \frac{1}{8}\|\Delta^2 P\|^2 + C\|\nabla\Delta P\|^2 + C, \tag{4.62}$$

and

$$\left|2Re\int_\Omega r_1 P\Delta^3\overline{P}\nabla W dx\right|$$

$$= \left|2Re\int_\Omega (\Delta P\nabla W + 2\Delta W\nabla P + P\nabla\Delta W)\Delta^2\overline{P}dx\right|$$

$$\leqslant 2(\|\Delta P\|\|\nabla W\|_\infty + 2\|\Delta W\|\|\nabla P\|_\infty + \|\nabla\Delta W\|\|P\|_\infty)\|\Delta^2 P\|$$

$$\leqslant \frac{1}{8}\|\Delta^2 P\|^2 + C\|\nabla\Delta P\|^{\frac{n}{2}}\|P\|^{\frac{4-n}{2}} + C\|\nabla\Delta P\|^2 + C$$

$$\leqslant \frac{1}{8}\|\Delta^2 P\|^2 + C\|\nabla\Delta P\|^2 + C, \tag{4.63}$$

for the remaining terms, using the Gagliardo-Nirenberg inequality and the previous estimates, we obtain

$$\left|2\|\nabla\Delta^2 W\|^2 + 2Re\int_\Omega gr_2 P\Delta^3\overline{P}\nabla\Delta W dx\right|$$

$$= 2\|\nabla\Delta^2 W\|^2 + |2Re\int_\Omega gr_2(\nabla\Delta^2 WP$$

$$+ 2\Delta^2 W\nabla P + \nabla\Delta W\Delta P)\Delta^2\overline{P}dx|$$

$$\leqslant 2\|\nabla\Delta^2 W\|^2 + 2g|r_2|(\|\nabla\Delta^2 W\|\|P\|_\infty$$

$$+ 2\|\nabla P\|\|\Delta^2 W\|_\infty + \|\nabla\Delta W\|_\infty\|\Delta P\|)\|\Delta^2 P\|$$

$$\leqslant \frac{1}{8}\|\Delta^2 P\|^2 + 2\|\nabla\Delta^2 W\|^2 + C\|\Delta^3 W\|^{\frac{5}{3}}\|W\|^{\frac{1}{3}}$$

$$+ C\|\Delta^3 W\|^{\frac{4+n}{4}}\|\Delta W\|^{\frac{4-n}{4}} + C\|\nabla\Delta^2 W\|^{\frac{4+n}{4}}\|\nabla W\|^{\frac{4-n}{4}}$$

$$\leqslant \frac{1}{8}\|\Delta^2 P\|^2 + g\|\Delta^3 W\|^2 + C\|\Delta^2 W\|^2 + C. \tag{4.64}$$

Combing these above estimates yields that

$$\frac{d}{dt}(\|\nabla\Delta P\|^2 + \|\Delta^2 W\|^2) \leqslant C(\|\nabla\Delta P\|^2 + \|\Delta^2 W\|^2) + C. \tag{4.65}$$

Therefore, from the above Lemmas, we get the following Lemma.

Lemma 4.10 Let $P_0(x) \in H^3_{per}(\Omega)$, $Q_0(x) \in H^5_{per}(\Omega)$ and suppose that $\sigma > 1$ and Ω is a bounded domain with $\partial\Omega$ in C^m. Then

$$\|P\|_{H^3} + \|Q\|_{H^5} \leqslant C, \tag{4.66}$$

where C is a positive constant.

5 The local solutions and global solutions

In this section, we will obtain the existence and uniqueness of the local solutions and global solutions for the periodic initial value problem (1.1)-(1.4). From the Lemmas in section 3, we deduce our main result:

Theorem 5.1 (local existence) Assume $P_0(x) \in H^3_{per}(\Omega), Q_0(x) \in H^5_{per}(\Omega)$ and the parameter σ, μ, ν satisfy the assumption (A1) and (A2). Then there exist local solutions $P(x,t)$ and $Q(x,t)$ to the periodic initial value problem (1.1)-(1.4), satisfying

$$P(x,t) \in C(0,t_0); H^3_{per}(\Omega)), \quad Q(x,t) \in C(0,t_0); H^5_{per}(\Omega)),$$

where t_0 depends on $\|P_0\|_{H^3_{per}}$ and $\|Q_0\|_{H^5_{per}}$.

Finally, we are able to deduce from this local existence theorem combined with a priori estimates that the solutions exist globally in time.

Theorem 5.2 (global existence) Assume $P_0(x) \in H^3_{per}(\Omega), Q_0(x) \in H^5_{per}(\Omega)$ and the parameter σ, μ, ν satisfy the assumption (A1) and (A2). Then there exist global solutions $P(x,t)$ and $Q(x,t)$ to the periodic initial value problem (1.1)-(1.4), satisfying

$$P(x,t) \in C(0,\infty); H^3_{per}(\Omega)), \quad Q(x,t) \in C(0,\infty); H^5_{per}(\Omega)),$$

where the periodic initial value problem (1.1)-(1.4).

Proof From Theorem 5.1, we know that there exist local solutions $P(x,t)$ and $Q(x,t)$ to the periodic initial value problem (1.1)-(1.4) and the existence time t_0 depends on $\|P_0(x)\|_{H^3_{per}}$ and $\|Q_0(x)\|_{H^5_{per}}$. According to the priori estimates in Section 4 and the so-called continuity method, we complete the proof.

References

[1] A. Golovin, B. Matkowsky, A. Bayliss. Coupled KS-CGL and coupled Burgers-CGL equations for flames governed by a sequential reaction. Physica D, (1999) 253-298.

[2] F.A. Williams, F. Williams. Combustion Thoery. Benjamin Cunmmings: Menlo Park, 1985.

[3] J. Pelaez, A. Linan. Structure and stability of flames with two sequential reactions. SIAM J. Appl. Math., 45 (1985) 503-522.

[4] J. Pelaez. Stability of premixed flames with two thin reaction layers. SIAM J. Appl. Math., 47 (1987) 781-799.

[5] A.A. Nepomnyashchy. Order parameter equations for long wavelength instabilities. Physica D, 86 (1995) 90-95.

[6] G.I. Sivashinsky. Nonlinear analysis of hydrodynamic instability in laminar flames I. Derivation of basic equations. Acta Astro., 4(11-12) (1977) 1177-1206.

[7] D.L. Li, B.L. Guo, X.H. Liu. Existence of global solution for complex Ginzburg Landau equation in three dimensions. Appl. Math. J. Chinese Univ. Ser. A, 19(4) (2004) 409-416.

[8] D.L. Li, B.L. Guo, X.H. Liu. Regularity of the attractor for 3-D complex Ginzburg-Landau equation. Acta Math. Appl. Sin., English Series, 27(2) (2011) 289-302.

[9] J.M. Ghidaglia, B. H'eron. Dimension of the attractor associated to the Ginzburg-Landau equation. Physica D: Nonlinear Phenomena, 28(3) (1987) 282-304.

[10] B.L. Guo. The existence and nonexistence of a global smooth solution for the initial value problem of generalized Kuramoto-Sivashinsky type equations. J. Math. Res. Exposition, 11(1) (1991) 57-70.

[11] B.L. Guo. The nonlinear Galerkin methods for the generalized Kuramoto-Sivashinsky type equations. Adv. Math., 22(2) (1993) 182-184.

[12] C.M. Postlethwaite, M. Silber. Spatial and temporal feedback control of traveling wave so- lutions of the two-dimensional complex Ginzburg-Landau equation. Physica D: Nonlinear Phenomena, 236(1) (2007) 65-74.

[13] J.A. Sherratt, M.J. Smith, J.D.M. Rademacher. Patterns of sources and sinks in the complex Ginzburg-Landau equation with zero linear dispersion. SIAM J. Appl. Dyn. Syst., 9(3) (2010)883-918.

[14] G.G. Doronin, N.A. Larkin. Kuramoto-Sivashinsky model for a dusty medium. Math. Meth. Appl. Sci., 26(3) (2003) 179-192.

[15] T. MacKenzie, A.J. Roberts. Accurately model the Kuramoto-Sivashinsky dynamics with holistic discretization. SIAM J. Appl. Dyn. Syst., 5(3) (2006) 365-402.

[16] A.A. Golovin, A.A. Nepomnyashchy, B.J. Matkowsky. Traveling and spiral waves for sequential flames with translation symmetry: Coupled CGL-Burgers equations. Physica D, 160 (2001) 1-28.

[17] B.L Guo, X.L. Wu. Global Existence and Nonlinear Stability for the Coupled CGL-Burgers Equations for Sequential flames in R^N. arXiv:1310.4326 [math.AP] (2013).

[18] C.G Guo, S.M. Mei, B. Guo. Global smooth solutions of the Gebeealized KS-CGL equations for flames governed by a sequential reaction. COMMUN. MATH. SCI., 12(8) (2014) 1457-1474.

[19] L.C. Evans. Partial Differential Equations. American Mathematical Society, 1988.
[20] A. Friedman. Partial Differential Equations. Reinhart and Winston, 1969.
[21] A. Pazy. Semigroups of linear operators and applications to partial differential equations. Berlin: Springer, 1998.

On the Global Well-posedness of the Fractional Klein-Gordon-Schrödinger System with Rough Initial Data*

Huang Chunyan (黄春妍), Guo Boling (郭柏灵),
Huang Daiwen (黄代文) and Li Qiaoxin (李巧欣)

Abstract In this paper, we investigate the low regularity local and global well-posedness of the Cauchy problem for the coupled Klein-Gordon-Schrödinger system with fractional Laplacian in the Schrödinger equation in $\mathbf{R}^{1+1}$. We use Bourgain space method to study this problem and prove that this system is locally well-posed for Schrödinger data in H^{s_1} and wave data in $H^{s_2} \times H^{s_2-1}$ for $3/4 - \alpha < s_1 \leqslant 0$ and $-1/2 < s_2 < 3/2$, where α is the fractional power of Laplacian which satisfies $3/4 < \alpha \leqslant 1$. Based on this local well-posedness result, we also obtain the global well-posedness of this system for $s_1 = 0$ and $-1/2 < s_2 < 1/2$ by using the conservation law for the L^2 norm of u.

Keywords Klein-Gordon-Schrödinger system; fractional Laplacian; Bourgain space; low regularity

1 Introduction

We consider the Cauchy problem to the following coupled system of Schrödinger equation with fractional Laplacian and Klein-Gordon equation(FKGS for short) through Yukawa coupling in one space dimension:

$$\begin{cases} \mathrm{i}u_t + (-\Delta)^\alpha u = u\phi, \\ \phi_{tt} - \Delta\phi + \phi = |u|^2, \\ u(0) = u_0, \\ \phi(0) = \phi_0, \phi_t(0) = \phi_1, \end{cases} \quad (1.1)$$

---

* Sci China Math, 2016, 59(7): 1345–1366. DOI:10.1007/s11425-016-5133-6.

where u and ϕ are complex and real valued unknown functions of $(t,x) \in \mathbf{R} \times \mathbf{R}$ respectively. $\alpha > 0$ is a real number and $(-\Delta)^\alpha$ denotes the fractional Laplacian which is defined as $\widehat{(-\Delta)^\alpha u}(t,\xi) = |\xi|^{2\alpha}\hat{u}(t,\xi)$, here $\hat{\ }$ is the Fourier transform respect to x. We are interested in the local and global well-posedness results on system (1.1) with initial data

$$(u_0, \phi_0, \phi_1) \in H^{s_1} \times H^{s_2} \times H^{s_2-1}$$

for s_1 and s_2 as small as possible.

When $\alpha = 1$, (1.1) reduces to the standard Klein-Gordon Schrödinger system (KGS) which has been studied extensively by many mathematicians in [3, 6, 7, 9, 10, 12–14, 16–18]. KGS system is a classical model that describes the interaction of scalar nucleons interacting with neutral scalar meson. The nucleons are described by the complex scalar field u and the meson by the real scalar field ϕ. Pecher [16] showed the local well-posedness for KGS in three dimension with data $(u_0, \phi_0, \phi_1) \in L^2 \times L^2 \times H^{-1}$ based on estimates given by Ginibre, Tsutsumi and Velo [11] for the Zakharov system. Tzirakis [18] proved that the KGS system in one, two and three dimensions has a global solution below the energy space by using I-method introduced in [8] and setting up mixed type Strichartz estimates for the solutions of Schrödinger and Klein-Gordon equations respectively. They obtained a polynomial in time bound for the norms of the solution. Colliander, Holmer and Tzirakis [7] proved the global well-posedness for KGS in three dimension with initial data $(u_0, \phi_0, \phi_1) \in L^2 \times L^2 \times H^{-1}$ by using the conservation law for the L^2 norm of u and controlling the growth of ϕ via the estimates in the local theory. This is also true for two and one dimension by similar arguments. Recently, Pecher [17] obtained some new sharp well-posedness results for the KGS system. In $2d$, it was proved that the KGS system is local well-posed for $s_1 = -1/4+$, $s_2 = -1/2$ and ill-posed for $s_1 < -1/4, s_2 < -1/2$. In $3d$, global well-posedness was obtained for $s_1 \geqslant 0$ and $s_1 - 1/2 \leqslant s_2 \leqslant s_1 + 1$. These results are based on the sharp local well-posedness in two dimension by Bejenaru, Herr, Holmer and Tataru [1] as well as the three dimensional estimates by Bejenaru and Herr [2] for the Zakharov system.

To the knowledge of the authors, there are few works on FKGS system (1.1) with general α. For this system, we have the following conservation laws:

$$M(u) := \int_{\mathbf{R}} |u(x,t)|^2 dx = M(u_0), \qquad (1.2)$$

$$E(u,\phi,\phi_t) := \int_{\mathbf{R}} \phi_t^2 + |\nabla\phi|^2 + \phi^2 + 2|D^\alpha u|^2 + 2\phi|u|^2 dx = E(u_0,\phi_0,\phi_1). \quad (1.3)$$

So the energy space is $H^\alpha \times H^1 \times L^2$. By sobolev embedding theorem together with the conservation laws (1.2) and (1.3), it is easy to verify that the Cauchy problem (1.1) is globally well-posed in the energy space if the initial data $(u_0,\phi_0,\phi_1) \in H^\alpha(\mathbf{R}) \times H^1(\mathbf{R}) \times L^2(\mathbf{R})$. The purpose of this paper is to study the low-regularity well-posedness for (1.1). The main results of this paper are the following two theorems:

Theorem 1.1 Let $3/4 < \alpha \leqslant 1$. Assume that the initial data $(u_0,\phi_0,\phi_1) \in H^{s_1}(\mathbf{R}) \times H^{s_2}(\mathbf{R}) \times H^{s_2-1}(\mathbf{R})$. If s_1, s_2 satisfy

$$3/4 - \alpha < s_1 \leqslant 0, \quad (1.4)$$

$$-1/2 < s_2 < 3/2. \quad (1.5)$$

Then the Cauchy problem (1.1) is locally well-posed.

Remark 1.1 The condition supposed in (1.4) allows for the initial data with negative regularity. For instance, when $\alpha = 5/6$, $s_1 = -1/20$, and $s_2 = -1/3$ satisfies the condition.

Based on the local well-posedness result in Theorem 1.1 and the conservation of L^2 norm of u, we can prove the following:

Theorem 1.2 Let $3/4 < \alpha \leqslant 1$ and the initial data $(u_0,\phi_0,\phi_1) \in L^2(\mathbf{R}) \times H^{s_2}(\mathbf{R}) \times H^{s_2-1}(\mathbf{R})$. Then the Cauchy problem (1.1) is globally well-posed under the assumption

$$-1/2 < s_2 < 1/2. \quad (1.6)$$

Moreover, for any $T > 0$ and any initial data in $L^2 \times H^{s_2} \times H^{s_2-1}$, the solution $(u(t),\phi(t),\phi_t(t))$ satisfies:

$$\|u\|_{L^2} = \|u_0\|_{L^2}$$

and

$$\sup_{0 \leqslant t \leqslant T} (\|\phi(t)\|_{H^{s_2}} + \|\partial_t\phi(t)\|_{H^{s_2-1}}) \leqslant C(1+T)\max\{\|u_0\|_{L^2}, \|\phi_0\|_{H^{s_2}} + \|\phi_1\|_{H^{s_2-1}}\}.$$

Remark 1.2 When $\alpha = 1$, Theorem 1.2 gives better result than [18] in one space dimension where they proved that if $s > 1/2$, for any $T > 0$ and any initial data in $H^s \times H^s \times H^{s-1}$, the KGS system is global well-posed and

$$\sup_{0 \leqslant t \leqslant T} (\|u(t)\|_{H^s} + \|\phi(t)\|_{H^s} + \|\partial_t\phi(t)\|_{H^{s-1}}) \leqslant C(1+T)^{\frac{1-s}{3s-1.5-\varepsilon}},$$

in which the constant depends only on the $H^s \times H^s \times H^{s-1}$ norm of the initial data. We can see that to obtain global well-posedness result of the KGS system, the regularity condition we imposed on the initial data of the system is much lower than that of [18].

Remark 1.3 For the proof of global well-posedness, we only use the conservation of L^2 norm of u and didn't use the energy conservation law. Hence, if we replace $u\phi$ and $|u|^2$ in system (1.1) by $-u\phi$ and $-|u|^2$, we can obtain the same global well-posedness result.

Now we sketch the approach of the proof. Let $\langle D \rangle = \sqrt{I-\Delta}$ and $v = \phi - i\langle D \rangle^{-1}\phi_t$. Then system (1.1) is equivalent to the following one order system:

$$\begin{cases} iu_t + (-\Delta)^\alpha u = \frac{1}{2}(uv + u\bar{v}), \\ iv_t + \langle D \rangle v = \langle D \rangle^{-1}(|u|^2), \\ u(0) = u_0, \ v(0) = v_0. \end{cases} \quad (1.7)$$

We rewrite system (1.7) as the following integral equations:

$$u(t) = S_\alpha(t)u_0 + \frac{1}{2}\int_0^t S_\alpha(t-\tau)(uv + u\bar{v})(\tau)d\tau, \quad (1.8)$$

$$v(t) = W(t)v_0 + \int_0^t W(t-\tau)\langle D \rangle^{-1}(|u|^2)(\tau)d\tau, \quad (1.9)$$

where $S_\alpha(t) := e^{it(-\Delta)^\alpha} = \mathscr{F}^{-1}e^{it|\xi|^{2\alpha}}\mathscr{F}$ and $W(t) := e^{it\langle D \rangle} = \mathscr{F}^{-1}e^{it\sqrt{1+|\xi|^2}}\mathscr{F}$.

We will use the Bourgain-space method [4, 5, 19] to study the well-posedness for this equivalent one order system (1.7). The local well-posednes of system (1.7) usually follows from a contraction mapping argument. We choose the resolution space to be a closed ball in the product Bourgain space of functions defined on the space-time slab $[0, T_1] \times \mathbf{R}$:

$$\mathcal{D} = \{(u, v) \in X^{s_1, b} \times Y^{s_2, b} : \|(u, v)\|_{X^{s_1, b} \times Y^{s_2, b}} \leqslant 4C \max\{\|u_0\|_{H^{s_1}}, \|v_0\|_{H^{s_2}}\}\},$$

and verify that the operator Λ defined in (4.1) and (4.2) is a contractive map on $\mathcal{D}$ which implies the existence of solution to (1.7) on $[0, T_1]$.

Since the linear solutions of (1.7) with time localization have good control in Bourgain spaces (see Lemma 2.1), our main task is to set up the following three bilinear estimates in relevant Bourgain spaces:

$$\|uv\|_{X^{s_1, b-1}} \leqslant C\|u\|_{X^{s_1, b'}}\|v\|_{Y^{s_2, b'}},$$

$$\|u\bar{v}\|_{X^{s_1,b-1}} \leqslant C\|u\|_{X^{s_1,b'}}\|v\|_{Y^{s_2,b'}}.$$

$$\|\langle D\rangle^{-1}(u\bar{v})\|_{Y^{s_2,b-1}} \leqslant C\|u\|_{X^{s_1,b'}}\|v\|_{X^{s_1,b'}}.$$

These bilinear estimates can be treated systematically adopting the $[k:\mathbb{Z}]$ multiplier idea and reduced to estimate a trilinear functional (3.22) with similar structure which is done in Lemma 3.1. With the bilinear estimates listed above and choosing the length of the time interval $[0, T_1]$ be small enough, we can prove that system (1.7) is locally well-posed on $[0, T_1]$ where

$$T_1 \sim \max\{\|u_0\|_{L^2}, \|v_0\|_{H^{s_2}}\}^{-1}.$$

Inspired by some ideas in [7], using the conservation of L^2 norm of u, we can verify that the length of time interval T_1 has a uniform lower bound and hence we can iterate the local well-posedness argument with fixed step and finally obtain the global well-posedness of system (1.7).

2 Preliminaries

We first recall some notations and basic facts that will be used in this paper. $C \geqslant 1$, $c \leqslant 1$ will denote universal positive constants which can be different at different places. $X \lesssim Y$ (for $X, Y > 0$) means that $X \leqslant CY$. We use $X \sim Y$ to denote that there exist positive constants c, C satisfying $cY \leqslant X \leqslant CY$. $a+ := a+\varepsilon$ denotes a number slightly larger than a. For three real numbers k_1, k_2, k_3, we will use k_{max}, k_{med} and k_{min} to denote the maximum, medium and minimum of k_1, k_2, k_3. In general, let $s, b \in \mathbb{R}$ and $h : \mathbb{R}^n \to \mathbb{R}$ be a continuous function, the Bourgain space $X_h^{s,b}(\mathbb{R} \times \mathbb{R}^n)$ is defined as the closure of the Schwartz functions $\mathcal{S}(\mathbb{R} \times \mathbb{R}^n)$ under the norm

$$\|u\|_{X_h^{s,b}} = \|\langle\xi\rangle^s\langle\tau - h(\xi)\rangle^b \widehat{u}(\xi,\tau)\|_{L_\tau^2 L_\xi^2},$$

where $\langle\xi\rangle = (1+|\xi|^2)^{1/2}$ and $\widehat{u} = \mathscr{F}u$ denotes Fourier transform with respect to t and x variable of u. $\mathscr{F}_x u$ denotes the Fourier transform only on x. We denote $h(D)$ to be the Fourier multiplier:

$$h(D)u = \int e^{ix\xi} h(\xi) \mathscr{F}_x u(\xi) d\xi.$$

The first equation in the system (1.7) corresponds to dispersion relation $h_1(\xi) = |\xi|^{2\alpha}$ and the second equation in (1.7) is with dispersion relation $h_2(\xi) = (1+|\xi|^2)^{1/2}$.

It is well known that there holds the following standard linear estimates in Bourgain spaces with time localization(see [19, 20]):

Lemma 2.1 (linear estimates) Let $\eta(t) \in \mathcal{S}(\mathbf{R})$ be a Schwartz cutoff function in time. If $b \in (1/2, 1]$, $s \in \mathbf{R}$, $\phi \in H^s(\mathbf{R}^n)$ and $f \in X_h^{s,b-1}$ then

$$\|\eta(t)e^{ith(D)}\phi\|_{X_h^{s,b}} \leqslant C\|\phi\|_{H^s}, \tag{2.1}$$

$$\|\eta(t)\int_0^t e^{i(t-s)h(D)}f(s)ds\|_{X_h^{s,b}} \leqslant C\|f\|_{X_h^{s,b-1}}. \tag{2.2}$$

Lemma 2.2 (scaling property, see [20]) Assume that $s \in \mathbf{R}, 1/2 < b < b' < 1$, $0 < \delta \leqslant 1$, we have:

$$\|\eta_\delta(t)f\|_{X_h^{s,b-1}} \leqslant C\delta^{b'-b}\|f\|_{X_h^{s,b'-1}}, \tag{2.3}$$

where $\eta_\delta(t) = \eta(t/\delta)$.

In this paper, we will use the particular Bourgain spaces $X^{s,b}, \bar{X}^{s,b}, Y^{s,b}, \bar{Y}^{s,b}$ which are defined by the completion of the Schwartz functions $\mathcal{S}(\mathbf{R} \times \mathbf{R}^n)$ under the following norms:

$$\|u\|_{X^{s,b}} = \|\langle\xi\rangle^s \langle\tau - |\xi|^\alpha\rangle^b \widehat{u}(\xi,\tau)\|_{L_\tau^2 L_\xi^2}, \tag{2.4}$$

$$\|u\|_{\bar{X}^{s,b}} = \|\langle\xi\rangle^s \langle\tau + |\xi|^\alpha\rangle^b \widehat{u}(\xi,\tau)\|_{L_\tau^2 L_\xi^2}, \tag{2.5}$$

$$\|u\|_{Y^{s,b}} = \|\langle\xi\rangle^s \langle\tau - \langle\xi\rangle\rangle^b \widehat{u}(\xi,\tau)\|_{L_\tau^2 L_\xi^2}, \tag{2.6}$$

$$\|u\|_{\bar{Y}^{s,b}} = \|\langle\xi\rangle^s \langle\tau + \langle\xi\rangle\rangle^b \widehat{u}(\xi,\tau)\|_{L_\tau^2 L_\xi^2}. \tag{2.7}$$

We remark that in general Bourgain spaces $X_h^{s,b}$ are not conjugate invariant. It is easy to check that there holds the complex conjugation relationship between the norms defined above:

Lemma 2.3 For any $s, b \in \mathbf{R}$, we have

$$\|\bar{u}\|_{X^{s,b}} = \|u\|_{\bar{X}^{s,b}}, \quad \|\bar{u}\|_{Y^{s,b}} = \|u\|_{\bar{Y}^{s,b}}. \tag{2.8}$$

3 Bilinear estimates

In this section, we will prove the following two type bilinear estimates which are crucial to the proof of our main Theorem 1.1.

Proposition 3.1 Let $3/4 < \alpha \leqslant 1$. Assume that s_1, s_2 satisfies

$$1/2 - \alpha < s_1 \leqslant 0, \tag{3.1}$$

$$s_2 > -1/2. \tag{3.2}$$

Then there exist $b, b' \in (1/2, 1)$ with $0 < b - b' < 1/2$, such that

$$\|uv\|_{X^{s_1,b-1}} \leqslant C\|u\|_{X^{s_1,b'}}\|v\|_{Y^{s_2,b'}} \tag{3.3}$$

$$\|u\bar{v}\|_{X^{s_1,b-1}} \leqslant C\|u\|_{X^{s_1,b'}}\|v\|_{Y^{s_2,b'}}. \tag{3.4}$$

The second type bilinear estimate is:

Proposition 3.2 Let $3/4 < \alpha \leqslant 1$. Suppose that s_1, s_2 satisfies

$$3/4 - \alpha < s_1 \leqslant 0, \tag{3.5}$$

$$s_2 < 3/2. \tag{3.6}$$

Then there exist $b, b' \in (1/2, 1)$ with $0 < b - b' < 1/2$ satisfying the following estimate

$$\|\langle D\rangle^{-1}(u\bar{v})\|_{Y^{s_2,b-1}} \leqslant C\|u\|_{X^{s_1,b'}}\|v\|_{X^{s_1,b'}}. \tag{3.7}$$

The bilinear estimates in Propositions 3.1 and 3.2 can be proved in a similar framework. By the definition of spaces, (3.3) is equivalent to

$$\|\langle\tau - |\xi|^{2\alpha}\rangle^{b-1}\langle\xi\rangle^{s_1}\widehat{u} * \widehat{v}(\xi,\tau)\|_{L^2_{\xi,\tau}}$$
$$\lesssim \|\langle\tau - |\xi|^{2\alpha}\rangle^{b'}\langle\xi\rangle^{s_1}\widehat{u}\|_{L^2_{\xi,\tau}}\|\langle\tau - \langle\xi\rangle\rangle^{b'}\langle\xi\rangle^{s_2}\widehat{v}\|_{L^2_{\xi,\tau}},$$

which is further equivalent to show that

$$\|\langle\tau - |\xi|^{2\alpha}\rangle^{b-1}\langle\xi\rangle^{s_1}(\langle\tau - |\xi|^{2\alpha}\rangle^{-b'}\langle\xi\rangle^{-s_1}\widehat{u}) * (\langle\tau - \langle\xi\rangle\rangle^{-b'}\langle\xi\rangle^{-s_2}\widehat{v})\|_{L^2_{\xi,\tau}} \tag{3.8}$$

is bounded by $\|\widehat{u}\|_{L^2_{\xi,\tau}}\|\widehat{v}\|_{L^2_{\xi,\tau}}$. We will use dyadic decomposition to prove this estimate. Let $\{\varphi_k\}_{k=0}^{\infty}$ to be the nonhomogeneous dyadic functions satisfying

$$\sum_{k=0}^{\infty}\varphi_k(\xi) = 1, \quad \text{supp}\varphi_k \subset I_k,$$

where

$$I_k = [2^{k-1}, 2^{k+1}] \quad \text{for } k \geqslant 1; \quad I_0 = [-2, 2].$$

We decompose the functions $\langle \tau - |\xi|^{2\alpha} \rangle$, $\tau - \langle \xi \rangle$ and $\langle \xi \rangle^s$ in (3.8) into dyadic functions and obtain that

$$(3.8) \lesssim \sum_{k_i \in \mathbb{Z}_+} \sum_{j_i \in \mathbb{Z}_+} \frac{2^{j_3(b-1)} 2^{k_3 s_1}}{2^{k_1 s_1} 2^{k_2 s_2} 2^{j_1 b'} 2^{j_2 b'}} \|1_{D^S_{k_3,j_3}} \cdot (u^S_{k_1,j_1} * v^K_{k_2,j_2})\|_{L^2_{\xi,\tau}}, \qquad (3.9)$$

where $u^S_{k_1,j_1} = \hat{u}(\xi,\tau)\varphi_{k_1}(\xi)\varphi_{j_1}(\tau - |\xi|^{2\alpha})$ is a function supported in $D^S_{k_1,j_1}$, $v^K_{k_2,j_2} = \hat{v}(\xi,\tau)\varphi_{k_2}(\xi)\varphi_{j_2}(\tau - \langle\xi\rangle)$ is supported in $D^K_{k_2,j_2}$ and $1_{D^S_{k_3,j_3}} = \varphi_{k_3}(\xi)\varphi_{j_3}(\tau - |\xi|^{2\alpha})$ is supported in $D^S_{k_3,j_3}$. Here we denote

$$D^S_{k,j} = \{(\xi,\tau) : \xi \in I_k, \tau - |\xi|^{2\alpha} \in I_j\},$$
$$D^K_{k,j} = \{(\xi,\tau) : \xi \in I_k, \tau - \langle\xi\rangle \in I_j\}.$$

So the problem is reduced to estimate $\|1_{D^S_{k_3,j_3}} \cdot (u^S_{k_1,j_1} * v^K_{k_2,j_2})\|_{L^2_{\xi,\tau}}$. If we can prove that for any $f \in L^2_{\xi,\tau}$,

$$\int f(\xi,\tau) 1_{D^S_{k_3,j_3}} \cdot (u^S_{k_1,j_1} * v^K_{k_2,j_2}) d\xi d\tau = \int f^S_{k_3,j_3}(u^S_{k_1,j_1} * v^K_{k_2,j_2}) d\xi d\tau$$
$$\leqslant C(k_i,j_i) \|f^S_{k_3,j_3}\|_{L^2_{\xi,\tau}} \|u^S_{k_1,j_1}\|_{L^2_{\xi,\tau}} \|v^K_{k_2,j_2}\|_{L^2_{\xi,\tau}}, \qquad (3.10)$$

where we write $f^S_{k_3,j_3} = f(\xi,\tau)1_{D^S_{k_3,j_3}}$, then by duality,

$$\|1_{D^S_{k_3,j_3}} \cdot (u^S_{k_1,j_1} * v^K_{k_2,j_2})\|_{L^2_{\xi,\tau}} \leqslant C(k_i,j_i) \|u^S_{k_1,j_1}\|_{L^2_{\xi,\tau}} \|v^K_{k_2,j_2}\|_{L^2_{\xi,\tau}}. \qquad (3.11)$$

Replace this estimate in (3.9), we can bound (3.8) by $\|\hat{u}\|_{L^2_{\xi,\tau}} \|\hat{v}\|_{L^2_{\xi,\tau}}$ under certain conditions on s_1, s_2, b, b' and α. So the key task is to estimate the integral:

$$J_1 = \int f^S_{k_3,j_3}(u^S_{k_1,j_1} * v^K_{k_2,j_2}) d\xi d\tau$$
$$= \int f^S_{k_3,j_3}(\xi_1 + \xi_2, \tau_1 + \tau_2) u^S_{k_1,j_1}(\xi_1,\tau_1) v^K_{k_2,j_2}(\xi_2,\tau_2) d\xi_1 d\xi_2 d\tau_1 d\tau_2. \qquad (3.12)$$

For simplicity, let $f_1(\xi,\tau) = u^S_{k_1,j_1}(\xi, \tau + |\xi|^{2\alpha})$, $f_2(\xi,\tau) = v^K_{k_2,j_2}(\xi, \tau + \langle\xi\rangle)$ and $f_3(\xi,\tau) = f^S_{k_3,j_3}(\xi, \tau + |\xi|^{2\alpha})$, then f_i is supported in

$$D_i = \{(\xi,\tau) : |\xi| \sim 2^{k_i}, |\tau| \sim 2^{j_i}\}.$$

The integral J_1 can be rewritten as the trilinear functional:

$$J_1(f_1,f_2,f_3) = \int f_1(\xi_1,\tau_1) f_2(\xi_2,\tau_2) f_3(\xi_1+\xi_2, \tau_1+\tau_2 + H_1(\xi_1,\xi_2)) d\xi_1 d\xi_2 d\tau_1 d\tau_2, \qquad (3.13)$$

where
$$H_1(\xi_1,\xi_2) = |\xi_1|^{2\alpha} + \langle \xi_2 \rangle - |\xi_1+\xi_2|^{2\alpha}. \tag{3.14}$$

In parallel, the proof of (3.7) can be transferred to bound a integral with similar structure. If we can prove
$$\|\langle D \rangle^{-1}(uv)\|_{Y^{s_2,b-1}} \lesssim \|u\|_{X^{s_1,b'}} \|\bar{v}\|_{X^{s_1,b'}}, \tag{3.15}$$

then clearly we obtain (3.7). Note that $\|\bar{v}\|_{X^{s_1,b'}} = \|v\|_{\bar{X}^{s_1,b'}}$, then it sufficient to prove
$$\|\langle D \rangle^{-1}(uv)\|_{Y^{s_2,b-1}} \lesssim \|u\|_{X^{s_1,b'}} \|v\|_{\bar{X}^{s_1,b'}}. \tag{3.16}$$

Thus by the definition of the Bourgain spaces, it is equivalent to prove
$$\|\langle \tau - \langle \xi \rangle \rangle^{b-1} \langle \xi \rangle^{s_2-1} \widehat{u} * \widehat{v}(\xi,\tau)\|_{L^2_{\xi,\tau}}$$
$$\lesssim \|\langle \tau - |\xi|^{2\alpha} \rangle^{b'} \langle \xi \rangle^{s_1} \widehat{u}\|_{L^2_{\xi,\tau}} \|\langle \tau + |\xi|^{2\alpha} \rangle^{b'} \langle \xi \rangle^{s_1} \widehat{v}\|_{L^2_{\xi,\tau}},$$

which is further equivalent to show that
$$\|\langle \tau - \langle \xi \rangle \rangle^{b-1} \langle \xi \rangle^{s_2-1}(\langle \tau - |\xi|^{2\alpha} \rangle^{-b'} \langle \xi \rangle^{-s_1} \widehat{u}) * (\langle \tau + |\xi|^{2\alpha} \rangle^{-b'} \langle \xi \rangle^{-s_1} \widehat{v})\|_{L^2_{\xi,\tau}} \tag{3.17}$$

is bounded by $\|\widehat{u}\|_{L^2_{\xi,\tau}} \|\widehat{v}\|_{L^2_{\xi,\tau}}$. Again by dyadic decomposition, we get
$$(3.17) \lesssim \sum_{k_i \in \mathbb{Z}_+} \sum_{j_i \in \mathbb{Z}_+} \frac{2^{j_3(b-1)} 2^{k_3(s_2-1)}}{2^{k_1 s_1} 2^{k_2 s_1} 2^{j_1 b'} 2^{j_2 b'}} \|1_{D^K_{k_3,j_3}} \cdot (u^S_{k_1,j_1} * v^{\bar{S}}_{k_2,j_2})\|_2, \tag{3.18}$$

where $u^S_{k_1,j_1} = \widehat{u}(\xi,\tau)\varphi_{k_1}(\xi)\varphi_{j_1}(\tau - |\xi|^{2\alpha})$ is a function supported in $D^S_{k_1,j_1}$, $v^{\bar{S}}_{k_2,j_2} = \widehat{v}(\xi,\tau)\varphi_{k_2}(\xi)\varphi_{j_2}(\tau + |\xi|^{2\alpha})$ is supported in $\bar{D}^S_{k_2,j_2}$ and $1_{D^K_{k_3,j_3}} = \varphi_{k_3}(\xi)\varphi_{j_3}(\tau - \langle \xi \rangle)$ is supported in $D^K_{k_3,j_3}$. Here we denote
$$\bar{D}^S_{k,j} = \{(\xi,\tau) : \xi \in I_k, \tau + |\xi|^{2\alpha} \in I_j\}.$$

By duality, the estimation of $\|1_{D^K_{k_3,j_3}} \cdot (u^S_{k_1,j_1} * v^{\bar{S}}_{k_2,j_2})\|_2$ reduces to estimate the following integral:
$$J_2 = \int f^K_{k_3,j_3}(u^S_{k_1,j_1} * v^{\bar{S}}_{k_2,j_2}) d\xi d\tau$$
$$= \int f^K_{k_3,j_3}(\xi_1+\xi_2, \tau_1+\tau_2) u^S_{k_1,j_1}(\xi_1,\tau_1) v^{\bar{S}}_{k_2,j_2}(\xi_2,\tau_2) d\xi_1 d\xi_2 d\tau_1 d\tau_2. \tag{3.19}$$

Let $f_1(\xi, \tau) = u^S_{k_1,j_1}(\xi, \tau + |\xi|^{2\alpha})$, $f_2(\xi, \tau) = v^{\bar{S}}_{k_2,j_2}(\xi, \tau - |\xi|^{2\alpha})$, and $f_3(\xi, \tau) = f^K_{k_3,j_3}(\xi, \tau + \langle\xi\rangle)$, then f_i is supported in

$$D_i = \{(\xi, \tau) : |\xi| \sim 2^{k_i}, |\tau| \sim 2^{j_i}\}.$$

The integral J_2 in (3.19) can be rewritten as the following trilinear operator:

$$J_2(f_1, f_2, f_3) = \int f_1(\xi_1, \tau_1) f_2(\xi_2, \tau_2) f_3(\xi_1 + \xi_2, \tau_1 + \tau_2 + H_2(\xi_1, \xi_2)) d\xi_1 d\xi_2 d\tau_1 d\tau_2, \quad (3.20)$$

where

$$H_2(\xi_1, \xi_2) = |\xi_1|^{2\alpha} - |\xi_2|^{2\alpha} - \langle\xi_1 + \xi_2\rangle. \quad (3.21)$$

From the discussion above, we see that the proof of the bilinear estimates in Proposition 3.1 and 3.2 relies heavily on suitable upper bounds for (3.13) and (3.20). We will handle them in a unified way. For $l = 1, 2$, let

$$J_l = \int f_1(\xi_1, \tau_1) f_2(\xi_2, \tau_2) f_3(\xi_1 + \xi_2, \tau_1 + \tau_2 + H_l(\xi_1, \xi_2)) d\xi_1 d\xi_2 d\tau_1 d\tau_2, \quad (3.22)$$

where $H_1(\xi_1, \xi_2)$ and $H_2(\xi_1, \xi_2)$ are defined as in (3.14) and (3.21). There holds the following estimates:

Lemma 3.1 *Let $k_i, j_i \in \mathbb{Z}_+$, $f_i(\xi, \tau) \in L^2(\mathbf{R}^2)$ be non-negative functions with $\|f_i\|_2 \leqslant 1$, and is supported in $\{|\xi| \sim 2^{k_i}, |\tau| \lesssim 2^{j_i}\}$, $i = 1, 2, 3$. Then (1) For all $k_i, j_i \in \mathbb{Z}_+$, we have for $l = 1, 2$*

$$J_l \lesssim 2^{j_{min}/2} 2^{k_{min}/2}. \quad (3.23)$$

(2) *If $2^{k_{min}} \ll 2^{k_{med}} \sim 2^{k_{max}}$, there are following three cases:*

(2.1) *When $2^{k_1} \ll 2^{k_2} \sim 2^{k_3}$, J_1 satisfying (3.37) $\sim$ (3.39) and J_2 satisfying (3.40) $\sim$ (3.42).*

(2.2) *When $2^{k_3} \ll 2^{k_1} \sim 2^{k_2}$, J_1 satisfying (3.43) $\sim$ (3.45) and J_2 satisfying (3.46) $\sim$ (3.48).*

(2.3) *When $2^{k_2} \ll 2^{k_1} \sim 2^{k_3}$, J_1 satisfying (3.50) $\sim$ (3.52) and J_2 satisfying (3.53) $\sim$ (3.55).*

(3) *If $2^{k_{min}} \sim 2^{k_{med}} \sim 2^{k_{max}}$, then for $l = 1, 2$,*

$$J_l \lesssim 2^{(j_1+j_2+j_3)/2} 2^{-j_{max}/2} 2^{-k_1(2\alpha-1)/2}.$$

Remark 3.1 *From the support property of f_i, one can easily check that*

$$J_l(f_1, f_2, f_3) = 0$$

unless

$$2^{k_{med}} \sim 2^{k_{max}} \tag{3.24}$$

and

$$2^{j_{max}} \sim \max\{2^{j_{med}}, |H_l|\}. \tag{3.25}$$

(3.25) tells us that $|H_l| \lesssim 2^{j_{max}}$ which means that $\log_2 |H_l|$ gives a near lower bound for j_{max}. This observation is crucial to the proof of the bilinear estimates in Propositions 3.1 and 3.2.

proof (1) We only prove the bound for J_1 since the estimate for J_2 is similar. From (3.12), we see that

$$J_1 = \int_{\mathbf{R}^4} f^S_{k_3,j_3}(\xi_1 + \xi_2, \tau_1 + \tau_2) u^S_{k_1,j_1}(\xi_1, \tau_1) v^K_{k_2,j_2}(\xi_2, \tau_2) d\xi_1 d\xi_2 d\tau_1 d\tau_2. \tag{3.26}$$

The proof can be separated into the following several cases:

Case 1.1 $k_1 = k_{min}, j_1 = j_{min}$. By first using Cauchy-Schwarz inequality in ξ_2, τ_2 variable,

$$J_1 \lesssim \int_{\mathbf{R}^2} |u^S_{k_1,j_1}(\xi_1, \tau_1)| \|f^S_{k_3,j_3}\|_{L^2_{\xi_2,\tau_2}} \|v^K_{k_2,j_2}\|_{L^2_{\xi_2,\tau_2}} d\xi_1 d\tau_1,$$

then taking Cauchy-Schwarz in ξ_1 and τ_1 together with the support property of $u^S_{k_1,j_1}$, we have

$$J_1 \lesssim 2^{k_1/2} 2^{j_1/2} \|u^S_{k_1,j_1}\|_{L^2_{\xi_1,\tau_1}} \|f^S_{k_3,j_3}\|_{L^2_{\xi_2,\tau_2}} \|v^K_{k_2,j_2}\|_{L^2_{\xi_2,\tau_2}} \lesssim 2^{k_1/2} 2^{j_1/2}.$$

Case 1.2 $k_1 = k_{min}, j_2 = j_{min}$. We first use Cauchy-Schwarz inequality in τ_1 variable,

$$J_1 \lesssim \int_{\mathbf{R}^3} \|f^S_{k_3,j_3}(\xi_1+\xi_2,\cdot)\|_{L^2_{\tau_1}} \|u^S_{k_1,j_1}(\xi_1,\cdot)\|_{L^2_{\tau_1}} |v^K_{k_2,j_2}(\xi_2,\tau_2)| d\xi_1 d\xi_2 d\tau_2,$$

then use Cauchy-Schwarz in τ_2 together with the support property of $v^K_{k_2,j_2}$,

$$J_1 \lesssim 2^{j_2/2} \int_{\mathbf{R}^2} \|f^S_{k_3,j_3}(\xi_1+\xi_2,\cdot)\|_{L^2_{\tau_1}} \|u^S_{k_1,j_1}(\xi_1,\cdot)\|_{L^2_{\tau_1}} \|v^K_{k_2,j_2}(\xi_2,\cdot)\|_{L^2_{\tau_2}} d\xi_1 d\xi_2,$$

then use Cauchy-Schwarz in ξ_2, we get

$$J_1 \lesssim 2^{j_2/2} \int_{\mathbf{R}^1} \|f^S_{k_3,j_3}(\cdot,\cdot)\|_{L^2_{\xi_2,\tau_1}} \|u^S_{k_1,j_1}(\xi_1,\cdot)\|_{L^2_{\tau_1}} \|v^K_{k_2,j_2}(\cdot,\cdot)\|_{L^2_{\xi_2,\tau_2}} d\xi_1,$$

last by using Cauchy-Schwarz in ξ_1 together with the support property of $u^S_{k_1,j_1}$, we obtain

$$J_1 \lesssim 2^{k_1/2} 2^{j_2/2} \|u^S_{k_1,j_1}\|_{L^2_{\xi_1,\tau_1}} \|f^S_{k_3,j_3}\|_{L^2_{\xi_2,\tau_1}} \|v^K_{k_2,j_2}\|_{L^2_{\xi_2,\tau_2}} \lesssim 2^{k_1/2} 2^{j_2/2}.$$

The other cases can be verified similarly and we omit the details.

(2) There are no symmetries between k_i, j_i. So we consider it case by case.

Case 2.1 $2^{k_1} \ll 2^{k_2} \sim 2^{k_3}$.

Using Cauchy-Schwarz inequality in ξ_1, ξ_2 for (3.22), we get

$$J_l \lesssim \int \|f_1(\xi_1,\tau_1) f_2(\xi_2,\tau_2)\|_{L^2_{\xi_1,\xi_2}} \|f_3(\xi_1+\xi_2, \tau_1+\tau_2+H_l(\xi_1,\xi_2))\|_{L^2_{\xi_1,\xi_2}} d\tau_1 d\tau_2. \quad (3.27)$$

To estimate $\|f_3(\xi_1+\xi_2, \tau_1+\tau_2+H_l(\xi_1,\xi_2))\|_{L^2_{\xi_1,\xi_2}}$, we make a change of variable

$$\eta_1 = \xi_1+\xi_2, \quad \eta_2 = \tau_1+\tau_2+H_l(\xi_1,\xi_2),$$

and from the fact that in the support of f_i,

$$|\det(\frac{\partial(\eta_1,\eta_2)}{\partial(\xi_1,\xi_2)})| = |\partial_{\xi_2} H_l(\xi_1,\xi_2) - \partial_{\xi_1} H_l(\xi_1,\xi_2)|$$

$$= \begin{cases} \left|\frac{\xi_2}{\langle\xi_2\rangle} - 2\alpha\right| \xi_1|^{2\alpha-1} \frac{\xi_1}{|\xi_1|}\right| \sim 2^{k_1(2\alpha-1)}, & \text{if } 2^{k_1} \gg 1, \ l=1, \\ \left|2\alpha|\xi_2|^{2\alpha-1}\frac{\xi_2}{|\xi_2|} + 2\alpha|\xi_1|^{2\alpha-1}\frac{\xi_1}{|\xi_1|}\right| \sim 2^{k_2(2\alpha-1)}, & l=2. \end{cases}$$

Then

$$\|f_3(\xi_1+\xi_2, \tau_1+\tau_2+H_l(\xi_1,\xi_2))\|^2_{L^2_{\xi_1,\xi_2}} = \int |f_3(\eta_1,\eta_2)|^2 \det(\frac{\partial(\xi_1,\xi_2)}{\partial(\eta_1,\eta_2)}) d\eta_1 d\eta_2$$

$$\sim \begin{cases} 2^{-k_1(2\alpha-1)} \|f_3(\eta_1,\eta_2)\|^2_{L^2_{\eta_1,\eta_2}}, & l=1; \\ 2^{-k_2(2\alpha-1)} \|f_3(\eta_1,\eta_2)\|^2_{L^2_{\eta_1,\eta_2}}, & l=2. \end{cases}$$

Replace these estimates in (3.27) and applying Cauchy-Schwartz inequality in τ_1 and τ_2, we obtain that

$$J_1 \lesssim 2^{j_1/2} 2^{j_2/2} 2^{-k_1(2\alpha-1)/2}, \quad \text{if } 2^{k_1} \gg 1, \quad (3.28)$$

$$J_2 \lesssim 2^{j_1/2} 2^{j_2/2} 2^{-k_2(2\alpha-1)/2}. \quad (3.29)$$

These estimates will be used for $j_3 = j_{max}$. We may rewrite J_l as

$$J_l = \int f_1(\xi_1,\tau_1) f_2(\xi_2-\xi_1, \tau_2-\tau_1-H_l(\xi_1,\xi_2-\xi_1)) f_3(\xi_2,\tau_2) d\xi_1 d\xi_2 d\tau_1 d\tau_2.$$

Similarly, we proceed as above, first by Cauchy-Schwarz inequality in ξ_1, ξ_2, we have

$$J_l \lesssim \int \|f_1(\xi_1, \tau_1) f_3(\xi_2, \tau_2)\|_{L^2_{\xi_1,\xi_2}} \|f_2(\xi_2 - \xi_1, \tau_2 - \tau_1 - H_l(\xi_1, \xi_2 - \xi_1))\|_{L^2_{\xi_1,\xi_2}} d\tau_1 d\tau_2, \tag{3.30}$$

and then making a change of variable

$$\eta_1 = \xi_2 - \xi_1, \quad \eta_2 = \tau_2 - \tau_1 - H_l(\xi_1, \xi_2 - \xi_1).$$

From the fact that in the support of f_i,

$$\left|\det\left(\frac{\partial(\eta_1, \eta_2)}{\partial(\xi_1, \xi_2)}\right)\right| = |(\partial_1 H_l)(\xi_1, \xi_2 - \xi_1)|$$

$$= \begin{cases} |2\alpha|\xi_1|^{2\alpha-2}\xi_1 - 2\alpha|\xi_2|^{2\alpha-2}\xi_2| \sim 2^{k_2(2\alpha-1)}, & l = 1, \\ \left|2\alpha|\xi_1|^{2\alpha-2}\xi_1 - \dfrac{\xi_2}{\langle\xi_2\rangle}\right| \sim 2^{k_1(2\alpha-1)}, & \text{if } 2^{k_1} \gg 1, \quad l = 2. \end{cases}$$

We conclude that

$$\|f_2(\xi_2 - \xi_1, \tau_2 - \tau_1 - H_l(\xi_1, \xi_2 - \xi_1))\|^2_{L^2_{\xi_1,\xi_2}} = \int |f_2(\eta_1, \eta_2)|^2 \det\left(\frac{\partial(\xi_1, \xi_2)}{\partial(\eta_1, \eta_2)}\right) d\eta_1 d\eta_2$$

$$\sim \begin{cases} 2^{-k_2(2\alpha-1)} \|f_2(\eta_1, \eta_2)\|^2_{L^2_{\eta_1,\eta_2}}, & l = 1; \\ 2^{-k_1(2\alpha-1)} \|f_2(\eta_1, \eta_2)\|^2_{L^2_{\eta_1,\eta_2}}, & l = 2. \end{cases}$$

Replace these estimates in (3.30) and apply Cauchy-Schwartz inequality in τ_1 and τ_2 (we remark that here $|\tau_2| \lesssim 2^{j_3}$), we get

$$J_1 \lesssim 2^{j_1/2} 2^{j_3/2} 2^{-k_2(2\alpha-1)/2}, \tag{3.31}$$

$$J_2 \lesssim 2^{j_1/2} 2^{j_3/2} 2^{-k_1(2\alpha-1)/2}, \quad \text{if } 2^{k_1} \gg 1, \tag{3.32}$$

which will be used for $j_2 = j_{max}$.

We may also rewrite J_l as

$$J_l = \int f_1(\xi_1 - \xi_2, \tau_1 - \tau_2 - H_l(\xi_1 - \xi_2, \xi_2)) f_2(\xi_2, \tau_2) f_3(\xi_1, \tau_1) d\xi_1 d\xi_2 d\tau_1 d\tau_2. \tag{3.33}$$

Using Cauchy-Schwarz inequality in ξ_1, ξ_2 for (3.33), we get

$$J_l \lesssim \int \|f_1(\xi_1 - \xi_2, \tau_1 - \tau_2 - H_l(\xi_1 - \xi_2, \xi_2))\|_{L^2_{\xi_1,\xi_2}} \|f_2(\xi_2, \tau_2) f_3(\xi_1, \tau_1)\|_{L^2_{\xi_1,\xi_2}} d\tau_1 d\tau_2. \tag{3.34}$$

Similarly, we make the change of variables

$$\eta_1 = \xi_1 - \xi_2, \quad \eta_2 = \tau_1 - \tau_2 - H_l(\xi_1 - \xi_2, \xi_2).$$

From the fact that in the support of f_i (noting that $|\xi_1| \sim 2^{k_3}$ and $|\xi_2| \sim 2^{k_2}$ in this case),

$$\left|\det\left(\frac{\partial(\eta_1,\eta_2)}{\partial(\xi_1,\xi_2)}\right)\right| = |(\partial_2 H_l)(\xi_1-\xi_2,\xi_2)|$$

$$= \begin{cases} \left|\frac{\xi_2}{\langle\xi_2\rangle} - 2\alpha\big|\xi_1\big|^{2\alpha-1}\frac{\xi_1}{|\xi_1|}\right| \sim 2^{k_3(2\alpha-1)}, & l=1, \\ \left|\frac{\xi_1}{\langle\xi_1\rangle} + 2\alpha|\xi_2|^{2\alpha-1}\frac{\xi_2}{|\xi_2|}\right| \sim 2^{k_3(2\alpha-1)}, & l=2. \end{cases}$$

We conclude that

$$\|f_1(\xi_1-\xi_2,\tau_1-\tau_2-H_l(\xi_1-\xi_2,\xi_2))\|_{L^2_{\xi_1,\xi_2}}^2 = \int |f_1(\eta_1,\eta_2)|^2 \det\left(\frac{\partial(\xi_1,\xi_2)}{\partial(\eta_1,\eta_2)}\right) d\eta_1 d\eta_2$$

$$\sim \begin{cases} 2^{-k_3(2\alpha-1)}\|f_1(\eta_1,\eta_2)\|_{L^2_{\eta_1,\eta_2}}^2, & l=1; \\ 2^{-k_3(2\alpha-1)}\|f_1(\eta_1,\eta_2)\|_{L^2_{\eta_1,\eta_2}}^2, & l=2. \end{cases}$$

Putting these estimates into (3.34) and using Cauchy-Schwartz in τ_1 and τ_2, noting that here $|\tau_1| \lesssim 2^{j_3}$, we have

$$J_1 \lesssim 2^{j_2/2} 2^{j_3/2} 2^{-k_3(2\alpha-1)/2}, \tag{3.35}$$

$$J_2 \lesssim 2^{j_2/2} 2^{j_3/2} 2^{-k_3(2\alpha-1)/2}. \tag{3.36}$$

These bounds will be used for $j_1 = j_{max}$. In all, we get that

$$J_1 \lesssim 2^{j_1/2} 2^{j_2/2} 2^{-k_1(2\alpha-1)/2}, \quad \text{if } 2^{k_1} \gg 1, \tag{3.37}$$

$$J_1 \lesssim 2^{j_1/2} 2^{j_3/2} 2^{-k_2(2\alpha-1)/2}, \tag{3.38}$$

$$J_1 \lesssim 2^{j_2/2} 2^{j_3/2} 2^{-k_3(2\alpha-1)/2}, \tag{3.39}$$

and

$$J_2 \lesssim 2^{j_1/2} 2^{j_2/2} 2^{-k_2(2\alpha-1)/2}, \tag{3.40}$$

$$J_2 \lesssim 2^{j_1/2} 2^{j_3/2} 2^{-k_1(2\alpha-1)/2}, \quad \text{if } 2^{k_1} \gg 1, \tag{3.41}$$

$$J_2 \lesssim 2^{j_2/2} 2^{j_3/2} 2^{-k_3(2\alpha-1)/2}. \tag{3.42}$$

Case 2.2 $2^{k_3} \ll 2^{k_1} \sim 2^{k_2}$.

Under this case, we use the same idea as in the proof of Case 2.1. Checking the determinant in the support of f_i, we get the following estimates:

$$J_1 \lesssim 2^{j_1/2} 2^{j_2/2} 2^{-k_1(2\alpha-1)/2}, \tag{3.43}$$

$$J_1 \lesssim 2^{j_1/2} 2^{j_3/2} 2^{-k_1(2\alpha-1)/2}, \tag{3.44}$$

$$J_1 \lesssim 2^{j_2/2} 2^{j_3/2} 2^{-k_3(2\alpha-1)/2}, \text{ if } 2^{k_3} \gg 1, \tag{3.45}$$

and

$$J_2 \lesssim 2^{j_1/2} 2^{j_2/2} 2^{-k_1(\alpha-1)} 2^{-k_3/2}, \tag{3.46}$$

$$J_2 \lesssim 2^{j_1/2} 2^{j_3/2} 2^{-k_1(2\alpha-1)/2}, \tag{3.47}$$

$$J_2 \lesssim 2^{j_2/2} 2^{j_3/2} 2^{-k_1(2\alpha-1)/2}. \tag{3.48}$$

Here we only explain the proof for (3.46) where the determinant checking is a little bit different. In this case we need to write J_2 as

$$J_2 = \int f_1(\xi_1, \tau_1) f_2(\xi_2, \tau_2) f_3(\xi_1 + \xi_2, \tau_1 + \tau_2 + H_2(\xi_1, \xi_2)) d\xi_1 d\xi_2 d\tau_1 d\tau_2, \tag{3.49}$$

which means that $|\xi_1| \sim 2^{k_1}$, $|\xi_2| \sim 2^{k_2}$, and $|\xi_1 + \xi_2| \sim 2^{k_3}$. Since $2^{k_3} \ll 2^{k_1} \sim 2^{k_2}$, this implies that ξ_1 and ξ_2 have opposite sign.

From the calculation above, we see that

$$\left| \det\left(\frac{\partial(\eta_1, \eta_2)}{\partial(\xi_1, \xi_2)} \right) \right| = |\partial_{\xi_2} H_2(\xi_1, \xi_2) - \partial_{\xi_1} H_2(\xi_1, \xi_2)| = |2\alpha|\xi_2|^{2\alpha-2}\xi_2 + 2\alpha|\xi_1|^{2\alpha-2}\xi_1|$$

$$= |2\alpha|\xi_2|^{2\alpha-2}\xi_2 - 2\alpha| -\xi_1|^{2\alpha-2}(-\xi_1)| \sim |\xi_*|^{2\alpha-2}|\xi_1 + \xi_2|,$$

where in the last step we use the mean value theorem. Owing to ξ_1 and ξ_2 have opposite sign and $|\xi_1| \sim 2^{k_1}$, $|\xi_2| \sim 2^{k_2} \sim 2^{k_1}$, we conclude that $|\xi_*| \sim 2^{k_1}$ and hence

$$\left| \det\left(\frac{\partial(\eta_1, \eta_2)}{\partial(\xi_1, \xi_2)} \right) \right| \sim 2^{2k_1(\alpha-1)} 2^{k_3},$$

which implies the bound (3.46) by applying Cauchy-Schwarz as above. For other estimates, the method is quite similar, we skip the details.

Case 2.3 $2^{k_2} \ll 2^{k_1} \sim 2^{k_3}$.

In this case, we repeat the proof of Case 2.1. Checking the determinant in the support of f_i and using Cauchy-Schwarz inequality, we obtain the following estimates:

$$J_1 \lesssim 2^{j_1/2} 2^{j_2/2} 2^{-k_1(2\alpha-1)/2}, \tag{3.50}$$

$$J_1 \lesssim 2^{j_1/2} 2^{j_3/2} 2^{-k_1(\alpha-1)} 2^{-k_2/2}, \text{ if } 2^{k_2} \gg 1, \tag{3.51}$$

$$J_1 \lesssim 2^{j_2/2} 2^{j_3/2} 2^{-k_3(2\alpha-1)/2}, \tag{3.52}$$

and

$$J_2 \lesssim 2^{j_1/2} 2^{j_2/2} 2^{-k_1(2\alpha-1)/2}, \tag{3.53}$$

$$J_2 \lesssim 2^{j_1/2} 2^{j_3/2} 2^{-k_1(2\alpha-1)/2}, \tag{3.54}$$

$$J_2 \lesssim 2^{j_2/2} 2^{j_3/2} 2^{-k_2(2\alpha-1)/2}, \text{ if } 2^{k_2} \gg 1. \tag{3.55}$$

For the estimate (3.51), we write J_1 as

$$J_1 = \int f_1(\xi_1, \tau_1) f_2(\xi_2 - \xi_1, \tau_2 - \tau_1 - H_1(\xi_1, \xi_2 - \xi_1)) f_3(\xi_2, \tau_2) d\xi_1 d\xi_2 d\tau_1 d\tau_2.$$

By the calculation in Case 2.1, we see that

$$\left| \det \left(\frac{\partial(\eta_1, \eta_2)}{\partial(\xi_1, \xi_2)} \right) \right| = |(\partial_1 H_1)(\xi_1, \xi_2 - \xi_1)|$$
$$= |2\alpha|\xi_1|^{2\alpha-2}\xi_1 - 2\alpha|\xi_2|^{2\alpha-2}\xi_2| \sim |\xi_*|^{2\alpha-2}|\xi_1 - \xi_2|,$$

where in the last step we use the mean value theorem. From the support property of f_i, we see that $|\xi_1| \sim 2^{k_1}$, $|\xi_2| \sim 2^{k_3}$ and $|\xi_2 - \xi_1| \sim 2^{k_2}$, therefore ξ_1 and ξ_2 have the same sign which tells us that $|\xi_*| \sim 2^{k_1}$. Hence

$$\left| \det \left(\frac{\partial(\eta_1, \eta_2)}{\partial(\xi_1, \xi_2)} \right) \right| \sim 2^{2(\alpha-1)k_1} 2^{k_2},$$

which together with Cauchy-Schwarz inequality implies (3.51) as desired. The proof for other estimates are similar and we omit the details.

(3) Since in this case $2^{k_1} \sim 2^{k_2} \sim 2^{k_3}$, by the proof of part (2), in the support of f_i, we get

$$J_l \lesssim 2^{(j_1+j_2+j_3)/2} 2^{-j_{max}/2} 2^{-k_1(2\alpha-1)/2}. \tag{3.56}$$

Remark 3.2 *From the proof of the above lemma, we see that if $H_1(\xi_1, \xi_2)$ is replaced by*

$$\widetilde{H}_1(\xi_1, \xi_2) = |\xi_1|^{2\alpha} - \langle \xi_2 \rangle - |\xi_1 + \xi_2|^{2\alpha}, \tag{3.57}$$

then for

$$\widetilde{J}_1 = \int f_1(\xi_1, \tau_1) f_2(\xi_2, \tau_2) f_3(\xi_1 + \xi_2, \tau_1 + \tau_2 + \widetilde{H}_1(\xi_1, \xi_2)) d\xi_1 d\xi_2 d\tau_1 d\tau_2,$$

we have the same estimates as J_1.

Proof of Proposition 3.1 From the analysis above, we now bound (3.9) case by case. We divide the summation over k_i into following cases.

Case 1 $2^{k_{max}} \lesssim 1$.

This case is trivial by using (3.23).

Case 2 $2^{k_{min}} \sim 2^{k_{max}} \gg 1$.

In this case, from the support property of f_i, we have $|\xi_1| \sim |\xi_2| \sim |\xi_1+\xi_2| \sim 2^{k_1}$. By using mean value theorem, we see that

$$|H_1(\xi_1,\xi_2)| = ||\xi_1|^{2\alpha} + \langle\xi_2\rangle - |\xi_1+\xi_2|^{2\alpha}|$$
$$= C|\xi_*|^{2\alpha-1}|\xi_2| + \langle\xi_2\rangle,$$

where $|\xi_*|$ is a value between $|\xi_1|$ and $|\xi_1+\xi_2|$. Hence

$$|H_1(\xi_1,\xi_2)| \sim 2^{2k_1\alpha},$$

which tells us that $j_{max} \gtrsim 2k_1\alpha$. By (3.9) and (3.56) we get

$$(3.8) \lesssim \sum_{k_i\in\mathbb{Z}_+}\sum_{j_i\in\mathbb{Z}_+} \frac{2^{j_3(b-1)}2^{k_3s_1}2^{(j_1+j_2+j_3)/2}}{2^{k_1s_1}2^{k_2s_2}2^{j_1b'}2^{j_2b'}}2^{-j_{max}/2}2^{-k_1(2\alpha-1)/2}\|u_{k_1,j_1}^S\|_2\|v_{k_2,j_2}^K\|_2.$$

If $j_1 = j_{max}$, then we get

$$(3.8) \lesssim \sum_{k_i\in\mathbb{Z}_+}\sum_{j_i\in\mathbb{Z}_+} \frac{2^{j_3(b-1)}2^{(j_2+j_3)/2}}{2^{k_2s_2}2^{j_1b'}2^{j_2b'}}2^{-k_1(2\alpha-1)/2}\|u_{k_1,j_1}^S\|_2\|v_{k_2,j_2}^K\|_2$$
$$\lesssim \sum_{k_i\in\mathbb{Z}_+}\sum_{j_i\in\mathbb{Z}_+} 2^{j_3(b-1/2)}2^{j_2(1/2-b')}2^{-j_1b'}2^{-k_1(s_2+\alpha-1/2)}\|u_{k_1,j_1}^S\|_2\|v_{k_2,j_2}^K\|_2. \quad (3.58)$$

Suppose that $b, b' > 1/2$, then

$$(3.8) \lesssim \sum_{k_1\in\mathbb{Z}_+}\sum_{j_1\in\mathbb{Z}_+} 2^{j_1(b-1/2)}2^{-j_1b'}2^{-k_1(s_2+\alpha-1/2)}\|u_{k_1,j_1}^S\|_2\|v_{k_2,j_2}^K\|_2$$
$$= \sum_{k_1\in\mathbb{Z}_+}\sum_{j_1\in\mathbb{Z}_+} 2^{j_1(b-b'-1/2)}2^{-k_1(s_2+\alpha-1/2)}\|u_{k_1,j_1}^S\|_2\|v_{k_2,j_2}^K\|_2. \quad (3.59)$$

Noting that $j_1 = j_{max} \gtrsim 2k_1\alpha$, suppose that $b - b' - 1/2 < 0$, then

$$\sum_{j_1\in\mathbb{Z}_+} 2^{j_1(b-b'-1/2)} \lesssim 2^{2k_1\alpha(b-b'-1/2)}.$$

Hence

$$(3.8) \lesssim \sum_{k_1\in\mathbb{Z}_+} 2^{2k_1\alpha(b-b'-1/2)}2^{-k_1(s_2+\alpha-1/2)}\|u_{k_1,j_1}^S\|_2\|v_{k_2,j_2}^K\|_2 \lesssim \|\hat{u}\|_2\|\hat{v}\|_2$$

provided that

$$b, b' > 1/2, \quad b - b' < 1/2, \quad 2\alpha(b-b') + 1/2 < 2\alpha + s_2. \quad (3.60)$$

If $j_2 = j_{max}$, this case is identical to the case $j_1 = j_{max}$. If $j_3 = j_{max}$, then $j_3 \gtrsim 2k_1\alpha$ and

$$(3.8) \lesssim \sum_{k_i \in \mathbb{Z}_+} \sum_{j_i \in \mathbb{Z}_+} \frac{2^{j_3(b-1)} 2^{(j_1+j_2)/2}}{2^{k_2 s_2} 2^{j_1 b'} 2^{j_2 b'}} 2^{-k_1(2\alpha-1)/2} \|u^S_{k_1,j_1}\|_2 \|v^K_{k_2,j_2}\|_2$$

$$\lesssim \sum_{k_1 \in \mathbb{Z}_+} \frac{2^{2k_1\alpha(b-1)}}{2^{k_1 s_2}} 2^{-k_1(2\alpha-1)/2} \|u^S_{k_1,j_1}\|_2 \|v^K_{k_2,j_2}\|_2 \lesssim \|\hat{u}\|_2 \|\hat{v}\|_2$$

provided that

$$b < 1, \quad b' > 1/2, \quad \alpha(2b-3) + 1/2 < s_2. \tag{3.61}$$

Case 3 $2^{k_{min}} \ll 2^{k_{med}} \sim 2^{k_{max}}, 2^{k_{max}} \gg 1$.

Case 3.1 $2^{k_1} \ll 2^{k_2} \sim 2^{k_3}$.

In this case, we have

$$|H_1(\xi_1, \xi_2)| \sim 2^{2k_2\alpha} \sim 2^{2k_3\alpha}.$$

If $2^{k_1} \lesssim 1$, then we get by (3.9) and (3.23) that

$$(3.8) \lesssim \sum_{k_i \in \mathbb{Z}_+} \sum_{j_i \in \mathbb{Z}_+} \frac{2^{j_3(b-1)} 2^{k_3 s_1}}{2^{k_1 s_1} 2^{k_2 s_2} 2^{j_1 b'} 2^{j_2 b'}} 2^{j_{min}/2} 2^{k_1/2} \|u^S_{k_1,j_1}\|_2 \|v^K_{k_2,j_2}\|_2. \tag{3.62}$$

Without loss of generality, we may assume that now $j_3 = j_{max}$ since it is the worst case. Then $j_3 \gtrsim 2k_3\alpha$ and

$$(3.8) \lesssim \sum_{k_i \in \mathbb{Z}_+} 2^{2k_3\alpha(b-1)} 2^{k_3(s_1-s_2)} 2^{k_1(1/2-s_1)} \|u^S_{k_1,j_1}\|_2 \|v^K_{k_2,j_2}\|_2$$

$$\lesssim \sum_{k_3 \in \mathbb{Z}_+} 2^{2k_3\alpha(b-1)} 2^{k_3(s_1-s_2)} \|u^S_{k_1,j_1}\|_2 \|v^K_{k_2,j_2}\|_2 \lesssim \|\hat{u}\|_2 \|\hat{v}\|_2$$

provided that

$$2\alpha(b-1) + s_1 < s_2. \tag{3.63}$$

Now we assume $2^{k_1} \gg 1$. If $j_3 = j_{max}$, then by (3.9) and (3.37) we get

$$(3.8) \lesssim \sum_{k_i \in \mathbb{Z}_+} \sum_{j_i \in \mathbb{Z}_+} \frac{2^{j_3(b-1)} 2^{k_3 s_1} 2^{(j_1+j_2)/2}}{2^{k_1 s_1} 2^{k_2 s_2} 2^{j_1 b'} 2^{j_2 b'}} 2^{-k_1(2\alpha-1)/2} \|u^S_{k_1,j_1}\|_2 \|v^K_{k_2,j_2}\|_2$$

$$\lesssim \sum_{k_i \in \mathbb{Z}_+} 2^{2k_3\alpha(b-1)} 2^{k_3(s_1-s_2)} 2^{-k_1(\alpha-1/2+s_1)} \|u^S_{k_1,j_1}\|_2 \|v^K_{k_2,j_2}\|_2 \lesssim \|\hat{u}\|_2 \|\hat{v}\|_2$$

provided that

$$2\alpha(b-1) + 1/2 < s_2 + \alpha, \quad \alpha + s_1 \leqslant 1/2; \text{ or} \tag{3.64}$$

$$2\alpha(b-1) + s_1 < s_2, \quad \alpha + s_1 > 1/2. \tag{3.65}$$

If $j_2 = j_{max}$, then by (3.9) and (3.38) we get

$$(3.8) \lesssim \sum_{k_i \in \mathbb{Z}_+} \sum_{j_i \in \mathbb{Z}_+} \frac{2^{j_3(b-1)} 2^{k_3 s_1} 2^{(j_1+j_3)/2}}{2^{k_1 s_1} 2^{k_2 s_2} 2^{j_1 b'} 2^{j_2 b'}} 2^{-k_2(2\alpha-1)/2} \|u^S_{k_1,j_1}\|_2 \|v^K_{k_2,j_2}\|_2$$

$$\lesssim \sum_{k_i \in \mathbb{Z}_+} 2^{2k_3\alpha(b-b'-1/2)} 2^{k_3(s_1-s_2-\alpha+1/2)} 2^{-k_1 s_1} \|u^S_{k_1,j_1}\|_2 \|v^K_{k_2,j_2}\|_2 \lesssim \|\hat{u}\|_2 \|\hat{v}\|_2$$

provided that

$$2\alpha(b - b' - 1) + 1/2 < s_2, \quad s_1 \leqslant 0; \tag{3.66}$$

or

$$2\alpha(b - b' - 1) + s_1 + 1/2 < s_2, \quad s_1 > 0. \tag{3.67}$$

If $j_1 = j_{max}$, then it is identical to the case $j_2 = j_{max}$.

Case 3.2 $2^{k_3} \ll 2^{k_1} \sim 2^{k_2}$.

In this case, we have

$$|H_1(\xi_1, \xi_2)| \sim 2^{2k_1\alpha} \sim 2^{2k_2\alpha}.$$

If $2^{k_3} \lesssim 1$, then we get by (3.9) and (3.23) that

$$(3.8) \lesssim \sum_{k_i \in \mathbb{Z}_+} \sum_{j_i \in \mathbb{Z}_+} \frac{2^{j_3(b-1)} 2^{k_3 s_1}}{2^{k_1 s_1} 2^{k_2 s_2} 2^{j_1 b'} 2^{j_2 b'}} 2^{j_{min}/2} 2^{k_3/2} \|u^S_{k_1,j_1}\|_2 \|v^K_{k_2,j_2}\|_2$$

$$\lesssim \sum_{k_i \in \mathbb{Z}_+} 2^{2k_1\alpha(b-1)} 2^{k_3(s_1+1/2)} 2^{-k_1(s_1+s_2)} \|u^S_{k_1,j_1}\|_2 \|v^K_{k_2,j_2}\|_2$$

$$\lesssim \sum_{k_1 \in \mathbb{Z}_+} 2^{2k_1\alpha(b-1)} 2^{-k_1(s_1+s_2)} \|u^S_{k_1,j_1}\|_2 \|v^K_{k_2,j_2}\|_2 \lesssim \|\hat{u}\|_2 \|\hat{v}\|_2$$

provided that

$$b < 1, \quad b' > 1/2, \quad 2\alpha(b-1) < s_1 + s_2. \tag{3.68}$$

Now we assume $2^{k_3} \gg 1$. If $j_3 = j_{max}$, then from (3.9) and (3.43), we have

$$(3.8) \lesssim \sum_{k_i \in \mathbb{Z}_+} \sum_{j_i \in \mathbb{Z}_+} \frac{2^{j_3(b-1)} 2^{k_3 s_1} 2^{(j_1+j_2)/2}}{2^{k_1 s_1} 2^{k_2 s_2} 2^{j_1 b'} 2^{j_2 b'}} 2^{-k_1(2\alpha-1)/2} \|u^S_{k_1,j_1}\|_2 \|v^K_{k_2,j_2}\|_2$$

$$\lesssim \sum_{k_i \in \mathbb{Z}_+} 2^{2k_1\alpha(b-1)} 2^{k_3 s_1} 2^{-k_1(\alpha - 1/2 + s_1 + s_2)} \|u_{k_1,j_1}^S\|_2 \|v_{k_2,j_2}^K\|_2 \lesssim \|\hat{u}\|_2 \|\hat{v}\|_2$$

provided that
$$2\alpha(b - 3/2) < s_2 - 1/2, \quad s_1 \geqslant 0; \tag{3.69}$$

or
$$2\alpha(b - 3/2) < s_1 + s_2 - 1/2, \quad s_1 \leqslant 0. \tag{3.70}$$

If $j_2 = j_{max}$, then by (3.9) and (3.44), we obtain that

$$(3.8) \lesssim \sum_{k_i \in \mathbb{Z}_+} \sum_{j_i \in \mathbb{Z}_+} \frac{2^{j_3(b-1)} 2^{k_3 s_1} 2^{(j_1+j_3)/2}}{2^{k_1 s_1} 2^{k_2 s_2} 2^{j_1 b'} 2^{j_2 b'}} 2^{-k_2(2\alpha-1)/2} \|u_{k_1,j_1}^S\|_2 \|v_{k_2,j_2}^K\|_2$$
$$\lesssim \sum_{k_i \in \mathbb{Z}_+} 2^{2k_2\alpha(b-b'-1/2)} 2^{k_3 s_1} 2^{k_2(-s_1-s_2-\alpha+1/2)} \|u_{k_1,j_1}^S\|_2 \|v_{k_2,j_2}^K\|_2 \lesssim \|\hat{u}\|_2 \|\hat{v}\|_2$$

provided that
$$2\alpha(b - b' - 1) + 1/2 < s_2, \quad s_1 \geqslant 0; \tag{3.71}$$

or
$$2\alpha(b - b' - 1) + 1/2 < s_1 + s_2, \quad s_1 \leqslant 0. \tag{3.72}$$

If $j_1 = j_{max}$, then by (3.9) and (3.45) we get

$$(3.8) \lesssim \sum_{k_i \in \mathbb{Z}_+} \sum_{j_i \in \mathbb{Z}_+} \frac{2^{j_3(b-1)} 2^{k_3 s_1} 2^{(j_2+j_3)/2}}{2^{k_1 s_1} 2^{k_2 s_2} 2^{j_1 b'} 2^{j_2 b'}} 2^{-k_3(2\alpha-1)/2} \|u_{k_1,j_1}^S\|_2 \|v_{k_2,j_2}^K\|_2$$
$$\lesssim \sum_{k_i \in \mathbb{Z}_+} 2^{2k_2\alpha(b-b'-1/2)} 2^{k_3(s_1-\alpha+1/2)} 2^{k_2(-s_1-s_2)} \|u_{k_1,j_1}^S\|_2 \|v_{k_2,j_2}^K\|_2 \lesssim \|\hat{u}\|_2 \|\hat{v}\|_2$$

provided that
$$2\alpha(b - b' - 1) + 1/2 < s_2, \quad s_1 + 1/2 \geqslant \alpha; \tag{3.73}$$

or
$$2\alpha(b - b' - 1/2) < s_1 + s_2, \quad s_1 + 1/2 < \alpha. \tag{3.74}$$

Case 3.3 $2^{k_2} \ll 2^{k_1} \sim 2^{k_3}$.

First, we assume $2^{k_2} \lesssim 1$. We get by (3.9) and (3.23) that

$$(3.8) \lesssim \sum_{k_i \in \mathbb{Z}_+} \sum_{j_i \in \mathbb{Z}_+} \frac{2^{j_3(b-1)} 2^{k_3 s_1}}{2^{j_1 b'} 2^{j_2 b'} 2^{k_1 s_1} 2^{k_2 s_2}} 2^{j_{min}/2} 2^{k_2/2} \|u_{k_1,j_1}^S\|_2 \|v_{k_2,j_2}^K\|_2$$
$$\lesssim \|\hat{u}\|_2 \|\hat{v}\|_2.$$

Now we assume $2^{k_2} \gg 1$. In this case, by applying mean value theorem as above, we have
$$|H_1(\xi_1,\xi_2)| \sim 2^{k_2} 2^{k_3(2\alpha-1)}.$$

If $j_3 = j_{max}$, then from (3.9) and (3.50) we see that

$$(3.8) \lesssim \sum_{k_i \in \mathbb{Z}_+} \sum_{j_i \in \mathbb{Z}_+} \frac{2^{j_3(b-1)} 2^{(j_1+j_2)/2}}{2^{k_2 s_2} 2^{j_1 b'} 2^{j_2 b'}} 2^{-k_1(2\alpha-1)/2} \|u_{k_1,j_1}^S\|_2 \|v_{k_2,j_2}^K\|_2$$

$$\lesssim \sum_{k_i \in \mathbb{Z}_+} 2^{(2\alpha-1)k_1(b-3/2)} 2^{k_2(b-1-s_2)} \|u_{k_1,j_1}^S\|_2 \|v_{k_2,j_2}^K\|_2 \lesssim \|\hat u\|_2 \|\hat v\|_2$$

provided that
$$2\alpha b + 1/2 < 3\alpha + s_2, \quad b \geqslant 1 + s_2; \quad (3.75)$$

or
$$b < 3/2, \quad b < 1 + s_2. \quad (3.76)$$

If $j_2 = j_{max}$, then by (3.9) and (3.51) we get

$$(3.8) \lesssim \sum_{k_i \in \mathbb{Z}_+} \sum_{j_i \in \mathbb{Z}_+} \frac{2^{j_3(b-1)} 2^{k_3 s_1} 2^{(j_1+j_3)/2}}{2^{k_1 s_1} 2^{k_2 s_2} 2^{j_1 b'} 2^{j_2 b'}} 2^{-k_1(2\alpha-2)/2} 2^{-k_2/2} \|u_{k_1,j_1}^S\|_2 \|v_{k_2,j_2}^K\|_2$$

$$\lesssim \sum_{k_i \in \mathbb{Z}_+} 2^{k_1(2\alpha-1)(b-b'-1)+k_1/2} 2^{k_2(b-b'-1-s_2)} \|u_{k_1,j_1}^S\|_2 \|v_{k_2,j_2}^K\|_2 \lesssim \|\hat u\|_2 \|\hat v\|_2$$

provided that
$$2\alpha(b-b'-1) + 1/2 < s_2, \quad b-b' \geqslant 1 + s_2; \quad (3.77)$$

or
$$(2\alpha-1)(b-b'-1) + 1/2 < 0, \quad b-b' < 1 + s_2. \quad (3.78)$$

If $j_1 = j_{max}$, then by (3.9) and (3.52) we get

$$(3.8) \lesssim \sum_{k_i \in \mathbb{Z}_+} \sum_{j_i \in \mathbb{Z}_+} \frac{2^{j_3(b-1)} 2^{k_3 s_1} 2^{(j_2+j_3)/2}}{2^{k_1 s_1} 2^{k_2 s_2} 2^{j_1 b'} 2^{j_2 b'}} 2^{-k_3(2\alpha-1)/2} \|u_{k_1,j_1}^S\|_2 \|v_{k_2,j_2}^K\|_2$$

$$\lesssim \sum_{k_i \in \mathbb{Z}_+} 2^{k_1(2\alpha-1)(b-b'-1)} 2^{k_2(b-b'-1/2-s_2)} \|u_{k_1,j_1}^S\|_2 \|v_{k_2,j_2}^K\|_2 \lesssim \|\hat u\|_2 \|\hat v\|_2$$

provided that
$$2\alpha(b-b'-1) + 1/2 < s_2, \quad b-b' \geqslant 1/2 + s_2; \quad (3.79)$$

or
$$b-b' < 1, \quad b-b' < 1/2 + s_2. \quad (3.80)$$

Since we are interested in the low regularity problem, therefore we take the intersection of conditions (3.60), (3.61), (3.63), (3.65), (3.66), (3.68), (3.70), (3.72), (3.74), (3.76), (3.78) and (3.80), then we obtain the assumptions in Proposition 3.1 which completes the proof of (3.3) in the proposition.

For (3.4), it can be verified similarly. We now sketch the proof. To show (3.4), it sufficient to prove

$$\|uv\|_{X^{s_1,b-1}} \lesssim \|u\|_{X^{s_1,b'}} \|\bar{v}\|_{Y^{s_2,b'}} = \|u\|_{X^{s_1,b'}} \|v\|_{\bar{Y}^{s_2,b'}}.$$

By the definition of spaces, we need to show that

$$\|\langle \tau - |\xi|^{2\alpha} \rangle^{b-1} \langle \xi \rangle^{s_1} \widehat{u} * \widehat{v}(\xi,\tau)\|_{L^2_{\xi,\tau}}$$
$$\lesssim \|\langle \tau - |\xi|^{2\alpha} \rangle^{b'} \langle \xi \rangle^{s_1} \widehat{u}\|_{L^2_{\xi,\tau}} \|\langle \tau + \langle \xi \rangle \rangle^{b'} \langle \xi \rangle^{s_2} \widehat{v}\|_{L^2_{\xi,\tau}}$$

which is equivalent to prove that

$$\|\langle \tau - |\xi|^{2\alpha} \rangle^{b-1} \langle \xi \rangle^{s_1} (\langle \tau - |\xi|^{2\alpha} \rangle^{-b'} \langle \xi \rangle^{-s_1} \widehat{u}) * (\langle \tau + \langle \xi \rangle \rangle^{-b'} \langle \xi \rangle^{-s_2} \widehat{v})\|_{L^2_{\xi,\tau}} \quad (3.81)$$

is bounded by $\lesssim \|\widehat{u}\|_{L^2_{\xi,\tau}} \|\widehat{v}\|_{L^2_{\xi,\tau}}$. Again by dyadic decomposition,

$$(3.81) \lesssim \sum_{k_i \in \mathbb{Z}_+} \sum_{j_i \in \mathbb{Z}_+} \frac{2^{j_3(b-1)} 2^{k_3 s_1}}{2^{k_1 s_1} 2^{k_2 s_2} 2^{j_1 b'} 2^{j_2 b'}} \|1_{D^S_{k_3,j_3}} \cdot (u^S_{k_1,j_1} * v^K_{k_2,j_2})\|_{L^2_{\xi,\tau}},$$

where $u^S_{k_1,j_1}$ is supported in $D^S_{k_1,j_1}$, $v^K_{k_2,j_2} = \widehat{v}(\xi,\tau)\varphi_{k_2}(\xi)\varphi_{j_2}(\tau + \langle \xi \rangle)$ is supported in $\bar{D}^K_{k_2,j_2} = \{(\xi,\tau) : \xi \in I_{k_2}, \tau + \langle \xi \rangle \in I_{j_2}\}$ and $1_{D^S_{k_3,j_3}}$ is supported in $D^S_{k_3,j_3}$. So the problem is reduced to estimate $\|1_{D^S_{k_3,j_3}} \cdot (u^S_{k_1,j_1} * v^K_{k_2,j_2})\|_{L^2_{\xi,\tau}}$ which is further reduced to estimate the integral:

$$\widetilde{J}_1 = \int f^S_{k_3,j_3}(u^S_{k_1,j_1} * v^K_{k_2,j_2}) d\xi d\tau$$
$$= \int f^S_{k_3,j_3}(\xi_1 + \xi_2, \tau_1 + \tau_2) u^S_{k_1,j_1}(\xi_1,\tau_1) v^K_{k_2,j_2}(\xi_2,\tau_2) d\xi_1 d\xi_2 d\tau_1 d\tau_2.$$

By taking transformation $f_1(\xi,\tau) = u^S_{k_1,j_1}(\xi, \tau + |\xi|^{2\alpha})$, $f_2(\xi,\tau) = v^K_{k_2,j_2}(\xi, \tau - \langle \xi \rangle)$, and $f_3(\xi,\tau) = f^S_{k_3,j_3}(\xi, \tau + |\xi|^{2\alpha})$, $\widetilde{J}_1$ can be rewritten as:

$$\widetilde{J}_1 = \int f_1(\xi_1,\tau_1) f_2(\xi_2,\tau_2) f_3(\xi_1 + \xi_2, \tau_1 + \tau_2 + \widetilde{H}_1(\xi_1,\xi_2)) d\xi_1 d\xi_2 d\tau_1 d\tau_2,$$

where $\widetilde{H}_1$ is defined in (3.57). From Remark 3.2, we see that $\widetilde{J}_1$ satisfies the same estimate as J_1. So repeat the proof for (3.3), we obtain (3.4) as desired.

Proof of Proposition 3.2 Recall that from (3.18), we have

$$(3.17) \lesssim \sum_{k_i \in \mathbb{Z}_+} \sum_{j_i \in \mathbb{Z}_+} \frac{2^{j_3(b-1)} 2^{k_3(s_2-1)}}{2^{k_1 s_1} 2^{k_2 s_1} 2^{j_1 b'} 2^{j_2 b'}} \|1_{D_{k_3,j_3}^K} \cdot (u_{k_1,j_1}^S * v_{k_2,j_2}^{\bar{S}})\|_2, \quad (3.82)$$

We will bound (3.82) case by case. We divide the summation over k_i into following cases.

Case 1 $2^{k_{max}} \lesssim 1$. This case is trivial by using estimate (3.23).

Case 2 $2^{k_{min}} \sim 2^{k_{max}} \gg 1$. In this case, we have

$$|H_2(\xi_1, \xi_2)| \sim 2^{2k_1 \alpha}.$$

By (3.18) and (3.56), we get

$$(3.17)$$
$$\lesssim \sum_{k_i \in \mathbb{Z}_+} \sum_{j_i \in \mathbb{Z}_+} \frac{2^{j_3(b-1)} 2^{k_3(s_2-1)} 2^{(j_1+j_2+j_3)/2}}{2^{k_1 s_1} 2^{k_2 s_1} 2^{j_1 b'} 2^{j_2 b'}} 2^{-j_{max}/2} 2^{-k_1(2\alpha-1)/2} \|u_{k_1,j_1}^S\|_2 \|v_{k_2,j_2}^{\bar{S}}\|_2.$$

If $j_1 = j_{max}$, then we get

$$(3.17) \lesssim \sum_{k_1 \in \mathbb{Z}_+} 2^{2k_1 \alpha(b-b'-1/2)} 2^{k_1(s_2-2s_1-\alpha-1/2)} \|u_{k_1,j_1}^S\|_2 \|v_{k_2,j_2}^{\bar{S}}\|_2 \lesssim \|\hat{u}\|_2 \|\hat{v}\|_2$$

provided that

$$b, b' > 1/2, \quad b - b' < 1/2, \quad 2\alpha(b - b' - 1) + s_2 < 2s_1 + 1/2. \quad (3.83)$$

If $j_2 = j_{max}$, this case is identical to the case $j_1 = j_{max}$. If $j_3 = j_{max}$, then

$$(3.17) \lesssim \sum_{k_1 \in \mathbb{Z}_+} 2^{2k_1 \alpha(b-1)} 2^{k_1(s_2-2s_1-\alpha-1/2)} \|u_{k_1,j_1}^S\|_2 \|v_{k_2,j_2}^{\bar{S}}\|_2 \lesssim \|\hat{u}\|_2 \|\hat{v}\|_2$$

provided that

$$b < 1, \quad b' > 1/2, \quad \alpha(2b - 3) + s_2 < 2s_1 + 1/2. \quad (3.84)$$

Case 3 $2^{k_{min}} \ll 2^{k_{med}} \sim 2^{k_{max}}$, $2^{k_{max}} \gg 1$.

Case 3.1 $2^{k_1} \ll 2^{k_2} \sim 2^{k_3}$.

In this case, we have

$$|H_2(\xi_1, \xi_2)| \sim 2^{2k_2 \alpha}.$$

If $2^{k_1} \lesssim 1$, then we get by (3.18) and (3.23),

$$(3.17) \lesssim \sum_{k_i \in \mathbb{Z}_+} \sum_{j_i \in \mathbb{Z}_+} \frac{2^{j_3(b-1)} 2^{k_3(s_2-1)}}{2^{k_1 s_1} 2^{k_2 s_1} 2^{j_1 b'} 2^{j_2 b'}} 2^{j_{min}/2} 2^{k_1/2} \|u_{k_1,j_1}^S\|_2 \|v_{k_2,j_2}^{\bar{S}}\|_2$$

$$\lesssim \sum_{k_i \in \mathbb{Z}_+} 2^{2k_3 \alpha(b-1)} 2^{k_3(s_2-s_1-1)} 2^{k_1(1/2-s_1)} \|u_{k_1,j_1}^S\|_2 \|v_{k_2,j_2}^{\bar{S}}\|_2$$

$$\lesssim \sum_{k_i \in \mathbb{Z}_+} 2^{2k_3 \alpha(b-1)} 2^{k_3(s_2-s_1-1)} \|u_{k_1,j_1}^S\|_2 \|v_{k_2,j_2}^{\bar{S}}\|_2 \lesssim \|\hat{u}\|_2 \|\hat{v}\|_2$$

provided that

$$2\alpha(b-1) + s_2 < s_1 + 1. \tag{3.85}$$

Now we assume $2^{k_1} \gg 1$. If $j_3 = j_{max}$, then by (3.18) and (3.40), we have

$$(3.17) \lesssim \sum_{k_i \in \mathbb{Z}_+} \sum_{j_i \in \mathbb{Z}_+} \frac{2^{j_3(b-1)} 2^{k_3(s_2-1)} 2^{(j_1+j_2)/2}}{2^{k_1 s_1} 2^{k_2 s_1} 2^{j_1 b'} 2^{j_2 b'}} 2^{-k_2(2\alpha-1)/2} \|u_{k_1,j_1}^S\|_2 \|v_{k_2,j_2}^{\bar{S}}\|_2$$

$$\lesssim \sum_{k_i \in \mathbb{Z}_+} 2^{2k_3 \alpha(b-1)} 2^{k_3(s_2-s_1-\alpha-1/2)} 2^{-k_1 s_1} \|u_{k_1,j_1}^S\|_2 \|v_{k_2,j_2}^{\bar{S}}\|_2 \lesssim \|\hat{u}\|_2 \|\hat{v}\|_2$$

provided that

$$2\alpha(b-1) + s_2 < 2s_1 + \alpha + 1/2, \quad s_1 \leqslant 0; \tag{3.86}$$

or

$$2\alpha(b-1) + s_2 < s_1 + \alpha + 1/2, \quad s_1 > 0. \tag{3.87}$$

If $j_2 = j_{max}$, then by (3.18) and (3.41), we get

$$(3.17) \lesssim \sum_{k_i \in \mathbb{Z}_+} \sum_{j_i \in \mathbb{Z}_+} \frac{2^{j_3(b-1)} 2^{k_3(s_2-1)} 2^{(j_1+j_3)/2}}{2^{k_1 s_1} 2^{k_2 s_1} 2^{j_1 b'} 2^{j_2 b'}} 2^{-k_1(2\alpha-1)/2} \|u_{k_1,j_1}^S\|_2 \|v_{k_2,j_2}^{\bar{S}}\|_2$$

$$\lesssim \sum_{k_i \in \mathbb{Z}_+} 2^{2k_3 \alpha(b-b'-1/2)} 2^{k_3(s_2-s_1-1)} 2^{-k_1(\alpha-1/2+s_1)} \|u_{k_1,j_1}^S\|_2 \|v_{k_2,j_2}^{\bar{S}}\|_2 \lesssim \|\hat{u}\|_2 \|\hat{v}\|_2$$

provided that

$$2\alpha(b - b' - 1/2) + s_2 < 2s_1 + \alpha + 1/2, \quad \alpha + s_1 \leqslant 1/2; \tag{3.88}$$

or

$$2\alpha(b - b' - 1/2) + s_2 < s_1 + 1, \quad \alpha + s_1 > 1/2. \tag{3.89}$$

If $j_1 = j_{max}$, then by (3.18) and (3.42), we obtain that

$$(3.17) \lesssim \sum_{k_i \in \mathbb{Z}_+} \sum_{j_i \in \mathbb{Z}_+} \frac{2^{j_3(b-1)} 2^{k_3(s_2-1)} 2^{(j_2+j_3)/2}}{2^{k_1 s_1} 2^{k_2 s_1} 2^{j_1 b'} 2^{j_2 b'}} 2^{-k_3(2\alpha-1)/2} \|u_{k_1,j_1}^S\|_2 \|v_{k_2,j_2}^{\bar{S}}\|_2$$

$$\lesssim \sum_{k_i \in \mathbb{Z}_+} 2^{2k_3\alpha(b-b'-1/2)} 2^{k_3(s_2-s_1-\alpha-1/2)} 2^{-k_1 s_1} \|u^S_{k_1,j_1}\|_2 \|v^{\bar{S}}_{k_2,j_2}\|_2 \lesssim \|\hat{u}\|_2 \|\hat{v}\|_2$$

provided that

$$2\alpha(b-b'-1/2) + s_2 < 2s_1 + \alpha + 1/2, \quad s_1 \leqslant 0; \tag{3.90}$$

or

$$2\alpha(b-b'-1/2) + s_2 < s_1 + \alpha + 1/2, \quad s_1 > 0. \tag{3.91}$$

Case 3.2 $2^{k_3} \ll 2^{k_1} \sim 2^{k_2}$.

First, we assume $2^{k_3} \lesssim 1$. We decompose further the low frequency as before and assume $k_3 \in \mathbb{Z}$. If $2^{k_3} 2^{k_2(2\alpha-1)} \gg 1$, then we have

$$|H_2(\xi_1, \xi_2)| \sim 2^{k_3} 2^{k_2(2\alpha-1)}.$$

We get by (3.18) and (3.23) that

$$(3.17)$$
$$\lesssim \sum_{k_1,k_2 \in \mathbb{Z}_+} \sum_{k_3 : 2^{k_3} \lesssim 2^{-k_2(2\alpha-1)}} \sum_{j_i \in \mathbb{Z}_+} \frac{2^{j_3(b-1)}}{2^{2k_1 s_1} 2^{j_1 b'} 2^{j_2 b'}} 2^{j_{min}/2} 2^{k_3/2} \|u^S_{k_1,j_1}\|_2 \|v^{\bar{S}}_{k_2,j_2}\|_2$$
$$+ \sum_{k_1,k_2 \in \mathbb{Z}_+} \sum_{k_3 : 2^{k_3} \gg 2^{-k_2(2\alpha-1)}} \sum_{j_i \in \mathbb{Z}_+} \frac{2^{j_3(b-1)}}{2^{2k_1 s_1} 2^{j_1 b'} 2^{j_2 b'}} 2^{j_{min}/2} 2^{k_3/2} \|u^S_{k_1,j_1}\|_2 \|v^{\bar{S}}_{k_2,j_2}\|_2$$
$$= I + II.$$

We may assume that $j_3 = j_{max}$ in I and II since it is the worst case. Since

$$I \lesssim \sum_{k_1,k_2 \in \mathbb{Z}_+} \sum_{j_i \in \mathbb{Z}_+} \frac{2^{j_3(b-1)}}{2^{2k_1 s_1} 2^{j_1 b'} 2^{j_2 b'}} 2^{j_{min}/2} 2^{-k_2(\alpha-1/2)} \|u^S_{k_1,j_1}\|_2 \|v^{\bar{S}}_{k_2,j_2}\|_2$$
$$\lesssim \sum_{k_1 \in \mathbb{Z}_+} 2^{k_1(-2s_1-\alpha+1/2)} \|u^S_{k_1,j_1}\|_2 \|v^{\bar{S}}_{k_2,j_2}\|_2$$

and

$$II \lesssim \sum_{k_1,k_2 \in \mathbb{Z}_+} \sum_{k_3 : 2^{k_3} \gg 2^{-k_2(2\alpha-1)}} \frac{2^{[k_3+k_2(2\alpha-1)](b-1)}}{2^{2k_1 s_1}} 2^{k_3/2} \|u^S_{k_1,j_1}\|_2 \|v^{\bar{S}}_{k_2,j_2}\|_2$$
$$\lesssim \sum_{k_1,k_2 \in \mathbb{Z}_+} 2^{k_1[(2\alpha-1)(b-1)-2s_1]} \|u^S_{k_1,j_1}\|_2 \|v^{\bar{S}}_{k_2,j_2}\|_2$$

then

$$(3.17) \lesssim \|\hat{u}\|_2 \|\hat{v}\|_2$$

under the assumption that
$$1/2 < 2s_1 + \alpha, \tag{3.92}$$
and
$$(2\alpha - 1)(b-1) < 2s_1. \tag{3.93}$$

Now we assume $2^{k_3} \gg 1$. In this case, we have
$$|H_2(\xi_1, \xi_2)| \sim 2^{k_2(2\alpha-1)}2^{k_3}.$$

If $j_3 = j_{max}$, then from (3.18) and (3.46), we see that
$$(3.17) \lesssim \sum_{k_i \in \mathbb{Z}_+} \sum_{j_i \in \mathbb{Z}_+} \frac{2^{j_3(b-1)}2^{k_3(s_2-1)}2^{(j_1+j_2)/2}}{2^{k_1 s_1}2^{k_2 s_1}2^{j_1 b'}2^{j_2 b'}} 2^{-k_1(\alpha-1)}2^{-k_3/2}\|u_{k_1,j_1}^S\|_2\|v_{k_2,j_2}^{\bar{S}}\|_2$$
$$\lesssim \sum_{k_i \in \mathbb{Z}_+} 2^{k_2(2\alpha-1)(b-1)}2^{k_3(b+s_2-5/2)}2^{-k_2(\alpha-1+2s_1)}\|u_{k_1,j_1}^S\|_2\|v_{k_2,j_2}^{\bar{S}}\|_2 \lesssim \|\hat{u}\|_2\|\hat{v}\|_2$$
provided that
$$(2\alpha-1)(b-1)+b+s_2 < 2s_1+\alpha+3/2, \quad b+s_2 \geqslant 5/2; \tag{3.94}$$
or
$$(2\alpha-1)(b-1) < 2s_1+\alpha-1, \quad b+s_2 < 5/2. \tag{3.95}$$

If $j_2 = j_{max}$, then by (3.18) and (3.47), we obtain
$$(3.17) \lesssim \sum_{k_i \in \mathbb{Z}_+} \sum_{j_i \in \mathbb{Z}_+} \frac{2^{j_3(b-1)}2^{k_3(s_2-1)}2^{(j_1+j_3)/2}}{2^{k_1 s_1}2^{k_2 s_1}2^{j_1 b'}2^{j_2 b'}} 2^{-k_2(2\alpha-1)/2}\|u_{k_1,j_1}^S\|_2\|v_{k_2,j_2}^{\bar{S}}\|_2$$
$$\lesssim \sum_{k_i \in \mathbb{Z}_+} 2^{k_2(2\alpha-1)(b-b'-1/2)}2^{k_3(s_2-3/2+b-b')}2^{k_2(-2s_1-\alpha+1/2)}\|u_{k_1,j_1}^S\|_2\|v_{k_2,j_2}^{\bar{S}}\|_2$$
$$\lesssim \|\hat{u}\|_2\|\hat{v}\|_2$$
provided that
$$(2\alpha-1)(b-b'-1/2)+s_2+b-b' < 2s_1+\alpha+1, \quad s_2+b-b' \geqslant 3/2; \tag{3.96}$$
or
$$(2\alpha-1)(b-b'-1/2) < 2s_1+\alpha-1/2, \quad s_2+b-b' < 3/2. \tag{3.97}$$

If $j_1 = j_{max}$, then this case is identical to the case $j_2 = j_{max}$.

Case 3.3 $2^{k_2} \ll 2^{k_1} \sim 2^{k_3}$.

This case is identical to the Case 3.1: $2^{k_1} \ll 2^{k_2} \sim 2^{k_3}$ and we skip the detailed proof. Taking the intersection of conditions (3.83)–(3.86), (3.89), (3.90), (3.92), (3.93), (3.95) and (3.97), we obtain the assumptions in Proposition 3.2 which finishes the proof.

4 Well-posedness for FKGS system

In this part, we prove the main Theorems 1.1 and 1.2.

Proof of Theorem 1.1 We define the following operators:

$$\Lambda^1_{u_0}(u,v) = \psi(t)S_\alpha(t)u_0 + C\psi(t)\int_0^t S_\alpha(t-\tau)\psi_T(\tau)(uv + u\bar{v})(\tau)d\tau, \tag{4.1}$$

$$\Lambda^2_{v_0}(u,v) = \psi(t)W(t)v_0 + C\psi(t)\int_0^t W(t-\tau)\psi_T(\tau)\langle D\rangle^{-1}(|u|^2)(\tau)d\tau, \tag{4.2}$$

then $\Lambda : (u,v) \to (\Lambda^1_{u_0}, \Lambda^2_{v_0})$. Here $\psi(t)$ is a smooth function supported on $[-2, 2]$ and equals 1 on $[-1, 1]$. Let $\psi_\delta = \psi(\cdot/\delta)$ for any $\delta > 0$. We choose the resolution space as:

$$\mathcal{D} = \{(u,v) \in X^{s_1,b} \times Y^{s_2,b} : \|(u,v)\|_{X^{s_1,b} \times Y^{s_2,b}} \leqslant 4C\max\{\|u_0\|_{H^{s_1}}, \|v_0\|_{H^{s_2}}\}\}, \tag{4.3}$$

where the product norm $\|(u,v)\|_{X^{s,b} \times Y^{s,b}} := \|u\|_{X^{s,b}} + \|v\|_{Y^{s,b}}$. We will prove that Λ is a contractive map on $\mathcal{D}$ if T is sufficiently small. Using Lemmas 2.1, 2.3, Propositions 3.1 and 3.2, for $1/2 < b < \tilde{b} < 1$ we have

$$\|\Lambda^1_{u_0}(u,v)\|_{X^{s_1,b}} \leqslant \|\psi(t)S_\alpha(t)u_0\|_{X^{s_1,b}}$$
$$+ C\|\psi(t)\int_0^t S_\alpha(t-\tau)\psi_T(\tau)(uv+u\bar{v})(\tau)d\tau\|_{X^{s_1,b}},$$
$$\leqslant C\|u_0\|_{H^{s_1}} + C\|\psi_T(\tau)(uv+u\bar{v})(\tau)\|_{X^{s_1,b-1}}$$
$$\leqslant C\|u_0\|_{H^{s_1}} + CT^{\tilde{b}-b}\|uv+u\bar{v}\|_{X^{s_1,\tilde{b}-1}}$$
$$\leqslant C\|u_0\|_{H^{s_1}} + CT^{\tilde{b}-b}\|u\|_{X^{s_1,b}}\|v\|_{Y^{s_2,b}}$$
$$\leqslant C\|u_0\|_{H^{s_1}} + 16C^3T^{\tilde{b}-b}\max\{\|u_0\|_{H^{s_1}}, \|v_0\|_{H^{s_2}}\}^2, \tag{4.4}$$

and similarly

$$\|\Lambda^2_{v_0}(u,v)\|_{Y^{s_2,b}} \leqslant \|\psi(t)W(t)v_0\|_{Y^{s_2,b}}$$
$$+ C\|\psi(t)\int_0^t W(t-\tau)\psi_T(\tau)\langle D\rangle^{-1}(|u|^2)(\tau)d\tau\|_{Y^{s_2,b}},$$
$$\leqslant C\|v_0\|_{H^{s_2}} + C\|\psi_T(\tau)u\bar{u}\|_{Y^{s_2,b-1}}$$
$$\leqslant C\|v_0\|_{H^{s_2}} + CT^{\tilde{b}-b}\|u\bar{u}\|_{Y^{s_2,\tilde{b}-1}}$$
$$\leqslant C\|v_0\|_{H^{s_2}} + CT^{\tilde{b}-b}\|u\|_{X^{s_1,b}}\|u\|_{X^{s_1,b}}$$
$$\leqslant C\|v_0\|_{H^{s_2}} + 16C^3T^{\tilde{b}-b}\max\{\|u_0\|_{H^{s_1}}, \|v_0\|_{H^{s_2}}\}^2. \tag{4.5}$$

We choose T sufficiently small satisfying $16C^2 T^{\widetilde{b}-b} \max\{\|u_0\|_{H^{s_1}}, \|v_0\|_{H^{s_2}}\} < 1/2$, then Λ maps $\mathcal{D}$ to $\mathcal{D}$. Let $(u_1, v_1), (u_2, v_2) \in \mathcal{D}$, by the calculation above, we see that

$$\|\Lambda(u_1, v_1) - \Lambda(u_2, v_2)\|_{X^{s_1,b} \times Y^{s_2,b}} < 1/2 \|(u_1 - u_2, v_1 - v_2)\|_{X^{s_1,b} \times Y^{s_2,b}},$$

hence Λ is a contraction mapping on $\mathcal{D}$ and there exists a unique $(u(t), v(t)) \in \mathcal{D}$ satisfying

$$u(t) = \psi(t) S_\alpha(t) u_0 + C\psi(t) \int_0^t S_\alpha(t-\tau) \psi_T(\tau)(uv + u\bar{v})(\tau) d\tau, \qquad (4.6)$$

$$v(t) = \psi(t) W(t) v_0 + C\psi(t) \int_0^t W(t-\tau) \psi_T(\tau) \langle D \rangle^{-1}(|u|^2)(\tau) d\tau. \qquad (4.7)$$

When $t \in [-T, T]$, $(u(t), v(t))$ given by (4.6) and (4.7) is a solution to system (1.7) and $(u, v) \in X_T^{s_1,b} \times Y_T^{s_2,b}$ where $X_T^{s,b}$ is defined as:

$$\|u\|_{X_T^{s,b}} = \inf\{\|\widetilde{u}\|_{X^{s,b}} : \widetilde{u}(t) = u(t), t \in [-T, T]\}.$$

Now we prove (u, v) is a unique solution to system (1.7) in $X_T^{s_1,b} \times Y_T^{s_2,b}$ by following some ideas in [15]. Let (u_1, v_1) and $(u_2, v_2) \in X_T^{s_1,b} \times Y_T^{s_2,b}$ be two solutions to system (1.7) with the same initial data (u_0, v_0), we will prove that $(u_1, v_1) = (u_2, v_2)$ on $[-T, T]$. From symmetry, we only need to prove $(u_1, v_1) = (u_2, v_2)$ on $[0, T]$. Let $\delta > 0$ to be a small data which will be chosen later, define for $i = 1, 2$:

$$\widetilde{u}_i = \begin{cases} u_i(t), & t \in [0, \delta], \\ u_i(2\delta - t), & t \in [\delta, 2\delta], \\ u_0, & \text{otherwise}, \end{cases}$$

and

$$\widetilde{v}_i = \begin{cases} v_i(t), & t \in [0, \delta], \\ v_i(2\delta - t), & t \in [\delta, 2\delta], \\ v_0, & \text{otherwise}. \end{cases}$$

Then $t \to (\widetilde{u}_i(t), \widetilde{v}_i(t))$ is continuous and $(\widetilde{u}_1(t) - \widetilde{u}_2(t), \widetilde{v}_1(t) - \widetilde{v}_2(t)) = (0, 0)$ for $t \in \mathbf{R} \setminus [0, 2\delta]$. Since (u_1, v_1) and (u_2, v_2) are two solutions to system (1.7), then for $t \in [0, \delta]$:

$$u_1(t) - u_2(t) = C\psi(t) \int_0^t S_\alpha(t-\tau) \psi_\delta(\tau)(u_1 v_1 + u_1 \bar{v}_1 - u_2 v_2 - u_2 \bar{v}_2)(\tau) d\tau, \qquad (4.8)$$

$$= C\psi(t)\int_0^t S_\alpha(t-\tau)\psi_\delta(\tau)[(\tilde{u}_1-\tilde{u}_2)(v_1+\bar{v}_1)+u_2(\tilde{v}_1-\tilde{v}_2)+u_2(\overline{\tilde{v}_1}-\overline{\tilde{v}_2})](\tau)d\tau, \tag{4.9}$$

$$v_1(t)-v_2(t)$$
$$= C\psi(t)\int_0^t W(t-\tau)\psi_\delta(\tau)\langle D\rangle^{-1}(|u_1|^2-|u_2|^2))(\tau)(\tau)d\tau,$$
$$= C\psi(t)\int_0^t W(t-\tau)\psi_\delta(\tau)\langle D\rangle^{-1}[u_1(\overline{\tilde{u}_1}-\overline{\tilde{u}_2})+(\tilde{u}_1-\tilde{u}_2)(\bar{u}_2)](\tau)d\tau. \tag{4.10}$$

For $T>0$, define $X_{T+}^{s,b}$:

$$\|u\|_{X_{T+}^{s,b}}=\inf\{\|\tilde{u}\|_{X^{s,b}}:\tilde{u}(t)=u(t),t\in[0,T]\}.$$

Then

$$\|u_1-u_2\|_{X_{\delta+}^{s_1,b}}\leqslant\|u_1-u_2\|_{X^{s_1,b}}$$
$$\leqslant C\delta^{\tilde{b}-b}(\|\tilde{u}_1-\tilde{u}_2\|_{X^{s_1,b}}\|v_1\|_{Y_T^{s_2,b}}+\|\tilde{v}_1-\tilde{v}_2\|_{Y^{s_2,b}}\|u_2\|_{X_T^{s_1,b}}), \tag{4.11}$$

and

$$\|v_1-v_2\|_{Y_{\delta+}^{s_2,b}}\leqslant\|v_1-v_2\|_{Y^{s_2,b}}$$
$$\leqslant C\delta^{\tilde{b}-b}\|\tilde{u}_1-\tilde{u}_2\|_{X^{s_1,b}}(\|u_1\|_{X_T^{s_1,b}}+\|u_2\|_{X_T^{s_1,b}}), \tag{4.12}$$

where $1/2<b<\tilde{b}<1$ as in the proof of existence. By the construction above and applying interpolation:

$$\|\tilde{u}_1-\tilde{u}_2\|_{X^{s_1,b}}\leqslant 4\|u_1-u_2\|_{X_{\delta+}^{s_1,b}},\quad \|\tilde{v}_1-\tilde{v}_2\|_{Y^{s_2,b}}\leqslant 4\|v_1-v_2\|_{Y_{\delta+}^{s_2,b}}. \tag{4.13}$$

Hence,

$$\|u_1-u_2\|_{X_{\delta+}^{s_1,b}}+\|v_1-v_2\|_{Y_{\delta+}^{s_2,b}}$$
$$\leqslant 2C\delta^{\tilde{b}-b}[\|u_1-u_2\|_{X_{\delta+}^{s_1,b}}(\|v_1\|_{Y_T^{s_2,b}}+\|u_1\|_{X_T^{s_1,b}}+\|u_2\|_{X_T^{s_1,b}})$$
$$+\|v_1-v_2\|_{Y_{\delta+}^{s_2,b}}\|u_2\|_{X_T^{s_1,b}}].$$

Choose δ small such that $2C\delta^{\tilde{b}-b}\max\{\|v_1\|_{Y_T^{s_2,b}},\|u_1\|_{X_T^{s_1,b}},\|u_2\|_{X_T^{s_1,b}}\}<1/8$, then $\|u_1-u_2\|_{X_{\delta+}^{s_1,b}}+\|v_1-v_2\|_{Y_{\delta+}^{s_2,b}}<1/2(\|u_1-u_2\|_{X_{\delta+}^{s_1,b}}+\|v_1-v_2\|_{Y_{\delta+}^{s_2,b}})$ which implies the uniqueness of the solution in $X_{\delta+}^{s_1,b} \times Y_{\delta+}^{s_2,b}$. Repeating the procedure, we get the uniqueness in $X_T^{s_1,b} \times Y_T^{s_2,b}$. The proof for Theorem 1.1 is completed.

Proof of Theorem 1.2 From the proof of Theorem 1.1, we see that system (1.7) is locally well-posed if the initial data $(u_0, v_0) \in L^2 \times H^{s_2}$ and the maximum existence time for the solution $T_1 \sim \max\{\|u_0\|_{L^2}, \|v_0\|_{H^{s_2}}\}^{-1}$. To extend the local solution to be a global one, we intend to show that the existence time T_1 has a positive lower bound. Namely, we should give a prior bound for $\|u(t)\|_{L^2}$ and $\|v(t)\|_{H^{s_2}}$. Since the L^2 norm of u is conserved, we only need to control $\|v\|_{H^{s_2}}$. Note that

$$\|v(t)\|_{H^{s_2}} \lesssim \|v_0\|_{H^{s_2}} + \|\langle D \rangle^{-1}(|u|^2)\|_{L_t^1 H_x^{s_2}}$$
$$\lesssim \|v_0\|_{H^{s_2}} + \||u|^2\|_{L_t^1 H_x^{s_2-1}}$$
$$\lesssim \|v_0\|_{H^{s_2}} + \||u|^2\|_{L_t^1 L_x^1},$$

where in the last inequality, we use the embedding $L_x^1 \hookrightarrow H^{s_2-1}$ if $s_2 < 1/2$. Hence for any $T > 0$,

$$\|v(t)\|_{L_t^\infty H^{s_2}([0,T] \times \mathbf{R})} \lesssim \|v_0\|_{H^{s_2}} + T\||u|^2\|_{L_t^\infty L_x^1}$$
$$\lesssim \|v_0\|_{H^{s_2}} + T\|u\|_{L_t^\infty L_x^2}^2$$
$$\lesssim \|v_0\|_{H^{s_2}} + T\|u_0\|_{L_x^2}^2,$$

which implies that we can repeat the local well-posedness procedure with time steps of equal length (depends only on $\|v_0\|_{H^{s_2}}$ and $\|u_0\|_{L^2}$). This proves the global well-posedness for system (1.7) with data in $L^2 \times H^{s_2}$ under the assumption that $s_2 < 1/2$.

Acknowledgements The first author is supported by National Natural Science Foundation of China (Grant No.11201498). We are grateful to the anonymous referees for their valuable comments and suggestions.

References

[1] Bejenaru I, Herr S, Holmer J, et al.. On the 2D Zakharov system with L^2 Schrödinger data[J]. Nonlinearity, 2009, 22: 1063-1089.
[2] Bejenaru I, Herr S. Convolutions of singular measures and applications to the Zakharov systems[J]. J. Funct. Analysis, 2011, 261: 478-506.
[3] Baillon J B, Chadam J M. The cauchy problem for the coupled Schrödinger-Klein-Gordon equations[C]. in: G. M. de La Penha, Mediros L A (Eds), Contemporary Developments in Continuum Mechanics and Partial Differential Equations. North-Holland, Amsterdam, 1978: 37-44.

[4] Bourgain J. Fourier transform restriction phenomena for certain lattice subseets and applications to nonlinear evolution equations I. Schrödinger equations[J]. Geom. Funct. Anal., 1993, 3: 107-156.

[5] Bourgain J. Fourier transform restriction phenomena for certain lattice subseets and applications to nonlinear evolution equations II. The KdV-equation[J]. Geom. Funct. Anal., 1993, 3: 209-262.

[6] Biler P. Attractors for the system of Schrödinger and Klein-Gordon equations with Yukawa coupling[J]. Siam J. Math. Anal., 1990, 21: 1190-1212.

[7] Colliander J, Holmer J, Tzirakis N. Low regularity global well-posedness for the Zakharov and Klein-Gordon-Schrödinger systems[J]. Transact. AMS, 2008, 360: 4619-4638.

[8] Colliander J, Keel M, Staffilani G, et al.. Global well-posedness for Schrödinger equations with derivative[J]. Siam J. Math. Anal., 2001, 33: 649-669.

[9] Fukuda I, Tsutsumi M. On coupled Klein-Gordon-Schrödinger equations I[J]. Bull. Sci. Engrg. Res. Lab. Waseda Univ., 1975, 69: 51-62.

[10] Fukuda I, Tsutsumi M. On coupled Klein-Gordon-Schrödinger equations II[J]. J. Math. Anal. Appl., 1978, 66: 358-378.

[11] Ginibre J, Tsutsumi Y, Velo G. On the Cauchy problem for the Zakharov system[J]. J. Funct. Analysis, 1997, 151: 384-436.

[12] Guo B, Miao C. Asymptotic behavior of coupled Klein-Gordon-Schödinger equations[J]. Sci. China Scr. A, 1995, 25: 705-714.

[13] Guo B, Li Y. Attractors for Klein-Gordon-Schödinger equations in $\mathbf{R}^3$[J]. J. Diff. Equa., 1997, 136: 356-377.

[14] Hayashi N, Wahl W. On the golobal strong solution of coupled Klein-Gordon-Schrödinger equations[J]. J. Math. Soc. Japan, 1987, 39:489-497.

[15] Molinet L, Ribaud F. On the low regularity of the Korteweg-de Vries-Burgers equation[J]. Int. Math. Res. Not., 2002, 37: 1979-2005.

[16] Pecher H. Global solutions of the Klein-Gordon-Schrödinger system with rough data[J]. Diff. Int. Equations, 2004, 17: 179-214.

[17] Pecher H. Some new well-posedness results for the Klein-Gordon-Schrödinger system[J]. Diff. Int. Equations, 2012, 25: 117-142.

[18] Tzirakis N. The Cauchy problem for the Klein-Gordon-Schrödinger system in low dimensions below the energy space[J]. Commun. Part. Diff. Equat., 2005, 30: 605-641.

[19] Tao T. Nonlinear dispersive equations: local and global analysis[M]. CBMS Regional Conference Series in Mathematics, vol. 106. Washington D C: American Mathematical Soc., 2006.

[20] Wang B, Huo Z, Hao C, Guo Z. Harmonic analysis methods for nonlinear evolution equations[M]. World Scientific, 2011.

Global Existence and Semiclassical Limit for Quantum Hydrodynamic Equations with Viscosity and Heat Conduction*

Pu Xueke (蒲学科) and Guo Boling (郭柏灵)

Abstract The hydrodynamic equations with quantum effects are studied in this paper. First we establish the global existence of smooth solutions with small initial data and then in the second part, we establish the convergence of the solutions of the quantum hydrodynamic equations to those of the classical hydrodynamic equations. The energy equation is considered in this paper, which added new difficulties to the energy estimates, especially to the selection of the appropriate Sobolev spaces.

Keywords quantum hydrodynamics; global solutions; semiclassical limit

1 Introduction

The hydrodynamic equations and related models with quantum effects are extensively studied in recent two decades. In these models, the quantum effects is included into the classical hydrodynamic equations by incorporating the first quantum corrections of $O(\hbar^2)$, where $\hbar$ is the Planck constant. One of the main applications of the quantum hydrodynamic equations is as a simplified but not a simplistic approach for quantum plasmas. In particular, the nonlinear aspects of quantum plasmas of quantum plasmas are much more accessible using a fluid description, in comparasion with kinetic theory. One may see the recent monograph of Haas [8] for many physics backgrounds and mathematical derivation of many interesting models. Many other applications of the quantum hydrodynamic equations consisting of analyzing the flow the electrons in quantum semiconductor devices in nano-size [7], where quantum effects like particle tunnelling through potential barriers and built-up in quantum wells,

---

*Kinetic and Related Models, 2016, 9(1): 165–191. DOI: 10.3934/krm.2016.9.165

can not be simulated by classical hydrodynamic model. Similar macroscopic quantum models are also used in many other physical fields such as superfluid and superconductivity [6].

Let us first consider the following classical hydrodynamic equations in conservation form, describing the motion of the electrons in plasmas by omitting the electric potential

$$\begin{cases} \dfrac{\partial n}{\partial t} + \dfrac{1}{m}\dfrac{\partial \Pi_i}{\partial x_i} = 0, & (1.1a) \\ \dfrac{\partial \Pi_j}{\partial t} + \dfrac{\partial}{\partial x_i}(u_i \Pi_j - P_{ij}) = 0, & (1.1b) \\ \dfrac{\partial W}{\partial t} + \dfrac{\partial}{\partial x_i}(u_i W - u_j P_{ij} + q_i) = 0, & (1.1c) \end{cases}$$

where n is the density, m is the effective electron mass, $u = (u_1, u_2, u_3)$ is the velocity, Π_j is the momentum density, P_{ij} is the stress tensor, W is the energy density and q is the heat flux. In this system, repeated indices are summed over under the Einstein convention. This system also emerges from descriptions of the motion of the electrons in semiconductor devices, with the electrical potential and the relaxation omitted.

As in the classical hydrodynamic equations, the quantum conservation laws have the same form as their classical counterparts. However, to close the moment expansion at the third order, we define the above quantities Π_i, P_{ij} and W in terms of the density n, the velocity u and the temperature T. As usual, the heat flux is assumed to obey the Fourier law $q = -\kappa \nabla T$ and the momentum density is defined by $\Pi_i = mnu_i$, where m is the electron mass and u the velocity. The symmetric stress tensor P_{ij} and the energy density W are defined, with quantum corrections, by

$$P_{ij} = -nT\delta_{ij} + \frac{\hbar^2 n}{12m}\frac{\partial^2}{\partial x_i \partial x_j}\log n + O(\hbar^4)$$

and

$$W = \frac{3}{2}nT + \frac{1}{2}mn|u|^2 - \frac{\hbar^2 n}{24m}\Delta \log n + O(\hbar^4),$$

respectively, where $\hbar$ is the Planck constant, and is very small compared to macro quantities.

As far as the quantum corrections are concerned, the quantum correction to the energy density was first derived by Wigner [25] for thermodynamic equilibrium, and the quantum correction to the stress tensor was proposed by Ancona and Tiersten [2] and Ancona and Iafrate [1] on the Wigner formalism. See also [7] for derivation of the system (1.1) by a moment expansion of the Wigner-Boltzmann equation and an expansion of the thermal equilibrium Wigner distribution function to $O(\hbar^2)$, leading to the expression for Π and W above. We also remark the quantum correction term is closely related to the quantum Bohm potential [4]

$$Q(n) = -\frac{\hbar^2}{2m}\frac{\Delta\sqrt{n}}{\sqrt{n}},$$

where n is the charge density. It relates to the quantum correction term in P_{ij} with

$$-n\nabla Q(n) = \frac{\hbar^2}{4m}\text{div}(n\nabla^2 \log n) = \frac{\hbar^2}{4m}\Delta\nabla\rho - \frac{\hbar^2}{m}\text{div}(\nabla\sqrt{n}\otimes\sqrt{n}).$$

For the system (1.1), there is no dissipation in the second equation. Given n and T, the second equation if hyperbolic, and generally we can not expect global smooth solutions for this system. In this paper, we consider the following viscous system by taking into account the stress tensor $\mathbb{S}$,

$$\begin{cases} \partial_t n + \nabla\cdot(nu) = 0, & (1.2\text{a}) \\ \partial_t u + u\cdot\nabla u + \dfrac{1}{mn}\nabla(nT) - \dfrac{\hbar^2}{12m^2 n}\text{div}\{n(\nabla\otimes\nabla)\log n\} = \dfrac{1}{mn}\text{div}\mathbb{S}, & (1.2\text{b}) \\ \partial_t T + u\cdot\nabla T + \dfrac{2}{3}T\nabla\cdot u - \dfrac{2}{3n}\nabla\cdot(\kappa\nabla T) + \dfrac{\hbar^2}{36mn}\nabla\cdot(n\Delta u) \\ \qquad = \dfrac{2}{3mn}\{\nabla\cdot(u\mathbb{S}) - u\cdot\text{div}\mathbb{S}\}. & (1.2\text{c}) \end{cases}$$

Here, $\mathbb{S}$ is the stress tensor defined by

$$\mathbb{S} = \mu(\nabla u + (\nabla u)^T) + \lambda(\text{div}\, u)\mathbb{I},$$

where $\mathbb{I}$ is the $d\times d$ identity matrix, $\mu > 0$ and λ are the primary coefficients of viscosity and the second coefficients of viscosity, respectively, satisfying $2\mu + 3\lambda > 0$. Without quantum corrections (i.e., setting $\hbar = 0$), this system is exactly the classical hydrodynamic equations studied in the seminal paper of Matsumura and Nishida [20].

Although important, there is little result on the system (1.2) to the best of our knowledge. But there does exist a large amount of work for system very similar to (1.2). These work comes from two main origins. The first one is from the quantum correction to various hydrodynamic equations, especially in semiconductors and in plasmas. Gardner [7] derived the full 3D quantum hydrodynamic model by a moment expansion of the Wigner-Boltzmann equation. Hsiao and Li [14] reviewed the recent progress on well-posedness, stability analysis, and small scaling limits for the (bi-polar) quantum hydrodynamic models, where the interested readers may find many useful references therein. Jungel [11] proved global existence of weak solutions for the isentropic case. See also [13, 17–19].

The other one, being equally important, emerges from the study of the compressible fluid models of Korteweg type, which are usually used to describe the motion of compressible fluids with capillarity effect of materials. See Korteweg [16] and the pioneering work of Dunn and Serrin [5]. The reference list can be very long, and we only mention a few of them. Hattori and Li [9, 10] considered the local and global existence of smooth solutions for for the fluid model of Korteweg type for small initial data. Wang and Tan [24] studied the optimal decay for the compressible fluid model of Korteweg type. Recently, Bian, Yao and Zhu [3] studied the global existence of small smooth solutions and the vanishing capillarity limit of this model. Jungel *et al* [12] showed a combined incompressible and vanishing capillarity limit in the barotropic compressible Navier-Stokes equations for smooth solutions.

Almost all of the above mentioned results considered the isothermal case, studying only the continuity equation and the momentum equation, or with electric potential described by a Poisson equation. To the best of our knowledge, there is no mathematical studies for the full quantum hydrodynamic system (1.1). The system (1.2) is itself interesting, since the energy equation also includes the quantum effects through the energy density W, which brings new features into this system. This makes it different from the previous known results, to be precisely stated in the following.

The aim of this paper is two fold. On one hand, we show the global existence of smooth solutions for (1.2) with fixed constant $\hbar > 0$ when the initial data is small near the constant stationary solution $(n, u, T) = (1, 0, 1)$. To be precise, we denote the perturbation by $(\rho, u, \theta) = (n - 1, u, T - 1)$ and transform the (1.2) into (2.1). The result is then stated in Theorem 2.2 for (2.1), where the estimate is stated in terms of the planck constant $\hbar$, and we can see clearly how the quantum corrections affect the estimate. On the other hand, since (1.2) modifies the classical hydrodynamic equations to a macro-micro level in the sense that it incorporates the (micro) quantum corrections, it is expected that as the Planck constant $\hbar \to 0$, the solution of the system (1.2) converges to that of the classical hydrodynamic equations. This limit is rigorously studied in this paper and stated in Theorem 2.3. In particular, algebraic convergence rate is given in terms of $\hbar$.

Among others, one of the main novelties is the selection of the Sobolev space like $(\rho, u, \theta) \in H^{k+2} \times H^{k+1} \times H^k$. The underlying reason lies in the fact that the quantum effects in the energy density involves higher order derivatives of the velocity, and hence we cannot seek solutions in the same Sobolev spaces for θ and u.

This paper is organized as follows. In Sect. 2, we present some preliminaries. We translate the system (1.2) into a convenient form and state the main results in this paper. In Sect. 3, we give the *a priori* estimates, and then prove Theorem (2.2) (existence result) at the end of this section. Finally, in Sect. 4, we prove Theorem (2.3), by showing the convergence of the solutions of the quantum hydrodynamic equations (2.1) to that of the classical hydrodynamic equations (2.3). The algebraic convergence rate is also given in terms of the Planck constant $\hbar$.

Notations. Throughout, C denotes some generic constant independent of time $t > 0$ and the Planck constant $\hbar > 0$. Let $p \in [1, \infty]$, L^p denotes the usual Lebesgue space with norm $\|\cdot\|_{L^p}$. When $p = 2$, it is usually write $\|\cdot\| = \|\cdot\|_{L^p}$, omitting the subscript. Let H^k denote the Sobolev space of the measurable functions whose generalized derivatives up to k^{th} order belong to L^2 with norm $\|\cdot\|_{H^k} = (\sum_{j=0}^k \|D^j \cdot \|^2)^{1/2}$. $\dot{H}^{s,p}$ denotes the homogeneous Sobolev spaces and $[A, B] = AB - BA$ denotes the commutator of A and B.

2 Preliminaries and main results

In this section, we reformulate the system (1.2) in convenient variables. First we take $(n, u, T) = (1, 0, 1)$ to be a constant solution to (1.2) and consider

$$\rho = n - 1, \quad u = u, \quad \theta = T - 1.$$

In these unknowns, with $m = 1$, the (1.2) transforms into

$$\begin{cases} \partial_t \rho + u \cdot \nabla \rho + (1+\rho)\mathrm{div}u = 0, & (2.1\mathrm{a}) \\ \partial_t u - \dfrac{\mu}{\rho+1}\Delta u - \dfrac{\mu+\lambda}{\rho+1}\nabla \mathrm{div}u = -u\cdot \nabla u - \nabla \theta - \dfrac{\theta+1}{\rho+1}\nabla \rho \\ \quad + \dfrac{\hbar^2}{12}\dfrac{\Delta \nabla \rho}{\rho+1} - \dfrac{\hbar^2}{3}\dfrac{\mathrm{div}(\nabla \sqrt{\rho+1} \otimes \nabla \sqrt{\rho+1})}{\rho+1}, & (2.1\mathrm{b}) \\ \partial_t \theta - \dfrac{2\kappa}{3(1+\rho)}\Delta \theta = -u\cdot \nabla \theta - \dfrac{2}{3}(\theta+1)\nabla \cdot u \\ \quad + \dfrac{\hbar^2}{36(1+\rho)}\mathrm{div}((1+\rho)\Delta u) + \dfrac{2}{3(1+\rho)}\left\{\dfrac{\mu}{2}|\nabla u + (\nabla u)^T|^2 + \lambda(\mathrm{div}u)^2\right\}, & (2.1\mathrm{c}) \end{cases}$$

with initial data

$$(\rho, u, \theta)(0, x) = (\rho_0, u_0, \theta_0)(x) = (n_0 - 1, u_0, T_0 - 1)(x).$$

We first state the local-in-time existence of smooth solutions to (2.1). To be precise, we first set

$$|||(\rho, u, \theta)|||_0^2 := \|(\rho, u, \theta)\|_{L^2}^2 + \|\nabla \rho\|_{L^2}^2 + \|(\hbar\nabla\rho, \hbar\nabla u)\|_{L^2}^2 + \|\hbar^2 \Delta \rho\|_{L^2}^2,$$

$$|||(\rho, u, \theta)|||_k^2 := |||(\rho, u, \theta)|||_{k-1}^2 + |||(\nabla^k \rho, \nabla^k u, \nabla^k \theta)|||_0^2,$$

and

$$\mathcal{E}_k(0, T) = \left\{ (\rho, u, \theta)(t, \cdot) : \sup_{t \in [0, T]} |||(\rho, u, \theta)(t)|||_k^2 < \infty \right\}.$$

Theorem 2.1 (local existence) *For any initial data such that $n_0 \geqslant \delta > 0$ is satisfied and $(\rho_0 = n_0 - 1, u_0, \theta_0) \in H^{k+2} \times H^{k+1} \times H^k$ ($k \geqslant 3$), there exists some $T > 0$ such that the Cauchy problem (2.1) has a unique solution (ρ, u, θ) in $[0, T]$ such that $(\rho, u, \theta) \in \mathcal{E}_k(0, T)$ and*

$$|||(\rho, u, \theta)(t)|||_k^2 \leqslant C_k |||(\rho_0, u_0, \theta_0)|||_k^2.$$

This theorem can be proved in a similar fashion as in [9] by the dual argument and iteration techniques, and hence omitted for brevity.

Now, we consider the global existence of smooth solutions. Let $T > 0$, we set

$$E = \sup_{0 \leqslant t \leqslant T} |||(\rho, u, \theta)(t, \cdot)|||_3. \tag{2.2}$$

One of the main purpose is to show the following

Theorem 2.2 (global existence) *Suppose the initial data*

$$(\rho, u, \theta)(0) \in H^5 \times H^4 \times H^3$$

and set $E_0 := |||(\rho, u, \theta)(0)|||_3 < \infty$. There exists some $\hbar_0 > 0$, $\varepsilon_0 > 0$, $\nu_0 > 0$ and $C_0 < \infty$, such that if $E_0 < \varepsilon_0$ and $\hbar < \hbar_0$, then there exists a unique global in time solution $(\rho, u, \theta) \in \mathcal{E}_3(0, T)$ of the Cauchy problem (2.1) for any $t \in (0, \infty)$, and the following estimate hold

$$|||(\rho, u, \theta)(t)|||_3^2 + \nu_0 \int_0^t \sum_{k=1}^4 \|\nabla^k(\rho, u, \theta, \hbar\nabla u, \hbar\nabla\rho)(s)\|^2 ds \leqslant C_0 |||(\rho, u, \theta)(0)|||_3^2,$$

where the constants $\nu_0 > 0$ and $C_0 > 0$ are independent of time t and $\hbar$.

Formally, as $\hbar \to 0$, (2.1) tends to the following classical hydrodynamic equations for (ρ^0, u^0, θ^0) (studied in [20])

$$\begin{cases} \partial_t \rho^0 + u^0 \cdot \nabla \rho^0 + (1+\rho^0)\mathrm{div} u^0 = 0, & (2.3a) \\ \partial_t u^0 - \dfrac{\mu}{\rho^0+1}\Delta u^0 - \dfrac{\mu+\lambda}{\rho^0+1}\nabla \mathrm{div} u^0 = -u^0 \cdot \nabla u^0 - \nabla \theta^0 - \dfrac{\theta^0+1}{\rho^0+1}\nabla \rho^0, & (2.3b) \\ \partial_t \theta^0 - \dfrac{2\kappa}{3(1+\rho^0)}\Delta \theta^0 = -u^0 \cdot \nabla \theta^0 - \dfrac{2}{3}(\theta^0+1)\nabla \cdot u^0 \\ \qquad + \dfrac{2}{3(1+\rho^0)}\left\{\dfrac{\mu}{2}|\nabla u^0 + (\nabla u^0)^T|^2 + \lambda(\mathrm{div} u^0)^2\right\}. & (2.3c) \end{cases}$$

The convergence result is stated in the following

Theorem 2.3 (semiclassical limit) Let $(\rho^\hbar, u^\hbar, \theta^\hbar)$ be the solution of (2.1) and (ρ^0, u^0, θ^0) be the solution of (2.3) with the same initial data $(\rho_0, u_0, \theta_0) \in H^5 \times H^4 \times H^3$. Then for all fixed time $T \in (0, \infty)$, we have the algebraic convergence

$$\sup_{t \in [0,T]} \|(\rho^\hbar - \rho^0, u^\hbar - u^0, \theta^\hbar - \theta^0)\|_{H^1}^2 \leqslant \left[c_2 e^{c_1 T}/c_1\right] \hbar^4$$

and

$$\sup_{t \in [0,T]} \|(\rho^\hbar - \rho^0, u^\hbar - u^0, \theta^\hbar - \theta^0)\|_{H^2}^2 \leqslant \left[c_2 e^{c_1 T}/c_1\right] \hbar^2,$$

for some constant positive constants c_1 and c_2, independent of $\hbar$ and t.

The following three lemmas will be frequently used, and hence cited here for reader's convenience.

Lemma 2.1 (Gagliardo-Nirenberg [21]) Let $p, q, r \in [1, \infty]$ and $0 \leqslant i, j \leqslant l$ be integers, there exist some generic constants $\theta \in [0,1]$ and $C > 0$, such that

$$\|\nabla^j u\|_{L^p(\mathbb{R}^N)} \leqslant C \|\nabla^l u\|_{L^q(\mathbb{R}^N)}^\theta \|\nabla^i u\|_{L^r(\mathbb{R}^N)}^{1-\theta},$$

where

$$j - \frac{N}{p} = \theta\left(l - \frac{N}{q}\right) + (1-\theta)\left(i - \frac{N}{r}\right).$$

When $\theta = 1$, $l - j \neq N/q$.

Lemma 2.2 Let $g(\rho)$ and $g(\rho, \theta)$ be smooth functions of ρ and (ρ, θ), respectively, with bounded derivatives of any order, and $\|\rho\|_{L^\infty} < 1$. Then for any integer $m \geqslant 1$, we have

$$\|\nabla^m g(\rho)\|_{L^p} \leqslant C\|\nabla^m \rho\|_{L^p}, \quad \|\nabla^m g(\rho, \theta)\|_{L^p} \leqslant C\|\nabla^m \rho, \nabla^m \theta\|_{L^p}, \quad \forall p \in [1, \infty],$$

where C may depend on g and m. In particular,

$$\left\|\partial_x^\alpha \left(\frac{\theta+1}{1+\rho}\right)\right\|_{L^p} \leqslant C\|(\rho,\theta)\|_{\dot{H}^{|\alpha|,p}}, \quad \forall p \in [1,\infty].$$

Proof This can be proved in a similar fashion as in [24] and [3] making use of the Gagliardo-Nirenberg inequality, and hence omitted here for brevity. □

Lemma 2.3 (Kato-Ponce [15]) *Let α be any multi-index with $|\alpha| = k$ and $p \in (1,\infty)$. Then there exists some constant $C > 0$ such that*

$$\|\partial_x^\alpha(fg)\|_{L^p} \leqslant C\{\|f\|_{L^{p_1}}\|g\|_{\dot{H}^{s,p_2}} + \|f\|_{\dot{H}^{s,p_3}}\|g\|_{L^{p_4}}\},$$

$$\|[\partial_x^\alpha, f]g\|_{L^p} \leqslant C\{\|\nabla f\|_{L^{p_1}}\|g\|_{\dot{H}^{k-1,p_2}} + \|f\|_{\dot{H}^{k,p_3}}\|g\|_{L^{p_4}}\},$$

where $f, g \in \mathcal{S}$, the Schwartz class and $p_2, p_3 \in (1, +\infty)$ such that

$$\frac{1}{p} = \frac{1}{p_1} + \frac{1}{p_2} = \frac{1}{p_3} + \frac{1}{p_4}.$$

3 A priori estimates

In this section, we establish useful a priori estimates of the solutions to (2.1) and prove Theorem 2.2. First of all, we let the Planck constant $\hbar < 1$. To simplify the proof slightly, we assume that there exists a positive number $\varepsilon \ll 1$ such that

$$E = \sup_{t \in [0,T]} \||(\rho, u, \theta)(t)\||_3 \leqslant \varepsilon, \tag{3.1}$$

which together with Sobolev embedding, implies that

$$\sup_{t \in [0,T]} \|(\rho, u, \theta), \nabla(\rho, u, \theta), \nabla^2 \rho, \hbar\nabla^2(\rho, u), \hbar^2 \nabla^3 \rho\|_{L^\infty} \leqslant CE \leqslant C\varepsilon, \tag{3.2}$$

and from (2.1) the following

$$\|\partial_t \rho\|_{L^p} \leqslant C\|\nabla \cdot (u(1+\rho))\|_{L^p} \leqslant CE \leqslant C\varepsilon, \quad \forall 1 \leqslant p \leqslant \infty, \tag{3.3}$$

and

$$\|\partial_t \theta\|_{L^p} \leqslant \frac{4\kappa}{3}\|\Delta \theta\|_{L^p} + \|u \cdot \nabla \theta\|_{L^p} + \|\mathrm{div}\, u\|_{L^p} + \hbar^2\|\nabla((1+\rho)\Delta u)\|_{L^p} \\ + C\|\nabla u\|_{L^{2p}}^2 \leqslant CE \leqslant C\varepsilon, \quad \forall 1 \leqslant p \leqslant 6. \tag{3.4}$$

In particular, we choose ε small enough such that

$$\sup_{t \in [0,T]} \|(\rho, \theta)(t)\|_{L^\infty} \leqslant 1/2. \tag{3.5}$$

3.1 Basic estimates

Now, we consider the zeroth order estimate for the system (2.1). As in [20], we set

$$s = (1+\theta)/(1+\rho)^{2/3} - 1, \qquad (3.6)$$

and define a function $E^0(\rho, u, s)$ for $\rho, u = (u^1, u^2, u^3)$ and s by

$$E^0(\rho, u, s) = \frac{3R(1+s)}{2}\left((1+\rho)^{5/3} - 1 - \frac{5\rho}{3}\right) + \frac{(1+\rho)}{2}|u|^2 + Rs\rho + \frac{3R(1+\rho)s^2}{4}. \qquad (3.7)$$

The following lemma is proved in [20].

Lemma 3.1 *There exists constants $0 < \rho_2 \leqslant 1/2$ and $0 < C_1 \leqslant C_2 < \infty$ such that E^0 is positive definite, i.e.,*

$$\rho^2 + |u|^2 + \theta^2 \leqslant C_1 E^0 \leqslant C_2(\rho^2 + |u|^2 + \theta^2), \quad for\ |\rho| \leqslant \rho_2.$$

We first prove the zeroth order estimate in the following:

Proposition 3.1 *There exists a constant $\varepsilon_0 > 0$ such that if $E \leqslant \varepsilon \leqslant \varepsilon_0$, then the following a priori estimate holds for all $t \in [0, T]$,*

$$|||(\rho, u, \theta)(t)|||_0^2 + \nu_0 \int_0^t \|D(\rho, u, \theta)(s), (\hbar\Delta u, \hbar\Delta\rho)(s)\|^2 ds \leqslant C|||(\rho, u, \theta)(0)|||_0^2, \qquad (3.8)$$

where $\nu_0 > 0$, $C = C(\varepsilon_0)$ are independent of t.

The proof if postponed to the end of Section 3.1.

Lemma 3.2 *There exists $0 < \varepsilon_0 < 1$ and $h_0 > 0$ such that if $E \leqslant \varepsilon \leqslant \varepsilon_0$ and $\hbar \leqslant h_0$, then for a suitable $\beta > 0$, there holds*

$$\|(\rho, u, s)(t)\|^2 + \beta\|\nabla\rho(t)\|^2 + \nu_0\int_0^t \|\nabla(\rho, u, s)(\tau)\|^2 d\tau + \nu_0\int_0^t \|\hbar\Delta\rho(\tau)\|^2 d\tau$$

$$\leqslant C|||(\rho, u, \theta)(0)|||_0^2 + C\hbar^4 \int_0^t \|\Delta u\|^2 ds, \qquad (3.9)$$

for some constant $C > 0$ independent of t.

Proof Under the transform of (3.6), the system (2.1) is transformed into the following system for (ρ, u, s),

$$\partial_t \rho + u \cdot \nabla \rho + (1+\rho)\mathrm{div}\, u = 0, \tag{3.10a}$$

$$\partial_t u + \frac{\nabla((1+\rho)^\gamma(1+s))}{1+\rho} - \frac{\mu}{\rho+1}\Delta u - \frac{\mu+\lambda}{\rho+1}\nabla \mathrm{div}\, u + u \cdot \nabla u$$
$$= \frac{\hbar^2}{12}\frac{\Delta\nabla\rho}{\rho+1} - \frac{\hbar^2}{3}\frac{\mathrm{div}(\nabla\sqrt{\rho+1} \otimes \nabla\sqrt{\rho+1})}{\rho+1} =: g_\hbar, \tag{3.10b}$$

$$\partial_t s + u \cdot \nabla s - \frac{2\kappa}{3(1+\rho)}\mathrm{div}\left\{\frac{\nabla s}{1+\rho} + \frac{2(1+s)\nabla\rho}{3(1+\rho)^2}\right\}$$
$$-\kappa\gamma(\gamma-1)\left\{\frac{\nabla s}{(1+\rho)^2} + \frac{2(1+s)}{3(1+\rho^3)}\nabla\rho\right\}\nabla\rho$$
$$+\frac{2}{3(1+\rho)^\gamma}\left\{\frac{\mu}{2}|\nabla u + (\nabla u)^T|^2 + \lambda(\mathrm{div}\, u)^2\right\} \tag{3.10c}$$
$$= -\frac{\hbar^2}{36(1+\rho)^\gamma}\mathrm{div}((1+\rho)\Delta u) =: h_\hbar. \tag{3.10d}$$

Recall that $E^0(\rho, u, s)$ is given in (3.7). We compute

$$\partial_t E^0 = (1+\rho)u \cdot u_t + \left\{\frac{u^2}{2} + \frac{\gamma}{\gamma-1}(1 + s((1+\rho)^{\gamma-1} - 1)) + s\right.$$
$$\left. + \frac{1}{2(\gamma-1)}s^2\right\}\rho_t + \left\{\frac{1}{\gamma-1}((1+\rho)^\gamma - 1 - \gamma\rho) + \frac{(1+\rho)s}{\gamma-1}\right\}s_t$$
$$= \mathrm{div}\{\cdots\} - \mu|\nabla u|^2 - (\mu+\lambda)|\nabla \cdot u|^2 - 2\kappa(\gamma-1)\nabla s \cdot \nabla\rho$$
$$- \kappa(\gamma-1)^2|\nabla\rho|^2 - \kappa|\nabla s|^2 + O(E)|D(\rho, u, s)|^2$$
$$+ \underbrace{(1+\rho)u \cdot g_\hbar}_{I} - \underbrace{\left\{\frac{1}{\gamma-1}((1+\rho)^\gamma - 1 - \gamma\rho) + \frac{(1+\rho)s}{\gamma-1}\right\}h_\hbar}_{II}. \tag{3.11}$$

Now, we consider the integration in space of the last two terms I and II on the RHS of (3.11). For the first term I, by integration by parts, and using (3.10a), we obtain

$$\int I\, dx = -\frac{\hbar^2}{12}\int \nabla \cdot u \Delta\rho + \frac{\hbar^2}{12}\int \nabla u(\nabla\rho \otimes \nabla\rho/(1+\rho)). \tag{3.12}$$

The last term on the RHS is easy to be bounded by

$$\left|\frac{\hbar^2}{12}\int \nabla u(\nabla\rho \otimes \nabla\rho/(1+\rho))\right| \leq C\|\nabla u\|_{L^\infty}\|\nabla\rho\|^2 \leq C\hbar^2 E\|\nabla\rho\|^2.$$

For the first term on the RHS, we use (2.1a) to obtain

$$-\frac{\hbar^2}{12}\int \nabla\cdot u\Delta\rho = \frac{\hbar^2}{12}\int \nabla\nabla\cdot u\nabla\rho$$
$$= -\frac{\hbar^2}{12}\int \frac{1}{1+\rho}\nabla(\rho_t + u\cdot\nabla\rho)\nabla\rho$$
$$= -\frac{\hbar^2}{24}\frac{d}{dt}\int \frac{|\nabla\rho|^2}{1+\rho} - \frac{\hbar^2}{24}\int \frac{\rho_t|\nabla\rho|^2}{(1+\rho)^2} - \frac{\hbar^2}{24}\int \frac{\nabla(u\cdot\nabla\rho)}{1+\rho}\nabla\rho.$$

But from (2.1a), it is easy to know that

$$\|\rho_t\|_{L^\infty} \leqslant C\|\mathrm{div}(\rho u)\|_{L^\infty} \leqslant CE^2 \leqslant CE,$$

and by integration by parts,

$$\int \frac{\nabla(u\cdot\nabla\rho)}{1+\rho}\nabla\rho = -\int \frac{\partial_i u\cdot\nabla\rho}{1+\rho}\partial_i\rho - \int \frac{u_i\cdot\partial_i\nabla\rho}{1+\rho}\nabla\rho$$
$$= -\int \frac{\partial_i u\cdot\nabla\rho}{1+\rho}\partial_i\rho + \int \frac{\nabla\cdot u|\nabla\rho|^2}{1+\rho} - \int \frac{u\cdot\nabla\rho}{(1+\rho)^2}|\nabla\rho|^2$$
$$\leqslant C(\|\nabla u\|_{L^\infty} + \|u\|_\infty\|\nabla\rho\|_{L^\infty})\|\nabla\rho\|^2$$
$$\leqslant CE\|\nabla\rho\|^2.$$

Therefore it is easy to see from (3.12) that

$$\int I dx \geqslant -\frac{\hbar^2}{24}\frac{d}{dt}\int \frac{|\nabla\rho|^2}{1+\rho} - C\hbar^2 E\|D(\rho,u,s)\|^2.$$

On the other hand, for the second term II in (3.11), we have

$$\int II dx \leqslant \delta_0\|(\nabla\rho,\nabla s)\| + \frac{C\hbar^4}{\delta_0}\|\Delta u\|^2.$$

In addition to (3.11), we compute

$$\frac{\partial}{\partial t}\left\{\frac{1}{2}|\nabla\rho|^2 + \frac{(1+\rho)^2}{2\mu+\lambda}\nabla\rho\cdot u\right\}$$
$$= \left\{\nabla\rho + \frac{(1+\rho)^2}{2\mu+\lambda}u\right\}\cdot\nabla\rho_t + \frac{(1+\rho)^2}{2\mu+\lambda}\nabla\rho\cdot u_t + \frac{2(1+\rho)}{2\mu+\lambda}u\cdot\nabla\rho\rho_t$$
$$= \mathrm{div}\{\cdots\} - \frac{1}{2\mu+\lambda}\nabla\rho\cdot\nabla s - \frac{\gamma}{2\mu+\lambda}|\nabla\rho|^2$$
$$+ \frac{(\mathrm{div}u)^2}{2\mu+\lambda} + O(E)|D(\rho,u,s)|^2 + \underbrace{\frac{(1+\rho)^2}{2\mu+\lambda}\nabla\rho\cdot g_\hbar}_{J}. \qquad (3.13)$$

After integration in space we obtain for the last term,

$$\int J dx = \frac{\hbar^2}{12}\int \frac{(1+\rho)}{2\mu+\lambda}\nabla\rho\cdot\Delta\nabla\rho + \frac{\hbar^2}{3}\int \frac{(1+\rho)}{2\mu+\lambda}\nabla\rho\cdot\operatorname{div}(\nabla\sqrt{\rho+1}\otimes\nabla\sqrt{\rho+1})$$
$$=: J_1 + J_2.$$

For the term J_1, we have

$$J_1 = -\frac{\hbar^2}{12}\int \frac{(1+\rho)}{2\mu+\lambda}|\Delta\rho|^2 - \frac{\hbar^2}{12}\int \frac{(1+\rho)}{2\mu+\lambda}|\nabla\rho|^2\Delta\rho$$
$$\geq -\frac{\hbar^2}{24}\int \frac{(1+\rho)}{2\mu+\lambda}|\Delta\rho|^2 - C\hbar^2 E^2\|\nabla\rho\|^2.$$

For the term J_2, we have

$$J_2 \leq \frac{\hbar^2}{48}\int \frac{(1+\rho)}{2\mu+\lambda}|\Delta\rho|^2 + C\hbar^2 E^2\|\nabla\rho\|^2.$$

Multiplying (3.13) with a constant β and integration in space, and then add the resultant to (3.11) integrated in space, we obtain

$$\frac{d}{dt}\int E^0(\rho,u,s) + \frac{\hbar^2}{24}\frac{|\nabla\rho|^2}{1+\rho} + \beta\left(\frac{1}{2}|\nabla\rho|^2 + \frac{(1+\rho)^2}{2\mu+\lambda}\nabla\rho\cdot u\right)dx$$
$$+\int \mu|\nabla u|^2 + (\mu+\lambda)|\nabla\cdot u|^2 + \frac{4}{3}\kappa\nabla s\cdot\nabla\rho + \frac{4}{9}\kappa|\nabla\rho|^2 + \kappa|\nabla s|^2 dx$$
$$+\beta\int \frac{\nabla\rho\cdot\nabla s}{2\mu+\lambda} + \frac{5}{3}\frac{1}{2\mu+\lambda}|\nabla\rho|^2 - \frac{(\operatorname{div}u)^2}{2\mu+\lambda} + \frac{\hbar^2}{48}\frac{1+\rho}{2\mu+\lambda}|\Delta\rho|^2 dx$$
$$\leq O(E)\|D(\rho,u,s)\|^2 + \delta_0\|(\nabla\rho,\nabla s)\|^2 + \frac{C\hbar^4}{\delta_0}\|\Delta u\|^2. \qquad (3.14)$$

Note that as in [20], if we take β small such that

$$0 < \beta < \min\left\{\frac{(2\mu+\lambda)^2}{8(1+\rho_2)^4}, (\mu+\lambda)(2\mu+\lambda), 4\kappa(2\mu+\lambda)\right\},$$

where ρ_2 is given in Lemma 3.1, then

$$E^0(\rho,u,s) + \beta\left(\frac{1}{2}|\nabla\rho|^2 + \frac{(1+\rho)^2}{2\mu+\lambda}\nabla\rho\cdot u\right) \geq \frac{1}{8}\left(\rho^2 + \frac{7}{9}s^2 + |u|^2\right) + \frac{\beta}{4}|\nabla\rho|^2,$$

and

$$\mu|\nabla u|^2 + (\mu+\lambda)|\nabla\cdot u|^2 + \frac{4}{3}\kappa\nabla s\cdot\nabla\rho + \frac{4}{9}\kappa|\nabla\rho|^2 + \kappa|\nabla s|^2$$
$$+\beta\left(\frac{\nabla\rho\cdot\nabla s}{2\mu+\lambda} + \frac{5}{3}\frac{1}{2\mu+\lambda}|\nabla\rho|^2 - \frac{(\operatorname{div}u)^2}{2\mu+\lambda}\right) \geq \mu|\nabla u|^2 + \frac{5\kappa}{27}|\nabla s|^2 + \frac{5}{3}\kappa|\nabla\rho|^2.$$

Integrating in time over $[0, t]$ and taking δ_0 and E sufficiently small (say, $\delta_0 = 1/20$), we obtain

$$\int_{\mathbb{R}^3} \left\{ \frac{1}{8}(\rho^2 + \frac{\hbar^2}{36}|\nabla\rho|^2 + \frac{7}{9}s^2 + |u|^2) + \frac{\beta}{4}|\nabla\rho|^2 \right\} dx$$
$$+ \frac{1}{2}\int_0^t \int_{\mathbb{R}^3} \left\{ \mu|\nabla u|^2 + \frac{5\kappa}{27}|\nabla s|^2 + \frac{5}{3}\kappa|\nabla\rho|^2 + \frac{\hbar^2}{96}\frac{1}{2\mu+\lambda}|\Delta\rho|^2 \right\}$$
$$\leqslant \int_{\mathbb{R}^3} E^0(\rho, u, s) + \beta\left(\frac{1}{2}|\nabla\rho|^2 + \frac{(1+\rho)^2}{2\mu+\lambda}\nabla\rho \cdot u\right) dx \bigg|_{t=0} + C\hbar^4 \int_0^t \|\Delta u\|^2 ds$$
$$\leqslant C\||(\rho, u, \theta)(0)\|\|_0^2 + C\hbar^4 \int_0^t \|\Delta u\|^2 ds. \tag{3.15}$$

Now, properly choose the constant ν_0 and $C > 0$, we finish the proof. □

Lemma 3.3 *Under the same condition in Lemma 3.2, we have*

$$\hbar^2 \left\{ \|\nabla u(t)\|^2 + \hbar^2\|\Delta\rho(t)\|^2 \right\} + \int_0^t \|\hbar\Delta u(s), \hbar\nabla\nabla \cdot u(s)\|^2 ds$$
$$\leqslant C\hbar^2\|\nabla u_0, \hbar\Delta\rho_0\|^2 + C\hbar^2 E \int_0^t \|\nabla\rho\|^2 ds + \delta_1 \int_0^t \|\nabla\theta, \nabla\rho\|^2 ds + \frac{C\hbar^4}{\delta_1}\int_0^t \|\Delta u\|^2 ds, \tag{3.16}$$

for any positive constants $\delta_1 > 0$ and $t > 0$.

Proof We now take the inner product of (2.1b) with $-\hbar^2 \Delta u$ to obtain

$$\sum_{i=1}^3 L_i := \frac{\hbar^2}{2}\frac{d}{dt}\int |\nabla u|^2 + \mu\hbar^2 \int \frac{|\Delta u|^2}{1+\rho} + (\mu+\lambda)\hbar^2 \int \frac{(\nabla\nabla \cdot u)}{1+\rho} \cdot \Delta u$$
$$= \hbar^2 \int (u \cdot \nabla u)\Delta u + \hbar^2 \int \nabla\theta \cdot \Delta u + \hbar^2 \int \frac{\theta+1}{\rho+1}\nabla\rho \cdot \Delta u - \frac{\hbar^4}{12}\int \frac{\Delta\nabla\rho}{1+\rho}\Delta u$$
$$+ \frac{\hbar^4}{3}\int \frac{\operatorname{div}(\nabla\sqrt{\rho+1} \otimes \nabla\sqrt{\rho+1})}{1+\rho}\Delta u =: \sum_{i=1}^5 R_i. \tag{3.17}$$

For L_3, we have by integration by parts twice

$$L_3 = (\mu+\lambda)\hbar^2 \int \frac{|\nabla\nabla \cdot u|^2}{(1+\rho)}$$
$$- (\mu+\lambda)\hbar^2 \int \frac{\partial_i\rho}{(1+\rho)^2}\partial_k\nabla \cdot u\partial_k u^i + (\mu+\lambda)\hbar^2 \int \frac{\partial_k\rho}{(1+\rho)^2}\partial_i\nabla \cdot u\partial_k u^i$$
$$\geqslant (\mu+\lambda)\hbar^2 \int \frac{|\nabla\nabla \cdot u|^2}{(1+\rho)} - C\hbar^2 E\|\nabla\rho, \Delta u\|^2,$$

since $\|\nabla u\|_{L^\infty} \leqslant C\|u\|_{H^3} \leqslant CE$.

For R_1, after integration by parts twice, we obtain
$$R_1 = -\hbar^2 \int (u \cdot \nabla u) \Delta u - 2\hbar^2 \int \partial_k u^i \partial_k u^j \partial_i u^j + \hbar^2 \int \nabla \cdot u |\nabla u|^2,$$
which implies that
$$R_1 = -\hbar^2 \int \partial_k u^i \partial_k u^j \partial_i u^j + \frac{\hbar^2}{2} \int \nabla \cdot u |\nabla u|^2 \leqslant C\hbar^2 E \|\nabla u\|^2.$$

For R_2 and R_3, by Hölder inequality we obtain
$$R_2 + R_3 \leqslant \delta_1 (\|\nabla \theta, \nabla \rho\|^2 + \frac{C\hbar^4}{\delta_1} \|\Delta u\|^2, \quad \forall \delta_1 > 0.$$

For R_4, we have by integration by parts and (2.1a)
$$R_4 = -\frac{\hbar^4}{12} \int \frac{\Delta \rho \nabla \rho}{(1+\rho)^2} \cdot \Delta u + \frac{\hbar^4}{12} \int \frac{\Delta \rho}{1+\rho} \Delta \nabla \cdot u$$
$$= -\frac{\hbar^4}{12} \int \frac{\Delta \rho \nabla \rho}{(1+\rho)^2} \cdot \Delta u$$
$$\quad -\frac{\hbar^4}{12} \int \frac{\Delta \rho}{(1+\rho)^2} \{\partial_t \Delta \rho + [\Delta, 1+\rho] \mathrm{div} u + [\Delta, u]\nabla \rho + u \cdot \nabla \Delta \rho\} = \sum_{i=1}^{5} R_{4i}.$$

It is easy to show the following estimates
$$R_{41} \leqslant C\hbar^4 \|\nabla \rho\|_{L^\infty} \|\Delta \rho, \Delta u\|^2 \leqslant CE\hbar^2 \|\hbar \Delta \rho, \hbar \Delta u\|^2,$$
$$R_{42} = -\frac{\hbar^4}{24} \frac{d}{dt} \int \frac{|\Delta \rho|^2}{(1+\rho)^2} - \frac{\hbar^4}{12} \int \frac{\partial_t \rho}{(1+\rho)^3} |\Delta \rho|^2$$
$$\leqslant -\frac{\hbar^4}{24} \frac{d}{dt} \int \frac{|\Delta \rho|^2}{(1+\rho)^2} + CE\hbar^2 \|\hbar \Delta \rho\|^2,$$
thanks to (3.3) and
$$R_{43} + R_{44} \leqslant C\hbar^4 \|\Delta \rho\| (\|[\Delta, 1+\rho]\mathrm{div} u\| + \|[\Delta, u]\nabla \rho\|)$$
$$\leqslant C\hbar^4 \|\Delta \rho\| (\|\Delta u\| \|\nabla \rho\|_{L^\infty} + \|\Delta \rho\| \|\nabla u\|_{L^\infty})$$
$$\leqslant CE\hbar^2 \|\hbar \Delta \rho, \hbar \Delta u\|^2,$$
thanks to Lemma 2.3, and by integration by parts
$$R_{45} = \frac{\hbar^4}{12} \int \frac{\nabla \Delta \rho}{(1+\rho)^2} \cdot u \Delta \rho + \frac{\hbar^4}{12} \int \frac{|\Delta \rho|^2}{(1+\rho)^2} \mathrm{div} u - \frac{\hbar^4}{6} \int \frac{|\Delta \rho|^2}{(1+\rho)^3} u \cdot \nabla \rho$$
$$= \frac{\hbar^4}{24} \int \frac{|\Delta \rho|^2}{(1+\rho)^2} \mathrm{div} u - \frac{\hbar^4}{4} \int \frac{|\Delta \rho|^2}{(1+\rho)^3} u \cdot \nabla \rho$$
$$\leqslant CE\hbar^2 \|\hbar \Delta \rho\|^2.$$

Therefore, we obtain
$$R_4 \leqslant -\frac{\hbar^4}{24}\frac{d}{dt}\int \frac{|\Delta\rho|^2}{(1+\rho)^2} + CE\hbar^2\|\hbar\Delta\rho\|^2.$$

For the term R_5, it is easy to show that
$$R_5 \leqslant CE\hbar^2\|\hbar\Delta\rho\|^2.$$

Hence, putting all the estimates together, we have from (3.17) that
$$\frac{\hbar^2}{2}\frac{d}{dt}\int|\nabla u|^2 + \frac{\hbar^4}{24}\frac{d}{dt}\int\frac{|\Delta\rho|^2}{(1+\rho)^2} + \mu\hbar^2\int\frac{|\Delta u|^2}{1+\rho} + (\mu+\lambda)\hbar^2\int\frac{|\nabla\nabla\cdot u|^2}{(1+\rho)}$$
$$\leqslant C\hbar^2E\|\nabla\rho,\Delta u\|^2 + \delta_1\|\nabla(\rho,\theta)\|^2 + \frac{C\hbar^4}{\delta_1}\|\Delta u\|^2 + CE\hbar^2\|\hbar\Delta\rho,\hbar\Delta u\|^2.$$

Take ε_0 and $\hbar_0$ small, then for any $\varepsilon \leqslant \varepsilon_0$ and $\hbar \leqslant \hbar_0$, integration in time over $[0,t]$ yields the result for any positive constant $\delta_1 > 0$. $\square$

Proof of Proposition of 3.1 By (3.6) and Lemma 3.1, the left hand side of (3.9) is equivalent to the norm
$$\|(\rho,u,\theta)(t)\|^2 + \beta\|\nabla\rho(t)\|^2 + \nu_0\int_0^t\|(\nabla\rho,\nabla u,\nabla\theta,\hbar\Delta\rho)(s)\|^2 ds,$$

for some another suitable constant $\nu_0 > 0$. Hence (3.9) implies that
$$\|(\rho,u,\theta)(t)\|^2 + \beta\|\nabla\rho(t)\|^2 + \nu_0\int_0^t\|(\nabla\rho,\nabla u,\nabla\theta,\hbar\Delta\rho)(s)\|^2 ds$$
$$\leqslant C\||(\rho,u,\theta)(0)\|\|_0^2 + C\hbar^4\int_0^t\|\Delta u\|^2 ds. \tag{3.18}$$

Now, taking δ_1 and $\hbar_0$ in (3.16) sufficiently small, say, $\delta_1 = \nu_0/4$ and $\hbar_0^2 < \nu_0/4C\varepsilon_0$, we then obtain
$$\hbar^2\left\{\|\nabla u(t)\|^2 + \hbar^2\|\Delta\rho(t)\|^2\right\} + \int_0^t\|(\hbar\Delta u,\hbar\nabla\nabla\cdot u)(s)\|^2 ds$$
$$\leqslant C\hbar^2\|\nabla u_0,\hbar\Delta\rho_0\|^2 + \frac{\nu_0}{2}\int_0^t\|(\nabla\rho,\nabla\theta)\|^2 ds + \frac{4C\hbar^4}{\nu_0}\int_0^t\|\Delta u\|^2 ds. \tag{3.19}$$

Add (3.18) and (3.19) together, and then taking $\hbar_0$ even smaller such that $C\hbar_0^2 + 4C\hbar_0^2/\nu_0 < 1/2$, we then obtain
$$\||(\rho,u,\theta)(t)|\|^2 + \nu_0\int_0^t\|(\nabla\rho,\nabla u,\nabla\theta,\hbar\Delta\rho,\hbar\Delta u)(s)\|^2 ds \leqslant C\||(\rho,u,\theta)(0)\|\|_0^2, \tag{3.20}$$

for some positive constant $\nu_0 > 0$ depends only on μ, λ and κ. in particular, ν_0 and C are both independent of t. $\square$

3.2 Higher order estimates

In the following, we denote $\partial^\alpha = \partial_{x_1}^{\alpha_1}\partial_{x_2}^{\alpha_2}\partial_{x_3}^{\alpha_3}$ the partial differential derivative operator with multi-index $\alpha = (\alpha_1, \alpha_2, \alpha_3)$. For our purpose, $|\alpha| \leqslant 3$ suffices. We sometimes abuse the notation to use $\alpha \pm 1$ to stand for $\alpha \pm \beta$ for a multi-index with $|\beta| = 1$ and $\alpha \geqslant \beta$ in the case of $\alpha - \beta$. We will prove the following

Proposition 3.2 Let α be any multi-index with $1 \leqslant |\alpha| \leqslant 3$ and $s = |\alpha|$. There exist some constants $\varepsilon_0 > 0$ and $\hbar_0 > 0$ such that if $E \leqslant \varepsilon_0$ and $\hbar \leqslant \hbar_0$, then the following a priori estimate hold for all $t \in [0, T]$,

$$|||\partial^\alpha (\rho, u, \theta)(t)|||^2 + \nu_0 \int_0^t \|\partial^\alpha \nabla(\rho, u, \theta, \hbar\nabla\rho, \hbar\nabla u)(\tau)\|^2 d\tau$$
$$\leqslant C |||\partial^\alpha(\rho, u, \theta)(0)|||^2 + C \int_0^t \|\nabla(\rho, u, \theta)\|^2_{\dot{H}^{s-1}} d\tau + C \int_0^t \|\hbar\Delta(\rho, u)(\tau)\|^2_{\dot{H}^{s-1}} d\tau, \tag{3.21}$$

for some $\nu_0 > 0$ and $C = C(\varepsilon_0)$ independent of t.

This proposition is proved as a direct sequence of the following lemmas.

Lemma 3.4 Under the assumptions in Proposition 3.2, there exists some constants $\delta_0 < 1$ and $\varepsilon_0 < 1$ sufficiently small, such that

$$\|\partial^\alpha \theta(t)\|^2 + \kappa \int_0^t \|\partial^\alpha \nabla\theta(\tau)\|^2 d\tau$$
$$\leqslant \|\partial^\alpha \theta_0\|^2 + \delta_0(\mu + \lambda)\int_0^t \|\partial^\alpha \nabla \cdot u(\tau)\|^2 d\tau + \frac{C\hbar^4}{\delta_0 \kappa}\int_0^t \|\partial^\alpha \Delta u(\tau)\|^2 d\tau$$
$$+ \frac{C}{\delta_0} \int_0^t \|\nabla(\rho, u, \theta)(\tau)\|^2_{\dot{H}^{s-1}} d\tau + \frac{CE^2\hbar^2}{\delta_0\kappa}\int_0^t \|\Delta u(\tau)\|^2_{\dot{H}^{s-1}} d\tau, \tag{3.22}$$

for all $\delta \leqslant \delta_0$ and $E \leqslant \varepsilon \leqslant \varepsilon_0$, where C is independent of t.

Proof Applying ∂^α to (2.1c) and then taking inner product of the resultant with $\partial^\alpha \theta$ to obtain

$$\frac{1}{2}\frac{d}{dt}\|\partial^\alpha\theta\|^2 + \frac{2\kappa}{3}\int \frac{|\partial^\alpha \nabla\theta|^2}{(1+\rho)} = \frac{2\kappa}{3}\int \frac{\nabla\partial^\alpha\theta \cdot \nabla\rho}{(1+\rho)^2}\partial^\alpha\theta$$
$$- \int \partial^\alpha(u\cdot\nabla\theta)\partial^\alpha\theta - \frac{2}{3}\int \partial^\alpha\left((\theta+1)\nabla\cdot u\right)\partial^\alpha\theta + \frac{\hbar^2}{36}\int \partial^\alpha\left(\frac{\text{div}((1+\rho)\Delta u)}{(1+\rho)}\right)\partial^\alpha\theta$$
$$+ \frac{2}{3}\int \partial^\alpha \left(\frac{\mu|\nabla u + (\nabla u)^T|^2 + 2\lambda(\text{div} u)^2}{(1+\rho)}\right)\partial^\alpha\theta = \sum R_i.$$

For the first term R_1, we have
$$R_1 \leqslant C\kappa \|\partial^\alpha \nabla \theta\| \|\nabla \rho\|_{L^3} \|\partial^\alpha \theta\|_{L^6} \leqslant C\kappa E \|\partial^\alpha \nabla \theta\|^2.$$

For the term R_2, we have
$$\begin{aligned} R_2 &\leqslant \|\partial^\alpha(u \cdot \nabla \theta)\|_{L^{6/5}} \|\partial^\alpha \theta\|_{L^6} \\ &\leqslant (\|u\|_{L^3} \|\partial^\alpha \nabla \theta\|_{L^2} + \|\partial^\alpha u\|_{L^6} \|\nabla \theta\|_{L^{3/2}}) \|\nabla \partial^\alpha \theta\|_{L^2} \\ &\leqslant \delta_0 \kappa \|\nabla \partial^\alpha \theta\|_{L^2}^2 + \frac{CE^2}{\delta_0 \kappa} \|\nabla \partial^\alpha u\|_{L^2}^2, \end{aligned}$$

thanks to Lemma 2.3. For the term R_3, since $|\theta + 1| \leqslant 3/2$ by (3.3), we have
$$\begin{aligned} R_3 &\leqslant (\|\theta + 1\|_{L^\infty} \|\partial^\alpha \nabla \cdot u\|_{L^2} + \|\partial^\alpha \theta\|_{L^6} \|\nabla \cdot u\|_{L^3}) \|\partial^\alpha \theta\|_{L^2} \\ &\leqslant \delta_0 \kappa \|\nabla \partial^\alpha \theta\|_{L^2}^2 + \delta_0(\mu + \lambda) \|\partial^\alpha \nabla \cdot u\|_{L^2}^2 + \frac{C(1+E^2)}{\delta_0} \|\partial^\alpha \theta\|_{L^2}^2. \end{aligned}$$

For the term R_4, by integration by parts,
$$\begin{aligned} R_4 &= -\frac{\hbar^2}{36} \int \partial^{\alpha-1} \left(\frac{\operatorname{div}((1+\rho)\Delta u)}{(1+\rho)} \right) \partial^{\alpha+1} \theta \\ &\leqslant C\hbar^2 \|\partial^{\alpha+1} \theta\| \left\{ \|\partial^{\alpha-1} \operatorname{div}((1+\rho)\Delta u)\| + \left\| \left[\partial^{\alpha-1}, \frac{1}{1+\rho} \right] \operatorname{div}((1+\rho)\Delta u) \right\| \right\} \\ &\leqslant C\hbar^2 \|\partial^{\alpha+1} \theta\| \left\{ \|\partial^{\alpha-1} \operatorname{div} \Delta u\| + \|\Delta u\|_{L^\infty} \|\partial^{\alpha-1} \nabla \rho\|_{L^2} \right\} + C\hbar^2 \|\partial^{\alpha+1} \theta\| \\ &\quad \times \left\{ \left\| \nabla \left(\frac{1}{1+\rho} \right) \right\|_{L^\infty} \|\partial^{\alpha-2} \operatorname{div}((1+\rho)\Delta u)\|_{L^2} + \left\| \frac{1}{1+\rho} \right\|_{\dot{H}^{s-1,6}} \|\operatorname{div}((1+\rho)\Delta u)\|_{L^3} \right\}, \end{aligned}$$

where $s = |\alpha|$. Then making use of Lemmas 2.1, 2.2 and 2.3 and (3.1)-(3.5), one obtains
$$\begin{aligned} R_4 &\leqslant C\hbar^2 \|\partial^{\alpha+1} \theta\| (\|\Delta u\|_{\dot{H}^s} + \|\Delta u\|_{L^\infty} \|\rho\|_{\dot{H}^s}) \\ &\quad + C\hbar^2 \|\partial^{\alpha+1} \theta\| \|\nabla \rho\|_{L^\infty} (\|\Delta u\|_{\dot{H}^{s-1}} + \|\rho\|_{\dot{H}^{s-1,6}} \|\Delta u\|_{L^3}) \\ &\quad + C\hbar^2 \|\partial^{\alpha+1} \theta\| \|\rho\|_{\dot{H}^{s-1,6}} (\|\nabla \Delta u\|_{L^3} + \|\nabla \rho\|_{L^\infty} \|\Delta u\|_{L^3}) \\ &\leqslant \delta_0 \kappa \|\partial^{\alpha+1} \theta\|^2 + \frac{C\hbar^4}{\delta_0 \kappa} \|\Delta u\|_{\dot{H}^s}^2 + \frac{CE^2 \hbar^2}{\delta_0 \kappa} \|\rho\|_{\dot{H}^s}^2 + \frac{CE^2 \hbar^2}{\delta_0 \kappa} \|\Delta u\|_{\dot{H}^{s-1}}^2. \end{aligned}$$

Similar to R_4, we have for the term R_5 that
$$\begin{aligned} R_5 &= \frac{2}{3} \int \partial^\alpha \left(\frac{\mu |\nabla u + (\nabla u)^T|^2 + 2\lambda (\operatorname{div} u)^2}{(1+\rho)} \right) \partial^\alpha \theta \\ &\leqslant \delta_0 \kappa \|\partial^{\alpha+1} \theta\|^2 + \frac{CE^2}{\delta_0 \kappa} \|(\nabla \rho, \nabla u)\|_{\dot{H}^{s-1}}^2. \end{aligned}$$

Putting these estimates together, we obtain

$$\frac{1}{2}\frac{d}{dt}\|\partial^\alpha\theta\|^2 + \frac{2\kappa}{3}\int\frac{|\partial^\alpha\nabla\theta|^2}{(1+\rho)} \leqslant (C\kappa E + \delta_0\kappa)\|\partial^\alpha\nabla\theta\|^2 + \delta_0(\mu+\lambda)\|\partial^\alpha\nabla\cdot u\|_{L^2}^2$$
$$+ \frac{C}{\delta_0}\|\partial^\alpha\theta\|_{L^2}^2 + \frac{C\hbar^4}{\delta_0\kappa}\|\partial^\alpha\Delta u\|^2 + \frac{CE^2}{\delta_0\kappa}\|(\nabla\rho,\nabla u)\|_{\dot H^{s-1}}^2 + \frac{CE^2\hbar^2}{\delta_0\kappa}\|\Delta u\|_{\dot H^{s-1}}^2.$$

Integrating this inequality in time over $[0,t]$ and noting $1/2 < 1+\rho < 3/2$, we know that there exists some constants $\delta_0 < 1$ and $\varepsilon_0 < 1$ sufficiently small, such that (3.22) holds for all $\delta \leqslant \delta_0$ and $E \leqslant \varepsilon \leqslant \varepsilon_0$. $\square$

Lemma 3.5 *Under the assumptions in Proposition 3.2, there exists some constant $\varepsilon_0 < 1$ sufficiently small and $\hbar_0 < 1$, such that*

$$\|\partial^\alpha(\rho,u,\hbar\nabla\rho)(t)\|^2 + \nu_0\int_0^t \|\partial^\alpha\nabla u(\tau)\|^2 d\tau$$
$$\leqslant C\|\partial^\alpha(\rho,u,\hbar\nabla\rho)(0)\|^2 + CE\int_0^t\|\nabla(\rho,u)(\tau)\|_{\dot H^s}^2 d\tau + C\int_0^t\|\nabla(\rho,u,\theta)(\tau)\|_{\dot H^{s-1}}^2 d\tau,$$
(3.23)

for all $E \leqslant \varepsilon \leqslant \varepsilon_0$ and $\hbar < \hbar_0$, where C is independent of t.

Proof Applying ∂^α to (2.1b) and then taking inner product of the resultant with $\partial^\alpha u$, we obtain

$$\sum L_i = \frac{1}{2}\frac{d}{dt}\|\partial^\alpha u\|^2 - \mu\int\partial^\alpha\left(\frac{\Delta u}{\rho+1}\right)\partial^\alpha u - (\mu+\lambda)\int\partial^\alpha\left(\frac{\nabla\mathrm{div}\,u}{\rho+1}\right)\partial^\alpha u$$
$$= -\int\partial^\alpha(u\cdot\nabla u)\partial^\alpha u - \int\partial^\alpha\nabla\theta\partial^\alpha u - \int\partial^\alpha\left(\frac{\theta+1}{\rho+1}\nabla\rho\right)\partial^\alpha u$$
$$+ \frac{\hbar^2}{12}\int\partial^\alpha\left(\frac{\Delta\nabla\rho}{\rho+1}\right)\partial^\alpha u - \frac{\hbar^2}{3}\int\partial^\alpha\left(\frac{\mathrm{div}\{\cdots\}}{\rho+1}\right)\partial^\alpha u = \sum R_i. \quad (3.24)$$

Now, for the term L_2, we have by integration by parts

$$L_2 = -\mu\int\frac{\partial^\alpha\Delta u}{1+\rho}\partial^\alpha u - \mu\int\left[\partial^\alpha,\frac{1}{1+\rho}\right]\Delta u\partial^\alpha u$$
$$= \mu\int\frac{|\partial^\alpha\nabla u|^2}{1+\rho} - \mu\int\frac{\partial^\alpha\nabla u}{(1+\rho)^2}\nabla\rho\partial^\alpha u - \mu\int\left[\partial^\alpha,\frac{1}{1+\rho}\right]\Delta u\partial^\alpha u.$$

Invoking Lemmas 2.2 and 2.3, we obtain

$$\left\|\left[\partial^\alpha,\frac{1}{1+\rho}\right]\Delta u\right\|_{L^{6/5}} \leqslant C\|\nabla\rho\|_{L^3}\|\partial^{\alpha-1}\Delta u\|_{L^2} + C\|\Delta u\|_{L^3}\left\|\partial^\alpha\left(\frac{1}{1+\rho}\right)\right\|_{L^2}$$
$$\leqslant CE\|\partial^{\alpha-1}\Delta u\|_{L^2} + CE\|\nabla\rho\|_{\dot H^{s-1}}.$$

Hence

$$L_2 \geq \frac{2\mu}{3}\|\partial^\alpha \nabla u\|^2 - C\mu\|\partial^\alpha \nabla u\|_{L^2}\|\nabla\rho\|_{L^3}\|\partial^\alpha u\|_{L^6} - C\left\|\left[\partial^\alpha, \frac{1}{1+\rho}\right]\Delta u\right\|_{L^{6/5}}\|\partial^\alpha u\|_{L^6}$$

$$\geq \frac{2\mu}{3}\|\partial^\alpha \nabla u\|^2 - CE(1+\mu)\|\partial^\alpha \nabla u\|^2 - CE\|\nabla\rho\|_{\dot{H}^{s-1}}^2$$

$$\geq \frac{\mu}{2}\|\partial^\alpha \nabla u\|^2 - CE\|\nabla\rho\|_{\dot{H}^{s-1}}^2$$

by taking $E \leq \mu/6C(1+\mu)$. Similarly, we have for L_3 that

$$L_3 \geq \frac{\mu+\lambda}{2}\|\partial^\alpha \nabla \cdot u\|^2 - CE\|\nabla\rho\|_{\dot{H}^{s-1}}^2.$$

For the RHS term R_1, we have by integration by parts that

$$R_1 = \frac{1}{2}\int \nabla \cdot u|\partial^\alpha u|^2 - \int [\partial^\alpha, u]\nabla u \partial^\alpha u$$

$$\leq C\|\nabla u\|_{L^\infty}\|\partial^\alpha u\|^2 \leq CE\|\nabla u\|_{\dot{H}^{s-1}}^2.$$

For the term R_2, we have for any $\delta_0 > 0$ that

$$R_2 \leq \delta_0\mu\|\nabla \partial^\alpha u\|^2 + \frac{C}{\delta_0\mu}\|\nabla\theta\|_{\dot{H}^{s-1}}^2.$$

The term R_3 will be treated with much more effort, from which some good terms will appear. By integration by parts,

$$R_3 = -\int \frac{\theta+1}{\rho+1}\nabla\partial^\alpha\rho\partial^\alpha u - \int \partial^\alpha u\left[\partial^\alpha, \frac{\theta+1}{\rho+1}\right]\nabla\rho$$

$$= \int \frac{\theta+1}{\rho+1}\partial^\alpha\rho\partial^\alpha \mathrm{div}u + \int \nabla\left(\frac{\theta+1}{\rho+1}\right)\partial^\alpha\rho\partial^\alpha u - \int \partial^\alpha u\left[\partial^\alpha, \frac{\theta+1}{\rho+1}\right]\nabla\rho = \sum R_{3i}.$$

It is easy to show that

$$R_{32} \leq C\|\nabla(\theta,\rho)\|_{L^\infty}\|\partial^\alpha(\rho,u)\|^2 \leq CE\|\nabla(\rho,u)\|_{\dot{H}^{s-1}}^2,$$

and

$$R_{33} \leq C\|\partial^\alpha u\|\left(\left\|\nabla\left(\frac{1+\theta}{1+\rho}\right)\right\|_{L^\infty}\|\nabla\rho\|_{\dot{H}^{s-1}} + \|\nabla\rho\|_{L^\infty}\left\|\frac{1+\theta}{1+\rho}\right\|_{\dot{H}^s}\right)$$

$$\leq CE\|\nabla(\rho,u,\theta)\|_{\dot{H}^{s-1}}^2,$$

thanks to Lemma 2.2. Differentiating the continuity equation (2.1a) with ∂^α yields

$$(1+\rho)\partial^\alpha\mathrm{div}u = -\partial_t\partial^\alpha\rho - \partial^\alpha(u\cdot\nabla\rho) - [\partial^\alpha, 1+\rho]\mathrm{div}u,$$

which implies that
$$R_{31} = -\int \frac{\theta+1}{(1+\rho)^2} \partial^\alpha \rho \{\partial_t \partial^\alpha \rho + \partial^\alpha (u \cdot \nabla \rho) + [\partial^\alpha, 1+\rho] \text{div} u\} = \sum R_{31i}.$$

It is immediately from (3.3) and (3.4) with $p = 3/2$ that
$$\begin{aligned}R_{311} &= -\frac{1}{2}\frac{d}{dt}\int \frac{\theta+1}{(1+\rho)^2}|\partial^\alpha \rho|^2 + \frac{1}{2}\int \partial_t\left(\frac{\theta+1}{(1+\rho)^2}\right)|\partial^\alpha \rho|^2 \\ &\leqslant -\frac{1}{2}\frac{d}{dt}\int \frac{\theta+1}{(1+\rho)^2}|\partial^\alpha \rho|^2 + C\|\partial_t\theta, \partial_t\rho\|_{L^{3/2}}\|\partial^\alpha \rho\|_{L^6}^2 \\ &\leqslant -\frac{1}{2}\frac{d}{dt}\int \frac{\theta+1}{(1+\rho)^2}|\partial^\alpha \rho|^2 + CE\|\partial^\alpha \nabla\rho\|^2.\end{aligned}$$

For the term R_{312}, we have by integration by parts that
$$\begin{aligned}R_{312} &= \frac{1}{2}\int \text{div}\left(\frac{(\theta+1)u}{(1+\rho)^2}\right)|\partial^\alpha \rho|^2 - \int \frac{(\theta+1)}{(1+\rho)^2}\partial^\alpha \rho[\partial^\alpha, u]\nabla\rho \\ &\leqslant C\|\nabla(\rho,u,\theta)\|_{L^\infty}\|\partial^\alpha \rho\|^2 + \|\partial^\alpha \rho\|\left(\|\nabla u\|_{L^\infty}\|\nabla \rho\|_{\dot H^{s-1}}\|\nabla\rho\|_{L^\infty}\|u\|_{\dot H^s}\right) \\ &\leqslant CE\|\nabla(\rho,u)\|_{\dot H^{s-1}}^2.\end{aligned}$$

The same estimate hold for R_{313}. Combining all the estimates for R_3, we obtain
$$R_3 \leqslant -\frac{1}{2}\frac{d}{dt}\int \frac{\theta+1}{(1+\rho)^2}|\partial^\alpha \rho|^2 + CE\|\partial^\alpha \nabla\rho\|^2 + CE\|\nabla(\rho,u,\theta)\|_{\dot H^{s-1}}^2.$$

Now, we consider the estimate of R_4. By integration by parts, we obtain
$$\begin{aligned}R_4 &= \frac{\hbar^2}{12}\int \partial^\alpha \partial_i \partial_j \left(\frac{\partial_i \rho}{\rho+1}\right)\partial^\alpha u^j - \frac{\hbar^2}{12}\int \partial^\alpha \left(\left[\partial_i \partial_j, \frac{1}{\rho+1}\right]\partial_i \rho\right)\partial^\alpha u^j \\ &= \frac{\hbar^2}{12}\int \partial^\alpha \left(\frac{\nabla\rho}{\rho+1}\right)\partial^\alpha \nabla\text{div}u + \frac{\hbar^2}{12}\int \partial^{\alpha-1}\left(\left[\nabla^2, \frac{1}{\rho+1}\right]\nabla\rho\right)\partial^{\alpha+1}u \\ &= \frac{\hbar^2}{12}\int \frac{\partial^\alpha \nabla\rho}{\rho+1}\partial^\alpha \nabla\text{div}u + \frac{\hbar^2}{12}\int \left[\partial^\alpha, \frac{1}{\rho+1}\right]\nabla\rho\partial^\alpha \nabla\text{div}u \\ &\quad + \frac{\hbar^2}{12}\int \partial^{\alpha-1}\left(\left[\nabla^2, \frac{1}{\rho+1}\right]\nabla\rho\right)\partial^{\alpha+1}u =: \sum R_{4i}.\end{aligned}$$

Using the continuity equation (2.1a) and similar to the term R_{31}, it can be shown that
$$R_{41} = -\frac{\hbar^2}{12}\int \frac{\partial^\alpha \nabla\rho}{(1+\rho)^2}\{\partial_t \partial^\alpha \nabla\rho + \partial^\alpha \nabla(u \cdot \nabla\rho) + [\partial^\alpha \nabla, 1+\rho]\text{div}u\} = \sum R_{41i}.$$

For R_{411}, we obtain
$$\begin{aligned}R_{411} &= -\frac{\hbar^2}{24}\frac{d}{dt}\int \frac{|\partial^\alpha \nabla\rho|^2}{(1+\rho)^2} - \frac{\hbar^2}{12}\frac{d}{dt}\int \frac{\partial_t \rho|\partial^\alpha \nabla\rho|^2}{(1+\rho)^3} \\ &\leqslant -\frac{\hbar^2}{24}\frac{d}{dt}\int \frac{|\partial^\alpha \nabla\rho|^2}{(1+\rho)^2} + C\hbar^2 E\|\partial^\alpha \nabla\rho\|^2.\end{aligned}$$

By integration by parts,

$$R_{412} = -\frac{\hbar^2}{12}\int \frac{\partial^\alpha \nabla\rho}{(1+\rho)^2} u \cdot \partial^\alpha \nabla^2\rho - \frac{\hbar^2}{12}\int \frac{\partial^\alpha \nabla\rho}{(1+\rho)^2}[\partial^\alpha \nabla, u]\nabla\rho$$

$$= \frac{\hbar^2}{24}\int \mathrm{div}\left(\frac{u}{(1+\rho)^2}\right)|\partial^\alpha \nabla\rho|^2 - \frac{\hbar^2}{12}\int \frac{\partial^\alpha \nabla\rho}{(1+\rho)^2}[\partial^\alpha \nabla, u]\nabla\rho,$$

and hence by Lemma 2.3 and (3.2)

$$R_{412} \leqslant C\hbar^2 \|\nabla(\rho,u)\|_{L^\infty}\|\partial^\alpha \nabla\rho\|^2$$
$$+ C\hbar^2 \|\partial^\alpha \nabla\rho\|(\|\partial^\alpha \nabla\rho\|\|\nabla u\|_{L^\infty} + \|\nabla\rho\|_{L^\infty}\|\partial^\alpha \nabla u\|)$$
$$\leqslant C\hbar^2 E\|\partial^\alpha \nabla(\rho,u)\|^2.$$

Similarly, by Lemma 2.3 and (3.2),

$$R_{413} \leqslant C\hbar^2 E\|\partial^\alpha \nabla(\rho,u)\|^2.$$

For the term R_{42}, we have

$$R_{42} = -\frac{\hbar^2}{12}\int \nabla\left(\left[\partial^\alpha, \frac{1}{\rho+1}\right]\nabla\rho\right)\partial^\alpha \mathrm{div}\,u$$

$$= \frac{\hbar^2}{12}\int \left(\left[\partial^\alpha, \frac{\nabla\rho}{(\rho+1)^2}\right]\nabla\rho\right)\partial^\alpha \mathrm{div}\,u - \frac{\hbar^2}{12}\int \left[\partial^\alpha, \frac{1}{\rho+1}\right]\nabla^2\rho\,\partial^\alpha \mathrm{div}\,u.$$

But by the commutator estimates, we have

$$\left\|\left[\partial^\alpha, \frac{\nabla\rho}{(\rho+1)^2}\right]\nabla\rho\right\|_{L^2} \leqslant \|\nabla\rho\|_{\dot H^{s-1,6}}\left\|\frac{\nabla\rho}{(\rho+1)^2}\right\|_{L^3} + \|\nabla\rho\|_{L^\infty}\left\|\frac{\nabla\rho}{(\rho+1)^2}\right\|_{\dot H^s}$$
$$\leqslant CE\|\nabla\rho\|_{\dot H^s},$$

and

$$\left\|\left[\partial^\alpha, \frac{1}{\rho+1}\right]\nabla^2\rho\right\|_{L^2} \leqslant \|\nabla^2\rho\|_{\dot H^{s-1}}\left\|\nabla\left(\frac{1}{\rho+1}\right)\right\|_{L^\infty} + \|\nabla^2\rho\|_{L^3}\left\|\frac{1}{\rho+1}\right\|_{\dot H^{s,6}}$$
$$\leqslant CE\|\nabla\rho\|_{\dot H^s},$$

hence

$$R_{42} \leqslant C\hbar^2 E\|\nabla(\rho,u)\|_{\dot H^s}^2.$$

Similarly, for R_{43}, we obtain

$$R_{43} \leqslant C\hbar^2 E\|\nabla(\rho,u)\|_{\dot H^s}^2.$$

Putting all the estimates for R_4 together, we obtain
$$R_4 \leqslant -\frac{\hbar^2}{24}\frac{d}{dt}\int\frac{|\partial^\alpha\nabla\rho|^2}{(1+\rho)^2} + C\hbar^2 E\|\nabla(\rho,u)\|_{H^s}^2.$$

Finally, for R_5, it is easy to show
$$R_5 = \frac{\hbar^2}{12}\int \partial^{\alpha-1}\left(\frac{\text{div}\{\nabla\rho\otimes\nabla\rho/(1+\rho)\}}{\rho+1}\right)\partial^{\alpha+1}u$$
$$\leqslant C\hbar^2 E\|\partial^\alpha\nabla(\rho,u)\|^2 + C\hbar^2 E\|\nabla\rho\|_{H^{s-1}}^2.$$

Now, putting all these estimates for (3.24) together, and taking $\delta_0 = 1/4$, we obtain,
$$\frac{1}{2}\frac{d}{dt}\|\partial^\alpha u\|^2 + \frac{1}{2}\frac{d}{dt}\int\frac{\theta+1}{(1+\rho)^2}|\partial^\alpha\rho|^2 + \frac{\hbar^2}{24}\frac{d}{dt}\int\frac{|\partial^\alpha\nabla\rho|^2}{(1+\rho)^2}$$
$$+ \frac{\mu}{4}\|\partial^\alpha\nabla u\|^2 + \frac{\mu+\lambda}{2}\|\partial^\alpha\nabla\cdot u\|^2$$
$$\leqslant C\|\partial^\alpha\theta\|^2 + CE\|\partial^\alpha\nabla\rho\|^2 + CE\|\nabla(\rho,u,\theta)\|_{H^{s-1}}^2 + C\hbar^2 E\|\partial^\alpha\nabla u\|^2.$$

Integrating in time over $[0,t]$ completes the proof, thanks to $\hbar < 1$. □

Lemma 3.6 *Under the assumptions in Proposition 3.2, there exists some constant $\varepsilon_0 < 1$ sufficiently small and $\hbar_0 < 1$, such that*
$$\hbar^2\|\nabla\partial^\alpha(u,\rho,\hbar\nabla\rho)(t)\|^2 + \nu_0\hbar^2\int_0^t\|\Delta\partial^\alpha u(\tau)\|^2 d\tau$$
$$\leqslant C\hbar^2\|\nabla\partial^\alpha(u,\rho,\hbar\nabla\rho)(0)\|^2 + C\hbar^4 E\int_0^t\|\Delta\partial^\alpha u(\tau)\|^2 d\tau + C\hbar^2 E\int_0^t\|\Delta\partial^\alpha\rho(\tau)\|^2 d\tau$$
$$+ C\hbar^2\int_0^t\|\partial^\alpha\nabla\theta(\tau)\|^2 d\tau + C\hbar^2 E\int_0^t\|\nabla(\rho,u,\theta)(\tau)\|_{H^s}^2 d\tau, \qquad (3.25)$$

for all $E \leqslant \varepsilon \leqslant \varepsilon_0$ and $\hbar < \hbar_0$, where C is independent of t.

Proof Applying ∂^α to (2.1b) and then taking inner product of the resultant with $-\hbar^2\Delta\partial^\alpha u$, we obtain
$$\sum L_i = \frac{\hbar^2}{2}\frac{d}{dt}\|\nabla\partial^\alpha u\|^2 + \mu\hbar^2\int\partial^\alpha\left(\frac{\Delta u}{\rho+1}\right)\Delta\partial^\alpha u + (\mu+\lambda)\hbar^2\int\partial^\alpha\left(\frac{\nabla\text{div}u}{\rho+1}\right)\Delta\partial^\alpha u$$
$$= \hbar^2\int\partial^\alpha(u\cdot\nabla u)\Delta\partial^\alpha u + \hbar^2\int\partial^\alpha\nabla\theta\Delta\partial^\alpha u + \hbar^2\int\partial^\alpha\left(\frac{\theta+1}{\rho+1}\nabla\rho\right)\Delta\partial^\alpha u$$
$$- \frac{\hbar^4}{12}\int\partial^\alpha\left(\frac{\Delta\nabla\rho}{\rho+1}\right)\Delta\partial^\alpha u + \frac{\hbar^4}{3}\int\partial^\alpha\left(\frac{\text{div}\{\cdots\}}{\rho+1}\right)\Delta\partial^\alpha u = \sum R_i.$$

Now, for the term L_2, we have by integration by parts
$$L_2 = \mu\hbar^2\int\frac{|\Delta\partial^\alpha u|^2}{1+\rho} + \mu\hbar^2\int\left[\partial^\alpha, \frac{1}{1+\rho}\right]\Delta u\Delta\partial^\alpha u.$$

Since
$$\left\|\left[\partial^\alpha, \frac{1}{1+\rho}\right]\Delta u\right\|_{L^2} \leqslant C\|\nabla\rho\|_{L^\infty}\|\partial^{\alpha-1}\Delta u\|_{L^2} + C\|\Delta u\|_{L^3}\left\|\partial^\alpha\left(\frac{1}{1+\rho}\right)\right\|_{L^6}$$
$$\leqslant CE\|\Delta u\|_{\dot H^{s-1}} + C\|\Delta u\|_{H^1}\|\rho\|_{\dot H^{s,6}}$$
$$\leqslant CE\|(\rho,u)\|_{\dot H^{s+1}},$$

thanks to Lemma 2.2 and (2.1), we have
$$L_2 \geqslant \frac{\mu\hbar^2}{2}\|\Delta\partial^\alpha u\|^2 - C\hbar^2 E^2\|(\rho,u)\|^2_{\dot H^{s+1}}.$$

Similarly, for the L_3, we have
$$L_3 \geqslant \frac{(\mu+\lambda)\hbar^2}{2}\|\partial^\alpha\nabla\mathrm{div}u\|^2 - C\hbar^2 E^2\|(\rho,u)\|^2_{\dot H^{s+1}}.$$

For the RHS term R_1, we have by integration by parts twice that
$$R_1 = -\hbar^2\int\partial_j\partial^\alpha(u^i\cdot\partial_i u)\partial_j\partial^\alpha u$$
$$= -\hbar^2\int u^i\partial_j\partial^\alpha\partial_i u\cdot\partial_j\partial^\alpha u - \hbar^2\int[\partial_j\partial^\alpha, u^i]\partial_i u\partial_j\partial^\alpha u$$
$$= \frac{\hbar^2}{2}\int\partial_i u^i|\partial_j\partial^\alpha u|^2 - \hbar^2\int[\partial_j\partial^\alpha, u^i]\partial_i u\partial_j\partial^\alpha u \leqslant C\hbar^2 E\|u\|^2_{\dot H^{s+1}}.$$

For the term R_2, we have
$$R_2 = \hbar^2\int\partial^\alpha\nabla\theta\Delta\partial^\alpha u \leqslant \delta_0\mu\hbar^2\|\Delta\partial^\alpha u\|^2 + \frac{C\hbar^2}{\delta_0\mu}\|\partial^\alpha\nabla\theta\|^2.$$

By integration by parts,
$$R_3 = -\hbar^2\int\nabla\partial^\alpha\left(\frac{\theta+1}{\rho+1}\nabla\rho\right)\nabla\partial^\alpha u$$
$$= -\hbar^2\int\frac{\theta+1}{\rho+1}\nabla\partial^\alpha\nabla\rho\nabla\partial^\alpha u - \hbar^2\int\nabla\partial^\alpha u\left[\nabla\partial^\alpha, \frac{\theta+1}{\rho+1}\right]\nabla\rho$$
$$= \hbar^2\int\frac{\theta+1}{\rho+1}\nabla\partial^\alpha\rho\nabla\partial^\alpha\mathrm{div}u + \hbar^2\int\nabla\left(\frac{\theta+1}{\rho+1}\right)\nabla\partial^\alpha\rho\nabla\partial^\alpha u$$
$$- \hbar^2\int\nabla\partial^\alpha u\left[\nabla\partial^\alpha, \frac{\theta+1}{\rho+1}\right]\nabla\rho = \sum R_{3i}.$$

It is easy to show that
$$R_{32} \leqslant \hbar^2\|\nabla(\theta,\rho)\|_{L^\infty}\|\partial^\alpha\nabla(\rho,u)\|^2 \leqslant C\hbar^2 E\|\nabla\partial^\alpha(\rho,u)\|^2$$

and by Lemmas 2.2 and 2.3,

$$R_{33} \leqslant C\hbar^2 \|\nabla \partial^\alpha u\| \left\| \left[\nabla \partial^\alpha, \frac{\theta+1}{\rho+1}\right] \nabla \rho \right\|$$

$$\leqslant C\hbar^2 \|\nabla \partial^\alpha u\| (\|\nabla \rho\|_{\dot{H}^s} \|\nabla(\rho,\theta)\|_{L^\infty} + \|\nabla \rho\|_{L^\infty} \|(\rho,\theta)\|_{\dot{H}^{s+1}})$$

$$\leqslant C\hbar^2 E \|\nabla(\rho,u,\theta)\|_{\dot{H}^s}^2.$$

Differentiating the continuity equation (2.1a) with ∂^α and then inserting the resultant to R_{31}, we obtain

$$R_{31} = -\hbar^2 \int \frac{\theta+1}{(1+\rho)^2} \nabla \partial^\alpha \rho \{\partial_t \nabla \partial^\alpha \rho + \nabla \partial^\alpha (u \cdot \nabla \rho) + [\nabla \partial^\alpha, 1+\rho]\mathrm{div} u\} = \sum R_{31i}.$$

It is immediately that

$$R_{311} = -\frac{\hbar^2}{2} \frac{d}{dt} \int \frac{\theta+1}{(1+\rho)^2} |\nabla \partial^\alpha \rho|^2 + \frac{\hbar^2}{2} \int \partial_t \left(\frac{\theta+1}{(1+\rho)^2}\right) |\nabla \partial^\alpha \rho|^2$$

$$\leqslant -\frac{\hbar^2}{2} \frac{d}{dt} \int \frac{\theta+1}{(1+\rho)^2} |\nabla \partial^\alpha \rho|^2 + C\hbar^2 E \|\partial^\alpha \Delta \rho\|^2,$$

thanks to (3.3) and (3.4) again. For the term R_{312}, we have by integration by parts that

$$R_{312} = \frac{\hbar^2}{2} \int \mathrm{div}\left(\frac{(\theta+1)u}{(1+\rho)^2}\right) |\nabla \partial^\alpha \rho|^2 - \hbar^2 \int \frac{(\theta+1)}{(1+\rho)^2} \nabla \partial^\alpha \rho [\nabla \partial^\alpha, u] \nabla \rho$$

$$\leqslant C\hbar^2 E \|\nabla \partial^\alpha (\rho,u)\|^2.$$

The same estimate hold for R_{313}. Now, we consider the estimate of R_4. By integration by parts, we obtain

$$R_4 = -\frac{\hbar^4}{12} \int \frac{\Delta \partial^\alpha \nabla \rho}{\rho+1} \Delta \partial^\alpha u - \frac{\hbar^4}{12} \int \left[\partial^\alpha, \frac{1}{\rho+1}\right] \Delta \nabla \rho \Delta \partial^\alpha u$$

$$= \frac{\hbar^4}{12} \int \frac{\Delta \partial^\alpha \rho}{\rho+1} \Delta \partial^\alpha \mathrm{div} u - \frac{\hbar^4}{12} \int \frac{\Delta \partial^\alpha \rho \Delta \partial^\alpha u}{(\rho+1)^2} \nabla \rho - \frac{\hbar^4}{12} \int \left[\partial^\alpha, \frac{1}{\rho+1}\right] \Delta \nabla \rho \Delta \partial^\alpha u$$

$$= \sum R_{4i}.$$

For the last two terms, it can be shown that

$$R_{42} + R_{43} \leqslant C\hbar^4 \|\nabla \rho\|_\infty \|\Delta \partial^\alpha (\rho,u)\|^2 + C\hbar^4 \|\Delta \partial^\alpha u\| \|\Delta \nabla \rho\|_{L^3} \|\partial^\alpha (1/(1+\rho))\|_{L^6}$$

$$\leqslant C\hbar^4 E \|\Delta \partial^\alpha (\rho,u)\|^2 + C\hbar^4 E \|\nabla \rho\|_{\dot{H}^s}^2.$$

Using the continuity equation (2.1a) and similar to the term R_{31}, it can be shown that

$$R_{41} = -\frac{\hbar^4}{12} \int \frac{\Delta \partial^\alpha \rho}{(1+\rho)^2} \{\partial_t \partial^\alpha \Delta \rho + \partial^\alpha \Delta (u \cdot \nabla \rho) + [\partial^\alpha \Delta, 1+\rho]\mathrm{div} u\}$$

$$\leqslant -\frac{\hbar^4}{24} \frac{d}{dt} \int \frac{|\partial^\alpha \Delta \rho|^2}{(1+\rho)^2} + C\hbar^4 E \|\partial^\alpha \Delta(\rho,u)\|^2.$$

Finally, for R_5, we have

$$R_5 = \frac{\hbar^4}{12}\int \partial^\alpha \left(\frac{\mathrm{div}\{\nabla\rho \otimes \nabla\rho/(1+\rho)\}}{\rho+1}\right)\Delta\partial^\alpha u$$
$$\leqslant C\hbar^4 E\|\partial^\alpha \Delta(\rho,u)\|^2 + C\hbar^4 E\|\nabla\rho\|^2_{\dot{H}^s}.$$

Now, putting all these estimates together, and taking $\delta_0 = 1/4$, we obtain,

$$\frac{\hbar^2}{2}\frac{d}{dt}\|\nabla\partial^\alpha u\|^2 + \frac{\mu\hbar^2}{2}\|\Delta\partial^\alpha u\|^2 + \frac{(\mu+\lambda)\hbar^2}{2}\|\partial^\alpha \nabla \mathrm{div}u\|^2$$
$$+ \frac{\hbar^2}{2}\frac{d}{dt}\int \frac{\theta+1}{(1+\rho)^2}|\nabla\partial^\alpha\rho|^2 + \frac{\hbar^4}{24}\frac{d}{dt}\int \frac{|\partial^\alpha\Delta\rho|^2}{(1+\rho)^2}$$
$$\leqslant C\hbar^4 E\|\Delta\partial^\alpha u\|^2 + C\hbar^2 E\|\Delta\partial^\alpha\rho\|^2 + C\hbar^2\|\partial^\alpha\nabla\theta\|^2 + C\hbar^2 E\|\nabla(\rho,u,\theta)\|^2_{\dot{H}^s}.$$

Integrating in time over $[0,t]$ completes the proof. $\square$

Lemma 3.7 *Under the assumptions in Proposition* 3.2, *there exists some constant* $\varepsilon_0 < 1$ *sufficiently small and* $\hbar_0 < 1$, *such that*

$$(2\mu+\lambda)\beta\|\partial^\gamma\Delta\rho(t)\|^2 + \beta\int_0^t \|\partial^\gamma\Delta(\rho,\hbar\nabla\rho)(\tau)\|^2 d\tau$$
$$\leqslant C\beta\|\partial^\gamma\Delta\rho_0\|^2 + \beta\|\partial^\gamma\nabla u_0\|^2 + C\beta\|\partial^\gamma \mathrm{div}u(t)\|^2$$
$$+ C\beta\int_0^t \|\partial^\gamma\Delta\theta\|^2 d\tau + C\beta\int_0^t \|\partial^\alpha\nabla u\|^2_{L^2}d\tau + C\beta\int_0^t \|\nabla(\rho,u,\theta)\|^2_{\dot{H}^{s-1}}d\tau,$$

for any $\beta > 0$, $E \leqslant \varepsilon \leqslant \varepsilon_0$ *and* $\hbar \leqslant \hbar_0$, *where* C *depends only on* μ, λ *and some Sobolev constants. In particular,* C *does not depend on* $\hbar > 0$ *or* $t > 0$.

Proof Let γ be a multi-index such that $|\gamma| = |\alpha|-1$ and $\gamma \leqslant \alpha$. We apply ∂^γ to (2.1b) and then take the inner product of the resultant with $-\beta\partial^\gamma\nabla\Delta\rho$ to obtain

$$\sum_{i=1}^8 L_i = -\beta\int \partial^\gamma u_t \cdot \partial^\gamma\nabla\Delta\rho + \beta\mu\int \partial^\gamma\left(\frac{\Delta u}{1+\rho}\right)\partial^\gamma\nabla\Delta\rho$$
$$+ \beta(\mu+\lambda)\int \partial^\gamma\left(\frac{\nabla\mathrm{div}u}{1+\rho}\right)\partial^\gamma\nabla\Delta\rho - \beta\int \partial^\gamma(u\cdot\nabla u)\partial^\gamma\nabla\Delta\rho$$
$$- \beta\int \partial^\gamma\nabla\theta\partial^\gamma\nabla\Delta\rho - \beta\int \partial^\gamma\left(\frac{\theta+1}{1+\rho}\nabla\rho\right)\partial^\gamma\nabla\Delta\rho$$
$$+ \frac{\beta\hbar^2}{12}\int \partial^\gamma\left(\frac{\nabla\Delta\rho}{1+\rho}\right)\partial^\gamma\nabla\Delta\rho - \frac{\beta\hbar^2}{3}\int \partial^\gamma\left(\frac{\mathrm{div}\{\cdots\}}{1+\rho}\right)\partial^\gamma\nabla\Delta\rho = 0. \quad (3.26)$$

We first note that by Hölder inequality

$$L_7 = \frac{\beta\hbar^2}{12}\int\frac{|\partial^\gamma\nabla\Delta\rho|^2}{1+\rho} + \frac{\beta\hbar^2}{12}\int\left[\partial^\gamma,\frac{1}{1+\rho}\right]\nabla\Delta\rho\partial^\gamma\nabla\Delta\rho$$

$$\geq \frac{\beta\hbar^2}{18}\|\partial^\gamma\nabla\Delta\rho\|^2 - C\beta\hbar^2\|\partial^\gamma\nabla\Delta\rho\|$$

$$\cdot\left(\|\nabla\Delta\rho\|_{\dot{H}^{s-2}}\|\nabla\rho\|_{L^\infty} + \|\nabla\Delta\rho\|_{L^3}\left\|\partial^\gamma\left(\frac{1}{1+\rho}\right)\right\|_{L^6}\right)$$

$$\geq \frac{\beta\hbar^2}{36}\|\partial^\gamma\nabla\Delta\rho\|^2 - C\beta\hbar^2 E\|\nabla\rho\|_{\dot{H}^s}^2 - C\beta\hbar^2 E\|\nabla\rho\|_{\dot{H}^{s-1}}^2.$$

and by integration by parts and Hölder inequality

$$L_6 = \beta\int\frac{1+\theta}{1+\rho}\partial^\gamma\Delta\rho\partial^\gamma\Delta\rho + \beta\int\left[\partial^\gamma\nabla\cdot,\frac{1+\theta}{1+\rho}\right]\nabla\rho\partial^\gamma\Delta\rho$$

$$\geq \frac{\beta}{18}\|\partial^\gamma\Delta\rho\|^2 - C\beta\|\partial^\gamma\Delta\rho\|\left(\|\nabla\rho\|_{\dot{H}^{s-1}}\|\nabla(\rho,\theta)\|_{L^\infty} + \|\nabla\rho\|_{L^\infty}\left\|\partial^\gamma\nabla\left(\frac{1+\theta}{1+\rho}\right)\right\|\right)$$

$$\geq \frac{\beta}{36}\|\partial^\gamma\Delta\rho\|^2 - C\beta E^2\|\nabla(\rho,\theta)\|_{\dot{H}^{s-1}}^2.$$

For the term L_8, we have

$$|L_8| \leq \delta_1\beta\hbar^2\|\partial^\gamma\nabla\Delta\rho\|^2 + \frac{C\beta\hbar^2 E^2}{\delta_1}\|\nabla\rho\|_{\dot{H}^s}^2.$$

For the term L_5, we have by integration by parts

$$|L_5| = \left|\beta\int\partial^\gamma\Delta\theta\partial^\gamma\Delta\rho\right| \leq \beta\delta_1\|\partial^\gamma\Delta\rho\|^2 + \frac{\beta}{4\delta_1}\|\partial^\gamma\Delta\theta\|^2.$$

For the term L_4, we have by integration by parts

$$|L_4| = \left|\beta\int\nabla\partial^\gamma(u\cdot\nabla u)\partial^\gamma\Delta\rho\right|$$

$$\leq \beta\|\partial^\gamma\Delta\rho\|(\|u\|_{L^\infty}\|\nabla u\|_{\dot{H}^s} + \|\nabla u\|_{\dot{H}^{s-1}}\|\nabla u\|_{L^\infty})$$

$$\leq \beta\delta_1\|\partial^\gamma\Delta\rho\|^2 + \frac{C\beta E^2}{\delta_1}\|\nabla u\|_{\dot{H}^{s-1}}^2.$$

For the term L_3, we have by integration by parts,

$$(\mu+\lambda)^{-1}L_3 = -\beta\int\partial^\gamma\nabla\cdot\left(\frac{\nabla\operatorname{div}u}{1+\rho}\right)\partial^\gamma\Delta\rho$$

$$= -\beta\int\frac{\partial^\gamma\Delta\operatorname{div}u}{1+\rho}\partial^\gamma\Delta\rho - \beta\int\left[\partial^\gamma\nabla,\frac{1}{1+\rho}\right]\nabla\operatorname{div}u\partial^\gamma\Delta\rho$$

$$= -\beta\int\frac{\partial^\gamma\Delta((1+\rho)\operatorname{div}u)}{(1+\rho)^2}\partial^\gamma\Delta\rho + \beta\int\frac{[\partial^\gamma\Delta,1+\rho]\operatorname{div}u}{(1+\rho)^2}\partial^\gamma\Delta\rho$$

$$-\beta\int\left[\partial^\gamma\nabla,\frac{1}{1+\rho}\right]\nabla\operatorname{div}u\partial^\gamma\Delta\rho =: L_{31} + L_{32} + L_{33}.$$

For the first term L_{31}, using (2.1a), it is easy to show by integration by parts,

$$L_{31} = \frac{\beta}{2}\frac{d}{dt}\int \frac{|\partial^\gamma \Delta \rho|^2}{(1+\rho)^2} + \beta \int \frac{\rho_t|\partial^\gamma \Delta \rho|^2}{(1+\rho)^3} - \frac{\beta}{2}\int \mathrm{div}\left(\frac{u}{(1+\rho)^2}\right)|\partial^\gamma \Delta \rho|^2$$
$$+ \beta \int \frac{[\partial^\gamma \Delta, u]\nabla\rho}{(1+\rho)^2}\partial^\gamma \Delta \rho$$
$$\geqslant \frac{\beta}{2}\frac{d}{dt}\int \frac{|\partial^\gamma \Delta \rho|^2}{(1+\rho)^2} - C\beta E\|\partial^\gamma \Delta(\rho, u)\|^2,$$

thanks to (3.3) with $p = \infty$ and Lemma 2.3. By commutator estimates,

$$|L_{32}| \leqslant C\beta\|\partial^\gamma \Delta u\|(\|\nabla \rho\|_{L^\infty}\|\mathrm{div} u\|_{\dot{H}^s} + \|\partial^\gamma \Delta \rho\|\|\mathrm{div} u\|_{L^\infty})$$
$$\leqslant C\beta E\|\nabla(\rho, u)\|_{\dot{H}^s}^2,$$

and

$$|L_{33}| \leqslant C\beta\|\partial^\gamma \Delta \rho\|(\|\nabla \mathrm{div} u\|_{\dot{H}^{s-1}}\|\nabla \rho\|_{L^\infty} + \|\nabla \mathrm{div} u\|_{L^3}\|\rho\|_{\dot{H}^{s,6}})$$
$$\leqslant C\beta E\|\nabla(\rho, u)\|_{\dot{H}^s}^2.$$

The term L_2 can be treated similarly and will lead to

$$L_2 \geqslant \frac{\beta\mu}{2}\frac{d}{dt}\int \frac{|\partial^\gamma \Delta \rho|^2}{(1+\rho)^2} - C\beta E\|\nabla(\rho, u)\|_{\dot{H}^s}^2.$$

Now, we focus on L_1. Integrating L_1 in time over $[0, t]$, we have

$$\beta\int_0^t L_1(s)ds = \beta\int_0^t \int \partial^\gamma \mathrm{div} u_t \partial^\gamma \Delta \rho d\tau$$
$$= \beta\int_{\mathbb{R}^3} \partial^\gamma \mathrm{div} u \partial^\gamma \Delta \rho dx \Big|_0^t - \beta\int_0^t \int_{\mathbb{R}^3} \partial^\gamma \mathrm{div} u \partial^\gamma \Delta \rho_t dx d\tau$$
$$=: I(t) - I(0) - \beta\int_0^t L_{12}(s)ds.$$

By direct estimates, we have

$$|I(t)| \leqslant \delta_1 \beta\|\partial^\gamma \Delta \rho(t)\|^2 + \frac{\beta}{\delta_1}\|\partial^\gamma \mathrm{div} u(t)\|^2,$$
$$|I(0)| \leqslant \beta\|\partial^\gamma \Delta \rho_0\|^2 + \beta\|\partial^\gamma \nabla u_0\|^2.$$

For $\beta\int_0^t L_{12}(s)ds$, we have by integration by parts and (2.1a),

$$-\beta \int_0^t L_{12}(s)ds = -\beta \int_0^t \int_{\mathbb{R}^3} \nabla \partial^\gamma \mathrm{div} u \nabla \partial^\gamma (u \cdot \nabla \rho) dx ds$$

$$-\beta \int_0^t \int_{\mathbb{R}^3} \nabla \partial^\gamma \mathrm{div} u \nabla \partial^\gamma [(1+\rho)\mathrm{div} u] dx ds$$

$$\leqslant C\beta \int_0^t \|\nabla \partial^\gamma \mathrm{div} u\|(\|u\|_{L^\infty}\|\nabla \rho\|_{\dot{H}^s} + \|u\|_{\dot{H}^{s,6}}\|\nabla \rho\|_{L^3})d\tau$$

$$+ C\beta \int_0^t \|\nabla \partial^\gamma \mathrm{div} u\|(\|\mathrm{div} u\|_{\dot{H}^s} + \|\mathrm{div} u\|_{L^3}\|\rho\|_{\dot{H}^{s,6}})d\tau$$

$$\leqslant C\beta \int_0^t \|\nabla \partial^\gamma \mathrm{div} u\|^2 d\tau + C\beta E^2 \int_0^t \|\nabla(\rho, u)\|_{\dot{H}^s}^2 d\tau.$$

Now, we fix some δ_1 small, depending on μ and λ, such that $\delta_1 = \min\{(2\mu + \lambda)/18, 1/142\}$, where C is the constant appearing in the above estimates. Then for such δ_1, fix some small ε_0 such that $\varepsilon_0 < \min\{1, 1/(72C)\}$. By integrating (3.26) in time over $[0, t]$, and then taking $E < \varepsilon_0$ small, we complete the proof. □

Now, we prove Proposition 3.2.

Proof of Proposition 3.2 Adding the estimates in Lemmas 3.4-3.7 together. First, we take $\delta_0 = \nu_0/4(\mu + \lambda)$, and then $\hbar_0$ small such that $\hbar_0^2 = \min\{\delta_0 \nu_0 \kappa/4C, \kappa/4C, 1\}$, then choose β small such that $C\beta \leqslant C\beta_0 := \min\{\kappa/4, \nu_0/4, 1/4\}$, and then for such fixed β, we choose ε_0 small such that $C\varepsilon_0 \leqslant \min\{\nu_0/4, \kappa/4, \beta/4\}$, then there holds,

$$\|\partial^\alpha(\rho, u, \theta, \hbar\nabla\rho)(t)\|^2 + \hbar^2\|\nabla\partial^\alpha(u, \hbar\nabla\rho)(t)\|^2 + \beta\|\partial^\gamma \Delta\rho(t)\|^2$$

$$+ \nu_0 \int_0^t \|\partial^\alpha \nabla(u, \theta, \hbar\nabla u)(\tau)\|^2 d\tau + \beta \int_0^t \|\partial^\gamma \Delta(\rho, \hbar\nabla\rho)(\tau)\|^2 d\tau$$

$$\leqslant C\|\partial^\alpha(\rho, u, \theta, \hbar\nabla\rho)(0)\|^2 + C\hbar^2\|\nabla\partial^\alpha(u, \hbar\nabla\rho)(0)\|^2 + C\|\partial^\gamma \Delta\rho_0\|^2$$

$$+ C\int_0^t \|\nabla(\rho, u, \theta)\|_{\dot{H}^{s-1}}^2 + C\hbar^2 \int_0^t \|\nabla(\rho, u)(\tau)\|_{\dot{H}^s}^2 d\tau,$$

for all $E \leqslant \varepsilon_0$, where C is independent of t. Rephrasing this in the $||| \cdot |||$ norm, we obtain

$$|||\partial^\alpha(\rho, u, \theta)(t)|||^2 + \nu_0 \int_0^t \|\partial^\alpha \nabla(\rho, u, \theta, \hbar\nabla\rho, \hbar\nabla u)(\tau)\|^2 d\tau$$

$$\leqslant C|||\partial^\alpha(\rho, u, \theta)(0)|||^2 + C\int_0^t \|\nabla(\rho, u, \theta)\|_{\dot{H}^{s-1}}^2 + C\hbar^2 \int_0^t \|\nabla(\rho, u)(\tau)\|_{\dot{H}^s}^2 d\tau.$$

The proof is complete. □

Now, we are ready to obtain *a priori* estimate for the solution of (2.1).

Theorem 3.4 *Suppose that for some $T > 0$, $(\rho, u, \theta) \in \mathcal{E}_3(0, T)$ is a solution of (2.1) satisfying $E \leqslant \max_{0 \leqslant t \leqslant T} |||(\rho, u, \theta)(\tau)|||_3 \leqslant \varepsilon$. Then there exists some $\varepsilon_0 > 0$, $\nu_0 = \nu_0(\varepsilon_0) > 0$ and $C_0 = C_0(\varepsilon_0, \nu_0)$ such that*

$$|||(\rho, u, \theta)(t)|||_3^2 + \nu_0 \int_0^t \|D(\rho, u, \theta, \hbar\nabla\rho, \hbar\nabla u)(\tau)\|_{H^3}^2 d\tau \leqslant C |||(\rho, u, \theta)(0)|||_3^2,$$

for any $(\rho, u, \theta)(t)$ satisfying $E \leqslant \varepsilon_0$.

Proof Add (3.21) for $|\alpha| = 1$ to $\left(\dfrac{C}{\nu_0} + 1\right)$ times (3.8) to obtain

$$|||(\rho, u, \theta)(t)|||_1^2 + \nu_0 \int_0^t \|D(\rho, u, \theta, \hbar\nabla u, \hbar\nabla\rho)(s)\|_{H^1}^2 ds \leqslant C |||(\rho, u, \theta)(0)|||_1^2. \quad (3.27)$$

Then, we add (3.21) for $|\alpha| = 2$ to $\left(\dfrac{C}{\nu_0} + 1\right)$ times (3.27) to obtain

$$|||(\rho, u, \theta)(t)|||_2^2 + \nu_0 \int_0^t \|D(\rho, u, \theta, \hbar\nabla u, \hbar\nabla\rho)(s)\|_{H^2}^2 ds \leqslant C |||(\rho, u, \theta)(0)|||_2^2. \quad (3.28)$$

Again, we add (3.21) for $|\alpha| = 3$ to $\left(\dfrac{C}{\nu_0} + 1\right)$ times (3.28), to obtain

$$|||(\rho, u, \theta)(t)|||_3^2 + \nu_0 \int_0^t \|D(\rho, u, \theta, \hbar\nabla u, \hbar\nabla\rho)(s)\|_{H^3}^2 ds \leqslant C |||(\rho, u, \theta)(0)|||_3^2, \quad (3.29)$$

which is the desired estimate, completing the proof. $\square$

Now, we prove Theorem 2.2.

Proof of Theorem 2.2 The proof is easy by combining the estimates in Theorem 3.4, continuity method and the local existence result. $\square$

4 Proof of Theorem 2.3

The following estimate was obtained in [20]. Under the smallness assumption,

$$\|(\rho, u, \theta)(t)\|_3^2 + \nu_0 \int_0^t \|D\rho(\tau)\|_2^2 + \|D(u, \theta)(\tau)\|_3^2 d\tau \leqslant C_0 \|(\rho, u, \theta)(0)\|_3^2 \quad (4.1)$$

and the following density estimate

$$\|D^4\rho(t)\|^2 - C\|D^3 u(t)\|^2 + \nu_0 \int_0^t \|D^3\rho(\tau)\|^2 d\tau$$

$$\leqslant C\|\rho_0\|_4^2 + C\|u_0\|_3^2 + C \int_0^t \|D^4(u, \theta)(\tau)\|^2 + \|D(\rho, u, \theta)(\tau)\|_2^2 d\tau. \quad (4.2)$$

Adding (4.2) to $\left(C + \dfrac{C}{\nu_0} + 1\right)$ times (4.1), one obtains

$$\|(\rho, u, \theta)(t)\|_3^2 + \|D^4\rho(t)\|^2 + \nu_0 \int_0^t \|D(\rho, u, \theta)(\tau)\|_3^2 d\tau \leqslant C_0 \|(\rho, u, \theta)(0)\|_3^2 + C\|\rho_0\|_4^2. \tag{4.3}$$

Now we turn to prove Theorem 2.3. To clearly specify the dependence of the solution on the parameter $\hbar$, we denote $(\rho^\hbar, u^\hbar, \theta^\hbar)$ the solution of the system (2.1) and (ρ^0, u^0, θ^0) the solution to (2.3). First of all, from Theorem 2.2 and (4.3), we have the estimate

$$\|(\rho^\hbar, u^\hbar, \theta^\hbar)(t)\|_3^2 + \|D^4\rho^\hbar(t)\|^2 + \hbar^2\|(\rho^\hbar, u^\hbar)(t)\|_4^2 + \hbar^4\|\rho^\hbar(t)\|_5^2$$
$$\leqslant C(\|(\rho_0, u_0, \theta_0)(0)\|_3^2 + \|(\rho_0, u_0)(0)\|_4^2 + \|\rho_0\|_5^2), \tag{4.4}$$

and

$$\|(\rho^0, u^0, \theta^0)(t)\|_3^2 + \|D^4\rho^0(t)\|^2 \leqslant C(\|(\rho_0, u_0, \theta_0)\|_3^2 + C\|\rho_0\|_4^2). \tag{4.5}$$

Now, we let
$$N = \rho^\hbar - \rho^0, \quad U = u^\hbar - u^0, \quad \Theta = \theta^\hbar - \theta^0.$$

Then (N, U, Θ) satisfy

$$\begin{cases}
\partial_t N + \nabla((1 + \rho^0)U + N u^0) = 0, \tag{4.6a}\\[4pt]
\partial_t U - \dfrac{\mu}{\rho^\hbar + 1}\Delta U - \left(\dfrac{\mu}{\rho^\hbar + 1} - \dfrac{\mu}{\rho^0 + 1}\right)\Delta u^0 - \dfrac{\mu + \lambda}{\rho^\hbar + 1}\nabla \text{div} U \\[4pt]
\quad - \left(\dfrac{\mu + \lambda}{\rho^\hbar + 1} - \dfrac{\mu + \lambda}{\rho^0 + 1}\right)\nabla \text{div} u^0 = -u^\hbar \cdot \nabla U - U \cdot \nabla u^0 - \nabla\Theta \\[4pt]
\quad - \dfrac{\Theta}{\rho^\hbar + 1}\nabla\rho^\hbar - \dfrac{\theta^0 + 1}{\rho^0 + 1}\nabla N - \left(\dfrac{\theta^0 + 1}{\rho^\hbar + 1} - \dfrac{\theta^0 + 1}{\rho^0 + 1}\right)\nabla\rho^0 \\[4pt]
\quad + \dfrac{\hbar^2}{12}\dfrac{\Delta\nabla\rho^\hbar}{\rho^\hbar + 1} - \dfrac{\hbar^2}{3}\dfrac{\text{div}(\nabla\sqrt{\rho^\hbar + 1} \otimes \nabla\sqrt{\rho^\hbar + 1})}{\rho^\hbar + 1}, \tag{4.6b}\\[6pt]
\partial_t \Theta - \dfrac{2\kappa}{3(1 + \rho^\hbar)}\Delta\Theta - \dfrac{2\kappa}{3}\left(\dfrac{1}{1 + \rho^\hbar} - \dfrac{1}{1 + \rho^0}\right)\Delta\theta^0 = -u^0 \cdot \nabla\Theta - U \cdot \nabla\theta^0 \\[4pt]
\quad - \dfrac{2}{3}(\theta^\hbar + 1)\nabla \cdot U - \dfrac{2}{3}\Theta\nabla \cdot u^0 + \dfrac{\hbar^2}{36(1 + \rho^\hbar)}\text{div}((1 + \rho^\hbar)\Delta u^\hbar) \\[4pt]
\quad + \dfrac{2}{3(1 + \rho^\hbar)}\left\{\dfrac{\mu}{2}(\nabla(u^\hbar + u^0) + (\nabla(u^\hbar + u^0))^T)(\nabla U + (\nabla U)^T)\right. \\[4pt]
\quad \left. + \lambda(\text{div}(u^\hbar + u^0)\text{div} U\right\} \\[4pt]
\quad + \dfrac{2}{3}\left(\dfrac{1}{1 + \rho^\hbar} - \dfrac{1}{1 + \rho^0}\right)\left\{\dfrac{\mu}{2}|\nabla u^0 + (\nabla u^0)^T|^2 + \lambda(\text{div} u^0)^2\right\}. \tag{4.6c}
\end{cases}$$

Now, we multiply (4.6) with N, U and Θ, respectively, integrate the resultant over $\mathbb{R}^3$ and then sum them up to obtain an energy inequality. Among the many terms, we only treat the following three typical thems in the following. First, for the viscosity term, we have

$$-\int \frac{\mu}{\rho^\hbar + 1}\Delta U U = \int \frac{\mu}{\rho^\hbar + 1}|\nabla U|^2 - \int \frac{\mu \nabla \rho^\hbar}{(\rho^\hbar + 1)^2}\nabla U U$$
$$\geqslant \frac{\mu}{2}\int |\nabla U|^2 - C\|U\|^2,$$

where the constant C depends on μ and the H^3 norm of $\rho^\hbar$. Secondly, for the last term on the left of (4.6b), we have

$$-\int \left(\frac{\mu+\lambda}{\rho^\hbar+1} - \frac{\mu+\lambda}{\rho^0+1}\right)\nabla \mathrm{div} u^0 U = \int \left(\frac{(\mu+\lambda)N}{(\rho^\hbar+1)(\rho^0+1)}\right)\nabla \mathrm{div} u^0 U$$
$$\leqslant C\|\nabla \mathrm{div} u^0\|_{L^3}\|N\|_{L^2}\|U\|_{L^6}$$
$$\leqslant \frac{\mu}{8}\|\nabla U\|^2 + C\|(N,U)\|_{L^2}^2,$$

where the constant C depends on μ and the H^3 norm of $(\rho^\hbar, \rho^0, u^0)$. Thirdly, for the second to the last term on the RHS of (4.6b), we have

$$\frac{\hbar^2}{12}\int \frac{\Delta \nabla \rho^\hbar}{\rho^\hbar + 1} U \leqslant \frac{\mu}{8}\|U\|^2 + \frac{C\hbar^4}{\mu}\|\rho^\hbar\|_{H^2}^2 \leqslant \frac{\mu}{8}\|U\|^2 + C\hbar^4,$$

where the constant C may depend on μ and the H^2 norm of $\rho^\hbar$. The other terms, either depending linearly on the difference N, U or Θ, or depending on the small parameter $\hbar^2$, can be estimated similarly. Therefore, we finally obtain after long but standard estimates

$$\frac{1}{2}\frac{d}{dt}\|(N,U,\Theta)(t)\|_{L^2}^2 + \nu\|\nabla(U,\Theta)\|^2 \leqslant C\|(N,U,\Theta)(t)\|^2 + C\hbar^4, \qquad (4.7)$$

where ν depends on the parameters μ and κ, and C depends on the H^3 norm of $\|(\rho^\hbar, u^\hbar, \theta^\hbar)\|$.

Similarly, taking inner product with $\Delta(N, U, \Theta)$, one can obtain

$$\frac{1}{2}\frac{d}{dt}\|\nabla(N,U,\Theta)(t)\|_{L^2}^2 + \nu\|\Delta(U,\Theta)\|^2 \leqslant C\|\nabla(N,U,\Theta)(t)\|^2 + C\hbar^4, \qquad (4.8)$$

where ν also depends on the parameters μ and κ, and C depends on the H^3 norm of $\|(\rho^\hbar, u^\hbar, \theta^\hbar)\|$. In the derivation of this inequality, we have the following term,

$$\int \frac{\hbar^2}{36(1+\rho^\hbar)}\mathrm{div}((1+\rho^\hbar)\Delta u^\hbar)\Delta\Theta \leqslant \frac{\kappa}{4}\|\Delta\Theta\|^2 + C\hbar^4\|\mathrm{div}((1+\rho^\hbar)\Delta u^\hbar)\|^2$$
$$\leqslant \frac{\kappa}{4}\|\Delta\Theta\|^2 + C\hbar^4,$$

where C depends on κ and the H^3-norm of ρ and Θ. This procedure can not be proceeded into the higher norms, since the estimates then depends on the H^4-norm of $u^\hbar$.

Combining the two inequalities (4.7) and (4.8), we obtain

$$\frac{1}{2}\frac{d}{dt}\|(N,U,\Theta)(t)\|_{H^1}^2 + \nu\|\nabla(U,\Theta)\|_{H^1}^2 \leqslant c_1\|(N,U,\Theta)(t)\|_{H^1}^2 + c_2\hbar^4,$$

which implies, thanks to the Gronwall inequality, that

$$\|(N,U,\Theta)(t)\|_{H^1}^2 \leqslant \left[c_2 e^{c_1 t}/c_1\right]\hbar^4.$$

In particular, we note that c_1 and c_2 are independent of $\hbar$.

We also remark that we can improve the convergence to the H^2-norm of (N,U,Θ) at the price of losing the decay rate. To be precise, we take the inner product of the system (4.6) with $\Delta^2(N,U,\Theta)$ to obtain an energy inequality. Among the terms, we consider the typical term

$$\int \frac{\hbar^2 \operatorname{div}((1+\rho^\hbar)\Delta u^\hbar)}{36(1+\rho^\hbar)}\Delta^2\Theta = -\int \nabla\left(\frac{\hbar^2 \operatorname{div}((1+\rho^\hbar)\Delta u^\hbar)}{36(1+\rho^\hbar)}\right)\nabla\Delta\Theta$$

$$\leqslant \frac{\kappa}{4}\|\nabla\Delta\Theta\|^2 + \hbar^4\Big(\|\nabla\operatorname{div}\Delta u^\hbar\|_{L^2}^2 + \|\Delta\rho^\hbar\|_{L^6}^2\|\Delta u^\hbar\|_{L^3}^2$$

$$+ \|\nabla\rho^\hbar\|_{L^\infty}^2\|\operatorname{div}\Delta u^\hbar\|_{L^2}^2 + \|\nabla\rho^\hbar\|_{L^\infty}^4\|\Delta u^\hbar\|_{L^2}^2\Big)$$

$$\leqslant \frac{\kappa}{4}\|\nabla\Delta\Theta\|^2 + \hbar^2\Big(\|\hbar\nabla\operatorname{div}\Delta u^\hbar\|_{L^2}^2 + (1+\|\rho^\hbar\|_{H^3}^4)\|u^\hbar\|_{H^3}^2\Big)$$

$$\leqslant \frac{\kappa}{4}\|\nabla\Delta\Theta\|^2 + C\hbar^2,$$

thanks to the estimate in (4.4). The other terms, either depending linearly on (N,U,Θ) or depending on the Planck constant $\hbar$, can be treated similarly by making use of (4.4). Finally, we arrive at the inequality

$$\frac{1}{2}\frac{d}{dt}\|(N,U,\Theta)(t)\|_{H^2}^2 + \nu\|\nabla(U,\Theta)\|_{H^2}^2 \leqslant c_1\|(N,U,\Theta)(t)\|_{H^2}^2 + c_2\hbar^2,$$

which implies, thanks to the Gronwall inequality, that

$$\|(N,U,\Theta)(t)\|_{H^2}^2 \leqslant \left[c_2 e^{c_1 t}/c_1\right]\hbar^2.$$

In particular, c_1 and c_2 are independent of $\hbar$. This completes the proof.

References

[1] M.G. Ancona and G.J. Iafrate. Quantum correction to the equation of state of an electron gas in semiconductor. Phys. Rev. B, 39, (1989)9536-9540.

[2] M.G. Ancona and H.F. Tiersten. Macroscopic physics of the silicon inversion layer. Phys. Rev. B, 35, (1987)7959-7965.

[3] D. Bian, L. Yao and C. Zhu, Vanishing capillarity limit of the compressible fluid models of Korteweg type to the Navier-Stokes equations. SIAM J. Math. Anal., 46(2), (2014)1633-1650.

[4] D. Bohm. A suggested interpretation of the quantum theory in terms of "hidden" valuables: I; II. Phys. Rev., 85, (1952)166-179; 180-193.

[5] J.E. Dunn and J. Serrin. On the thermodynamics of interstitial working. Arch. Ration. Mech. Anal., 88, (1985)95-133.

[6] R. Feynman. Statistical Mechanics, a Set of Lectures. New York: W.A. Benjamin, 1972.

[7] C.L. Gardner. The quantum hydrodynamic model for semiconductor devices. SIAM J. Appl. Math., 54(2), (1994)409-427.

[8] F. Haas. Quantum plasmas: An hydrodynamic approach, Springer. New York, 2011.

[9] H. Hattori, D. Li. Solutions for two-dimensional system for materials of Korteweg type. SIAM J. Math. Anal., 25(2), (1994)85-98.

[10] H. Hattori, D. Li. Global solutions of a high dimensional system for Korteweg materials. J. Math. Anal. Appl., 198, (1996)84-97.

[11] A. Jungel. Global weak solutions to compressible Navier-Stokes equations for quantum fluids. SIAM J. Math. Anal., 42(3), (2010)1025-1045.

[12] A. Jungel, C.-K. Lin and K.-C. Wu. An asymptotic limit of a Navier-Stokes system with capillary effects, Comm. Math. Phys., 329, (2014)725-744.

[13] A. Jungel, J.-P. Milisic. Full compressible Navier-Stokes equations for quantum fluids: derivation and numerical solution. Kinet. Relat. Models, 4(3), (2011)785-807.

[14] L. Hsiao and H. Li. The well-posedness and asymptotics of multi-dimensional quantum hydrodynamics. Acta Math. Sci., 29B(3), (2009)552-568.

[15] T. Kato and G. Ponce. Commutator estimates and the Euler and Navier-Stokes equations, Comm. Pure Appl. Math., 41, (1988)891-907.

[16] D. Korteweg. Sur la forme que prennent les équations du mouvement des fluides si l'on tient compte des forces capillaires par des variations de densité. Arch. Néer. Sci. Exactes Sér, II 6, (1901)1-24.

[17] H. Li and C.K. Lin. Zero Debye length asymptotic of the quantum hydrodynamic model for semiconductors. Comm. Math. Phys., 256(1), (2005)195-212.

[18] H. Li and P. Marcati. Existence and asymptotic behavior of multi-dimensional quantum hydrodynamic model for semiconductors. Comm. Math. Phys., 245(20), (2004)215-247.

[19] H. Li and P. Markowich. A review of hydrodynamical models for semiconductors: asymptotic behavior. Bol. Soc. Brasil Mat., 32(3), (2001)321-342.

[20] A. Matsumura, T. Nishida. The initial value problem for the equations of motion of viscous and heat-conductive gases. J. Math. Kyoto Univ., 20-1, (1980)67-104.

[21] L. Nirenberg. On elliptic partial differential equations. Ann. Scuola Norm. Sup. Pisa, 13(3), (1959)115-162.

[22] X. Pu. Dispersive limit of the Euler-Poisson system in higher dimensions. SIAM J. Math. Anal., 45(2), (2013)834-878.

[23] X. Pu and B. Guo. Global existence and convergence rates of smooth solutions for the full compressible MHD equations. Z. Angew. Math. Phys., 64, (2013)519-538.

[24] Y. Wang and Z. Tan. Optimal decay rates for the compressible fluid model of Korteweg type. J. Math. Anal. Appl., 379, (2011)256-271.

[25] E. Wigner. On the quantum correction for thermodynamic equilibrium. Phys. Rev., 40, (1932)749-759.

Dynamics for a Generalized Incompressible Navier-Stokes Equations in R²*

Guo Boling (郭柏灵), Huang Daiwen (黄代文), Li Qiaoxin (李巧欣) and Sun Chunyou (孙春友)

Abstract In this paper, we consider the dynamics for a damped generalized incompressible Navier-Stokes equations defined on $\mathbb{R}^2$. The generalized Navier-Stokes equations here refer to the equations obtained by replacing the Laplacian in the classical Navier-Stokes equations by the more general operator $(-\Delta)^\alpha$ with $\alpha \in \left(\frac{1}{2}, 1\right)$. We prove the rate of dissipation of enstrophy vanishes as $\nu \to 0^+$, where ν is the viscosity parameter. Moreover, we prove the existence and finite dimensionality of a global attractor in $(H^1(\mathbf{R}^2))^2$ as $\nu > 0$ is kept fixed for the generalized Navier-Stokes equations.

Keywords generalized Navier-Stokes equations; inviscid limit; global attractor

1 Introduction

We consider the damped and driven generalized incompressible Navier-Stokes (GNS) equations in $\mathbf{R}^2$,

$$\begin{cases} \partial_t u + \nu(-\Delta)^\alpha u + u \cdot \nabla u + \gamma u + \nabla p = f, \\ \nabla \cdot u = 0, \\ u(x,0) = u_0(x), \end{cases} \tag{1.1}$$

where $\alpha \in \left(\frac{1}{2}, 1\right)$, $\gamma > 0$ is a fixed damping coefficient, and the coefficient $\nu > 0$ is a parameter that we will let vary. The force f is given and time-independent, and the initial velocity is divergence-free and belongs to $(H^1(\mathbf{R}^2))^2$. The fractional Laplacian

---

* Adv. Nonlinear Stud., 2016, 16(2): 249–272. DOI:10.1515/ans-2015-5018.

$\Lambda^{2\alpha} = (-\Delta)^{\alpha}$ is defined in terms of the Fourier transform

$$\widehat{\Lambda^{2\alpha}\phi}(\xi) = |\xi|^{2\alpha}\widehat{\phi}(\xi),$$

where $\widehat{\phi}(\xi) = \int_{\mathbb{R}^2} \phi(x)e^{-ix\cdot\xi}dx$ and $\Lambda = (-\Delta)^{\frac{1}{2}}$.

The fractional Laplacian operator appears in a wide class of physical systems and engineering problems, including Lévy flights, stochastic interfaces and anomalous diffusion problems. In fluid mechanics, the fractional Laplacian is often applied to describe many complicated phenomenons via partial differential equations. Equations (1.1) are generalizations of the classical Navier-Stokes equations.

When $\alpha = 1$, the GNS equations (1.1) reduce to the usual Navier-Stokes equations. One advantage of working with the GNS equations is that they allow simultaneous consideration of their solutions corresponding to a range of α's. The well-posedness and regularity about the GNS equations in different spaces have been studied by many authors, see, for example, Wu [22, 23], Li and Zhai [15], Wu and Fan [24], Zhang and Fang [25] and so on.

In this paper, we mainly consider the dynamics of the solutions of (1.1). More precisely, we accomplish two major goals. First, as $\nu \to 0^+$, in spirit of [5, 12], we consider the zero viscosity limit of long time averages of vorticity. Second, as $\nu > 0$ is kept fixed, we prove the existence and finite dimensionality of a compact global attractor for the dynamical systems generated by (1.1) in $(H^1(\mathbf{R}^2))^2$.

In Section 2, we recall the notation and properties that we will used throughout this paper. Especially, we give a simple proof about the existence and uniqueness of the solution for the GNS equations (1.1), see *Lemma* 2.1.

Principal substantive questions related to turbulence have been raised since the beginning of the twentieth century, a large number of empirical and heuristical results were derived, among them the works of Lamb [14] on addressing idealized inviscid flows, and Taylor [19] on viscous flows. Anomalous dissipation is important in revealing the turbulence, see [5, 6, 12, 13] for the details. Anomalous dissipation of energy in three-dimensional turbulence is one of the basic statements of physical theory, which is open mostly in mathematics. Recently, Constantin *et al.* have proved in [5] the absence of anomalous dissipation of enstrophy for two-dimensional forced damped Navier-Stokes equations, and in [6] the absence of anomalous dissipation of energy for forced surface quasi-geostrophic equations. In Section 3, following the blueprint in Constantin and Ramos [5], we prove the absence of anomalous dissipation of enstrophy for the two-dimensional GNS equations (1.1), that is, the following theorem.

Theorem 1.1 *Let $u_0 \in (L^2(\mathbf{R}^2))^2$ be divergence-free and $\nabla^\perp \cdot u_0 = \omega_0 \in L^1(\mathbf{R}^2) \cap L^\infty(\mathbf{R}^2)$ and $f \in W^{1,1}(\mathbf{R}^2) \cap W^{1,\infty}(\mathbf{R}^2)$. Let $S^{(\nu)}(t, \omega_0)$ be the vorticity of the solution of the damped and driven generalized Navier-Stokes equations. Then*

$$\lim_{\nu \to 0} \nu \left(\limsup_{t \to \infty} \frac{1}{t} \int_0^t \|\Lambda^\alpha S^{(\nu)}(s + t_0, \omega_0)\|_{L^2(\mathbf{R}^2)}^2 ds \right) = 0$$

holds for any $t_0 > 0$.

Theorem 1.1 shows that a same result as that in [5] holds when we relax the exponent $\alpha = 1$ to $\alpha \in \left(\frac{1}{2}, 1\right)$. However, it is unknown for the case $\alpha \leqslant \frac{1}{2}$.

The main notation we used in the third section is the so called stationary statistical solutions, which are measures in the phase space and naturally arise as long time average of solutions. From the pioneering works of [10–12], we know that the stationary statistical solutions are supported in the corresponding global attractor. Hence, in Section 4, we will concentrate on studying the asymptotic behaviors of solutions for system (1.1). More precisely, we consider the existence and finite dimensionality of a compact global attractor in $(H^1(\mathbf{R}^2))^2$ as $\nu > 0$ is kept fixed for the GNS equations (1.1). The main result in this section is the following theorem.

Theorem 1.2 *Let $\alpha \in \left(\frac{1}{2}, 1\right)$, $\nu, \gamma > 0$ and $f \in (W^{1,2}(\mathbf{R}^2))^2$ be time-independent. Then, the semigroup $\{S(t)\}_{t \geqslant 0}$ generated by the solutions of (1.1) has a unique global attractor $\mathscr{A}$ in V; that is, $\mathscr{A}$ is compact, invariant in V and attracts every bounded subset of V in the V-norm. Moreover, the fractal dimension of $\mathscr{A}$ is finite in V.*

Since $\nu > 0$ is kept fixed and (1.1) has certain smoothing effect, we can get easily some regularity estimates, which, combined with the *tail estimate* technique introduced in Wang [21], allow us to deduce the existence of a compact global attractor directly. Nevertheless, the finite dimensionality requires much more than just regularity, especially for the systems defined on unbounded domains that lack compact embeddings. In this part, we apply a criterion originally introduced in Chueshov and Lasiecka [3] for the hyperbolic equation, but with a combination of the idea of *l-trajectory* in Malek and Prazak [17], see *Theorem* 4.2 below. Note that, such method allows us to deduce the finite dimensionality of the attractor only via the energy estimates and tail estimates that are already established for the existence of the attractor.

2 Preliminaries

2.1 Notation and solution

Set
$$\mathcal{V} = \{u \in (C_c^\infty(\mathbf{R}^2))^2 \mid \operatorname{div} u = 0\}.$$

For each $s \in [0, 2]$, define H^s as the completion of $\mathcal{V}$ with respect to the norm $\|\cdot\|_s$, where, for any $u \in \mathcal{V}$,
$$\|u\|_s := \left(\int_{\mathbf{R}^2} (1+|\xi|^s)|\hat{u}(\xi)|^2 d\xi\right)^{\frac{1}{2}};$$

Especially, to simplify the notation we denote
$$H = H^0, \quad \|\cdot\| = \|\cdot\|_0 \text{ with inner products } (u,v) = \int_{\mathbf{R}^2} u \cdot v dx, \ u, v \in H^0,$$
$$V = H^1, \quad V' = \text{for the dual space of } V.$$

Then, the mathematical framework of (1.1) is now classical, and we consider the following weak formulation of (1.1): find
$$u \in L^\infty(0, T; V) \cap L^2(0, T; H^{2\alpha}) \text{ for all } T > 0, \tag{2.1}$$

such that
$$\frac{d}{dt}(u,v) + \nu(\Lambda^\alpha u, \Lambda^\alpha v) + \gamma(u,v) + b(u,u,v) = (f,v) \text{ for all } v \in V, \ t > 0 \tag{2.2}$$

and
$$u(x, 0) = u_0, \tag{2.3}$$

where $b : V \times V \times V \to \mathbf{R}$ is given by
$$b(u, v, w) = \sum_{i,j=1}^{2} \int_{\mathbf{R}^2} u_i \frac{\partial v_j}{\partial x_i} w_j dx.$$

We have the following result.

Lemma 2.1 Let $\nu, \gamma > 0$, $\alpha \in \left(\frac{1}{2}, 1\right)$ and $f \in (W^{1,2}(\mathbf{R}^2))^2$. Then for each initial data $u_0 \in V$ and any $T > 0$, there exists a unique $u \in L^\infty(0, T; V)$ such that (2.2) and (2.3) hold. Moreover, $u' \in L^2(0, T; H)$ and $u \in C([0, T]; V)$.

This existence and uniqueness indeed can be seen as a special result of Wu [22, 23]. Here, since we can use the vorticity equation in $\mathbf{R}^2$, we give a simple proof as follows.

Moreover, since our initial data and forcing term are nice, the definition about the solution given in (2.1)-(2.3) is different, e.g., compared with [22, 23].

Proof The existence can be obtained by the well-known viscosity solution method in [16]: consider the following equations, $\varepsilon > 0$,

$$\begin{cases} \partial_t u_\varepsilon + \nu(-\Delta)^\alpha u_\varepsilon + u_\varepsilon \cdot \nabla u_\varepsilon + \gamma u_\varepsilon + \nabla p_\varepsilon - \varepsilon \Delta u_\varepsilon = f, \\ \nabla \cdot u_\varepsilon = 0, \\ u_\varepsilon(x,0) = u_0(x). \end{cases} \quad (2.4)$$

By the classical result for Navier-Stokes equations we know that for each $\varepsilon > 0$ and $u_0 \in V$, there exists a unique solution $u_\varepsilon \in L^\infty(0,T;V) \cap L^2(0,T;H^2(\mathbf{R}^2))$ which satisfies $(2.4)_1$ almost everywhere in $\mathbf{R}^2 \times (0,T)$.

We will get a solution for the weak formulation (2.2) by taking $\varepsilon \to 0^+$.

For this, we need to show that $\{u_\varepsilon\}$ is bounded in V with the bounds independent of ε, which will imply $u \in L^\infty(0,T;V)$. Indeed, this can be done by considering the corresponding vorticity ω_ε (the curl of the incompressible two-dimensional velocity)

$$\omega_\varepsilon = \nabla^\perp \cdot u_\varepsilon = \partial_1 u_{\varepsilon 2} - \partial_2 u_{\varepsilon 1},$$

which obeys

$$\partial_t \omega_\varepsilon + u_\varepsilon \cdot \nabla \omega_\varepsilon + \gamma \omega_\varepsilon + \nu \Lambda^{2\alpha} \omega_\varepsilon - \varepsilon \Delta \omega_\varepsilon = g$$

with $g \in L^2(\mathbf{R}^2)$, $g = \nabla^\perp \cdot f$. It is easy to see that $\|\omega_\varepsilon(\cdot,t)\|_{L^2(\mathbf{R}^2)}$ can be controlled by the bounds which depend only on $\|\omega_\varepsilon(\cdot,0)\|_{L^2(\mathbf{R}^2)}$ and the coefficients in the equations. Then the uniformly (with respect to ε) boundedness of u_ε in V can be deduced immediately by noting $u_\varepsilon = K \star \omega_\varepsilon$ (e.g., see *Lemma* 3.1; where $K = \dfrac{1}{2\pi} \dfrac{x^\perp}{|x|^2}$ is the Biot-Savart kernel). Note that, here we used crucially the fact that $u_0 \in V$ and $f \in (W^{1,2}(\mathbf{R}^2))^2$.

To see $u \in L^2(0,T;H^{2\alpha})$, we multiply (1.1) (this can be justified by multiplying (2.4) with $\Lambda^{2\alpha} u_\varepsilon$ and then taking the limitation) by $\Lambda^{2\alpha} u$ and integrate in space to deduce that

$$\frac{d}{dt}\|\Lambda^\alpha u\|^2 + 2\gamma\|\Lambda^\alpha u\|^2 + 2\nu\|\Lambda^{2\alpha} u\|^2 \leq 2\int_{\mathbf{R}^2} |u \cdot \nabla u||\Lambda^{2\alpha} u| dx.$$

We will use the fact $u \in L^\infty(0,T;V)$ to deal with the nonlinear term as follows:

$$\int_{\mathbf{R}^2} |u \cdot \nabla u||\Lambda^{2\alpha} u| dx \leq \|\Lambda^{2\alpha} u\| \cdot \|\nabla u\|_{L^{\frac{2p}{p-2}}(\mathbf{R}^2)} \|u\|_{L^p(\mathbf{R}^2)}, \quad (2.5)$$

where $p \in (2, \infty)$ is large enough $\left(\text{e.g., } H^{(2\alpha-1)^-} \hookrightarrow L^{\frac{2p}{p-2}}(\mathbf{R}^2) \text{ since } \alpha \in \left(\frac{1}{2}, 1\right)\right)$ such that

$$\|\nabla u\|_{L^{\frac{2p}{p-2}}(\mathbf{R}^2)} \leqslant C_\eta \|u\| + \eta \|\Lambda^{2\alpha} u\| \tag{2.6}$$

for any $\eta > 0$; where $(2\alpha-1)^-$ denotes the number smaller than $2\alpha-1$ but approaching $2\alpha - 1$. Consequently, combining with $H^1 \hookrightarrow L^p(\mathbf{R}^2)$ for any $p \in [2, \infty)$, we have

$$\frac{d}{dt}\|\Lambda^\alpha u(t)\|^2 + 2\gamma\|\Lambda^\alpha u(t)\|^2 + 2\nu\|\Lambda^{2\alpha} u(t)\|^2$$
$$\leqslant 2\|\Lambda^{2\alpha} u(t)\|(C_\eta\|u(t)\| + \eta\|\Lambda^{2\alpha} u(t)\|)\|u\|_{L^\infty(0,T;V)}$$
$$\leqslant (3\eta\|\Lambda^{2\alpha} u(t)\|^2 + C'_\eta\|u(t)\|^2)\|u\|_{L^\infty(0,T;V)} \quad \text{for a.e. } t \in (0,T).$$

Hence, by taking η small enough such that $3\eta\|u\|_{L^\infty(0,T;V)} \leqslant \nu$, we integrate the above inequality over $(0,T)$ and obtain that

$$\nu \int_0^T \|\Lambda^{2\alpha} u(s)\|^2 ds \leqslant \|\Lambda^\alpha u(0)\|^2 + C'_\eta \|u\|_{L^\infty(0,T;V)} \int_0^T \|u(s)\|^2 ds,$$

which, combined with the fact $u(0) \in V$ and the a priori bounds about $\|u\|_{L^\infty(0,T;V)}$ again, implies that $u \in L^2(0,T; H^{2\alpha})$.

We can see from (1.1) that $u' \in L^2(0,T;H)$ directly after we obtained $u \in L^\infty(0,T;V) \cap L^2(0,T;H^{2\alpha})$. Then $u \in C([0,T);V)$ follows immediately.

The uniqueness is obvious by noticing the fact that the solution belongs to $L^\infty(0,T;V)$: let (u^i, p^i) be the solutions corresponding to initial data u_0^i, $i = 1, 2$, and write $w = u^1 - u^2$, then w solves the following equation:

$$\begin{cases} \partial_t w + \nu(-\Delta)^\alpha w + u^1 \cdot \nabla u^1 - u^2 \cdot \nabla u^2 + \gamma w + \nabla p^1 - \nabla p^2 = 0, \\ \nabla \cdot w = 0, \\ w(x,0) = u_0^1 - u_0^2. \end{cases} \tag{2.7}$$

Multiplying (2.7) by w and integrating in space, we get

$$\frac{d}{dt}\|w(t)\|^2 + 2\min\{\gamma,\nu\}\|w(t)\|^2_{H^\alpha} \leqslant 2|b(w, u^1, w)| \leqslant 2\|w(t)\|^2_{L^4(\mathbf{R}^2)}\|u^1(t)\|_V. \tag{2.8}$$

Then, noting that $u^1 \in L^\infty(0,T;V)$ and using the fact $\left(\text{since } \alpha > \frac{1}{2}\right)$

$$\|w(t)\|_{L^4(\mathbf{R}^2)} \leqslant C\|w(t)\|^r \|w(t)\|^{1-r}_{H^\alpha}$$

for some constant $r \in (0,1)$, we can deduce from (2.8) that

$$\frac{d}{dt}\|w(t)\|^2 \leqslant C_1 \|w(t)\|^2,$$

where the constant C_1 depends on $\|u^1\|_{L^\infty(0,T;V)}$ and γ, ν; which immediately implies that

$$\|w(t)\|^2 \leqslant e^{C_1 t}\|u_0^1 - u_0^2\|^2 \quad \text{for all } t \in [0,T). \tag{2.9}$$

This finishes the proof.

2.2 Useful properties

We will frequently use the following properties. For the proofs we refer to [7, 9].

Proposition 2.1 Let $0 < \alpha < 2, x \in \mathbf{R}^2$, and $\theta \in \mathcal{S}$, the Schwartz class, then

$$\Lambda^\alpha \theta(x) = C_\alpha P.V. \int_{\mathbf{R}^2} \frac{[\theta(x) - \theta(y)]}{|x-y|^{2+\alpha}} dy.$$

where $C_\alpha > 0$.

Proposition 2.2 (pointwise identity) Let $0 \leqslant \alpha \leqslant 2, x \in \mathbf{R}^2$ and $\theta \in \mathcal{S}$. Then

$$2\theta \Lambda^\alpha \theta(x) = \Lambda^\alpha(\theta^2(x)) + D_\alpha[\theta](x),$$

where

$$D_\alpha[\theta](x) = P.V. \int_{\mathbf{R}^2} \frac{[\theta(y) - \theta(x)]^2}{|x-y|^{2+\alpha}} dy \geqslant 0.$$

Proposition 2.3 (positivity lemma) Let $0 \leqslant \alpha \leqslant 2$, $x \in \mathbf{R}^2$, and $\theta, \Lambda^\alpha \theta \in L^p$ with $1 \leqslant p < \infty$. Then

$$\int_{\mathbf{R}^2} |\theta|^{p-2} \theta \Lambda^\alpha \theta \, dx \geqslant 0.$$

3 Inviscid limit

Throughout the section, we assume that $\alpha \in \left(\frac{1}{2}, 1\right)$ and $\gamma > 0$ are kept fixed, and that the force $f \in (W^{1,1}(\mathbf{R}^2) \cap W^{1,\infty}(\mathbf{R}^2))^2$ is given and time-independent. The results follow the idea in Constantin and Ramos [5], see also Foias et al. [10–12] for more details.

3.1 Preliminaries

We start with some properties of the solutions, which are the same as that in [5] for the usual damped and driven Navier-Stokes equations.

Theorem 3.1 Let $u_0 \in V$. Then the solution of (1.1) with initial datum u_0 exists for all time, is unique and satisfies the energy equation:

$$\frac{d}{2dt} \int_{\mathbf{R}^2} |u|^2 dx + \gamma \int_{\mathbf{R}^2} |u|^2 dx + \nu \int_{\mathbf{R}^2} |\Lambda^\alpha u|^2 dx = \int_{\mathbf{R}^2} f \cdot u \, dx. \tag{3.1}$$

The kinetic energy is bounded uniformly in time, with bounds independent of viscosity ν:

$$\|u(\cdot,t)\|_{L^2(\mathbf{R}^2)} \leqslant e^{-\gamma t}\left\{\|u_0\|_{L^2(\mathbf{R}^2)} - \frac{1}{\gamma^2}\|f\|_{L^2(\mathbf{R}^2)}\right\} + \frac{1}{\gamma^2}\|f\|_{L^2(\mathbf{R}^2)}. \quad (3.2)$$

The vorticity

$$\omega = \nabla^\perp \cdot u = \partial_1 u_2 - \partial_2 u_1$$

obeys

$$\partial_t \omega + u \cdot \nabla \omega + \gamma \omega + \nu \Lambda^{2\alpha} \omega = g \quad (3.3)$$

with $g \in L^1(\mathbf{R}^2) \cap L^\infty(\mathbf{R}^2)$, $g = \nabla^\perp \cdot f$. The map

$$[0, +\infty) \to L^2(\mathbf{R}^2), \quad t \to \omega(t)$$

is continuous. If the initial vorticity ω_0 is in $L^p(\mathbf{R}^2), p > 1$, then the p-enstrophy is bounded uniformly in time, with bounds independent of viscosity:

$$\|\omega(\cdot,t)\|_{L^p(\mathbf{R}^2)} \leqslant e^{-\gamma t}\left\{\|\omega_0\|_{L^p(\mathbf{R}^2)} - \frac{1}{\gamma^2}\|g\|_{L^p(\mathbf{R}^2)}\right\} + \frac{1}{\gamma^2}\|g\|_{L^p(\mathbf{R}^2)} \quad (3.4)$$

for $p \geqslant 1$. Moreover, the positive semi-orbit

$$O_+(\omega) = \{\omega = \omega(\cdot, t) \mid t \geqslant 0\} \subset L^2(\mathbf{R}^2)$$

is equi-integrable in $L^2(\mathbf{R}^2)$: for every $\epsilon > 0$, there exists $R > 0$ such that

$$\int_{|x| \geqslant R} |\omega(x,t)|^2 dx \leqslant \epsilon \quad (3.5)$$

holds for all $t \geqslant 0$, where the radius R depends on the coefficients γ, ν, α and $\omega(x,0)$.

Proof See *Lemma* 2.1 for the existence and uniqueness of solutions.

The energy equation (3.1) follows from the incompressibility of u and integration by parts. The bounds (3.2) and (3.4) follow from the positivity lemma and an application of the Gronwall inequality; see, for example, Section 4.

In the following, we only prove the equi-integrability (3.5). As in [5], we consider the function

$$Y_R(t) = \int_{\mathbf{R}^2} \chi\left(\frac{x}{R}\right) \omega^2(x,t) dx,$$

where $\chi(\cdot)$ is a nonnegative smooth function supported in $\left\{x \in \mathbf{R}^2 : |x| \geqslant \frac{1}{2}\right\}$ and identically equal to 1 for $|x| \geqslant 1$. We Multiply the equation (3.3) by $2\chi\left(\frac{x}{R}\right)\omega(x,t)$ and integrate in space. The only challenging term we encounter is

$$2\nu \int_{\mathbf{R}^2} \Lambda^{2\alpha} \omega(x) \chi\left(\frac{x}{R}\right) \omega(x,t) dx.$$

Using *Proposition* 2.2 we have

$$2\nu \int_{\mathbf{R}^2} \Lambda^{2\alpha}\omega(x,t)\chi\left(\frac{x}{R}\right)\omega(x,t)dx \geq \nu \int_{\mathbf{R}^2} \Lambda^{2\alpha}(\omega^2(x,t))\left(1-\left(1-\chi\left(\frac{x}{R}\right)\right)\right)dx$$

$$= -\nu \int_{\mathbf{R}^2} \omega^2(x,t)\Lambda^{2\alpha}\left(1-\chi\left(\frac{x}{R}\right)\right)dx,$$

where we have used

$$\int_{\mathbf{R}^2} \Lambda^{2\alpha}(\omega^2)dx = \widehat{\Lambda^{2\alpha}(\omega^2)}(0) = 0,$$

since $\alpha > 0$, and the fact that $\Lambda^{2\alpha}\left(1-\chi\left(\frac{x}{R}\right)\right)$ is well defined because $1-\chi\left(\frac{x}{R}\right) \in C_c^\infty(\mathbf{R}^2)$. Moreover, we have

$$1-\chi\left(\frac{x}{R}\right) =: \phi\left(\frac{x}{R}\right).$$

In view of

$$\Lambda^{2\alpha}\left(\phi\left(\frac{x}{R}\right)\right) = \frac{1}{R^{2\alpha}}(\Lambda^{2\alpha}\phi)\left(\frac{x}{R}\right),$$

and

$$|(\Lambda^{2\alpha}\phi)(x)| \leq \int_{\mathbf{R}^2} |\widehat{\Lambda^{2\alpha}\phi}(\xi)|d\xi = \int_{\mathbf{R}^2} |\xi|^{2\alpha}|\widehat{\phi}(\xi)|d\xi \leq C_0,$$

we have

$$2\nu \int_{\mathbf{R}^2} \Lambda^{2\alpha}\omega(x,t)\chi\left(\frac{x}{R}\right)\omega(x,t)dx \geq -\frac{C_0\nu}{R^{2\alpha}}\|\omega(\cdot,t)\|_{L^2(\mathbf{R}^2)}^2. \tag{3.6}$$

So we can obtain that

$$\frac{d}{dt}Y_R(t) + 2\gamma Y_R(t) \leq \frac{C}{R}\|u(\cdot,t)\|_{L^2(\mathbf{R}^2)}\|\omega(\cdot,t)\|_{L^4(\mathbf{R}^2)}^2 + \frac{C_0\nu}{R^{2\alpha}}\|\omega(\cdot,t)\|_{L^2(\mathbf{R}^2)}^2$$

$$+ C\left(Y_R(t)\int_{|x|\geq \frac{R}{2}}|g(x)|^2 dx\right)^{\frac{1}{2}},$$

that is,

$$Y_R(t) - e^{-\gamma t}Y_R(0) \leq \frac{C}{R}\max_{s\in[0,t]}\{\|u(\cdot,s)\|_{L^2(\mathbf{R}^2)}\} \cdot \int_0^t e^{\gamma(s-t)}\|\omega(\cdot,s)\|_{L^4(\mathbf{R}^2)}^2 ds$$

$$+ \frac{C_0\nu}{\gamma R^{2\alpha}}\max_{s\in[0,t]}\{\|\omega(\cdot,s)\|_{L^2(\mathbf{R}^2)}^2\} + \frac{C}{\gamma^2}\int_{|x|\geq \frac{R}{2}}|g(x)|^2 dx. \tag{3.7}$$

On the other hand, multiplying (3.3) by $\omega(x,t)$ and integrating in space, we obtain that

$$\frac{d}{dt}\int_{\mathbf{R}^2}|\omega(x,t)|^2 dx + 2\gamma \int_{\mathbf{R}^2}|\omega(x,t)|^2 dx + 2\nu \int_{\mathbf{R}^2}|\Lambda^\alpha \omega(x,t)|^2 dx$$

$$\leq 2\int_{\mathbf{R}^2}|\omega(x,t)||g(x)|dx,$$

which implies that, for any $t \geq 0$,

$$\|\omega(x,t)\|_{L^2(\mathbf{R}^2)}^2 + 2\nu \int_0^t e^{\gamma(s-t)} \|\Lambda^\alpha \omega(x,s)\|_{L^2(\mathbf{R}^2)}^2 ds \leq F < \infty,$$

with a positive constant F which is bounded in terms of γ, $\|g\|_{L^2(\mathbf{R}^2)}^2$ and $\|\omega(x,0)\|_{L^2(\mathbf{R}^2)}^2$ (but is independent of t). Hence, combining with the embedding $H^\alpha \hookrightarrow L^4(\mathbf{R}^2)$ for $\alpha > \frac{1}{2}$, we have

$$\int_0^t e^{\gamma(s-t)} \|\omega(x,s)\|_{L^4(\mathbf{R}^2)}^2 ds \leq C_{F,\nu} < \infty \quad \text{for any } t \geq 0. \tag{3.8}$$

Therefore, from (3.2), (3.4), (3.8) and the fact that $|g|^2$ is integrable, the right hand side of (3.7) will be arbitrary small if we take R large enough. Then the equi-integrability (3.5) follows from the fact that $Y_R(0)$ is small for large R.

Remark 3.1 *The vorticity equation (3.3) with a different draft term is a special case of the viscous surface quasi-geostrophic equation, which has been studied by many authors (e.g., see [1, 6-8] and the references therein). Especially, in [8], Dlotko, Kania and Sun obtained some approximation estimates that may be helpful in considering the inviscid limit problem (as what we do in this section for (1.1)) for the surface quasi-geostrophic equation.*

In the following, let $\gamma, \nu > 0$ be fixed. We give some a priori estimates about the vorticity $\omega(x,t)$.

We write $S^{(\nu)}(t,\omega_0)$ for the solution of the vorticity equation (3.3) at time $t \geq 0$ which started at time $t = 0$ from the initial data ω_0.

Theorem 3.2 *Let $\omega_0 \in L^2(\mathbf{R}^2)$, $g \in L^2(\mathbf{R}^2)$ and $u \in L^\infty(0,\infty; V)$. Then for any $t_0 > 0$, the positive semi-orbit*

$$O_+(t_0,\omega_0) = \{\omega(\cdot,t) \mid t \geq t_0\}$$

is relatively compact in $L^2(\mathbf{R}^2)$.

The proof of this theorem follows from the equi-integrability (3.5) and *Lemma 3.2* below.

Lemma 3.1 *Let $s \geq 0$ and $\Lambda^s \omega \in L^2(\mathbf{R}^2)$, then for $u = \frac{1}{2\pi} \frac{x^\perp}{|x|^2} \star \omega$, we have $\Lambda^{1+s} u \in (L^2(\mathbf{R}^2))^2$ and*

$$\|\Lambda^{1+s} u\|_{(L^2(\mathbf{R}^2))^2} \leq C \|\Lambda^s \omega\|_{L^2(\mathbf{R}^2)}.$$

Proof Note that $u = \dfrac{1}{2\pi}\dfrac{x^\perp}{|x|^2} \star \omega = \Lambda^{-1}\mathcal{R}^\perp \omega$, hence

$$\|\Lambda^{1+s}u\|_{(L^2(\mathbf{R}^2))^2} = \|\Lambda^s\mathcal{R}^\perp\omega\|_{(L^2(\mathbf{R}^2))^2} \leqslant C\|\Lambda^s\omega\|_{L^2(\mathbf{R}^2)},$$

where the constant C depends only on the L^2-bounds of the Riesz transform $\mathcal{R}$.

Lemma 3.2 Let $\omega_0 \in L^2(\mathbf{R}^2)$, $g \in L^2(\mathbf{R}^2)$ and $u \in L^\infty(0,\infty;V)$. Then the solution $\omega(\cdot,t)$ of (3.3) is uniformly (with respect to $t \geqslant t_0$) bounded in $W^{\alpha,2}(\mathbf{R}^2)$ for any $t_0 > 0$.

Proof At first, we multiply (3.3) by ω to deduce that

$$\int_0^{t_0}\int_{\mathbf{R}^2}|\Lambda^\alpha\omega(x,\tau)|^2 dx d\tau \leqslant \tilde{M}_0, \tag{3.9}$$

where the constant $\tilde{M}_0$ depends only on ν, $\|g\|_{L^2(\mathbf{R}^2)}$ and $\|\omega(\cdot,0)\|_{L^2(\mathbf{R}^2)}$.

Secondly, we multiply again the vorticity equation (3.3) by $\Lambda^{2\alpha}\omega$ and integrate in space, to get

$$\frac{d}{2dt}\int_{\mathbf{R}^2}|\Lambda^\alpha\omega|^2 dx + \int_{\mathbf{R}^2}(u\cdot\nabla\omega)\Lambda^{2\alpha}\omega dx + \gamma\int_{\mathbf{R}^2}|\Lambda^\alpha\omega|^2 dx + \nu\int_{\mathbf{R}^2}|\Lambda^{2\alpha}\omega|^2 dx$$
$$= \int_{\mathbf{R}^2}g\Lambda^{2\alpha}\omega dx.$$

The nonlinear term $\int_{\mathbf{R}^2}(u\cdot\nabla\omega)\Lambda^{2\alpha}\omega dx$ can be estimated as (2.5) and (2.6):

$$\left|\int_{\mathbf{R}^2}(u\cdot\nabla\omega)\Lambda^{2\alpha}\omega dx\right| \leqslant C\|u\|_V \cdot \|\omega\|_{L^2(\mathbf{R}^2)}^{1-r} \cdot \|\Lambda^{2\alpha}\omega\|_{L^2(\mathbf{R}^2)}^{1+r}$$

with some positive constant $r \in (0,1)$. Then, applying the Young inequality we obtain that

$$\frac{d}{dt}\int_{\mathbf{R}^2}|\Lambda^\alpha\omega|^2 dx + 2\gamma\int_{\mathbf{R}^2}|\Lambda^\alpha\omega|^2 dx \leqslant \tilde{M}_1, \tag{3.10}$$

where the constant $\tilde{M}_1$ depends on $\|u(\cdot,t)\|_V$, $\|\omega(\cdot,t)\|_{L^2(\mathbf{R}^2)}$, $\|g\|_{L^2(\mathbf{R}^2)}$ and ν.

Then, combining with (3.4), (3.9), (3.10) and *Lemma* 3.1, we can finish the proof by applying a Gronwall's inequality.

Moreover, we have the following a priori estimates.

Lemma 3.3 Let $\omega_0 \in L^2(\mathbf{R}^2) \cap L^\infty(\mathbf{R}^2)$, $g \in L^1(\mathbf{R}^2) \cap L^\infty(\mathbf{R}^2)$ and $u \in L^\infty(0,\infty;V)$. Then the solution $\omega(\cdot,t)$ of (3.3) satisfies

$$\omega \in L^\infty(\mathbf{R}^2 \times (t_0,\infty))$$

for any $t_0 > 0$.

Proof The proof is based on an idea of Caffarelli and Vasseur [1]. The details for our case are exactly as in [*Lemmas* 2.2 and 2.3 in [2]]. Note that our assumptions here are stronger than that required in [2].

3.2 Stationary statistical solutions and enstrophy balance

We introduce first the notation of stationary statistical solution for damped and driven generalized incompressible Navier-Stokes equations in the vorticity phase space, in spirit of [5, 10–12]. The solution is a Borel probability measure in $L^2(\mathbf{R}^2)$.

Definition 3.1 *A stationary statistical solution of damped and driven generalized incompressible Navier-Stokes equations in the vorticity phase space is a Borel probability measure $\mu^{(\nu)}$ on $L^2(\mathbf{R}^2)$ such that*

$$\int_{L^2(\mathbf{R}^2)} \|\omega\|^2_{H^\alpha(\mathbf{R}^2)} d\mu^{(\nu)}(\omega) < \infty \tag{3.11}$$

and

$$\int_{L^2(\mathbf{R}^2)} \langle u \cdot \nabla \omega + \gamma \omega - g, \Psi'(\omega) \rangle + \nu \langle \Lambda^{2\alpha} \omega, \Psi'(\omega) \rangle d\mu^{(\nu)}(\omega) = 0 \tag{3.12}$$

for any test functional $\Psi \in \mathcal{T}$, *with* $u = \dfrac{1}{2\pi} \dfrac{x^\perp}{|x|^2} \star \omega$, *where we used the notation* $\langle v, w \rangle = \displaystyle\int_{\mathbf{R}^2} v(x) w(x) dx$; *and*

$$\int_{E_1 \leqslant \|\omega\|_{L^2(\mathbf{R}^2)} \leqslant E_2} \left(\gamma \|\omega\|^2_{L^2(\mathbf{R}^2)} + \nu \|\Lambda^\alpha \omega\|^2_{L^2(\mathbf{R}^2)} - \langle g, \omega \rangle \right) d\mu^{(\nu)}(\omega) \leqslant 0 \tag{3.13}$$

for any $0 \leqslant E_1 \leqslant E_2$.

The class $\mathcal{T}$ of cylindrical test functions is given as follows.

Definition 3.2 [5, 12] *The class $\mathcal{T}$ of test functions is the set of functions* $\Psi : L^2(\mathbf{R}^2) \to \mathbf{R}$ *of the form*

$$\Psi(\omega) := \Psi_I(\omega) = \psi(\langle \omega, \mathbf{w_1} \rangle, \cdots, \langle \omega, \mathbf{w_m} \rangle),$$

or

$$\Psi(\omega) := \Psi_\varepsilon(\omega) = \psi(\langle \alpha_\varepsilon(\omega), \mathbf{w_1} \rangle, \cdots, \langle \alpha_\varepsilon(\omega), \mathbf{w_m} \rangle),$$

where ψ is a C^1 scalar valued function defined on $\mathbf{R}^m$, $m \in \mathbb{N}$, $\mathbf{w_1}, \cdots, \mathbf{w_m}$ belong to $C_0^2(\mathbf{R}^2)$ and

$$\alpha_\varepsilon(\omega) = J_\varepsilon \beta(J_\varepsilon(\omega)),$$

where $\beta \in C^3$ is a compactly supported function of one real variable, and J_ϵ is the convolution operator
$$J_\epsilon(\omega) = j_\epsilon \star \omega,$$
with $j \geqslant 0$ a fixed smooth, even function supported in $|z| \leqslant 1$ and with $\int_{\mathbf{R}^2} j(z) dz = 1$.

Remark 3.2 *We make mathematical sense of the conditions (3.11) and (3.13) in Definition 3.1: the function $\omega \mapsto \|\omega\|^2_{H^\alpha(\mathbf{R}^2)}$ is a Borel measurable in $L^2(\mathbf{R}^2)$ because it is the limit of a sequence of continuous functions $\omega \mapsto \|J_\epsilon \omega\|^2_{H^\alpha(\mathbf{R}^2)}$.*

Remark 3.3 *The support of any stationary statistical solution of damped and driven generalized incompressible Navier-Stokes equations is included in the ball:* $\text{supp } \mu^{(\nu)} \subset \left\{ \omega \in L^2(\mathbf{R}^2) : \|\omega\|_{L^2(\mathbf{R}^2)} \leqslant \frac{1}{\gamma} \|g\|_{L^2(\mathbf{R}^2)} \right\}$. *Indeed (note that the following proof is differs slightly from that in [5, 12]), set* $E = \{\omega \in L^2(\mathbf{R}^2) : E_1^2 \leqslant \|\omega\|^2_{L^2(\mathbf{R}^2)} \leqslant E_2^2\}$, *then from (3.13) we have*

$$\gamma \int_E \|\omega\|^2_{L^2(\mathbf{R}^2)} d\mu^\nu(\omega) \leqslant \|g\|_{L^2(\mathbf{R}^2)} \int_E \|\omega\|_{L^2(\mathbf{R}^2)} d\mu^\nu(\omega)$$
$$\leqslant \|g\|_{L^2(\mathbf{R}^2)} \left(\int_E \|\omega\|^2_{L^2(\mathbf{R}^2)} d\mu^\nu(\omega) \right)^{\frac{1}{2}} \cdot (mes(E))^{\frac{1}{2}}.$$

Therefore,
$$\int_E \|\omega\|^2_{L^2(\mathbf{R}^2)} d\mu^\nu(\omega) \leqslant \frac{\|g\|^2_{L^2(\mathbf{R}^2)}}{\gamma^2} mes(E),$$

which implies immediately that
$$\int_E \left(\|\omega\|^2_{L^2(\mathbf{R}^2)} - \frac{\|g\|^2_{L^2(\mathbf{R}^2)}}{\gamma^2} \right) d\mu^\nu(\omega) \leqslant 0.$$

Thus, we can deduce the support of μ^ν by taking $E_1 = \frac{\|g\|_{L^2(\mathbf{R}^2)}}{\gamma}$ in (3.13).

Remark 3.4 *The test functions Ψ in Definition 3.2 are locally bounded and weakly sequentially continuous in $L^2(\mathbf{R}^2)$.*

We can compute Ψ' for test functions $\Psi \in \mathcal{T}$ as follow (see [5]) :
$$\Psi'_I(\omega) = \sum_{j=1}^m \partial_j \psi(\langle \omega, \mathbf{w_1} \rangle, \cdots, \langle \omega, \mathbf{w_m} \rangle) \mathbf{w_j},$$

and
$$\Psi'_\epsilon(\omega) = \sum_{j=1}^m \partial_j \psi(\langle \alpha_\epsilon(\omega), \mathbf{w_1}\rangle, \cdots, \langle \alpha_\epsilon(\omega), \mathbf{w_m}\rangle)(\beta'(\omega_\epsilon)\mathbf{w_{j_\epsilon}})_\epsilon.$$

Next, we state some important properties.

Lemma 3.4 ([5]) *Let $\Psi \in \mathcal{T}$ and $\omega \in L^2(\mathbf{R}^2)$, then $\Psi'(\omega) \in C_0^2(\mathbf{R}^2)$. We consider*
$$F_i : L^2(\mathbf{R}^2) \to \mathbf{R},$$
i=1, 2, 3 given by
$$F_1(\omega) = \langle \gamma\omega - g, \Psi'(\omega)\rangle,$$
$$F_2(\omega) = \langle \Lambda^\alpha \omega, \Lambda^\alpha \Psi'(\omega)\rangle$$
and
$$F_3(\omega) = \langle u \cdot \nabla\omega, \Psi'(\omega)\rangle, \quad u = \frac{1}{2\pi}\frac{x^\perp}{|x|^2} \star \omega.$$
Then, these three maps are well defined for $\omega \in L^2(\mathbf{R}^2)$, locally bounded and weakly continuous in $L^2(\mathbf{R}^2)$.

Proof The maps $F_1(\cdot)$ and $F_3(\cdot)$ are exactly as in [5]. For $F_2(\cdot)$, we just need to follow the idea of [5]:

The fact that $F_2(\omega)$ is well defined follows from that $\Lambda^{2\alpha}\Psi'(\omega)$ is well defined (since $\Psi'(\omega) \in C_0^2(\mathbf{R}^2)$ by the fact that $\mathbf{w_j} \in C_0^2(\mathbf{R}^2)$) and
$$F_2(\omega) = \langle \Lambda^{2\alpha}\Psi'(\omega), \omega\rangle.$$

Concerning the weak continuity of F_2 : for Ψ_I, we have
$$\langle \Lambda^{2\alpha}\Psi'_I(\omega), \omega\rangle = \sum_{j=1}^m \partial_j \psi(\langle \omega, \mathbf{w_1}\rangle, \cdots, \langle \omega, \mathbf{w_m}\rangle)\langle \Lambda^{2\alpha}\mathbf{w_j}, \omega\rangle,$$
which is obviously a weakly continuous and locally bounded function of $\omega \in L^2(\mathbf{R}^2)$. In the case of Ψ_ϵ, we have
$$\langle \Lambda^{2\alpha}\Psi'_\epsilon(\omega), \omega\rangle = \sum_{j=1}^m \partial_j \psi(\langle \alpha_\epsilon(\omega), \mathbf{w_1}\rangle, \cdots, \langle \alpha_\epsilon(\omega), \mathbf{w_m}\rangle)\langle \Lambda^{2\alpha}(\beta'(\omega_\epsilon)\mathbf{w_{j_\epsilon}})_\epsilon, \omega\rangle$$
$$= \sum_{j=1}^m \partial_j \psi(\langle \alpha_\epsilon(\omega), \mathbf{w_1}\rangle, \cdots, \langle \alpha_\epsilon(\omega), \mathbf{w_m}\rangle)\langle \mathbf{w_{j_\epsilon}}, \beta'(\omega_\epsilon)\Lambda^{2\alpha}\omega_\epsilon\rangle.$$

If $\omega^i \rightharpoonup \omega$ in $L^2(\mathbf{R}^2)$, then $\omega^i_\epsilon \to \omega_\epsilon$ and $\Lambda^{2\alpha}\omega^i_\epsilon \to \Lambda^{2\alpha}\omega_\epsilon$ converge pointwise, and they are bounded. Consequently, $\beta'(\omega^i_\epsilon)\Lambda^{2\alpha}\omega^i_\epsilon$ converges pointwise and is uniformly bounded. Therefore we use the Lebesgue dominated convergence theorem and obtain that $F_2(\omega)$ is weakly continuous. It is also clear that

$$\|\beta'(\omega_\epsilon)\Lambda^{2\alpha}\omega_\epsilon\|_{L^2(\mathbf{R}^2)} \leqslant C_\epsilon\|\omega\|_{L^2(\mathbf{R}^2)}.$$

Thus, $F_2(\cdot)$ is locally bounded in $L^2(\mathbf{R}^2)$.

We define the notation of a renormalized stationary statistical solution of the Euler equation.

Definition 3.3 ([5]) *A Borel probability measure μ^0 on $L^2(\mathbf{R}^2)$ is a renormalized stationary statistical solution of the damped and driven Euler equation if*

$$\int_{L^2(\mathbf{R}^2)} \langle u\cdot\nabla\omega + \gamma\omega - g, \Psi'(\omega)\rangle d\mu^0(\omega) = 0 \tag{3.14}$$

for any test functional $\Psi \in \mathcal{T}$, where $u = \dfrac{1}{2\pi}\dfrac{x^\perp}{|x|^2}\star\omega$.

We say that a renormalized stationary statistical solution μ^0 of the Euler equation satisfies the enstrophy balance if

$$\int_{L^2(\mathbf{R}^2)} (\gamma\|\omega\|_{L^2(\mathbf{R}^2)}^2 - \langle g,\omega\rangle) d\mu^0(\omega) = 0 \tag{3.15}$$

holds.

Theorem 3.3 *Let $\mu^{(\nu)}$ be a sequence of stationary statistical solutions of damped and driven generalized incompressible Navier-Stokes equations in vorticity phase space, with $\nu \to 0$. There exist a subsequence, denoted also $\mu^{(\nu)}$ and a Borel probability measure μ^0 on $L^2(\mathbf{R}^2)$ such that*

$$\lim_{\nu\to 0}\int_{L^2(\mathbf{R}^2)} \Phi(\omega)d\mu^{(\nu)}(\omega) = \int_{L^2(\mathbf{R}^2)} \Phi(\omega)d\mu^0(\omega)$$

for all weakly continuous, locally bounded real-valued functions Φ. Furthermore the weak limit measure μ^0 is a renormalized stationary statistical solution of the damped and driven Euler equation.

Proof By *Remark* 3.3, the support of $\mu^{(\nu)}$ is included in

$$B = \left\{\omega \in L^2(\mathbf{R}^2) : \|\omega\|_{L^2(\mathbf{R}^2)} \leqslant \frac{1}{\gamma}\|g\|_{L^2(\mathbf{R}^2)}\right\}.$$

The set B endowed with the weak $L^2(\mathbf{R}^2)$ topology is a separable metrizable compact space. We apply Prokhorov's theorem. There exists a subsequence of $\mu^{(\nu)}$ that converges weakly to a Borel probability measure μ^0 in B. So the weak limit μ^0 is a Borel probability measure on B. Because B is convex and so weakly closed in $L^2(\mathbf{R}^2)$, we can extend the measure μ^0 to $L^2(\mathbf{R}^2)$ by setting $\mu^0(X) = \mu^0(X \cap B)$. We claim that μ^0 is a renormalized stationary statistical solution of the damped and driven Euler equation. In order to verify that μ^0 satisfies (3.14), take $\Psi \in \mathcal{T}$. Then for each $i = 1, 2, 3$, noting that $supp\, \mu^{(\nu)} \subset B$, we have

$$\lim_{\nu \to 0} \int_{L^2(\mathbf{R}^2)} F_i(\omega) d\mu^{(\nu)}(\omega) = \int_{L^2(\mathbf{R}^2)} F_i(\omega) d\mu^0(\omega)$$

in view of Lemma 3.4 (here the limitation $\lim_{\nu \to \infty}$ is taking for the subsequence which is weak convergence). In particular, the sequence $\int_{L^2(\mathbf{R}^2)} F_2(\omega) d\mu^{(\nu)}(\omega)$ is bounded and so

$$\lim_{\nu \to 0} \nu \int_{L^2(\mathbf{R}^2)} F_2(\omega) d\mu^{(\nu)}(\omega) = 0.$$

Because $\mu^{(\nu)}$ are stationary statistical solutions of damped and driven generalized incompressible Navier-Stokes equations, using (3.12) we have

$$\int_{L^2(\mathbf{R}^2)} (F_1(\omega) + F_3(\omega)) d\mu^{(\nu)}(\omega) = -\nu \int_{L^2(\mathbf{R}^2)} F_2(\omega) d\mu^{(\nu)}(\omega).$$

Passing to the limit $\nu \to 0$, we obtain

$$\int_{L^2(\mathbf{R}^2)} (F_1(\omega) + F_3(\omega)) d\mu^0(\omega) = 0.$$

That is, measure μ^0 satisfy condition (3.14), and therefore is a renormalized stationary statistical solution of the damped and driven Euler equation.

We consider the sets

$$B_p^\infty(r) = \{\omega \in B \mid \|\omega\|_{L^p(\mathbf{R}^2)} \leq r, \|\omega\|_{L^\infty(\mathbf{R}^2)} \leq r\}$$

defined for $r > 0$, $1 \leq p < 2$.

In exactly the same way as the proof of [Theorem 4.7, [5]], we can check that if we proved that the limitation μ^0 is a renormalized stationary statistical solution of the damped and driven Euler equation and μ^0 is supported in some bounded subset, then μ^0 must satisfy the enstrophy balance (3.15). Hence, similar to [Theorem 4.7, [5]], from *Theorem* 3.3 we also have the following result.

Theorem 3.4 *Let $\mu^{(\nu)}$ be a sequence of stationary statistical solutions of damped and driven generalized incompressible Navier-Stokes equations in vorticity phase space,* with $\nu \to 0$. Assume that there exist $1 < p < 2$ and $r > 0$ such that

$$supp\mu^{(\nu)} \subset B_p^\infty(r).$$

Then the limit μ^0 of any weakly convergent subsequence is a renormalized stationary statistical solution of the damped and driven Euler equation that is supported in set $B_p^\infty(r)$ and satisfies the enstrophy balance (3.15).

3.3 Long time averages and the inviscid limit

In this section we consider the stationary statistical solutions obtained as generalized (Banach) limits of long time averages of functionals of determined solutions of the damped and driven generalized incompressible Navier-Stokes equations. These stationary statistical solutions have enough properties to pass to the inviscid limit and are used to prove that the time averaged enstrophy dissipation vanishes in the zero viscosity limit. We start by recalling the concept of the generalized (Banach) limit (see for example [12]):

Definition 3.4 *A generalized limit (Banach limit) is a linear continuous functional*

$$Lim_{t\to\infty} : \mathcal{BC}([0,\infty)) \to \mathbf{R}$$

such that

1. *$Lim_{t\to\infty}(g) \geqslant 0$ for all $g \in \mathcal{BC}([0,\infty))$ with $g(s) \geqslant 0$ for all $s \geqslant 0$,*
2. *$Lim_{t\to\infty}(g) = \lim_{t\to\infty} g(t)$, whenever the usual limit exists.*

The space $\mathcal{BC}([0,\infty))$ is the Banach space of all bounded continuous real valued functions defined on $[0,\infty)$ endowed with the sup norm.

Remark 3.5 ([12]) *It can be shown that any generalized limit satisfies*

$$\liminf_{t\to\infty} g(t) \leqslant Lim_{t\to\infty}(g) \leqslant \limsup_{t\to\infty} g(t), \quad \text{for all } g \in \mathcal{BC}([0,\infty)).$$

Remark 3.6 ([12]) *Given a fixed $g_0 \in \mathcal{BC}([0,\infty))$ and a sequence $t_j \to \infty$ for which $\lim_{j\to\infty} g_0(t_j) = l$ exists, we can construct a generalized limit $Lim_{t\to\infty}$ satisfying $Lim_{t\to\infty}(g_0) = l$. This implies that one can choose a functional $Lim_{t\to\infty}$ so that $Lim_{t\to\infty} g_0 = \limsup_{t\to\infty} g_0(t)$.*

We now state the result about long time averages of the damped and driven generalized incompressible Navier-Stokes equations.

Theorem 3.5 *Let $u_0 \in (L^2(\mathbf{R}^2))^2$ and $\nabla^\perp \cdot u_0 = \omega_0 \in L^1(\mathbf{R}^2) \cap L^\infty(\mathbf{R}^2)$, $f \in W^{1,1}(\mathbf{R}^2) \cap W^{1,\infty}(\mathbf{R}^2)$, and $Lim_{t\to\infty}$ be a Banach limit. Given $t_0 > 0$. Then, μ^ν, defined by*

$$\int_{L^2(\mathbf{R}^2)} \Phi(\omega) d\mu^{(\nu)}(\omega) = Lim_{t\to\infty} \frac{1}{t} \int_0^t \Phi(S^{(\nu)}(s+t_0,\omega_0))ds, \tag{3.16}$$

is a stationary statistical solution of the damped and driven generalized incompressible Navier-Stokes equation in the vorticity phase space, where Φ is a continuous real functional on $L^2(\mathbf{R}^2)$; For any $p > 1$ there exists r depending only on γ, f, ω_0 but not on ν nor t_0 such that

$$supp\mu^{(\nu)} \subset B_p^\infty(r).$$

The following inequality holds:

$$\nu \int_{L^2(\mathbf{R}^2)} \|\Lambda^\alpha \omega\|_{L^2(\mathbf{R}^2)}^2 d\mu^{(\nu)}(\omega) \leq \int_{L^2(\mathbf{R}^2)} [\langle g, \omega \rangle - \gamma \|\omega\|_{L^2(\mathbf{R}^2)}^2] d\mu^{(\nu)}(\omega). \tag{3.17}$$

Proof From *Theorem 3.2*, the positive semi-orbit

$$O_+(t_0, \omega_0) = \{\omega(\cdot, t) = S^{(\mu)}(t, \omega_0) \mid t \geq t_0\}$$

is relatively compact in $L^2(\mathbf{R}^2)$. For any $\Phi \in C(L^2(\mathbf{R}^2))$, we have $\Phi \in C(\overline{O_+(t_0, \omega_0)})$ and the function $s \mapsto \Phi(S^{(\nu)}(s+t_0, \omega_0))$ is a continuous bounded function on $[0, \infty)$ and so is its time average on $[0, t]$. Thus we may apply the generalized limit $Lim_{t\to\infty}$ to it and define the functional:

$$\Phi \mapsto Lim_{t\to\infty} \frac{1}{t} \int_0^t \Phi(S^{(\nu)}(s+t_0, \omega_0))ds.$$

This functional is linear and nonnegative. Hence, applying the Riesz representation theorem on compact spaces, it follows that there exists a Borel measure $\mu^{(\nu)}$ representing it, that is (3.16) holds. The measure $\mu^{(\nu)}$ is supported on $\overline{O_+(t_0, \omega_0)}$, extend $\mu^{(\nu)}$ to $L^2(\mathbf{R}^2)$ given by $\mu^{(\nu)}(X) = \mu^{(\nu)}(X \cap \overline{O_+(t_0, \omega_0)})$, for any X Borelian in $L^2(\mathbf{R}^2)$. It follows from *Theorem 3.1* that the measure $\mu^{(\nu)}$ is supported in the set $B_p^\infty(r)$.

Take a test function $\Psi \in \mathcal{T}$. Then, noticing that $O_+(t_0, \omega_0)$ is precompact in $L^2(\mathbf{R}^2)$, we can calculate directly that

$$\int_{L^2(\mathbf{R}^2)} \langle u \cdot \nabla \omega + \gamma \omega + \nu \Lambda^{2\alpha} \omega, \Psi'(\omega) \rangle d\mu^{(\nu)}(\omega)$$

$$= Lim_{t\to\infty} \frac{1}{t} \int_0^t \frac{d}{ds} \Psi(S^{(\nu)}(s+t_0, \omega_0) ds = 0$$

holds, where the second equality is due to the boundedness of Ψ on $O_+(t_0,\omega_0)$. This verifies (3.12) of *Definition* 3.1.

In order to verify conditions (3.11) and (3.13) we take solution $\omega(t) = S^{(\nu)}(t,\omega_0)$ and mollify it, $\omega_\epsilon(t) = J_\epsilon(\omega(t))$.

We obtain from (3.3) that

$$\frac{d}{2dt}\|\omega_\epsilon(t)\|^2_{L^2(\mathbf{R}^2)} + \gamma\|\omega_\epsilon(t)\|^2_{L^2(\mathbf{R}^2)} + \nu\|\Lambda^\alpha\omega_\epsilon(t)\|^2_{L^2(\mathbf{R}^2)} - \langle J_\epsilon g, \omega_\epsilon(t)\rangle$$
$$= (\rho_\epsilon(u(t),\omega(t)), \nabla\omega_\epsilon(t)),$$

where we have used the identity in [4] with

$$\rho_\epsilon(u,\omega) = \int_{\mathbf{R}^2} j(z)\big(u(x-\epsilon z) - u(x)\big)\big(\omega(x-\epsilon z) - \omega(x)\big)dz - (u-u_\epsilon)(\omega-\omega_\epsilon).$$

Integrating in time we deduce

$$\frac{1}{t}\int_0^t [\gamma\|\omega_\epsilon(s+t_0)\|^2_{L^2(\mathbf{R}^2)} + \nu\|\Lambda^\alpha\omega_\epsilon(s+t_0)\|^2_{L^2(\mathbf{R}^2)} - \langle J_\epsilon g, \omega_\epsilon(s+t_0)\rangle]ds$$
$$= \frac{1}{2t}[\|\omega_\epsilon(t_0)\|^2_{L^2(\mathbf{R}^2)} - \|\omega_\epsilon(t+t_0)\|^2_{L^2(\mathbf{R}^2)}]$$
$$+ \frac{1}{t}\int_0^t \big(\rho_\epsilon(u(s+t_0),\omega(s+t_0)), \nabla\omega_\epsilon(s+t_0)\big)ds.$$

We apply $Lim_{t\to\infty}$ and from (3.16) we have that

$$\int_{L^2(\mathbf{R}^2)} [\gamma\|\omega_\epsilon\|^2_{L^2(\mathbf{R}^2)} + \nu\|\Lambda^\alpha\omega_\epsilon\|^2_{L^2(\mathbf{R}^2)} - \langle J_\epsilon g, \omega_\epsilon\rangle]d\mu^{(\nu)}(\omega)$$
$$= Lim_{t\to\infty}\frac{1}{t}\int_0^t \big(\rho_\epsilon(u(s+t_0),\omega(s+t_0)), \nabla\omega_\epsilon(s+t_0)\big)ds.$$

Similar to [5], noting that $\partial_i\omega_\epsilon(x) = \int_{\mathbf{R}^2} \partial_i j_\epsilon(z-x)\omega(z)dz$ $(i=1,2)$, we have

$$\|\nabla\omega_\epsilon(t)\|_{L^\infty(\mathbf{R}^2)} \leqslant C_j\frac{1}{\epsilon}\|\omega(t)\|_{L^\infty(\mathbf{R}^2)}$$

with the constant C_j depending only on the mollifier $j(\cdot)$. For $u = (u_1,u_2) = \frac{1}{2\pi}\frac{x^\perp}{|x|^2}\star\omega$, we have

$$\|u(\cdot - \epsilon z) - u(\cdot)\|_H \leqslant c_1\epsilon|z|(\|\nabla u_1\|_H + \|\nabla u_2\|_H) \leqslant c_2\epsilon|z|\|\omega\|_{L^2(\mathbf{R}^2)},$$

and obviously

$$u(x) - u_\epsilon(x) = \int_{\mathbf{R}^2} j(z)\big(u(x) - u(x-\epsilon z)\big)dz.$$

Therefore, we have

$$|(\rho_\epsilon(u(s+t_0),\omega(s+t_0)),\nabla\omega_\epsilon(s+t_0))|$$

$$\leqslant C_j \frac{1}{\epsilon}\|\omega(s+t_0)\|_{L^\infty(\mathbf{R}^2)} \left(\int_{\mathbf{R}^2} j(z)\|\delta_{\epsilon z}\omega(s+t_0)\|_{L^2(\mathbf{R}^2)} \cdot c_2 \epsilon |z| \|\omega(s+t_0)\|_{L^2(\mathbf{R}^2)} dz \right.$$

$$\left. + \|\omega(s+t_0) - \omega_\epsilon(s+t_0)\|_{L^2(\mathbf{R}^2)} \int_{\mathbf{R}^2} j(z) c_2 \epsilon |z| \|\omega(s+t_0)\|_{L^2(\mathbf{R}^2)} dz \right)$$

$$\leqslant M' \left(\int_{\mathbf{R}^2} j(z)\|\delta_{\epsilon z}\omega(s+t_0)\|_{L^2(\mathbf{R}^2)}|z| dz + \|\omega(s+t_0) - \omega_\epsilon(s+t_0)\|_{L^2(\mathbf{R}^2)} \right)$$

$$\leqslant 2M' \int_{\mathbf{R}^2} j(z)\|\delta_{\epsilon z}\omega(s+t_0)\|_{L^2(\mathbf{R}^2)} dz,$$

where $\delta_h \omega(x) = \omega(x) - \omega(x-h)$, and M' is a constant depends only on $\|\omega(s+t_0)\|_{L^\infty(\mathbf{R}^2)}\|\omega(s+t_0)\|_{L^2(\mathbf{R}^2)}$. Consequently, we have

$$\left| Lim_{t\to\infty} \frac{1}{t} \int_0^t (\rho_\epsilon(u(s+t_0),\omega(s+t_0)),\nabla\omega_\epsilon(s+t_0)) ds \right|$$

$$\leqslant M Lim_{t\to\infty} \frac{1}{t} \int_0^t \int_{\mathbf{R}^2} j(z)\|\delta_{\epsilon z}\omega(s+t_0)\|_{L^2(\mathbf{R}^2)} dz ds, \qquad (3.18)$$

where M is a bound on $\sup_{s\geqslant 0} \|\omega(s+t_0)\|_{L^\infty(\mathbf{R}^2)}\|\omega(s+t_0)\|_{L^2(\mathbf{R}^2)}$.

Note that $\overline{O_+(t_0,\omega_0)}$ is compact in $L^2(\mathbf{R}^2)$. Then for every small number $h > 0$ there exists $\epsilon > 0$ such that

$$\|\delta_{\epsilon z}\omega(s+t_0)\|_{L^2(\mathbf{R}^2)} \leqslant h$$

for all $s \geqslant 0$ and all z in the compact support of j. Therefore we have

$$\int_{L^2(\mathbf{R}^2)} [\gamma\|\omega_\epsilon\|_{L^2(\mathbf{R}^2)}^2 + \nu\|\Lambda^\alpha \omega_\epsilon\|_{L^2(\mathbf{R}^2)}^2 - \langle J_\epsilon g, \omega_\epsilon \rangle] d\mu^{(\nu)}(\omega) \leqslant h(\epsilon), \qquad (3.19)$$

where $0 \leqslant h(\epsilon)$, a function satisfying $\lim_{\epsilon \to 0} h(\epsilon) = 0$. We remove the mollifier. First we note that

$$\int_{L^2(\mathbf{R}^2)} (\gamma\|\omega\|_{L^2(\mathbf{R}^2)}^2 - \langle g, \omega \rangle) d\mu^{(\nu)}(\omega) = \lim_{\epsilon \to 0} \int_{L^2(\mathbf{R}^2)} (\gamma\|\omega_\epsilon\|_{L^2(\mathbf{R}^2)}^2 - \langle J_\epsilon g, \omega_\epsilon \rangle) d\mu^{(\nu)}(\omega)$$

holds trivially. This, together with (3.19), implies that

$$\nu \limsup_{\epsilon \to 0} \int_{L^2(\mathbf{R}^2)} \|\Lambda^\alpha \omega_\epsilon\|_{L^2(\mathbf{R}^2)}^2 d\mu^{(\nu)}(\omega) \leqslant - \int_{L^2(\mathbf{R}^2)} (\gamma\|\omega\|_{L^2(\mathbf{R}^2)}^2 - \langle g, \omega \rangle) d\mu^{(\nu)}(\omega).$$

By Fatou's lemma, we have

$$\nu \int_{L^2(\mathbf{R}^2)} \|\Lambda^\alpha \omega\|_{L^2(\mathbf{R}^2)}^2 d\mu^{(\nu)}(\omega) \leqslant - \int_{L^2(\mathbf{R}^2)} (\gamma\|\omega\|_{L^2(\mathbf{R}^2)}^2 - \langle g, \omega \rangle) d\mu^{(\nu)}(\omega),$$

which proves (3.11) and (3.17).

To verify (3.13), we take, similarly to [5], a smooth, nonnegative, compactly supported function $\chi'(y)$ defined for $y \geqslant 0$. Then $\chi(y) = \int_0^y \chi'(x)dx$ is bounded on $\mathbf{R}_+$ and

$$\frac{d}{dt}\chi(\|\omega_\epsilon(t)\|_{L^2(\mathbf{R}^2)}^2) = \chi'(\|\omega_\epsilon(t)\|_{L^2(\mathbf{R}^2)}^2)\frac{d}{dt}\|\omega_\epsilon(t)\|_{L^2(\mathbf{R}^2)}^2.$$

We proceed as above by taking time average and long time limit to obtain

$$\frac{1}{t}\int_0^t \chi'(\|\omega_\epsilon(t_0+s)\|_{L^2(\mathbf{R}^2)}^2)[\gamma\|\omega_\epsilon(s+t_0)\|_{L^2(\mathbf{R}^2)}^2 + \nu\|\Lambda^\alpha\omega_\epsilon(s+t_0)\|_{L^2(\mathbf{R}^2)}^2$$
$$- \langle J_\epsilon g, \omega_\epsilon(s+t_0)\rangle]ds$$
$$= \frac{1}{2t}[\|\chi(\omega_\epsilon(t_0))\|_{L^2(\mathbf{R}^2)}^2 - \|\chi(\omega_\epsilon(t+t_0))\|_{L^2(\mathbf{R}^2)}^2]$$
$$+ \frac{1}{t}\int_0^t \chi'(\|\omega_\epsilon(t_0+s)\|_{L^2(\mathbf{R}^2)}^2)\big(\rho_\epsilon(u(s+t_0),\omega(s+t_0)),\nabla\omega_\epsilon(s+t_0)\big)ds.$$

Noting that $\chi'(\cdot)$ is bounded, we have

$$\mid Lim_{t\to\infty}\frac{1}{t}\int_0^t \chi'(\|\omega_\epsilon(t_0+s)\|_{L^2(\mathbf{R}^2)}^2)\big(\rho_\epsilon(u(s+t_0),\omega(s+t_0)),\nabla\omega_\epsilon(s+t_0)\big)ds \mid$$
$$\leqslant c_1 M Lim_{t\to\infty}\frac{1}{t}\int_0^t \int_{\mathbf{R}^2} j(z)\|\delta_{\epsilon z}\omega(s+t_0)\|_{L^2(\mathbf{R}^2)}dzds,$$

where the constant c_1 is the bound of $\chi'(\cdot)$, and the constant M is the same as that in (3.18).

Hence, we can remove the mollifier as above and obtain

$$\int_{L^2(\mathbf{R}^2)} \chi'(\|\omega\|_{L^2(\mathbf{R}^2)}^2)\big(\nu\|\Lambda^\alpha\omega\|_{L^2(\mathbf{R}^2)}^2 + \gamma\|\omega\|_{L^2(\mathbf{R}^2)}^2 - \langle g,\omega\rangle\big)d\mu^{(\nu)}(\omega) \leqslant 0.$$

Taking $\chi'(y) \to 1_{[E_1^2, E_2^2]}$ pointwise with $0 \leqslant \chi'(y) \leqslant 2$ and using Fatou's lemma, we can deduce (3.13) of *Definition* 3.1. This concludes the proof of Theorem 3.5.

We are now ready to prove our first main result about the inviscid limit.

Proof of Theorem 1.1 We argue by contradiction and assume the conclusion were false. Then there exist $\delta > 0$ and a sequence $\nu_k \to 0$, and for each ν_k, there exists a sequence of time $t_j \to \infty$ such that

$$\frac{\nu_k}{t_j}\int_0^{t_j} \|\Lambda^\alpha S^{(\nu_k)}(s+t_0,\omega_0)\|_{L^2(\mathbf{R}^2)}^2 ds \geqslant \delta$$

holds for all t_j. From the energy estimates of (3.3), we have

$$\delta \leq \frac{\nu_k}{t_j} \int_0^{t_j} \|\Lambda^\alpha S^{(\nu_k)}(s+t_0,\omega_0)\|_{L^2(\mathbf{R}^2)}^2 ds$$

$$\leq \frac{1}{t_j} \int_0^{t_j} [-\gamma\|S^{(\nu_k)}(s+t_0,\omega_0)\|_{L^2(\mathbf{R}^2)}^2 + \langle g, S^{(\nu_k)}(s+t_0,\omega_0)\rangle] ds$$

$$+ \frac{1}{2t_j}[\|S^{(\nu_k)}(t_0,\omega_0)\|_{L^2(\mathbf{R}^2)}^2 - \|S^{(\nu_k)}(t+t_0,\omega_0)\|_{L^2(\mathbf{R}^2)}^2].$$

It follow that

$$\limsup_{t\to\infty} \frac{1}{t}\int_0^t [-\gamma\|S^{(\nu_k)}(s+t_0,\omega_0)\|_{L^2(\mathbf{R}^2)}^2 + \langle g, S^{(\nu_k)}(s+t_0,\omega_0)\rangle] ds \geq \delta.$$

By *Remark* 3.6, we can choose a generalized limit $Lim_{t\to\infty}$ such that

$$Lim_{t\to\infty} \frac{1}{t}\int_0^t [-\gamma\|S^{(\nu_k)}(s+t_0,\omega_0)\|_{L^2(\mathbf{R}^2)}^2 + \langle g, S^{(\nu_k)}(s+t_0,\omega_0)\rangle] ds$$

$$= \limsup_{t\to\infty} \frac{1}{t}\int_0^t [-\gamma\|S^{(\nu_k)}(s+t_0,\omega_0)\|_{L^2(\mathbf{R}^2)}^2 + \langle g, S^{(\nu_k)}(s+t_0,\omega_0)\rangle] ds.$$

Now, by *Theorem* 3.5, there exists a stationary statistical solution $\mu^{(\nu_k)}$ supported in $B_p^\infty(r)$ such that

$$\int_{L^2(\mathbf{R}^2)} \big(-\gamma\|\omega\|_{L^2(\mathbf{R}^2)}^2 + \langle g, \omega\rangle\big) d\mu^{(\nu_k)}(\omega) \geq \delta > 0. \qquad (3.20)$$

Passing to a weakly convergent subsequence, denoted again $\mu^{(\nu_k)}$, using *Theorems* 3.3 and 3.4, we find a renormalized stationary statistical solution μ^0 of the damped and driven Euler equation that satisfies enstrophy balance (3.15).

Because the function $\omega \mapsto \langle g,\omega\rangle$ is weakly continuous, we have

$$\lim_{k\to\infty} \int_{L^2(\mathbf{R}^2)} \langle g,\omega\rangle d\mu^{(\nu_k)}(\omega) = \int_{L^2(\mathbf{R}^2)} \langle g,\omega\rangle d\mu^0(\omega).$$

On the other hand, by Fatou lemma

$$\gamma \int_{L^2(\mathbf{R}^2)} \|\omega\|_{L^2(\mathbf{R}^2)}^2 d\mu^0(\omega) \leq \gamma \liminf_{k\to\infty} \int_{L^2(\mathbf{R}^2)} \|\omega\|_{L^2(\mathbf{R}^2)}^2 d\mu^{(\nu_k)}(\omega).$$

From (3.20) we obtain

$$\int_{L^2(\mathbf{R}^2)} (\gamma\|\omega\|_{L^2(\mathbf{R}^2)}^2 - \langle g,\omega\rangle) d\mu^0(\omega) \leq -\delta < 0,$$

contradicting energy dissipation balance (3.15). This concludes the proof of *Theorem* 1.1.

4 Global attractor for the GNS equations

Throughout the section, we assume that $\alpha \in \left(\dfrac{1}{2}, 1\right)$, $\nu, \gamma > 0$ are kept fixed, the force $f \in (W^{1,2}(\mathbf{R}^2))^2$ is fixed and time-independent.

4.1 Notation

We first recall in this subsection the notation about global attractor that we will use later; see [18, 20] for more details.

We consider a semigroup $\{S(t)\}_{t \geqslant 0}$ on a Banach space X, i.e. a family of mappings $S(t) : X \to X$, such that

$$S(0) = I_X \quad \text{and} \quad S(t+s) = S(t)S(s) \quad \text{for all } s, t \in [0, \infty) \text{ and } x \in X.$$

Definition 4.1 *Let $\{S(t)\}_{t \geqslant 0}$ be a semigroup on a Banach space X. A subset $\mathscr{A} \subset X$ is called a global attractor for the semigroup if $\mathscr{A}$ is compact in X and enjoys the following properties*:

(i) *$\mathscr{A}$ is an invariant set, i.e., $S(t)\mathscr{A} = \mathscr{A}$ for any $t \geqslant 0$;*

(ii) *$\mathscr{A}$ attracts all bounded sets of X, i.e., for any bounded subset B of X,*

$$\mathrm{dist}(S(t)B, \mathscr{A}) \to 0, \quad \text{as } t \to \infty,$$

where $\mathrm{dist}(A, B)$ is the Hausdorff semidistance of two sets A and B:

$$\mathrm{dist}(A, B) = \sup_{x \in A} \inf_{y \in B} \|x - y\|_X.$$

4.2 Global attractor

According to *Lemma* 2.1, we can define the operator semigroup $\{S(t)\}_{t \geqslant 0}$ on V as follows:

$$S(t)u_0 : \mathbf{R}^+ \times V \to V, \quad S(t)u_0 = u(t), \tag{4.1}$$

where $u(t)$ is the unique solution of (1.1) corresponding to the initial data $u_0 \in V$.

The main result of this subsection is to prove that the semigroup $\{S(t)\}_{t \geqslant 0}$ defined by (4.1) has a global attractor in the phase space V, that is:

Theorem 4.1 (existence) *Let $\alpha \in \left(\dfrac{1}{2}, 1\right)$, $\nu, \gamma > 0$ and $f \in (W^{1,2}(\mathbf{R}^2))^2$ be time-independent. Then, the semigroup $\{S(t)\}_{t \geqslant 0}$ generated by the solutions of (1.1) has a unique global attractor $\mathscr{A}$ in V.*

To prove *Theorem* 4.1, we need some dissipation estimates.

Lemma 4.1 *Under the assumptions of Theorem 4.1, there exists a subset $\mathscr{B} \subset V$, which is bounded in $H^{1+\alpha}$ and satisfies the following : for any bounded subset B of V, there exists a $t_B > 0$ which depends only on $\|B\|_V$ such that*

$$S(t)B \subset \mathscr{B} \quad \text{for all } t \geqslant t_B.$$

Proof We divide our proof into three steps.

Step 1. Multiplying equation (1.1) by u and integrating in space, we can deduce that

$$\frac{d}{dt}\|u\|^2 + 2\gamma\|u\|^2 + 2\nu\|\Lambda^\alpha u\|^2 \leqslant \|f\|\|u\|,$$

which implies that

$$\|u(t)\|^2 + 2\nu \int_0^t e^{\gamma(s-t)}\|\Lambda^\alpha u(s)\|^2 ds \leqslant e^{-\gamma t}\|u(0)\|^2 + \frac{1}{\gamma^2}\|f\|^2.$$

Hence, there exists a constant t_1 which depends only on $\|u(0)\|$ such that

$$\|u(t)\|^2 + 2\nu \int_0^t e^{\gamma(s-t)}\|\Lambda^\alpha u(s)\|^2 ds \leqslant \frac{1}{\gamma^2}\|f\|^2 + 1 := M_1 \quad \text{for all } t \geqslant t_1. \quad (4.2)$$

Step 2. Multiplying the vorticity equation (3.3) by ω and integrating over $\mathbf{R}^2$, we have that

$$\frac{d}{dt}\|\omega\|^2 + 2\gamma\|\omega\|^2 + 2\nu\|\Lambda^\alpha \omega\|^2 \leqslant \|g\|\|\omega\|,$$

which, similarly, implies that there exists a constant t_2 which depends only on $\|\omega(0)\|$, and so $\|u(0)\|_V$ such that

$$\|\omega(t)\|^2 + 2\nu \int_0^t e^{\gamma(s-t)}\|\Lambda^\alpha \omega(s)\|^2 ds \leqslant \frac{1}{\gamma^2}\|g\|^2 + 1 := M_2 \quad \text{for all } t \geqslant t_2. \quad (4.3)$$

It follows immediately from (4.2), (4.3) and *Lemma* 3.1 that

$$\|u(t)\|_V^2 = \|u(t)\|^2 + \|\Lambda u(t)\|^2 \leqslant M_1 + CM_2 := M_3 \quad \text{for all } t \geqslant t_1 + t_2,$$

where the constant C comes from *Lemma* 3.1. Moreover,

$$\int_t^{t+1} \|\Lambda^\alpha \omega(s)\|^2 ds \leqslant \frac{M_2}{2e\nu} \quad \text{for all } t \geqslant t_2. \quad (4.4)$$

Step 3. Multiplying the vorticity equation (3.3) by $\Lambda^{2\alpha}\omega$ and integrating over $\mathbf{R}^2$, we obtain, in much the same way as in the proof of *Lemma* 3.2, that

$$\frac{d}{dt}\|\Lambda^\alpha \omega(t)\|^2 + 2\gamma\|\Lambda^\alpha \omega(t)\|^2 \leqslant M_4 \quad \text{for all } t \geqslant t_1 + t_2,$$

where the positive constant M_4 depends only on M_2, M_3, ν and $\|g\|$. Combined with (4.4), this implies that

$$\|\Lambda^\alpha \omega(t)\|^2 \leqslant \frac{M_4}{2\gamma} + 1 \quad \text{for all } t \geqslant t_1 + t_2 + \frac{1}{2\gamma} \ln \frac{M_2}{2e\nu}.$$

Therefore, applying *Lemma* 3.1 again, we can finish our proof by setting

$$\mathscr{B} := \left\{ u \in H^{1+\alpha} \ : \ \|u\|^2_{H^{1+\alpha}} \leqslant M_1 + C^2 \left(\frac{M_4}{2\gamma} + 1 \right) \right\}$$

and

$$t_B = t_1 + t_2 + \frac{1}{2\gamma} \ln \frac{M_2}{2e\nu}.$$

Since the embedding $H^{1+\alpha} \hookrightarrow V$ is not compact, to deduce the necessary asymptotic compactness, we will use the *tail estimates* (see Wang [21]).

Lemma 4.2 *Under the assumptions of Theorem* 4.1, *for any* $\varepsilon > 0$ *and any bounded subset* $B \subset V$, *there exist* $T_B > 0$ *and* $K_B > 0$, *such that*

$$\int_{|x| \geqslant K_B} \|S(t)u_0\|^2 dx \leqslant \varepsilon \quad \text{for any } t \geqslant T_B \text{ and } u_0 \in B.$$

Proof The proof is standard and similar to the one of *Theorem* 3.1 (or see *Lemma* 4.5 below): taking $\chi(\cdot)$ to be a proper nonnegative smooth cut-off function and multiplying the equation (1.1) by $2\chi\left(\dfrac{x}{R}\right)u(x,t)$, we can finish the proof by applying the Gronwall inequality.

We are now ready to prove *Theorem* 4.1.

Proof of Theorem 4.1 *Lemma* 4.1 implies that $\{S(t)\}_{t \geqslant 0}$ has a bounded absorbing set $\mathscr{B}$ in V; *Lemmas* 4.1 and 4.2 imply that $\{S(t)\}_{t \geqslant 0}$ is asymptotical compact in V. The continuity with respect to initial data in $\mathscr{B}$ follows from (2.9) and interpolation. Hence, *Theorem* 4.1 follows from the standard criterion in [18, 20].

4.3 Finite dimensionality of the attractor

In this section we prove that the fractal dimension of the global attractor $\mathscr{A}$ obtained in *Theorem* 4.1 is finite in H^1. We recall that the fractal dimension $\dim_F(Z; X)$ of a compact set Z in space (topology) X is given by

$$\dim_F(Z; X) = \limsup_{r \to 0} \frac{\ln N_Z(r; X)}{-\ln r},$$

where $N_Z(r; X)$ is the minimal number of balls in X of radius r needed to cover Z.

From *Lemma* 4.1, we know that $\mathscr{A}$ is bounded in $H^{1+\alpha}$, consequently, $\mathscr{A}$ is compact in H^s for any $s \in [0,1]$, especially, $\mathscr{A}$ is closed in H^α.

For convenience, we denote by M the $H^{1+\alpha}$-bounds of $\mathscr{A}$,

$$M = \|\mathscr{A}\|_{H^{1+\alpha}}^2 = \sup_{y \in \mathscr{A}} \|y\|_{H^{1+\alpha}}^2 < \infty. \tag{4.5}$$

Let $u_0, v_0 \in \mathscr{A}$, and $u(t) = S(t)u_0$, $v(t) = S(t)v_0$ be the corresponding solutions of (1.1) respectively. Set $w(t) = u(t) - v(t)$; then $w(t)$ solves the following equations

$$\begin{cases} \partial_t w + \nu(-\Delta)^\alpha w + w \cdot \nabla u + v \cdot \nabla w + \gamma w + \nabla p_u - \nabla p_v = 0, \\ \nabla \cdot w = 0, \\ w(x,0) = u_0 - v_0. \end{cases} \tag{4.6}$$

Lemma 4.3 *There exists a positive constant l_1, which depends only on γ, ν and M, such that*

$$\|w(t)\|^2 \leqslant e^{l_1 t}\|u_0 - v_0\|^2 \quad \text{for all } t \geqslant 0. \tag{4.7}$$

Proof Multiplying (4.6) by w and integrating in space, we have

$$\frac{d}{dt}\|w(t)\|^2 + 2\gamma\|w(t)\|^2 + 2\nu\|\Lambda^\alpha\|^2 \leqslant 2\int_{\mathbf{R}^2} |w|^2 |\nabla u| dx;$$

in which,

$$\int_{\mathbf{R}^2} |w|^2 |\nabla u| dx \leqslant \|\nabla u\| \cdot \|w\|_{L^4(\mathbf{R}^2)}^2 \leqslant M\|w\|^{2r}\|w\|_{H^\alpha}^{2-2r}$$

$$\leqslant \frac{l_1}{2}\|w\|^2 + \frac{\min\{\gamma,\nu\}}{2}\|w\|_{H^\alpha}^2,$$

where, $r \in (0,1)$ and we have used the fact that $\alpha \in \left(\frac{1}{2}, 1\right)$ and the invariance of $\mathscr{A}$, and l_1 is a constant that depends only on γ, ν and M. Then, we have

$$\frac{d}{dt}\|w(t)\|^2 + \gamma\|w(t)\|^2 + \nu\|\Lambda^\alpha\|^2 \leqslant l_1\|w(t)\|^2, \tag{4.8}$$

thus (4.7) follows by an application of the Gronwall inequality.

Lemma 4.4 *There exists a positive constant l_2, which depends only on ν, γ and M, such that*

$$\|w(t)\|_{H^\alpha}^2 + \nu \int_0^t e^{\gamma(s-t)}\|w(s)\|_{H^{2\alpha}}^2 ds \leqslant e^{-\gamma t}\|w(0)\|_{H^\alpha}^2 + l_2 \int_0^t e^{\gamma(s-t)}\|w(s)\|^2 ds \tag{4.9}$$

for all $t \geqslant 0$.

Proof Multiplying (4.6) by $\Lambda^{2\alpha}w$ and integrating over $\mathbf{R}^2$, we have

$$\frac{1}{2}\frac{d}{dt}\|\Lambda^\alpha w(t)\|^2 + \gamma\|\Lambda^\alpha w(t)\|^2 + \nu\|\Lambda^{2\alpha}w\|^2$$
$$\leqslant \int_{\mathbf{R}^2} |w||\nabla u||\Lambda^{2\alpha}w|dx + \int_{\mathbf{R}^2} |v||\nabla w||\Lambda^{2\alpha}w|dx. \tag{4.10}$$

Applying the embedding $H^{1+\alpha} \hookrightarrow L^\infty(\mathbf{R}^2)$ and the interpolation, we have

$$\int_{\mathbf{R}^2} |v||\nabla w||\Lambda^{2\alpha}w|dx \leqslant M\|\nabla w\|\cdot\|\Lambda^{2\alpha}w\| \leqslant M\|w\|^r\cdot\|w\|_{H^{2\alpha}}^{2-r}$$
$$\leqslant C'_{M,\nu,\gamma}\|w\|^2 + \frac{\min\{\nu,\gamma\}}{4}\|w\|_{H^{2\alpha}}^2; \tag{4.11}$$

and similarly, applying the embedding $H^\alpha \hookrightarrow L^4(\mathbf{R}^2)$ and the interpolation, we have

$$\int_{\mathbf{R}^2} |w||\nabla u||\Lambda^{2\alpha}w|dx \leqslant \|w\|_{L^4(\mathbf{R}^2)}\|\nabla u\|_{L^4(\mathbf{R}^2)}\|\Lambda^{2\alpha}w\|$$
$$\leqslant M\|w\|_{L^4(\mathbf{R}^2)}\|\Lambda^{2\alpha}w\| \leqslant C''_{M,\nu,\gamma}\|w\|^2 + \frac{\min\{\nu,\gamma\}}{4}\|w\|_{H^{2\alpha}}^2. \tag{4.12}$$

Inserting (4.11) and (4.12) into (4.10), and combining with (4.8), we obtain that

$$\frac{d}{dt}\|w(t)\|_{H^\alpha}^2 + \gamma\|w(t)\|_{H^\alpha}^2 + \nu\|w(t)\|_{H^{2\alpha}}^2 \leqslant l_2\|w\|^2,$$

which, applying the Gronwall inequality, implies (4.9) immediately. Here the constant l_2 depends only on $l_1, \nu, \gamma, C'_{M,\nu,\gamma}$ and $C''_{M,\nu,\gamma}$.

We also need the following a priori estimates to overcome the difficulty arising from the unboundedness of the spatial domain.

Lemma 4.5 *There exist $t^* > 0$ and $k^* \gg 1$, such that*

$$\|w(t)\|_{H^\alpha}^2 + \int_0^t e^{\gamma(s-t)}\|w_t(s)\|^2 ds + \nu\int_0^t e^{\gamma(s-t)}\|w(s)\|_{H^{2\alpha}}^2 ds$$
$$\leqslant a(t)\|w(0)\|_{H^\alpha}^2 + E_w(t,k^*) \quad \text{for any } t \geqslant 0, \tag{4.13}$$

where $a(\cdot) : [0,\infty) \to [0,\infty)$ is continuous and satisfies

$$a(t^*) + a(2t^*) \leqslant \frac{1}{2}, \tag{4.14}$$

and

$$E_w(t,k) = l_2\left(1 + \frac{l_3}{\nu}\right)\int_0^t e^{\gamma(s-t)}\int_{|x|\leqslant 2k} |w(x,s)|^2 dx ds, \tag{4.15}$$

where the constant l_2 comes from Lemma 4.4, and l_3 is a constant that depends only on M, ν, γ.

Proof We divide our proof into three steps.

Step 1. Take $\chi(\cdot)$ to be a nonnegative smooth function supported in $\{x \in \mathbf{R}^2 : |x| \geqslant 1\}$ and identically equal to 1 for $|x| \geqslant 2$.

Multiplying the equation (4.6) by $2\chi\left(\frac{x}{k}\right) w(x,t)$ and integrating in space, we obtain that

$$\frac{d}{dt}\int_{\mathbf{R}^2} \chi\left(\frac{x}{k}\right) |w|^2 dx + 2\gamma \int_{\mathbf{R}^2} \chi\left(\frac{x}{k}\right) |w|^2 dx \leqslant \frac{C_0 \nu}{k^{2\alpha}} \|w\|^2 + 2\int_{\mathbf{R}^2} \chi\left(\frac{x}{k}\right) |w|^2 |\nabla u| dx, \quad (4.16)$$

where we have used an estimate similar as (3.6) for dealing with the fractional term $\Lambda^{2\alpha} w$ and the fact $b\left(v, w, \chi\left(\frac{x}{k}\right) w\right) = b\left(\chi\left(\frac{x}{k}\right) v, w, w\right) = 0$. At the same time, using again that $\alpha > \frac{1}{2}$, we have

$$\int_{\mathbf{R}^2} \chi\left(\frac{x}{k}\right) |w|^2 |\nabla u| dx \leqslant \left(\int_{|x| \geqslant k} |\nabla u|^2 dx\right)^{\frac{1}{2}} \|w\|^2_{L^4(\mathbf{R}^2)}$$

$$\leqslant c_0 \left(\int_{|x| \geqslant k} |\nabla u|^2 dx\right)^{\frac{1}{2}} \|w\|^2_{H^\alpha}, \quad (4.17)$$

where the constant c_0 is the embedding constant for $H^\alpha \hookrightarrow L^4(\mathbf{R}^2)$.

Consequently, from (4.16) and (4.17), as k is large enough such that $k^{2\alpha} \geqslant \frac{C_0 \nu}{\gamma}$, by applying the Gronwall inequality, we have that

$$\int_{|x| \geqslant 2k} |w(t)|^2 dx \leqslant \int_{\mathbf{R}^2} \chi\left(\frac{x}{k}\right) |w(t)|^2 dx$$

$$\leqslant e^{-\gamma t} \|w(0)\|^2 + 2c_0 \int_0^t e^{\gamma(s-t)} \left(\int_{|x| \geqslant k} |\nabla u(s)|^2 dx\right)^{\frac{1}{2}} \|w(s)\|^2_{H^\alpha} ds. \quad (4.18)$$

On the other hand, from (4.7) and (4.9), we have

$$\|w(t)\|^2_{H^\alpha} \leqslant e^{-\gamma t} \|w(0)\|^2_{H^\alpha} + \frac{l_2}{l_1} e^{l_1 t} \|w(0)\|^2 \leqslant \left(1 + \frac{l_2}{l_1} e^{l_1 t}\right) \|w(0)\|^2_{H^\alpha}. \quad (4.19)$$

We denote

$$I_k := \max_{v \in \mathscr{A}} \int_{|x| \geqslant k} |\nabla v(x)|^2 dx.$$

Then, combining with the invariance of $\mathscr{A}$, as $k^{2\alpha} \geqslant \frac{C_0 \nu}{\gamma}$, from (4.7), (4.18) and (4.19), we have that

$$\int_{|x| \geqslant 2k} |w(t)|^2 dx \leqslant e^{-\gamma t} \|w(0)\|^2 + \frac{2c_0}{\gamma} \left(1 + \frac{l_2}{l_1} e^{l_1 t}\right) I_k \|w(0)\|^2_{H^\alpha}. \quad (4.20)$$

Step 2. Denote by $\langle \cdot, \cdot \rangle$ the dual product between H and $H'(=H)$. Then, for any $\varphi \in H$, from the equation (4.6) we deduce that

$$|\langle w_t(s), \varphi \rangle| \leqslant (\gamma \|w(s)\| + \nu \|\Lambda^{2\alpha} w(s)\|) \|\varphi\|$$
$$+ (\|w(s)\|_{L^4(\mathbf{R}^2)} \|\nabla u(s)\|_{L^4(\mathbf{R}^2)} + \|v\|_{L^\infty(\mathbf{R}^2)} \|\nabla w(s)\|) \|\varphi\|,$$

as that for (4.11) and (4.12), we can deduce that

$$|\langle w_t(s), \varphi \rangle| \leqslant (\gamma \|w(s)\| + \nu \|\Lambda^{2\alpha} w(s)\|) \|\varphi\| + l_3' \|w(s)\|_{H^{2\alpha}} \|\varphi\|$$

for some constant l_3' that depends only on M. Then, combining with (4.9), we have

$$\int_0^t e^{\gamma(s-t)} \|w_t(s)\|^2 ds \leqslant C_{\gamma, \nu, l_3'} \int_0^t e^{\gamma(s-t)} \|w(s)\|_{H^{2\alpha}}^2 ds$$
$$:= l_3 \int_0^t e^{\gamma(s-t)} \|w(s)\|_{H^{2\alpha}}^2 ds. \tag{4.21}$$

Step 3. Now, returning to (4.9), we have that

$$\|w(t)\|_{H^\alpha}^2 + \nu \int_0^t e^{\gamma(s-t)} \|w(s)\|_{H^{2\alpha}}^2 ds \leqslant e^{-\gamma t} \|w(0)\|_{H^\alpha}^2$$
$$+ l_2 \int_0^t e^{\gamma(s-t)} \int_{|x| \leqslant 2k} |w(x,s)|^2 dx ds + l_2 \int_0^t e^{\gamma(s-t)} \int_{|x| \geqslant 2k} |w(x,s)|^2 dx ds. \tag{4.22}$$

We set

$$E'(t,k) = e^{-\gamma t} + l_2 t e^{-\gamma t} + \frac{2 l_2 c_0 t}{\gamma} \left(1 + \frac{l_2}{l_1} e^{l_1 t}\right) I_k \tag{4.23}$$

and

$$E(t,k) = \left(1 + \frac{l_3}{\nu}\right) E'(t,k).$$

Then, from (4.21), (4.22) and (4.20), we have

$$\|w(t)\|_{H^\alpha}^2 + \int_0^t e^{\gamma(s-t)} \|w_t(s)\|^2 ds + \nu \int_0^t e^{\gamma(s-t)} \|w(s)\|_{H^{2\alpha}}^2 ds$$
$$\leqslant E(t,k) \|w(0)\|_{H^\alpha}^2 + E_w(t,k) \quad \text{for all } t \geqslant 0. \tag{4.24}$$

Since $\mathscr{A}$ is compact in V, we have $I_k \to 0$ as $k \to \infty$; thus, in (4.23), we can first take t large enough, e.g., $t = t^* > 0$, and then take k large enough, e.g., $k = k^* \gg 1$, such that

$$E(t^*, k^*) \leqslant \frac{1}{4} \quad \text{and} \quad E(2t^*, k^*) \leqslant \frac{1}{4}.$$

Thus, we can finish the proof from (4.24) by taking $a(s) = E(s, k^*)$.

We are now ready to state and prove the main result of this subsection.

Theorem 4.2 (finite dimensionality) *The fractal dimension of the global attractor $\mathscr{A}$ (obtained in Theorem 4.1) is finite in H^1.*

Proof The proof is based on the idea of *l*-trajectories, see [17]; here we use also a criterion given in [3], namely [*Theorem* 2.15, [3]].

Let t^* and k^* be the constants given in *Lemma* 4.5.

Define the space $W \subset L^2(0, t^*; (L^2(\mathbf{R}^2))^2)$ as follows:

$$W = \{\phi \in L^2(0, t^*; (L^2(\mathbf{R}^2))^2)$$
$$: \int_0^{t^*} e^{\gamma(s-t^*)} \int_{\mathbf{R}^2} (|\Lambda^{2\alpha}\phi(x,s)|^2 + |\phi_t(x,s)|^2) dx ds < \infty\},$$

endowed with the norm

$$\|\phi\|_W^2 = \int_0^{t^*} e^{\gamma(s-t^*)} \left(\nu\|\phi(s)\|_{H^{2\alpha}}^2 + \|\phi_t(s)\|_{(L^2(\mathbf{R}^2))^2}^2\right) ds.$$

Then, $(W, \|\cdot\|_W)$ is a Banach space.

Define

$$X = H^\alpha \times W,$$

endowed with the norm

$$\|(y, z)\|_X^2 = \|y\|_{H^\alpha}^2 + \|z\|_W^2 \quad \text{for all } (y, z) \in X.$$

For the global attractor $\mathscr{A}$ given in *Theorem* 4.1, define

$$\mathscr{A}_{t^*} = \{(u_0, l(u_0)) : u_0 \in \mathscr{A}, \ l(u_0) = \{S(s)u_0 : s \in [0, t^*]\}\};$$

and define the mapping $\mathcal{L}$ on $\mathscr{A}_{t^*}$ as follows:

$$\mathcal{L} : \mathscr{A}_{t^*} \mapsto X,$$
$$\mathcal{L}(u_0, l(u_0)) = (S(t^*)u_0, l(S(t^*)u_0)) \quad \text{for all } (u_0, l(u_0)) \in \mathscr{A}_{t^*}.$$

From the invariance and compactness of $\mathscr{A}$ and Lemma 4.5, we have that $\mathscr{A}_{t^*}$ is a closed in X and $\mathcal{L}\mathscr{A}_{t^*} = \mathscr{A}_{t^*}$.

For any $(u_0, l(u_0)), (v_0, l(v_0)) \in \mathscr{A}_{t^*}$, we have that

$$\|\mathcal{L}(u_0, l(u_0)) - \mathcal{L}(v_0, l(v_0))\|_X^2$$
$$= \|S(t^*)u_0 - S(t^*)v_0\|_{H^\alpha}^2 + \int_0^{t^*} e^{\gamma(s-t^*)}(\nu\|S(s+t^*)u_0 - S(s+t^*)v_0\|_{H^{2\alpha}}^2$$
$$+ \|(S(s+t^*)u_0 - S(s+t^*)v_0)_t\|^2) ds$$

$$= \|S(t^*)u_0 - S(t^*)v_0\|_{H^\alpha}^2$$
$$+ \int_{t^*}^{2t^*} e^{\gamma(s-2t^*)}(\nu\|S(s)u_0 - S(s)v_0\|_{H^{2\alpha}}^2 + \|(S(s)u_0 - S(s)v_0)_t\|^2)ds$$
$$= I_1 + I_2;$$

where, from *Lemma* 4.5, we have

$$I_1 \leqslant a(t^*)\|u_0 - v_0\|_{H^\alpha}^2 + l_2\left(1 + \frac{l_3}{\nu}\right)\int_0^{t^*} e^{\gamma(s-t^*)}\int_{|x|\leqslant 2k^*} |S(s)u_0 - S(s)v_0|^2 dxds$$

and

$$I_2 \leqslant \int_0^{2t^*} e^{\gamma(s-2t^*)}(\nu\|S(s)u_0 - S(s)v_0\|_{H^{2\alpha}}^2 + \|(S(s)u_0 - S(s)v_0)_t\|^2)ds$$
$$\leqslant a(2t^*)\|u_0 - v_0\|_{H^\alpha}^2 + l_2\left(1 + \frac{l_3}{\nu}\right)\int_0^{2t^*} e^{\gamma(s-2t^*)}\int_{|x|\leqslant 2k^*}|S(s)u_0 - S(s)v_0|^2 dxds$$
$$\leqslant a(2t^*)\|u_0 - v_0\|_{H^\alpha}^2 + l_2\left(1+\frac{l_3}{\nu}\right)e^{-\gamma t^*}\int_0^{t^*} e^{\gamma(s-t^*)}\int_{|x|\leqslant 2k^*}|S(s)u_0 - S(s)v_0|^2 dxds$$
$$+ l_2\left(1 + \frac{l_3}{\nu}\right)\int_0^{t^*} e^{\gamma(s-t^*)}\int_{|x|\leqslant 2k^*}|S(t^*+s)u_0 - S(t^*+s)v_0|^2 dxds.$$

Therefore,

$$\|\mathcal{L}(u_0, l(u_0)) - \mathcal{L}(v_0, l(v_0))\|_X^2$$
$$\leqslant \big(a(t^*) + a(2t^*)\big)\|u_0 - v_0\|_{H^\alpha}^2$$
$$+ l_2\left(1 + \frac{l_3}{\nu}\right)(1 + e^{-\gamma t^*})\int_0^{t^*} e^{\gamma(s-t^*)}\int_{|x|\leqslant 2k^*}|S(s)u_0 - S(s)v_0|^2 dxds$$
$$+ l_2\left(1 + \frac{l_3}{\nu}\right)\int_0^{t^*} e^{\gamma(s-t^*)}\int_{|x|\leqslant 2k^*}|S(t^*+s)u_0 - S(t^*+s)v_0|^2 dxds$$
$$\leqslant \big(a(t^*) + a(2t^*)\big)\|(u_0, l(u_0)) - (v_0, l(v_0))\|_X^2$$
$$+ l_2\left(1 + \frac{l_3}{\nu}\right)(1 + e^{-\gamma t^*})\|(u_0, l(u_0)) - (v_0, l(v_0))\|_c^2$$
$$+ l_2\left(1 + \frac{l_3}{\nu}\right)\|\mathcal{L}(u_0, l(u_0)) - \mathcal{L}(v_0, l(v_0))\|_c^2, \tag{4.25}$$

where $\|\cdot\|_c$ is the compact seminorm on X defined as follows (a seminorm $\|\cdot\|_c$ on X is said to be compact iff for any bounded set $B \subset X$ there exists a sequence $\{\psi_n\} \subset B$ such that $\|\psi_n - \psi_m\|_c \to 0$ as $m, n \to \infty$):

$$\|(v_0, z(\cdot))\|_c^2 := \int_0^{t^*} e^{\gamma(s-t^*)}\int_{|x|\leqslant 2k^*}|z(x,s)|^2 dxds \quad \text{for all } (v_0, z(\cdot)) \in X.$$

At the same time, combining (4.7) with the first inequality of (4.25), it is easy to see the following Lipschitz continuity:

$$\|\mathcal{L}(u_0, l(u_0)) - \mathcal{L}(v_0, l(v_0))\|_X^2$$
$$\leqslant \left(a(t^*) + a(2t^*)\right)\|u_0 - v_0\|_{H^\alpha}^2$$
$$+ l_2\left(1 + \frac{l_3}{\nu}\right)(1 + e^{-\gamma t^*} + e^{l_1 t^*})\int_0^{t^*} e^{\gamma(s-t^*)} \int_{|x|\leqslant 2k^*} |S(s)u_0 - S(s)v_0|^2 dx ds$$
$$\leqslant L\|(u_0, l(u_0)) - (v_0, l(v_0))\|_X^2, \tag{4.26}$$

where the constant L depends only on l_i, γ, ν and t^*.

Note that $a(t^*) + a(2t^*) \leqslant \frac{1}{2}$. By (4.25) and (4.26), we have verified all conditions in [*Theorem* 2.15, [3]] for the mapping $\mathcal{L}$ on $\mathscr{A}_{t^*}$. Thus $\mathscr{A}_{t^*}$ is compact in X and has finite fractal dimension.

Finally, note that the canonical projection $P: X \to H^\alpha$ defined as $P(y, z) = y$ is obviously Lipschitz with Lipschitz constant 1 and $\mathscr{A} = P\mathscr{A}_{t^*}$. We thus have

$$\dim_F(\mathscr{A}; H^\alpha) \leqslant \dim_F(\mathscr{A}_{t^*}; X) < \infty,$$

and the finite dimensionality in H^1 follows immediately by (4.5) and interpolation.

Proof of Theorem 1.2 It is a direct result of *Theorems* 4.1 and 4.2.

5 Some remarks

For the vorticity equation (3.3), as the viscosity parameter $\nu > 0$ is kept fixed, under the same assumptions about the forcing term and initial data as that in Section 3, i.e., $g \in L^1(\mathbf{R}^2) \cap L^\infty(\mathbf{R}^2)$ and $\omega_0 \in L^2(\mathbf{R}^2)$, proceed as that in Section 4, we can also obtain a compact global attractor $\mathscr{A}^{(\nu)} \subset L^2(\mathbf{R}^2)$ for the corresponding semigroup $\{S^{(\nu)}(t)\}_{t\geqslant 0}$ in the phase space $L^2(\mathbf{R}^2)$. The necessary asymptotical compactness follows from *Lemma* 3.2 and the fact that the estimates in (3.7) depends only on the $L^2(\mathbf{R}^2)$-size of initial data (which will deduce the *tail estimate* for $\{S^{(\nu)}(t)\}_{t\geqslant 0}$ in $L^2(\mathbf{R}^2)$). Moreover, the stationary statistical solution $\mu^{(\nu)}$ obtained in *Theorem* 3.5 will support in $\mathscr{A}^{(\nu)}$ for each $\nu > 0$.

Concerning the fractal dimension of $\mathscr{A}$ in *Theorem* 4.2, we indeed (e.g., by checking the general criterions presented in [3, 17]) can give an upper bound which depends explicitly on the parameters in (4.13)-(4.15) (which depend on γ, ν and the size of the forcing term f).

The assumption $\alpha > \frac{1}{2}$ (used, e.g., *Lemma* 2.1 and subsequently) is essential in this paper. Hence, how to obtain the same results as that in Sections 3 and 4 for the case $\alpha = \frac{1}{2}$ would be more interesting and challenging.

Acknowledgements The second author acknowledges the support of the 973 Program (no. 2013CB834100) and the NSFC grants 91130005 and 11271052. The fourth author acknowledges the support of the NSFC grants 11471148 and 11522109.

References

[1] Caffarelli L, Vasseur A. Drift diffusion equations with fractional diffusion and the quasi-geostrophic equation[J]. Ann. Math., 2010, 171(2):1903-1930.

[2] Cheskidov A, Dai M. The existence of a global attractor for the forced critical surface Quasi-Geostrophic equation in L^2[J]. J. Math. Fluid Mech., 2018, 20(1):213-225.

[3] Chueshov I, Lasiecka I. Long-Time Behavior of Second Order Evolution Equations with Nonlinear Damping[M]. Memoirs of the American Mathematical Society, 195(912), 2008.

[4] Constantin P, E W, Titi E. Onsager's conjecture on the energy conservation for solutions of Euler's equation[J]. Commun. Math. Phys., 1994, 165:207-209.

[5] Constantin P, Ramos F. Inviscid limit for damped and driven incompressible Navier-Stokes equations in $\mathbf{R}^2$[J]. Commun. Math. Phys., 2007, 275:529-551.

[6] Constantin P, Tarfulea A, Vicol V. Absence of anomalous dissipation of energy in forced two dimensional fluid equations[J]. Arch. Ration. Mech. Anal., 2014, 212:875-903.

[7] Córdoba A, Córdoba D. A maximum principle applied to quasi-geostrophic equations[J]. Commun. Math. Phys., 2004, 249:511-528.

[8] Dlotko T, Kania M, Sun C. Quasi-geostrophic equation in $\mathbf{R}^2$[J]. J. Differential Equations, 2015, 259:531-561.

[9] Droniou J, Imbert C. Fractal first-order partial differential equations[J]. Arch. Rational Mech. Anal., 2006, 182:299-331.

[10] Foias C. Statistical study of Navier-Stokes equations. I[J]. Rend. Sem. Mat. Univ. Padova, 1972, 48:219-348.

[11] Foias C. Statistical study of Navier-Stokes equations. II[J]. Rend. Sem. Mat. Univ. Padova, 1973, 49:9-123.

[12] Foias C, Manley O, Rosa R,Temam R. Navier-Stokes Equations and Turbulence[M]. Encyclopedia of Mathematics and its applications, 83, Cambridge University Press, 2001.

[13] Frisch U. Turbulence-The legacy of A. N. Kolmogorov[M]. Cambridge University Press, Cambridge, 1995.

[14] Lamb H. Hydrodynamics[M]. 6th ed., Combridge University Press, 1957.

[15] Li P, Zhai Z. Well-posedness and regularity of generalized Navier-Stokes equations in some critical Q-spaces[J]. J. Functional Analysis, 2010, 259:2457-2519.

[16] Lions J. Quelques méthodes de résolution des problèmes aux limites non linéaires[M]. Dunod Gauthier-Villars, Paris, 1969.

[17] Malek J, Prazak D. Large time behavior via the method of l-trajectories[J]. J. Differential Equations, 2002, 181:243-279.

[18] Robinson J. Infinite-Dimensional Dynamical Systems: An Introduction to Dissipative Parabolic PDEs and the Theory of Global Attractors[M]. Cambridge University Press, 2001.

[19] Taylor G. Statistical theory of turbulence[J]. Proc. Roy. Soc. London Ser. A, 1935, 151:421-478.

[20] Temam R. Infinite-Dimensional Dynamical Systems in Mechanics and Physics[M]. Springer, New York, 1997.

[21] Wang B. Attractors for reaction-diffusion equations in unbounded domains[J]. Physica D, 1999, 128:41-52.

[22] Wu J. The generalized incompressible Navier-Stokes equations in Besov spaces[J]. Dyn. Partial Differ. Eq., 2004, 1:381-400.

[23] Wu J. Lower bounds for an integral involving fractional Laplacians and the generalized Navier-Stokes equations in Besov spaces[J]. Commun. Math. Phys., 2005, 263:803-831.

[24] Wu H, Fan J. Weak-strong uniqueness for the generalized Navier-Stokes equations[J]. Applied Math. Letters, 2012, 25:423-428.

[25] Zhang T, Fang D. Random data Cauchy theory for the generalized incompressible Navier-Stokes equations[J]. J. Math. Fluid Mech., 2012, 14:311-324.

Globally Smooth Solution and Blow-up Phenomenon for a Nonlinearly Coupled Schrödinger System Atomic Bose-Einstein Condensates*

Guo Boling (郭柏灵), Li Qiaoxin (李巧欣) and Wu Xinglong (吴兴龙)

Abstract In this paper, we study the nonlinearly coupled Schrödinger equations for atomic Bose-Einstein condensates. By using the Galërkin method and *a priori* estimates, the global existence of smooth solution is obtained. And under some conditions of the coefficients and p, the blow-up theorem is established.

Keywords the Galërkin method; locally smooth solution; globally smooth solution; *a priori* estimates; blow-up solution.

1 Introduction

In this paper we consider the following nonlinearly coupled Schrödinger system

$$\begin{cases} i\hbar u_t = \left(-\dfrac{\hbar^2}{2M}\Delta + \lambda_u|u|^2 + \lambda|v|^2 + g_{11}|u|^{2p} + g|u|^{p-1}|v|^{p+1}\right)u + \sqrt{2}\alpha \bar{u}v, \\ i\hbar v_t = \left(-\dfrac{\hbar^2}{4M}\Delta + \epsilon + \lambda_v|v|^2 + \lambda|u|^2 + g|u|^{p+1}|v|^{p-1} + g_{22}|v|^{2p}\right)v + \dfrac{\alpha}{\sqrt{2}}u^2, \end{cases} \quad (1.1)$$

with the initial condition and periodic boundary condition

$$u(x,0) = u_0(x), \quad v(x,0) = v_0(x), \quad x \in \Omega, \quad (1.2)$$

$$u(x+2L,t) = u(x,t), \quad v(x+2L,t) = v(x,t), \quad x \in \Omega, \quad t \geqslant 0, \quad (1.3)$$

where $\Delta = \dfrac{\partial^2}{\partial x^2}$, $i = \sqrt{-1}$, $L > 0$, $\Omega = (-L, L)$, $\hbar$ is Planck constant, $M > 0$ is the

---

* Commun. Math. Sci, 2016, 14(4): 1005−1021.

mass of a single atom, $\lambda_u, \lambda_v, \lambda$ represent the strengths of the atom-atom, molecule-molecule and atom-molecule interactions, respectively and ϵ, α are any real constants.

It was an eagerly anticipated event that the Ketterle's group found the Feschbach resonances in the inter-particle interactions of a dilute Bose-Einstein condensate of Na-atoms at MIT [22]. Since all quantities of interest in the atomic BECs crucially depend on the scattering length, a tunable interaction suggests very interesting studies of the many-body behavior of condensate systems. Then for the time evolution of the dilute single condensate system, the corresponding equation becomes the Gross-Pitaevskii equation which is the Dirac time-dependent variational scheme.

When $p = \hbar = M = 1$, the system (1.1) is similar to the following coupled Gross-Pitaevskii equtions

$$\begin{cases} iu_t = \left(-\dfrac{\nabla^2}{2} + V(x) + (\lambda_u|u|^2 + \lambda|v|^2)\right)u + \alpha v, \\ iv_t = \left(-\dfrac{\nabla^2}{2} + V(x) + \epsilon + (\lambda_v|v|^2 + \lambda|u|^2)\right)v + \alpha u. \end{cases} \quad (1.4)$$

Several authors are interested in the existence of solution for this problem. In 2011, Weizhu Bao and Yongyong Cai [4] proved the existence and uniqueness results for the ground states of the above coupled Gross-Pitaevskii equations and obtained the limiting behavior of the ground states with large parameters. Tingchun Wang and Xiaofei Zhao [23] also studied this problem. They proposed and analyzed the finite difference methods for solving (1.4) in two dimensions.

The single Gross-Pitaevskii equation has been considered by many authors. In 2003, Weizhu Bao, Dieter Jaksch, and Peter A. Markowich [6] obtained the numerical solution of the time-dependent Gross-Pitaevskii equation (TGPE) and in [5], Weizhu Bao, Weijun Tang proposed a new numerical method to compute the ground-state solution of trapped interacting Bose-Einstein condensation at zero or very low temperature by directly minimizing the energy functional via finite element approximation. Further discussion can be found in Refs [2, 3, 18, 19].

However, to our knowledge, the TGPE has not yet been fully studied. When $p = 1$, the system (1.1) reduce to

$$\begin{cases} i\hbar u_t = \left(-\dfrac{\hbar^2}{2M}\Delta + a|u|^2 + b|v|^2\right)u + \sqrt{2}\alpha\bar{u}v, \\ i\hbar v_t = \left(-\dfrac{\hbar^2}{4M}\Delta + \varepsilon + b|u|^2 + c|v|^2\right)v + \dfrac{\alpha}{\sqrt{2}}u^2, \end{cases} \quad (1.5)$$

these coupled non-linear equations replace the usual Gross-Pitaevskii equation that describes the time evolution of the dilute single condensate system [12, 17, 22]. In this paper, we consider the system (1.1) which is a more general problem than (1.5) as a mathematical model in nonlinear partial differential equations. And our aim is to obtain the globally smooth solution of system (1.1) and under some conditions to establish the blow-up theorem for the case of $p \geqslant 2$. The main difficulty is to establish certain delicate *a priori* estimates that govern our strategy to prove the existence of the smooth solution.

In [13]- [16], the initial value problem and the periodic boundary value problem were studied by Boling Guo for a class of systems of standard nonlinear Schrödinger equations. In this paper, we prove the existence and uniqueness of the global solution to the periodic boundary value problem for the nonlinearly coupled Schrödinger system (1.1) by using the Faedo-Galëkin method.

For the *a priori* estimates of the solution to the system (1.1)–(1.3) are unconcerned with the period L, we can derive the global smooth solution as $L \to \infty$, a.e. $x \in \mathbb{R}$. Theorem 1.2 is the global smooth solution to the periodic boundary value problem for the system (1.1) and Theorem 1.3 is the global smooth solution for $x \in \mathbb{R}$.

Before starting the main results, we review the notations and the calculus inequalities used in this paper.

To simplify the notation in this paper we shall denote by $\int U(x)dx$ the integration $\int_\Omega U(x)dx$, C is a generic constant and may assume different values in different formulates.

Let $L^m(\Omega), 1 \leqslant m \leqslant \infty$ be the classical Lebesgue space with the norm

$$\|u\|_m = \left(\int_\Omega |u|^m dx\right)^{\frac{1}{m}} \quad (1 \leqslant m < \infty),$$

$$\|u\|_\infty = ess.sup.\{|u(x)| : x \in \Omega\} \quad (m = \infty).$$

The usual L^2 inner product is $(u,v) = \int_\Omega u\bar{v}dx$, where $\bar{v}$ denotes the complex conjugate of v, and the norm of L^2 is $\|u\|_2 = \sqrt{(u,u)}$.

Denote $H^m(\Omega), m = 1, 2, \cdots$ be the Sobolev space of complex-valued functions with the norm

$$\|u\|_{H^m} = \left(\int_\Omega \sum_{|\alpha|\leqslant m} |D^\alpha u|^2 dx\right)^{\frac{1}{2}}.$$

Define $\Lambda = \{u \in H^1(\Omega) : |x|u \in L^2(\Omega)\}$ with norm $\|u\|_\Lambda^2 dx = \int_\Omega (|\nabla u|^2 + |x|^2|u|^2)dx$ and $\Lambda \times \Lambda$ is simply denoted by Λ^2.

The following auxiliary lemmas will be needed.

Lemma 1.1 (the Gagliardo-Nirenberg inequality) *Assuming $u \in L^q(\mathbb{R})$, $\partial_x^m u \in L^r(\mathbb{R})$, $1 \leqslant q, r \leqslant \infty$. Let p and α satisfy*

$$\frac{1}{p} = j + \alpha\left(\frac{1}{r} - m\right) + (1-\alpha)\frac{1}{q}; \quad \frac{j}{m} \leqslant \alpha \leqslant 1.$$

Then

$$\|\partial_x^j u\|_p \leqslant C(p,m,j,q,r)\|\partial_x^m u\|_r^\alpha \|u\|_q^{1-\alpha}. \tag{1.6}$$

In particular, as $m = 1, j = 0, p = 4, r = 2, q = 2$, we have

$$\|u\|_4^4 \leqslant C\|u_x\|_2^{\frac{1}{2}}\|u\|_2^3, \tag{1.7}$$

$$\|u\|_{2p+2}^{2p+2} \leqslant C\|u_x\|_2^p \|u\|_2^{p+2}. \tag{1.8}$$

Lemma 1.2 (the Gronwall inequality) *Let c be a constant, and $b(t), u(t)$ be nonnegative continuous functions in the interval $[0,T]$ satisfying*

$$u(t) \leqslant c + \int_0^t b(\tau)u(\tau)d\tau, \quad t \in [0,T].$$

Then $u(t)$ satisfies the estimate

$$u(t) \leqslant c\exp\left(\int_0^t b(\tau)d\tau\right), \quad \text{for} \quad t \in [0,T]. \tag{1.9}$$

Our main results are:

Theorem 1.1 *Let $p \geqslant 1, m > \frac{1}{2}, u_0(x) \in H_{per}^m(\Omega), v_0(x) \in H_{per}^m(\Omega)$. Then the system (1.1)–(1.3) has a locally smooth solution, which satisfies*

$$(u,v) \in L^\infty([0,T_0]; H_{per}^m(\Omega))^2. \tag{1.10}$$

Theorem 1.2 *Let $u_0(x) \in H_{per}^m(\Omega), v_0(x) \in H_{per}^m(\Omega)$, and $m > \frac{1}{2}$. Suppose one of the following conditions holds*

(i) $p \geqslant 2$, $g_{11}, g_{22} \geqslant 0$, $\begin{pmatrix} g_{11} & g \\ g & g_{22} \end{pmatrix}$ is positive define,

(ii) $1 \leqslant p < 2$,

then $\forall T > 0$, the system (1.1)–(1.3) has a uniquely global smooth solution

$$(u,v) \in L^\infty\left([0,T]; H_{per}^m(\Omega)\right)^2. \tag{1.11}$$

Theorem 1.3 Let $u_0(x) \in H^m(\mathbb{R}), v_0(x) \in H^m(\mathbb{R})$, and $m > \dfrac{1}{2}$. If one of the following conditions holds

(i) $p \geqslant 2$, $g_{11}, g_{22} \geqslant 0$, $\begin{pmatrix} g_{11} & g \\ g & g_{22} \end{pmatrix}$ is positive define,

(ii) $1 \leqslant p < 2$,

then the system (1.1)–(1.2) has a uniquely global smooth solution

$$(u,v) \in L^\infty_{loc}([0,\infty); H^m(\mathbb{R}))^2. \tag{1.12}$$

Theorem 1.4 Let $p \geqslant 2$ and $(u(t), v(t)) \in \Lambda^2$. If $g_{11}, g_{22} \geqslant 0$, $\begin{pmatrix} g_{11} & g \\ g & g_{22} \end{pmatrix}$ is negative definite, $\epsilon > \left|\dfrac{3\sqrt{2}}{8}\alpha\right|$, $\lambda_u > \left|\dfrac{3\sqrt{2}}{2}\alpha\right|$, $\hbar, \lambda_v, \lambda > 0$ and one of the following conditions holds

(i) $E_0 = E(u_0, v_0) < 0$,

(ii) $E_0 = 0$ and $\mathrm{Im} \int (x\overline{u_0}v_{x0} + x\overline{v_{x0}}u_0) < 0$,

(iii) $E_0 > 0$ and $\mathrm{Im} \int (x\overline{u_0}v_{x0} + x\overline{v_0}u_{x0}) < -\dfrac{2}{\hbar}\sqrt{h(0)E_0}$.

Then the solution $(u(t), v(t))$ of the system (1.1) blows up in a finite time, i.e., there is $T_* > 0$ such that

$$\lim_{t \to T_*^-} \int (|u_x|^2 + |v_x|^2)dx = +\infty. \tag{1.13}$$

Theorem 1.1. can be easily proved by Galërkin method(see, e.g., Ref. [11]). The detailed proof is omitted here.

2 The global existence of smooth solution

In this section, we give the demonstration of *a priori* estimates that guarantee the existence of the global smooth solution of the system (1.1)–(1.3). Also, we get the uniqueness of the solution.

Lemma 2.1 Let $u_0(x) \in L^2(\Omega), v_0(x) \in L^2(\Omega)$ and (u,v) be a locally smooth solution of the system (1.1) with initial data (u_0, v_0), then we have the identity

$$\|u(x,t)\|_2^2 + 2\|v(x,t)\|_2^2 \equiv \|u_0(x)\|_2^2 + 2\|v_0(x)\|_2^2. \tag{2.1}$$

Proof Taking the inner product for the first equation of the system (1.1) with $\overline{u}$ and the second equation with $\overline{v}$, respectively, and integrating the resulting equations with respect to x on Ω, and then taking the imaginary part of the resulting equations, we obtain

$$\begin{cases} \dfrac{\hbar}{2}\dfrac{d}{dt}\|u\|_2^2 = \sqrt{2}\alpha \mathrm{Im}\int (\overline{u})^2 v\, dx, \\ \dfrac{\hbar}{2}\dfrac{d}{dt}\|v\|_2^2 = \dfrac{\alpha}{\sqrt{2}}\mathrm{Im}\int u^2 \overline{v}\, dx. \end{cases} \quad (2.2)$$

Multiplying the second equation of the system (2.2) by 2 and then sum up the first equation, it follows that

$$\dfrac{\hbar}{2}\dfrac{d}{dt}\|u\|_2^2 + \hbar\dfrac{d}{dt}\|v\|_2^2 = 0,$$

which implies the identity (2.1).

Lemma 2.2 *Under the conditions of Lemma 2.1, $u_0 \in H_0^1(\Omega)$, $v_0 \in H_0^1(\Omega)$, $M > 0$ and one of the following conditions holds*

(i) $p \geq 2$, $g_{11}, g_{22} \geq 0$, $\begin{pmatrix} g_{11} & g \\ g & g_{22} \end{pmatrix}$ *is positive define,*

(ii) $1 \leq p < 2$,

we can get

$$\sup_{0 \leq t \leq T} (2\|u(\cdot,t)\|_{H^1} + \|v(\cdot,t)\|_{H^1}) \leq C, \quad \forall T > 0, \quad (2.3)$$

where C is a constant depending only on $\|u_0\|_{H^1}, \|v_0\|_{H^1}$.

Proof The inner product is taken to the first equation of the system (1.1) with $\overline{u}_t$ and the second equation with $\overline{v}_t$, and then integrating and taking the real part of the resulting equations, we get

$$\begin{cases} 0 = \dfrac{\hbar^2}{4M}\dfrac{d}{dt}\int |u_x|^2 dx + \dfrac{\lambda_u}{4}\dfrac{d}{dt}\int |u|^4 dx + \lambda\mathrm{Re}\int |v|^2 u\overline{u}_t dx + \sqrt{2}\alpha\mathrm{Re}\int \overline{u}v\overline{u}_t dx \\ \qquad + g_{11}\int |u|^{2p}u\overline{u}_t dx + g\int |u|^{p-1}|v|^{p+1}u\overline{u}_t dx, \\ 0 = \dfrac{\hbar^2}{8M}\dfrac{d}{dt}\int |v_x|^2 dx + \dfrac{\varepsilon}{2}\dfrac{d}{dt}\int |v|^2 dx + \dfrac{\lambda_v}{4}\dfrac{d}{dt}\int |v|^4 dx + \lambda\mathrm{Re}\int |u|^2 v\overline{v}_t dx \\ \qquad + \dfrac{\alpha}{\sqrt{2}}\mathrm{Re}\int u^2 \overline{v}_t dx + g\int |u|^{p+1}|v|^{p-1}v\overline{v}_t dx + g_{22}\int |v|^{2p}v\overline{v}_t dx. \end{cases} \quad (2.4)$$

Summing up the two equations of the system (2.4), we have

$$\frac{\hbar^2}{4M}\frac{d}{dt}\left(\int |u_x|^2 dx + \frac{1}{2}\int |v_x|^2 dx\right) + \frac{1}{4}\frac{d}{dt}\left(\lambda_u \int |u|^4 dx + \lambda_v \int |v|^4 dx\right)$$
$$+ \frac{\epsilon}{2}\frac{d}{dt}\int |v|^2 dx + \frac{\lambda}{2}\frac{d}{dt}\int |u|^2|v|^2 dx + \frac{\alpha}{\sqrt{2}}Re\frac{d}{dt}\int u^2 \bar{v} dx$$
$$+ \frac{1}{2p+2}\frac{d}{dt}\left(g_{11}\int |u|^{2p+2} dx + 2g\int |u|^{p+1}|v|^{p+1} dx + g_{22}\int |v|^{2p+2} dx\right) = 0.$$

Let
$$\text{I} := \frac{\hbar^2}{4M}\left(\int |u_x|^2 dx + \frac{1}{2}\int |v_x|^2 dx\right), \quad \text{II} := \frac{1}{4}\left(\lambda_u \int |u|^4 dx + \lambda_v \int |v|^4 dx\right),$$
$$\text{III} := \frac{\lambda}{2}\int |u|^2|v|^2 dx, \quad \text{IV} := \frac{\epsilon}{2}\int |v|^2 dx, \quad \text{V} := \frac{\alpha}{\sqrt{2}}Re\int u^2 \bar{v} dx,$$
$$\text{VI} := \frac{1}{2p+2}\left(g_{11}\int |u|^{2p+2} dx + 2g\int |u|^{p+1}|v|^{p+1} dx + g_{22}\int |v|^{2p+2} dx\right).$$

Then
$$E(t) = \text{I} + \text{II} + \text{III} + \text{IV} + \text{V} + \text{VI} \equiv E(0). \tag{2.5}$$

Applying Lemma 1.1 and the Young inequality, we have
$$\|u\|_4^4 \leqslant C\|u\|_2^3\|u_x\|_2 \leqslant \frac{C}{2}\left(\frac{1}{\delta^2}\|u\|_2^6 + \delta^2\|u_x\|_2^2\right), \tag{2.6}$$

$$\|v\|_4^4 \leqslant C\|v\|_2^3\|v_x\|_2 \leqslant \frac{C}{2}\left(\frac{1}{\delta^2}\|v\|_2^6 + \delta^2\|v_x\|_2^2\right). \tag{2.7}$$

Then we can bound the term II by
$$|\text{II}| \leqslant \frac{C}{8}\left(\frac{1}{\delta^2}(\lambda_u\|u\|_2^6 + \lambda_v\|v\|_2^6) + \delta^2(\lambda_u\|u_x\|_2^2 + \lambda_v\|v_x\|_2^2)\right). \tag{2.8}$$

For the term III, using the Hölder's inequality
$$\frac{\lambda}{2}\int |u|^2|v|^2 dx \leqslant \frac{\lambda}{4}\left(\|u\|_4^4 + \|v\|_4^4\right). \tag{2.9}$$

Combining the inequalities (2.6) and (2.7), the term III can be bounded by
$$|\text{III}| \leqslant \frac{C}{8}\left(\frac{1}{\delta^2}(\|u\|_2^6 + \|v\|_2^6) + \delta^2(\|u_x\|_2^2 + \|v_x\|_2^2)\right). \tag{2.10}$$

The term
$$\text{V} = \frac{\alpha}{\sqrt{2}}Re\int u^2 \bar{v} dx \leqslant \frac{\alpha}{\sqrt{2}}\int |u|^2|v| dx \leqslant C\|v\|_2\|u\|_4^2. \tag{2.11}$$

By applying the inequality (2.7) and Lemma 2.1 to yield
$$|\text{V}| \leqslant C\left(\frac{1}{\delta^2}\|u\|_2^6 + \delta^2\|u_x\|_2^2\right). \tag{2.12}$$

Using the estimates of the term II, III and V, we deduce

$$|\text{II}| + |\text{III}| + |\text{V}| \leq C\left(\frac{1}{\delta^2}(\|u\|_2^6 + \|v\|_2^6) + \delta^2(\|u_x\|_2^2 + \|v_x\|_2^2)\right). \tag{2.13}$$

In view of Lemma 2.1, it follows that

$$\text{IV} = \frac{\epsilon}{2}\|v\|_2^2 \leq C. \tag{2.14}$$

Next we need to estimate the term VI

$$\|u\|_{2p+2}^{2p+2} \leq C\|u_x\|_2^p \|u\|_2^{p+2}, \qquad \|v\|_{2p+2}^{2p+2} \leq C\|v_x\|_2^p \|v\|_2^{p+2}.$$

Using Hölder's inequality, we have

$$\int |u|^{p+1}|v|^{p+1} \leq C(\|u\|_{2p+2}^{2p+2} + \|v\|_{2p+2}^{2p+2})$$

$$\leq C\left(\frac{1}{\delta}\left(\|u\|_2^{(p+2)\frac{2}{2-p}} + \|v\|_2^{(p+2)\frac{2}{2-p}}\right) + \delta(\|u_x\|_2^2 + \|v_x\|_2^2)\right). \tag{2.15}$$

Therefore, the term VI can be bounded by

$$|\text{VI}| \leq C\left(\frac{1}{\delta}\left(\|u\|_2^{(p+2)\frac{2}{2-p}} + \|v\|_2^{(p+2)\frac{2}{2-p}}\right) + \delta(\|u_x\|_2^2 + \|v_x\|_2^2)\right). \tag{2.16}$$

For $1 \leq p < 2$, combining the estimates (2.13), (2.14) and (2.16)

$$\left(\frac{\hbar^2}{4M} - \delta\right)\left(\|u_x\|_2^2 + \frac{1}{2}\|v_x\|_2^2\right) \leq C,$$

a.e.,

$$2\|u_x\|_2^2 + \|v_x\|_2^2 \leq C.$$

For $p \geq 2$,

$$E := \text{I} + \text{II} + \text{III} + \text{IV} + \text{V} + \text{VI} \equiv C.$$

Note that the matrix $\begin{pmatrix} g_{11} & g \\ g & g_{22} \end{pmatrix} \geq 0$, consequently

$$\text{I} + \text{II} + \text{III} + \text{IV} + \text{V} \leq C.$$

Combining the estimates (2.13) and (2.14)

$$\text{I} - \delta\left(\|u_x\|_2^2 + \frac{1}{2}\|v_x\|_2^2\right) \leq C,$$

a.e.,

$$2\|u_x\|_2^2 + \|v_x\|_2^2 \leq C.$$

This completes the proof of Lemma 2.2.

Lemma 2.3 *Let T be any positive number, $u_0 \in H_0^2(\Omega), v_0 \in H_0^2(\Omega)$. Under the conditions of Lemma 2.1, we have*

$$\sup_{0 \leqslant t \leqslant T} (2\|u(\cdot,t)\|_{H^2} + \|v(\cdot,t)\|_{H^2}) \leqslant C, \quad \forall T > 0, \tag{2.17}$$

where the constant C depends only on T and $\|u_o\|_{H^2}, \|v_0\|_{H^2}$.

Proof Making the inner product of $\bar{u}_{xxxx}$ with the first equation of the system (1.1) and $\bar{v}_{xxxx}$ with the second equation and integrating the resulting equations with respect to x on Ω, and then taking the imaginary part of the resulting equations, we obtain

$$\frac{\hbar}{2}\frac{d}{dt}\|u_{xx}\|_2^2 = \lambda_u Im \int \left(u^2(\bar{u}_{xx})^2 + 2|u_x|^2 u\bar{u}_{xx} + 2|u|_x^2 u_x\bar{u}_{xx}\right) dx$$

$$+ \lambda Im \int \left(v_{xx}\bar{v}u\bar{u}_{xx} + \bar{v}_{xx}vu\bar{u}_{xx} + 2|v_x|^2 u\bar{u}_{xx} + 2|v|_x^2 u_x\bar{u}_{xx}\right) dx$$

$$+ g_{11} Im \int (|u|^{2p}u)_{xx}\bar{u}_{xx} dx + gIm \int (|u|^{p-1}|v|^{p+1}u)_{xx}\bar{u}_{xx} dx$$

$$+ \sqrt{2}\alpha Im \int \left(|u_{xx}|^2\bar{v} + \bar{v}_{xx}u\bar{u}_{xx} + 2(u_x\bar{v}_x)\bar{u}_{xx}\right) dx,$$

$$\frac{\hbar}{2}\frac{d}{dt}\|v_{xx}\|_2^2 = \lambda_v Im \int \left(v^2(\bar{v}_{xx})^2 + 2|v_x|^2 v\bar{v}_{xx} + 2|v|_x^2 v_x\bar{v}_{xx}\right) dx$$

$$+ \lambda Im \int \left(u_{xx}\bar{u}v\bar{v}_{xx} + \bar{u}_{xx}uv\bar{v}_{xx} + 2|u_x|^2 v\bar{v}_{xx} + 2(|u|_x^2 v_x)\bar{v}_{xx}\right) dx$$

$$+ g_{22} Im \int (|v|^{2p}v)_{xx}\bar{v}_{xx} dx + gIm \int (|v|^{p-1}|u|^{p+1}v)_{xx}\bar{v}_{xx} dx$$

$$+ \frac{2\alpha}{\sqrt{2}} Im \int (u_{xx}u\bar{v}_{xx} + 2u_x u_x)\bar{v}_{xx} dx.$$

Denote $\dfrac{\hbar}{2}\dfrac{d}{dt}\|u_{xx}\|_2^2 = \mathrm{I} + \mathrm{II} + \mathrm{III} + \mathrm{IV}$, where

$$\mathrm{I} \equiv \lambda_u Im \int \left(u^2(\bar{u}_{xx})^2 + 2|u_x|^2 u\bar{u}_{xx} + 2|u|_x^2 u_x\bar{u}_{xx}\right) dx,$$

$$\mathrm{II} \equiv \lambda Im \int \left(v_{xx}\bar{v}u\bar{u}_{xx} + \bar{v}_{xx}vu\bar{u}_{xx} + 2|v_x|^2 u\bar{u}_{xx} + 2|v|_x^2 u_x\bar{u}_{xx}\right) dx,$$

$$\mathrm{III} \equiv g_{11} Im \int (|u|^{2p}u)_{xx}\bar{u}_{xx} dx + gIm \int (|u|^{p-1}|v|^{p+1}u)_{xx}\bar{u}_{xx} dx,$$

$$\mathrm{IV} \equiv \sqrt{2}\alpha Im \int \left(|u_{xx}|^2\bar{v} + \bar{v}_{xx}u\bar{u}_{xx} + 2u_x\bar{v}_x\bar{u}_{xx}\right) dx.$$

And we denote $\dfrac{\hbar}{2}\dfrac{d}{dt}\|v_{xx}\|_2^2 = \text{V} + \text{VI} + \text{VII} + \text{VIII}$, where

$$\text{V} \equiv \lambda_v Im \int \left(v^2(\bar{v}_{xx})^2 + 2|v_x|^2 v\bar{v}_{xx} + 2|v|_x^2 v\bar{v}_{xx}\right) dx,$$

$$\text{VI} \equiv \lambda Im \int \left(u_{xx}\bar{u}v\bar{v}_{xx} + \bar{u}_{xx}uv\bar{v}_{xx} + 2|u_x|^2 v\bar{v}_{xx} + 2|u|_x^2 v\bar{v}_{xx}\right) dx,$$

$$\text{VII} \equiv g_{22} Im \int (|v|^{2p}v)_{xx}\bar{v}_{xx}dx + gIm \int (|v|^{p-1}|u|^{p+1}v)_{xx}\bar{v}_{xx}dx,$$

$$\text{VIII} \equiv \dfrac{2\alpha}{\sqrt{2}} Im \int (u_{xx}u\bar{v}_{xx} + 2u_x u_x \bar{v}_{xx}) dx.$$

Firstly, we estimate the term I.

By using Sobolev embedding theorem and the Hölder's inequality, we have

$$|\text{I}| \leqslant C\|u_{xx}\|_2^2 + C_1\|u_x\|_4^2\|u_{xx}\|_2.$$

Applying the inequality (1.7) and Lemma 2.2,

$$C_1\|u_x\|_4^2\|u_{xx}\|_2 \leqslant C_2\|u_{xx}\|_2^{\frac{1}{2}}\|u_x\|_2^{\frac{3}{2}}\|u_{xx}\|_2 \leqslant C_3\|u_{xx}\|_2^{\frac{3}{2}} \leqslant C_4\|u_{xx}\|_2^2 + C_5.$$

So

$$|\text{I}| \leqslant C_1\|u_{xx}\|_2^2 + C_2. \tag{2.18}$$

Next, we will deal with the term II.

Applying the Sobolev embedding theorem and the Hölder inequality

$$|\text{II}| \leqslant C_1(\|u_{xx}\|_2^2 + \|v_{xx}\|_2^2) + C_2\|u_{xx}\|_2\|v_x\|_4^2 + C_3\|v_x\|_4\|u_x\|_4\|u_{xx}\|_2$$

Applying the inequality (1.7) and Lemma 2.2,

$$C_2\|u_{xx}\|_2\|v_x\|_4^2 + C_3\|v_x\|_4\|u_x\|_4\|u_{xx}\|_2$$
$$\leqslant C_4\|u_{xx}\|_2\|v_{xx}\|_2^{\frac{1}{2}}\|v_x\|_2^{\frac{3}{2}} + C_5\|v_{xx}\|_2^{\frac{1}{4}}\|u_{xx}\|_2^{\frac{5}{4}}$$
$$\leqslant C(\|u_{xx}\|_2^2 + \|v_{xx}\|_2^2 + 1).$$

Therefore

$$|\text{II}| \leqslant C(\|v_{xx}\|_2^2 + \|u_{xx}\|_2^2 + 1). \tag{2.19}$$

For the term III

$$g_{11}Im\int(|u|^{2p}u)_{xx}\overline{u}_{xx}dx$$

$$=g_{11}Im\int(|u|^{2p})_{xx}u\overline{u}_{xx}dx+2g_{11}Im\int|u|^{2p}_{x}u_{x}\overline{u}_{xx}dx$$

$$=g_{11}Im\bigg\{\int 2p|u|^{2p-2}|u_x|^2u\overline{u}_{xx}+p|u|^{2p-2}(\overline{u}_{xx})^2u^2$$

$$+p(p-1)|u|^{2p-4}(|u|^2)_x^2u\overline{u}_{xx}dx\bigg\}+2g_{11}Im\int|u|^{2p}_{x}u_x\overline{u}_{xx}dx.$$

Using the Sobolev embedding theorem and Hölder's inequality, we have

$$|g_{11}Im\int(|u|^{2p}u)_{xx}\overline{u}_{xx}dx|\leqslant C(1+\|u_{xx}\|_2^2).$$

Similarly,

$$gIm\int(|u|^{p-1}|v|^{p+1}u)_{xx}\overline{u}_{xx}dx$$

$$=gIm\left(\int|u|^{p-1}_{xx}|v|^{p+1}u\overline{u}_{xx}dx+\int|v|^{p+1}_{xx}|u|^{p-1}u\overline{u}_{xx}dx\right)$$

$$+2gIm\left(\int|u|^{p-1}_x|v|^{p+1}_xudx+\int|u|^{p-1}|v|^{p+1}_xu_xdx+\int|u|^{p-1}_x|v|^{p+1}u_xdx\right)$$

$$=gIm\left(\int(p-1)|u|^{p-3}|u_{xx}|^2|v|^{p+1}u\overline{u}_{xx}dx+\int\frac{p-1}{2}|u|^{p-3}(\overline{u}_{xx})^2u^2|v|^{p+1}dx\right)$$

$$+gIm\left(\int\frac{p-1}{2}\frac{p-3}{2}|u|^{p-5}(|u|_x)^2|v|^{p+1}u\overline{u}_{xx}dx+\int(|v|^{p+1})_{xx}|u|^{p-1}u\overline{u}_{xx}dx\right)$$

$$+2gIm(\int|u|^{p-1}_x|v|^{p+1}_xudx+\int|u|^{p-1}|v|^{p+1}_xu_xdx+\int|u|^{p-1}_x|v|^{p+1}u_xdx)$$

$$\leqslant C(1+\|v_{xx}\|_2^2+\|u_{xx}\|_2^2).$$

Therefore

$$|\text{III}|\leqslant C(1+\|v_{xx}\|_2^2+\|u_{xx}\|_2^2). \qquad (2.20)$$

For the term IV, we can immediately get the estimate

$$|\text{IV}|\leqslant C(\|v_{xx}\|_2^2+\|u_{xx}\|_2^2+1). \qquad (2.21)$$

Comparing the term I with the term V, II with VI, we can get

$$|\text{V}|\leqslant C(\|v_{xx}\|_2^2+1),\quad|\text{VI}|\leqslant C(\|v_{xx}\|_2^2+\|u_{xx}\|_2^2+1). \qquad (2.22)$$

And comparing III and VII, we can get

$$|\text{VII}|\leqslant C(1+\|v_{xx}\|_2^2+\|u_{xx}\|_2^2). \qquad (2.23)$$

For the term VIII, we have

$$|\text{VIII}| \leqslant C(\|u_{xx}\|_2^2 + \|v_{xx}\|_2^2 + 1). \qquad (2.24)$$

Applying (2.18) – (2.24), we can obtain the following estimate

$$\frac{d}{dt}(\|u_{xx}\|_2^2 + \|v_{xx}\|_2^2) \leqslant C(\|u_{xx}\|_2^2 + \|v_{xx}\|_2^2 + 1).$$

Using the Gronwall inequality

$$\|u_{xx}\|_2^2 + \|v_{xx}\|_2^2 \leqslant C.$$

This completes the proof of Lemma 2.3.

Lemma 2.4 *Let $m \geqslant 0$ be any integer number. Under the conditions of Lemma 2.1, we have*

$$\sup_{0 \leqslant t \leqslant T}(2\|u(\cdot,t)\|_{H^m} + \|v(\cdot,t)\|_{H^m}) \leqslant C, \quad \forall T > 0, \qquad (2.25)$$

where the constant C depends only on T and $\|u_0\|_{H^m}, \|v_0\|_{H^m}$.

Proof This lemma is proved by mathematical induction as follows. When $m = 0, 1, 2$, according to Lemmas 2.1, 2.2, 2.3, the inequality (2.25) is held.

Suppose that (2.25) is valid for $m \leqslant k$. We will prove that (2.25) holds for $m = k+1$.

Making the inner product of $D^{2(k+1)}\bar{u}$ with the first equation of the system (1.1) and $D^{2(k+1)}\bar{v}$ with the second equation, and then integrating and taking the imaginary part of the resulting equations, we get

$$\frac{\hbar}{2}\frac{d}{dt}\|D^{k+1}u\|_2^2$$

$$= Im\left(\lambda_u \int D^{k+1}(|u|^2 u) D^{k+1}\bar{u}dx + \lambda \int D^{k+1}(|v|^2 u) D^{k+1}\bar{u}dx\right)$$

$$+ Im(g_{11} \int D^{k+1}(|u|^{2p}u) D^{k+1}\bar{u}dx + g \int D^{k+1}(|u|^{p-1}|v|^{p+1}u) D^{k+1}\bar{u}dx$$

$$+ \sqrt{2}\alpha Im \int D^{k+1}(\bar{u}v) D^{k+1}\bar{u}dx$$

$$:= I + II + III,$$

$$\frac{\hbar}{2}\frac{d}{dt}\|D^{k+1}v\|_2^2$$

$$= Im\left(\lambda_v \int D^{k+1}(|v|^2 v) D^{k+1}\overline{v} dx + \lambda \int D^{k+1}(|u|^2 v) D^{k+1}\overline{v} dx\right)$$

$$+ Im\left(g_{22}\int D^{k+1}(|v|^{2p}v) D^{k+1}\overline{v} dx + g\int D^{k+1}(|v|^{p-1}|u|^{p+1}v) D^{k+1}\overline{v} dx\right)$$

$$+ \frac{\alpha}{\sqrt{2}} Im \int D^{k+1}(u^2) D^{k+1}\overline{v} dx$$

$$:= IV + V + VI.$$

By the normal computation, we can obtain

$$\int D^{k+1}(|u|^2 u) D^{k+1}\overline{u} dx$$

$$= \int (D^{k+1}|u|^2) u D^{k+1}\overline{u} + C_1 \int (D^k|u|^2) Du D^{k+1}\overline{u} dx$$

$$+ \cdots + C_k \int (D|u|^2) D^k u D^{k+1}\overline{u} dx + \int |u|^2 D^{k+1} u D^{k+1}\overline{u} dx. \tag{2.26}$$

Usimg the induction assumption for $m \leqslant k$, $\|u\|_{H^m} + \|v\|_{H^m} \leqslant C$, so when $k=3$, we can get $\|u\|_{H^3} \leqslant C$, $\|v\|_{H^3} \leqslant C$. Using the Sobolev's embedding theorem, $\|D^2 u\|_\infty \leqslant C$, $\|D^2 v\|_\infty \leqslant C$. Therefore

$$\lambda_u Im \int D^{k+1}(|u|^2 u) D^{k+1}\overline{u} dx \leqslant C_1 \|D^{k+1}u\|_2^2 + C_2, \tag{2.27}$$

Applying the same computation

$$\lambda Im \int D^{k+1}(|v|^2 u) D^{k+1}\overline{u} dx \leqslant C_1 \|D^{k+1}v\|_2^2 + C_2 \|D^{k+1}u\|_2^2 + C_3. \tag{2.28}$$

Therefore

$$|I| \leqslant C_1 \|D^{k+1}v\|_2^2 + C_2 \|D^{k+1}u\|_2^2 + C_3. \tag{2.29}$$

Also by the normal computation, we can obtain

$$\int D^{k+1}(|u|^{2p}u) D^{k+1}\overline{u} dx$$

$$= \int (D^{k+1}|u|^{2p}) u D^{k+1}\overline{u} + C_1 \int (D^k|u|^{2p}) Du D^{k+1}\overline{u} dx$$

$$+ \cdots + C_k \int (D|u|^{2p}) D^k u D^{k+1}\overline{u} dx + \int |u|^{2p} D^{k+1} u D^{k+1}\overline{u} dx,$$

so

$$g_{11} Im \int D^{k+1}(|u|^{2p}u) D^{k+1}\overline{u} dx \leqslant C_1 \|D^{k+1}u\|_2^2 + C_2.$$

Using the same computation

$$g Im \int D^{k+1}(|u|^{p-1}|v|^{p+1}u) D^{k+1}\overline{u} dx \leqslant C_1 \|D^{k+1}v\|_2^2 + C_2 \|D^{k+1}u\|_2^2 + C_3.$$

Hence
$$|\text{II}| \leq C_1 \|D^{k+1}v\|_2^2 + C_2 \|D^{k+1}u\|_2^2 + C_3. \tag{2.30}$$

Applying the induction computation, we have
$$\int (D^{k+1}\bar{u}v) D^{k+1}\bar{u} dx$$
$$= \int D^{k+1}\bar{u} v D^{k+1}\bar{u} dx + C_1 \int D^m \bar{u} Dv D^{k+1}\bar{u} dx$$
$$+ \cdots + C_k \int D\bar{u} D^k v D^{k+1}\bar{u} dx + \int \bar{u} D^{k+1} v D^{k+1}\bar{u} dx.$$

So we have the estimate of the term III
$$|\text{III}| \leq C_1 \|D^{k+1}u\|_2^2 + C_2 \|D^{k+1}v\|_2^2 + C_3. \tag{2.31}$$

Comparing the term I with the term IV, and II with V, we get
$$|\text{IV}| \leq C_1 \|D^{k+1}v\|_2^2 + C_2 \|D^{k+1}u\|_2^2 + C_3, \tag{2.32}$$
$$|\text{V}| \leq C_1 \|D^{k+1}v\|_2^2 + C_2 \|D^{k+1}u\|_2^2 + C_3. \tag{2.33}$$

For the term VI, by direct computation, we have
$$\int D^{k+1} u^2 D^{k+1}\bar{u} dx$$
$$= \int D^{k+1} u u D^{k+1}\bar{u} dx + C_1 \int D^k u Du D^{k+1}\bar{u} dx$$
$$+ \cdots + C_k \int Du D^k u D^{k+1}\bar{u} dx + \int u D^{k+1} u D^{k+1}\bar{u} dx.$$

Usimg the induction assumption and the Sobolev embedding theorem
$$|\text{VI}| \leq C_1 \|D^{k+1}u\|_2^2 + C_2. \tag{2.34}$$

Comparing (2.29)–(2.34), we have
$$\frac{d}{dt}(\|D^{k+1}u\|_2^2 + \|D^{k+1}v\|_2^2) \leq C(\|D^{k+1}u\|_2^2 + \|D^{k+1}v\|_2^2 + 1).$$

Using the Gronwall inequality
$$\|D^{k+1}u\|_2^2 + \|D^{k+1}v\|_2^2 \leq C.$$

This completes the proof of Lemma 2.4.

Finally, we prove the uniqueness of the solution to the system (1.1)–(1.3) in the following.

Let $(u_1, v_1), (u_2, v_2)$ be two solutions which satisfy the system (1.1)–(1.3), then $(s = u_1 - u_2, m = v_1 - v_2)$ satisfies

$$\begin{cases} i\hbar s_t = -\dfrac{\hbar^2 \nabla^2}{2M} s + \lambda_u(|u_1|^2 u_1 - |u_2|^2 u_2) + \lambda(|v_1|^2 u_1 - |v_2|^2 u_2) \\ \qquad + g_{11}(|u_1|^{2p} u_1 - |u_2|^{2p} u_2) + g(|v_1|^{p+1}|u_1|^{p-1} u_1 - |v_2|^{p+1}|u_2|^{p-1} u_2) \\ \qquad + \sqrt{2}\alpha(\overline{u_1} v_1 - \overline{u_2} v_2), \\ i\hbar m_t = -\dfrac{\hbar^2 \nabla^2}{4M} m + \epsilon m + \lambda_v(|v_1|^2 v_1 - |v_2|^2 v_2) + \lambda(|u_1|^2 v_1 - |u_2|^2 v_2) \\ \qquad + g_{22}(|v_1|^{2p} v_1 - |v_2|^{2p} v_2) + g(|u_1|^{p+1}|v_1|^{p-1} v_1 - |u_2|^{p+1}|v_2|^{p-1} v_2) \\ \qquad + \dfrac{\alpha}{\sqrt{2}}(u_1^2 - u_2^2), \end{cases}$$
(2.35)

$$s(0) = 0, \quad m(0) = 0.$$

Taking the inner product of the first equation of the system (2.35) with $\bar{s}$ and the second equation with $\bar{m}$, considering the imaginary part of the resulting equations, we obtain:

$$\dfrac{\hbar}{2} \dfrac{d}{dt} \|s\|_2^2$$
$$= \lambda_u Im \int (|u_1|^2 u_1 - |u_2|^2 u_2) \bar{s} dx + \lambda Im \int (|v_1|^2 u_1 - |v_2|^2 u_2) \bar{s} dx$$
$$+ g_{11} Im \int (|u_1|^{2p} u_1 - |u_2|^{2p} u_2) \bar{s} dx + \sqrt{2}\alpha Im \int (\overline{u_1} v_1 - \overline{u_2} v_2) \bar{s} dx$$
$$+ g Im \int (|v_1|^{p+1}|u_1|^{p-1} u_1 - |v_2|^{p+1}|u_2|^{p-1} u_2) \bar{s} dx,$$

$$\dfrac{\hbar}{2} \dfrac{d}{dt} \|m\|_2^2$$
$$= \lambda_v Im \int (|v_1|^2 v_1 - |v_2|^2 v_2) \bar{m} dx + \lambda Im \int (|u_1|^2 v_1 - |u_2|^2 v_2) \bar{m} dx$$
$$+ g_{22} Im \int (|v_1|^{2p} v_1 - |v_2|^{2p} v_2) \bar{m} dx + \dfrac{\alpha}{\sqrt{2}} Im \int (u_1^2 - u_2^2) \bar{m} dx$$
$$+ g Im \int (|u_1|^{p+1}|v_1|^{p-1} v_1 - |u_2|^{p+1}|v_2|^{p-1} v_2) \bar{m} dx.$$

But

$$\lambda_u Im \int (|u_1|^2 u_1 - |u_2|^2 u_2) \bar{s} dx$$

$$\leqslant C\int |(|u_1|^2 s\bar{s} + (|u_1|^2 - |u_2|^2)u_2\bar{s})|dx$$
$$\leqslant C_1\|s\|_2^2.$$

Similarly
$$g_{11}Im\int (|u_1|^{2p}u_1 - |u_2|^{2p}u_2)\bar{s}dx \leqslant C\|s\|_2^2,$$
$$\lambda_v Im\int (|v_1|^2 v_1 - |v_2|^2 v_2)\overline{m}dx \leqslant C\|m\|_2^2,$$
$$g_{22}Im\int (|v_1|^{2p}v_1 - |v_2|^{2p}v_2)\overline{m}dx \leqslant C\|m\|_2^2.$$

And
$$\lambda Im\int (|v_1|^2 u_1 - |v_2|^2 u_2)\bar{s}dx$$
$$\leqslant C\int |(|v_1|^2 s\bar{s} + (|v_1|^2 - |v_2|^2)u_2\bar{s})|dx$$
$$\leqslant C(\|s\|_2^2 + \|m\|_2^2),$$
$$gIm\int (|v_1|^{p+1}|u_1|^{p-1}u_1 - |v_2|^{p+1}|u_2|^{p-1}u_2)\bar{s}dx \leqslant C(\|s\|_2^2 + \|m\|_2^2),$$
$$gIm\int (|u_1|^{p+1}|v_1|^{p-1}v_1 - |u_2|^{p+1}|v_2|^{p-1}v_2)\overline{m}dx \leqslant C(\|s\|_2^2 + \|m\|_2^2),$$
$$\int (\overline{u_1}v_1 - \overline{u_2}v_2)\bar{s}dx = \int (\overline{u_1}v_1 - \overline{u_1}v_2 + \overline{u_1}v_2 - \overline{u_2}v_2)\bar{s}dx$$
$$\leqslant C_1\int (m\bar{s} + |s|^2)dx \leqslant C_2(\|m\|_2^2 + \|s\|_2^2),$$
$$\int (u_1^2 - u_2^2)\overline{m}dx = \int ((u_1^2 - u_1 u_2) + (u_1 u_2 - u_2^2))\overline{m}dx$$
$$\leqslant C_1\int s\overline{m}dx \leqslant C_2(\|s\|_2^2 + \|m\|_2^2).$$

By the above inequalities, one can easily check that obtain
$$\frac{d}{dt}(\|s\|_2^2 + \|m\|_2^2) \leqslant C(\|s\|_2^2 + \|m\|_2^2).$$

Applying the Gronwall inequality, we get $s = 0, m = 0$. Thus the uniqueness is obtained.

Remark 2.1 By virtue of the local smooth solution, the *a priori* estimates and the continuous extension theorem, we obtain the global smooth solution of the period initial value problem (1.1)–(1.3). So that, Theorem 1.2 is obtained.

Remark 2.2 All the above estimates are unconcerned with the period L and only depend on the norm of initial data. Therefore, by using the *a priori* estimates of the solution to the system (1.1)–(1.3) for L, as in Ref. [25], we derive the global smooth solution as $L \to \infty$. So that, Theorem 1.3. is obtained.

3 Blow-up phenomenon of the solution

In this section, we give some conditions on the existence of blow-up solutions of the system (1.1). Define $h(t) = \int |x|^2(|u(t)|^2 + 2|v(t)|^2)dx$.

Theorem 3.1 Let $p \geqslant 2, (u_0, v_0) \in \Lambda^2$ and $(u(t), v(t)) \in C([0, T_0), \Lambda^2)$ be the solution of the system (1.1). Define $h(t) = \int |x|^2(|u(t)|^2 + 2|v(t)|^2)dx$. Then $h(t)$ is well defined for $t \in [0, T_0)$. Moreover,

$$h'(t) = \frac{2\hbar}{M} Im \int (\overline{u}xu_x + \overline{v}xv_x)dx, \tag{3.36}$$

$$\begin{aligned}h''(t) =& \frac{2\hbar^2}{M^2} \int \left(|u_x|^2 + \frac{1}{2}|v_x|^2\right) dx + \frac{1}{M}\left(\lambda_u \int |u|^4 dx + \lambda_v \int |v|^4 dx + 2\lambda \int |u|^2|v|^2 dx\right) \\&+ \frac{\sqrt{2}\alpha}{M} Re \int u^2 \overline{v} dx + \frac{2p}{M(p+1)}\left(g_{11}\int |u|^{2p+2} dx + g_{22}\int |v|^{2p+2} dx\right. \\&\left. + 2g \int |u|^{p+1}|v|^{p+1} dx\right).\end{aligned} \tag{3.37}$$

Proof We only prove (3.36) and (3.37) formally.

$$\begin{aligned}h'(t) &= 2Re \int |x|^2(\overline{u}u_t + 2\overline{v}v_t)dx \\&= -\frac{\hbar}{M} Im \int |x|^2 (\overline{u}u_{xx} + \overline{v}v_{xx})dx \\&= \frac{2\hbar}{M} Im \int (\overline{u}xu_x + \overline{v}xv_x)dx,\end{aligned}$$

$$\begin{aligned}h''(t) =& -\frac{2\hbar}{M} Im \int (u_t(\overline{u} + 2x\overline{u}_x) + v_t(\overline{v} + 2x\overline{v}_x))dx \\=& -\frac{2}{M} Re \int \left(\left(\frac{\hbar^2 \partial^2}{2M\partial_{xx}} - \lambda_u|u|^2 - \lambda|v|^2 - g_{11}|u|^{2p} - g|u|^{p-1}|v|^{p+1}\right)u \right.\\&\left. - \sqrt{2}\alpha \overline{u}v\right)(\overline{u} + 2x\overline{u}_x)dx + \int \left(\left(\frac{\hbar^2 \partial^2}{4M\partial_{xx}} - \epsilon - \lambda_v|v|^2 - \lambda|u|^2\right.\right.\\&\left.\left. - g|u|^{p+1}|v|^{p-1} - g_{22}|v|^{2p}\right)v - \frac{\alpha}{\sqrt{2}}u^2\right)(\overline{v} + 2x\overline{v}_x)dx.\end{aligned}$$

Note that
$$\frac{\hbar^2}{2M}\mathrm{Re}\int \bar{u}u_{xx}dx = -\frac{\hbar^2}{2M}\int |u_x|^2 dx, \quad \frac{\hbar^2}{2M}\mathrm{Re}\int 2u_{xx}x\bar{u}_x dx = -\frac{\hbar^2}{2M}\int |u_x|^2 dx.$$

Then
$$\frac{\hbar^2}{2M}\mathrm{Re}\int u_{xx}(\bar{u}+2x\bar{u}_x)dx = -\frac{\hbar^2}{M}\int |u_x|^2 dx.$$

Similarly
$$\frac{\hbar^2}{4M}\mathrm{Re}\int v_{xx}(\bar{v}+2x\bar{v}_x)dx = -\frac{\hbar^2}{2M}\int |v_x|^2 dx,$$

$$\lambda_u \mathrm{Re}\int |u|^2 u(\bar{u}+2x\bar{u}_x)dx = \frac{\lambda_u}{2}\int |u|^4 dx,$$

$$\lambda \mathrm{Re}\int |v|^2 u(\bar{u}+2x\bar{u}_x)dx = \lambda \int (|u|^2|v|^2 + |v|^2 x|u|_x^2)dx,$$

$$g_{11}\mathrm{Re}\int |u|^{2p}u(\bar{u}+2x\bar{u}_x)dx = g_{11}\int \left(|u|^{2p+2} - \frac{1}{p+1}|u|^{2p+2}\right)dx,$$

$$g\mathrm{Re}\int |u|^{p-1}|v|^{p+1}u(\bar{u}+2x\bar{u}_x)dx = g\int \left(|u|^{p+1}|v|^{p+1} + \frac{2}{p+1}|v|^{p+1}x|u|_x^{p+1}\right)dx,$$

$$\lambda_v \mathrm{Re}\int |v|^2 v(\bar{v}+2x\bar{v}_x)dx = \frac{\lambda_v}{2}\int |v|^4 dx,$$

$$\lambda \mathrm{Re}\int |u|^2 v(\bar{v}+2x\bar{v}_x)dx = \lambda \int (|v|^2|u|^2 + |u|^2 x|v|_x^2)dx,$$

$$\mathrm{Re}\int \epsilon v(\bar{v}+2x\bar{v}_x)dx = \epsilon \int (|v|^2 - |v|^2)dx = 0,$$

$$g\mathrm{Re}\int |u|^{p+1}|v|^{p-1}v(\bar{v}+2x\bar{v}_x)dx = g\int \left(|v|^{p+1}|u|^{p+1} + \frac{2}{p+1}|u|^{p+1}x|v|_x^{p+1}\right)dx,$$

$$g_{22}\mathrm{Re}\int |v|^{2p}v(\bar{v}+2x\bar{v}_x)dx = g_{22}\int \left(|v|^{2p+2} - \frac{1}{p+1}|v|^{2p+2}\right)dx.$$

Therefore
$$h''(t) = \frac{2\hbar^2}{M^2}\int \left(|u_x|^2 + \frac{1}{2}|v_x|^2\right)dx + \frac{1}{M}\left(\lambda_u \int |u|^4 dx + \lambda_v \int |v|^4 dx + 2\lambda \int |u|^2|v|^2 dx\right)$$
$$+ \frac{\sqrt{2}\alpha}{M}\mathrm{Re}\int u^2\bar{v}dx + \frac{2p}{M(p+1)}\left(g_{11}\int |u|^{2p+2}dx\right.$$
$$\left. + g_{22}\int |v|^{2p+2}dx + 2g\int |u|^{p+1}|v|^{p+1}dx\right).$$

This completes the proof of Theorem 3.1.

Proof of Theorem 1.4 Assuming that the solutions exists globally in time, we obtain from Theorem 3.1 that

$$h''(t) = \frac{8}{M}E(t) - \frac{4\epsilon}{M}\int |v|^2 dx - \frac{1}{M}\left(\lambda_u \int |u|^4 dx + \lambda_v \int |v|^4\right.$$
$$+ 2\lambda \int |u|^2|v|^2 dx\bigg) - \frac{3\sqrt{2}\alpha}{M}\mathrm{Re}\int u^2 \bar{v} dx - \frac{2(2-p)}{(p+1)M}\left(g_{11}\int |u|^{2p+2}dx\right.$$
$$+ g_{22}\int |v|^{2p+2}dx + 2g\int |u|^{p+1}|v|^{p+1}dx\bigg).$$

Using the Hölder inequality, we have

$$-\frac{3\sqrt{2}\alpha}{M}\mathrm{Re}\int u^2\bar{v}dx \leqslant \left|\frac{3\sqrt{2}\alpha}{2M}\right|\left(\int |u|^4 dx + \int |v|^2 dx\right).$$

So

$$h''(t) \leqslant \frac{8}{M}E(t) - \left(\frac{4\epsilon}{M} - \left|\frac{3\sqrt{2}\alpha}{2M}\right|\right)\int |v|^2 dx$$
$$- \frac{1}{M}\left(\lambda_u - \left|\frac{3\sqrt{2}\alpha}{2M}\right|\right)\int |u|^4 dx - \frac{\lambda_v}{M}\int |v|^4 dx - \frac{2\lambda}{M}\int |u|^2|v|^2 dx$$
$$- \frac{2(2-p)}{(p+1)M}\left(\int (g_{11}|u|^{2p+2} + g_{22}|v|^{2p+2} + 2g|u|^{p+1}|v|^{p+1})dx\right).$$

Because $M > 0$, $\left(4\epsilon - \left|\frac{3\sqrt{2}\alpha}{2}\right|\right) > 0$, $\left(\lambda_u - \left|\frac{3\sqrt{2}\alpha}{2M}\right|\right) > 0$, $\lambda_v, \lambda > 0$, we deduce

$$h''(t) \leqslant \frac{8}{M}E(t) - \frac{2(2-p)}{(p+1)M}\left(g_{11}\int |u|^{2p+2}dx + g_{22}\int |v|^{2p+2}dx + 2g\int |u|^{p+1}|v|^{p+1}dx\right).$$

Since $p \geqslant 2$, $g_{11} > 0$ and $\begin{pmatrix} g_{11} & g \\ g & g_{22} \end{pmatrix}$ is negative definite, it follows that

$$h''(t) \leqslant \frac{8}{M}E(t) = \frac{8}{M}E(0). \tag{3.38}$$

By a classical analysis, we have that

$$h(t) = h(0) + h'(0)t + \int_0^t (t-\tau)h''(\tau)d\tau, \quad 0 \leqslant t < +\infty. \tag{3.39}$$

It follows that

$$h(t) \leqslant h(0) + h'(0)t + \frac{4}{M}E_0 t^2, \quad 0 \leqslant t < +\infty.$$

Moreover, $h(t)$ is a nonnegative function,

$$h(0) = \int |x|^2 (|u_0|^2 + 2|v_0|^2) dx$$

and

$$h'(0) = \frac{2\hbar}{M} Im \int (x\overline{u_0} u_{0x} + x\overline{v_0} v_{0x}) dx. \tag{3.40}$$

In the following, we discuss this theorem through three cases.

(a) If (i) holds, then from (3.38) we have $h''(t) \leq \frac{8}{M} E(0) < 0$. Then $h(t)$ is a concave function of t which implies that there exists $T^* < \infty$ such that $\lim_{t \to T^*} h(t) = 0$.

(b) If (ii) holds, then from (3.38) and (3.40) we have $h''(t) \leq \frac{8}{M} E(0) = 0$ and $h'(0) < 0$. Thus there exists $T^* < \infty$ such that $\lim_{t \to T^*} h(t) = 0$.

(c) If (iii) holds, then let

$$f(t) = h(0) + h'(0)t + \frac{4}{M} E_0 t^2,$$

we have $\left(h'(0)^2 - \frac{16}{M} h(0) E_0 \right) \geq 0$. Thus there exists at least t_1 such that

$$f(t_1) = h(0) + h'(0)t_1 + \frac{4}{M} E_0 t_1^2 = 0. \tag{3.41}$$

Therefore, from (3.39) and (3.41), there exists $T_* < \infty$ such that $\lim_{t \to T_*} h(t) = 0$.

By (a), (b) and (c), we can get $\lim_{t \to T_*} h(t) = 0$, which together with (2.1) leads to a contradiction. So the maximal existence time T of the solution (u, v) to the system (1.1)–(1.3) is finite.

This completes the proof of Theorem 1.4.

References

[1] Adams R A and Fournier J F. Sobolev Spaces. Academic Press, 2009.

[2] Adhikari S K and Muruganandam P. Bose-Einstein condensation dynamics from the numerical solution of the Gross-Pitaevskii equation. Journal. Phys., B. **35**, (2002).

[3] Adhikari S K. Numrical study of the spherically sysmmetric Gross-Pitaevskii equation in two space dimensions. Journal. Phys. Rev., E **65**, 2937-2944 (2000).

[4] Bao W Z and Cai Y Y. Ground States of Two-component Bose-Einstein Condensates with an Internal Atomic Josephson Junction. East Asian Journal on Applied Mathematics, **1**, 49-81 (2011).

[5] Bao W Z and Tang W J. Ground-state solution of Bose-Einstein condensate by directly minimizing the energy functiona. Journal of Computational Physics, **187**, 230-254 (2003).

[6] Bao W Z, Jaksch D and Markowich P A. Numerical solution of the Gross-Pitaevskii equation for Boes-Einstein condensation. Journal of Computational Physics, **187**, 318-342 (2003).

[7] Bao W Z, Jin S and Markowich P A. On time-splitting spectral approximations for the Schrödinger equation in the semiclassical regime. Journal of Computational Physics, **175**, 487-524 (2002).

[8] Chen J Q and Guo B L. Blow-up profile to the solutions of two-coupled Schrödinger equations. Journal of Mathematical Physics, **50**, 023505 (2009).

[9] Chen S H and Guo B L. Classical solutions of time-dependent Ginzburg-Landau theory for atomic Fermi gases near the BCS-BEC crossover. Journal of Differential Equations, **251**, 1415-1427 (2011).

[10] Ho T L. Spintor Bose condensates in optical traps. Journal. Phys. Rev., **81**, (1998).

[11] Ginibre J and Velo G. The global Cauchy problem for the nonlinear Schrödinger equation revisited. Ann. Inst. H. Poincaré Anal. Non Linéaire, **2** 309-327 (1985).

[12] Gross E P. Structure of a quantized vortex in boson systems. Nuovo Cimento, **20** 454-477 (1961).

[13] Guo B L. The global solution for some systems of nonlinear Schrödinger equations. Proc. of DD-1 Symposium, **3** 1227-1246(1980).

[14] Guo B L. The initial and periodic value problem of one class nonlinear Schrödinger equations describing excitons in molecular crystals. Acta Mathematica Scientia, **2 (3)** (1982) 269-276.

[15] Guo B L. The initial value problems and periodic boundary value problem of one class of higher order multi-dimensional nonlinear Schrödinger equations. Chinese Science Bulletin, **6** (1982) 324-327.

[16] Guo B L. Nonlinear Evolution Equations, Shanghai Scientific and Technological Education Publishing House. Shanghai, 1985 (in Chinese).

[17] Kerman A and Tommasini P. Gaussian Time-Dependent Variational Principle for Bosons I. Uniform Case. Ann. Phys. (NY), **260**, 250-274 (1997).

[18] Lee M D, Morgan S A, Davis M J and Burnett K. Energy-dependent scattering and the Gross-Pitaevskii equation in two-dimensional Bose-Einstein condensates, Journal. Phys. Rev. A., **65**, (2002).

[19] Pérez-García V M, Michinel H J, Cirac I, Lewenstein M and Zoller P. Dynamics of Bose-Einstein condensates: Variational solutions of the Gross-Pitaevskii equations. Journal. Phys. Rev. A., **56**, (1997).

[20] Pitaevskii, L P, Vortex lines in an imperfect Bose gas. Sov. Phys-JETP, **13** (1961).

[21] Ruprecht P A, Burnett K, Dodd R J, Clark C W, Dodd R J and Clark C W. Collective excitations of atomic Bose-Einstein Condensates. Journal. Phys. Rev., **77**, (1996).

[22] Timmermans E, Tommasini P, Hussein M and Kerman A. Feshbach resonances in atomic Bose-Einstein condensates. Physics Reports, **315**, 199-230 (1999).

[23] Wang T C and Zhao X F. Optimal l^∞ error estimates of finite difference methods for the coupled Gross-Pitaevskii equations in high dimensions. Science China Mathematics, **57**, (2014) doi: 10.1007/s11425-014-4773-7.

[24] Zhang J J, Guo C X and Guo B L. On the Cauchy problem for the magnetic Zakharov system. Monatsh Mathematics, **170**, 89-111 (2013).

[25] Zhou Y and Guo B L. Periodic boundary problem and initial value problem for the generalized Korteweg-de Vries systems of higher order. Acta Math. Sinica, **27**, 154- 176 (1984) in Chinese.

High-order Rogue Wave Solutions for the Coupled Nonlinear Schrödinger Equations-II*

Zhao Lichen (赵立臣), Guo Boling (郭柏灵),
and Ling Liming (凌黎明)

Abstract We study on dynamics of high-order rogue wave in two-component coupled nonlinear Schrödinger equations(CNLSE). Based on the generalized Darboux transformation and formal series method, we obtain the high-order rogue wave solution without the special limitation on the wave vectors. As an application, we exhibit the first, second-order rogue wave solution and the superposition of them by computer plotting. We find the distribution patterns for vector rogue waves are much more abundant than the ones for scalar rogue waves, and also different from the ones obtained with the constrain conditions on background fields. The results further enrich and deep our realization on rogue wave excitation dynamics in such diverse fields as Bose-Einstein condensates, nonlinear fibers, and superfluids.

Keywords darboux transformation; CNLSE; rogue wave.

1 Introduction

Vector rogue wave(RW) has been paid much attention, since the components in nonlinear systems are usually more than two and the dynamics of vector RW demonstrate many striking dynamics in contrast to the ones in scalar system. Firstly, there are some new excitation patterns for vector RW, in contrast to the well-known eye-shaped one in scalar system [1–4]. For example, dark RW was presented numerically [5] and analytically [6–8] for two-component coupled systems. Four-petaled

---

* J. Math Phys., 57, 043508 (2016).

RW was reported recently in three-component coupled systems [9–11], and even two-component coupled system [12]. Secondly, the number of RW in temporal-spatial distribution plane is also different from the ones in scalar systems. We have demonstrated that two or four fundamental RWs can emerge on the distribution plane [6, 13], which is absent for scalar systems [14–18].

It should be noted that the two or four RW patterns were derived under some certain constrains on background fields [6, 13]. The relative wave vector for the two background fields should satisfy a certain relation with the background amplitudes. Then, are the constrain conditions essential for these new distribution patterns? We revisit on high-order RWs by developing the deriving method. Furthermore, the RW patterns for four RW pattern or six RW pattern are all eye-shaped one for the coupled model with constrain conditions. Considering the four-petaled RW and anti-eye-shaped RW can exist in the system, one could expect that these different pattern can coexist for some special superposition cases. Therefore, it is essential to find out how they can be superposed and the properties of the superposition.

In this paper, we present a method to derive high-order RW solution with releasing the constrains conditions on background fields. We find that two or four fundamental RWs can emerge for the superposition of first-order RW with itself or the second-order vector RW in the coupled system without the constrain conditions before. Especially, one four-petaled RW can coexist with three eye-shaped ones and other different pattern combination can exist in the coupled system, in contrast to the four eye-shaped ones coexisting case reported before. Moreover, six RWs can emerge on the distribution for the superposition of two second-order vector RWs, which can be constituted of different fundamental RW patterns. For example, three four-petaled and three eye-shaped ones can coexist on the distribution plane. These results are different from the ones in our previously reported results [13], and could further enrich our knowledge on RW excitation dynamics in many different coupled systems.

2 Generalized Darboux transformation and rogue wave formula

In this section, we firstly recall some basic knowledge of generalized Darboux transformation. We consider the following focusing coupled Schrödinger equation (CNLSE)

$$iq_{1,t} + \frac{1}{2}q_{1,xx} + (|q_1|^2 + |q_2|^2)q_1 = 0,$$
$$iq_{2,t} + \frac{1}{2}q_{2,xx} + (|q_1|^2 + |q_2|^2)q_2 = 0.$$
(2.1)

The CNLSE can be used to describe evolution of localized waves in a two-mode nonlinear fiber, two-component Bose-Einstein condensate, and other coupled nonlinear systems [21, 22]. It admits the following Lax pair:

$$\Phi_x = U(\lambda; Q)\Phi,$$
$$\Phi_t = V(\lambda; Q)\Phi,$$
(2.2)

where

$$U(\lambda; Q) = i\lambda(\sigma_3 + I_3) + Q,$$
$$V(\lambda; Q) = i\lambda^2(\sigma_3 + I_3) + \lambda Q + \frac{i}{2}\sigma_3(Q^2 - Q_x) + ic\, I_3,$$
$$Q = \begin{pmatrix} 0 & i\bar{q}_1 & i\bar{q}_2 \\ iq_1 & 0 & 0 \\ iq_2 & 0 & 0 \end{pmatrix}, \quad \sigma_3 = \operatorname{diag}(1, -1, -1),$$

I_3 is a 3×3 identity matrix and the symbol overbar represents complex conjugation. The parameter c is a real constant. The compatibility condition $\Phi_{xt} = \Phi_{tx}$ gives the CNLSE (2.1).

We can convert the system (2.2) into a new system

$$\Phi[1]_x = U(\lambda; Q[1])\Phi[1],$$
$$\Phi[1]_t = V(\lambda; Q[1])\Phi[1],$$
(2.3)

by the following elementary Darboux transformation [19, 20],

$$\Phi[1] = T\Phi, \quad T = I + \frac{\bar{\lambda}_1 - \lambda_1}{\lambda - \bar{\lambda}_1}\frac{\Phi_1\Phi_1^\dagger}{\Phi_1^\dagger\Phi_1},$$
$$Q[1] = Q + i(\bar{\lambda}_1 - \lambda_1)[P_1, \sigma_3],$$
(2.4)

where Φ_1 is a special solution for system (2.1) at $\lambda = \lambda_1$. By the way, if the solution Φ_1 is a nonzero solution, the transformation is nonsingular about $(x, t) \in \mathbb{R}^2$. If there exists $\Phi_1^\dagger \Phi_1|_{x=x_0, t=t_0} = 0$, then $\Phi_1(x_0, t_0) = 0$. By the existence and uniqueness theorem of ODE, we can obtain $\Phi(x, t) = 0$. This is a contradiction. Thus any nonzero solutions could keep the non-singularity of DT. It follows that the solutions of CNLSE are non-singularity.

To obtain the general RW solutions for CNLSE, we choose the general plane wave solution

$$q_1[0] = a_1 \exp \theta_1, \quad q_2[0] = a_2 \exp \theta_2, \tag{2.5}$$

where $\theta_1 = i\left[b_1 x + \left(a_1^2 + a_2^2 - \dfrac{1}{2}b_1^2\right)t\right]$, $\theta_2 = i\left[b_2 x + \left(a_1^2 + a_2^2 - \dfrac{1}{2}b_2^2\right)t\right]$. Since CNLSE equation possesses Galileo symmetry, then we can set $b_2 = -b_1$. It has been shown that the relative wave vector $2b_1$ can induce the RW pattern transition [12]. We firstly investigate the fundamental solution of Lax pair with the general plane wave solution to develop a new method for deriving new RW solutions. Substituting seed solution (2.5) into equation (2.2), we can obtain a vector solution of the Lax pair with $c = a_1^2 + a_2^2$,

$$\Phi(\lambda) = D \begin{bmatrix} \exp \omega \\ \dfrac{a_1 \exp \omega}{\chi + b_1} \\ \dfrac{a_2 \exp \omega}{\chi - b_1} \end{bmatrix},$$

where

$$D = \text{diag}\left(1, e^{\theta_1}, e^{\theta_2}\right),$$

$$\omega = i\left[\chi x + \dfrac{1}{2}\chi^2 t\right],$$

and χ is a multiple root for the following cubic equation

$$\xi^3 - 2\lambda \xi^2 - \left(a_1^2 + a_2^2 + b_1^2\right)\xi + (a_1^2 - a_2^2)b_1 + 2b_1^2 \lambda = 0. \tag{2.6}$$

If the above cubic equation (2.6) possesses a multiple root, there exist rogue wave solutions. In other words, the spectral parameter λ_i must be satisfied the following quartic equation (discriminant equation for equation (2.6)):

$$b_1^2 \lambda^4 + \dfrac{1}{2}\left(a_1^2 - a_2^2\right)b_1 \lambda^3 - \left[\dfrac{1}{2}b_1^4 - \dfrac{5}{4}\left(a_1^2 + a_2^2\right)b_1^2 - \dfrac{1}{16}\left(a_1^2 + a_2^2\right)^2\right]\lambda^2$$

$$+ \dfrac{9b_1\left(a_1^2 + a_2^2 - 2b_1^2\right)\left(a_1^2 - a_2^2\right)}{16}\lambda + \dfrac{1}{16}b_1^6 + \dfrac{3}{16}\left(a_1^2 + a_2^2\right)b_1^4 + \dfrac{1}{16}\left(a_1^2 + a_2^2\right)^3$$

$$- \dfrac{3\left(5a_1^2 - a_2^2\right)\left(a_1^2 - 5a_2^2\right)b_1^2}{64} = 0. \tag{2.7}$$

It is readily to see that the discriminant for quartic equation (2.7) is

$$\Delta = \dfrac{a_1^2 a_2^2 b_1^2}{16384}\left[(a_1^2 + a_2^2 + 4b_1^2)^3 - 27a_1^2 a_2^2(4b_1^2)\right] \geqslant 0.$$

When $\Delta = 0$, this is the degenerate case, which have been researched in the previous work [13]. We found that four or six RWs can exist in the coupled system with some additional constrains on relative wave vector and background amplitudes. In this work, we consider the non-dengenerate case $\Delta > 0$ which can relax these constrain conditions, and try to find if there are some new patterns for RW excitations in the coupled system. In this case, the discriminant (2.7) possesses two pairs of conjugation complex roots for the fixed parameters a_1, a_2 and b_1 (In general, when $\Delta > 0$, the quantic equation (2.7) could possesses four real roots. But in this case, we can prove that the quartic equation (2.7) merely has two pairs of conjugate complex roots with a similar way given in [23]). We denote two pairs of conjugate complex roots as λ_i and $\bar{\lambda}_i (i = 1, 2)$. The corresponding double roots for equation (2.6) are χ_i and $\bar{\chi}_i$ respectively. Since the dynamics of RW solutions with conjugate roots are similar just with RW location difference, we just consider λ_i and χ_i cases.

In this paper, we use the formal series to tackle with high-order rational solutions for CNLSE. Since the Kadan formula is complicated to derive for the matrix whose eigenvalue equation is a high-order one, and it is not easy to derive high-order RW solution from roots of Kadan form, we would like to replace the Kadan formula with the asymptotical series. We have the following lemmas:

Lemma 2.1 *The formal series*

$$\lambda_i(\epsilon_i) = \lambda_i + d_i \epsilon_i^2, \quad d_i = \frac{2\lambda_i - 3\chi_i}{2b_1^2 - 2\chi_i^2},$$
$$\chi_i(\epsilon_i) = \sum_{j=0}^{\infty} \chi_i^{[j]} \epsilon_i^j, \quad \chi_i^{[0]} = \chi_i, \tag{2.8}$$

satisfy the cubic equation (2.6). The parameter ϵ_i is small complex one. The parameters $\chi_i^{[1]} = 1$ and $\chi_i^{[j \geqslant 2]}$ can be determined recursively

$$\chi_i^{[j-1]} = \left[\sum_{\substack{m+n+k=j \\ 0 \leqslant m,n,k \leqslant j-2}} \frac{-\chi_i^{[m]} \chi_i^{[n]} \chi_i^{[k]}}{2(3\chi_i - \lambda_i)} + \sum_{\substack{m+n=j \\ 0 \leqslant m,n \leqslant j-2}} \frac{\lambda_i \chi_i^{[m]} \chi_i^{[n]}}{(3\chi_i - \lambda_i)} + \sum_{\substack{m+n=j-2 \\ 0 \leqslant m,n}} \frac{d_i \chi_i^{[m]} \chi_i^{[n]}}{(3\chi_i - \lambda_i)} \right],$$
$$j \geqslant 3.$$

Specially, we have

$$\chi_i^{[2]} = \frac{(b_1^2 - 4\lambda_i \chi_i + 5\chi_i^2)}{2(2\lambda_i - 3\chi_i)(b_1^2 - \chi_i^2)},$$

$$\chi_i^{[3]} = \frac{2\chi_i^4 + 4\left(a_1^2 + a_2^2 + 3b_1^2\right)\chi_i^2 + 8b_1\left(a_1^2 - a_2^2\right)\lambda_1 + 3(a_1^2 + a_2^2)^2 + 2b_1^4}{8\left(b_1^2 - \chi_i^2\right)^2 \left(2\lambda_i - 3\chi_i\right)^2}.$$

It follows that

$$\omega_i = i\left[\chi_i x + \frac{1}{2}\chi_i^2 t\right] = \sum_{k=0}^{\infty} X_i^{[k]} \epsilon_i^k,$$

where

$$X_i^{[k]} = i\left(\chi_i^{[k]} x + \frac{1}{2}\sum_{j=0}^{k} \chi_i^{[j]} \chi_i^{[k-j]} t\right).$$

Particularly, we can know that the first three terms for $X_i^{[K]}$ are

$$X_i^{[1]} = i\left(x + \chi_i t\right),$$
$$X_i^{[2]} = i\left[\chi_i^{[2]} x + \left(\chi_i^{[2]}\chi_i + \frac{1}{2}\right)t\right],$$
$$X_i^{[3]} = i\left[\chi_i^{[3]} x + \left(\chi_i^{[3]}\chi_i + \chi_i^{[2]}\right)t\right].$$

Moreover, based on the elementary Schur polynomials we have the expansion

Lemma 2.2

$$\exp\left(\sum_{k=1}^{\infty} X_i^{[k]} \epsilon_i^k\right) = \sum_{j=0}^{\infty} S_i^{[j]} \epsilon_i^j,$$

where $S_i^{[j]}$ is

$$S_i^{[j]} = \sum_{\sum_{k=0}^{m} kl_k = j} \frac{(X_i^{[1]})^{l_1}(X_i^{[2]})^{l_2}\cdots(X_i^{[m]})^{l_m}}{l_1! l_2! \cdots l_m!}.$$

Specially

$$S_i^{[0]} = 1, \quad S_i^{[1]} = X_i^{[1]},$$
$$S_i^{[2]} = \frac{1}{2}(X_i^{[1]})^2 + X_i^{[1]},$$
$$S_i^{[3]} = X_i^{[3]} + X_i^{[1]} X_i^{[2]} + \frac{1}{6}(X_i^{[1]})^3.$$

On the other hand, we need the following series expansion

Lemma 2.3

$$\frac{1}{\chi_i(\epsilon_i) \pm b_1} = \sum_{k=0}^{\infty} \mu_{i,\pm}^{[k]} \epsilon_i^k,$$

where

$$\mu_{i,\pm}^{[0]} = \frac{1}{(\pm b_1 + \chi_i)},$$

$$\mu_{i,\pm}^{[k]} = \frac{-1}{(\pm b_1 + \chi_i)} \sum_{j=0}^{k-1} \mu_{i,\pm}^{[j]} \chi_i^{[k-j]}, \quad j \geqslant 1.$$

Specially

$$\mu_{i,\pm}^{[1]} = \frac{-1}{(\pm b_1 + \chi_i)^2},$$

$$\mu_{i,\pm}^{[2]} = -\frac{(\pm b_1 + \chi_i)\chi_i^{[2]} - 1}{(\pm b_1 + \chi_i)^3},$$

$$\mu_{i,\pm}^{[3]} = -\frac{(\pm b_1 + \chi_i)^2 \chi_i^{[3]} - 2(\pm b_1 + \chi_i)\chi_i^{[2]} + 1}{(\pm b_1 + \chi_i)^4}.$$

Finally, we have the asymptotic series for fundamental solution

$$\Phi_i(\epsilon_i) = \sum_{j=0}^{\infty} \Phi_i^{[j]} \epsilon_i^j, \tag{2.9}$$

where

$$\Phi_i^{[j]} = D \begin{bmatrix} S_i^{[j]} \\ \sum_{m=0}^{j} \mu_{i,+}^{[m]} S_i^{[j-m]} \\ \sum_{m=0}^{j} \mu_{i,-}^{[m]} S_i^{[j-m]} \end{bmatrix},$$

which solves the following differential equations

$$\begin{aligned}\Phi_x &= U(\lambda_i + d_i \epsilon_i^2; Q[0])\Phi, \\ \Phi_t &= V(\lambda_i + d_i \epsilon_i^2; Q[0])\Phi,\end{aligned} \tag{2.10}$$

where

$$Q[0] = \begin{pmatrix} 0 & ia_1 \exp[-\theta_1] & ia_2 \exp[-\theta_2] \\ ia_1 \exp[\theta_1] & 0 & 0 \\ ia_2 \exp[\theta_2] & 0 & 0 \end{pmatrix}.$$

Using a simple symmetry $\epsilon_i \to -\epsilon_i$, we can obtain another asymptotic series $\Phi_i(-\epsilon_i)$ for fundamental solution. Based on the solution $\Phi_i(\pm \epsilon_i)$, we can construct the high-order RW solution. We choose the asymptotic series as

$$\begin{aligned}\Theta_i &\equiv \frac{\Phi_i(\epsilon_i) + \Phi_i(-\epsilon_i)}{2} = \sum_{j=0}^{\infty} \Phi_i^{[2j]} \epsilon_i^{2j}, \\ \Xi_i &\equiv \frac{\Phi_i(\epsilon_i) - \Phi_i(-\epsilon_i)}{2\epsilon_i} = \sum_{j=0}^{\infty} \Phi_i^{[2j+1]} \epsilon_i^{2j}.\end{aligned} \tag{2.11}$$

To introduce some free parameters for high-order RW solutions which can be used to vary the RW pattern distribution, we choose the special solution

$$\widehat{\Psi}_i = \Xi_i + \alpha_i(\epsilon_i)\Theta_i$$
$$= \left(\frac{\epsilon_i\alpha_i(\epsilon_i)+1}{2\epsilon_i}\right)\Phi_i(\epsilon_i) + \left(\frac{\epsilon_i\alpha_i(\epsilon_i)-1}{2\epsilon_i}\right)\Phi_i(-\epsilon_i),$$

where

$$\alpha_i(\epsilon_i) = \sum_{k=0}^{\infty} \alpha_i^{[k]}\epsilon_i^{2k}.$$

To represent the solution with a compact form, we replace the parameter $\epsilon_i\alpha_i(\epsilon_i)+1$ with $\exp[\epsilon_i\alpha_i(\epsilon_i)]$. That is

$$\Psi_j(\epsilon_j) = \left(\frac{\exp[\epsilon_i\alpha_i(\epsilon_i)]\Phi_i(\epsilon_i) - \exp[-\epsilon_i\alpha_i(\epsilon_i)]\Phi_i(-\epsilon_i)}{2\epsilon_i}\right).$$

Moreover,

$$\Psi_j(\epsilon_j) = \sum_{m=1}^{\infty} \Psi_j^{[m-1]}\epsilon_j^{2(m-1)}, \quad \Psi_j^{[m-1]} = \Phi_j^{[2m-1]} + \alpha_j^{[m-1]}.$$

We expand the series

$$\frac{1}{\lambda_j(\epsilon_j) - \overline{\lambda_i(\epsilon_i)}} = \sum_{m=0}^{+\infty} \frac{(\overline{d_i\epsilon_i}^2 - d_j\epsilon_j^2)^m}{(\lambda_j - \overline{\lambda_i})^{m+1}},$$

and

$$\langle\Psi_i(\epsilon_i), \Psi_j(\epsilon_j)\rangle = \sum_{m=0}^{+\infty}\left[\sum_{n=0}^{m}\langle\Psi_i^{[m-n]}, \Psi_j^{[n]}\rangle\epsilon_j^{2n}\overline{\epsilon_i}^{2(m-n)}\right],$$

where the inner product is defined as

$$\langle\Psi_i(\epsilon_i), \Psi_j(\epsilon_j)\rangle \equiv [\Psi_i(\epsilon_i)]^\dagger \Psi_j(\epsilon_j).$$

By directly calculations, we can obtain that

$$\frac{\langle\Psi_i(\epsilon_i), \Psi_j(\epsilon_j)\rangle}{\lambda_j(\epsilon_j) - \overline{\lambda_i(\epsilon_i)}} = \left[\sum_{m=0}^{+\infty} \frac{(\overline{d_i\epsilon_i}^2 - d_j\epsilon_j^2)^m}{(\lambda_j - \overline{\lambda_i})^{m+1}}\right]\sum_{m=0}^{+\infty}\left[\sum_{n=0}^{m}\langle\Psi_i^{[m-n]}, \Psi_j^{[n]}\rangle\epsilon_j^{2n}\overline{\epsilon_i}^{2(m-n)}\right]$$

$$= \sum_{k=0}^{+\infty}\sum_{l=0}^{k}\left[\left(\sum_{m=0}^{l}\binom{l}{m}\frac{(\overline{d_i})^{l-m}(-d_j)^m}{(\lambda_j - \overline{\lambda_i})^{l+1}}\epsilon_j^{2m}\overline{\epsilon_i}^{2(l-m)}\right)\right.$$
$$\left.\cdot\left(\sum_{n=0}^{k-l}\langle\Psi_i^{[k-l-n]}, \Psi_j^{[n]}\rangle\epsilon_j^{2n}\overline{\epsilon_i}^{2(k-l-n)}\right)\right]$$

$$= \sum_{k=0}^{+\infty}\sum_{s=0}^{k}\left[\sum_{l=0}^{k}\sum_{\substack{0\leqslant m\leqslant l,\\ 0\leqslant n\leqslant k-l\\ m+n=s}}\binom{l}{m}\frac{(\overline{d_i})^{l-m}(-d_j)^m}{(\lambda_j - \overline{\lambda_i})^{l+1}}\langle\Psi_i^{[k-l-n]}, \Psi_j^{[n]}\rangle\right]\epsilon_j^{2s}\overline{\epsilon_i}^{2(k-s)}.$$

Lemma 2.4 *Thus we can obtain*

$$\frac{\langle \Psi_i(\epsilon_i), \Psi_j(\epsilon_j)\rangle}{\lambda_j(\epsilon_j) - \overline{\lambda_i(\epsilon_i)}} = \sum_{r,t=1}^{+\infty,+\infty} M_{i,j}^{[r,t]} \epsilon_j^{2(t-1)} \overline{\epsilon}_i^{2(r-1)},$$

where

$$M_{i,j}^{[r,t]} = \sum_{l=0}^{r+t-2} \sum_{\substack{m+n=t-1 \\ 0 \leqslant n \leqslant r+t-l-2, \\ 0 \leqslant m \leqslant l}} \binom{l}{m} \frac{(\overline{d_i})^{l-m}(-d_j)^m}{(\lambda_j - \overline{\lambda_i})^{l+1}} \cdot \langle \Psi_i^{[r+t-l-n-2]}, \Psi_j^{[n]}\rangle.$$

Specially, we have

$$M_{i,j}^{[1,1]} = \frac{\langle \Psi_i^{[0]}, \Psi_j^{[0]}\rangle}{\lambda_j - \overline{\lambda_i}},$$

$$M_{i,j}^{[1,2]} = \frac{1}{\lambda_j - \overline{\lambda_i}}\left(\langle \Psi_i^{[0]}, \Psi_j^{[1]}\rangle - \frac{\langle \Psi_i^{[0]}, \Psi_j^{[0]}\rangle d_j}{\lambda_j - \overline{\lambda_i}}\right),$$

$$M_{i,j}^{[2,1]} = \frac{1}{\lambda_j - \overline{\lambda_i}}\left(\langle \Psi_i^{[1]}, \Psi_j^{[0]}\rangle + \frac{\langle \Psi_i^{[0]}, \Psi_j^{[0]}\rangle \overline{d_i}}{\lambda_j - \overline{\lambda_i}}\right),$$

$$M_{i,j}^{[2,2]} = \left[-\frac{\left(\langle \Psi_i^{[0]}, \Psi_j^{[0]}\rangle d_j + \langle \Psi_i^{[0]}, \Psi_j^{[1]}\rangle (-\lambda_j + \overline{\lambda_i})\right)\overline{d_i}}{(\lambda_j - \overline{\lambda_i})^2}\right.$$
$$+ \frac{\left(-\langle \Psi_i^{[0]}, \Psi_j^{[0]}\rangle \overline{d_i} + \langle \Psi_i^{[1]}, \Psi_j^{[0]}\rangle (-\lambda_j + \overline{\lambda_i})\right)d_j}{(\lambda_j - \overline{\lambda_i})^2}$$
$$\left.+\langle \Psi_i^{[1]}, \Psi_j^{[1]}\rangle\right]\frac{1}{\lambda_j - \overline{\lambda_i}}.$$

To obtain the high-order RW solution, we merely need to take limit $\epsilon_j \to 0$.

Theorem 2.1 *Therefore the general RW solutions can be represented as*

$$Q[N] = Q[0] + \mathrm{i}[P, \sigma_3], \quad P = -XM^{-1}X^\dagger,$$
$$M = \begin{bmatrix} M_{1,1} & M_{1,2} \\ M_{2,1} & M_{2,2} \end{bmatrix},$$

where

$$M_{1,1} = \left(M_{1,1}^{[r,t]}\right)_{1 \leqslant r,t \leqslant N_1}, \quad M_{2,2} = \left(M_{2,2}^{[r,t]}\right)_{1 \leqslant r,t \leqslant N_2},$$

$$M_{1,2} = \left(M_{1,2}^{[r,t]}\right)_{\substack{1\leqslant r\leqslant N_1;\\ 1\leqslant t\leqslant N_2}}, \quad M_{1,2} = \left(M_{1,2}^{[r,t]}\right)_{\substack{1\leqslant r\leqslant N_2;\\ 1\leqslant t\leqslant N_1}},$$

$$X = \begin{bmatrix} X_1 \\ X_2 \\ X_3 \end{bmatrix} = \left[\Psi_1^{[1]}, \cdots, \Psi_1^{[N_1]}, \Psi_2^{[1]}, \cdots, \Psi_2^{[N_2]}\right].$$

By simple linear algebra, we can represent above formulas as

$$\begin{aligned} q_1[N] &= a_1 \left[\frac{\det(M_1)}{\det(M)}\right] \exp\theta_1, \\ q_2[N] &= a_2 \left[\frac{\det(M_2)}{\det(M)}\right] \exp\theta_2, \end{aligned} \quad (2.12)$$

where

$$M_1 = M - 2Y_2^\dagger X_1, \quad M_2 = M - 2Y_3^\dagger X_1,$$

$$Y_2 = \frac{X_2 e^{-\theta_1}}{a_1} \text{ and } Y_3 = \frac{X_3 e^{-\theta_2}}{a_2}.$$

The generalized form can be used to derive rogue wave solution with arbitrary order without the constrain conditions on background fields and the derivative with spectral parameter. The solution formulas are given by a purely algebraic way. Especially, high-order RW with different fundamental patterns can be obtained, in contrast to the ones reported before [13, 24, 25].

3 Exact rogue wave solutions and their dynamics

We find that many new patterns for RW excitation can exist in the coupled system without the constrain conditions before. In what following, we demonstrate the pattern dynamics of the first-order rogue wave solution, second-order rogue wave solution, and superposition of them respectively.

a) Fundamental rogue wave

The first-order RW solution can be given directly by the above formulas (2.12). The solution can be presented as follows by some simplifications

$$q_1[1] = a_1 \left[1 + \frac{-2ir_1}{\chi_1 + b_1} \frac{(x+p_1t)^2 + r_1^2 t^2 + \dfrac{i(x+p_1t - ir_1 t)}{p_1 + b_1 + ir_1}}{\left(x + p_1 t + \dfrac{1}{2r_1}\right)^2 + r_1^2 t^2 + \dfrac{1}{4r_1^2}}\right] e^{\theta_1},$$

$$q_2[1] = a_2 \left[1 + \frac{-2ir_1}{\chi_1 - b_1} \frac{(x+p_1t)^2 + r_1^2 t^2 + \dfrac{i(x+p_1t-ir_1t)}{p_1 - b_1 + ir_1}}{\left(x + p_1 t + \dfrac{1}{2r_1}\right)^2 + r_1^2 t^2 + \dfrac{1}{4r_1^2}} \right] e^{\theta_2},$$

where $p_1 = \text{Re}(\chi_1)$, $r_1 = \text{Im}(\chi_1)$ and χ_1 is a double root for cubic equation (2.6). We find that there are three different types of rogue wave solution:

- If $\dfrac{(p_1 \pm b_1)^2}{r_1^2} \geqslant 3$, then the rogue wave is called anti-eye-shaped rogue wave (or dark RW).
- If $\dfrac{1}{3} < \dfrac{(p_1 \pm b_1)^2}{r_1^2} < 3$, then the rogue wave is four-petaled rogue wave.
- If $\dfrac{(p_1 \pm b_1)^2}{r_1^2} \leqslant \dfrac{1}{3}$, then the rogue wave is called eye-shaped rogue wave (or bright RW).

Similar patterns for RW have been demonstrated in [12]. The explicit conditions for the RW transition were not given there. Additionally, breathers with these different pattern units were derived in a three-component coupled system [26]. The value of $\dfrac{(p_1 \pm b_1)^2}{r_1^2}$ can be used to make judgment on the RW pattern in the two components conveniently based on the above results. Explicitly, $\dfrac{(p_1 + b_1)^2}{r_1^2}$ can be used to clarify the RW pattern in the first component, and $\dfrac{(p_1 - b_1)^2}{r_1^2}$ is used to clarify RW pattern for the second component. When the multi RWs interact with each other, there will be many different patterns for which it is hard to know which fundamental RWs constitute them. These criterions can be used to clarify the fundamental RW pattern conveniently.

For the general case, since the roots for cubic equation and quartic equation are presented with the radical solution very complexly, a simple way to replace these radical solution is using the numerical roots. The above forms can be used to investigate dynamics of them directly. Especially, we can investigate RW solutions analytically with $a_1 = a_2$ case under which the RW pattern can be clarified explicitly. Since CNLSE possesses the scaling symmetry, we can set $a_1 = a_2 = 1$ without losing generality in this case. Setting $b_1 = \dfrac{1}{2}\sqrt{1 - \gamma_1^2}$, then we can have

$$\lambda_1 = \frac{i\delta_+(3+\gamma_1)}{4(1+\gamma_1)}, \quad \chi_1 = \frac{i\delta_+}{2},$$

$$\lambda_2 = \frac{i\delta_-(3-\gamma_1)}{4(1-\gamma_1)}, \quad \chi_2 = \frac{i\delta_-}{2},$$

where

$$\delta_\pm = \sqrt{(1\pm\gamma_1)(3\pm\gamma_1)}.$$

If $0 < b_1 < 1/2$, we can obtain the RW with velocity equals to zero. RW patterns in the two components are identical with each other in this case. And the types of RW are determined by the value $\frac{1-\gamma_1}{3+\gamma_1}$ and $\frac{1+\gamma_1}{3-\gamma_1}$ respectively. It is readily to see that there can be both eye-shaped or four-petaled RW exists in the two components for this case. It is impossible for anti-eye-shaped RW to exist correspondingly in both components for the coupled model, since $\frac{1-\gamma_1}{3+\gamma_1}$ and $\frac{1+\gamma_1}{3-\gamma_1}$ can not be larger than 2 for $0 < b_1 < 1/2$ (namely $|\gamma_1| < 1$). However, the anti-eye-shaped RW can exist in both components for the coupled model with negative nonlinear terms [27].

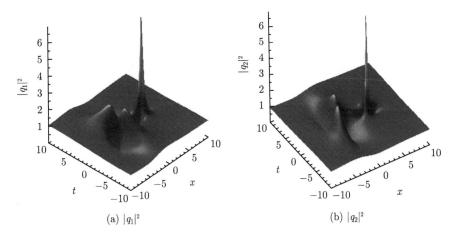

Fig. 1 The coexistence of eye-shaped rogue wave and four-petaled one in each component. Parameters: $b_1 = \frac{2}{5}, \gamma_1 = \frac{3}{5}, a_1 = a_2 = 1, \alpha_1^{[0]} = \frac{17}{5}, \alpha_2^{[0]} = \frac{\sqrt{6}}{5} - 1$ and $N_1 = N_2 = 1$

If $b_1 > 1/2$, then we introduce

$$k_1 = \frac{4\kappa}{1+\kappa} + \frac{i\sqrt{\kappa(\kappa^2 - 14\kappa + 1)}}{1+\kappa},$$

$$\gamma_1 = \frac{1}{2}\left(k_1 + \frac{1}{k_1}\right) - 2, \quad b_1 = \frac{\kappa^2 - 6\kappa + 1}{4(1+\kappa)\sqrt{\kappa}}$$

$$\delta_+ = \frac{1}{2}\left(k_1 - \frac{1}{k_1}\right), \quad \delta_- = \overline{\delta_+},$$

where $\kappa \in (0, 7 - 4\sqrt{3}) \cup (7 + 4\sqrt{3}, \infty)$. Furthermore, we have

$$p_1 = -\frac{\sqrt{\kappa^2 - 14\kappa + 1}}{4\sqrt{\kappa}}, \quad p_2 = -p_1,$$
$$r_1 = \frac{\kappa - 1}{1 + \kappa}, \quad r_2 = r_1. \tag{3.1}$$

We can verify that $\frac{(p_1 + b_1)^2}{r_1^2} < 1/3$ for eye-shaped RW and $\frac{(p_1 - b_1)^2}{r_1^2} > 1/3$ for four-petaled RW or anti-eye-shaped one. And the types for RW pattern in q_1 and q_2 components are different.

b) Superposition for fundamental rogue wave solutions

Superposition for fundamental RW solutions here is refer to nonlinear superposition for two fundamental RW solutions with different spectral parameter setting for the double root χ. This will make it possible to obtain RW with different patterns coexisting, since the RW pattern is determined by the value of χ and relative wave vector parameter b_1. It has been shown above that the RW pattern for two components can be clarified to two cases for different b_1 values. Firstly, we give two special cases which show that there are two different types of dynamic behavior. Similar coexistence of RW with different patterns were also demonstrated in other coupled system with some certain constrains on background fields [26]. It should be emphasized that there is no constrains on background field here. We show that eye-shaped RW can coexist with a four-petaled one or an anti-eye-shaped one. Nextly, we discuss them explicitly.

When $0 < b_1 < 1/2$ and $a_1 = a_2 = 1$, we can obtain the two RWs with different patterns in each component. One pattern is a four-petaled structure, and the other is an eye-shaped structure. The types for two RWs in components q_1 and q_2 are consistence (Fig. 1). In this case, it is impossible to obtain an anti-eye-shaped RW coexist with one eye-shaped RW, since anti-eye-shaped RW can not exist correspondingly in both components for the coupled model as discussed above.

When $b_1 > 1/2$ and $a_1 = a_2 = 1$, we can obtain the two RW solution which admits one anti-eye-shaped RW or a four-petaled RW and an eye-shaped one in each component. The types for two rogue waves in components q_1 and q_2 are different. As an example, we show one case for anti-eye-shaped RW coexisting with eye-shaped RW in Fig. 2. It is seen that at the same location, one anti-eye-shaped RW in one component corresponds to an eye-shaped RW in the other component. It should be noted that the case for a four-petaled RW coexisting with an eye-shaped RW is distinctive from the ones in Fig. 1.

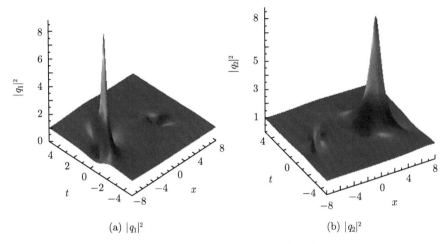

Fig. 2 The coexistence of eye-shaped rogue wave and anti-eye-shaped one in each component. Parameters: $\kappa = \dfrac{1}{25}$, $b_1 = \dfrac{119}{130}$, $a_1 = a_2 = 1$, $\alpha_1^{[0]} = -\dfrac{76}{13}$, $\alpha_2^{[0]} = \dfrac{28}{13}$, $N_1 = N_2 = 1$

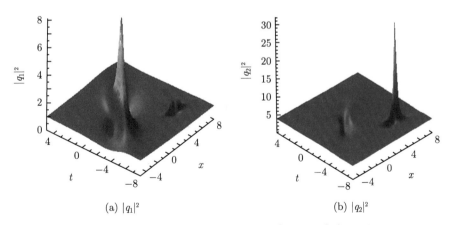

Fig. 3 The coexistence of eye-shaped rogue wave and four-petaled one in component q_1, and eye-shaped rogue wave with an anti-eye-shaped one in component q_2. $1/3 < \dfrac{(p_1+b_1)^2}{r_1^2} \approx 0.8744365601 < 3$ four-petaled rogue wave, $0 < \dfrac{(p_1-b_1)^2}{r_1^2} \approx 0.0045075247 < 1/3$ bright rogue wave, $\dfrac{(p_2+b_1)^2}{r_2^2} \approx 0.03465434964 < 1/3$ bright rogue wave, $\dfrac{(p_2-b_1)^2}{r_2^2} \approx 6.722765218 > 3$ dark rogue wave

For the general case $a_1 \neq a_2$, we merely give a special example to show its dynamics. For instance, we choose the parameters
$$b_1 = 1, \quad a_1 = 1, \quad a_2 = 2, \quad N_1 = N_2 = 1,$$
$$\lambda_1 = \frac{3 - i\sqrt{207 + 48\sqrt{3}}}{8}, \quad \lambda_2 = \frac{3 - i\sqrt{207 - 48\sqrt{3}}}{8},$$
$$\chi_1 = \frac{\sqrt{3}}{2} - i\frac{6 + \sqrt{3}}{66}\sqrt{207 + 48\sqrt{3}}, \quad \alpha_1^{[0]} = 10,$$
$$\chi_2 = -\frac{\sqrt{3}}{2} + i\frac{\sqrt{3} - 6}{66}\sqrt{207 - 48\sqrt{3}}, \quad \alpha_2^{[0]} = 0,$$
then we obtain the figure (Fig. 3). One eye-shaped RW and a four-petaled RW coexist in one component, and an anti-eye-shaped RW with an eye-shaped RW coexist in the other component. These would provide us more interesting excitation patterns for two RW case in coupled system. It should be noted that the above patterns all contains eye-shaped RW. Can an anti-eye-shaped RW coexist with a four-petaled one? It has been shown that this pattern can emerge in one component in a three-component coupled system [26]. We are not sure whether this pattern can emerge in this two-component coupled system.

c) Second-order rogue wave solution

As the second-order RW in scalar system [14–18], the second-order RW solution here is still obtained by superposition of fundamental RW solutions with the same spectral parameter. Therefore, one can obtain three fundamental RWs with identical pattern in each component for the second-order RW solution, which is similar to the ones in scalar case. But the three fundamental RWs can be three eye-shaped ones, three anti-eye-shaped ones, and three four-petaled ones, in contrast three eye-shaped ones in scalar system. By choosing parameters, we can obtain the different types of RW solution. The three RWs can be superposed together, and construct symmetric structure as the second-order RW with highest peak for scalar NLS [14–18]. However, it is usually very complicated to obtain the symmetric structure since there are much more parameters than the one for scalar NLS. Based on the solution form presented here, we can obtain them more easily. We show them by three categories.

With $b_1 < 1/2$ and $a_1 = a_2 = 1$, it is possible to obtain three eye-shaped RWs or four-petaled RWs in both components. As an example, we show dynamics of the second-order RW with highest peak with $b_1 = 2/5$ in (Fig. 4). Then, which three fundamental RW pattern constitute the pattern in Fig. 4? The values of parameters b_1 and χ can be used to analyze the patterns in Fig. 4 based on the above criterions for clarifying fundamental RW. It is proven that the patterns in the two components are both superimposed by three four-petaled RWs.

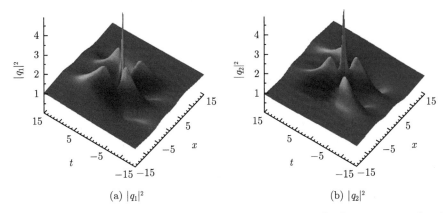

(a) $|q_1|^2$ (b) $|q_2|^2$

Fig. 4 The superposition of three four-petaled rogue waves in both components for the second-order rogue wave solution. The parameters are: $b_1 = \frac{2}{5}$, $a_1 = a_2 = 1$, $\lambda_1 = \frac{3i}{5}\sqrt{6}$, $\chi_1 = \frac{i}{5}\sqrt{6}$, $\alpha_1^{[0]} = 0$, $\alpha_1^{[1]} = 0$, $N_1 = 2$, $N_2 = 0$

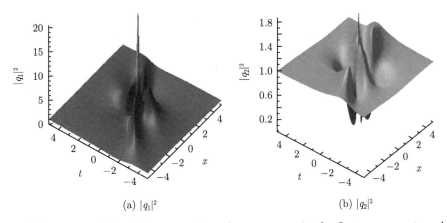

(a) $|q_1|^2$ (b) $|q_2|^2$

Fig. 5 The superposition of three eye-shaped rogue waves in the first component, and three anti-eye-shaped rogue waves in the second component. The parameters are: $b_1 = \frac{119}{130}$, $\lambda_1 = -\frac{69}{2380}\sqrt{69} - \frac{1728}{1547}i$, $\chi_1 = -\frac{1}{10}\sqrt{69} - \frac{12}{13}i$, $\alpha_1^{[0]} = 0$, $\alpha_1^{[1]} = 0$, $N_1 = 2$, $N_2 = 0$

The second case is with $b_1 > 1/2$ and $a_1 = a_2 = 1$. In this case, the patterns for RW in two components are different. As an example, we show the dynamics of them with $b_1 = 119/130$ in Fig. 5. It is seen that the patterns in two components are indeed different. Then, we can also clarify which fundamental RW pattern superpose the patterns in Fig. 5. For the patterns in Fig. 5, $b_1 = \dfrac{119}{130}$, $p_1 = -\dfrac{\sqrt{69}}{10}$, and $r_1 = -\dfrac{12}{13}$. From the value of $\dfrac{(p_1 \pm b_1)^2}{r_1^2}$, we can know that the pattern in q_1 is superimposed by three eye-shaped RWs, and the pattern in q_2 is constituted of three anti-eye-shaped ones. It is shown that three eye-shaped RWs admits the largest peak value among these three different pattern superposition ways, and the anti-eye-shaped ones admits the lowest value (compare the peak values in Fig. s4 and 5).

The third case is the general case with $a_1 \neq a_2$, $b_1 \neq 0$. The dynamics of RW can be investigated by second-order solution. The patterns of them are superposed by three eye-shaped ones (similar to Fig. 5(a)), or three four-petaled ones (similar to Fig. 4(a) or (b)), or three anti-eye-shaped ones (similar to Fig. 5(b)). Therefore, we do not show the figures.

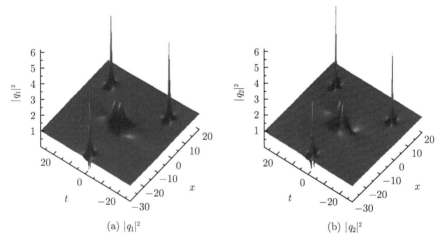

(a) $|q_1|^2$ (b) $|q_2|^2$

Fig. 6 The coexistence of one four-petaled RW and three eye-shaped ones. Parameters: $b_1 = \dfrac{2}{5}$, $a_1 = a_2 = 1$, $\lambda_1 = \dfrac{27i}{20}$, $\lambda_2 = \dfrac{3\sqrt{6}i}{5}$, $\chi_1 = \dfrac{6i}{5}$, $\chi_2 = \dfrac{\sqrt{6}i}{5}$, $\alpha_1^{[0]} = \dfrac{12}{5}i$, $\alpha_1^{[1]} = 5000i$, $\alpha_2^{[0]} = \dfrac{2\sqrt{6}}{5}i$, $N_1 = 2$ and $N_2 = 1$

d) Superposition for the second-order rogue wave solution

From the superposition of these different types RW solutions, we can obtain some novel dynamic behavior. We give them by two different categories. The first case is the interaction between the first-order RW and second-order RW. The second case is the interaction between two second-order RWs.

The superposition for the interaction between the first-order RW and second-order RW can demonstrate cases that there are four RWs on the temporal-spatial distribution plane. Especially, the four RWs can admit many different patterns. It is possible to obtain at least $2+2+2+2 = 8$ different patterns, such as three eye-shaped ones with one anti-eye-shaped RW in both components, three eye-shaped ones with a four-petaled RW in both components(which includes two distinctive cases: RW patterns in the two components at the same locations are identical or different), the combined ones in each component(corresponding to Fig. 3 case), and the inverse cases of them. For example, we show one case that one four-petaled RW coexist with three eye-shaped RWs in Fig. 6. All these patterns are in contrast to the four eye-shaped ones obtained before [13]. The nonlinear interactions among them can induce many other different patterns based on these 8 fundamental pattern combinations.

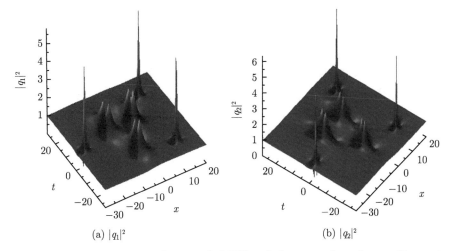

Fig. 7 The coexistence of three four-petaled RW and three eye-shaped ones. Parameters: $b_1 = \frac{2}{5}$, $a_1 = a_2 = 1$, $\lambda_1 = \frac{27i}{20}$, $\lambda_2 = \frac{3\sqrt{6}i}{5}$, $\chi_1 = \frac{6i}{5}$, $\chi_2 = \frac{\sqrt{6}i}{5}$, $\alpha_1^{[0]} = \frac{12}{5}i$, $\alpha_1^{[1]} = 5000i$, $\alpha_2^{[0]} = \frac{2\sqrt{6}}{5}i$, $\alpha_2^{[1]} = 100$, $N_1 = N_2 = 2$

Moreover, we can observe the superposition of two second-order RW solutions. There are six RWs in the distribution plane. The patterns of them can be different, in contrast to the six eye-shaped ones obtained with constrain conditions on background fields [13]. The six RWs just possess two types of fundamental RW patterns and each type pattern is occupied by three RWs. We find that it is possible to obtain at least 4 different patterns, mainly including three cases corresponding to Fig. s1-3 and one case similar to Fig. 1, but with RW patterns in the two components at the same locations are different. As an example, we show one case that three four-petaled RWs and three eye-shaped ones coexist in Fig. 7. Many different superposition patterns can be obtained by varying the parameters based on these 4 fundamental pattern combinations.

4 Conclusion

In this paper, we present a method to derive high-order RW for coupled NLS model with no constrain conditions on background fields. It is demonstrated that coexistence of RWs with different patterns can emerge in the coupled system. For example, one four-petaled RW and three eye-shaped ones can coexist, in contrast to the four eye-shaped ones reported before. Three four-petaled RW and three eye-shaped ones can constitute some new patterns for six fundamental ones case. There are many other different cases for the coexistence of RWs with different patterns. The superposition of different order RW solution can demonstrate more complex dynamics for RW excitation in the coupled systems. These RW pattern excitations are all in contrast to the ones in scalar NLS equation and the previous results obtained in coupled NLS systems. Since our formula is given by an algebraic form, it is possible to investigate dynamics of any order RW solution (for instance the explicit figure for 20-th or higher order RW) by computer soft. These results will further enrich our realization on RW dynamics in many different nonlinear physical systems. The method provided in this paper can be extended to derive the high-order RW solutions for coupled Hirota equations and even N-component NLS system or N-wave system with a general case.

Note added. Very recently, Chen and Mihalache presented fundamental and second-order RW solutions in the same coupled model [28]. Some of their results are overlapped with ours. Optical dark RW and baseband modulational instability in vector NLS described systems were observed or demonstrated in experiments [29, 30], which bring many possibilities to observe the theoretical results here for RW excitation patterns.

Acknowledgments This work is supported by National Natural Science Foundation of China (Contact No. 11401221, 11405129) and Fundamental Research Funds for the Central Universities (Contact No. 2014ZB0034).

References

[1] V Ruban, Y. Kodama, M. Ruderma, et al..*Rogue waves towards a unifying concept?: Discussions and debates.* Eur. Phys. Journ. Special Topics, 185, 5-15 (2010).

[2] N. Akhmediev and E. Pelinovsky. *Editorial-Introductory remarks on "Discussion & Debate: Rogue Waves-Towards a Unifying Concept?"* Eur. Phys. J. Special Topics, 185, 1 (2010).

[3] C. Kharif and E. Pelinovsky. *Physical mechanisms of the rogue wave phenomenon.* Eur. J. Mech. B/Fluids, 22, 603 (2003).

[4] E. Pelinovsky and C. Kharif. Extreme Ocean. Waves. Berlin: Springer, 2008.

[5] Y.V. Bludov, V.V. Konotop, and N. Akhmediev. *Vector rogue waves in binary mixtures of Bose-Einstein condensates.* Eur. Phys. J. Special Topics, 185, 169 (2010).

[6] L.C. Zhao and J. Liu. *Localized nonlinear waves in a two-mode nonlinear fiber.* J. Opt. Soc. Am. B, 29, 3119-3127 (2012).

[7] S.H. Chen, L.Y. Song.*Rogue waves in coupled Hirota systems.* Phys. Rev. E, 87, 032910 (2013).

[8] F. Baronio, M. Conforti, A. Degasperis, S. Lombardo, M. Onorato, and S. Wabnitz. *Vector rogue waves and baseband modulation instability in the defocusing regime.* Phys. Rev. Lett., 113, 034101 (2014).

[9] L.C. Zhao and J. Liu. *Rogue-wave solutions of a three-component coupled nonlinear Schrödinger equation.* Phys. Rev. E, 87, 013201 (2013).

[10] F. Baronio, M. Conforti, A. Degasperis, and S. Lombardo. *Rogue waves emerging from the resonant interaction of three waves.* Phys. Rev. Lett., 111, 114101 (2013).

[11] S. Chen, X. M. Cai, P. Grelu, J. M. Soto-Crespo, S. Wabnitz, and F. Baronio. *Complementary optical rogue waves in parametric three-wave mixing.* Opt. Express, 24, 5886 (2016).

[12] L.C. Zhao, G.G. Xin, Z.Y. Yang. *Rogue-wave pattern transition induced by relative frequency.* Phys. Rev. E, 90, 022918 (2014).

[13] L.M. Ling, B.L. Guo, L.C. Zhao. *High-order rogue waves in vector nonlinear Schrödinger equations.* Phys. Rev. E, 89, 041201(R) (2014).

[14] Y. Ohta, J.K. Yang. *General high-order rogue waves and their dynamics in the nonlinear Schrödinger equation.* Proc. R. Soc. A, 468, 1716-1740 (2012).

[15] B.L. Guo, L.M. Ling, Q. P. Liu. *Nonlinear Schrödinger equation: generalized Darboux transformation and rogue wave solutions.* Phys. Rev. E, 85, 026607 (2012); B.L. Guo, L.L. Ling and Q. P. Liu, *High-Order Solutions and Generalized Darboux Transformations of Derivative Nonlinear Schrödinger Equations.* Stud. Appl. Math. 130, 317-344 (2013).

[16] J.S. He, H.R. Zhang, L.H. Wang, K. Porsezian and A.S. Fokas. *Generating mechanism for higher-order rogue waves.* Phys. Rev. E, 87, 052914 (2013).

[17] L.M. Ling, L.C. Zhao. *Simple determinant representation for rogue waves of the nonlinear Schrödinger equation.* Phys. Rev. E, 88, 043201 (2013).

[18] D.J. Kedziora, A. Ankiewicz, and N. Akhmediev. *Classifying the hierarchy of nonlinear-Schrödinger-equation rogue-wave solutions.* Phys. Rev. E, 88, 013207 (2013).

[19] V. B. Matveev and M. A. Salle. *Darboux transformations and solitons.* Springer, 1991.

[20] C.H. Gu, H.S. Hu and Z.X. Zhou. *Darboux transformations in integrable systems: theory and their applications to geometry.* Springer, 2006.

[21] F. Baronio, A. Degasperis, M. Conforti, and S. Wabnitz. *Solutions of the vector nonlinear Schrödinger equations: evidence for deterministic rogue waves.* Phys. Rev. Lett., 109, 044102 2012.

[22] B.L. Guo and L. M. Ling. *Rogue wave, breathers and bright-dark-rogue solutions for the coupled Schrödinger equations,* Chin. Phys. Lett., 28, 110202 (2011).

[23] L.M. Ling, L.C. Zhao and B. Guo. *Darboux transformation and classification of solution for mixed coupled nonlinear Schrödinger equations.* arXiv preprint arXiv:1407.5194.

[24] B.G. Zhai, W.G. Zhang, X.L. Wang and H.Q. Zhang. *Multi-rogue waves and rational solutions of the coupled nonlinear Schrödinger equations.* Nonlinear Analysis: Real World Applications, 14, 14-27 (2013).

[25] G. Mu, Z. Qin and R. Grimshaw. *Dynamics of Rogue Waves on a Multisoliton Background in a Vector Nonlinear Schroödinger Equation.* SIAM J. Appl. Math., 75: 1-20 (2015).

[26] C. Liu, Z.Y. Yang, L.C. Zhao, W.L. Yang. *Vector breathers and the inelastic interaction in a three-mode nonlinear optical fiber.* Phys. Rev. A 89, 055803 (2014).

[27] J.H. Li, H.N. Chan, K.S. Chiang and K.W. Chow. *Breathers and 'black' rogue waves of coupled nonlinear Schrödinger equations with dispersion and nonlinearity of opposite signs.* Commun. Nonlinear Sci. Numer. Simulat., 28, 28-38 (2015).

[28] S. Chen, and D. Mihalache. *Vector rogue waves in the Manakov system: diversity and compossibility,* J. Phys. A: Math. Theor., 48, 215202 (2015).

[29] B. Frisquet, B. Kibler, Ph. Morin, F. Baronio, M. Conforti, G. Millot, and S.Wabnitz. *Optical dark rogue waves.* Sci. Rep. 6, 20785 (2016).

[30] B. Frisquet, B. Kibler, J. Fatome, P. Morin, F. Baronio, M. Conforti, G. Millot,and S. Wabnitz. *Polarization modulation instability in a Manakov fiber system.* Phys. Rev. A, 92, 053854 (2015).

A Second-order Finite Difference Method for Two-dimensional Fractional Percolation Equations*

Guo Boling (郭柏灵), Xu Qiang (徐强), and Zhu Ailing (朱爱玲)

Abstract A finite difference method which is second-order accurate in time and in space is proposed for two-dimensional fractional percolation equations. Using the Fourier transform, a general approximation for the mixed fractional derivatives is analyzed. An approach based on the classical Crank-Nicolson scheme combined with the Richardson extrapolation is used to obtain temporally and spatially second-order accurate numerical estimates. Consistency, stability and convergence of the method are established. Numerical experiments illustrating the effectiveness of the theoretical analysis are provided.

Keywords fractional percolation equation; second-order finite difference method; Crank-Nicolson scheme; Richardson extrapolation; stability analysis

1 Introduction

In this paper, we are concerned with the development of finite difference methods for the two-dimensional fractional percolation problem which seeks unknown pressure function $p(x,y,t)$ satisfying

$$\frac{\partial p}{\partial t} = \frac{\partial^{\beta_1}}{\partial x^{\beta_1}}\left(k_x(x,y)\frac{\partial^{\alpha_1} p}{\partial x^{\alpha_1}}\right) + \frac{\partial^{\beta_2}}{\partial y^{\beta_2}}\left(k_y(x,y)\frac{\partial^{\alpha_2} p}{\partial y^{\alpha_2}}\right) + f(x,y,t),$$
$$(x,y) \in \Omega, \quad 0 < t \leq T, \tag{1.1}$$

$$p(x,y,0) = p_0(x,y), \quad (x,y) \in \Omega, \tag{1.2}$$

$$p(x_L,y,t) = p(x,y_L,t) = 0, \quad p(x_R,y,t) = v_x(y,t), \quad p(x,y_R,t) = v_y(x,t),$$

---

* Commun. Comput. Phys., 2016, 19(3): 733–757. DOI:10.4208/cicp.011214.140715a.

$$(x,y) \in \Omega, \quad 0 \leqslant t \leqslant T, \tag{1.3}$$

where $\Omega = \{(x,y) | x_L \leqslant x \leqslant x_R, y_L \leqslant y \leqslant y_R\}, 0 < \alpha_1, \alpha_2 < 1, 0 < \beta_1, \beta_2 \leqslant 1$, $f(x,y,t)$ is the source term, positive $k_x(x,y)$ and $k_y(x,y)$ are percolation coefficients along the x and y direction, respectively.

The fractional partial derivatives in (1.1) are defined in the Riemann-Liouville form. Generally, for any $\gamma > 0$, the Riemann-Liouville fractional partial derivatives $\dfrac{\partial^\gamma w(x,y)}{\partial x^\gamma}$ and $\dfrac{\partial^\gamma w(x,y)}{\partial y^\gamma}$ of order γ are defined by [28, 32, 35]

$$\frac{\partial^\gamma w(x,y)}{\partial x^\gamma} = \frac{1}{\Gamma(n-\gamma)} \frac{\partial^n}{\partial x^n} \int_{x_L}^{x} \frac{w(\xi,y)}{(x-\xi)^{\gamma+1-n}} \, d\xi \tag{1.4}$$

and

$$\frac{\partial^\gamma w(x,y)}{\partial y^\gamma} = \frac{1}{\Gamma(n-\gamma)} \frac{\partial^n}{\partial y^n} \int_{y_L}^{y} \frac{w(x,\eta)}{(y-\eta)^{\gamma+1-n}} \, d\eta, \tag{1.5}$$

where n is an integer such that $n-1 < \gamma \leqslant n$. If γ is an integer, then the above definitions give the standard integer partial derivatives.

The percolation equations have been applied successfully in groundwater hydraulics, groundwater dynamics and fluid dynamics in porous media [7, 31, 40]. Under the assumptions of seepage flow continuity and the traditional Darcy's law, the traditional percolation equation [24, 34] for two-dimensional seepage flow in porous media is just the special case of equation (1.1) that $\alpha_i = \beta_i = 1$ for $i = 1, 2$. In view of the limitations of these two assumptions above, He [12] proposed the modified Darcy's law with Riemann-Liouville fractional derivatives

$$q_x = k_x(x,y) \frac{\partial^{\alpha_1} p}{\partial x^{\alpha_1}}, \qquad q_y = k_y(x,y) \frac{\partial^{\alpha_2} p}{\partial y^{\alpha_2}}, \qquad 0 < \alpha_1, \alpha_2 < 1, \tag{1.6}$$

as a generalization of Darcy's law for realistically describing the movement of solute in a non-homogeneous porous medium. Furthermore, considering the fact that the seepage flow is neither continued nor rigid body motion, He employed the fractional differential operators $\dfrac{\partial^{\beta_1}}{\partial x^{\beta_1}}$ and $\dfrac{\partial^{\beta_2}}{\partial y^{\beta_2}}$, where $0 < \beta_1, \beta_2 \leqslant 1$ in the percolation equation, and then the fractional percolation model (1.1) was obtained, which is the focus of this paper.

As analytic solutions of most fractional differential equations cannot be obtained explicitly, numerical methods become major ways and a wide variety of techniques have been developed, including finite difference methods [25–27, 29, 37–39, 49–51], finite element methods [2, 8–10, 16], finite volume methods [42, 44, 45, 48], spectral methods [15, 17, 47], and mesh-free methods [11, 20]. Recently, Liu et al. [21] proposed a first-order alternating direction implicit scheme for the three-dimensional non-continued seepage flow in uniform media and a second-order method which combined modified Douglas scheme with Richardson extrapolation for the three-dimensional continued seepage flow in non-uniform media. Chen et al. [5] developed an implicit finite difference method for the initial-boundary value problem of one-dimensional fractional percolation equation with left-sided mixed Riemann-Liouville fractional derivative, and they [6] considered an alternating direction implicit difference method for the two-dimensional case by a similar technique. The numerical simulation of variable-order fractional percolation equation in non-homogeneous porous media is discussed by Chen et al. in [4]. The numerical methods in [4–6] were proved to be first-order accurate in time and in space.

For solving high dimensional fractional differential equations, some numerical methods have been considered by many authors. Liu et al. [22] developed an implicit finite difference scheme for the two-dimensional fractional FitzHugh-Nagumo monodomain model, and they proved that the numerical method is unconditionally stable and converges linearly. A spatially second-order accurate semi-alternating direction method for the two-dimensional fractional FitzHugh-Nagumo monodomain model on an approximate irregular domain is proposed in [23]. Yu et al. [46] presented an alternating direction method for the space and time fractional Bloch-Torrey equation in three-dimensions, and a spatially second-order accurate implicit finite difference method with fractional centered derivative approximating the Riesz fractional derivative for the same problem is derived by Song et al. in [36].

Moreover, for fractional diffusion equations, a second-order finite difference method based on the classical Crank-Nicolson scheme and the Richardson extrapolation is developed by Tadjeran et al. in [39]. A second-order method which combines the alternating direction implicit approach with the Crank-Nicolson discretization and the Richardson extrapolation for the two-dimensional fractional diffusion equations is studied in [38]. By analyzing the Toeplitz-like structure of the full coefficient matrix, Basu and Wang [1] proposed a fast second-order finite difference method for fractional diffusion equations, which reduces the computational work from $O(N^3)$ to $O(N \log^2 N)$ per time step and reduces the memory requirement from $O(N^2)$ to $O(N)$. Lin et al. [18] developed two preconditioned iterative methods to solve the relevant linear systems of a second-order finite difference scheme for fractional diffusion equations, which require $O(N \log N)$ operations per iteration and $O(N)$ storage. Yang et al. [45] established a finite volume scheme with preconditioned Lanczos method as an attractive and high-efficiency approach for the two-dimensional space fractional reaction-diffusion equations. Tian et al. [41] developed a class of second-order weighted and shifted Grünwald difference operators for Riemann-Liouville fractional derivatives and then proposed the corresponding unconditionally stable difference schemes for the fractional diffusion equations with constant coefficients in one and two dimensions. A quasi-compact difference scheme which is third-order accurate in space is established for the fractional diffusion equations by Zhou et al. [50]. Liu et al. [19] developed an alternating direction implicit method for the two-dimensional Riesz space fractional diffusion equations with a nonlinear reaction term. Zeng et al. [47] proposed a Crank-Nicolson alternating direction implicit Galerkin-Legendre spectral method for the two-dimensional Riesz space fractional nonlinear reaction-diffusion equations. A second-order finite difference-spectral method for the fractional diffusion equations is considered by Huang et al. [13]. However, second-order accurate numerical methods for two-dimensional fractional percolation equations are still limited. This motivates us to examine a numerical approach which combines the classical Crank-Nicolson scheme with the Richardson extrapolation for the initial-boundary value problem (1.1) (1.2) (1.3).

The rest of the paper is organized as follows. In Section 2 we establish a general approximation for the mixed fractional derivatives in (1.1). In Section 3 we present a Crank-Nicolson scheme and consider its consistency. In Section 4 we prove stability and convergence of the scheme. In Section 5 we improve the order of convergence of the Crank-Nicolson scheme by the Richardson extrapolation. In Section 6 we carry out numerical experiments to evaluate the performance of the method. Finally, we draw our conclusions in Section 7.

2 Approximation for the mixed fractional derivatives

We begin with some definitions and properties.

Definition 2.1 (see [28, 32, 35]) *For any $\gamma > 0$, the Liouville fractional integral $\mathrm{I}^{n-\gamma} u(x)$ of order $n - \gamma$ is defined by*

$$\mathrm{I}^{n-\gamma} u(x) = \frac{1}{\Gamma(n-\gamma)} \int_{-\infty}^{x} \frac{u(\xi)}{(x-\xi)^{\gamma+1-n}} \, d\xi \qquad (2.1)$$

and the Liouville fractional derivative $D^\gamma u(x)$ of order γ is defined by

$$D^\gamma u(x) = \frac{d^n}{dx^n} I^{n-\gamma} u(x), \tag{2.2}$$

where n is an integer such that $n - 1 < \gamma \leqslant n$.

For any positive integer r, let $W^{r,1}(\mathbf{R})$ denote the collection of functions in $C^r(\mathbf{R})$ whose derivatives up to order r belong to $L^1(\mathbf{R})$. For any $u(x) \in L^1(\mathbf{R})$, we use the definition of the Fourier transform as

$$\mathcal{F}[u(x)](\omega) = \hat{u}(\omega) := \int_{\mathbf{R}} e^{i\omega x} u(x)\, dx, \tag{2.3}$$

where i is the imaginary unit. Three useful properties of the Fourier transform are listed in the following lemma.

Lemma 2.1 (see [35])

(1) If $u(x) \in L^1(\mathbf{R})$ and $h \in \mathbf{R}$, then

$$\mathcal{F}[u(x-h)](\omega) = e^{ih\omega} \hat{u}(\omega). \tag{2.4}$$

(2) If $u(x), v(x), u(x)v(x) \in L^1(\mathbf{R})$, then

$$\mathcal{F}[u(x)v(x)](\omega) = \frac{1}{2\pi} \hat{u}(\omega) * \hat{v}(\omega) = \frac{1}{2\pi} \int_{\mathbf{R}} \hat{u}(\omega - \tau)\hat{v}(\tau)\, d\tau. \tag{2.5}$$

(3) If $\gamma > 0$, n is an integer such that $n - 1 < \gamma \leqslant n$, $u(x) \in W^{n,1}(\mathbf{R})$, then

$$\mathcal{F}[D^\gamma u(x)](\omega) = (-i\omega)^\gamma \hat{u}(\omega). \tag{2.6}$$

Under the assumption that $u(x_L) = 0$, the Riemman-Liouville mixed fractional derivative $\frac{\partial^\beta}{\partial x^\beta}\left(k(x)\frac{\partial^\alpha}{\partial x^\alpha}\right)$ and $D^\beta(k(x)D^\alpha u(x))$ in the Liouville form are equivalent in view of the zero-extension of $u(x)$ for $x < x_L$. Notice the Dirichlet boundary condition (1.3) of the fractional percolation problem, it is sufficient for us to study a general approximation for the mixed fractional derivative $D^\beta(k(x)D^\alpha u(x))$.

For functions $k(x)$ and $u(x)$, we define the operator $\Delta^{\alpha,\beta}_{h,q,\tilde{q}}$ as follow:

$$\Delta^{\alpha,\beta}_{h,q,\tilde{q}}(k(x), u(x)) = \frac{1}{h^{\alpha+\beta}} \sum_{l=0}^{\infty} \sum_{j=0}^{\infty} g_l^{(\beta)} g_j^{(\alpha)} k(x - (l - \tilde{q})h) u(x - (j + l - q)h), \tag{2.7}$$

where $0 < \alpha < 1$, $0 < \beta \leqslant 1$, h is the grid size, q and $\tilde{q}$ are any real numbers, $g_j^{(\alpha)}$ and $g_l^{(\beta)}$ are the alternating fractional binomial coefficients related to α and β, respectively, e.g.,

$$\begin{cases} g_0^{(\alpha)} = 1, \\ g_j^{(\alpha)} = \dfrac{(-1)^j}{j!}\alpha(\alpha-1)\cdots(\alpha-j+1), \quad j=1,2,3,\cdots. \end{cases} \quad (2.8)$$

It has been proved (see Theorem 1 in [5]) that $\Delta_{h,q,0}^{\alpha,\beta}(k(x),u(x))$ is a first-order accurate approximation for $D^\beta(k(x)D^\alpha u(x))$, i.e.,

$$\Delta_{h,q,0}^{\alpha,\beta}(k(x),u(x)) = D^\beta(k(x)D^\alpha u(x)) + O(h) \quad (2.9)$$

uniformly in $x \in \mathbf{R}$ as $h \to 0$. With the help of the Fourier transform, we have a more general expansion of $\Delta_{h,q,\tilde{q}}^{\alpha,\beta}(k(x),u(x))$ for any real numbers q and $\tilde{q}$.

Theorem 2.1 *Assume that $0 < \alpha < 1$, $0 < \beta \leqslant 1$, n is a positive integer, $k(x) \in W^{n+3,1}(\mathbf{R})$, $u(x) \in W^{2n+3,1}(\mathbf{R})$ and $I^{1-\alpha}u(x) \in W^{2n,1}(\mathbf{R})$, then for any q, $\tilde{q} \in \mathbf{R}$, there exist some constants $a_l^{(\alpha)}$ and $a_l^{(\beta)}$ independent of $h, k(x)$ and $u(x)$, such that*

$$\Delta_{h,q,\tilde{q}}^{\alpha,\beta}(k(x),u(x))$$
$$= D^\beta(k(x)D^\alpha u(x)) + \sum_{j=1}^{n-1}\sum_{l=0}^{j} a_l^{(\alpha)} a_{j-l}^{(\beta)} D^{\beta+j-l}(k(x)D^{\alpha+l}u(x)) h^j + O(h^n) \quad (2.10)$$

uniformly in $x \in \mathbf{R}$ as $h \to 0$.

To prove Theorem 2.1, we need the following lemmas.

Lemma 2.2 *Under the assumption of Theorem 2.1, for positive integers m_k and m_u which satisfy $m_k < n+2$ and $m_k + m_u < 2n+2$, there exists a constant $C(m_k, m_u)$ such that*

$$\int_{\mathbf{R}}\int_{\mathbf{R}} \left| \hat{k}(\omega-\tau)\hat{u}(\tau)(-i\tau)^{m_u}(-i\omega)^{m_k} \right| d\tau d\omega \leqslant C(m_k, m_u), \quad (2.11)$$

where $\hat{k}(\omega)$ and $\hat{u}(\omega)$ are the Fourier transform of $k(x)$ and $u(x)$, respectively.

Proof Since $k(x) \in W^{n+3,1}(\mathbf{R})$ and $u(x) \in W^{2n+3,1}(\mathbf{R})$, according to the Riemman-Lebesgue Lemma, we have

$$|\hat{k}(\omega)| \leqslant C_k(1+|\omega|)^{-(n+3)} \quad (2.12)$$

and

$$|\hat{u}(\omega)| \leqslant C_u(1+|\omega|)^{-(2n+3)} \quad (2.13)$$

for all $\omega \in \mathbf{R}$, where C_k and C_u are constants. Hence $w^{m_k}\hat{k}(\omega)$, $w^{m_u}\hat{u}(\omega)$ and $w^{m_k+m_u}\hat{u}(\omega)$ belong to $L^1(\mathbf{R})$. Let

$$C(m_k,m_u)=2^{m_k-1}\left(\|\hat{k}(\omega)\|_{L^1}\|w^{m_k+m_u}\hat{u}(\omega)\|_{L^1}+\|w^{m_k}\hat{k}(\omega)\|_{L^1}\|w^{m_u}\hat{u}(\omega)\|_{L^1}\right), \quad (2.14)$$

then we have

$$\int_{\mathbf{R}}\int_{\mathbf{R}}\left|\hat{k}(\omega-\tau)\hat{u}(\tau)(-\mathrm{i}\tau)^{m_u}(-\mathrm{i}\omega)^{m_k}\right|d\tau d\omega$$
$$\leqslant \int_{\mathbf{R}}|\hat{u}(\tau)||\tau|^{m_u}d\tau\int_{\mathbf{R}}|\hat{k}(\omega-\tau)||\omega|^{m_k}d\omega$$
$$\leqslant 2^{m_k-1}\int_{\mathbf{R}}|\hat{u}(\tau)||\tau|^{m_u}d\tau\int_{\mathbf{R}}|\hat{k}(\omega-\tau)|(|\tau|^{m_k}+|\omega-\tau|^{m_k})d\omega$$
$$=C(m_k,m_u). \quad (2.15)$$

The proof of Lemma 2.2 is completed. $\square$

Lemma 2.3 *Under the assumption of Theorem 2.1, for $0 \leqslant j, l \leqslant n-1$, we have*

$$\mathcal{F}^{-1}\left(\frac{1}{2\pi}\int_{\mathbf{R}}\hat{k}(\omega-\tau)\hat{u}(\tau)(-\mathrm{i}\tau)^{\alpha+l}(-\mathrm{i}\omega)^{\beta+j}d\tau\right)=\mathrm{D}^{\beta+j}\left(k(x)\mathrm{D}^{\alpha+l}u(x)\right). \quad (2.16)$$

Proof Since $\mathrm{D}^{\alpha+l}u(x)=\dfrac{d^{l+1}}{dx^{l+1}}\mathrm{I}^{1-\alpha}u(x)\in W^{2n-l-1,1}(\mathbf{R})$ and $k(x)\in W^{n+3,1}(\mathbf{R})$, it is easy to verify that

$$\mathcal{F}\left[\mathrm{D}^{\beta+j}\left(k(x)\mathrm{D}^{\alpha+l}u(x)\right)\right](\omega)$$
$$=(-\mathrm{i}\omega)^{\beta+j}\mathcal{F}\left[k(x)\mathrm{D}^{\alpha+l}u(x)\right](\omega)$$
$$=(-\mathrm{i}\omega)^{\beta+j}\frac{1}{2\pi}\left(\hat{k}(\omega)*\mathcal{F}\left[\mathrm{D}^{\alpha+l}u(x)\right](\omega)\right)$$
$$=(-\mathrm{i}\omega)^{\beta+j}\frac{1}{2\pi}\left(\hat{k}(\omega)*(-\mathrm{i}\omega)^{\alpha+l}\hat{u}(\omega)\right)$$
$$=\frac{1}{2\pi}\int_{\mathbf{R}}\hat{k}(\omega-\tau)\hat{u}(\tau)(-\mathrm{i}\tau)^{\alpha+l}(-\mathrm{i}\omega)^{\beta+j}d\tau. \quad (2.17)$$

By Lemma 2.2, we have

$$\int_{\mathbf{R}}\hat{k}(\omega-\tau)\hat{u}(\tau)(-\mathrm{i}\tau)^{\alpha+l}(-\mathrm{i}\omega)^{\beta+j}d\tau\in L^1(\mathbf{R}),$$

then (2.16) follows. $\square$

We are now turning to the proof of Theorem 2.1.

Proof of Theorem 2.1 First, we consider a decomposition of the Fourier transform of $\Delta_{h,q,\tilde{q}}^{\alpha,\beta}(k(x), u(x))$. Note that

$$(1-z)^\gamma = \sum_{j=0}^{\infty} g_j^{(\gamma)} z^\gamma \tag{2.18}$$

converges absolutely for $|z| \leqslant 1$, where $\gamma \in \{\alpha, \beta\}$. Hence $\Delta_{h,q,\tilde{q}}^{\alpha,\beta}(k(x), u(x)) \in L^1(\mathbf{R})$ and therefore

$$\mathcal{F}\left[\Delta_{h,q,\tilde{q}}^{\alpha,\beta}(k(x), u(x))\right](\omega)$$

$$= \frac{1}{h^{\alpha+\beta}} \sum_{l=0}^{\infty} g_l^{(\beta)} \mathcal{F}\left[\sum_{j=0}^{\infty} g_j^{(\alpha)} k(x - (l-\tilde{q})h) u(x - (j+l-q)h)\right](\omega)$$

$$= \frac{1}{h^{\alpha+\beta}} \sum_{l=0}^{\infty} g_l^{(\beta)} \frac{1}{2\pi} \left\{\mathcal{F}[k(x - (l-\tilde{q})h)](\omega) * \sum_{j=0}^{\infty} g_j^{(\alpha)} \mathcal{F}[u(x - (j+l-q)h)](\omega)\right\}$$

$$= \frac{1}{h^{\alpha+\beta}} \sum_{l=0}^{\infty} g_l^{(\beta)} \frac{1}{2\pi} \left\{\left[e^{i\omega(l-\tilde{q})h} \hat{k}(\omega)\right] * \sum_{j=0}^{\infty} g_j^{(\alpha)} e^{i\omega(j+l-q)h} \hat{u}(\omega)\right\}$$

$$= \frac{1}{h^{\alpha+\beta}} \sum_{l=0}^{\infty} g_l^{(\beta)} \frac{1}{2\pi} \int_{\mathbf{R}} e^{i(\omega-\tau)(l-\tilde{q})h} \hat{k}(\omega - \tau) \sum_{j=0}^{\infty} g_j^{(\alpha)} e^{i\tau(j+l-q)h} \hat{u}(\tau) \, d\tau$$

$$= \frac{1}{h^{\beta}} \sum_{l=0}^{\infty} g_l^{(\beta)} e^{i\omega(l-\tilde{q})h} \frac{1}{2\pi} \int_{\mathbf{R}} \frac{1}{h^{\alpha}} \sum_{j=0}^{\infty} g_j^{(\alpha)} e^{i\tau(j-q+\tilde{q})h} \hat{k}(\omega - \tau) \hat{u}(\tau) \, d\tau$$

$$= \frac{1}{h^{\beta}} (1 - e^{i\omega h})^{\beta} e^{-i\omega\tilde{q}h} \frac{1}{2\pi} \int_{\mathbf{R}} \frac{1}{h^{\alpha}} (1 - e^{i\tau h})^{\alpha} e^{-i\tau(q-\tilde{q})h} \hat{k}(\omega - \tau) \hat{u}(\tau) \, d\tau$$

$$= (-i\omega)^{\beta} \left(\frac{1 - e^{i\omega h}}{-i\omega h}\right)^{\beta} \frac{e^{-i\omega\tilde{q}h}}{2\pi} \int_{\mathbf{R}} (-i\tau)^{\alpha} \left(\frac{1 - e^{i\tau h}}{-i\tau h}\right)^{\alpha} e^{-i\tau(q-\tilde{q})h} \hat{k}(\omega - \tau) \hat{u}(\tau) \, d\tau. \tag{2.19}$$

Define

$$W_{\gamma,\varsigma}(z) = \left(\frac{1 - e^{-z}}{z}\right)^{\gamma} e^{\varsigma z}, \tag{2.20}$$

for $\varsigma \in \mathbf{R}$ and $\gamma \in \{\alpha, \beta\}$. Then the previous expression (2.19) may be rewritten as

$$\mathcal{F}\left[\Delta_{h,q,\tilde{q}}^{\alpha,\beta}(k(x), u(x))\right](\omega)$$

$$= \frac{1}{2\pi} \int_{\mathbf{R}} \hat{k}(\omega - \tau) \hat{u}(\tau) (-i\tau)^{\alpha} (-i\omega)^{\beta} W_{\alpha, q-\tilde{q}}(-i\tau h) W_{\beta, \tilde{q}}(-i\omega h) \, d\tau. \tag{2.21}$$

Since $W_{\alpha, q-\tilde{q}}(z)$ and $W_{\beta, \tilde{q}}(z)$ are analytic in some neighborhood of the origin, we have the power series expansions

$$W_{\alpha,q-\tilde{q}}(z) = \sum_{l=0}^{\infty} a_l^{(\alpha)} z^l \tag{2.22}$$

and

$$W_{\beta,\tilde{q}}(z) = \sum_{l=0}^{\infty} a_l^{(\beta)} z^l, \tag{2.23}$$

which converge absolutely for all $|z| \leqslant R$ with some $R > 0$. Note that $a_0^{(\alpha)} = a_0^{(\beta)} = 1$, and that $a_1^{(\alpha)} = q - \tilde{q} - \frac{\alpha}{2}$, $a_1^{(\beta)} = \tilde{q} - \frac{\beta}{2}$. For $x \in \mathbf{R}$, let

$$R_n^{(\alpha)}(x) = W_{\alpha,q-\tilde{q}}(-\mathrm{i}x) - \sum_{l=0}^{n-1} a_l^{(\alpha)}(-\mathrm{i}x)^l, \tag{2.24}$$

$$R_n^{(\beta)}(x) = W_{\beta,\tilde{q}}(-\mathrm{i}x) - \sum_{l=0}^{n-1} a_l^{(\beta)}(-\mathrm{i}x)^l. \tag{2.25}$$

According to Theorem 3.1 in [39], there exist positive constants C_1 and C_2 such that

$$|R_n^{(\alpha)}(x)| \leqslant C_1 |x|^n \tag{2.26}$$

and

$$|R_n^{(\beta)}(x)| \leqslant C_2 |x|^n \tag{2.27}$$

uniformly in $x \in \mathbf{R}$. Thus,

$$W_{\alpha,q-\tilde{q}}(-\mathrm{i}\tau h) W_{\beta,\tilde{q}}(-\mathrm{i}\omega h)$$
$$= \sum_{j=0}^{n-1} a_j^{(\alpha)} (-\mathrm{i}\tau h)^j \sum_{l=0}^{n-1} a_l^{(\beta)} (-\mathrm{i}\omega h)^l + R_n^{(\beta)}(\omega h) \sum_{j=0}^{n-1} a_j^{(\alpha)} (-\mathrm{i}\tau h)^j$$
$$+ R_n^{(\alpha)}(\tau h) \sum_{l=0}^{n-1} a_l^{(\beta)} (-\mathrm{i}\omega h)^l + R_n^{(\alpha)}(\tau h) R_n^{(\beta)}(\omega h)$$
$$= \sum_{j=0}^{n-1} \sum_{l=0}^{j} a_l^{(\alpha)} a_{j-l}^{(\beta)} (-\mathrm{i}\tau h)^l (-\mathrm{i}\omega h)^{j-l} + \sum_{j=1}^{n-1} \sum_{l=n-j}^{n-1} a_j^{(\alpha)} a_l^{(\beta)} (-\mathrm{i}\tau h)^j (-\mathrm{i}\omega h)^l$$
$$+ R_n^{(\beta)}(\omega h) \sum_{j=0}^{n-1} a_j^{(\alpha)} (-\mathrm{i}\tau h)^j + R_n^{(\alpha)}(\tau h) \sum_{l=0}^{n-1} a_l^{(\beta)} (-\mathrm{i}\omega h)^l + R_n^{(\alpha)}(\tau h) R_n^{(\beta)}(\omega h)$$
$$=: \phi_1(\tau,\omega,h) + \phi_2(\tau,\omega,h) + \phi_3(\tau,\omega,h) + \phi_4(\tau,\omega,h) + \phi_5(\tau,\omega,h). \tag{2.28}$$

Substitute (2.28) into (2.21), we obtain

$$\mathcal{F}\left[\Delta_{h,q,\tilde{q}}^{\alpha,\beta}(k(x),u(x))\right](\omega)$$

$$= \sum_{s=1}^{5} \frac{1}{2\pi} \int_{\mathbf{R}} \hat{k}(\omega - \tau)\hat{u}(\tau)(-i\tau)^{\alpha}(-i\omega)^{\beta}\phi_s(\tau,\omega,h)d\tau$$

$$=: \sum_{s=1}^{5} \hat{\psi}_s(\omega, h). \tag{2.29}$$

Next, we analyze the Fourier inversion of $\hat{\psi}_s(\omega,h)$ on the right hand side of (2.29). By Lemma 2.2, the Fourier inversion of $\hat{\psi}_1(\omega,h)$ may be estimated as

$$\left|\mathcal{F}^{-1}\left[\hat{\psi}_1(\omega,h)\right](x)\right|$$

$$= \left|\frac{1}{2\pi}\int_{\mathbf{R}} e^{-i\omega x}\hat{\psi}_1(\omega,h)\,d\omega\right|$$

$$= \left|\frac{1}{4\pi^2}\int_{\mathbf{R}} e^{-i\omega x}\sum_{j=0}^{n-1}\sum_{l=0}^{j} a_l^{(\alpha)}a_{j-l}^{(\beta)}h^j \int_{\mathbf{R}} \hat{k}(\omega-\tau)\hat{u}(\tau)(-i\tau)^{\alpha+l}(-i\omega)^{\beta+j-l}\,d\tau d\omega\right|$$

$$\leqslant \frac{1}{4\pi^2}\sum_{j=0}^{n-1}\sum_{l=0}^{j}\left|a_l^{(\alpha)}\right|\left|a_{j-l}^{(\beta)}\right|h^j \int_{\mathbf{R}}\int_{\mathbf{R}}\left|\hat{k}(\omega-\tau)\hat{u}(\tau)(-i\tau)^{\alpha+l}(-i\omega)^{\beta+j-l}\right|d\tau d\omega$$

$$\leqslant \frac{1}{4\pi^2}\sum_{j=0}^{n-1}\sum_{l=0}^{j}\left|a_l^{(\alpha)}\right|\left|a_{j-l}^{(\beta)}\right|C(\alpha+l,\beta+j-l)h^j. \tag{2.30}$$

Similarly, due to (2.26) and (2.27), we have that

$$\left|\mathcal{F}^{-1}\left[\hat{\psi}_2(\omega,h)\right](x)\right| \leqslant \frac{1}{4\pi^2}\sum_{j=1}^{n-1}\sum_{l=n-j}^{n-1}\left|a_j^{(\alpha)}\right|\left|a_l^{(\beta)}\right|C(\alpha+j,\beta+l)h^{j+l}, \tag{2.31}$$

$$\left|\mathcal{F}^{-1}\left[\hat{\psi}_3(h,\omega)\right](x)\right| \leqslant \frac{C_2}{4\pi^2}\sum_{j=0}^{n-1}\left|a_j^{(\alpha)}\right|C(\alpha+j,\beta+n)h^{j+n}, \tag{2.32}$$

$$\left|\mathcal{F}^{-1}\left[\hat{\psi}_4(h,\omega)\right](x)\right| \leqslant \frac{C_1}{4\pi^2}\sum_{l=0}^{n-1}\left|a_l^{(\beta)}\right|C(\alpha+n,\beta+l)h^{n+l}, \tag{2.33}$$

and

$$\left|\mathcal{F}^{-1}\left[\hat{\psi}_5(h,\omega)\right](x)\right| \leqslant \frac{C_1 C_2}{4\pi^2}C(\alpha+n,\beta+n)h^{2n} \tag{2.34}$$

uniformly in $x \in \mathbf{R}$, respectively.

Finally, we compute the Fourier inversion of $\mathcal{F}\left[\Delta_{h,q,\tilde{q}}^{\alpha,\beta}(k(x),u(x))\right](\omega)$ to obtain the expansion (2.10). Using Lemma 2.3, we have

$$\mathcal{F}^{-1}\left[\hat{\psi}_1(\omega,h)\right](x)$$

$$= \sum_{j=0}^{n-1} \sum_{l=0}^{j} a_l^{(\alpha)} a_{j-l}^{(\beta)} \mathcal{F}^{-1} \left[\frac{1}{2\pi} \int_{\mathbf{R}} \hat{k}(\omega - \tau) \hat{u}(\tau)(-i\tau)^{\alpha+l}(-i\omega)^{\beta+j-l} d\tau \right] h^j$$

$$= \sum_{j=0}^{n-1} \sum_{l=0}^{j} a_l^{(\alpha)} a_{j-l}^{(\beta)} D^{\beta+j-l} \left(k(x) D^{\alpha+l} u(x) \right) h^j. \tag{2.35}$$

Thus we conclude that

$$\Delta_{h,q,\tilde{q}}^{\alpha,\beta}(k(x), u(x))$$

$$= \sum_{j=0}^{n-1} \sum_{l=0}^{j} a_l^{(\alpha)} a_{j-l}^{(\beta)} D^{\beta+j-l} \left(k(x) D^{\alpha+l} u(x) \right) h^j + \sum_{s=2}^{5} \mathcal{F}^{-1} \left[\hat{\psi}_s(\omega, h) \right](x)$$

$$= D^{\beta} \left(k(x) D^{\alpha} u(x) \right) + \sum_{j=1}^{n-1} \sum_{l=0}^{j} a_l^{(\alpha)} a_{j-l}^{(\beta)} D^{\beta+j-l} \left(k(x) D^{\alpha+l} u(x) \right) h^j + O(h^n) \tag{2.36}$$

uniformly in $x \in \mathbf{R}$ as $h \to 0$. This completes the proof. $\square$

3 A Crank-Nicolson scheme and its consistency

For the numerical approximation scheme, define $t_m = m\Delta t$ to be the integration time $0 \leqslant t_m \leqslant T$, h_x is the grid size in x-direction, $h_x = (x_R - x_L)/(N_x + 1)$, with $x_i = x_L + ih_x$ for $i = 0, \cdots, N_x + 1$; h_y is the grid size in y-direction, $h_y = (y_R - y_L)/(N_y + 1)$, with $y_j = y_L + jh_y$ for $j = 0, \cdots, N_y + 1$. Denote by $P_{i,j}^m$ the exact solution p at the mesh point (x_i, y_j, t_m). Let $p_{i,j}^m$ to be the numerical approximation to $P_{i,j}^m$ and define $k_{i,j}^x = k_x(x_i, y_j)$, $k_{i,j}^y = k_y(x_i, y_j)$, $f_{i,j}^m = f(x_i, y_j, t_m)$.

According to Theorem 2.1 and the Dirichlet boundary condition $p(x_L, y, t) = 0$ for $y_L \leqslant y \leqslant y_R$, $0 < t \leqslant T$, the mixed fractional derivative with respect to x in (1.1) can be discretized with $q = \tilde{q} = 1$ and $n = 2$ as:

$$\frac{\partial^{\beta_1}}{\partial x^{\beta_1}} \left(k_x(x,y) \frac{\partial^{\alpha_1} p}{\partial x^{\alpha_1}} \right) \bigg|_{(x_i, y_j, t_m)}$$

$$= \frac{1}{h_x^{\alpha_1 + \beta_1}} \sum_{l=0}^{i} g_l^{(\beta_1)} k_{i-l+1,j}^x \sum_{s=0}^{i-l} g_s^{(\alpha_1)} P_{i+1-s-l,j}^m + \zeta_x(x_i, y_j, t_m) h_x + O(h_x^2), \tag{3.1}$$

for $i = 1, \cdots, N_x$, $j = 1, \cdots, N_y$, $m > 0$, where

$$\zeta_x(x, y, t)$$

$$= \left(\frac{\beta_1}{2} - 1 \right) \frac{\partial^{\beta_1 + 1}}{\partial x^{\beta_1 + 1}} \left(k_x(x,y) \frac{\partial^{\alpha_1} p}{\partial x^{\alpha_1}} \right) + \frac{\alpha_1}{2} \frac{\partial^{\beta_1}}{\partial x^{\beta_1}} \left(k_x(x,y) \frac{\partial^{\alpha_1 + 1} p}{\partial x^{\alpha_1 + 1}} \right). \tag{3.2}$$

Similarly, the mixed fractional derivative with respect to y in (1.1) can be discretized as

$$\frac{\partial^{\beta_2}}{\partial y^{\beta_2}}\left(k_y(x,y)\frac{\partial^{\alpha_2}p}{\partial y^{\alpha_2}}\right)\bigg|_{(x_i,y_j,t_m)}$$
$$=\frac{1}{h_y^{\alpha_2+\beta_2}}\sum_{l=0}^{j}g_l^{(\beta_2)}k_{i,j-l+1}^{y}\sum_{s=0}^{j-l}g_s^{(\alpha_2)}P_{i,j+1-s-l}^{m}+\zeta_y(x_i,y_j,t_m)h_y+O(h_y^2), \quad (3.3)$$

for $i=1,\cdots,N_x$, $j=1,\cdots,N_y$, $m>0$, where

$$\zeta_y(x,y,t)$$
$$=\left(\frac{\beta_2}{2}-1\right)\frac{\partial^{\beta_2+1}}{\partial y^{\beta_2+1}}\left(k_y(x,y)\frac{\partial^{\alpha_2}p}{\partial y^{\alpha_2}}\right)+\frac{\alpha_2}{2}\frac{\partial^{\beta_2}}{\partial y^{\beta_2}}\left(k_y(x,y)\frac{\partial^{\alpha_2+1}p}{\partial y^{\alpha_2+1}}\right). \quad (3.4)$$

Substitute (3.1) and (3.3) into the fractional percolation equation (1.1) to get the Crank-Nicolson type numerical scheme, then the resulting two-dimensional implicit finite difference equations may be expressed as follows:

$$\frac{p_{i,j}^m-p_{i,j}^{m-1}}{\Delta t}=\frac{1}{2}\left(\delta_x p_{i,j}^m+\delta_x p_{i,j}^{m-1}+\delta_y p_{i,j}^m+\delta_y p_{i,j}^{m-1}\right)+\frac{1}{2}\left(f_{i,j}^m+f_{i,j}^{m-1}\right),$$
$$i=1,\cdots,N_x, \quad j=1,\cdots,N_y, \quad m>0, \quad (3.5)$$

$$p_{0,j}^m=0, \quad p_{N_x+1,j}^m=v_x(y_j,t_m), \quad j=0,\cdots,N_y+1, \quad m>0,$$
$$p_{i,0}^m=0, \quad p_{i,N_y+1}^m=v_y(x_i,t_m), \quad i=0,\cdots,N_x+1, \quad m>0, \quad (3.6)$$
$$p_{i,j}^0=p_0(x_i,y_j), \quad i=1,\cdots,N_x, \quad j=1,\cdots,N_y, \quad (3.7)$$

where the operators δ_x and δ_y are defined as

$$\delta_x p_{i,j}^m=\frac{1}{h_x^{\alpha_1+\beta_1}}\sum_{l=0}^{i}g_l^{(\beta_1)}k_{i-l+1,j}^x\sum_{s=0}^{i-l}g_s^{(\alpha_1)}p_{i+1-s-l,j}^m, \quad (3.8)$$

$$\delta_y p_{i,j}^m=\frac{1}{h_y^{\alpha_2+\beta_2}}\sum_{l=0}^{j}g_l^{(\beta_2)}k_{i,j-l+1}^y\sum_{s=0}^{j-l}g_s^{(\alpha_2)}p_{i,j+1-s-l}^m. \quad (3.9)$$

By exchanging the order of summations above, (3.8) and (3.9) can be rewritten as

$$\delta_x p_{i,j}^m=\frac{1}{h_x^{\alpha_1+\beta_1}}\sum_{s=1}^{i+1}\left[\sum_{l=0}^{i+1-s}g_l^{(\beta_1)}g_{i+1-s-l}^{(\alpha_1)}k_{i-l+1,j}^x\right]p_{s,j}^m, \quad (3.10)$$

$$\delta_y p_{i,j}^m=\frac{1}{h_y^{\alpha_2+\beta_2}}\sum_{s=1}^{j+1}\left[\sum_{l=0}^{j+1-s}g_l^{(\beta_2)}g_{j+1-s-l}^{(\alpha_2)}k_{i,j-l+1}^y\right]p_{i,s}^m. \quad (3.11)$$

Denote the local truncation error by $R_{i,j}^m$. Notice the fact that the center divided difference in t provides an $O((\Delta t)^2)$ temporal error component. It follows from (3.1) and (3.3) that

$$R_{i,j}^m = \frac{P_{i,j}^m - P_{i,j}^{m-1}}{\Delta t} - \frac{1}{2}\left(\delta_x P_{i,j}^m + \delta_x P_{i,j}^{m-1} + \delta_y P_{i,j}^m + \delta_y P_{i,j}^{m-1}\right) - \frac{1}{2}\left(f_{i,j}^m + f_{i,j}^{m-1}\right)$$

$$= -\frac{(\Delta t)^2}{12}\frac{\partial^3 P}{\partial t^3}\bigg|_{(x_i,y_j,t_{m-\frac{1}{2}})} + O((\Delta t)^3) + \frac{h_x}{2}(\zeta_x(x_i,y_j,t_m) + \zeta_x(x_i,y_j,t_{m-1}))$$

$$+ \frac{h_y}{2}(\zeta_y(x_i,y_j,t_m) + \zeta_y(x_i,y_j,t_{m-1})) + O(h_x^2) + O(h_y^2),$$

$$= O((\Delta t)^2) + O(h_x) + O(h_y), \tag{3.12}$$

where $i = 1, \cdots, N_x$, $j = 1, \cdots, N_y$, $m = 1, \cdots, M$. This implies that the Crank-Nicolson scheme defined by (3.5) is consistent.

4 Stability and convergence

Let $r_1 = \Delta t/(h_x^{\alpha_1+\beta_1})$, $r_2 = \Delta t/(h_y^{\alpha_2+\beta_2})$, $\mathbf{p}^m$ and $\mathbf{f}^m$ be the N-dimensional vectors defined by

$$\mathbf{p}^m := \left[p_{1,1}^m, \cdots, p_{N_x,1}^m, p_{1,2}^m, \cdots, p_{N_x,2}^m, \cdots, p_{1,N_y}^m, \cdots, p_{N_x,N_y}^m\right]^T, \tag{4.1}$$

$$\mathbf{f}^m := \left[f_{1,1}^m, \cdots, f_{N_x-1,1}^m, \tilde{f}_{N_x,1}^m, f_{1,2}^m, \cdots, f_{N_x-1,2}^m, \tilde{f}_{N_x,2}^m, \cdots,\right.$$

$$\left. f_{1,N_y-1}^m, \cdots, f_{N_x-1,N_y-1}^m, \tilde{f}_{N_x,N_y-1}^m, \tilde{f}_{1,N_y}^m, \cdots, \tilde{f}_{N_x-1,N_y-1}^m, \tilde{f}_{N_x,N_y}^m\right]^T, \tag{4.2}$$

where

$$\tilde{f}_{N_x,j}^m = f_{N_x,j}^m + \frac{1}{h_x^{\alpha_1+\beta_1}}k_{N_x+1,j}^x p_{N_x+1,j}^m, \quad 1 \leqslant j \leqslant N_y - 1,$$

$$\tilde{f}_{i,N_y}^m = f_{i,N_y}^m + \frac{1}{h_y^{\alpha_2+\beta_2}}k_{i,N_y+1}^y p_{i,N_y+1}^m, \quad 1 \leqslant i \leqslant N_x - 1,$$

$$\tilde{f}_{N_x,N_y}^m = f_{N_x,N_y}^m + \frac{1}{h_x^{\alpha_1+\beta_1}}k_{N_x+1,N_y}^x p_{N_x+1,N_y}^m + \frac{1}{h_y^{\alpha_2+\beta_2}}k_{N_x,N_y+1}^y p_{N_x,N_y+1}^m.$$

Let $\mathbf{I}$ and $\mathbf{A}$ be the N-by-N identity and stiffness matrices, respectively. Then the finite difference equation is expressed in the matrix form

$$\left(\mathbf{I} - \frac{1}{2}\mathbf{A}\right)\mathbf{p}^m = \left(\mathbf{I} + \frac{1}{2}\mathbf{A}\right)\mathbf{p}^{m-1} + \frac{\Delta t}{2}(\mathbf{f}^m + \mathbf{f}^{m-1}). \tag{4.3}$$

The N-by-N stiffness matrix $\mathbf{A}$ in (4.3) can be expressed as an N_y-by-N_y block matrix with each N_x-by-N_x block $A_{j,\tilde{s}}$ defined by

$$(A_{j,j})_{i,s} := r_1 \sum_{l=0}^{i+1-s} g_l^{(\beta_1)} g_{i+1-s-l}^{(\alpha_1)} k_{i-l+1,j}^x, \quad s \leqslant i-1,$$

$$(A_{j,j})_{i,i} := r_1 \sum_{l=0}^{1} g_l^{(\beta_1)} g_{1-l}^{(\alpha_1)} k_{i-l+1,j}^x + r_2 \sum_{l=0}^{1} g_l^{(\beta_2)} g_{1-l}^{(\alpha_2)} k_{i,j-l+1}^y,$$

$$(A_{j,j})_{i,i+1} := r_1 k_{i+1,j}^x,$$

$$(A_{j,j})_{i,s} := 0, \quad s > i+1,$$

$$(A_{j,\tilde{s}})_{i,s} := \delta_{i,s} r_2 \sum_{l=0}^{j+1-\tilde{s}} g_l^{(\beta_2)} g_{j+1-\tilde{s}-l}^{(\alpha_2)} k_{i,j-l+1}^y, \quad \tilde{s} \leqslant j-1,$$

$$(A_{j,j+1})_{i,s} := \delta_{i,s} r_2 k_{i,j+1}^y,$$

$$(A_{j,\tilde{s}})_{i,s} := 0, \quad \tilde{s} > j+1,$$

where $\delta_{i,s} = 1$ if $i = s$ or 0 otherwise.

To discuss the convergence of the numerical method, we define

$$e_{i,j}^m = P_{i,j}^m - p_{i,j}^m$$

with corresponding vector

$$\mathbf{e}^m = \left[e_{1,1}^m, \cdots, e_{N_x,1}^m, e_{1,2}^m, \cdots, e_{N_x,2}^m, \cdots, e_{1,N_y}^m, \cdots, e_{N_x,N_y}^m\right]^{\mathrm{T}}, \quad (4.4)$$

and denote the local truncation error vector by

$$\mathbf{R}^m = \left[R_{1,1}^m, \cdots, R_{N_x,1}^m, R_{1,2}^m, \cdots, R_{N_x,2}^m, \cdots, R_{1,N_y}^m, \cdots, R_{N_x,N_y}^m\right]^{\mathrm{T}}. \quad (4.5)$$

From (3.5) and (3.12), we have the error equation

$$\left(\mathbf{I} - \frac{1}{2}\mathbf{A}\right)\mathbf{e}^m = \left(\mathbf{I} + \frac{1}{2}\mathbf{A}\right)\mathbf{e}^{m-1} + \Delta t \mathbf{R}^m, \quad 1 \leqslant m \leqslant M, \quad (4.6)$$

and $\mathbf{e}^0 = \mathbf{0}$.

For any N-dimensional vector

$$\mathbf{u} = \left[u_{1,1}, \cdots, u_{N_x,1}, u_{1,2}, \cdots, u_{N_x,2}, \cdots, u_{1,N_y}, \cdots, u_{N_x,N_y}\right]^{\mathrm{T}},$$

let the discrete L^2 norm of $\mathbf{u}$ in this paper be

$$\|\mathbf{u}\|_2 = \left(h_x h_y \sum_{i=1}^{N_x} \sum_{j=1}^{N_y} |u_{i,j}|^2\right)^{\frac{1}{2}}. \quad (4.7)$$

We summarize some useful properties of the alternating fractional binomial coefficient $g_j^{(\alpha)}$ in the following lemma [3, 5, 25, 26, 43], and establish the stability of the Crank-Nicolson scheme under the assumption of continuity of seepage flow ($\beta_1 = \beta_2 = 1$).

Lemma 4.1 *Let $g_j^{(\alpha)}$ be defined in (2.8). We have*

$$\begin{cases} \sum_{j=0}^{\infty} g_j^{(\alpha)} = 0, \quad g_0^{(\alpha)} = 1, \quad g_1^{(\alpha)} = -\alpha, \quad \text{for } \alpha > 0; \\ g_1^{(\alpha)} < g_2^{(\alpha)} < \cdots < 0, \quad \text{for } 0 < \alpha < 1; \\ g_2^{(\alpha)} > g_3^{(\alpha)} > \cdots > 0, \quad \text{for } 1 < \alpha < 2; \\ \sum_{j=0}^{n} g_j^{(\beta)} g_{n-j}^{(\alpha)} = g_n^{(\alpha+\beta)}, \quad n = 0, 1, \cdots, \quad \text{for } \alpha, \beta > 0. \end{cases} \quad (4.8)$$

Theorem 4.1 *If $\beta_1 = \beta_2 = 1$, $k_x(x,y)$ decreases monotonically with respect to x and $k_y(x,y)$ decreases monotonically with respect to y in Ω, then the Crank-Nicolson scheme defined by (4.3) is unconditionally stable.*

Proof Our first goal is to show that the eigenvalues of the matrix $\mathbf{A}$ have negative real parts. For $\beta = 1$, we have $g_0^{(\beta)} = 1$, $g_1^{(\beta)} = -1$ and $g_j^{(\beta)} = 0 \, (j \geqslant 2)$. Then the entries of matrix $\mathbf{A}$ may be rewritten as

$$(A_{j,j})_{i,s} := r_1 \left(g_{i+1-s}^{(\alpha_1)} k_{i+1,j}^x - g_{i-s}^{(\alpha_1)} k_{i,j}^x \right), \quad s \leqslant i-1,$$

$$(A_{j,j})_{i,i} := r_1 \left(g_1^{(\alpha_1)} k_{i+1,j}^x - g_0^{(\alpha_1)} k_{i,j}^x \right) + r_2 \left(g_1^{(\alpha_2)} k_{i,j+1}^y - g_0^{(\alpha_2)} k_{i,j}^y \right),$$

$$(A_{j,j})_{i,i+1} := r_1 k_{i+1,j}^x,$$

$$(A_{j,j})_{i,s} := 0, \quad s > i+1,$$

$$(A_{j,\tilde{s}})_{i,s} := \delta_{i,s} r_2 \left(g_{j+1-\tilde{s}}^{(\alpha_2)} k_{i,j+1}^y - g_{j-\tilde{s}}^{(\alpha_2)} k_{i,j}^y \right), \quad \tilde{s} \leqslant j-1,$$

$$(A_{j,j+1})_{i,s} := \delta_{i,s} r_2 k_{i,j+1}^y,$$

$$(A_{j,\tilde{s}})_{i,s} := 0, \quad \tilde{s} > j+1.$$

Due to the monotonicity of $k_x(x,y)$ and $k_y(x,y)$, we have $0 < k_{i+1,j}^x < k_{i,j}^x$ and $0 < k_{i,j+1}^y < k_{i,j}^y$ for $i = 1, \cdots, N_x$, $j = 1, \cdots, N_y$. From Lemma 4.1, we have

$$\begin{cases} g_{i+1-s}^{(\alpha_1)} k_{i+1}^x - g_{i-s}^{(\alpha_1)} k_{i,j}^x > 0, \quad s \leqslant i-1, \\ g_{j+1-\tilde{s}}^{(\alpha_2)} k_{i,j+1}^y - g_{j-\tilde{s}}^{(\alpha_2)} k_{i,j}^y > 0, \quad \tilde{s} \leqslant j-1. \end{cases} \quad (4.9)$$

Therefore $(A_{j,j})_{i,s}, (A_{j,\tilde{s}})_{i,i} \geqslant 0$, for $i \neq s$, $j \neq \tilde{s}$, and

$$(A_{j,j})_{i,i} = -r_1(\alpha_1 k_{i+1,j}^x + k_{i,j}^x) - r_2(\alpha_2 k_{i,j+1}^y + k_{i,j}^y) < 0. \quad (4.10)$$

According to the Gerschgorin theorem (see [14]), the eigenvalues of the matrix $\mathbf{A}$ lie in the union of disks centered at $(A_{j,j})_{i,i}$ with radius

$$\sum_{s=1,s\neq i}^{N_x} |((A_{j,j})_{i,s})| + \sum_{\tilde{s}=1,\tilde{s}\neq j}^{N_y} \sum_{s=1}^{N_x} |(A_{j,\tilde{s}})_{i,s}|.$$

We estimate

$$\sum_{s=1,s\neq i}^{N_x} |((A_{j,j})_{i,s})| + \sum_{\tilde{s}=1,\tilde{s}\neq j}^{N_y} \sum_{s=1}^{N_x} |(A_{j,\tilde{s}})_{i,s}|$$

$$= \sum_{s=1,s\neq i}^{N_x} |((A_{j,j})_{i,s})| + \sum_{\tilde{s}=1,\tilde{s}\neq j}^{N_y} |(A_{j,\tilde{s}})_{i,i}|$$

$$= r_1 \left(\sum_{s=0,s\neq 1}^{i} g_s^{(\alpha_1)} k_{i+1,j}^x - \sum_{s=1}^{i-1} g_s^{(\alpha_1)} k_{i,j}^x \right) + r_2 \left(\sum_{\tilde{s}=0,\tilde{s}\neq 1}^{j} g_{\tilde{s}}^{(\alpha_2)} k_{i,j+1}^y - \sum_{\tilde{s}=1}^{j-1} g_{\tilde{s}}^{(\alpha_2)} k_{i,j}^y \right)$$

$$< r_1 \left(-g_1^{(\alpha_1)} k_{i+1,j}^x - \sum_{s=1,s\neq i}^{\infty} g_s^{(\alpha_1)} k_{i,j}^x \right) + r_2 \left(-g_1^{(\alpha_2)} k_{i,j+1}^y - \sum_{\tilde{s}=1,\tilde{s}\neq j}^{\infty} g_{\tilde{s}}^{(\alpha_2)} k_{i,j}^y \right)$$

$$< r_1 \left(-g_1^{(\alpha_1)} k_{i+1,j}^x + g_0^{(\alpha_1)} k_{i,j}^x \right) + r_2 \left(-g_1^{(\alpha_2)} k_{i,j+1}^y + g_0^{(\alpha_2)} k_{i,j}^y \right)$$

$$= -(A_{j,j})_{i,i}.$$

This implies that all of the Greschgorin disks of matrix $\mathbf{A}$ are within the left half of the complex plane. Thus, the eigenvalues of the matrix $\mathbf{A}$ have negative real-parts.

Next, λ is an eigenvalue of matrix $\mathbf{A}$ if and only if $1 - \lambda/2$ is an eigenvalue of the matrix $\left(\mathbf{I} - \frac{1}{2}\mathbf{A}\right)$, if and only if $(1 + \lambda/2)/(1 - \lambda/2)$ is an eigenvalue of the matrix $\left(\mathbf{I} - \frac{1}{2}\mathbf{A}\right)^{-1} \left(\mathbf{I} + \frac{1}{2}\mathbf{A}\right)$. We observe that the first part of this statement implies that all the eigenvalues of the matrix $\left(\mathbf{I} - \frac{1}{2}\mathbf{A}\right)$ have a magnitude larger than 1, and thus this matrix is invertible. Furthermore, since the real part of λ is negative, it is easy to see that

$$\left| \frac{1 + \frac{\lambda}{2}}{1 - \frac{\lambda}{2}} \right| < 1.$$

Therefore, the spectral radius of the system matrix $\left(\mathbf{I} - \frac{1}{2}\mathbf{A}\right)^{-1} \left(\mathbf{I} + \frac{1}{2}\mathbf{A}\right)$ is less than one. Thus, the Crank-Nicolson scheme (4.3) is unconditionally stable. □

The stability of the Crank-Nicolson scheme is discussed in Theorem 4.2 under the assumption that $k_x(x,y)$ and $k_y(x,y)$ are constant functions in Ω.

Theorem 4.2 *If $1 < \alpha_l + \beta_l < 2$ for $l = 1, 2$, $k_x(x,y) = K_x$ and $k_y(x,y) = K_y$ for $(x,y) \in \Omega$, where K_x and K_y are positive constants, then the Crank-Nicolson scheme defined by (4.3) is unconditionally stable, and there exists a constant C independent of Δt, h_x and h_y such that*

$$\|\mathbf{e}^m\|_2 \leqslant C((\Delta t)^2 + h_x + h_y), \tag{4.11}$$

for $1 \leqslant m \leqslant M$.

Proof From Lemma 4.1, the entries of matrix $\mathbf{A}$ may be rewritten as

$$\begin{aligned}
(A_{j,j})_{i,s} &:= r_1 g_{i+1-s}^{(\alpha_1+\beta_1)} K_x, \quad s \leqslant i-1, \\
(A_{j,j})_{i,i} &:= r_1 g_1^{(\alpha_1+\beta_1)} K_x + r_2 g_1^{(\alpha_2+\beta_2)} K_y, \\
(A_{j,j})_{i,i+1} &:= r_1 K_x, \\
(A_{j,j})_{i,s} &:= 0, \quad s > i+1, \\
(A_{j,\tilde{s}})_{i,s} &:= \delta_{i,s} r_2 g_{j+1-\tilde{s}}^{(\alpha_2+\beta_2)} K_y, \quad \tilde{s} \leqslant j-1, \\
(A_{j,j+1})_{i,s} &:= \delta_{i,s} r_2 K_y, \\
(A_{j,\tilde{s}})_{i,s} &:= 0, \quad \tilde{s} > j+1.
\end{aligned} \tag{4.12}$$

For $1 < \alpha_l + \beta_l < 2$ $(l=1,2)$, we have $g_1^{(\alpha_l+\beta_l)} = -(\alpha_l+\beta_l)$, $g_s^{(\alpha_l+\beta_l)} > 0$ $(s \neq 1)$ and $\sum_{s=0}^n g_s^{(\alpha_l+\beta_l)} < 0$, where n is a positive integer. By the Gerschgorin theorem, the eigenvalues of the matrix $\mathbf{A}$ lie in the union of disks centered at $(A_{j,j})_{i,i}$ with radius

$$\begin{aligned}
&\sum_{s=1,s\neq i}^{N_x} |(A_{j,j})_{i,s}| + \sum_{\tilde{s}=1,\tilde{s}\neq j}^{N_y} \sum_{s=1}^{N_x} |(A_{j,\tilde{s}})_{i,s}| \\
&= r_1 \sum_{s=0,s\neq 1}^{i} g_s^{(\alpha_1+\beta_1)} K_x + r_2 \sum_{\tilde{s}=0,\tilde{s}\neq 1}^{j} g_{\tilde{s}}^{(\alpha_2+\beta_2)} K_y \\
&< -\left(r_1 g_1^{(\alpha_1+\beta_1)} K_x + r_2 g_1^{(\alpha_2+\beta_2)} K_y\right) \\
&= -(A_{j,j})_{i,i}.
\end{aligned} \tag{4.13}$$

Hence every eigenvalue of the matrix $\mathbf{A}$ has negative real-part. Similar to the proof of Theorem 4.1, we can conclude the results of Theorem 4.1.

We now consider the convergence of the Crank-Nicolson scheme (4.3). From (4.12), we have an inequality similar to (4.13) as follow

$$\sum_{s=1,s\neq i}^{N_x} |(A_{j,j})_{s,i}| + \sum_{\tilde{s}=1,\tilde{s}\neq j}^{N_y} \sum_{s=1}^{N_x} |(A_{\tilde{s},j})_{s,i}|$$

$$= r_1 \sum_{s=0,s\neq 1}^{i} g_s^{(\alpha_1+\beta_1)} K_x + r_2 \sum_{\tilde{s}=0,\tilde{s}\neq 1}^{j} g_{\tilde{s}}^{(\alpha_2+\beta_2)} K_y$$

$$< -\left(r_1 g_1^{(\alpha_1+\beta_1)} K_x + r_2 g_1^{(\alpha_2+\beta_2)} K_y\right)$$

$$= -(A_{j,j})_{i,i}. \tag{4.14}$$

Combining (4.13) and (4.14) and using the Gerschgorin theorem, we have that the eigenvalues of the matrix $\mathbf{A} + \mathbf{A}^T$ are negative. Since $\mathbf{A} + \mathbf{A}^T$ is symmetric, it is negative definite. Thus we know

$$\left\|\left(\mathbf{I} - \frac{1}{2}\mathbf{A}\right)\mathbf{e}^m\right\|_2^2 = \|\mathbf{e}^m\|_2^2 - \frac{h_x h_y}{2}(\mathbf{e}^m)^T(\mathbf{A}+\mathbf{A}^T)\mathbf{e}^m + \frac{1}{4}\|\mathbf{A}\mathbf{e}^m\|_2^2$$

$$\geqslant \|\mathbf{e}^m\|_2^2 + \frac{1}{4}\|\mathbf{A}\mathbf{e}^m\|_2^2, \tag{4.15}$$

$$\left\|\left(\mathbf{I} + \frac{1}{2}\mathbf{A}\right)\mathbf{e}^{m-1}\right\|_2^2 = \|\mathbf{e}^{m-1}\|_2^2 + \frac{h_x h_y}{2}(\mathbf{e}^{m-1})^T(\mathbf{A}+\mathbf{A}^T)\mathbf{e}^{m-1} + \frac{1}{4}\|\mathbf{A}\mathbf{e}^{m-1}\|_2^2$$

$$\leqslant \|\mathbf{e}^{m-1}\|_2^2 + \frac{1}{4}\|\mathbf{A}\mathbf{e}^{m-1}\|_2^2, \tag{4.16}$$

for $1 \leqslant m \leqslant M$. Taking the discrete L^2 norm on the both sides of (4.6) implies

$$\left\|\left(\mathbf{I} - \frac{1}{2}\mathbf{A}\right)\mathbf{e}^m\right\|_2 \leqslant \left\|\left(\mathbf{I} + \frac{1}{2}\mathbf{A}\right)\mathbf{e}^{m-1}\right\|_2 + \Delta t\|\mathbf{R}^m\|_2, \quad 1 \leqslant m \leqslant M. \tag{4.17}$$

Together with (4.15) and (4.16), we have

$$\sqrt{\|\mathbf{e}^m\|_2^2 + \frac{1}{4}\|\mathbf{A}\mathbf{e}^m\|_2^2} \leqslant \sqrt{\|\mathbf{e}^{m-1}\|_2^2 + \frac{1}{4}\|\mathbf{A}\mathbf{e}^{m-1}\|_2^2} + \Delta t\|\mathbf{R}^m\|_2. \tag{4.18}$$

Summing up (4.18) from 1 to m leads to

$$\sqrt{\|\mathbf{e}^m\|_2^2 + \frac{1}{4}\|\mathbf{A}\mathbf{e}^m\|^2} \leqslant \Delta t \sum_{l=1}^{m} \|\mathbf{R}^l\|_2, \quad 1 \leqslant m \leqslant M,$$

where $\mathbf{e}^0 = \mathbf{0}$ is considered. Noticing that

$$|R_{i,j}^m| \leqslant c((\Delta t)^2 + h_x + h_y),$$

for $1 \leqslant i \leqslant N_x$, $1 \leqslant j \leqslant N_y$, $1 \leqslant m \leqslant M$, we obtain

$$\|e^m\|_2 \leqslant cm\Delta t((\Delta t)^2 + h_x + h_y) \leqslant C((\Delta t)^2 + h_x + h_y).$$

This completes the proof. □

For the case $1 < \alpha_l + \beta_l < 2$ ($l = 1, 2$), where the functions $k_x(x, y)$ and $k_y(x, y)$ are not constants in Ω, some numerical experiments show that the Crank-Nicolson scheme (4.3) is also unconditionally stable and convergent. We provide evidence in Table 2 for Example 1 in Section 6.

5 Improving the order of convergence by extrapolation

The stability of the Crank-Nicolson scheme was shown under two assumptions in Section , respectively. The method is consistent with a local truncation error which is $O((\Delta t)^2) + O(h_x) + O(h_y)$. By the fact that the errors in the approximations of the mixed fractional derivatives in (1.1) with respect to x and y have the form $\zeta_x(x, y, t)h_x + O(h_x^2)$ and $\zeta_y(x, y, t)h_y + O(h_y^2)$, respectively, where $\zeta_x(x, y, t)$ and $\zeta_y(x, y, t)$ do not depend on the grid size h_x or h_y, it follows that the Richardson extrapolation method [14] can be used to obtain a solution with local truncation error $O((\Delta t)^2) + O(h_x^2) + O(h_y^2)$. More specifically, we may compute two Crank-Nicolson solutions $p_{\Delta t, h_x, h_y}(x, y, t_m)$ and $p_{\Delta t, \frac{h_x}{2}, \frac{h_y}{2}}(x, y, t_m)$ for the grid sizes (h_x, h_y) and $\left(\frac{h_x}{2}, \frac{h_y}{2}\right)$ with the same Δt at each time step respectively, and then compute the extrapolation solution by

$$p^{ex}_{\Delta t, h_x, h_y}(x, y, t_m) = 2p_{\Delta t, \frac{h_x}{2}, \frac{h_y}{2}}(x, y, t_m) - p_{\Delta t, h_x, h_y}(x, y, t_m) \tag{5.1}$$

to obtain the $O((\Delta t)^2) + O(h_x^2) + O(h_y^2)$ accuracy, where (x, y) is a grid point on the coarse spatial grid. In other words, $(x, y) = (x_i, y_j)$ on the coarse grid of size (h_x, h_y), while $(x, y) = (x_{2i}, y_{2j})$ on the fine grid of size $(h_x/2, h_y/2)$.

6 Numerical examples

In this section, we carry out numerical experiments to demonstrate the effectiveness of the second-order accurate finite difference method. For a fixed time step Δt, we use $p_{\Delta t, h_x, h_y}$, $p_{\Delta t, \frac{h_x}{2}, \frac{h_y}{2}}$ and $p^{ex}_{\Delta t, h_x, h_y}$ to denote the Crank-Nicolson solution with grid size (h_x, h_y), the Crank-Nicolson solution with grid size $(h_x/2, h_y/2)$, and the corresponding extrapolated solution generated by (5.1), respectively. In the numerical experiments, we compute the discrete infinity norms of the error $p - p_{\Delta t, h_x, h_y}$, $p - p_{\Delta t, \frac{h_x}{2}, \frac{h_y}{2}}$ and $p - p^{ex}_{\Delta t, h_x, h_y}$ at time $T = M\Delta t$ as

$$E^{(\infty)}_{\Delta t, h_x, h_y} = \max_{1 \leqslant i \leqslant N_x, 1 \leqslant j \leqslant N_y} |P^M_{i,j} - p_{\Delta t, h_x, h_y}(x_i, y_j, t_M)|, \quad (6.1)$$

$$E^{(\infty)}_{\Delta t, \frac{h_x}{2}, \frac{h_y}{2}} = \max_{1 \leqslant i \leqslant 2N_x+1, 1 \leqslant j \leqslant 2N_y+1} |P^M_{i,j} - p_{\Delta t, \frac{h_x}{2}, \frac{h_y}{2}}(x_i, y_j, t_M)|, \quad (6.2)$$

$$E^{ex,(\infty)}_{\Delta t, h_x, h_y} = \max_{1 \leqslant i \leqslant N_x, 1 \leqslant j \leqslant N_y} |P^M_{i,j} - p^{ex}_{\Delta t, h_x, h_y}(x_i, y_j, t_M)|, \quad (6.3)$$

and the discrete L^2 norms of the errors as

$$E^{(2)}_{\Delta t, h_x, h_y} = \left(h_x h_y \sum_{i=1}^{N_x} \sum_{j=1}^{N_y} |P^M_{i,j} - p_{\Delta t, h_x, h_y}(x_i, y_j, t_M)|^2 \right)^{\frac{1}{2}}, \quad (6.4)$$

$$E^{(2)}_{\Delta t, \frac{h_x}{2}, \frac{h_y}{2}} = \left(\frac{h_x h_y}{4} \sum_{i=1}^{2N_x+1} \sum_{j=1}^{2N_y+1} |P^M_{i,j} - p_{\Delta t, \frac{h_x}{2}, \frac{h_y}{2}}(x_i, y_j, t_M)|^2 \right)^{\frac{1}{2}}, \quad (6.5)$$

$$E^{ex,(2)}_{\Delta t, h_x, h_y} = \left(h_x h_y \sum_{i=1}^{N_x} \sum_{j=1}^{N_y} |P^M_{i,j} - p^{ex}_{\Delta t, h_x, h_y}(x_i, y_j, t_M)|^2 \right)^{\frac{1}{2}}. \quad (6.6)$$

Let

$$G^{(\kappa)}_{\Delta t, h_x, h_y} = \log_2 \frac{E^{(\kappa)}_{2\Delta t, 2h_x, 2h_y}}{E^{(\kappa)}_{\Delta t, h_x, h_y}}, \quad G^{ex,(\kappa)}_{\Delta t, h_x, h_y} = \log_2 \frac{E^{ex,(\kappa)}_{2\Delta t, 2h_x, 2h_y}}{E^{ex,(\kappa)}_{\Delta t, h_x, h_y}}, \quad \kappa \in \{2, \infty\}$$

reflect the convergence order of the Crank-Nicolson method and extrapolated Crank-Nicolson method, respectively.

The algorithm was implemented in Fortran 90 on a Thinkpad E420 laptop with configuration: Intel(R) Core(TM) i3-2350M CPU 2.30GHZ and 4GB RAM. All computations were performed in double precision.

Example 1   We consider the two-dimensional fractional percolation equation (1.1) with the following data [6]. The spatial domain is $\Omega = \{(x,y) | 0 \leqslant x \leqslant 1, 0 \leqslant y \leqslant 1\}$. The percolation coefficients are $k_x(x,y) = 2 - x^2$ and $k_y(x,y) = 2 - y^2$, which are decreases monotonically with respect to x and y, respectively. The source term $f(x, y, t)$ is given by

$$f(x, y, t)$$

$$= -\frac{\Gamma(3)y^2 e^{-t}}{\Gamma(3-\alpha_1)} \left(\frac{2\Gamma(3-\alpha_1)x^{2-\alpha_1-\beta_1}}{\Gamma(3-\alpha_1-\beta_1)} - \frac{\Gamma(5-\alpha_1)x^{4-\alpha_1-\beta_1}}{\Gamma(5-\alpha_1-\beta_1)} \right)$$

$$-\frac{\Gamma(3)x^2 e^{-t}}{\Gamma(3-\alpha_2)} \left(\frac{2\Gamma(3-\alpha_2)y^{2-\alpha_2-\beta_2}}{\Gamma(3-\alpha_2-\beta_2)} - \frac{\Gamma(5-\alpha_2)y^{4-\alpha_2-\beta_2}}{\Gamma(5-\alpha_2-\beta_2)} \right) - x^2 y^2 e^{-t}.$$

Table 1 Error behavior for the Crank-Nicolson scheme and the extrapolated Crank-Nicolson scheme for Example 1 with $\alpha_1 = 0.5$, $\beta_1 = 1$, $\alpha_2 = 0.5$, $\beta_2 = 1$, $\Delta t = h_x = h_y$ at time $T = 1$

Δt	$E^{(\infty)}_{\Delta t, h_x, h_y}$	$G^{(\infty)}_{\Delta t, h_x, h_y}$	$E^{(\infty)}_{\Delta t, \frac{h_x}{2}, \frac{h_y}{2}}$	$G^{(\infty)}_{\Delta t, \frac{h_x}{2}, \frac{h_y}{2}}$	$E^{ex,(\infty)}_{\Delta t, h_x, h_y}$	$G^{ex,(\infty)}_{\Delta t, h_x, h_y}$
2^{-3}	8.813e-3	—	4.620e-3	—	7.981e-4	—
2^{-4}	4.584e-3	0.943	2.338e-3	0.982	1.792e-4	2.154
2^{-5}	2.351e-3	0.963	1.186e-3	0.978	3.399e-5	2.398
2^{-6}	1.190e-3	0.982	5.974e-4	0.990	8.488e-6	2.001
2^{-7}	5.983e-4	0.993	2.997e-4	0.995	2.124e-6	1.998
Δt	$E^{(2)}_{\Delta t, h_x, h_y}$	$G^{(2)}_{\Delta t, h_x, h_y}$	$E^{(2)}_{\Delta t, \frac{h_x}{2}, \frac{h_y}{2}}$	$G^{(2)}_{\Delta t, \frac{h_x}{2}, \frac{h_y}{2}}$	$E^{ex,(2)}_{\Delta t, h_x, h_y}$	$G^{ex,(2)}_{\Delta t, h_x, h_y}$
2^{-3}	4.124e-3	—	2.132e-3	—	1.941e-4	—
2^{-4}	2.174e-3	0.923	1.099e-3	0.956	4.101e-5	2.242
2^{-5}	1.110e-3	0.969	5.576e-4	0.978	9.969e-6	2.040
2^{-6}	5.604e-4	0.986	2.807e-4	0.990	2.521e-6	1.983
2^{-7}	2.830e-4	0.986	1.414e-4	0.989	6.302e-7	2.000

Table 2 Error behavior for the Crank-Nicolson scheme and the extrapolated Crank-Nicolson scheme for Example 1 with $\alpha_1 = 0.7$, $\beta_1 = 0.8$, $\alpha_2 = 0.6$, $\beta_2 = 0.9$, $\Delta t = h_x = h_y$ at time $T = 1$

Δt	$E^{(\infty)}_{\Delta t, h_x, h_y}$	$G^{(\infty)}_{\Delta t, h_x, h_y}$	$E^{(\infty)}_{\Delta t, \frac{h_x}{2}, \frac{h_y}{2}}$	$G^{(\infty)}_{\Delta t, \frac{h_x}{2}, \frac{h_y}{2}}$	$E^{ex,(\infty)}_{\Delta t, h_x, h_y}$	$G^{ex,(\infty)}_{\Delta t, h_x, h_y}$
2^{-3}	9.243e-3	—	4.820e-3	—	7.846e-4	—
2^{-4}	4.796e-3	0.946	2.449e-3	0.977	1.770e-4	2.148
2^{-5}	2.465e-3	0.960	1.240e-3	0.982	3.275e-5	2.434
2^{-6}	1.244e-3	0.986	6.245e-4	0.990	8.205e-6	1.997
2^{-7}	6.256e-4	0.993	3.133e-4	0.995	2.051e-6	1.999
Δt	$E^{(2)}_{\Delta t, h_x, h_y}$	$G^{(2)}_{\Delta t, h_x, h_y}$	$E^{(2)}_{\Delta t, \frac{h_x}{2}, \frac{h_y}{2}}$	$G^{(2)}_{\Delta t, \frac{h_x}{2}, \frac{h_y}{2}}$	$E^{ex,(2)}_{\Delta t, h_x, h_y}$	$G^{ex,(2)}_{\Delta t, h_x, h_y}$
2^{-3}	4.394e-3	—	2.261e-3	—	1.902e-4	—
2^{-4}	2.310e-3	0.927	1.165e-3	0.956	4.021e-5	2.241
2^{-5}	1.178e-3	0.971	5.908e-4	0.979	9.913e-6	2.020
2^{-6}	5.940e-4	0.987	2.974e-4	0.990	2.535e-6	1.967
2^{-7}	2.951e-4	1.009	1.503e-4	0.984	6.327e-7	2.002

Table 3 Error behavior for the Crank-Nicolson scheme and the extrapolated Crank-Nicolson scheme for Example 2 with $\alpha_1 = 0.85$, $\beta_1 = 0.85$, $\alpha_2 = 0.75$, $\beta_2 = 0.95$, $\Delta t = h_x = h_y$ at time $T = 1$

Δt	$E^{(\infty)}_{\Delta t,h_x,h_y}$	$G^{(\infty)}_{\Delta t,h_x,h_y}$	$E^{(\infty)}_{\Delta t,\frac{h_x}{2},\frac{h_y}{2}}$	$G^{(\infty)}_{\Delta t,\frac{h_x}{2},\frac{h_y}{2}}$	$E^{ex,(\infty)}_{\Delta t,h_x,h_y}$	$G^{ex,(\infty)}_{\Delta t,h_x,h_y}$
2^{-3}	2.737e-4	—	1.263e-4	—	8.042e-5	—
2^{-4}	1.180e-4	1.213	6.261e-5	1.012	2.080e-5	1.951
2^{-5}	6.198e-5	0.929	3.243e-5	0.948	5.262e-6	1.983
2^{-6}	3.226e-5	0.942	1.651e-5	0.974	1.317e-6	1.998
2^{-7}	1.646e-5	0.970	8.329e-6	0.987	3.384e-7	1.961
Δt	$E^{(2)}_{\Delta t,h_x,h_y}$	$G^{(2)}_{\Delta t,h_x,h_y}$	$E^{(2)}_{\Delta t,\frac{h_x}{2},\frac{h_y}{2}}$	$G^{(2)}_{\Delta t,\frac{h_x}{2},\frac{h_y}{2}}$	$E^{ex,(2)}_{\Delta t,h_x,h_y}$	$G^{ex,(2)}_{\Delta t,h_x,h_y}$
2^{-3}	1.352e-4	—	6.340e-5	—	4.457e-5	—
2^{-4}	6.209e-5	1.122	3.042e-5	1.059	1.029e-5	2.114
2^{-5}	3.036e-5	1.032	1.510e-5	1.010	2.499e-6	2.041
2^{-6}	1.509e-5	1.008	7.540e-6	1.001	6.182e-7	2.015
2^{-7}	7.652e-6	0.979	3.804e-6	0.987	1.624e-7	1.928

The exact solution of this problem is

$$p(x,y,t) = e^{-t}x^2y^2,$$

which satisfies the initial conditon

$$p_0(x,y) = x^2y^2,$$

and the Dirichlet boundary condtions

$$p(0,y,t) = p(x,0,t) = 0, \quad p(1,y,t) = e^{-t}y^2, \quad p(x,1,t) = e^{-t}x^2, \quad (x,y) \in \Omega.$$

This example problem is solved to time $T = 1$.

Example 2 In this example, we study the case for which the percolation coefficients are constants. Consider the fractional percolation equation (1.1) on the finite domain $\Omega = \{(x,y)|0 \leqslant x \leqslant 1, 0 \leqslant y \leqslant 1\}$, with the percolation coefficients

$$k_x(x,y) = k_y(x,y) = 1,$$

the source term

$$\begin{aligned}f(x,y,t) = &-e^{1-t}\left(x^2(1-x)^2y^2(1-y)^2\right.\\ &+ y^2(1-y)^2\left(\frac{\Gamma(3)x^{2-\alpha_1-\beta_1}}{\Gamma(3-\alpha_1-\beta_1)} - \frac{2\Gamma(4)x^{3-\alpha_1-\beta_1}}{\Gamma(4-\alpha_1-\beta_1)} + \frac{\Gamma(5)x^{4-\alpha_1-\beta_1}}{\Gamma(5-\alpha_1-\beta_1)}\right)\end{aligned}$$

$$+x^2(1-x)^2\left(\frac{\Gamma(3)y^{2-\alpha_2-\beta_2}}{\Gamma(3-\alpha_2-\beta_2)}-\frac{2\Gamma(4)y^{3-\alpha_2-\beta_2}}{\Gamma(4-\alpha_2-\beta_2)}+\frac{\Gamma(5)y^{4-\alpha_2-\beta_2}}{\Gamma(5-\alpha_2-\beta_2)}\right)\right).$$

the initial condition

$$p_0(x,y) = ex^2(1-x)^2 y^2(1-y)^2$$

for $(x,y) \in \Omega$ and the homogeneous Dirichlet boundary conditions. The exact solution to the corresponding fractional percolation equation is

$$p(x,y,t) = e^{1-t} x^2(1-x)^2 y^2(1-y)^2.$$

This example problem is solved to time $T = 1$.

We present the discrete infinity norms and discrete L^2 norms of the errors for the numerical solutions of the Crank-Nicolson method and the extrapolated Crank-Nicolson method in Table 1, for Example 1 with $\alpha_1 = 0.5$, $\beta_1 = 1$, $\alpha_2 = 0.5$, $\beta_2 = 1$. Table 2 reports the error behavior for Example 1 with $\alpha_1 = 0.7$, $\beta_1 = 0.8$, $\alpha_2 = 0.6$, $\beta_2 = 0.9$, and Table 3 reports that for Example 2 with $\alpha_1 = 0.85$, $\beta_1 = 0.85$, $\alpha_2 = 0.75$, $\beta_2 = 0.95$.

As can be seen, when the time step Δt is fixed, the extrapolated solution performs well than both the Crank-Nicolson solution on the coarse grid of size (h_x, h_y) and that on the fine grid of size $(h_x/2, h_y/2)$. From the numerical results under the parameter setting in Table 2 for Example 1, we observe that the convergence condition of the Crank-Nicolson scheme can be weaken. In Table 1-3, the behavior of the maximum error is almost linear when the Crank-Nicolson method is used, this is because the second-order convergence in time t is masked by the linear order of convergence in space. More importantly, the seventh column in each Table shows that the convergence of the extrapolated Crank-Nicolson method is of order $O((\Delta t)^2) + O(h_x^2) + O(h_y^2)$, which supports our theoretical analysis.

7 Conclusions

In this paper, we have presented a general approximation for the mixed fractional derivatives and developed a second-order accurate method which combines the Crank-Nicolson scheme with the Richardson extrapolation for the two-dimensional fractional percolation equations. We discussed the consistency and stability of the Crank-Nicolson scheme and proved relevant theorems. Thanks to the expansion of the operator which is used in approximating the mixed fractional spatial derivatives, the convergence order was improved from $O(\Delta t)+O(h_x)+O(h_y)$ to $O((\Delta t)^2)+O(h_x^2)+O(h_y^2)$ by employing the Richardson extrapolation. Numerical experiments verified the accuracy and efficiency of the extrapolated Crank-Nicolson method.

Acknowledgements The authors would like to express the sincere thanks to the referees for their valuable comments and suggestions which helped to improve the original paper. The authors are also grateful to Dr. Yongqiang Lyu from Shandong Normal University for his helpful suggestions. This work was supported by the fund of National Natural Science (11171193), the fund of Natural Science of Shandong Province (ZR2011MA016) and a Project of Shandong Province Science and Technology Development Program(2012GGB01198).

References

[1] T.S. Basu, H. Wang. A fast second-order finite difference method for space-fractional diffusion equations [J]. Int. J. Numer. Anal. Model., 9 (2012), 658-666.

[2] K. Burrage, N. Hale, D. Kay. An efficient implicit FEM scheme for fractional-in-space reaction-diffusion equations [J]. SIAM J. Sci. Comput., 34 (4) (2012), A2145-A2172.

[3] C.-M. Chen, F. Liu, K. Burrage. Finite difference methods and a fourier analysis for the fractional reaction-subdiffusion equation [J]. Appl. Math. Comput., 198 (2008), 754-769.

[4] S. Chen, F. Liu, K. Burrage. Numerical simulation of a new two-dimensional variable-order fractional percolation equation in non-homogeneous porous media [J]. Comput.Math. Appl., 67(9) (2013), 1673-1681.

[5] S. Chen, F. Liu, I. Turner, V. Anh. A novel implicit finite difference methods for the one dimensional fractional percolation equation [J]. Numer. Algor., 56 (2011), 517-535.

[6] S. Chen, F. Liu, I. Turner, V. Anh. An implicit numerical method for the two-dimensional fractional percolation equation [J]. Appl. Math. Comput., 219 (2013), 4322-4331.

[7] H. Chou, B. Lee, C. Chen. The transient infiltration process for seepage flow from cracks [A]. Advances in Subsurface Flow and Transport: Eastern and Western Approaches III [C], 2006.

[8] V.J. Ervin, J.P. Roop. Variational formulation for the stationary fractional advection dispersion equation [J]. Numer. Methods Part. Different. Equat., 22 (2005), 558-576.

[9] V.J. Ervin, J.P. Roop. Variational solution of fractional advection dispersion equations on bounded domains in $\mathbb{R}^d$ [J]. Numer. Methods Part. Different. Equat., 23 (2007), 256-281.

[10] V.J. Ervin, N. Heuer, J.P. Roop. Numerical approximation of a time dependent, nonlinear, space-fractional diffusion equation [J]. SIAM J. Numer. Anal., 45 (2) (2007), 572-591.

[11] Y.T. Gu, P. Zhuang, Q. Liu. An advanced meshless method for time fractional diffusion equation [J]. Int. J. Comput. Methods, 8 (2011), 653-665.

[12] J.-H. He. Approximate analytical solution for seepage flow with fractional derivatives in porous media [J]. Comput. Methods Appl. Mech. Eng., 167 (1998), 57-68.

[13] J. Huang, N. Nie, Y. Tang. A second order finite difference-spectral method for space fractional diffusion equations [J]. Science China Mathematics, Nov (2013), 1-15.

[14] E. Isaacson, H.B. Keller. Analysis of Numerical Methods [M]. Wiley, New York, 1966.

[15] X. Li, C. Xu. Existence and uniqueness of the weak solution of the space-time fractional diffusion equation and a spectral method approximation [J]. Commun. Comput. Phys., 8 (2010), 1016-1051.

[16] C. Li, Z. Zhao, Y. Chen. Numerical approximation of nonlinear fractional differential equations with subdiffusion and superdiffusion [J]. Comput. Math. Appl., 62(3) (2011), 855-875.

[17] Y. Lin, X. Li, C. Xu. Finite dfierence/specrtal approximations for the fractional cable equation [J]. Math. Comp., 80 (2011), 1369-1396.

[18] F.-R. Lin, S.-W. Yang, X.-Q. Jin. Preconditioned iterative methods for fractional diffusion equation [J]. J. Comput. Phys., 256 (2014), 109-117.

[19] F. Liu, S. Chen, I. Turner, K. Burrage, V. Anh. Numerical simulation for two-dimensional Riesz space fractional diffusion equations with a nonlinear reaction term [J]. Cent. Eur. J. Phys., 11(10) (2013), 1221-1232.

[20] Q. Liu, Y.T. Gu, P. Zhuang, F. Liu, Y.F. Nie. An implicit RBF meshless approach for time fractional diffusion equations [J]. Comput. Mech., 48 (2011), 1-12.

[21] Q. Liu, F. Liu, I. Turner, V. Anh. Numerical simulation for the 3D seepage flow with fractional derivatives in porous media [J]. IMA J. Appl. Math., 74 (2009), 201-229.

[22] F. Liu, I. Turner, V. Anh, Q. Yang, K. Burrage. A numerical method for the fractional Fitzhugh-Nagumo monodomain model [J]. ANZIAM J., 54 (2013), 608-629.

[23] F. Liu, P. Zhuang, I. Turner, V. Anh, K. Burrage. A semi-alternating direction method for a 2-D fractional FitzHugh-Nagumo monodomain model on an approximate irregular domain [J]. J. Comput. Phys., 293 (2015), 252-263.

[24] Z.-J. Luo, Y.-Y. Zhang, Y.-X. Wu. Finite element numerical simulation of three-dimensional seepage control for deep foundation pit dewatering [J]. J. Hydrodyn., Ser. B, 20(5) (2008), 596-602.

[25] M.M. Meerschaert, C. Tadjeran. Finite difference approximations for fractional advection dispersion flow equations [J]. J. Comput. Appl. Math., 172 (2004), 65-77.

[26] M.M. Meerschaert, C. Tadjeran. Finite difference approximations for two-sided space-fractional partial differential equations [J]. Appl. Numer. Math., 56 (2006), 80-90.

[27] M.M. Meerschaert, H.P. Scheffler, C. Tadjeran. Finite difference methods for two-dimensional fractional dispersion equation [J]. J. Comput. Phys., 211 (2006), 249-261.

[28] K. Miller, B. Ross. An Introduction to the Fractional Calculus and Fractional Differential Equations [M]. Wiley, New York, 1993.

[29] S. Momani, A.A. Rqayiq, D. Baleanu. A nonstandard finite difference scheme for two-sided space-fractional partial differential equations [J]. Internat. J. Bifur. Chaos Appl. Sci. Engrg., 22(4) (2012), 1-5.

[30] J. Ochoa-Tapia, F. Valdes-Parada, J. alvares-Ramirez. A fractional-order Darcy's law [J]. Phy. A, 374(1) (2007), 1-14.

[31] N. Petford, M.A. Koenders. Seepage flow and consolidation in a deforming porous medium [A]. Geophys. Res. Abstracts [C], 5 (2003), 13329.

[32] I. Podlubny. Fractional Differential Equations [M]. Academic Press, New York, 1999.

[33] R.D. Richtmyer, K.W. Morton. Difference Methods for Initial-Value Problems [M]. Krieger Publishing, Malabar, FL, 1994.

[34] K.R. Rushton, S.C. Redshaw. Seepage and groundwater flow [M]. Brisbane, Australia: Wiley-Interscience Publication, 1979.

[35] S. Samko, A. Kilbas, O. Marichev. Fractional Integrals and Derivatives: Theory and Applications [M]. Gordon and Breach Science Publishers, 1993.

[36] J. Song, Q. Yu, F. Liu, I. Turner. A spatially second-order accurate implicit numerical method for the space and time fractional Bloch-Torrey equation [J]. Numer. Algorithms, 66(4) (2014), 911-932.

[37] N. Sweilam, M. Khader, A. Nagy. Numerical solution of two-sided space-fractional wave equation using finite difference method [J]. J. Comput. Appl. Math., 235 (2011), 2832-2841.

[38] C. Tadjeran, M.M. Meerschaert. A second-order accurate numerical method for the two-dimensional fractional diffusion equation [J]. J. Comput. Phys., 220 (2007), 813-823.

[39] C. Tadjeran, M.M. Meerschaert, H.P. Scheffler. A second-order accurate numerical approximation for the fractional diffusion equation [J]. J. Comput. Phys., 213 (2006), 205-213.

[40] N.I. Thusyanthan, S.P.G. Madabhushi. Scaling of seepage flow velocity in centrifuge models [J]. Acta Gastroenterologica Latinoamericana, 38(2) (2003), 105-115.

[41] W. Tian, H. Zhou, W. Deng. A class of second order difference approximations for solving space fractional diffusion equations [J]. Math. Comp., 84(294) (2015), 1703-1727.

[42] H. Wang, N. Du. A superfast-preconditioned iterative method for steady-state space-fractional diffusion equations [J]. J. Comput. Phys., 240 (2013), 49-57.

[43] H. Wang, K. Wang, T. Sircar. A direct $O(N \log^2 N)$ finite difference method for fractional diffusion equations [J]. J. Comput. Phys., 229 (2010), 8095-8104.

[44] Q. Yang, T. Moroney, K. Burrage, I. Turner, F. Liu. Novel numerical methods for time-space fractional reaction diffusion equations in two dimensions [J]. Aust. New Zealand Ind. Appl. Math. J., 52 (2011), C395-C409.

[45] Q. Yang, I. Turner, T. Moroney, F. Liu. A finite volume scheme with preconditioned Lanczos method for two-dimensional space-fractional reaction-diffusion equations [J]. Applied Mathematical Modelling, 38(15-16) (2014), 3755-3762.

[46] Q. Yu, F. Liu, I. Turner, K. Burrage. A computationally effective alternating direction method for the space and time fractional Bloch-Torrey equation in 3-D [J]. Appl. Math. Comput., 219(8) (2012), 4082-4095.

[47] F. Zeng, F. Liu, C. Li, K. Burrage, I. Turner, V. Anh. A Crank-Nicolson ADI spectral method for the two-dimensional Riesz space fractional nonlinear reaction-diffusion equation [J]. SIAM J. Numer. Anal., 52(6) (2014), 2599-2622.

[48] X. Zhang, J.W. Crawford, L.K. Deeks, M.I. Stutter, A.G. Bengough, I.M. Young. A mass balance based numerical method for the fractional advection-dispersion equation: Theory and application [J]. Water Resour. Res., 41 (2005), 1-10.

[49] Y. Zhang, Z. Sun, X. Zhao. Compact alternating direction implicit scheme for the two-dimensional fractional diffusion-wave equation [J]. SIAM J. Numer. Anal., 50(3) (2012), 1535-1555.

[50] H. Zhou, W. Tian, W. Deng. Quasi-compact finite difference schemes for space fractional diffusion equations [J]. J. Sci. Comput., 56(1) (2013), 45-66.

[51] P. Zhuang, F. Liu, V. Anh, I. Turner. Numerical methods for the variable-order fractional advection-diffusion equation with a nonlinear source term [J]. SIAM J. Numer. Anal., 47(3) (2009), 1760-1781.

Classical Solutions of General Ginzburg-Landau Equations*

Chen Shuhong (陈淑红), Guo Boling (郭柏灵)

Abstract In this paper, we prove the existence of global classical solutions to time-dependent Ginzburg-Landau(TDGL) equations. By the properties of Besov and Sobolev spaces, together with the energy method, we establish the global existence and uniqueness of classical solutions to the initial boundary value problem for time-dependent Ginzburg-Landau equations.

Keywords classical solution; time-dependent Ginzburg-Landau theory; Besov space; energy method.

1 Introduction

This paper deals with the existence of global classical solutions to initial boundary value problems for the general coupled time-dependent Ginzburg-Landau equations:

$$-idu_t = \left(-\frac{dg^2+1}{U} + a\right)u + g\left[a + d(2\nu - 2\mu)\right]\varphi_B + \frac{c}{4m}\Delta u + \frac{g}{4m}(c-d)\Delta\varphi_B$$
$$-b|u + g\varphi_B|^{2\sigma}(u + g\varphi_B), \tag{1.1}$$

$$i\varphi_{Bt} = -\frac{g}{U}u + (2\nu - 2\mu)\varphi_B - \frac{1}{4m}\Delta\varphi_B, \tag{1.2}$$

$$u(x,0) = u_0(x), \quad \varphi_B(x,0) = \varphi_{B0}(x), \quad x \in \Omega, \tag{1.3}$$

$$u(x,t) = 0, \quad \varphi_B(x,t) = 0, \quad \text{on } [0,\infty) \times \partial\Omega; \tag{1.4}$$

and the definition domain of Laplacian Δ is chosen as:

$$D(\Delta) = H^2(\Omega) \cap H_0^1(\Omega),$$

---

*ACTA MATHEMATICA SCIENTIA Ser.B, 2016, 36(3): 717–732.

where Ω is a bounded domain in R^n with Lipschitzian boundary. The parameters a, b, c, d and m, U, g, ν, μ are all coupling coefficients, $t \geqslant 0$, σ is a positive integer and d is generally complex.

The main purpose of this paper is to prove the global existence of classical solutions to above problem. Our work is motivated by the paper of Chen and Guo [6] where they studied the equations (1.1) and (1.2) with the index $\sigma = 1$ and established the global existence of classical solutions. The equations with a cubic nonlinearity (i.e. $\sigma = 1$) are first obtained by Machida-Koyama [13], which is a coupled time dependent Ginzburg-Landau theory for superfluid atomic Fermi gases describing the BCS-BEC crossover near the Feshbach resonance from the Fermion-Boson model (double-channel model).

The Ginzburg-Landau equation has proved fruitful for illustrating the connections between infinite dimensional dynamics and finite dimensional dynamical systems [8,9,11,21,22]. To our best knowledge, however, there are few mathematical analysis results on nonlinear problems from the Fermion-Boson model in the literature, due to the fact that in general, passage from a single-channel model to a double-channel model is rather difficult. And the works of double-channel model are few, even less for coupled time dependent Ginzburg-Landau equations from the Fermion-Boson model. Though some results have been analyzed from the physical point of view [10, 17].

Recently, Chen and Guo [3-7] developed the weak solution theory of the problem (1.1)-(1.4) and established class solutions to the equations (1.1)-(1.2) with cubic nonlinearity [7]. A natural question is to ask: do similar results hold in the problem (1.1)-(1.4)? In fact, under certain condition, the global strong solutions of initial data problem (1.1)-(1.4) can be found.

This is the main work that will be solved in present paper. Though in the classical solution theory, the admissible parameter values a, c, d, m, U and the dimension n are interrelated. Just as the nonlinear Schrödinger equation the higher the dimension, the harder the problem is [2]. The difficulty lies in how to obtain a priori estimates of solutions with higher derivatives. What make things worse is that the approaches in both [16] and [18] could not be applied to the current paper, since the crossover between the BCS and BEC states can bring some technical difficulties. This arbitrariness contrasts sharply with the theory of weak solutions which have established without any restriction on the dimension n [10]. Further more, a priori estimates even in the case of nonlinear wave equations [12, 14, 15, 20] require some restrictions on the nonlinear term, which cause that one can obtain the desired result just in the case of $n \leqslant 11$.

In this paper, we will establish the global existence of classical solutions to the initial value problem (1.1)-(1.4) in various spatial dimensions. For general positive number σ one cannot expect to obtain solutions in $C((0,T], C^3(\Omega))$ because unbounded singularities may be introduced at zeros of $u(x,t) + g\varphi_B(x,t)$ upon further differentiation of

$$\nabla(|u+g\varphi_B|^{2\sigma}(u+g\varphi_B)) = (\sigma+1)|u+g\varphi_B|^{2\sigma}\nabla(u+g\varphi_B)+\sigma|u+g\varphi_B|^{2\sigma-2}\nabla(\overline{u+g\varphi_B}).$$

However, additional regularity can be gained in some cases. For example, when σ is a positive integer the nonlinearity is a polynomial in $(u+g\varphi_B)$ and $(\overline{u+g\varphi_B})$ and one can freely differentiate $|u+g\varphi_B|^{2\sigma}(u+g\varphi_B)$ without introducing any singularities. Whence, continuing with the bootstrapping argument, it can be shown that for every t in $(0,T]$ the solution has at least one more spatial derivative than it had initially. But then repeating this argument implies that the solution is in $C((0,T], C^\infty(\Omega))$. Moreover, because the equation relates temporal derivatives to spatial derivatives, the solution must posses all temporal derivatives too and is therefore a smooth (C^∞) solution of the time-dependent Ginzburg-Landau equations (1.1)-(1.2) so long as it is a weak solution. More precisely, we have proved the following result.

Let $d = d_r + id_i$ and $|d|^2 = d_r^2 + d_i^2$, we can state the main result of the paper as following:

Theorem 1.1 Let Ω be a bounded domain in R^n, with Lipschitzian boundary. And assume that $U > 0, c > 0, d_i > 0, m > 0, b > 0$, and σ is a positive integer. Let

$$n_0 = \max\left\{2, \left[\frac{n}{2}\right]+1\right\}, \quad q_0 = \max\left\{2, \frac{2\sigma n n_0 + 2(\sigma-1)n}{2+2\sigma n_0 - n}\right\}.$$

(i) Let the parameter d_i, d_r be restricted by the condition

$$|d_r| < \frac{2d_i\sqrt{q_0-1}}{q_0-2}.$$

Then, for any initial value $u_0 \in H^{k,2}, \varphi_{B0} \in H^{k,2}$ with $k \geqslant n_0$ the initial boundary value problem (1.1)-(1.4) has a unique classical solution

$$u(x,t) \in C^0\left([0,\infty); W^{k,2}(\Omega)\right) \cap C^1\left([0,\infty); W^{k-2,2}(\Omega)\right),$$

$$\varphi_B(x,t) \in C^0\left([0,\infty); W^{k,2}(\Omega)\right) \cap C^1\left([0,\infty); W^{k-2,2}(\Omega)\right).$$

(ii) If $\sigma n \geqslant 3$ and the parameter d is restricted by the condition
$$|d_r| < \frac{2d_i\sqrt{\sigma n - 1}}{\sigma n - 2},$$
then, the initial boundary value problem (1.1)-(1.4) with C^∞ initial data has a unique global classical solution
$$u(x,t) \in C^0\left([0,\infty); W^{k,2}(\Omega)\right) \cap C^1\left([0,\infty); W^{k-2,2}(\Omega)\right),$$
$$\varphi_B(x,t) \in C^0\left([0,\infty); W^{k,2}(\Omega)\right) \cap C^1\left([0,\infty); W^{k-2,2}(\Omega)\right).$$

For all of the above cases there exists constants C_k that are dependent only on the shape of the domain Ω, the parameters of the equations (1.1) and (1.2), independent of the solutions $u(x,t)$ and $\varphi_B(x,t)$, such that
$$\limsup_{t \to \infty} \|u(x,t)\|_{k,2} < C_k, \quad k \in N,$$
$$\limsup_{t \to \infty} \|\varphi_B(x,t)\|_{k,2} < C_k, \quad k \in N.$$

Remark 1.1 For $\sigma n = 1$ and $\sigma n = 2$ the Ginzburg-Landau equations (1.1)-(1.2) with $D(\Delta_\Omega)$ initial data possesses a unique global solution without any restriction on d expect for $d_i \geqslant 0$.

Theorem 1.2 Under the conditions of theorem 1.1. Then for every $u_0(x) \in W^{\infty,2}(\Omega)$, $\varphi_{B0}(x) \in W^{\infty,2}(\Omega), t \geqslant 0$, the initial problem (1.1)-(1.4) has a unique classical global solution
$$u(x,t) \in C^\infty\left((0,\infty) \times \Omega\right), \quad \varphi_B(x,t) \in C^\infty\left((0,\infty) \times \Omega\right).$$

2 Preliminary

In this section, we induce some necessary preliminaries. At first, We mention some known results on $W^{k,p}, H^{s,p}$ and $B^s_{p,q}$-spaces which will be useful in the subsequent sections [1], [19]:

Lemma 2.1 Let $\Omega \subset R^n$ be an arbitrary (bounded or unbounded) domain.

(a) For $s \in N \cup \{0\}$ and $1 < p < \infty$, then $\|\cdot\|^*_{s,p}$ is an equivalent norm to $\|\cdot\|_{s,p}$, i.e., it holds that
$$H^{s,p}(\Omega) = W^{s,p}(\Omega).$$

(b) Let $s,t \in R, t < s$ and $1 < p \leqslant q < \infty$. Provided that $\dfrac{1}{q} \geqslant \dfrac{1}{p} - \dfrac{s-t}{n}$, it holds that
$$H^{s,p}(\Omega) \hookrightarrow H^{t,q}(\Omega).$$

(c) Let $s \in R, 1 < p < \infty$ with $s > n/p$. Then,
$$H^{s,p}(\Omega) \hookrightarrow L^\infty(\Omega).$$

(d) Let $s \in R$ and $2 \leqslant p < \infty$. Then,
$$B^s_{p,2}(\Omega) \hookrightarrow H^{s,p}(\Omega) \hookrightarrow B^s_{p,p}(\Omega).$$

(e) Let $s \in R, 1 < p < \infty$ and $1 < q \leqslant \infty$. For any $\varepsilon > 0$,
$$B^{s+\varepsilon}_{p,\infty}(\Omega) \hookrightarrow B^s_{p,q}(\Omega) \hookrightarrow B^{s-\varepsilon}_{p,1}(\Omega),$$
and
$$H^{s+\varepsilon}_p(\Omega) \hookrightarrow B^s_{p,q}(\Omega) \hookrightarrow H^{s-\varepsilon}_p(\Omega).$$

Lemma 2.2 Let $\alpha_1, \alpha_2, \cdots, \alpha_N$ be multi-indices with $|\alpha_j| \leqslant k_j (j = 1, \cdots, N)$, $k_j \in R_+$ and $k_0 \in R_+$. Let $1 < p, q_j < \infty (i = 0, 1, \cdots, N)$ and $\beta \in R_+$ satisfying the relation
$$\sum_{a_j > 0} a_j < \frac{1}{p} < \sum_{j=1}^N \frac{1}{q_j},$$
where
$$a_0 = \beta \left(\frac{1}{q_0} - \frac{k_0}{n} \right), \quad a_j = \frac{1}{q_j} - \frac{k_j - |\alpha_j|}{n}, \quad j = 1, \cdots, N.$$
Then, we have
$$\left\| |u_0|^\beta D^{\alpha_1} u_1 \cdots D^{\alpha_N} u_N \right\|_p \leqslant C \|u_0\|^\beta_{k_0, q_0} \prod_{j=1}^N \|u_j\|_{k_j, q_j}.$$

Proof By Hölder inequality,
$$\left\| |u_0|^\beta D^{\alpha_1} u_1 \cdots D^{\alpha_N} u_N \right\|_p$$
$$\leqslant \left(\int |u_0|^{\frac{\beta}{a_0}} dx \right)^{a_0} \left(\int |D^{\alpha_1} u_1|^{\frac{1}{a_1}} dx \right)^{a_1} \cdots \left(\int |D^{\alpha_N} u_N|^{\frac{1}{a_N}} dx \right)^{a_N}$$
$$\leqslant C \|u_0\|^\beta_{k_0, q_0} \prod_{j=1}^N \|u_j\|_{k_j, q_j},$$
for $\sum_{a_j > 0} < \frac{1}{p}$ or $\sum_{j=1}^N a_j \leqslant \frac{1}{p}$ if all a_j are positive.

In fact, according to the assumptions above it is possible to choose $1 < p_1, \cdots, < p_N \leqslant +\infty$ in such a way that $\sum_{j=1}^N \frac{1}{p_j} = 1$, $\frac{1}{q_j} \geqslant \frac{1}{pp_j} \geqslant a_j$, if $a_j > 0$, and $\frac{1}{q_j} \geqslant \frac{1}{pp_j} > a_j$, if $a_j \leqslant 0$.

By the Sobolev imbedding theorem this implies $H^{k_j,q_j} \subset H^{|\alpha_j|,pp_j}(j=1,\cdots,N)$. Therefore from the Hölder inequality it follows that

$$\left\|\prod_{j=1}^{N}|D^{\alpha_j}u_j|\right\|_{L^p(R^n)} \leq \prod_{j=1}^{N}\|D^{\alpha_j}u_j\|_{L^{pp_j}(R^n)} \leq C_0\prod_{j=1}^{N}\|u_j\|_{k_j,q_j}.$$

Here and in what follows, the symbol C denotes a generic positive constant independent from the function being estimated. Other possible dependence of the constants C has to be determined from the context where it appears.

Then we mention $L^p - L^{p'}$ estimates for the solution of the following initial value problem:

$$iu_t - \Delta u = 0, \quad u(x,0) = v(x).$$

Lemma 2.3 ([15]) Let $1 < p \leq 2, \dfrac{1}{p}+\dfrac{1}{p'} = 1, t \in R$ and $1 \leq q < \infty$. Then one has

$$\left\|F^{-1}exp(i|\xi|^2 t)\hat{v}\right\|_{B^s_{p',q}} \leq C|t|^{\frac{n}{2}\left(1-\frac{2}{p}\right)}\|v\|_{B^s_{p,q}}$$

for all $v \in B^s_{p,q}$. And

$$\left\|F^{-1}exp(i|\xi|^2 t)\hat{v}\right\|_{s,p'} \leq C|t|^{\frac{n}{2}\left(1-\frac{2}{p}\right)}\|v\|_{s,p},$$

for all $v \in H^{s,p}$.

In the sequel we will use the following two inequalities frequently.

(1) Let $1 < p < \infty$. Then, for any function $u \in C^2(\bar{\Omega})$ such that

$$\int_{\partial\Omega}|u|^{p-2}\bar{u}\frac{\partial u}{\partial n}dS = 0,$$

one has the inequality [12]

$$|\text{Im}\langle\Delta u,|u|^{p-2}u\rangle| \leq \frac{|p-2|}{2\sqrt{p-1}}\text{Re}\langle-\Delta u,|u|^{p-2}u\rangle. \tag{2.1}$$

As usual, n stands for the outer unit normal to the boundary $\partial\Omega$.

As a consequence, we deduce

$$\text{Re}\left[(1+ir)\langle-\Delta u,|u|^{p-2}u\rangle\right] \geq 0, \quad \text{for} \quad |r| \leq \frac{2\sqrt{p-1}}{|p-2|}. \tag{2.2}$$

(2) Gagliardo-Nirenberg inequalities [12], [19]:

$$\|u\|_{k,p} \leq C\|u\|_{s,q}^{\theta}\|u\|_{r}^{1-\theta}, \tag{2.3}$$

where $1 \leqslant p, q, r < \infty, k/s \leqslant \theta \leqslant 1$, and $k - (n/p) = \theta(s - (n/q)) - (1-\theta)n/r$.

$$\|u\|_{C^\tau} \leqslant C\|u\|_{s,q}^\theta \|u\|_r^{1-\theta}, \tag{2.4}$$

with $0 \leqslant \theta \leqslant 1$ and $\tau \leqslant \theta(s - (n/q)) - (1-\theta)n/r$.

Now for $\alpha \in (0,1)$ we define $(-\Delta)^\alpha f$ to be the function w whose Fourier transform satisfying

$$\widehat{w}(y) = |y|^{2\alpha} \widehat{f}(y).$$

By extension we then obtain the operator $(-\Delta)^\alpha$. This is again a nonnegative operator. We note that the definition domain of $(-\Delta)^\alpha$ is a closed linear subspace of the fractional Sobolev space $H^{2\alpha}(\Omega)$. Since $\nabla^2 u = \Delta u$ for any scalar function u, we find that

$$\|\nabla^k u\|_2 = \|(-\Delta_\Omega)^{k/2} u\|_2, \quad u \in D((-\Delta_\Omega)^{k/2}).$$

and the following property:

Fix $u \in D((-\Delta)^\alpha) \cap D((-\Delta)^\beta)$. Then the mapping $0 \leqslant \alpha \mapsto \|(-\Delta)^\alpha u\|_2$ is log convex, i.e.

$$\|(-\Delta)^{\tau\alpha + (1-\tau)\beta} u\|_2 \leqslant \|(-\Delta)^\alpha u\|_2^\tau \cdot \|(-\Delta)^\beta u\|_2^{1-\tau}, \quad \tau \in [0,1]. \tag{2.5}$$

Lemma 2.4 Let Ω be a bounded domain in R^n (any n) with Lipschitzian boundary $\partial \Omega$.

For a positive integer σ and every integer $k > n/2$ and real number $q \in [1, \infty]$ there exists a constant $C_1(n, k, q)$ such that the following inequalities hold true:

(i) For $u \in H^k(\Omega)$ with $\tau = \dfrac{\left(k - \dfrac{n}{2}\right) + \dfrac{(2\sigma+1)n}{q}}{\left(k - \dfrac{n}{2}\right) + \dfrac{n}{q}}$, it holds that

$$\|\nabla^k(|u|^{2\sigma} u)\|_2 \leqslant C_1(k, n, q, 2\sigma+1) \|u\|_{k,2}^\tau \cdot \|u\|_q^{(2\sigma+1)-\tau}; \tag{2.6}$$

(ii) For $u, v \in H^k(\Omega)$, it holds that

$$\|\nabla^k(|u|^{2\sigma} u - |v|^{2\sigma} v)\|_2 \leqslant C_3(k, n, 2\sigma+1) \cdot \left(\|u\|_{k,2}^{2\sigma} + \|v\|_{k,2}^{2\sigma}\right) \|u - v\|_{k,2}. \tag{2.7}$$

Proof By writing $|u|^{2\sigma} u$ as $\underbrace{u\bar{u} \cdots u\bar{u}}_{2\sigma} u$ for a positive integer σ and using the Leibnitz formula gives

$$\nabla^k(|u|^{2\sigma} u) = \sum_{\substack{k_1, k_2, \cdots, k_{2\sigma+1} \\ k_1 + k_2 + \cdots + k_{2\sigma+1} = k}} \frac{k!}{k_1! k_2! \cdots k_{2\sigma+1}!} \nabla^{k_1} u \vee \nabla^{k_2} \bar{u} \vee \cdots \vee \nabla^{k_{2\sigma}} \bar{u} \vee \nabla^{k_{2\sigma+1}} u,$$

$$\tag{2.8}$$

where " ∨ " denotes the symmetric tensor outer product that acts on a symmetric k-tensor by symmetrizing the $(k+l)$-tensor that is their usual tensor outer product. From definition of norm it is easily checked that

$$\left|\nabla^k(|u|^{2\sigma}u)\right| \leqslant \sum_{\substack{k_1,k_2,\cdots,k_{2\sigma+1} \\ k_1+k_2+\cdots+k_{2\sigma+1}=k}} \frac{k!}{k_1!k_2!\cdots k_{2\sigma+1}!} \left|\nabla^{k_1}u\right|\left|\nabla^{k_2}u\right|\cdots\left|\nabla^{k_{2\sigma}}u\right|\left|\nabla^{k_{2\sigma+1}}u\right|. \tag{2.9}$$

By applying a triple Hölder estimate on each term of this sum we have

$$\left\|\nabla^k(|u|^{2\sigma}u)\right\|_2$$
$$\leqslant \sum_{\substack{k_1,k_2,\cdots,k_{2\sigma+1} \\ k_1+k_2+\cdots+k_{2\sigma+1}=k}} \frac{k!}{k_1!k_2!\cdots k_{2\sigma+1}!} \left\|\nabla^{k_1}u\right\|_{\frac{2k}{k_1}} \left\|\nabla^{k_2}u\right\|_{\frac{2k}{k_2}} \cdots \left\|\nabla^{k_{2\sigma}}u\right\|_{\frac{2k}{k_{2\sigma}}}$$
$$\cdot \left\|\nabla^{k_{2\sigma+1}}u\right\|_{\frac{2k}{k_{2\sigma+1}}}. \tag{2.10}$$

A Gagliardo-Nirenberg interpolation of each of the norms in this sum shows that

$$\left\|\nabla^{k_1}u\right\|_{\frac{2k}{k_1}} \leqslant C\|u\|_{k,2}^{\theta_1}\|u\|_q^{1-\theta_1}, \quad \text{for } \theta_1\left(k-\frac{n}{2}+\frac{n}{q}\right) = k_1 - \frac{nk_1}{2k} + \frac{n}{2};$$

$$\left\|\nabla^{k_2}u\right\|_{\frac{2k}{k_2}} \leqslant C\|u\|_{k,2}^{\theta_2}\|u\|_q^{1-\theta_2}, \quad \text{for } \theta_2\left(k-\frac{n}{2}+\frac{n}{q}\right) = k_1 - \frac{nk_2}{2k} + \frac{n}{2};$$

$$\cdots\cdots$$

$$\left\|\nabla^{k_{2\sigma+1}}u\right\|_{\frac{2k}{k_{2\sigma+1}}} \leqslant C\|u\|_{k,2}^{\theta_{2\sigma+1}}\|u\|_q^{1-\theta_{2\sigma+1}}, \quad \text{for } \theta_{2\sigma+1}\left(k-\frac{n}{2}+\frac{n}{q}\right) = k_{2\sigma+1} - \frac{nk_{2\sigma+1}}{2k} + \frac{n}{2}.$$

Since $k_1 + k_2 + \cdots + k_{2\sigma+1} = k$, we have

$$(\theta_1 + \theta_2 + \cdots + \theta_{2\sigma+1})\left(k - \frac{n}{2} + \frac{n}{q}\right) = k - \frac{n}{2} + \frac{n(2\sigma+1)}{q}.$$

Therefore, by letting

$$\tau = \theta_1 + \theta_2 + \cdots + \theta_{2\sigma+1},$$

we obtain that

$$\left\|\nabla^k(|u|^{2\sigma}u)\right\|_2 \leqslant C_1(n,k,q)\|u\|_{k,2}^\tau \|u\|_q^{(2\sigma+1)-\tau},$$

where

$$\tau = \frac{k - \dfrac{n}{2} + \dfrac{n(2\sigma+1)}{q}}{k - \dfrac{n}{2} + \dfrac{n}{q}}.$$

Hence, inequality (2.6) holds ture.

To prove inequality (2.7) we let $w = u - v$ and observe that

$$a^n - b^n = (a-b)(a^{n-1} + a^{n-2}b + \cdots + ab^{n-2} + b^{n-1}),$$

to find

$$\begin{aligned}
&|u|^{2\sigma}u - |v|^{2\sigma}v \\
&= |u|^{2\sigma}w + v(|u|^{2\sigma} - |v|^{2\sigma}) \\
&= |u|^{2\sigma}w + \overline{w}uv(|u|^{2(\sigma-1)} + |u|^{2(\sigma-2)}|v|^2 + \cdots + |v|^{(2\sigma-1)}).
\end{aligned}$$

Thus,

$$\begin{aligned}
&|\nabla^k(|u|^{2\sigma}u - |v|^{2\sigma}v)| \\
&\leqslant \sum_{k_1+k_2+\cdots+k_{2\sigma+1}} \frac{k!}{k_1!k_2!\cdots k_{2\sigma+1}!} \left(|\nabla^{k_1}u| + |\nabla^{k_1}v|\right)\left(|\nabla^{k_2}u| + |\nabla^{k_2}v|\right) \\
&\quad \cdots \cdot \left(|\nabla^{k_{2\sigma}}u| + |\nabla^{k_{2\sigma}}v|\right) \cdot |\nabla^{k_{2\sigma+1}}w|.
\end{aligned}$$

Using Hölder's inequality yields

$$\begin{aligned}
&\left\|\nabla^k(|u|^{2\sigma}u - |v|^{2\sigma}v)\right\| \\
&\leqslant \sum_{k_1+k_2+\cdots+k_{2\sigma+1}} \frac{k!}{k_1!k_2!\cdots k_{2\sigma+1}!} \left(\left\|\nabla^{k_1}u\right\|_{\frac{2k}{k_1}} + \left\|\nabla^{k_1}v\right\|_{\frac{2k}{k_1}}\right)\left(\left\|\nabla^{k_2}u\right\|_{\frac{2k}{k_2}} + \left\|\nabla^{k_2}v\right\|_{\frac{2k}{k_2}}\right) \\
&\quad \cdots \cdot \left(\left\|\nabla^{k_{2\sigma}}u\right\|_{\frac{2k}{k_{2\sigma}}} + \left\|\nabla^{k_{2\sigma}}v\right\|_{\frac{2k}{k_{2\sigma}}}\right) \cdot \left\|\nabla^{k_{2\sigma+1}}w\right\|_{\frac{2k}{k_{2\sigma+1}}}.
\end{aligned}$$

As seen above, by a Gagliardo-Nirenberg interpolation and Young's inequality, we have

$$\begin{aligned}
&\left(\left\|\nabla^{k_i}u\right\|_{\frac{2k}{k_i}} + \left\|\nabla^{k_i}v\right\|_{\frac{2k}{k_i}}\right) \\
&\leqslant C\left(\|u\|_{k,2}^{r_i} \cdot \|u\|_q^{1-r_i} + \|v\|_{k,2}^{r_i} \cdot \|v\|_q^{1-r_i}\right) \\
&\leqslant C_2\left(\|u\|_{k,2} + \|u\|_q + \|v\|_{k,2} + \|v\|_q\right),
\end{aligned}$$

for $0 \leqslant i \leqslant k$ and

$$\left\|\nabla^{k_{2\sigma+1}}w\right\|_{\frac{2k}{k_{2\sigma+1}}} \leqslant C\left(\|w\|_{k,2}^{r_{2\sigma+1}} \cdot \|w\|_q^{1-r_{2\sigma+1}}\right).$$

Therefore, with

$$I = C_2\left(\|u\|_{k,2} + \|u\|_q + \|v\|_{k,2} + \|v\|_q\right),$$

and

$$\gamma_0 = \frac{\frac{1}{k}\left(1 - \frac{n}{2}\right)}{k - \frac{n}{2} + \frac{n}{q}}, \quad \gamma_1 = \frac{\frac{n}{q}}{k - \frac{n}{2} + \frac{n}{q}},$$

we get
$$r_{2\sigma+1} = k_{2\sigma+1}\gamma_0 + \gamma_1$$

and
$$\begin{aligned}
&\left\|\nabla^k\left(|u|^{2\sigma}u - |v|^{2\sigma}v\right)\right\|_2 \\
&\leqslant C_1 I^{2\sigma} \sum_{k_1+\cdots+k_{2\sigma+1}} \frac{k!}{k_1!\cdots k_{2\sigma+1}!} \|w\|_{k,2}^{r_{2\sigma+1}} \cdot \|w\|_q^{1-r_{2\sigma+1}} \\
&\leqslant C_1 I^{2\sigma} \|w\|_q \left(\frac{\|w\|_{k,2}}{\|w\|_q}\right)^{\gamma_1} \sum_{k_1+\cdots+k_{2\sigma+1}} \frac{k!}{k_1!\cdots k_{2\sigma+1}!} \left(\frac{\|w\|_{k,2}}{\|w\|_q}\right)^{k_{2\sigma+1}\gamma_0} \\
&\leqslant C_1 I^{2\sigma} \|w\|_q \left(\frac{\|w\|_{k,2}}{\|w\|_q}\right)^{\gamma_1} \left[C_2 + \left(\frac{\|w\|_{k,2}}{\|w\|_q}\right)^{\gamma_0}\right]^k \\
&\leqslant C_1 I^{2\sigma} \|w\|_{k,2}^{\gamma_1} \cdot \left[C_2\|w\|_q + \|w\|_{k,2}^{\gamma_0}\right]^k \\
&\leqslant C_3 I^{2\sigma} \left(\|w\|_q + \|w\|_{k,2}\right).
\end{aligned}$$

For the last two inequalities we have used the relation
$$k\gamma_0 + \gamma_1 = 1.$$

By taking $q = 2$ we obtain finally that
$$\left\|\nabla^k\left(|u|^{2\sigma}u - |v|^{2\sigma}v\right)\right\|_2 \leqslant C_3(k, n, 2\sigma+1)\left(\|u\|_{k,2}^{2\sigma} + \|v\|_{k,2}^{2\sigma}\right) \cdot \|w\|_{k,2}.$$

This finishes the proof of lemma 2.4.

3 Local Existence of Smooth Solutions

In this section, we would establish the existence of local solution of (1.1)-(1.4). At first, we may therefore conclude the initial boundary value problem (1.1)-(1.4) as that:

$$(u+g\varphi_B)_t = i\left(\frac{a}{d} - \frac{1}{dU}\right)(u+g\varphi_B) + \frac{ig}{dU}\varphi_B + \frac{ic}{4md}\Delta(u+g\varphi_B)$$
$$-\frac{ib}{d}|u+g\varphi_B|^{2\sigma}(u+g\varphi_B), \tag{3.1}$$

$$\varphi_{Bt} = \frac{ig}{U}u - i(2\nu - 2\mu)\varphi_B + \frac{i}{4m}\Delta\varphi_B, \tag{3.2}$$

$$u(x,0) = u_0(x), \tag{3.3}$$

$$\varphi_B(x,0) = \varphi_{B0}(x). \tag{3.4}$$

Assume that $\vec{u} = \begin{pmatrix} u(x,t) + g\varphi_B(x,t) \\ \varphi_B(x,t) \end{pmatrix}$, then the initial value problem becomes:

$$\vec{u}_t = (A_1 + A_2)\vec{u} + (A_3 + iA_4)\Delta\vec{u} + J(\vec{u}), \tag{3.5}$$

where

$$A_1 = \begin{pmatrix} \frac{(aU-1)d_i}{|d|^2 U} & \frac{gd_i}{|d|^2 U} \\ 0 & 0 \end{pmatrix}, \quad A_2 = \begin{pmatrix} \frac{(aU-1)d_r}{|d|^2 U} & \frac{gd_r}{|d|^2 U} \\ \frac{g^2}{U} & -2\nu + 2\mu - \frac{g^2}{U} \end{pmatrix},$$

$$A_3 = \begin{pmatrix} \frac{cd_i}{4m|d|^2} & 0 \\ 0 & 0 \end{pmatrix}, \quad A_4 = \begin{pmatrix} \frac{cd_r}{4m|d|^2} & 0 \\ 0 & \frac{1}{4m} \end{pmatrix},$$

and

$$J(\vec{u}) = \begin{pmatrix} -\frac{ib}{d}|u + g\varphi_B|^{2\sigma}(u + g\varphi_B) \\ 0 \end{pmatrix}.$$

Obviously, A_3 is a nonnegative define matrixes and the operator $-\Delta$ is a positive self-adjoint operator on $L^2(\Omega)$.

As is standard way, we study the initial boundary value problem via the corresponding integral equation

$$\vec{u}(t) = e^{[(A_3\Delta + A_1) + i(A_4\Delta + A_2)]t}\vec{u}_0 + \int_0^t e^{[(A_3\Delta + A_1) + i(A_4\Delta + A_2)](t-\tau)} J(\vec{u}(x,\tau))d\tau. \tag{3.6}$$

Theorem 3.1 Let $k \in N, k > \frac{n}{2}$ and $k \geqslant 2, U > 0, b > 0, m > 0, d_i > 0$. Then for every $\vec{u}_0 \in D((-\Delta_\Omega)^{k/2})$, there exists a unique solution $\vec{u} = \vec{u}(x,t)$ to the initial value problem (1.1)-(1.4) in the class

$$\vec{u} \in C^0([0,T^*]; W^{k,2}(\Omega)) \cap C^1([0,T^*]; W^{k-2,2}(\Omega)), \tag{3.7}$$

where T^* has a lower bound dependent only on $\|\vec{u}_0\|_{k,2}$.

Furthermore, we have the following alternative: $(\forall \vec{u}_0 \in D((-\Delta_\Omega)^{k/2}))$

1) There exists a $T^{**} > 0$ such that (1.1)-(1.4) has a unique solution $\vec{u}(t,x)$ in the class

$$\vec{u}(x,t) \in C^0([0,T^{**}]; W^{k,2}(\Omega)) \cap C^1([0,T^{**}]; W^{k-2,2}(\Omega))$$

and

$$\limsup_{t \to T^{**}} \|\vec{u}(x,t)\|_{[\frac{n}{2}]+1,2} = +\infty.$$

2) (1.1)-(1.4) has a global solution $\vec{u}(x,t)$ in the class

$$\vec{u}(x,t) \in C^0([0,+\infty); W^{k,2}(\Omega)) \cap C^1([0,+\infty); W^{k-2,2}(\Omega)).$$

Proof The first part is standard. Indeed, we can apply Banach's fixed point theorem to the operator

$$\widetilde{T}: \vec{u}(x,t) \to \widetilde{T}(\vec{u}(x,t)) = e^{[(A_3\Delta+A_1)+i(A_4\Delta+A_2)]t}\vec{u}_0$$
$$+ \int_0^t e^{[(A_3\Delta+A_1)+i(A_4\Delta+A_2)](t-\tau)} J(\vec{u}(x,\tau))d\tau, \quad (3.8)$$

in the Banach space $X = C^0([0,T_1], W^{k,2}(\Omega))$, which is locally Lipschitz continuous by virtue of lemma 2.4. And letting

$$T_1 = \frac{1}{2\left[\frac{|b|}{|d|}C_3(2(\|\vec{u}_0\|_{k,2}+1)^2)\right](\max\{1, e^{(A_3\Delta+A_1)T_1}\}\|\vec{u}_0\|_{k,2}+1)},$$

and

$$\|\vec{u}(x,t)\|_X, \|\vec{v}(x,t)\|_X \leqslant C = \max\{1, e^{(A_3\Delta+A_1)T_1}\}\|\vec{u}_0\|_{k,2}+1.$$

Since the operator $-\Delta$ is a nonnegative operator, then

$$\|\widetilde{T}(\vec{u}(x,t))\|_X \leqslant e^{(A_3\Delta+A_1)T_1}\|\vec{u}_0\|_X + \int_0^{T_1}\|J(\vec{u}(\tau))\|_X d\tau$$
$$\leqslant \max\{1, e^{(A_3\Delta+A_1)T_1}\}\|\vec{u}_0\|_{k,2} + \int_0^{T_1}\|J(\vec{u}(\tau))\|_{k,2}d\tau$$
$$\leqslant C - 1 + \|J(\vec{u}(\tau))\|_X T_1$$
$$\leqslant C - 1 + \frac{|b|}{|d|}C_3(\|\vec{u}_0\|_{k,2}^2)\|\vec{u}\|_{k,2}T_1$$
$$< C,$$

and

$$\|\widetilde{T}(\vec{u}(x,t)) - \widetilde{T}(\vec{v}(x,t))\|_X \leqslant \int_0^T \|J(\vec{u}(x,\tau)) - J(\vec{v}(x,\tau))\|_X d\tau$$
$$\leqslant \frac{|b|}{|d|}C_3(\|\vec{u}_0\|_{k,2}+\|\vec{v}_0\|_{k,2})\|\vec{u}-\vec{v}\|_X T_1$$
$$\leqslant \frac{1}{2}\|\vec{u}-\vec{v}\|_X. \quad (3.9)$$

Then, by contraction mapping principle, We obtain a unique solution of (3.6) (and hence, of (1.1)-(1.4)) in the class

$$\vec{u}(x,t) \in C^0([0,T_1]; W^{k,2}(\Omega)) \cap C^1([0,T_1]; W^{k-2,2}(\Omega))$$

For the second part, we will show that if for any $T > 0$,

$$\sup_{0 \leqslant t \leqslant T} \|\vec{u}\|_{[\frac{n}{2}]+1,2} < +\infty, \tag{3.10}$$

then the solution exists globally, that is,

$$\vec{u}(x,t) \in C^0([0,+\infty); W^{k,2}(\Omega)) \cap C^1([0,+\infty); W^{k-2,2}(\Omega)).$$

Let T be the supermum of t such that (3.1)-(3.4) has a solution in the class

$$\vec{u}(x,t) \in C^0([0,t); W^{k,2}(\Omega)) \cap C^1([0,t); W^{k-2,2}(\Omega)).$$

If $k = \left[\frac{n}{2}\right]+1$, there is no need to discuss. Assume that $k > \left[\frac{n}{2}\right]+1$ and $T < +\infty$. In virtue of lemma 2.3 and lemma 2.4, we have

$$\|\vec{u}(x,t)\|_{k,2} \leqslant C\|\vec{u}_0\|_{k,2} + C_4 \int_0^t \|\vec{u}(\tau)\|_{k,2} d\tau, \tag{3.11}$$

$\forall t \in [0, T)$.

From (3.10) and (3.11), we obtain inductively

$$\limsup_{t \to T} \|\vec{u}(t)\|_{k,2} \leqslant C_k.$$

Then, $\lim_{t \to T} \|\vec{u}(x,t)\|_{k,2} = \|\vec{u}(x,T)\|_{k,2}$ exists in view of (3.6). Since the local solution exists for a time interval depending only on $\|\vec{u}_0\|_{k,2}$, the solution $\vec{u}$ can be continued beyond T. This contradicts the definition of T. Thus, $T = +\infty$. This completes the proof.

4 Global Existence of Smooth Solutions

The second assertion of theorem 3.1 in the previous section assures that in order to prove the existence of global smooth solutions in suffices to show that the $H^{[\frac{n}{2}]+1,2}$-norm of the solution is bounded for every finite interval in R_+. And we begin with some priori estimates:

Lemma 4.1 Assume that $U > 0, b > 0, c > 0, m > 0$ and let $q \geqslant 2$. If

$$|d_r| \leqslant \frac{2d_i\sqrt{q-1}}{q-2},$$

then

$$\|u + g\varphi_B\|_q^q + \|\varphi_B\|_q^q \leqslant C_7(u_0, \varphi_{B0}, T), \tag{4.1}$$

Classical Solutions of General Ginzburg-Landau Equations

$$\|\varphi_B\|_q^q \leqslant C_7(u_0, \varphi_{B0}, T), \qquad (4.2)$$

$$\|u\|_q^q \leqslant C_7(u_0, \varphi_{B0}, T). \qquad (4.3)$$

Proof Multiplying (3.1) by $|u+g\varphi_B|^{q-2}(u+g\varphi_B)$ and taking the inner products, one obtain the real part of the result inequality:

$$\frac{1}{q}\frac{d}{dt}\|u+g\varphi_B\|_q^q$$
$$= Re\left[\left(\frac{cd_i}{4m|d|^2}+i\frac{cd_r}{4m|d|^2}\right)\int \Delta(u+g\varphi_B)|u+g\varphi_B|^{q-2}\overline{(u+g\varphi_B)}dx\right]$$
$$+\left(\frac{ad_i}{|d|^2}-\frac{d_i}{|d|^2 U}\right)\|u+g\varphi_B\|_q^q + Re\left[\int \frac{ig}{dU}\varphi_B|u+g\varphi_B|^{q-2}\overline{(u+g\varphi_B)}dx\right]$$
$$-\frac{bd_i}{|d|^2}\|u+g\varphi_B\|_{q+2\sigma}^{q+2\sigma}. \qquad (4.4)$$

Notice that

$$|d_r| \leqslant \frac{2d_i\sqrt{q-1}}{q-2},$$

then, it follows from (2.4),

$$Re\left[\left(\frac{cd_i}{4m|d|^2}+i\frac{cd_r}{4m|d|^2}\right)\int \Delta(u+g\varphi_B)|u+g\varphi_B|^{q-2}\overline{(u+g\varphi_B)}dx\right]$$
$$=\frac{cd_i}{4m|d|^2}Re\left(\int \Delta(u+g\varphi_B)|u+g\varphi_B|^{q-2}\overline{(u+g\varphi_B)}dx\right)$$
$$-\frac{cd_r}{4m|d|^2}Im\left(\int \Delta(u+g\varphi_B)|u+g\varphi_B|^{q-2}\overline{(u+g\varphi_B)}dx\right)$$
$$\leqslant -\frac{cd_i}{4m|d|^2}\cdot\frac{2\sqrt{q-1}}{q-2}\left|\int \Delta(u+g\varphi_B)|u+g\varphi_B|^{q-2}\overline{(u+g\varphi_B)}dx\right|$$
$$-\frac{cd_r}{4m|d|^2}Im\left(\int \Delta(u+g\varphi_B)|u+g\varphi_B|^{q-2}\overline{(u+g\varphi_B)}dx\right)$$
$$\leqslant -\frac{cd_i}{4m|d|^2}\cdot\frac{2\sqrt{q-1}}{q-2}\left|\int \Delta(u+g\varphi_B)|u+g\varphi_B|^{q-2}\overline{(u+g\varphi_B)}dx\right|$$
$$+\left|\frac{cd_r}{4m|d|^2}Im\left(\int \Delta(u+g\varphi_B)|u+g\varphi_B|^{q-2}\overline{(u+g\varphi_B)}dx\right)\right|$$
$$=\left(\left|\frac{cd_r}{4m|d|^2}\right|-\frac{cd_i}{4m|d|^2}\cdot\frac{2\sqrt{q-1}}{q-2}\right)\left|\int \Delta(u+g\varphi_B)|u+g\varphi_B|^{q-2}\overline{(u+g\varphi_B)}dx\right|$$
$$\leqslant 0. \qquad (4.5)$$

Thus, under the condition $bd_i \geqslant 0$, (4.5) and Young's inequality, (4.4) becomes

$$\frac{d}{dt}\|u+g\varphi_B\|_q^q \leqslant q\left(\frac{ad_i}{|d|^2}-\frac{d_i}{|d|^2 U}+\frac{|g|}{|d|U}\right)\|u+g\varphi_B\|_q^q + \frac{q|g|}{|d|U}\|\varphi_B\|_q^q. \qquad (4.6)$$

Using Fourier transform, (1.2) can be reformed into

$$\varphi_B(x,t) = e^{iHt}\varphi_{B0} + \frac{ig}{U}\int_0^t e^{iH(t-\tau)}(u+g\varphi_B)d\tau - i\left(\frac{g^2}{U}+2\nu-2\mu\right)\int_0^t e^{iH(t-\tau)}\varphi_B d\tau, \tag{4.7}$$

where

$$e^{iHt} = F^{-1}\left(e^{-\frac{i}{4m}\lambda^2 t}F\right).$$

Let $q \geqslant 2$ and $\dfrac{1}{q}+\dfrac{1}{q'} = 1$, then we can find that $1 \leqslant q' \leqslant 2$. Lemma 2.1 (b), (d) and (e), give

$$\|e^{iHt}\varphi_{B0}\|_{B^0_{q',2}} \leqslant C\|e^{iHt}\varphi_{B0}\|_{q',2} \leqslant C\|e^{iHt}\varphi_{B0}\|_{k,2} = C\|\varphi_{B0}\|_{k,2} \leqslant C, \tag{4.8}$$

provided $k > q'$.

We apply lemma 2.3 to the integral equation (4.7) to obtain:

$$\|\varphi_B(x,t)\|^q_{B^0_{q,2}} \leqslant C + C\int_0^t (t-s)^\gamma \|u+g\varphi_B\|^q_{B^0_{q',2}} d\tau + C\int_0^t (t-s)^\gamma \|\varphi_B\|^q_{B^0_{q',2}} d\tau, \tag{4.9}$$

where $0 > \gamma = \dfrac{qn}{2}\left(1-\dfrac{2}{q'}\right) > -\dfrac{qn}{2}$. By lemma 2.1 (d), we have

$$\|\varphi_B(x,t)\|^q_q \leqslant C + C\int_0^t \|u+g\varphi_B\|^{q'}_{q'} d\tau + C\int_0^t \|\varphi_B\|^{q'}_{q'} d\tau. \tag{4.10}$$

Noting that $q' \leqslant q$, then by Hölder's inequality, we have

$$\|\varphi_B(x,t)\|^q_q \leqslant C + C\int_0^t \|u+g\varphi_B\|^q_q d\tau + C\int_0^t \|\varphi_B\|^q_q d\tau. \tag{4.11}$$

We see that (4.6) implies

$$\|u+g\varphi_B\|^q_q \leqslant q\left(\frac{ad_i}{|d|^2} - \frac{d_i}{|d|^2 U} + \frac{|g|}{|d|U}\right)\int_0^t \|u+g\varphi_B\|^q_q d\tau$$
$$+ \frac{q|g|}{|d|U}\int_0^t \|\varphi_B\|^q_q d\tau + \|u_0+g\varphi_{B0}\|^q_q. \tag{4.12}$$

The former two inequalities mean that

$$\|u+g\varphi_B\|^q_q + \|\varphi_B(x,t)\|^q_q \leqslant C_5 + C_6\int_0^t \left(\|u+g\varphi_B\|^q_q + \|\varphi_B\|^q_q\right) d\tau. \tag{4.13}$$

By Gronwall's inequality, we obtain

$$\|u+g\varphi_B\|^q_q + \|\varphi_B(x,t)\|^q_q \leqslant C_7(u_0, \varphi_{B0}, T). \tag{4.14}$$

Thus, we get the desired results immediately. This completes the proof of lemma 4.1.

Lemma 4.2 Under the conditions of lemma 4.1, if the L^q-norm of the solutions $u(x,t)$ and $\varphi_B(x,t)$ are bounded in $[0,T)$ for some $q \in [1,\infty)$ such that

$$q > \frac{2\sigma k n + 2(\sigma-1)n}{2+2k-n},$$

then the H^k-norm of the solutions $u(x,t)$ and $\varphi_B(x,t)$ are bounded in $[0,T)$.

Proof Multiplying (3.1) by $\Delta^k(u+g\varphi_B)$ and taking the inner products, and taking the real part of the result inequality to obtain:

$$\frac{1}{2}\frac{d}{dt}\|\nabla^k(u+g\varphi_B)\|_2^2$$
$$= -\frac{cd_i}{4m|d|^2}\|\nabla^{k+1}(u+g\varphi_B)\|_2^2 + \left(\frac{ad_i}{|d|^2} - \frac{d_i}{|d|^2 U}\right)\|\nabla^k(u+g\varphi_B)\|_2^2$$
$$- \mathrm{Re}\left(\int \frac{ig}{dU}\nabla^k\varphi_B \cdot \nabla^k(u+g\varphi_B)dx\right)$$
$$- \mathrm{Re}\left[\left(\frac{bd_i}{|d|^2} + i\frac{bd_r}{|d|^2}\right)\int |u+g\varphi_B|^{2\sigma}(u+g\varphi_B)\Delta^k\overline{(u+g\varphi_B)}dx\right]$$
$$\leqslant -\frac{cd_i}{4m|d|^2}\|\nabla^{k+1}(u+g\varphi_B)\|_2^2 + \left(\frac{ad_i}{|d|^2} - \frac{d_i}{|d|^2 U} + \frac{|g|}{2|d|U}\right)\|\nabla^k(u+g\varphi_B)\|_2^2$$
$$+ \frac{|g|}{2|d|U}\|\nabla^k\varphi_B\|_2^2$$
$$- \mathrm{Re}\left[\left(\frac{bd_i}{|d|^2} + i\frac{bd_r}{|d|^2}\right)\int |u+g\varphi_B|^{2\sigma}(u+g\varphi_B)\nabla^{(k+1)+(k-1)}\overline{(u+g\varphi_B)}dx\right].$$
(4.15)

Using Schwarz's inequality, we have

$$\int |u+g\varphi_B|^{2\sigma}(u+g\varphi_B)\nabla^{(k+1)+(k-1)}\overline{(u+g\varphi_B)}dx$$
$$= (-1)^{k+1}\int \nabla^{k+1}\left(|u+g\varphi_B|^{2\sigma}(u+g\varphi_B)\right) \cdot \nabla^{k-1}\overline{(u+g\varphi_B)}dx$$
$$\leqslant \left(\int \left|\nabla^{k+1}\left(|u+g\varphi_B|^{2\sigma}(u+g\varphi_B)\right)\right|^2 dx\right)^{\frac{1}{2}}\left(\int \left|\nabla^{k-1}(u+g\varphi_B)\right|^2 dx\right)^{\frac{1}{2}}$$
$$= \|\nabla^{k+1}\left(|u+g\varphi_B|^{2\sigma}(u+g\varphi_B)\right)\|_2 \cdot \|\nabla^{k-1}(u+g\varphi_B)\|_2.$$
(4.16)

For the first term we use (2.1) to find that

$$\|\nabla^{k+1}\left(|u+g\varphi_B|^{2\sigma}(u+g\varphi_B)\right)\|_2$$
$$\leqslant C_1(k+1,n,q)\|u+g\varphi_B\|_{k+1,2}^\tau \cdot \|u+g\varphi_B\|_q^{(2\sigma+1)-\tau},$$
(4.17)

where
$$\tau = \frac{\left(k+1-\dfrac{n}{2}\right)+\dfrac{n(2\sigma+1)}{q}}{\left(k+1-\dfrac{n}{2}\right)+\dfrac{n}{q}}. \tag{4.18}$$

The number q is just the one given in the lemma 4.1 such that the L^q-norm $\|u+g\varphi_B\|_q$ of the solution is bounded in $[0,T)$.

For the second term we use the log convexity (2.5) to find
$$\begin{aligned}\|\nabla^{k-1}(u+g\varphi_B)\|_2 &= \|(-\Delta)^{\frac{1}{2}(k+1)\cdot\frac{k-1}{k+1}+\frac{2}{k+2}\cdot 0}(u+g\varphi_B)\|_2 \\ &\leqslant \|\nabla^{k+1}(u+g\varphi_B)\|_2^{\frac{k-1}{k+1}}\cdot\|u+g\varphi_B\|_2^{\frac{2}{k+1}}.\end{aligned} \tag{4.19}$$

We find from above and lemma 4.1 for $q=2$,
$$\begin{aligned}&\left|\int |u+g\varphi_B|^{2\sigma}(u+g\varphi_B)\nabla^{(k+1)+(k-1)}\overline{(u+g\varphi_B)}dx\right| \\ &\leqslant C_1(k+1,n,q)\|u+g\varphi_B\|_{k+1,2}^\tau \\ &\quad\cdot\|u+g\varphi_B\|_q^{2\sigma+1-\tau}\|\nabla^{k+1}(u+g\varphi_B)\|_2^{\frac{k-1}{k+1}}\|u+g\varphi_B\|_2^{\frac{2}{k+1}} \\ &\leqslant C\|\nabla^{k+1}(u+g\varphi_B)\|_2^{2\cdot\frac{k-1}{2k+2}}\left(\|\nabla^{k+1}(u+g\varphi_B)\|_2^2+\|u+g\varphi_B\|_2^2\right)^{\frac{\tau}{2}} \\ &\leqslant C\left(\|\nabla^{k+1}(u+g\varphi_B)\|_2^{2\gamma}+C_8^\tau\|\nabla^{k+1}(u+g\varphi_B)\|_2^{\frac{k-1}{k+1}}\right),\end{aligned} \tag{4.20}$$

where
$$\gamma = \frac{k-1}{2k+2}+\frac{\tau}{2}<1, \tag{4.21}$$
which means that
$$q > \frac{2\sigma kn+2(\sigma-1)n}{2k+2-n}. \tag{4.22}$$

From (4.15), (4.21), lemma 4.1 for $q=2$, and by Young's inequality, we have
$$\frac{d}{dt}\|\nabla^k(u+g\varphi_B)\|_2^2 \leqslant 2\left(\frac{ad_i}{|d|^2}-\frac{d_i}{|d|U}+\frac{|g|}{2|d|U}\right)\|\nabla^k(u+g\varphi_B)\|_2^2+\frac{|g|}{2|d|U}\|\nabla^k\varphi_B\|_2^2. \tag{4.23}$$

Since the H^{k+1}-norm $\|\cdot\|_{k+1,2}$ is equivalent to the norm
$$\left(\|\nabla^{k+1}\|_2^2+\|\cdot\|_2^2\right)^{\frac{1}{2}} = \|\cdot\|_{k+1,2}, \tag{4.24}$$

we have
$$\|u+g\varphi_B\|_{k,2}^2 \leqslant C_9\int_0^t\|u+g\varphi_B\|_{k,2}^2 d\tau + C_{10}\int_0^t\|\varphi_B\|_{k,2}^2 d\tau. \tag{4.25}$$

And from (4.7), (4.8) and lemma 2.1 and lemma 2.3,

$$\|\varphi_B\|_{k,2}^2 \leqslant C + \frac{g^2}{U^2}\int_0^t \|u+g\varphi_B\|_{k,2}^2 d\tau + \left|\frac{g^2}{U}+2\nu-2\mu\right|^2 \int_0^t \|\varphi_B\|_{k,2}^2 d\tau. \quad (4.26)$$

Then, by (4.25) and (4.26), we have

$$\|u+g\varphi_B\|_{k,2}^2 + \|\varphi_B\|_{k,2}^2 \leqslant C_{11} + C_{12}\int_0^t \left(\|u+g\varphi_B\|_{k,2}^2 + \|\varphi_B\|_{k,2}^2\right) d\tau.$$

Using Gronwall's inequlity,

$$\|u+g\varphi_B\|_{k,2}^2 + \|\varphi_B\|_{k,2}^2 \leqslant C_{13}(u_0,\varphi_{B0},T). \quad (4.27)$$

Thus,

$$\|u+g\varphi_B\|_{k,2}^2 \leqslant C_{13}(u_0,\varphi_{B0},T), \quad (4.28)$$

$$\|\varphi_B\|_{k,2}^2 \leqslant C_{13}(u_0,\varphi_{B0},T), \quad (4.29)$$

$$\|u(x,t)\|_{k,2}^2 \leqslant C_{13}(u_0,\varphi_{B0},T). \quad (4.30)$$

This finishes the proof of lemma 4.2.

Proof of theorem 1.1 Let us examine when the condition in lemma 4.2 will be satisfied, i.e., when the bounded of the L^q-norm of the solution with

$$q > \frac{2\sigma kn + 2(\sigma-1)}{2+2k-n}$$

will be guaranteed. It appears that we should consider all k with $k \geqslant n_0$. However, for any $k \geqslant n_0$ we can divided the H^k-estimate of the solutions in two pieces: We first establish the bounded of the H^{n_0}-norm of the solutions and consequently the global existence of the solutions. Then, by Sobolev's imbedding we claim that the L^∞-norm of the solutions are bounded as well. This implies that for general $k \geqslant n_0$ we can take $q = \infty$ so that the condition $q > \dfrac{2\sigma kn + 2(\sigma-1)n}{2+2k-n}$ holds true and thus we finally conclude from lemma 4.2 that the H^k-norm of the solutions is bounded.

According to this discussion, in order to apply lemma 4.2 we need to find condition so that the bounded of the L^q-norm of the solutions with

$$q > \frac{2n\sigma n_0 + 2(\sigma-1)n_0}{2+2\sigma n_0 - n} > \frac{2\sigma kn + 2(\sigma-1)n}{2+2k-n}, \quad (4.31)$$

will be guaranteed. Actually, only when the term in the right side of (4.31) is greater than 2, then the requirement on q becomes a real need. Thus, in general we need to consider q satisfying

$$q > q_0 = \max\left\{2, \frac{2\sigma nn_0 + 2(\sigma-1)n_0}{2+2\sigma n_0 - n}\right\}. \quad (4.32)$$

For this purpose we appeal to lemma 4.1 which says that the L^q-norm ($q \geqslant 2$) of the solution will be bounded if the parameter d_i, d_r are chosen in such a way that $|d_r| \leqslant \dfrac{2d_i\sqrt{q-1}}{q-2}$. Thus, if the parameter d_i, d_r, are chosen in such a way that

$$|d_r| \leqslant \frac{2d_i\sqrt{q_0-1}}{q_0-2}. \tag{4.33}$$

then (4.32) will be satisfied for some $q > q_0$. It is quite comfortable to see that $q_0 = 2$ for $\sigma n = 2$ and thus condition (4.31) becomes superfluous if $\sigma n \leqslant 2$.

If in the above procedure we begin with C^∞ data, then what we need is to find condition on d_i, d_r such that the bounded of the L^q-norm of the solutions can be guaranteed for some $q > \dfrac{2\sigma kn + 2(\sigma-1)n}{2+2k-n}$ with any one $k > \dfrac{n}{2}$. Note that $\dfrac{2\sigma kn + 2(\sigma-1)n}{2+2k-n}$ converges to σn as $k \to \infty$. Thus, the bottom rung for such a q is $q = \sigma n$. Hence, in this case the condition (4.33) should be read as

$$|d_r| \leqslant \frac{2d_i\sqrt{\sigma n-1}}{\sigma n-2}, \quad (\sigma n \geqslant 3) \tag{4.34}$$

We collect these results as theorem 1.1. This completes the proof of theorem 1.1.

As we have observed above, lemma 4.2 implies that, in particular, for $\sigma n \leqslant 2$ the time-dependent Ginzburg-Landau equations with the initial data in H^k possesses a unique global classical solution whose H^2-norm is bounded, without any restriction on the parameters d_r and d_i of the equations except for $d_i \geqslant 0$.

From theorem 1.1, we obtain theorem 1.2 immediately.

Remark If the boundary condition (1.4) is replaced by the Neumenn boundary condition $\dfrac{\partial u}{\partial n} = 0, \dfrac{\partial \varphi_B}{\partial n} = 0$, on $[0,\infty) \times \partial\Omega$, where n is the unit outward normal on the boundary, or the space-periodicity with $\Omega = (0, L_1) \times (0, L_2) \times \cdots \times (0, L_n)$ and u, φ_B are Ω − periodic, the main results theorem 1.1 and theorem 1.2 are also hold.

Certainly, the corresponding definition domains of the Laplacian Δ are chosen as follows: $D(\Delta_N) = \left\{u, \varphi_B \in H^2(\Omega) : \dfrac{\partial u}{\partial n} = 0, \dfrac{\partial \varphi_B}{\partial n} = 0, \text{ on } \partial\Omega\right\}$, for the Neumann boundary conditions, and $D(\Delta_{per}) = H^2_{per}(\Omega)$ for the periodic boundary conditions.

Acknowledgments The paper is supported by National Natural Science Foundation of China (Nos: 11201415, 11571159); Program for New Century Excellent Talents in Fujian Province University (No: JA14191). And the authors specially thanks the referee for his/her carefully reading of the manuscript and for giving valuable suggestions.

References

[1] Bergh J, Löfström J. Interpolation Spaces. Berlin-Heidelberg-New York: Springer, 1976.

[2] Brenner P, Wahl WV. Global classical solutions of nonlinear wave equations. Math. Z., 1981, 176: 87-121.

[3] Chen S, Guo B. On the cauchy problem of the Ginzburg-Landau equations for atomic Fermi gases near the BCS-BEC crossover. J. Partial Differ. Equ., 2009, 22(3): 218-133.

[4] Chen S, Guo B. Solution theory of the coupled time-dependent Ginzburg-Landau equations. Int. J. Dyn. Syst. Differ. Equ., 2009, 2(1-2): 1-20.

[5] Chen S, Guo B. Existence of the weak solution of coupled time-dependent Ginzburg-Landau equations. J. Math. Phys., 2010, 51(3): 033507.

[6] Chen S, Guo B. Classical solutions of time-dependent Ginzburg-Landau theory for atomic Fermi gases near the BCS-BEC crossover. J. Diff. Equa., 2011, 251(6): 1415-1427.

[7] Chen S, Guo B. The existence of global solution of the Ginzburg-Landau equations for atomic Fermi gases nears the BCS-BC crossover (in Chinese). Acta Math Sci., 2011, 31A(5): 1359-1368.

[8] de Gennes PG. Superconductivity of Metals and Alloys (Addisom-Wesley, Reading, MA,1998); Abrikosov AA. Fundamentals of the Theory of Metals (New York: Elsevier-Science Ltd., 1988).

[9] Deoring CR, Gibbon JD, Levermore CD. Weak and strong solutions of the complex Ginzburg-Landau equation. Physica D., 1994, 71: 285-318.

[10] Drechsler M, Zwerger W. Crossover from BCS-superconductivity to Bose-condensation. Ann. Phys., 1992, 504(1): 15-23.

[11] Hayashi N. Classical solutions of nonlinear Schrödinger equations. Manuscripta Math., 1986, 55: 171-190.

[12] Henry D. Geometric theory of semilinear parabolic equations. Lect. Notes in Math. 840. Springer-Verlag, 1981.

[13] Machida M, Koyama T. Time-dependent Ginzburg-Landau theory for atomic Fermi gases near the BCS-BEC crossover. Phy. Rev. A., 2006, 74: 033603.

[14] Pecher H. L^p-Abschätzungen und klassische Lösungen für nichtlineare Wellengleichungen. I. Math. Z. 1976, 150: 159-183 and II, Manuscripta Math., 1977, 20: 227-244.

[15] Pecher H, Wahl W.V. Time dependent nonlinear Schrödinger equations. Manuscripta Math., 1979, 27: 125-157.

[16] Reed M. Abstract non-linear wave equations. Lecture Notes in Mathematics 507. Berlin-Heidelberg-New York: Springer, 1976.

[17] Sa de Melo C A R, Randeria M, Engelbrecht JR. Crossover from BCS to Bose superconductivity: transition temperature and time-dependent Ginzburg-Landau theory. Phys. Rev. Lett. 1993, 71: 3202-3205.

[18] Huang S, Peter Takáč. Global smooth solutions of the complex Ginzburg-Landau equation and their Dynamical properties. Discrete and continuous Dynamical Systems., 1999, 5(4): 825-848.

[19] Triebel H. Interpolation theory, function spaces, differential operators. Amsterdam-New York-Oxford: North Holland, 1978.

[20] Tsutsumi M, Hayashi N. Classical solutions of nonlinear Schrödinger equations in higher dimensions. Math. Z., 1981, 177: 217-234.

[21] Wang X, Gao F. Moderate deviations from hydrodynamic limit of a Ginzburg-Landau model. Acta Math Sci., 2006, 26(4): 691-701.

[20] Wen H, Ding S. Vortex dynamics of the anisotropic Ginzburg-Landau equation. Acta Math Sci., 2010, 30(3): 949-962.

Implicit Finite Difference Method for Fractional Percolation Equation with Dirichlet and Fractional Boundary Conditions*

Guo Boling (郭柏灵), Xu Qiang (徐强) and Yin Zhe (尹哲)

Abstract An implicit finite difference method is developed for a one-dimensional fractional percolation equation (FPE) with the Dirichlet and fractional boundary conditions. The stability and convergence are discussed for two special cases, i.e., a continued seepage flow with a monotone percolation coefficient and a seepage flow with the fractional Neumann boundary condition. The accuracy and efficiency of the method are checked with two numerical examples.

Keywords fractional percolation equation; Riemann-Liouville derivative; fractional boundary condition; finite difference method; stability and convergence; Toeplitz matrix

1 Introduction

Fractional differential equations have been attracted considerable attention in recent years, since they have applications in physics [1], geology [2, 3], biology [4], chemistry [5], and even finance [6]. As there are very few cases of fractional differential equations in which the explicit analytical solutions are available, numerical methods become major ways and then have been developed intensively [7–18]. In recent years, some efficient numerical methods have been considered by many authors. Zheng et al. [19] derived a high order space-time spectral method which employs the Jacobi polynomials for the temporal discretization and Fourier-like basis functions for the spatial discretization for the time fractional Fokker-Planck initial-boundary value problem. Shen et al. [20] studied finite difference methods for the nonlinear fractional diffusion/wave diffusion equations with variable order time-fractional derivative operator. Liu et al. [21] developed a novel weighted fractional finite volume method based on the nodal basis functions for the two-sided space fractional diffusion equation. They pointed out that this method can be extended to two-dimensional and three-dimensional problems with complex regions. Feng et al. [22] proposed a fractional finite volume method for the two-sided space fractional diffusion equation based on the nodal basis functions. They proved that the derived numerical scheme is unconditionally stable and convergent. Liu et al. [23] considered a meshless scheme based on radial basis functions for a fractal mobile/immobile transport model. Liu et al. [24] developed two meshless schemes based on the point interpolation method for the one-dimensional space fractional diffusion equation. Feng et al. [25] proposed a Crank-Nicolson scheme for the one-dimensional space fractional diffusion equation with variable coefficient. They proved that the numerical scheme is unconditionally stable and convergent with second-order accuracy. Liu et al. [26] derived a novel alternating direction implicit method for the two-dimensional Riesz space fractional diffusion equation with a nonlinear reaction term. Yang et al. [27] established a finite volume scheme with preconditioned Lanczos method as an attractive and high-efficiency approach for the two-dimensional space fractional reaction-diffusion equations.

The fractional percolation equation, which is a mathematical model of seepage flow problems in groundwater and fluid dynamics in porous media [28–30], is a partial differential equation obtained from the traditional percolation equation [31] by replacing the space integer derivatives by space fractional derivatives. A three-dimensional fractional percolation equation is proposed by He [32] as

$$\frac{1}{v}\frac{\partial p}{\partial t} = \frac{\partial^{\beta_1}}{\partial_+ x^{\beta_1}}\left(K_x \frac{\partial^{\alpha_1} p}{\partial_+ x^{\alpha_1}}\right) + \frac{\partial^{\beta_2}}{\partial_+ y^{\beta_2}}\left(K_y \frac{\partial^{\alpha_2} p}{\partial_+ y^{\alpha_2}}\right) + \frac{\partial^{\beta_3}}{\partial_+ z^{\beta_3}}\left(K_z \frac{\partial^{\alpha_3} p}{\partial_+ z^{\alpha_3}}\right)$$
$$+ f(x,y,z,t), \quad (x,y,z) \in \Omega, \tag{1.1}$$

where $0 < \alpha_i < 1$, $0 < \beta_i \leqslant 1$ for $i = 1, 2, 3$; $p = p(x, y, z, t)$ is the pressure; K_x, K_y, K_z are the percolation coefficients along the x, y and z directions, respectively; $f(x,y,z,t)$ is the source term; v is the velocity, Ω is the percolation domain.

The space fractional partial derivatives in (1.1) are defined in the Riemman-Liouville form. Generally, for any $\mu > 0$, the Riemann-Liouville fractional derivative

$\dfrac{d^\mu u(x)}{d_+ x^\mu}$ of order μ is defined by [33–35]

$$\frac{d^\mu u(x)}{d_+ x^\mu} = \frac{1}{\Gamma(n-\mu)} \frac{d^n}{dx^n} \int_0^x \frac{u(\xi)}{(x-\xi)^{\mu+1-n}} d\xi, \tag{1.2}$$

where n is an integer such that $n-1 < \mu \leqslant n$. If μ is an integer, then (1.2) gives the standard integer derivative.

In the fractional percolation equation (1.1), we denote the rate of fluid mass flux $\mathbf{q} = (q_x, q_y, q_z)$, where

$$q_x = K_x \frac{\partial^{\alpha_1} p}{\partial_+ x^{\alpha_1}}, \quad q_y = K_y \frac{\partial^{\alpha_2} p}{\partial_+ y^{\alpha_2}}, \quad q_z = K_z \frac{\partial^{\alpha_3} p}{\partial_+ z^{\alpha_3}}. \tag{1.3}$$

It should be pointed out that the expression (1.3) derives from the modified Darcy law which is proposed by He [32] and Ochoa-Tapia [36] for describing the movement of solute in a non-homogeneous porous medium. Notice that the continued seepage flow leads to $\beta_i = 1$ ($i = 1, 2, 3$) and equation (1.1) models a rigid body motion while $\beta_i = 0$ ($i = 1, 2, 3$). However, the reality is that the seepage flow is neither continuous nor rigid, hence $0 < \beta_i < 1$ is more general. Under this consideration, He [32] proposed an approximate analytical solution for equation (1.1) by the variational iteration method. In the numerical aspect, Chen et al. [37] developed an implicit finite difference method for the one-dimensional fractional percolation equation, and they [38] considered an alternating direction implicit difference method for the two-dimensional case. The numerical simulation of the two-dimensional variable-order fractional percolation equation is discussed [39]. Liu et al. [40] proposed two finite difference methods for the three-dimensional non-continued seepage flow problem with constant percolation coefficients and continued seepage flow problem with variable percolation coefficients, respectively.

In Refs. [37–40], the fractional percolation equations are studied with Dirichlet boundary conditions. As we all know, in the seepage flow problems, the Robin boundary condition

$$(\gamma p + \mathbf{q} \cdot \mathbf{n})|_{\partial \Omega} = v_r(x, y, z, t)$$

represents the linear replenishment or percolation of the fluid mass flux on the boundary $\partial \Omega$, where $\gamma > 0$ and $\mathbf{n}$ is the outward unit normal; and the Neumann boundary condition

$$\mathbf{q} \cdot \mathbf{n}|_{\partial \Omega} = v_r(x, y, z, t)$$

is used to describe the variation of the fluid mass flux **q** on $\partial\Omega$. Notice the modified Darcy law (1.3), if we consider the corresponding Robin or Neumann boundary condition for the fractional percolation equation (1.1), then the fractional derivative of pressure p are employed in the boundary expressions which may be characterized as fractional boundary conditions.

Recently, Jia and Wang [41] considered the space fractional diffusion equations with fractional derivative boundary conditions and developed fast implicit finite difference methods for both steady-state and time-dependent problems. To our knowledge, study on numerical computation of fractional percolation equations with fractional derivative boundary conditions is still limited. This motivates us to examine an implicit finite difference method for the one-dimensional fractional percolation equation with fractional Robin/Neumann boundary conditions in this paper.

The rest of the paper is organized as follows. In Section 2 we construct an implicit finite difference method and analyze its consistency. In Section 3 we study the unconditionally stability and convergence of the method in two cases. In Section 4 we carry out numerical experiments to verify the accuracy and efficiency of the proposed scheme. Finally, we draw our conclusions in Section 5.

2 Implicit finite difference method and its consistency

In this section, we discuss the following simplified one-dimensional fractional percolation problem with left-sided mixed fractional spatial derivative:

$$\frac{\partial p(x,t)}{\partial t} = \frac{\partial^\beta}{\partial_+ x^\beta}\left(k(x)\frac{\partial^\alpha p(x,t)}{\partial_+ x^\alpha}\right) + f(x,t), \quad 0 < x < L, \quad 0 < t \leqslant T, \qquad (2.1)$$

$$p(0,t) = 0, \quad \gamma p(L,t) + \left(k(x)\frac{\partial^\alpha p(x,t)}{\partial_+ x^\alpha}\right)\bigg|_{x=L} = v(t), \quad 0 < t \leqslant T, \qquad (2.2)$$

$$p(x,0) = p_0(x), \quad 0 \leqslant x \leqslant L, \qquad (2.3)$$

where $0 < \alpha < 1$, $0 < \beta \leqslant 1$, $0 < k_{\min} \leqslant k(x) \leqslant k_{\max}$. $\gamma = 0$ corresponds to a fractional Neumann boundary condition and $\gamma > 0$ corresponds to a fractional Robin boundary condition.

Let $h = L/N$ and $\Delta t = T/M$ be the spatial mesh-width and time step, with N and M being positive integers. We define a spatial and temporal partition $x_i = ih$ for $i = 0, 1, \cdots, N$ and $t_m = m\Delta t$ for $m = 0, 1, \cdots, M$. Let $k_i = k(x_i)$, $f_i^m = f(x_i, t_m)$, $v^m = v(t_m)$. Denote the exact and numerical solutions at the mesh point (x_i, t_m)

by $P_i^m = p(x_i, t_m)$ and p_i^m, respectively. The initial and the Dirichlet boundary conditions are set by $p_i^0 = p_0(x_i)$ and $p_0^m = 0$, respectively.

To approximate the left-sided mixed fractional spatial derivative, Chen [27] defined the operator $\Delta_{h,q}^{\alpha,\beta}$ as

$$\Delta_{h,q}^{\alpha,\beta}(k,p)(x,t) = \frac{1}{h^{\alpha+\beta}} \sum_{l=0}^{\infty} \sum_{j=0}^{\infty} g_l^{(\beta)} g_j^{(\alpha)} k(x-lh) p(x-(j+l-q)h, t),$$

where q is a nonnegative integer and $g_j^{(\alpha)} = (-1)^j \binom{\alpha}{j}$ with $\binom{\alpha}{j}$ being the fractional binomial coefficients. It is proved that [27]

$$\frac{\partial^\beta}{\partial_+ x^\beta}\left(k(x)\frac{\partial^\alpha p(x,t)}{\partial_+ x^\alpha}\right) = \Delta_{h,q}^{\alpha,\beta}(k,p)(x,t) + O(h),$$

uniformly in $x \in (0, L)$ as $h \to 0$ under the assumption of the homogenous Dirichlet boundary condition on $x = 0$. Thus, for $q = 1$, the left-sided mixed fractional spatial derivative in equation (2.1) can be approximated as

$$\frac{\partial^\beta}{\partial_+ x^\beta}\left(k(x)\frac{\partial^\alpha p(x,t)}{\partial_+ x^\alpha}\right)\bigg|_{(x_i,t_m)}$$
$$= \frac{1}{h^{\alpha+\beta}} \sum_{l=0}^{i} g_l^{(\beta)} k_{i-l} \sum_{j=0}^{i-l} g_j^{(\alpha)} P_{i+1-j-l}^m + O(h), \qquad (2.4)$$

for $1 \leq i \leq N-1$, $1 \leq m \leq M$. Using the standard left Grünwald-Letnikov fractional derivative cite10 to approximate the left-sided fractional derivative of order α in the fractional boundary condition on $x = L$, we have

$$\frac{\partial^\alpha p(x,t)}{\partial_+ x^\alpha}\bigg|_{(L,t_m)} = \frac{1}{h^\alpha} \sum_{j=0}^{N-1} g_j^{(\alpha)} P_{N-j}^m + O(h), \qquad (2.5)$$

for $1 \leq m \leq M$. We employ the backward Euler scheme to approximate the first order time derivative

$$\frac{\partial p}{\partial t}\bigg|_{(x_i,t_m)} = \frac{P_i^m - P_i^{m-1}}{\Delta t} + O(\Delta t), \qquad (2.6)$$

and present an implicit finite difference scheme as follow:

$$\frac{p_i^m - p_i^{m-1}}{\Delta t} = \frac{1}{h^{\alpha+\beta}} \sum_{l=0}^{i} g_l^{(\beta)} k_{i-l} \sum_{j=0}^{i-l} g_j^{(\alpha)} p_{i+1-j-l}^m + f_i^m, \quad 1 \leq i \leq N-1, \qquad (2.7)$$

$$p_0^m = 0, \quad \gamma p_N^m + \frac{k_N}{h^\alpha} \sum_{j=0}^{N-1} g_j^{(\alpha)} p_{N-j}^m = v^m. \tag{2.8}$$

Denote the local truncation error by R_i^m for $1 \leqslant i \leqslant N$, thanks to (2.4)–(2.6), we have that

$$R_i^m = \frac{P_i^m - P_i^{m-1}}{\Delta t} - \frac{1}{h^{\alpha+\beta}} \sum_{l=0}^{i} g_l^{(\beta)} k_{i-l} \sum_{j=0}^{i-l} g_j^{(\alpha)} P_{i+1-j-l}^m - f_i^m$$
$$= O(\Delta t) + O(h), \quad 1 \leqslant i \leqslant N-1, \tag{2.9}$$

and

$$R_N^m = \gamma P_N^m + \frac{k_N}{h^\alpha} \sum_{j=0}^{N-1} g_j^{(\alpha)} P_{N-j}^m - v^m = O(h). \tag{2.10}$$

This implies the consistency of the implicit finite difference scheme above both on the interior nodes and the right boundary node.

3 Stability and convergence analysis

In this section, we discuss the stability and convergence of the numerical method under two special cases: continued seepage flow ($\beta = 1$) with monotone percolation coefficient $k(x)$ and seepage flow with fractional Neumann boundary condition ($\gamma = 0$).

Let $r = \Delta t / h^{\alpha+\beta}$ be the mesh ratio, we rewrite (2.7) by exchanging the order of summation as follow

$$p_i^m - r \sum_{j=1}^{i+1} \sum_{l=0}^{i+1-j} g_l^{(\beta)} g_{i+1-j-l}^{(\alpha)} k_{i-l} p_j^m = p_i^{m-1} + \Delta t f_i^m, \quad 1 \leqslant i \leqslant N-1.$$

Denote column vectors
$$\mathbf{p}^m = (p_1^m, p_2^m, \cdots, p_N^m)^{\mathrm{T}},$$
$$\mathbf{q}^{m-1} = \left(p_1^{m-1}, p_2^{m-1}, \cdots, p_{N-1}^{m-1}, 0\right)^{\mathrm{T}}$$

and
$$\mathbf{f}^m = \left(\Delta t f_1^m, \Delta t f_2^m, \cdots, \Delta t f_{N-1}^m, h^\alpha v^m\right)^{\mathrm{T}},$$

for $1 \leqslant m \leqslant M$, then the matrix form of the implicit finite difference scheme (2.7)–(2.8) can be expressed as

$$\mathbf{A}\mathbf{p}^m = \mathbf{q}^{m-1} + \mathbf{f}^m, \quad 1 \leqslant m \leqslant M, \tag{3.1}$$

where the coefficient matrix $\mathbf{A} = [a_{i,j}]_{i,j=1}^{N}$ and its entries are

$$a_{i,j} = \begin{cases} -r\sum_{l=0}^{i+1-j} g_l^{(\beta)} g_{i+1-j-l}^{(\alpha)} k_{i-l}, & \text{for } 1 \leqslant j \leqslant i-1,\ 1 \leqslant i \leqslant N-1; \\ 1 - r\left(g_0^{(\beta)} g_1^{(\alpha)} k_i + g_1^{(\beta)} g_0^{(\alpha)} k_{i-1}\right), & \text{for } 1 \leqslant j = i \leqslant N-1; \\ -r g_0^{(\beta)} g_0^{(\alpha)} k_i, & \text{for } j = i+1,\ 1 \leqslant i \leqslant N-1; \\ 0, & \text{for } i+2 \leqslant j \leqslant N,\ 1 \leqslant i \leqslant N-2; \\ g_{N-j}^{(\alpha)} k_N, & \text{for } 1 \leqslant j \leqslant N-1,\ i = N; \\ h^{\alpha}\gamma + g_0^{(\alpha)} k_N, & \text{for } i = j = N. \end{cases}$$ (3.2)

Some properties of the alternating fractional binomial coefficients $g_j^{(\alpha)}$ and $g_l^{(\beta)}$ in (3.2) are listed in the following lemma [34].

Lemma 3.1 *Let μ, μ_1, μ_2 be positive real numbers and integer $n \geqslant 1$. We have*

(1) $g_0^{(\mu)} = 1$, $g_j^{(\mu)} = \left(1 - \dfrac{\mu+1}{j}\right) g_{j-1}^{(\mu)}$ *for* $j \geqslant 1$.

(2) $g_1^{(\mu)} < g_2^{(\mu)} < \cdots < 0$, $\sum_{j=0}^{n} g_j^{(\mu)} > 0$ *for* $0 < \mu < 1$.

(3) $g_2^{(\mu)} > g_3^{(\mu)} > \cdots > 0$, $\sum_{j=0}^{n} g_j^{(\mu)} < 0$ *for* $1 < \mu < 2$.

(4) $\sum_{j=0}^{n} g_j^{(\mu)} = (-1)^n \binom{\mu-1}{n}$.

(5) $\sum_{j=0}^{n} g_j^{(\mu_1)} g_{n-j}^{(\mu_2)} = g_n^{(\mu_1+\mu_2)}$.

To discuss the stability of the numerical method, we denote by $\tilde{p}_i^m$ ($1 \leqslant i \leqslant N$, $1 \leqslant m \leqslant M$) the approximate solution of the difference scheme with the initial condition $\tilde{p}_i^0$ ($1 \leqslant i \leqslant N$), and define $\varepsilon_i^m = p_i^m - \tilde{p}_i^m$, $\varepsilon^m = (\varepsilon_1^m, \varepsilon_2^m, \cdots, \varepsilon_N^m)^{\mathrm{T}}$, $\|\varepsilon^m\|_{\infty} = \max\limits_{1 \leqslant i \leqslant N} |\varepsilon_i^m|$. For the convergence analysis, we define $e_i^m = P_i^m - p_i^m$ ($1 \leqslant i \leqslant N$, $0 \leqslant m \leqslant M$) with $\mathbf{e}^m = (e_1^m, e_2^m, \cdots, e_N^m)^{\mathrm{T}}$ and $\|\mathbf{e}^m\|_{\infty} = \max\limits_{1 \leqslant i \leqslant N} |e_i^m|$. From the definition of the finite difference scheme, we have

$$\varepsilon_i^m - r \sum_{j=1}^{i+1} \sum_{l=0}^{i+1-j} g_l^{(\beta)} g_{i+1-j-l}^{(\alpha)} k_{i-l} \varepsilon_j^m = \varepsilon_i^{m-1}, \quad 1 \leqslant i \leqslant N-1, \tag{3.3}$$

$$\left(h^{\alpha}\gamma + g_0^{(\alpha)} k_N\right) \varepsilon_N^m + k_N \sum_{j=1}^{N-1} g_j^{(\alpha)} \varepsilon_{N-j}^m = 0, \tag{3.4}$$

with $\varepsilon_0^m = 0$, $m = 1, 2, \cdots, M$. Recalling the local truncation error (2.9) and (2.10), we have

$$e_i^m - r \sum_{j=1}^{i+1} \sum_{l=0}^{i+1-j} g_l^{(\beta)} g_{i+1-j-l}^{(\alpha)} k_{i-l} e_j^m = e_i^{m-1} + \Delta t R_i^m, \quad 1 \leqslant i \leqslant N-1, \quad (3.5)$$

$$\left(h^\alpha \gamma + g_0^{(\alpha)} k_N\right) e_N^m + k_N \sum_{j=1}^{N-1} g_j^{(\alpha)} e_{N-j}^m = h^\alpha R_N^m, \quad (3.6)$$

with $e_0^m = 0$, $m = 1, 2, \cdots, M$ and $e_i^0 = 0$, $i = 0, 1, \cdots, N$.

Now we give the following stability and convergence theorem under the assumption of the continuity of seepage flow ($\beta = 1$).

Theorem 3.1 *If $\beta = 1$, $\gamma \geqslant 0$, $k(x)$ decreases monotonically with respect to x in $[0, L]$, then the implicit finite difference scheme (3.1) is unconditionally stable, and there exists a positive constant C independent of Δt and h, such that*

$$\|\mathbf{e}^m\|_\infty \leqslant C(\Delta t + h), \quad 1 \leqslant m \leqslant M. \quad (3.7)$$

Proof For $\beta = 1$, we have $g_0^{(\beta)} = 1$, $g_1^{(\beta)} = -1$ and $g_j^{(\beta)} = 0$ ($j \geqslant 2$). The entries in the first $(N-1)$ rows of the matrix $\mathbf{A}$ can be rewritten as

$$a_{i,j} = \begin{cases} -r\left(g_{i+1-j}^{(\alpha)} k_i - g_{i-j}^{(\alpha)} k_{i-1}\right), & \text{for } 1 \leqslant j \leqslant i-1; \\ 1 - r\left(g_1^{(\alpha)} k_i - g_0^{(\alpha)} k_{i-1}\right), & \text{for } j = i; \\ -r g_0^{(\alpha)} k_i, & \text{for } j = i+1; \\ 0, & \text{for } i+2 \leqslant j \leqslant N, \end{cases} \quad (3.8)$$

where $1 \leqslant i \leqslant N-1$. By Lemma 3.1, we have

$$\sum_{j=1}^{N-1} \left|g_j^{(\alpha)}\right| = -\sum_{j=1}^{N-1} g_j^{(\alpha)} < g_0^{(\alpha)}. \quad (3.9)$$

Taking the monotonicity of $k(x)$ into consideration, for $1 \leqslant i \leqslant N-1$, we get

$$a_{i,j} \leqslant 0, \quad j \neq i, \quad (3.10)$$

and

$$\sum_{j=1}^{N} a_{i,j} = 1 - r\left(\sum_{j=0}^{i} g_j^{(\alpha)} k_i - \sum_{j=0}^{i-1} g_j^{(\alpha)} k_{i-1}\right) > 1. \quad (3.11)$$

It follows from (3.4) that

$$|\varepsilon_N^m| = \frac{\left|k_N \sum_{j=1}^{N-1} g_j^{(\alpha)} \varepsilon_{N-j}^m\right|}{\left|h^\alpha \gamma + g_0^{(\alpha)} k_N\right|} \leqslant \frac{k_N \sum_{j=1}^{N-1} \left|g_j^{(\alpha)}\right| \left|\varepsilon_{N-j}^m\right|}{h^\alpha \gamma + g_0^{(\alpha)} k_N} < \max_{1\leqslant i\leqslant N-1} |\varepsilon_i^m|. \qquad (3.12)$$

Therefore, we suppose that $\|\varepsilon^m\|_\infty = |\varepsilon_{i_0}^m|$ where $1 \leqslant i_0 \leqslant N-1$, then

$$\|\varepsilon^m\|_\infty = |\varepsilon_{i_0}^m| < \sum_{j=1}^{N} a_{i_0,j} |\varepsilon_{i_0}^m| \leqslant |a_{i_0,i_0} \varepsilon_{i_0}^m| - \sum_{j=1,j\neq i_0}^{N} |a_{i_0,j}||\varepsilon_j^m|$$

$$\leqslant |a_{i_0,i_0} \varepsilon_{i_0}^m| - \left|\sum_{j=1,j\neq i_0}^{N} a_{i_0,j} \varepsilon_j^m\right| \leqslant \left|\sum_{j=1}^{N} a_{i_0,j} \varepsilon_j^m\right| = |\varepsilon_{i_0}^{m-1}| \leqslant \|\varepsilon^{m-1}\|_\infty, \qquad (3.13)$$

which implies that $\|\varepsilon^m\|_\infty < \|\varepsilon^0\|_\infty$ for $1 \leqslant m \leqslant M$. Thus the numerical method defined by (3.1) is unconditionally stable with respect to the initial data.

For $1 \leqslant m \leqslant M$, if $\|\mathbf{e}^m\|_\infty = |e_N^m| \geqslant \max_{1\leqslant i\leqslant N-1} |e_i^m|$, it follows from (3.6) that

$$\left(h^\alpha \gamma + k_N \sum_{j=0}^{N-1} g_j^{(\alpha)}\right) |e_N^m|$$

$$= \left(h^\alpha \gamma + g_0^{(\alpha)} k_N\right) |e_N^m| - k_N \sum_{j=1}^{N-1} \left|g_j^{(\alpha)}\right| |e_N^m|$$

$$\leqslant \left(h^\alpha \gamma + g_0^{(\alpha)} k_N\right) |e_N^m| - k_N \sum_{j=1}^{N-1} \left|g_j^{(\alpha)}\right| |e_{N-j}^m|$$

$$\leqslant \left(h^\alpha \gamma + g_0^{(\alpha)} k_N\right) |e_N^m| - \left|k_N \sum_{j=1}^{N-1} g_j^{(\alpha)} e_{N-j}^m\right|$$

$$\leqslant \left|\left(h^\alpha \gamma + g_0^{(\alpha)} k_N\right) e_N^m + k_N \sum_{j=1}^{N-1} g_{N-j}^{(\alpha)} e_j^m\right|$$

$$= h^\alpha |R_N^m|. \qquad (3.14)$$

Thanks to Lemma 3.1 and Stirling formula for the Gamma function [25], we have

$$\left(\sum_{j=0}^{N-1} g_j^{(\alpha)}\right)^{-1} = (-1)^{N-1} \binom{\alpha-1}{N-1}^{-1} = \frac{\Gamma(N)}{\Gamma(1-\alpha)\Gamma(N-\alpha)} = O(N^\alpha), \qquad (3.15)$$

as $N \to \infty$. Combining (3.14) and (3.15), we get

$$|e_N^m| \leqslant \left(k_N \sum_{j=0}^{N-1} g_j^{(\alpha)}\right)^{-1} h^\alpha |R_N^m| \leqslant C_1 h. \tag{3.16}$$

If $\|\mathbf{e}^m\|_\infty = |e_{i_0}^m| > |e_N^m|$ for $1 \leqslant i_0 \leqslant N-1$, then

$$|e_{i_0}^m| < \sum_{j=1}^{N} a_{i_0,j}|e_{i_0}^m| \leqslant |a_{i_0,i_0} e_{i_0}^m| - \sum_{j=1, j\neq i_0}^{N} |a_{i_0,j}||e_j^m|$$

$$\leqslant |a_{i_0,i_0} e_{i_0}^m| - \left|\sum_{j=1, j\neq i_0}^{N} a_{i_0,j} e_j^m\right| \leqslant \left|\sum_{j=1}^{N} a_{i_0,j} e_j^m\right|$$

$$= |e_{i_0}^{m-1} + \Delta t R_{i_0}^m| \leqslant \max_{1 \leqslant i \leqslant N-1} |e_i^{m-1}| + \Delta t C_2(\Delta t + h). \tag{3.17}$$

By induction, we can finally obtain

$$\|\mathbf{e}^m\|_\infty \leqslant C_1 h + m\Delta t C_2(\Delta t + h) \leqslant C(\Delta t + h). \tag{3.18}$$

This completes the proof. □

Before proceeding to the case with the fractional Neumann boundary condition ($\gamma = 0$), let us consider a equivalent form of the finite difference scheme (3.1). We express the coefficient matrix $\mathbf{A}$ defined by (3.2) in a block form

$$\mathbf{A} = \begin{bmatrix} \mathbf{A}_{N-1,N-1} & \mathbf{A}_{N-1,N} \\ \mathbf{A}_{N,N-1}^{\mathrm{T}} & a_{N,N} \end{bmatrix}, \tag{3.19}$$

where $\mathbf{A}_{N-1,N-1}$ is the $(N-1)$-by-$(N-1)$ matrix that consists of the first $N-1$ rows and the first $N-1$ columns of $\mathbf{A}$, $\mathbf{A}_{N-1,N}$ is an $(N-1)$-dimensional column vector that consists of the first $N-1$ entries in the N-th column of $\mathbf{A}$, and $\mathbf{A}_{N,N-1}^{\mathrm{T}}$ is an $(N-1)$-dimensional column vector that consists of the first $N-1$ entries in the N-th row of $\mathbf{A}$. It is simple to show that the matrix $\mathbf{A}_{N-1,N-1}$ has the following decomposition

$$\mathbf{A}_{N-1,N-1} = \mathbf{I} - r\left(\mathbf{A}^{(\beta)} \mathrm{diag}(\mathbf{k}) \mathbf{A}^{(\alpha)} + k_{N-1}\mathbf{c}_{N-1}(\mathbf{g}^{(\alpha)})^{\mathrm{T}}\right), \tag{3.20}$$

where $\mathbf{I}$ is the identity matrix of order $N-1$, $\mathbf{c}_{N-1}$ is the $(N-1)$-th column of $\mathbf{I}$, $\mathbf{k} = (k_0, k_1, \cdots, k_{N-2})^{\mathrm{T}}$, $\mathbf{g}^{(\alpha)} = (g_{N-1}^{(\alpha)}, g_{N-2}^{(\alpha)}, \cdots, g_1^{(\alpha)})^{\mathrm{T}}$, $\mathbf{A}^{(\alpha)}$ and $\mathbf{A}^{(\beta)}$ are $(N-1)$-

by-$(N-1)$ Toeplitz matrices [42] of the form

$$\mathbf{A}^{(\alpha)} = \begin{bmatrix} g_0^{(\alpha)} & 0 & 0 & \cdots & \cdots & 0 \\ g_1^{(\alpha)} & g_0^{(\alpha)} & 0 & \cdots & \cdots & 0 \\ g_2^{(\alpha)} & g_1^{(\alpha)} & g_0^{(\alpha)} & \cdots & \cdots & 0 \\ \vdots & \vdots & \vdots & & & \vdots \\ g_{N-3}^{(\alpha)} & g_{N-4}^{(\alpha)} & g_{N-5}^{(\alpha)} & \cdots & & 0 \\ g_{N-2}^{(\alpha)} & g_{N-3}^{(\alpha)} & g_{N-4}^{(\alpha)} & \cdots & g_1^{(\alpha)} & g_0^{(\alpha)} \end{bmatrix}, \tag{3.21}$$

$$\mathbf{A}^{(\beta)} = \begin{bmatrix} g_1^{(\beta)} & g_0^{(\beta)} & 0 & \cdots & 0 & 0 \\ g_2^{(\beta)} & g_1^{(\beta)} & g_0^{(\beta)} & \cdots & & 0 \\ g_3^{(\beta)} & g_2^{(\beta)} & g_1^{(\beta)} & \cdots & & 0 \\ \vdots & \vdots & \vdots & & & \vdots \\ g_{N-2}^{(\beta)} & g_{N-3}^{(\beta)} & g_{N-4}^{(\beta)} & \cdots & g_1^{(\beta)} & g_0^{(\beta)} \\ g_{N-1}^{(\beta)} & g_{N-2}^{(\beta)} & g_{N-3}^{(\beta)} & \cdots & g_2^{(\beta)} & g_1^{(\beta)} \end{bmatrix}. \tag{3.22}$$

According to the definition (3.2), we can write $\mathbf{A}_{N-1,N}$ and $\mathbf{A}_{N,N-1}^{\mathrm{T}}$ as

$$\mathbf{A}_{N-1,N} = -rk_{N-1}g_0^{(\alpha)}\mathbf{c}_{N-1}, \tag{3.23}$$

$$\mathbf{A}_{N,N-1}^{\mathrm{T}} = k_N(\mathbf{g}^{(\alpha)})^{\mathrm{T}}. \tag{3.24}$$

Let $\omega_N = rk_{N-1}/k_N$. Notice that $a_{N,N} = h^\alpha \gamma + k_N g_0^{(\alpha)}$. Adding ω_N times the N-th equation to the $(N-1)$-th equation in the linear system (3.1), we get a equivalent form of (3.1)

$$\bar{\mathbf{A}}\mathbf{p}^m = \mathbf{q}^{m-1} + \bar{\mathbf{f}}^m, \quad 1 \leqslant m \leqslant M, \tag{3.25}$$

where

$$\bar{\mathbf{f}}^m = \left(\Delta t f_1^m, \Delta t f_2^m, \cdots, \Delta t f_{N-2}^m, \Delta t f_{N-1}^m + \omega_N h^\alpha v^m, h^\alpha v^m\right)^{\mathrm{T}}, \tag{3.26}$$

$$\bar{\mathbf{A}} = \begin{bmatrix} \bar{\mathbf{A}}_{N-1,N-1} & \bar{\mathbf{A}}_{N-1,N} \\ \mathbf{A}_{N,N-1}^{\mathrm{T}} & a_{N,N} \end{bmatrix}, \tag{3.27}$$

$$\bar{\mathbf{A}}_{N-1,N-1} = \mathbf{I} - r\left(\mathbf{A}^{(\beta)}\mathrm{diag}(\mathbf{k})\mathbf{A}^{(\alpha)}\right), \tag{3.28}$$

$$\bar{\mathbf{A}}_{N-1,N} = \omega_N h^\alpha \gamma \mathbf{c}_{N-1}. \tag{3.29}$$

For the stability of the numerical method under the assumption that $\gamma = 0$, we have the following theorem.

Lemma 3.2 *If $1 < \alpha + \beta < 2$, $\gamma = 0$, then the implicit finite difference scheme (3.1) is unconditionally stable.*

Proof Let $\zeta^m = (\varepsilon_1^m, \varepsilon_2^m, \cdots, \varepsilon_{N-1}^m)^{\mathrm{T}}$ for $1 \leqslant m \leqslant M$. Notice that $\gamma = 0$ leads to $\bar{\mathbf{A}}_{N-1,N} = \mathbf{0}$. Then we can write equations (3.3) and (3.4) into the matrix form

$$\bar{\mathbf{A}}_{N-1,N-1}\zeta^m = \zeta^{m-1}, \tag{3.30}$$

$$\mathbf{A}_{N,N-1}^{\mathrm{T}}\zeta^m + a_{N,N}\varepsilon_N^m = 0. \tag{3.31}$$

Similar to the inequality (3.12) in Theorem 3.1, we have from (3.31) that

$$|\varepsilon_N^m| < \max_{1 \leqslant i \leqslant N-1} |\varepsilon_i^m|. \tag{3.32}$$

This means $\|\varepsilon^m\|_\infty = \max_{1 \leqslant i \leqslant N-1} |\varepsilon_i^m| = \|\zeta^m\|_\infty$. Since $\zeta^m = \left(\bar{\mathbf{A}}_{N-1,N-1}\right)^{-m} \zeta^0$, it suffices to prove that the spectral radius of the matrix $\left(\bar{\mathbf{A}}_{N-1,N-1}\right)^{-1}$ is less that one.

Our first goal is to show that the eigenvalues of the matrix $\mathbf{A}^{(\alpha)}\mathbf{A}^{(\beta)}\mathrm{diag}(\mathbf{k})$ have negative real parts. Denote $\mathbf{A}^{(\alpha)}\mathbf{A}^{(\beta)}\mathrm{diag}(\mathbf{k}) = [b_{i,j}]_{i,j=1}^{N-1}$. It follows from Lemma 3.1 that the entries $b_{i,j}$ have the form

$$b_{i,j} = \begin{cases} \left(g_i^{(\alpha+\beta)} - g_i^{(\alpha)}\right) k_0, & \text{for } 1 \leqslant i \leqslant N-1, j = 1; \\ 0, & \text{for } 1 \leqslant i \leqslant j-2, 2 \leqslant j \leqslant N-1; \\ g_{i+1-j}^{(\alpha+\beta)} k_{j-1}, & \text{for } j-1 \leqslant i \leqslant N-1, 2 \leqslant j \leqslant N-1. \end{cases} \tag{3.33}$$

According to the Gerschgorin theorem [43], the eigenvalues of matrix

$$\mathbf{A}^{(\alpha)}\mathbf{A}^{(\beta)}\mathrm{diag}(\mathbf{k})$$

lie in the union of disks centered at $b_{j,j}$ with radius $\sum_{i=1}^{N-1} |b_{i,j}|$. We observe from Lemma 3.1 that $b_{j,j} < 0$ and $b_{i,j} \geqslant 0$ for all $i \neq j$. For $2 \leqslant j \leqslant N-1$, we have

$$b_{j,j} + \sum_{i=1,i\neq j}^{N-1} |b_{i,j}| = k_{j-1} \sum_{i=j-1}^{N-1} g_{i+1-j}^{(\alpha+\beta)} = k_{j-1} \sum_{l=0}^{N-j} g_l^{(\alpha+\beta)} < 0. \tag{3.34}$$

For $j = 1$, we compute

$$b_{1,1} + \sum_{i=2}^{N-1} |b_{i,1}| = k_0 \sum_{i=1}^{N-1} \left(g_i^{(\alpha+\beta)} - g_i^{(\alpha)}\right)$$

$$= k_0 \left(\sum_{i=0}^{N-1} g_i^{(\alpha+\beta)} - \sum_{i=0}^{N-1} g_i^{(\alpha)}\right) < 0. \tag{3.35}$$

The two inequalities (3.34) and (3.35) imply that all of the Greschgorin disks of matrix $\mathbf{A}^{(\alpha)}\mathbf{A}^{(\beta)}\mathrm{diag}(\mathbf{k})$ are within the left half of the complex plane. Thus, the eigenvalues of the matrix $\mathbf{A}^{(\alpha)}\mathbf{A}^{(\beta)}\mathrm{diag}(\mathbf{k})$ have negative real parts.

Next, by the fact that

$$\bar{\mathbf{A}}_{N-1,N-1} = \left(\mathbf{A}^{(\alpha)}\right)^{-1}\left(\mathbf{I} - r\left(\mathbf{A}^{(\alpha)}\mathbf{A}^{(\beta)}\mathrm{diag}(\mathbf{k})\right)\right)\mathbf{A}^{(\alpha)}, \qquad (3.36)$$

it follows that $\bar{\mathbf{A}}_{N-1,N-1}$ has the same eigenvalues with $\mathbf{I} - r\left(\mathbf{A}^{(\alpha)}\mathbf{A}^{(\beta)}\mathrm{diag}(\mathbf{k})\right)$. Thus each eigenvalue of $\bar{\mathbf{A}}_{N-1,N-1}$ has a magnitude larger that 1. Therefore, the spectral radius of the matrix $\left(\bar{\mathbf{A}}_{N-1,N-1}\right)^{-1}$ is less than 1. This completes the proof. □

Remark 3.1 *Under the assumption of Theorem 3.2, although a convergence estimate such as (3.7) is not given, some numerical experiments in Section 4 show that the implicit finite difference scheme defined by (3.1) is also temporally and spatially first-order accurate.*

4 Numerical examples

We present two numerical examples for solving initial-boundary value problem of the fractional percolation equations (2.1)–(2.3) by the implicit finite difference method given in Section 2. Denote $\|e_h^M\|_\infty$ the maximum error of the numerical solution at time $t = T$ for $\Delta t = h$. Let $G_h = \log_2\left(\|e_{2h}^M\|_\infty / \|e_h^{2M}\|_\infty\right)$ reflect the convergence order of the method. All numerical experiments are run in MATLAB on a Thinkpad E420 laptop with configuration: Intel(R) Core(TM) i3-2350M CPU 2.30GHZ and 4GB RAM.

Example 4.1 We solve the initial-boundary problem (2.1) with the following data. The spatial domain is $[0, L] = [0, 1]$ and the time interval is $[0, T] = [0, 1]$. The percolation coefficient is

$$k(x) = 2 - x^2.$$

It is obvious that $k(x)$ decreases monotonically with respect to x in $[0, 1]$. The exact solution of this problem is

$$p(x, t) = e^{-t}x^2(1-x)^2,$$

which satisfies the homogenous Dirichlet boundary condition on $x = 0$. The initial condition is

$$p_0(x) = x^2(1-x)^2.$$

The source term $f(x,t)$ and function $v(t)$ may be obtained by substituting the exact solution into (2.1) and its fractional Robin boundary condition (2.2) with $\gamma = 1$ by using the formula

$$\frac{\mathrm{d}^\mu}{\mathrm{d}_+ x^\mu} x^\nu = \frac{\Gamma(\nu+1)}{\Gamma(\nu+1-\mu)} x^{\nu-\mu}$$

for the Riemann-Liouville fractional derivative.

Example 4.2 In this example, set the spatial domain $[0, L] = [0, 1]$, the time interval $[0, T] = [0, 1]$ and the percolation coefficient

$$k(x) = -x^2 + x + 1.$$

The fractional boundary condition and the initial condition are set by

$$v(t) = \mathrm{e}^{-t} \left(\frac{\Gamma(3)}{\Gamma(3-\alpha)} + \gamma \right)$$

and

$$p_0(x) = x^2,$$

respectively. The source term is

$$f(x,t) = \mathrm{e}^{-t} \left(\frac{\Gamma(3) x^{2-\alpha-\beta}}{\Gamma(3-\alpha)} \left(\frac{\Gamma(5-\alpha) x^2}{\Gamma(5-\alpha-\beta)} - \frac{\Gamma(4-\alpha) x}{\Gamma(4-\alpha-\beta)} - \frac{\Gamma(3-\alpha)}{\Gamma(3-\alpha-\beta)} \right) - x^2 \right)$$

with $\alpha, \beta \in (0, 1)$. The exact solution of this problem is

$$p(x,t) = \mathrm{e}^{-t} x^2.$$

In our numerical experiments, we consider three different α in each case, respectively. Table 1 shows the maximum errors at time $T = 1$ for Example 4.1 with $\gamma = 1$ and $\beta = 1.0$, it illustrates the stability and $O(\Delta t) + O(h)$ convergence rate of the implicit finite difference scheme (3.1) proved in Theorem 3.1. The error data obtained at time $T = 1$ for Example 2 with $\gamma = 0$ and $\beta = 0.8$ is reported in Table 2, which reflects the stability of the numerical method and supports Theorem 3.2. In Table 3, the error data at time $T = 1$ for Example 2 with $\gamma = 1$ and $\beta = 0.8$ is given. It is observed from Table 2–3 that the numerical method is also temporally and spatially first-order accurate for non-continued seepage flow with fractional Neumann or Robin boundary condition. This implies that the stability and convergence condition in Theorem 3.1 and 3.2 can be weaken.

Table 1 Error behavior for the implicit finite difference scheme (3.11) for Example 4.1 at time $T = 1$ with $\gamma = 1$ and $\beta = 1.0$

$\Delta t = h$	$\alpha = 0.6$		$\alpha = 0.7$		$\alpha = 0.8$	
	$\|e_h^M\|_\infty$	G_h	$\|e_h^M\|_\infty$	G_h	$\|e_h^M\|_\infty$	G_h
2^{-6}	8.8564e-4	—	1.1144e-3	—	1.5178e-3	—
2^{-7}	4.5088e-4	0.973	5.8279e-4	0.935	7.7182e-4	0.975
2^{-8}	2.2753e-4	0.986	2.9408e-4	0.986	3.8922e-4	0.987
2^{-9}	1.1429e-4	0.993	1.4773e-4	0.993	1.9545e-4	0.993
2^{-10}	5.7284e-5	0.996	7.4040e-5	0.996	9.7942e-5	0.996

Table 2 Error behavior for the implicit finite difference scheme (3.11) for Example 4.2 at time $T = 1$ with $\gamma = 0$ and $\beta = 0.8$

$\Delta t = h$	$\alpha = 0.6$		$\alpha = 0.7$		$\alpha = 0.8$	
	$\|e_h^M\|_\infty$	G_h	$\|e_h^M\|_\infty$	G_h	$\|e_h^M\|_\infty$	G_h
2^{-6}	1.1443e-2	—	1.2077e-2	—	1.2846e-2	—
2^{-7}	5.7790e-3	0.985	6.0984e-3	0.985	6.4855e-3	0.986
2^{-8}	2.9040e-3	0.992	3.0643e-3	0.992	3.2584e-3	0.993
2^{-9}	1.4556e-3	0.996	1.5359e-3	0.996	1.6332e-3	0.996
2^{-10}	7.2873e-4	0.998	7.6892e-4	0.998	8.1759e-4	0.998

Table 3 Error behavior for the implicit finite difference scheme (3.11) for Example 4.2 at time $T = 1$ with $\gamma = 1$ and $\beta = 0.8$

$\Delta t = h$	$\alpha = 0.6$		$\alpha = 0.7$		$\alpha = 0.8$	
	$\|e_h^M\|_\infty$	G_h	$\|e_h^M\|_\infty$	G_h	$\|e_h^M\|_\infty$	G_h
2^{-6}	7.0591e-3	—	6.7768e-3	—	6.6690e-3	—
2^{-7}	3.5726e-3	0.982	3.4303e-3	0.982	3.3867e-3	0.977
2^{-8}	1.7972e-3	0.991	1.7257e-3	0.991	1.7038e-3	0.991
2^{-9}	9.0138e-4	0.995	8.6553e-4	0.995	8.5459e-4	0.995
2^{-10}	4.5138e-4	0.997	4.3343e-4	0.997	4.2796e-4	0.997

5   Conclusions

In this paper, an implicit finite difference method is developed for the one-dimensional fractional percolation equation with Dirichlet and fractional boundary conditions. We proved the unconditionally stability and convergence of the method under two cases: continued seepage flow with monotone percolation coefficient and seepage flow with fractional Neumann boundary condition. Numerical experiments illustrate the practicability of the numerical scheme. Numerical methods for the two or three dimensional fractional percolation equations will be considered in our future work.

Acknowledgements The authors would like to express their sincere thanks to the referees for their valuable comments and suggestions which helped to improve the original paper. The authors are also grateful to Dr. Yongqiang Lyu from Shandong Normal University for his helpful suggestions.

References

[1] Sokolov I. M, Klafter J, Blumen A. Fractional kinetics [J]. Physics Today, 55(11) (2002), 48-54.

[2] Benson D. A, Wheatcraft S. W, Meerschaert M. M. Application of a fractional advection-dispersion equation [J]. Water Resources Research, 36(6) (2000), 1403-1412.

[3] Benson D. A, Wheatcraft S. W, Meerschaert M. M. The fractional-order governing equation of Lévy motion [J]. Water Resources Research, 36(6) (2000), 1413-1423.

[4] Magin R. L. Fractional Calculus in Bioengineering [M]. Begell House Publishers, (2006).

[5] Kirchner J. W, Feng X, Neal C. Fractal stream chemistry and its implications for contaminant transport in catchments [J]. Nature, 403(6769) (2000), 524-527.

[6] Raberto M, Scalas E, Mainardi F. Waiting-times and returns in high-frequency financial data: an empirical study [J]. Phys. A, 314(1) (2002), 749-755.

[7] Liu F, Anh V, Turner I. Numerical solution of the space fractional Fokker-Planck Equation [J]. J. Comput. Appl. Math., 166(1) (2004), 209-219.

[8] Ervin V. J, Roop J. P. Variational formulation for the stationary fractional advection dispersion equation [J]. Numer. Methods Partial Differential Equations, 22(3) (2005), 558-576.

[9] Ervin V. J, Heuer N, Roop J. P. Numerical approximation of a time dependent, nonlinear, space-fractional diffusion equation [J]. SIAM J. Numer. Anal., 45(2) (2007), 572-591.

[10] Meerschaert M. M, Tadjeran C. Finite difference approximations for fractional advection-dispersion flow equations [J]. J. Comput. Appl. Math., 172(1) (2004), 65-77.

[11] Meerschaert M. M, Scheffler H. P, Tadjeran C. Finite difference methods for two-dimensional fractional dispersion equation [J]. J. Comput. Phys., 211(1) (2006), 249-261.

[12] Li X, Xu C. Existence and uniqueness of the week solution of the space-time fractional diffusion equation and a spectral method approximation [J]. Commun. Comput. Phys., 8(5) (2010), 1016-1051.

[13] Lin Y, Li X, Xu C. Finite dfiference/specrtal approximations for the fractional cable equation [J]. Math. Comp., 80(275) (2011), 1369-1396.

[14] Chen C. M, Liu F, Burrage K. Finite difference methods and a fourier analysis for the fractional reaction-subdiffusion equation [J]. Appl. Math. Comput., 198(2) (2008),

754-769.

[15] Liu F, Zhuang P, Burrage K. Numerical methods and analysis for a class of fractional advection-dispersion models [J]. Comput.Math. Appl., 64(10) (2012), 2990-3007.

[16] Shen S, Liu F, Anh V, Turner I, Chen J. A characteristic difference method for the variable-order fractional advection-diffusion equation [J]. J. Appl. Math. Comput., 42(1-2) (2013), 371-386.

[17] Liu F, Meerschaert M. M, McGough R. J, Zhuang P, Liu Q. Numerical methods for solving the multi-term time-fractional wave-diffusion equation [J]. Fract. Calc. Appl. Anal., 16(1) (2013), 9-25

[18] Hao Z. P, Sun Z. Z, Cao W. R. A fourth-order approximation of fractional derivatives with its applications [J]. J. Comput. Phys., 281 (2015), 787-805.

[19] Zheng M, Liu F, Turner I, Anh V. A novel high order space-time spectral method for the time fractional Fokker-Planck equation [J]. SIAM J. Sci. Comput., 37(2) (2015), A701-A724.

[20] Shen S, Liu F, Liu Q, Anh V. Numerical simulation of anomalous infiltration in porous media [J]. Numer. Algorithms, 68(3) (2015), 443-454.

[21] Liu F, Zhuang P, Turner I, Burrage K, Anh V. A new fractional finite volume method for solving the fractional diffusion equation [J]. Appl. Math. Model., 38(15) (2014), 3871-3878.

[22] Feng L.B, Zhuang P, Liu F, Turner I. Stability and convergence of a new finite volume method for a two-sided space-fractional diffusion equation [J] Appl. Math. Comput., 257 (2015), 52-65.

[23] Liu Q, Liu F, Turner I, Anh V, Gu Y. A RBF meshless approach for modeling a fractal mobile/immobile transport model [J]. Appl. Math. Comput., 224 (2014), 336-347.

[24] Liu Q, Liu F, Gu Y, Zhuang P, Chen J, Turner I. A meshless method based on point interpolation method (PIM) for the space fractional diffusion equation [J]. Appl. Math. Comput., 256 (2015), 930-938.

[25] Feng L, Zhuang P, Liu F, Turner I, Yang Q. Second-order approximation for the space fractional diffusion equation with variable coefficient [J]. Progr. Fract. Differ. Appl., 1 (2015), 23-35.

[26] Liu F, Chen S, Turner I, Burrage K, Anh V. Numerical simulation for two-dimensional Riesz space fractional diffusion equations with a nonlinear reaction term [J]. Cent. Eur. J. Phys., 11(10) (2013), 1221-1232.

[27] Yang Q, Turner I, Moroney T, Liu F. A finite volume scheme with preconditioned Lanczos method for two-dimensional space-fractional reaction-diffusion equations [J]. Appl. Math. Model., 38(15-16) (2014), 3755-3762.

[28] Petford N, Koenders, M. A. Seepage flow and consolidation in a deforming porous medium [J]. Geophys. Res. Abstracts, 5 (2003), 13329.

[29] Thusyanthan N. I, Madabhushi S. P. G. Scaling of seepage flow velocity in centrifuge

models [J]. Acta Gastroenterologica Latinoamericana, 38(2) (2003), 105-115.

[30] Chou H, Lee B, Chen C. The transient infiltration process for seepage flow from cracks [A]. Western Pacific Meeting, Advances in Subsurface Flow and Transport: Eastern and Western Approaches III [C], Eos, Transactions American Geophysical Union, (2006).

[31] Bear J. Dynamics of fluids in porous media [M]. Elsevier, New York, (1972), 184-186.

[32] He J. H. Approximate analytical solution for seepage flow with fractional derivatives in porous media [J]. Comput.Methods Appl. Mech. Engrg., 167(1) (1998), 57-68.

[33] Miller K. S, Ross B. An Introduction to the Fractional Calculus and Fractional Differential Equations [M]. Wiley, New York, (1993).

[34] Samko S. G, Kilbas A. A, Marichev O. I. Fractional Integrals and Derivatives: Theory and Applications [M]. Gordon Breach, Yverdon, (1993).

[35] Podlubny I, Fractional Differential Equations [M]. Academic Press, New York, (1999).

[36] Ochoa-Tapia J. A, Valdes-Parada F. J, Alvarez-Ramirez J. A fractional-order Darcy's law [J]. Phys. A, 374(1) (2007), 1-14.

[37] Chen S, Liu F, Anh V. A novel implicit finite difference methods for the one dimensional fractional percolation equation [J]. Numer. Algorithms, 56(4) (2011), 517-535.

[38] Chen S, Liu F, Turner I, Anh V. An implicit numerical method for the two-dimensional fractional percolation equation [J]. Appl. Math. Comput., 219(9) (2013), 4322-4331.

[39] Chen S, Liu F, Burrage K. Numerical simulation of a new two-dimensional variable-order fractional percolation equation in non-homogeneous porous media [J]. Comput.Math. Appl., 68(12) (2014), 2133-2141.

[40] Liu Q, Liu F, Turner I, Anh V. Numerical simulation for the 3D seepage flow with fractional derivatives in porous media [J]. IMA J. Appl. Math., 74 (2009), 201-229.

[41] Jia J, Wang H. Fast finite difference methods for space-fractional diffusion equations with fractional derivative boundary conditions [J]. J. Comput. Phys., 293 (2015), 359-369.

[42] Gray R. M. Toeplitz and circulant matrices: a review [J]. Commun. Inform. Theory, 2(3) (2006), 155-239.

[43] Isaacson E, Keller H. B. Analysis of Numerical Methods [J]. Wiley, New York, (1966).

Coupling Model for Unsteady MHD Flow of Generalized Maxwell Fluid with Radiation Thermal Transform*

Liu Yaqing (刘亚轻) and Guo Boling (郭柏灵)

Abstract This paper introduces a new model for Fourier law of heat conduction with time-fractional order to generalized Maxwell fluid. The flow is influenced by magnetic field, radiation heat and heat source. Fractional calculus approach is used to establish the constitutive relationship coupling model of viscoelastic fluid. We used the Laplace transform and solving ordinary differential equations with matrix form to obtain the velocity and temperature in Laplace domain. In order to obtain solutions from Laplace space back to the original space, we have used the numerical inversion of the Laplace transform. According to the results and graphs, some new theory has been constructed. Some comparisons for associated parameters and the corresponding flow and heat transfer characteristics are presented to analyze in detail.

Keywords Maxwell fluid; fractional derivative; radiation heat; heat source; Laplace transform

1 Introduction

Non-Newtonian fluids have been a famous topic of research for their diverse use in many industrial processes, such as polymer solutions, blood and heavy oils. These fluids have been modeled in a number of diverse manners with their constitutive equations varying greatly in complexity, among which the viscoelastic Maxwell fluid model has been studied widely[1-3] . Maxwell fluid has had some success in describing polymeric liquids, it being more amenable to analysis and more importantly experimental.

Fractional calculus has been used successfully for describing the behaviors of viscoelastic fluid[4–5]. The first application of fractional derivatives was given by Abel who applied fractional calculus in the solution of an integral equation. Bagley and Torvik [6] showed that fractional calculus models of viscoelastic material were in harmony with the molecular theory and obtained the fractional differential equation of order 1/2. The relaxation and retardation functions were determined for the four-parameter Maxwell model by Friedrich [7]. Song and Jiang [8] used the fractional calculus to analyze the experiment data of viscoelastic gum and obtained satisfactory results. In general, the constitutive equations for generalized non-Newtonian fluids are modified from the well known fluid models by replacing the time derivative of an integer order with the so called Riemann-Liouville fractional calculus operators. Qi and Xu [9–10] considered Stokes' first problem and some unsteady unidirectional flows for a viscoelastic fluid with the generalized Oldroyd-B model. Fetecau et al.[11–14] investigated some accelerated flows of generalized Oldroyd-B fluid. Khan et al. [15] studied the MHD flow of a generalized Oldroyd-B fluid with modified Darcy's law in a circular pipe. Zheng and Liu[16–17] considered some MHD flow of generalized viscoelastic fluid. Shen et al.[18–19] studied decay of vortex velocity and diffusion of temperature in a generalized second grade fluid and Reyleigh-Stokes problem for a heated generalized second grade fluid with fractional derivative. Recently, some new energy constitutive equation models have been proposed by Ezzat [20] and Qi and Liu [21] by taking the time fractional derivative into account and the associated flow and heat transfer characteristics are analyzed. Xu and Jiang [22] proposed a fractional dual-phase-lag model and the corresponding bioheat transfer equation. The inverse problem of estimating the fractional model parameters was also studied based on the non-linear least square method. Jiang and Qi [23] built a new fractional thermal wave model of the bioheat transfer and used it to examine the heat transfer in biological tissue.

Motivated by the above mentioned works, this paper considers a new model for Fourier law of heat conduction with time-fractional order to generalized Maxwell fluid. The flow induced by a moving plane and influenced by magnetic field, radiation heat and heat source. We used the Laplace transform and solving ordinary differential equations with matrix form to obtain the velocity and temperature in Laplace domain. In order to obtain solutions from Laplace space back to the original space, we have used the numerical inversion of the Laplace transform. Some graphs are presented for various parameters and the corresponding flow and heat transfer characteristics are discussed.

2 Governing equations

The governing equations for an incompressible fluid are given by

$$\nabla \mathbf{V} = 0, \quad \rho \frac{d\mathbf{V}}{dt} = \nabla \mathbf{T} + \rho \mathbf{b}, \tag{2.1}$$

where $\mathbf{T}$ is the Cauchy stress tensor, $\mathbf{V}$ is the velocity vector, ρ is the constant density of the fluid, $\mathbf{b}$ is the body force field.

We consider the fluid is permeated by an imposed magnetic field B_0 which acts in the positive y-coordinate. The Lorentz force has one component in the x-direction is

$$F_x = -\sigma_0 B_0^2 u - \sigma_0 k_0 B_0 \frac{\partial T}{\partial y}, \tag{2.2}$$

where σ_0 is the electrical conductivity, k_0 is the Seebeck coefficient, T is the absolute temperature, which is related with y and t.

Assuming the velocity and shear stress are of the form:

$$\mathbf{V} = u(y,t)\mathbf{i}, \quad \mathbf{S} = \mathbf{S}(y,t), \tag{2.3}$$

where u is the velocity and $\mathbf{i}$ is the unit vectors in the x-direction. Taking account of the initial condition $\mathbf{S}(y,0) = 0$ and in the absence of pressure gradient in the x-direction, the motion equation of the generalized Maxwell fluid with modified Ohm's law is

$$(1 + \lambda D_t^\alpha)\tau(y,t) = \mu \partial_y u(y,t), \tag{2.4}$$

$$(1 + \lambda D_t^\alpha)\frac{\partial u}{\partial t} = \nu \frac{\partial^2 u}{\partial y^2} - \frac{\sigma_0 B_0^2}{\rho}(1 + \lambda D_t^\alpha)u - \frac{\sigma_0 k_0 B_0}{\rho}(1 + \lambda D_t^\alpha)\frac{\partial T}{\partial y}, \tag{2.5}$$

where $\tau(y,t) = S_{xy}(y,t)$ is the shear stress that is different of zero, $\nu = \mu/\rho$ is the kinematic viscosity, λ is material constants, D_t^α is the Riemann-Liouville fractional differential operator of α order with respect to t defined as[4-5]

$$D_t^\alpha f(t) = \frac{1}{\Gamma(1-\alpha)} \frac{d}{dt} \int \frac{f(\tau)}{(t-\tau)^\alpha} d\tau, \quad 0 \leqslant \alpha \leqslant 1, \tag{2.6}$$

where $\Gamma(.)$ is the Gamma function.

The energy equation in terms of the heat conduction vector **q** is

$$\rho c_p \frac{D}{Dt} T = -\nabla \cdot \mathbf{q} + \Phi, \tag{2.7}$$

where c_p is the specific heat of a fluid at constant pressure, Φ is the internal heat and the operator D/Dt defined as

$$\frac{D}{Dt} = \frac{\partial}{\partial t} + (\mathbf{V} \cdot \nabla). \tag{2.8}$$

The generalized Fourier law of heat conduction including the current density effect and the radiation heat effect q_r be considered as

$$(1 + \lambda_r D_t^\beta)\mathbf{q} = -k_T \nabla T + \Pi J - q_r, \tag{2.9}$$

where k_T is the thermal conductivity, λ_r is relaxation time, Π is Peltier coefficient, J is the conduction current density vector.

The generalized energy equation[16,24–25] with time fractional derivative of order $(0 < \beta \leqslant 1)$ is

$$\rho c_p (1 + \lambda_r D_t^\beta)\frac{\partial T}{\partial t} = k_T \frac{\partial^2}{\partial y^2} + \nabla \cdot \Pi J + \nabla q_r. \tag{2.10}$$

Consider the Lorentz force F_x and heat source $Q(y,t)$, the energy equation is

$$\rho c_p (1 + \lambda_r D_t^\beta)\frac{\partial T}{\partial t} = (k_T + \sigma_0 \pi_0 k_0 B_0)\frac{\partial^2 T}{\partial y^2} + \sigma_0 \pi_0 B_0^2 \frac{\partial u}{\partial y} - \frac{\partial q_r}{\partial y} + (1 + \lambda_r D_t^\beta)Q, \tag{2.11}$$

where π_0 is the Peltier coefficient at T_0 (is a reference temperature).

Using the Rosseland approximation for radiation, the radiation heat flux is simplified as[26–27]

$$q_r = -\frac{4\sigma^*}{3k^*}\frac{\partial T^4}{\partial y}, \tag{2.12}$$

where σ^* and k^* are the Stefan-Boltzmann constant and the mean absorption coefficient, respectively. We assume that the temperature difference within the flow is sufficiently small such as that the term T^4 can be expressed as a linear function of temperature. Hence, expanding T^4 in a Taylor series about a free stream temperature T_∞ and neglecting higher-order terms we get

$$T^4 \cong 4T_\infty^3 T - 3T_\infty^4. \tag{2.13}$$

It should be noted that the above radiation transfer pertains to an optically thick model.

In view of Eqs.(2.12) and (2.13), Eq.(2.11) reduces to

$$\rho c_p (1 + \lambda_r D_t^\beta) \frac{\partial T}{\partial t} = k_T \left[1 + \frac{\sigma_0 \pi_0 k_0 B_0}{k_T} + \frac{4}{3N_R} \right] \frac{\partial^2 T}{\partial y^2} + \sigma_0 \pi_0 B_0^2 \frac{\partial u}{\partial y} + (1 + \lambda_r D_t^\beta) Q, \tag{2.14}$$

where $N_R = \dfrac{k^* K_T}{4\sigma^* T_\infty^3}$.

3 Flow due to uniform accelerated plate

Consider an incompressible fractional Maxwell fluid lying over an infinitely extended plate which is situated in the (x, z) plane. Initially, the fluid is at rest and at time $t = 0^+$, the infinite plate begins to slide in its own plane with a uniform acceleration Ut (U is a constant). Let T_w denotes temperature of the plate for $t > 0$, and suppose the temperature of the fluid at the moment $t = 0$ is T_∞ and there is a plane distribution of continuous heat sources located at the plane $y = 0$. By the influence of shear, the fluid above the plate is gradually set in motion. So the governing equations will be the form Eqs. (2.5) and (2.14).

To simplify the algebra, we introduced the following non-dimensional quantities:

$$u^* = \frac{u}{(U\nu)^{\frac{1}{3}}}, \quad y^* = \frac{y(U\nu)^{\frac{1}{3}}}{\nu}, \quad t^* = \frac{t(U\nu)^{\frac{2}{3}}}{\nu}, \quad \lambda^* = \lambda \left(\frac{(U\nu)^{\frac{2}{3}}}{\nu} \right)^\alpha,$$

$$\lambda_r^* = \lambda_r \left(\frac{(U\nu)^{\frac{2}{3}}}{\nu} \right)^\beta, \quad M^* = \frac{\sigma_0 B_0^2 (U\nu)^{\frac{1}{3}}}{\rho U}, \quad K_0^* = \frac{\sigma_0 k_0 B_0 (T_w - T_\infty)}{\rho (U\nu)^{\frac{2}{3}}},$$

$$T^* = \frac{T - T_\infty}{T_w - T_\infty}, \quad P_r^* = \frac{c_p \mu}{k_T}, \quad \Pi_0^* = \frac{\sigma_0 \pi_0 B_0^2 \nu}{(T_w - T_\infty) k_T}, \quad Z^* = \frac{\sigma_0 \pi_0 k_0 B_0}{k_T},$$

$$Q^* = \frac{\nu^2 Q}{k_T (U\nu)^{\frac{2}{3}} (T_w - T_\infty)}. \tag{3.1}$$

Dimensionless governing equations can be given (for brevity the dimensionless mark "*" is omitted here)

$$(1 + \lambda D_t^\alpha) \frac{\partial u}{\partial t} = \frac{\partial^2 u}{\partial y^2} - M(1 + \lambda D_t^\alpha) u - K_0 (1 + \lambda D_t^\alpha) \frac{\partial T}{\partial y}, \tag{3.2}$$

$$P_r (1 + \lambda_r D_t^\beta) \frac{\partial T}{\partial t} = \left[1 + Z + \frac{4}{3N_R} \right] \frac{\partial^2 T}{\partial y^2} + \Pi_0 \frac{\partial u}{\partial y} + (1 + \lambda_r D_t^\beta) Q. \tag{3.3}$$

In there, we consider a heat source of the form $Q = Q_0\delta(y)H(t)$ (Q_0 is a constant). Initially, the dimensionless velocity is $u(y,0) = 0$ and temperature is $T(y,0) = 0$. In order to solve the above problem, we use the Laplace transform principle of sequential fractional derivatives on both side of Eqs.(3.2)-(3.3), and writing the resulting equations in matrix form results in[28–34]

$$\frac{d}{dy}\begin{pmatrix} \overline{\theta}(y,s) \\ \overline{u}(y,s) \\ \overline{\theta}'(y,s) \\ \overline{u}'(y,s) \end{pmatrix} = \begin{pmatrix} 0 & 0 & 1 & 0 \\ 0 & 0 & 0 & 1 \\ C & 0 & 0 & -D \\ 0 & (s+M)(1+\lambda s^\alpha) & K_0(1+\lambda s^\alpha) & 0 \end{pmatrix} \begin{pmatrix} \overline{\theta}(y,s) \\ \overline{u}(y,s) \\ \overline{\theta}'(y,s) \\ \overline{u}'(y,s) \end{pmatrix} - \overline{Q}\gamma \begin{pmatrix} 0 \\ 0 \\ 1 \\ 0 \end{pmatrix}, \qquad (3.4)$$

where s is a transform parameter,

$$C = \frac{P_r S(1+\lambda_r s^\beta)}{1+Z+\dfrac{4}{3N_R}}, \quad D = \frac{\Pi_0}{1+Z+\dfrac{4}{3N_R}}, \quad \gamma = \frac{1+\lambda_r s^\beta}{1+Z+\dfrac{4}{3N_R}}.$$

Eq.(3.4) can be written in constricted form as

$$\overline{\mathbf{G}}'(y,s) = \mathbf{A}(s)\overline{\mathbf{G}}(y,s) + B(y,s) \qquad (3.5)$$

where $\overline{\mathbf{G}}(y,s)$ denotes the state vector in the transform domain whose components consist of the transformed temperature, velocity as well as their gradients.

We shall use the well known Cayley-Hamilton theorem to find the form of the matrix $\exp(\mathbf{A}(s)y)$. The characteristics equation of the matrix $\mathbf{A}(s)$ can be written as

$$k^4 - [(s+M)(1+\lambda s^\alpha) - K_0 D(1+\lambda s^\alpha) + C]k^2 + C(s+M)(1+\lambda s^\alpha) = 0. \qquad (3.6)$$

The roots of this equation $\pm k_1$ and $\pm k_2$, satisfy the relations

$$k_1^2 + k_2^2 = (s+M)(1+\lambda s^\alpha) - K_0 D(1+\lambda s^\alpha) + C, \qquad (3.7)$$

$$k_1^2 \cdot k_2^2 = C(s+M)(1+\lambda s^\alpha). \qquad (3.8)$$

The Taylor series expansion of the matrix exponential in $\exp(\mathbf{A}(s)y)$ has the form

$$\exp(\mathbf{A}(s)y) = \sum_{n=0}^{\infty} \frac{[\mathbf{A}(s)y]^n}{n!}. \qquad (3.9)$$

Using the Cayley-Hamilton theorem, we can express $\mathbf{A}^4$ and all higher power of $\mathbf{A}$ can be expressed in terms of $\mathbf{A}^3$, $\mathbf{A}^2$, $\mathbf{A}$ and $\mathbf{I}$, the unit matrix of order 4. The matrix exponential can now be written in the form

$$\exp(\mathbf{A}(s)y) = a_0(y,s)\mathbf{I} + a_1(y,s)\mathbf{A}(s) + a_2(y,s)\mathbf{A}^2(s) + a_3(y,s)\mathbf{A}^3(s), \quad (3.10)$$

where a_0, a_1, a_2, a_3 are coefficients depending on and s. This leading to the system of equations

$$\exp(\pm k_1 y) = a_0 \pm a_1 k_1 + a_2 k_1^2 \pm a_3 k_1^3, \quad (3.11)$$

$$\exp(\pm k_2 y) = a_0 \pm a_1 k_2 + a_2 k_2^2 \pm a_3 k_2^3. \quad (3.12)$$

The solution of the above system is given by

$$a_0 = \frac{k_1^2 \cosh k_2 y - k_2^2 \cosh k_1 y}{k_1^2 - k_2^2}, \quad a_1 = \frac{k_1^3 \sinh k_2 y - k_2^3 \sinh k_1 y}{k_1 k_2 (k_1^2 - k_2^2)},$$

$$a_2 = \frac{\cosh k_1 y - \cosh k_2 y}{k_1^2 - k_2^2}, \quad a_3 = \frac{k_2 \sinh k_1 y - k_1 \sinh k_2 y}{k_1 k_2 (k_1^2 - k_2^2)}.$$

Hence, we have

$$\exp(\mathbf{A}(s)y) = L(y,s) = [l_{ij}(y,s)], \quad i,j = 1,2,3,4, \quad (3.13)$$

where the elements $l_{ij}(y,s)$ are given in Appendix.

In the actual physical problem the space is divided into two regions accordingly as $y \geqslant 0$ and $y \leqslant 0$. Inside the region $0 \leqslant y < \infty$, the positive exponential terms, not bounded at infinity. Thus for $y \geqslant 0$, we replace each $\sinh ky$ and $\cosh ky$ by $-\exp(-ky)/2$ and $\exp(-ky)/2$ respectively. In the region $y \leqslant 0$ the negative exponential are suppressed instead.

The formal solution of Eqs.(3.5) can be written in the form

$$\overline{\mathbf{G}}(y,s) = \exp(\mathbf{A}(s)y)\left[\overline{\mathbf{G}}(0,s) + \int_0^y \exp(-A(s)z)B(z,s)dz\right], \quad (3.14)$$

where $\overline{\mathbf{G}}(0,s) = \begin{pmatrix} \overline{\theta}(0,s) \\ \overline{u}(0,s) \\ \overline{\theta}'(0,s) \\ \overline{u}'(0,s) \end{pmatrix}$. Using the integral prosperities of the Dirac delta function, we obtain

$$\overline{\mathbf{G}}(y,s) = \exp(\mathbf{A}(s)y)\left[\overline{\mathbf{G}}(0,s) + \xi(s)\right]. \quad (3.15)$$

For the heat source Q located in $y = 0$, the temperature is a symmetric of y while the velocity is anti-symmetric. Then

$$\xi(s) = -\frac{Q_0\gamma}{4s}\begin{pmatrix} \dfrac{k_1k_2 + (s+M)(1+\lambda s^\alpha)}{k_1k_2(k_1+k_2)} \\ 0 \\ 1 \\ \dfrac{K_0(1+\lambda s^\alpha)}{k_1+k_2} \end{pmatrix}.$$

Initially, the plate is at rest, at $t > 0$ the plate moving with velocity t, we have

$$u(0,t) = t \text{ or } \bar{u}(0,s) = \frac{1}{s^2}. \tag{3.16}$$

For the heat source located in $y = 0$ and the temperature is a symmetric of y, then

$$q(0,t) = \frac{Q_0}{2}H(t) \text{ or } \bar{q}(0,s) = \frac{Q_0}{2s}. \tag{3.17}$$

Using Fourier's law of heat conduction in the non-dimensional form, namely

$$\bar{q}(0,s) = -\frac{\left(1+Z+\dfrac{4}{3N_R}\right)\overline{T}' + \Pi_0\bar{u}}{1+\lambda_r s^\beta}. \tag{3.18}$$

We obtain the condition

$$\bar{\theta}'(0,s) = -\frac{Q_0\gamma}{2s} - \frac{D}{s^2} = -\frac{Q_0\gamma s + 2D}{2s^2}. \tag{3.19}$$

Eqs.(3.16) and (3.19) are two components of the initial state vector. To obtain the remaining two components, we substitute on both sides of Eq.(3.15) to obtain the following two solutions

$$\bar{\theta}(0,s) = \frac{Q_0\gamma s[k_1k_2 + (s+M)(1+\lambda s^\alpha)] + 2Dk_1k_2}{2s^2k_1k_2(k_1+k_2)}, \tag{3.20}$$

$$\bar{u}'(0,s) = \frac{2(C - k_1^2 - k_2^2 - k_1k_2) + Q_0\gamma s K_0(1+\lambda s^\alpha)}{2s^2(k_1+k_2)}. \tag{3.21}$$

Substituting the above in the right-hand side of Eq.(3.15), we obtain

$$\bar{\theta}(y,s) = \frac{Q_0\gamma}{2s(k_1^2-k_2^2)}\left[k_1 - \frac{(s+M)(1+\lambda s^\alpha)}{k_1} + \frac{2Dk_1}{sQ_0\gamma}\right]\exp(-k_1 y)$$
$$+ \frac{Q_0\gamma}{2s(k_1^2-k_2^2)}\left[-k_2 - \frac{(s+M)(1+\lambda s^\alpha)}{k_2} + \frac{2Dk_2}{sQ_0\gamma}\right]\exp(-k_2 y), \tag{3.22}$$

$$\bar{u}(y,s) = \frac{1}{2s(k_1^2 - k_2^2)} \left[-Q_0\gamma K_0(1 + \lambda s^\alpha) + \frac{2(k_1^2 - C)}{s} \right] \exp(-k_1 y)$$
$$+ \frac{1}{2s(k_1^2 - k_2^2)} \left[Q_0\gamma K_0(1 + \lambda s^\alpha) - \frac{2(k_2^2 - C)}{s} \right] \exp(-k_2 y). \quad (3.23)$$

Those complete the solution in the Laplace domain.

In order to invert the Laplace transform in the above equations, we adopt a numerical inversion method based on a Fourier series expansion [35]:

$$f(t) = \frac{e^{vt}}{T} \left[-\frac{1}{2} Re\{F(v)\} + \sum_{k=0}^{N} \left\{ Re \left\{ F\left(v + i\frac{k\pi}{T}\right) \right\} \cos\left(\frac{k\pi}{T} t\right) \right.\right.$$
$$\left.\left. - Im \left\{ F\left(v + i\frac{k\pi}{T}\right) \right\} \times \sin\left(\frac{k\pi}{T}\right) \right\} \right], \quad 0 \leqslant t \leqslant 2T, \quad (3.24)$$

where $F(t)$ is the Laplace transform of $f(t)$, $v > 0$ is an arbitrary constant, N is sufficiently large integer. The numerical results of velocity field and temperature field are plotted in Figs.1-10.

4 Flow due to variable accelerated plate

Here, the fluid and the plate are considered at rest for $t \leqslant 0$ and the infinitely extend plate starts moving in its own plane with variable accelerating Vt^2 (V is a constant). The governing equations, the initial and a part of the boundary conditions are the same. Instead of the initial condition $(3.16)_1$ as follow:

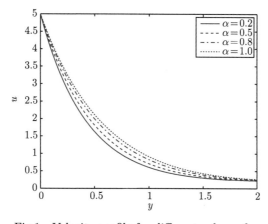

Fig.1 Velocity profile for different values of α

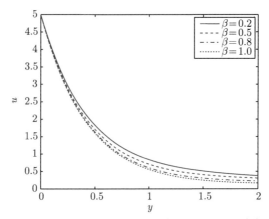

Fig.2 Velocity profile for different values of β

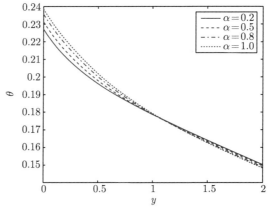

Fig.3 Temperature profile for different values of α

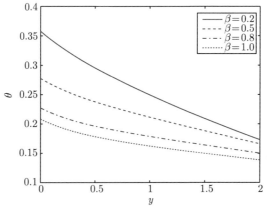

Fig.4 Temperature profile for different values of β

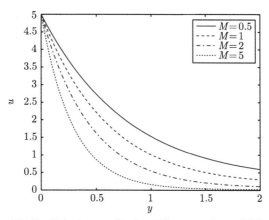

Fig.5 Velocity profile for different values of M

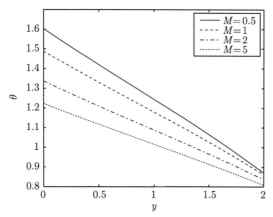

Fig.6 Temperature profile for different values of M

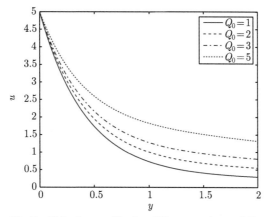

Fig.7 Velocity profile for different values of Q_0

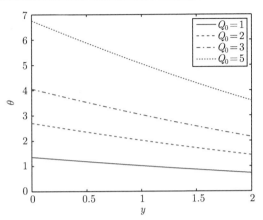

Fig.8 Temperature profile for different values of Q_0

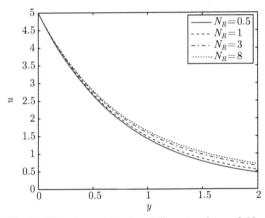

Fig.9 Velocity profile for different values of N_R

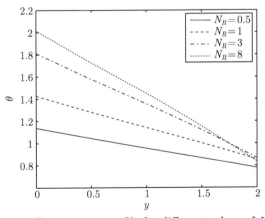

Fig.10 Temperature profile for different values of N_R

$$u(0,t) = Vt^2, \quad t > 0. \tag{4.1}$$

Here the non-dimensional quantities are:

$$u^* = \frac{u}{(V\nu^2)^{\frac{1}{5}}}, \quad y^* = \frac{y(V\nu^2)^{\frac{1}{5}}}{\nu}, \quad t^* = \frac{t(V\nu^2)^{\frac{2}{5}}}{\nu}, \quad \lambda^* = \lambda\left(\frac{(V\nu^2)^{\frac{2}{5}}}{\nu}\right)^\alpha,$$

$$\lambda_r^* = \lambda_r\left(\frac{(V\nu^2)^{\frac{2}{5}}}{\nu}\right)^\beta, \quad M^* = \frac{\sigma_0 B_0^2 (V\nu^2)^{\frac{1}{5}}}{\rho V}, \quad K_0^* = \frac{\sigma_0 k_0 B_0 (T_w - T_\infty)}{\rho(V\nu^2)^{\frac{2}{5}}},$$

$$T^* = \frac{T - T_\infty}{T_w - T_\infty}, \quad P_r^* = \frac{c_p \mu}{k_T}, \quad \Pi_0^* = \frac{\sigma_0 \pi_0 B_0^2 \nu}{(T_w - T_\infty) k_T}, \quad Z^* = \frac{\sigma_0 \pi_0 k_0 B_0}{k_T},$$

$$Q^* = \frac{\nu^2 Q}{k_T (V\nu^2)^{\frac{2}{5}} (T_w - T_\infty)}. \tag{4.2}$$

The non-dimensional governing equations are Eqs.(3.2)-(3.3), and the initial condition of velocity field in Laplace domain is $\bar{u}(0,s) = 2/s^3$.

Adopting a similar procedure as before, we have

$$\bar{\theta}'(0,s) = -\frac{Q_0\gamma}{2s} - \frac{2D}{s^3} = -\frac{Q_0\gamma s + 4D}{2s^3}, \tag{4.3}$$

$$\bar{\theta}(0,s) = \frac{Q_0\gamma s^2 [k_1 k_2 + (s+M)(1+\lambda s^\alpha)] + 4Dk_1 k_2}{2s^3 k_1 k_2 (k_1 + k_2)}, \tag{4.4}$$

$$\bar{u}'(0,s) = \frac{4(C - k_1^2 - k_2^2 - k_1 k_2) + Q_0\gamma s^2 K_0(1+\lambda s^\alpha)}{2s^3(k_1 + k_2)}. \tag{4.5}$$

We obtain the expression for the velocity field and temperature field in the transform plane are

$$\bar{\theta}(y,s) = \frac{Q_0\gamma}{2s(k_1^2 - k_2^2)}\left[k_1 - \frac{(s+M)(1+\lambda s^\alpha)}{k_1} + \frac{4Dk_1}{s^2 Q_0\gamma}\right]\exp(-k_1 y)$$
$$+ \frac{Q_0\gamma}{2s(k_1^2 - k_2^2)}\left[-k_2 - \frac{(s+M)(1+\lambda s^\alpha)}{k_2} + \frac{4Dk_2}{s^2 Q_0\gamma}\right]\exp(-k_2 y), \tag{4.6}$$

$$\bar{u}(y,s) = \frac{1}{2s(k_1^2 - k_2^2)}\left[-Q_0\gamma K_0(1+\lambda s^\alpha) + \frac{4(k_1^2 - C)}{s^2}\right]\exp(-k_1 y)$$
$$+ \frac{1}{2s(k_1^2 - k_2^2)}\left[Q_0\gamma K_0(1+\lambda s^\alpha) - \frac{4(k_2^2 - C)}{s^2}\right]\exp(-k_2 y). \tag{4.7}$$

Those complete the solution in the Laplace domain. Now we obtain the solution in the original space by doing the inverse of the Laplace transform numerically using Eq.(3.24) [35] to give us in Figs.11-20.

5 Results and Discussion

In section 3-4, we considered a flow of Maxwell fluid with uniform acceleration boundary condition and variable accelerating boundary condition, respectively.

Graphical results show the velocity profiles and the temperature profiles for different values of M, N_R, α, β, Q_0.

Figs.1-2 show the velocity changes with the fractional parameters α and β in time $t = 5$. It is clearly seen that the smaller the values of α and β, the more steadily the velocity changes. The boundary layer region around the surface is dependent on the fractional order α, the fractional parameter β has quite the opposite effect to that of α. Figs. 3-4 are the temperature profiles for fractional parameters α and β, which have the same effect with velocity profile. Figs.5-6 depict the influence of magnetic body for velocity field and temperature field. As it was to be expected, the magnetic force M is favorable to the velocity decays, and the temperature profiles have a marked decrease with increase in M. This is due to the fact that applied transverse magnetic field produces a drag in the form of Lorentz force thereby decreasing the magnitude of velocity. Figs.7-10 demonstrate the influence of N_R and Q_0 for velocity field and temperature field. It is clear that near the plate there is an increase in velocity and temperature with increasing the radiation heat effect N_R and heat source Q_0. The increase of the temperature results in the velocity increasing. For radiation heat is a radiating process,the radiation parameter increasing, the temperature decreases rapidly in y-direction. Figs. 11-20 are plotted to demonstrate the effects of different parameters on the velocity and temperature fields by using the solution obtained in Section 4. For the plane is moving with variable accelerating Vt^2, with the time increasing, the velocity and temperature are increasing remarkably.

6 Conclusions

The purpose of this paper is to introduce a new model for Fourier law of heat conduction with time-fractional order to generalized Maxwell fluid. The motion equation and energy equation are coupling with each other. We used Laplace transform and numerical inverse Laplace transform to analyze the velocity field and temperature field. The effect of increasing the magnetic force M is to decrease the velocity and temperature but the effect of heat source Q_0 have the opposite effect, and increasing radiation heat N_R can decreases the temperature rapidly. This fluid is related with magnetic field, radiation heat effect and heat source, which play an important role in engineering, petroleum industries and heat transform.

Acknowledgements The work is supported by the China Postdoctoral Science

Foundation (No.2015M580069).

Appendix A

The elements of the matrix $L(y, s)$,

$$l_{11} = \frac{(k_1^2 - C)\cosh(k_2 y) - (k_2^2 - C)\cosh(k_1 y)}{k_1^2 - k_2^2},$$

$$l_{12} = -D(s+M)(1+\lambda s^\alpha)\left[\frac{k_2 \sinh(k_1 y) - k_1 \sinh(k_2 y)}{k_1 k_2 (k_1^2 - k_2^2)}\right],$$

$$l_{13} = \frac{k_2[k_1^2 - (s+M)(1+\lambda s^\alpha)]\sinh(k_1 y) - k_1[k_2^2 - (s+M)(1+\lambda s^\alpha)]\sinh(k_2 y)}{k_1 k_2 (k_1^2 - k_2^2)},$$

$$l_{14} = -D\left[\frac{\cosh(k_1 y) - \cosh(k_2 y)}{k_1^2 - k_2^2}\right],$$

$$l_{21} = K_0 C(1+\lambda s^\alpha)\left[\frac{k_2 \sinh(k_1 y) - k_1 \sinh(k_2 y)}{k_1 k_2 (k_1^2 - k_2^2)}\right],$$

$$l_{22} = \frac{[k_1^2 - (s+M)(1+\lambda s^\alpha)]\cosh(k_2 y) - [k_2^2 - (s+M)(1+\lambda s^\alpha)]\cosh(k_1 y)}{k_1^2 - k_2^2},$$

$$l_{23} = K_0(1+\lambda s^\alpha)\left[\frac{\cosh(k_1 y) - \cosh(k_2 y)}{k_1^2 - k_2^2}\right],$$

$$l_{24} = \frac{k_2(k_1^2 - C)\sinh(k_1 y) - k_1(k_2^2 - C)\sinh(k_2 y)}{k_1 k_2 (k_1^2 - k_2^2)},$$

$$l_{31} = C\left[\frac{k_2[k_1^2 - (s+M)(1+\lambda s^\alpha)]\sinh(k_1 y) - k_1[k_2^2 - (s+M)(1+\lambda s^\alpha)]\sinh(k_2 y)}{k_1 k_2 (k_1^2 - k_2^2)}\right],$$

$$l_{32} = -D(s+M)(1+\lambda s^\alpha)\left[\frac{\cosh(k_1 y) - \cosh(k_2 y)}{k_1^2 - k_2^2}\right],$$

$$l_{33} = \frac{[k_1^2 - (s+M)(1+\lambda s^\alpha)]\cosh(k_1 y) - [k_2^2 - (s+M)(1+\lambda s^\alpha)]\cosh(k_2 y)}{k_1^2 - k_2^2},$$

$$l_{34} = -D\left[\frac{k_1 \sinh(k_1 y) - k_2 \sinh(k_2 y)}{k_1^2 - k_2^2}\right],$$

$$l_{41} = K_0 C(1+\lambda s^\alpha)\left[\frac{\cosh(k_1 y) - \cosh(k_2 y)}{k_1^2 - k_2^2}\right],$$

$$l_{42} = (s+M)(1+\lambda s^\alpha)\left[\frac{k_2(k_1^2 - C)\sinh(k_1 y) - k_1(k_2^2 - C)\sinh(k_2 y)}{k_1 k_2 (k_1^2 - k_2^2)}\right],$$

$$l_{43} = K_0(1+\lambda s^\alpha)\left[\frac{k_1 \sinh(k_1 y) - k_2 \sinh(k_2 y)}{k_1^2 - k_2^2}\right],$$

$$l_{44} = \frac{(k_1^2 - C)\cosh(k_1 y) - (k_2^2 - C)\cosh(k_2 y)}{k_1^2 - k_2^2}.$$

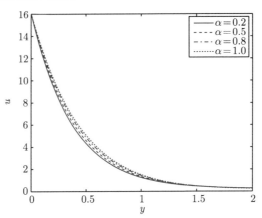

Fig.11 Velocity profile for different values of α

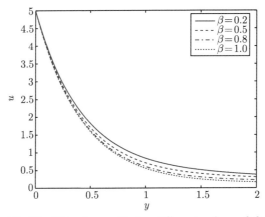

Fig.12 Velocity profile for different values of β

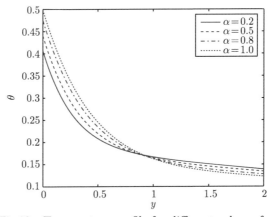

Fig.13 Temperature profile for different values of α

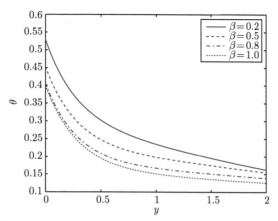

Fig.14　Temperature profile for different values of β

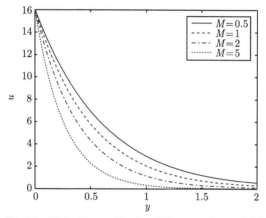

Fig.15　Velocity profile for different values of M

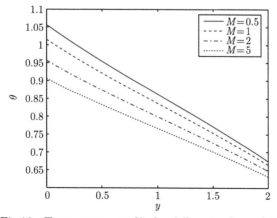

Fig.16　Temperature profile for different values of M

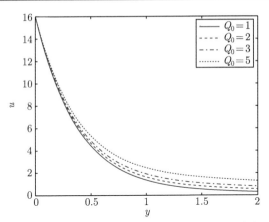

Fig.17　Velocity profile for different values of Q_0

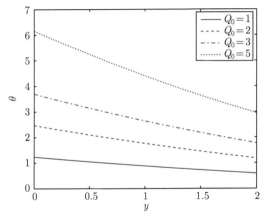

Fig.18　Temperature profile for different values of Q_0

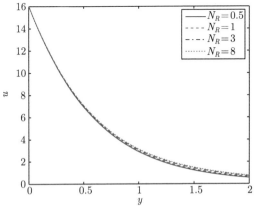

Fig.19　Velocity profile for different values of N_R

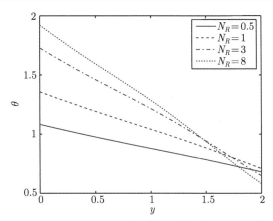

Fig.20 Temperature profile for different values of N_R

References

[1] Fetecau C, Athar M, Fetecau C. Unsteady flow of a generalized Maxwell fluid with fractional derivative due to a constantly accelerating plate[J]. Comput. Math. Appl., 2009, 57: 596-603.

[2] Hayat T, Sajid M. Homotopy analysis of MHD boundary layer flow of an upper-convected Maxwell fluid[J]. Int. J. Engrg. Sci., 2007, 45: 393-401.

[3] Tan W, Masuoka T. Stability analysis of a Maxwell fluid in a porous medium heated from below[J]. Phys. Lett. A, 2007, 360: 454-460.

[4] Podlubny I. Fractional Differential Equations. New York: Academic Press, 1999.

[5] Guo B, Pu X, Huang F. Fractional partial differential equations and their numerical solutions[M]. Beijing: Science Press, 2011.

[6] Bagley R, Torvik P. A theoretical basis for the application of fractional calculus to viscoelasticity[J]. J. Rheol. 1983, 27: 201-210.

[7] Friedrich C. Relaxation and retardation functions of the Maxwell model with fractional derivatives[J]. Rheol. Acta, 1991, 30: 151-158.

[8] Song D, Jiang T. Study on the constitutive equation with fractional derivative for the viscoelastic fluids Modified Jeffreys model and its application[J]. Rheol. Acta, 1998, 27: 512-517.

[9] Qi H, Xu M. Stokes' first problem for a viscoelastic fluid with the generalized Oldroyd-B model[J]. Acta Mech. Sin., 2007, 23: 463-469.

[10] Qi H, Xu M. Some unsteady unidirectional flows of a generalized Oldroyd-B fluid with fractional derivative[J]. Appl. Math. Model., 2009, 33: 4184-4191.

[11] Fetecau C, Prasad S. Rajagopal K. A note on the flow induced by a constantly accelerating plate in an Oldroyd-B fluid[J]. Appl. Math. Model., 2007, 31: 647-654.

[12] Fetecau Con, Fetecau Cor, Kamran M, Vieru D. Exact solutions for the flow of a generalized Oldroyd-B fluid induced by a constantlyaccelerating plate between two side walls perpendicular to the plate[J]. J. Non-Newton. Fluid Mech., 2009, 156: 189-201.

[13] Vieru D, Fetecau Cor, Fetecau C. Flow of a generalized Oldroyd-B fluid due to a constantly accelerating plate[J]. Appl. Math. Comput., 2008, 201: 834-842.

[14] Fetecau Cor, Athar M, Fetecau C. Unsteady flow of a generalized Maxwell fluid with fractional derivative due to a constantly accelerating plate[J]. Comput. Math. Appl., 2009, 57: 596-603.

[15] Khan M, Hayat T, Asghar S. Exact solution for MHD flow of a generalized Oldroyd-B fluid with modified Darcy's law[J]. Int. J. Engrg. Sci., 2006, 44: 333-339.

[16] Zheng L, Liu Y, Zhang X. A New Model for Plastic-Viscoelastic Magnetohydrodynamic (MHD) Flow with Radiation Thermal Transfer[J]. Int. J. Nonlinear Sci. Numer. Simul., 2013, 14: 435-441.

[17] Zheng L, Liu Y, Zhang X. Exact solutions for MHD flow of generalized Oldroyd-B fluid due to an infinite accelerating plate[J]. Math. Comput. Model., 2011, 54: 780-788.

[18] Shen F, Tan W, Zhao Y, Masuoka T. Decay of vortex velocity and diffusion of temperature in a generalized second grade fluid[J]. Appl. Math. Mech., 2004, 25: 1151-1159.

[19] Shen F, Tan W , Zhao Y, Masuoka T. The Reyleigh-stokes problem for a heated generalized second grade fluid with fractional derivative model[J]. Nonlinear Analysis RWA., 2006, 7: 1072-1080.

[20] Ezzat M. Thermoelectric MHD non-Newtonian fluid with fractional derivative heat transfer[J]. Physica B, 2010, 405: 4188-4194.

[21] Qi H. Liu J. Time-fractional radial diffusion in hollow geometries[J]. Meccanica, 2010, 45: 577-583.

[22] Jiang X, Qi H. Thermal wave model of binoheat transfer with modified Riemann-Liouville fractional derivative[J]. J. Phys. A: Math. Theor., 2012, 45: 1-11.

[23] Xu H, Jiang X. Time fractional deal-phase-lag heat conduction equation[J]. Chin. Phys. B, 2015, 24: 1-7.

[24] Jiang X, Xu M, Qi H. The fractional diffusion model with an absorption term and modified Fick's law for non-local transport processes[J]. Nonlinear Anal. RWA., 2010, 11: 262-269.

[25] Tan W, Masuoka T. Stokes' first problem for a second grade fluid in a porous half-space with heated boundary[J]. Int. J. Non-Linear Mech., 2005, 40: 515-522.

[26] EI-Aziz M. Radiation effect on the flow and heat transfer over an unsteady stretching sheet[J]. Int. Commun. Heat Mass Transfer, 2009, 36: 521-524.

[27] Cortell R. Effects of viscous dissipation and radiation on the thermal boundary layer over a nonlinearly stretching sheet[J]. Phys. Lett. A, 2008, 371: 631-636.

[28] Ezzat M, El-Bary A, Ezzat S. Combined heat and mass transfer for unsteady MHD

flow of perfect conducting micropolar fluid with thermal relaxation[J]. Energy Convers. Manag., 2011, 52: 934-945.

[29] Ezzat M, El-Karamany A. Magnetothermoelasticity with two relaxation times in conducting medium with variable electrical and thermal conductivity[J]. Appl. Math. Comput., 2003, 142: 449-467.

[30] Ezzat M, Othman M, El-Karamany A. State space approach to generalized thermo-viscoelasticity with two relaxation times[J]. Int. J. Eng. Sci., 2002, 40: 283-302.

[31] Ezzat M, El-Karamany A, Samaan A. The dependence of the modulus of elasticity on reference temperature in generalized thermoelasticity with thermal relaxation[J]. Appl. Math. Comput., 2004, 147: 169-189.

[32] Ezzat M. The relaxation effects of the volume properties of electrically conducting viscoelastic material[J]. Mat. Sci. Eng. B, 2006, 130: 11-23.

[33] Ezzat M, El-Karamany A. The relaxation effects of the volume properties of viscoelastic material in generalized thermoelasticity[J]. Int. J. Eng. Sci., 2003, 41: 2281-2298.

[34] El-Karamany A, Ezzat M. Thermal shock problem in generalized thermo-viscoelasticty under four theories[J]. Int. J. Eng. Sci., 2004, 42: 649-671.

[35] Honig G, Hirdes U. A method for the numerical inversion of Laplace transforms[J]. J. Comput. appl. Math., 1984, 10: 113-132.

Global Well-posedness for the Periodic Novikov Equation with Cubic Nonlinearity*

Wu Xinglong (吴兴龙) and Guo Boling (郭柏灵)

Abstract This paper is devoted to the study of the Cauchy problem to the periodic Novikov equation. Firstly, the local well–posedness for the equation is established. Secondly, we give the precise blow–up criterion, conservation laws and prove that the equation has global strong solutions in time, if the initial potential does not change sign on $\mathbb{R}$. Thirdly, with the initial potential satisfying the sign conditions, we show the existence of global weak solutions in time. Moreover, the uniqueness of global solution is addressed.

Keywords the periodic Novikov equation; well-posedness; blow–up scenario; conservation laws; global solutions; peakon solutions

1 Introduction

We consider the Cauchy problem of the following periodic Novikov equation with cubic nonlinearity

$$\begin{cases} u_t - u_{txx} + 4u^2 u_x = 3uu_x u_{xx} + u^2 u_{xxx}, & t > 0,\ x \in \mathbb{R}, \\ u(0, x) = u_0(x), & x \in \mathbb{R}, \\ u(t, x) = u(t, x+1), & t > 0,\ x \in \mathbb{R}, \end{cases} \quad (1.1)$$

which [18] arises as a zero curvature equation $F_t - G_x + [F, G] = 0$, Eq.(1.1) is the compatibility condition to the linear system

$$\Psi_x = F\Psi, \quad \Psi_t = G\Psi,$$

---

* Applicable Analysis, 2016, 95, 405–425. DOI:10.1080/00036811.2015.1005611.

where $m = u - u_{xx}$,

$$F = \begin{pmatrix} 0 & m\lambda & 1 \\ 0 & 0 & m\lambda \\ 1 & 0 & 0 \end{pmatrix}, G = \begin{pmatrix} \frac{1}{3\lambda^2} - uu_x & \frac{u_x}{\lambda} - u^2 m\lambda & u_x^2 \\ \frac{u}{\lambda} & -\frac{2}{3\lambda^2} & -\frac{u_x}{\lambda} - u^2 m\lambda \\ -u^2 & \frac{u}{\lambda} & \frac{1}{3\lambda^2} + uu_x \end{pmatrix}.$$

By the perturbative symmetry approach, Eq.(1.1) was derived by Novikov in a symmetry classification of nonlocal PDEs with cubic nonlinearity [23,25]. By virtue of a scalar Lax pair, he subsequently proves that the equation is integrable. If we introduce a new dependent potential function m, then Eq.(1.1) is given by

$$m_t + u^2 m_x + 3uu_x m = 0, \quad m = u - u_{xx}.$$

Using an asymptotic approximation to the Hamiltonian for the Green–Naghdi equations in shallow water theory. Camassa and Holm discovered [2] the following equation

$$m_t + um_x + 2u_x m = 0, \quad m = u - u_{xx}. \qquad (1.2)$$

To begin with, the Camassa–Holm (CH) equation approximates unidirectional fluid flow in Euler's equations at the next order beyond the KdV equation. It has a bi-Hamiltonian structure [15] and is completely integrable [5], and with a Lax pair based on a linear spectral problem of second order. Also, there are smooth soliton solutions of Eq.(1.2) on a non-zero constant background [3]. The Camassa-Holm equation has attracted a lot of interests in the past seventeen years for various reasons, cf. [6,7,9].

One might wonder whether the Camassa-Holm equation is the only integrable shallow water equation whose dispersionless version has weak solitons. In 1999, by an asymptotic integrability approach to isolate integrable third order equations, Degasperis and Procesi [13] discovered the following Degasperis–Procesi (DP) equation

$$m_t + um_x + 3u_x m = 0, \quad m = u - u_{xx}, \qquad (1.3)$$

which can also be regarded as a model for nonlinear shallow water dynamics [16]. Degasperis, Holm and Hone [12] derive the formal integrability of Eq.(1.3) by constructing a Lax pair. Eq.(1.3) [12] has a bi-Hamiltonian structure and an infinite sequence of conserved quantities, and admits exact peakon solutions.

Despite the three equations are similar in form, it should be emphasized that physics background of these equations are truly different. One of the important features of DP equation is that it has not only peakon solitons [12], i.e. solutions at the form $u(t,x) = ce^{-|x-ct|}$ and periodic peakon solutions [30]

$$u(t,x) = c\frac{\cosh\left(x - ct - [x - ct] - \frac{1}{2}\right)}{\sinh\frac{1}{2}}, \quad c \in \mathbb{R} \text{ and } c \neq 0.$$

Moreover it have shock peakons [4, 21], the form is

$$u(t,x) = -\frac{1}{t+k}\text{sgn}(x)e^{-|x|}, \quad k > 0,$$

and the periodic shock peakons [14].

Similar to the CH equation, the periodic Novikov equation has a bi-Hamiltonian structure, an infinite sequence of conserved quantities, and admits exact peakon solutions, i.e., solutions at the form

$$u(t,x) = \pm\sqrt{c}\frac{\cosh\left(x - ct - [x - ct] - \frac{1}{2}\right)}{\sinh\frac{1}{2}}, \quad c > 0.$$

Because of the Novikov equation itself is not symmetrical, i.e. $(u,x) \nrightarrow (-u,-x)$ and the cubic nonlinearity, thus some results of Eq.(1.1) are truly different with the CH and DP equation [17, 18, 27]. Recently, some authors study the Cauchy problem of the Novikov equation on the line and on the circle in [26–28].

The remainder of the paper is organized as follows. In section 2, we simply present local well-posedness of Eq.(1.1). In Section 3, we derive two conservation laws, the precise blow-up scenario and the strong solution to Eq.(1.1), which is unique and exists globally in time, provided the initial potential satisfies certain sign conditions. In Section 4, as in [10, 27, 30], with initial potential satisfying certain sign conditions, we derive the existence and uniqueness of global weak solutions.

Notation: In this paper, we introduce some notations. If A is an unbounded operator, $D(A)$ denotes the domain of the operator A, $[A, B] = AB - BA$ denotes the commutator of the linear operators A and B, $\|\cdot\|_X$ denotes the norm of the Banach space X. For convenience, let $\|\cdot\|_{L^p}, \|\cdot\|_s$ and $(\cdot,\cdot)_s$ denote the norm of $L^p(\mathbb{S})$ space, $H^s(\mathbb{S})$ space and inner product of $H^s(\mathbb{S})$, respectively, for $1 \leqslant p < \infty$, $s \in \mathbb{R}$. Let X and Y be Hilbert spaces such that Y is continuity and densely embedded in X and let $Q : Y \to X$ be a topological isomorphism. $L(Y,X)$ denotes the space of all bounded linear operators from Y to X ($L(X)$, if $X = Y$). Moreover, let $\langle \cdot, \cdot \rangle$ denote the duality paring between $H^1(\mathbb{S})$ and $H^{-1}(\mathbb{S})$. Let $M(\mathbb{S})$ be the space of Radon measures on $\mathbb{S}$ with bounded total variation and $M^+(\mathbb{S})$ be the subset of positive Radon measures. Furthermore, $BV(\mathbb{S})$ stands for the space of functions with bounded variation, $\mathbb{V}(f)$ is the total variation of $f \in BV(\mathbb{S})$.

2 Local well-posedness in time

In the subsection, we shall apply Kato's theory to establish the local well-posedness of the Cauchy problem to Eq.(1.1) in $H^s(\mathbb{S}), s > \frac{3}{2}$ with $\mathbb{S} = \mathbb{R}/\mathbb{Z}$ (the circle of unite length).

By computation, with $y = u - u_{xx}$, Eq.(1.1) becomes the following form of a quasi-linear evolution equation of hyperbolic type

$$\begin{cases} y_t + u^2 y_x + 3u u_x y = 0, \\ u(t, x) = u(t, x + 1), & t > 0, \; x \in \mathbb{R}, \\ y(0, x) = u_0(x) - u_{xx}(0, x), & x \in \mathbb{R}. \end{cases} \quad (2.1)$$

Note that if $G(x) := \dfrac{\cosh\left(x - [x] - \frac{1}{2}\right)}{2 \sinh\left(\frac{1}{2}\right)}$, where $[x]$ stands for the integer part of $x \in \mathbb{R}$, then for all the $f \in L^2(\mathbb{S})$, we have $(1 - \partial_x^2)^{-1} f = G * f$, i.e. $G * y = u$, where we denote by $*$ the convolution. By the Green formulation, we can rewrite Eq.(2.1) as follows

$$\begin{cases} u_t + u^2 u_x + G * (3 u u_x u_{xx} + 2 u_x^3 + 3 u^2 u_x) = 0, & t > 0, \; x \in \mathbb{R}, \\ u(0, x) = u_0(x), & x \in \mathbb{R}, \\ u(t, x) = u(t, x + 1), & t > 0, \; x \in \mathbb{R}. \end{cases} \quad (2.2)$$

Or in the equivalent form

$$\begin{cases} u_t + u^2 u_x = -(1 - \partial_x^2)^{-1} \left(\partial_x \left(\frac{3}{2} u u_x^2 + u^3 \right) + \frac{1}{2} u_x^3 \right), \\ u(0, x) = u_0(x), & x \in \mathbb{R}, \\ u(t, x) = u(t, x + 1), & t > 0, \; x \in \mathbb{R}. \end{cases} \quad (2.3)$$

Theorem 2.1 *Assume that the initial data $u_0 \in H^s(\mathbb{S})$, $s > \frac{3}{2}$. Then there exist a time $T = T(\|u_0\|_s)$ and a unique solution u to Eq.(1.1) (or Eq.(2.3)), which satisfies*

$$u = u(\cdot, u_0) \in C([0, T]; H^s(\mathbb{S})) \cap C^1([0, T]; H^{s-1}(\mathbb{S})).$$

Moreover, the solution u depends continuously on the initial data u_0, i.e. the mapping $u_0 \to u(\cdot, u_0) : H^s(\mathbb{S}) \to \mathcal{C}([0,T]; H^s(\mathbb{S})) \cap \mathcal{C}^1([0,T]; H^{s-1}(\mathbb{S}))$ is continuous.

Proof For $u \in H^s(\mathbb{S}), s > \frac{3}{2}$, define the operator $A(u) = u^2 \partial_x$. Analogous to Lemma 2.6 [27], we have that the operator $A(u)$ is quasi-m-accretive, uniformly on bounded sets in $H^{s-1}(\mathbb{S})$. It is easily check that $A(u) \in L(H^s(\mathbb{S}), H^{s-1}(\mathbb{S}))$. Moreover,

$$\|(A(y) - A(z))w\|_{s-1} \leqslant a_1 \|y-z\|_{s-1} \|w\|_s, \quad y, z, w \in H^s(\mathbb{S}).$$

Let $B(u) = [\Lambda, u^2 \partial_x] \Lambda^{-1}$, where $\Lambda = (1 - \partial_x)^{\frac{1}{2}}$. Similar to Lemma 2.8 [27], we can deduce that $B(u) \in L(H^{s-1}(\mathbb{S}))$ and satisfies

$$\|(B(y) - B(z))w\|_{s-1} \leqslant a_2 \|y-z\|_s \|w\|_{s-1}, \quad y, z \in H^s(\mathbb{S}), w \in H^{s-1}(\mathbb{S}).$$

Introducing the function $f(u) = -(1 - \partial_x^2)^{-1} \left(\partial_x \left(\frac{3}{2} u u_x^2 + u^3 \right) + \frac{1}{2} u_x^3 \right), X = H^{s-1}$ $(\mathbb{S}), Y = H^s(\mathbb{S})$, in view of Lemma 2.9 of [27], f is bounded on bounded sets in $H^s(\mathbb{S})$, and satisfies

$$\|g(y) - g(z)\|_s \leqslant a_3 \|y-z\|_s, \quad y, z \in H^s(\mathbb{S}),$$

$$\|g(y) - g(z)\|_{s-1} \leqslant a_4 \|y-z\|_{s-1}, \quad y, z \in H^s(\mathbb{S}).$$

Set $Y = H^s(\mathbb{S}), X = H^{s-1}(\mathbb{S})$, and $Q = (1 - \partial_x^2)^{\frac{1}{2}}$. Obviously, Q is an isomorphism of Y onto X. Thanks to Kato's theory for abstract quasilinear evolution equation of hyperbolic type [19], we can establish the local well-posedness of Eq.(2.3) in $H^s(\mathbb{S}), s > \frac{3}{2}$, then the solution u belongs to

$$\mathcal{C}([0,T]; H^s(\mathbb{S})) \cap \mathcal{C}^1([0,T]; H^{s-1}(\mathbb{S})).$$

This concludes the proof of Theorem 2.1. □

Theorem 2.2 Let $u_0 \in H^s(\mathbb{S}), s > \frac{3}{2}$. Then T in Theorem 2.1 may be chosen independent of s in the following sense. If $u = u(\cdot, u_0) \in \mathcal{C}([0,T]; H^s(\mathbb{S})) \cap \mathcal{C}^1([0,T]; H^{s-1}(\mathbb{S}))$ is a solution to Eq.(1.1) (or Eq.(2.3)), and if $u_0 \in H^{s_1}(\mathbb{S})$ for some $s_1 \neq s, s_1 > \frac{3}{2}$, then the solution $u \in \mathcal{C}([0,T]; H^{s_1}(\mathbb{S})) \cap \mathcal{C}^1([0,T]; H^{s_1-1}(\mathbb{S}))$ and with the same T. In particular, if $u_0 \in H^\infty(\mathbb{S}) = \bigcap_{s \geqslant 0} H^s(\mathbb{S})$, then the solution $u \in \mathcal{C}([0,T]; H^\infty(\mathbb{S}))$.

Proof In fact, the proof of Theorem 2.2 is similar to the one of Theorem 2.3 of [27]. Thus we omit it here. □

3 Conservation laws, Blow–up scenario and Global existence

In this section, we shall begin deriving conservation laws for strong solutions to Eq.(2.3). Using these conservation laws, we obtain a blow–up scenario. Then we establish two global existence theorems.

For the convenience of the readers, we first recall the following useful lemmas.

Lemma 3.1 [20] *Assume that $s > 0$. Then we have*

$$\|[\Lambda^s, g]f\|_{L^2} \leqslant c(\|\partial_x g\|_{L^\infty}\|\Lambda^{s-1}f\|_{L^2} + \|\Lambda^s g\|_{L^2}\|f\|_{L^\infty}),$$

where c is constant depending only on s.

Lemma 3.2 [11] *Assume that $F \in C^{m+2}(\mathbb{R})$ with $F(0) = 0$. Then for every $\frac{1}{2} < s \leqslant m$, we have*

$$\|F(u)\|_s \leqslant \tilde{F}(\|u\|_{L^\infty})\|u\|_s, \quad u \in H^s(\mathbb{S}),$$

where $\tilde{F}$ is a monotone increasing function depending only on F and s.

Lemma 3.3 [20] *Assume that $s > 0$. Then $H^s(\mathbb{S}) \cap L^\infty(\mathbb{S})$ is an algebra. Moreover*

$$\|fg\|_s \leqslant c(\|f\|_{L^\infty}\|g\|_s + \|f\|_s\|g\|_{L^\infty}),$$

where c is a constant depending only on s.

Lemma 3.4 [7] *Let $T > 0$ and $u \in C^1([0,T); H^2(\mathbb{S}))$. Then for every $t \in [0, T)$, there exist at least one pair points $\xi(t), \zeta(t) \in \mathbb{S}$, such that*

$$m(t) = \inf_{x \in \mathbb{S}} u(t, x) = u(t, \xi(t)), \quad M(t) = \sup_{x \in \mathbb{S}} u(t, x) = u(t, \zeta(t)),$$

and $m(t), M(t)$ are absolutely continuous in $[0, T)$. Moreover,

$$\frac{dm(t)}{dt} = u_t(t, \xi(t)), \quad \frac{dM(t)}{dt} = u_t(t, \zeta(t)), \quad a.e. \text{ on } [0, T).$$

Lemma 3.5 [8] *There exists a constant $k > 0$ such that*

$$\max_{x \in [0,1]} u^2(x) \leqslant k\|u\|_1^2, \quad u \in H^r(\mathbb{S}), r > \frac{5}{2}.$$

The best possible constant lies within the range $\left(1, \frac{13}{12}\right].$

Theorem 3.1   Let $u_0 \in H^s(\mathbb{S}), s > \dfrac{3}{2}$, and $y_0^{1/3} \in L^2(\mathbb{S})$. Then as long as the solution $u(t,x)$ to Eq.(1.1) given by Theorem 2.2 exists, we obtain

$$\int_{\mathbb{S}}(u^2(t,x)+u_x^2(t,x))dx = \int_{\mathbb{S}}(u_0^2+u_{0,x}^2)dx,$$

$$\int_{\mathbb{S}} y^{\frac{2}{3}}(t,x)dx = \int_{\mathbb{S}} y_0^{\frac{2}{3}}(x)dx,$$

where $u_0 = u(0,x), u_{0,x} = u_x(0,x)$, and $y = u - u_{xx}$. Moreover, we have

$$|u(t,x)| \leqslant \sqrt{k}\|u_0\|_1, \quad k \in \left(1, \dfrac{13}{12}\right].$$

Proof   In view of Theorem 2.2 and a simple density argument, it suffices to consider the case $s = 3$. Multiply Eq.(1.1) by u, we have

$$uu_t - uu_{txx} + 4u^3 u_x = 3u^2 u_x u_{xx} + u^3 u_{xxx}.$$

Integrating by parts on $\mathbb{S}$,

$$\dfrac{1}{2}\dfrac{d}{dt}\int_{\mathbb{S}}(u^2+u_x^2)dx = \int_{\mathbb{S}}(-4u^3 u_x + 3u^2 u_x u_{xx} + u^3 u_{xxx})dx$$
$$= \int_{\mathbb{S}}(3u^2 u_x u_{xx} - 3u^2 u_x u_{xx})dx$$
$$= 0.$$

Thus we can deduce that

$$\int_{\mathbb{S}}(u^2(t,x)+u_x^2(t,x))dx = \int_{\mathbb{S}}(u_0^2+u_{0,x}^2)dx.$$

In view of the above conservation law and Lemma 3.5, we have

$$|u(t,x)| \leqslant \sqrt{k}\|u\|_1 = \sqrt{k}\|u_0\|_1, \quad k \in \left(1, \dfrac{13}{12}\right].$$

Note that Eq.(2.1) can be written as

$$(y^{\frac{2}{3}})_t + (y^{\frac{2}{3}}u)_x = 0.$$

In view of $y_0^{1/3} \in L^2(\mathbb{S})$, it follows that

$$\int_{\mathbb{S}} y^{\frac{2}{3}}(t,x)dx = \int_{\mathbb{S}} y_0^{\frac{2}{3}}(x)dx.$$

This completes the proof of Theorem 3.1.   □

Theorem 3.2 Let the initial data $u_0 \in H^r(\mathbb{S}), r > \dfrac{3}{2}$. If T is the maximal existence time of corresponding solution $u(t, x)$ to Eq.(1.1) with the initial data u_0, then the $H^r(\mathbb{S})$-norm of $u(t, x)$ to Eq.(1.1) (or (2.3)) blows up on $[0, T)$ if and only if

$$\varlimsup_{t \uparrow T} \|u_x(t, x)\|_{L^\infty} = \infty.$$

Proof Let $u(t, x)$ be the solution to Eq.(1.1) with the initial data $u_0 \in H^r(\mathbb{S}), r > \dfrac{3}{2}$, which is guaranteed by Theorem 2.2.

If $\varlimsup_{t \uparrow T} \|u_x(t, x)\|_{L^\infty} = \infty$, by Sobolev's embedding theorem, we obtain the solution $u(t, x)$ will blow up in finite time.

Next, applying the operator Λ^r to Eq.(2.3), multiplying by $\Lambda^r u$, and integrating by parts on $\mathbb{S}$, we obtain

$$\frac{d}{dt}(u, u)_r = -2(u^2 u_x, u)_r + 2(f(u), u)_r, \tag{3.1}$$

where $f(u) = -(1 - \partial_x^2)^{-1}\left(\partial_x\left(\dfrac{3}{2}u u_x^2 + u^3\right) + \dfrac{1}{2}u_x^3\right).$

Suppose there exists a $M > 0$, such that $\varlimsup_{t \uparrow T}\|u_x(t, x)\|_{L^\infty} \leqslant M$. In view of Theorem 3.1, we have $\|u\|_{L^\infty} \leqslant \sqrt{\dfrac{13}{12}}\|u_0\|_1$. Then we have

$$\begin{aligned}
|(u^2 u_x, u)_r| &= |(\Lambda^r(u^2 u_x), \Lambda^r u)_0| \\
&= |([\Lambda^r, u^2]u_x, \Lambda^r u)_0 + (u^2 \Lambda^r u_x, \Lambda^r u)_0| \\
&\leqslant \|[\Lambda^r, u^2]u_x\|_{L^2}\|\Lambda^r u\|_{L^2} + \|u u_x\|_{L^\infty}\|\Lambda^r u\|_{L^2}^2 \\
&\leqslant c\left(\|(u^2)_x\|_{L^\infty}\|\Lambda^{r-1}u_x\|_{L^2} + \|\Lambda^r u^2\|_{L^2}\|u_x\|_{L^\infty}\right)\|u\|_r + cM\|u\|_r^2 \\
&\leqslant c\left(\|u u_x\|_{L^\infty}\|u\|_r + M\|u^2\|_r\right)\|u\|_r + cM\|u\|_r^2 \\
&\leqslant cM\|u\|_r^2 \leqslant c\|u\|_r^2,
\end{aligned} \tag{3.2}$$

where we applied Lemma 3.1 with $s = r$, Lemma 3.2 with $F(u) = u^2$ and $s = r$.

Note that $H^s(\mathbb{R}), s > \dfrac{1}{2}$ is a Banach algebra, we deduce that

$$\begin{aligned}
\|u u_x^2 + u^3\|_{r-1} &\leqslant \|[\Lambda^{r-1}, u]u_x^2 + u\Lambda^{r-1}u_x^2\|_{L^2} + \|u^3\|_{r-1} \\
&\leqslant c(\|u_x\|_{L^\infty}\|\Lambda^{r-1}u_x^2\|_{L^2} + \|\Lambda^{r-1}u\|_{L^2}\|u_x^2\|_{L^\infty}) + c\|u_x^2\|_{r-1} + \|u^3\|_{r-1} \\
&\leqslant c(M\|u_x^2\|_{r-1} + M^2\|u\|_{r-1}) + c\|u_x^2\|_{r-1} + \|u^3\|_{r-1} \\
&\leqslant c(\|u_x\|_{r-1} + \|u\|_{r-1}) \leqslant c\|u\|_r,
\end{aligned}$$

where we applied Lemma 3.1 with $s = r - 1$, Lemma 3.2 with $F(u) = u^3, u_x^2$, and $s = r - 1$, Lemma 3.3 with $s = r - 1$.

On the other hand, in view of the above relation, we estimate the second term of the right hand side of Eq. (3.1) as follows

$$\begin{aligned}(f(u), u)_r &= \left(-(1-\partial_x^2)^{-1}\left(\partial_x(\frac{3}{2}uu_x^2 + u^3) + \frac{1}{2}u_x^3\right), u\right)_r \\ &\leqslant c\|u\|_r \left(\frac{3}{2}\|uu_x^2 + u^3\|_{r-1} + \frac{1}{2}\|u_x^3\|_{r-2}\right) \\ &\leqslant c\|u\|_r(c\|u\|_r + \|u_x^3\|_{r-1}) \\ &\leqslant c\|u\|_r^2,\end{aligned} \quad (3.3)$$

where we applied Lemma 3.2 with $F(u) = u_x^3$ and $s = r - 1$.

Combining (3.1), (3.2) with (3.3), we obtain

$$\frac{d}{dt}\|u\|_r^2 \leqslant c\|u\|_r^2.$$

Therefore, using Gronwall's lemma to the above inequality, we have

$$\|u(t)\|_r^2 \leqslant \|u_0\|_r^2 \exp(ct).$$

This completes the proof of Theorem 3.2. □

Consider the following differential equation

$$\begin{cases} q_t = u^2(t, q), & t > 0, \ x \in \mathbb{R}, \\ q(0, x) = x, & x \in \mathbb{R}. \end{cases} \quad (3.4)$$

Applying classical results in the theory of ordinary differential equations, one can obtain the following useful result on the above initial value problem.

Lemma 3.6 *Assume $u_0 \in H^s(\mathbb{S}), s \geqslant 3$, and T be the maximal existence time of the corresponding solution $u(t, x)$ to Eq.(2.3) with the initial data u_0. Then Eq.(3.4) has a unique solution $q \in C^1([0, T) \times \mathbb{S}, \mathbb{S})$. Moreover, the map $q(t, \cdot)$ is an increasing diffeomorphism of $\mathbb{R}$ with*

$$q_x = \exp\left(\int_0^t 2uu_x(s, q(s, x))ds\right), \quad \forall (t, x) \in [0, T) \times \mathbb{S}.$$

Furthermore, letting $y = u - u_{xx}$, we have

$$y(t, q)q_x^{\frac{3}{2}}(t, x) = y_0(x), \quad \forall (t, x) \in [0, T) \times \mathbb{S}.$$

Proof   First, for fixed $x \in \mathbb{S}$, we can deal with an ordinary differential equation as follows. By the Sobolev's embedding theorem we have that $u \in \mathcal{C}^1([0,T) \times \mathbb{S}, \mathbb{S})$. Therefore, by the theory of ordinary differential equations [1] to yield the the first assertion. From Eq.(3.4) we obtain

$$\begin{cases} \dfrac{d}{dt} q_x = 2u(t,q)u_x(t,q)q_x, & t > 0, \ x \in \mathbb{S}, \\ q_x(0,x) = 1, & x \in \mathbb{S}. \end{cases}$$

Thus the solution is given by

$$q_x = \exp\left(\int_0^t 2uu_x(s, q(s,x)) ds \right) > 0, \quad \forall (t,x) \in [0,T) \times \mathbb{S}.$$

Applying Sobolev's embedding theorem and Lemma 3.4, we can find a constant $K > 0$ such that

$$q_x(t,x) \geqslant e^{-tK}, \quad (t,x) \in [0,T) \times \mathbb{S}.$$

Next, differentiating the left-hand side of the above equation with respect to the time variable t, and applying Eq.(2.1) and (3.4), we have

$$\begin{aligned} \frac{d}{dt}\left(y(t,q) q_x^{\frac{3}{2}}(t,x) \right) &= (y_t + y_x q_t) q_x^{\frac{3}{2}} + \frac{3}{2} y q_x^{\frac{1}{2}} q_{xt} \\ &= (y_t + y_x q_t + 3uu_x y) q_x^{\frac{3}{2}} \\ &= (y_t + y_x u^2 + 3uu_x y) q_x^{\frac{3}{2}} \\ &= 0. \end{aligned}$$

Thus we obtain

$$y(t,q) q_x^{\frac{3}{2}}(t,x) = y_0(x), \quad \forall (t,x) \in [0,T) \times \mathbb{S}.$$

This completes the proof of Lemma 3.6.   □

By virtue of Lemma 3.6, we can get another blow-up scenario of Eq.(2.1).

Theorem 3.3   *Assume that $u_0 \in H^s(\mathbb{S})$, $s > \dfrac{3}{2}$. If T is the maximal existence time of corresponding solution $u(t,x)$ to Eq.(2.1) with the initial data u_0, then the $H^s(\mathbb{S})$-norm of $u(t,x)$ to Eq.(2.1) (or (1.1)) blows up on $[0,T)$ if and only if*

$$\liminf_{t \uparrow T} \{ \inf_{x \in \mathbb{S}} uu_x(t,x) \} = -\infty.$$

Proof   Let $u(t,x)$ be the solution to Eq.(1.1) with the initial data $u_0 \in H^r(\mathbb{S})$, $r > \dfrac{3}{2}$, which is guaranteed by Theorem 2.2.

Multiplying the first equation of Eq.(2.1) by y and integration by parts, we end up with

$$\frac{1}{2}\frac{d}{dt}\int_S y^2(t,x)dx = -2\int_S uu_x y^2(t,x)dx. \tag{3.5}$$

Assume there exists a $M > 0$ such that $uu_x \geq -M$. One can easily check that

$$\frac{d}{dt}\int_S y^2(t,x)dx \leq 4M\int_S y^2(t,x)dx.$$

Applying Gronwal's inequality yields

$$\|u(t,\cdot)\|_2^2 = \int_S y^2(t,x)dx \leq e^{4Mt}\int_S y_0^2(x)dx$$
$$= e^{4Mt}\|u_0\|_2^2. \tag{3.6}$$

Differentiating the first equation of Eq.(2.1) with respect to x, after multiplying it by y_x and integration by parts, we can get

$$\frac{1}{2}\frac{d}{dt}\int_S y_x^2(t,x)dx = -4\int_S uu_x y^2(t,x)dx - 3\int_S (u_x^2 + uu_{xx})yy_x(t,x)dx. \tag{3.7}$$

Note that

$$\|u_x\|_{L^\infty} \leq \|u_x\|_1 \leq \|u\|_2 \leq e^{2Mt}\|u_0\|_2,$$

$$\|(q_x)^{-1}\|_{L^\infty} = \left\|\exp\left(\int_0^t -2uu_x(s,q(s,x))ds\right)\right\|_{L^\infty}$$
$$\leq \exp(ce^{2Mt}\|u_0\|_2)$$

and

$$\|y(t,x)\|_{L^\infty} = \|y(t,q)\|_{L^\infty} = \left\|y_0(x)q_x^{-\frac{3}{2}}\right\|_{L^\infty}$$
$$\leq \|y_0\|_{L^\infty}\exp(ce^{2Mt}\|u_0\|_2).$$

Combining (3.5) with (3.7), in view of the above inequality, we obtain

$$\frac{d}{dt}\int_S (y^2 + y_x^2)dx$$
$$= -4\int_S uu_x(y^2 + 2y_x^2)dx - 6\int_S (u_x^2 + uu_{xx})yy_x dx$$
$$\leq 8M\int_S (y^2 + y_x^2)dx + 3e^{4Mt}\|u_0\|_2^2\int_S (y^2 + y_x^2)dx$$
$$+ \frac{3}{2}\int_S \|y\|_{L^\infty}^2(u^2 + u_x^2 + u_{xx}^2 + u_{xxx}^2)dx$$

$$\leqslant \left(8M + 3e^{4Mt}\|u_0\|_2^2 + \frac{3}{2}\|y_0\|_{L^\infty}^2 \exp(2ce^{2Mt}\|u_0\|_2)\right) \int_{\mathbb{S}} (y^2 + y_x^2) dx. \qquad (3.8)$$

Applying Gronwall's inequality to (3.8), we obtain

$$\|u\|_3^2 \leqslant \int_{\mathbb{S}} (y^2 + y_x^2) dx$$

$$\leqslant \left(8M + 3e^{4Mt}\|u_0\|_2^2 + \frac{3}{2}\|y_0\|_{L^\infty}^2 \exp(2ce^{2Mt}\|u_0\|_2)\right) \int_{\mathbb{S}} (y_0^2 + y_{0,x}^2) dx$$

$$\leqslant (16M + 6e^{4Mt}\|u_0\|_2^2 + 3\|y_0\|_{L^\infty}^2 \exp(2ce^{2Mt}\|u_0\|_2)) \|u_0\|_3^2. \qquad (3.9)$$

Similarly, while $s > 3$, we can prove that the $H^s(\mathbb{S})$-norm of $u(t,x)$ is bounded.

As $\liminf_{t \uparrow T} \{\inf_{x \in \mathbb{S}} uu_x(t,x)\} = -\infty$, by Sobolev's embedding theorem. One can easily get that the solution $u(t,x)$ will blow up in finite time. This completes the proof of Theorem 3.3. □

Remark 3.1 Let T is the maximal existence time of corresponding solution $u(t,x)$ to Eq.(1.1) with the initial data u_0. While the solution $u(t,x)$ blows up, Theorem 3.3 yields that

$$\liminf_{t \uparrow T} \{\inf_{x \in \mathbb{S}} uu_x(t,x)\} = -\infty,$$

thanks to Theorem 3.1, we get $\|u\|_{L^\infty}$ is bounded. This yields

$$\overline{\lim}_{t \uparrow T} \|u_x(t,x)\|_{L^\infty} = \infty.$$

Next, in view of the blow-up scenario, Theorem 3.1 and Lemma 3.6, we will establish the following global existence theorems.

Theorem 3.4 Let $u_0 \in H^s(\mathbb{S})$, $s > \frac{3}{2}$. Assume $y_0 = (u_0 - u_{0,xx})$ is non-negative, and $y_0^{1/3} \in L^2(\mathbb{S})$. Then Eq.(1.1) (or Eq.(2.2)) has a unique global strong solution

$$u = u(\cdot, u_0) \in \mathcal{C}([0,\infty); H^s(\mathbb{S})) \cap \mathcal{C}^1([0,\infty); H^{s-1}(\mathbb{S})).$$

Define $y(t,\cdot) := u(t,\cdot) - u_{xx}(t,\cdot)$. Then $E(u) = \int_{\mathbb{S}} u^2 + u_x^2 \, dx$ and $F(u) = \int_{\mathbb{S}} y^{\frac{2}{3}} dx$ are two conservation laws and we have for all $t \in \mathbb{R}_+$,

(i) $y(t,\cdot) \geqslant 0, u(t,\cdot) \geqslant 0$, and $|u_x(t,\cdot)| \leqslant u(t,\cdot)$ on $\mathbb{S}$,
(ii) $\|u(t,\cdot)\|_{L^1} \leqslant \|y(t,\cdot)\|_{L^1}, \|u(t,\cdot)\|_{L^\infty} \leqslant \sqrt{k}\|u(t,\cdot)\|_1 = \sqrt{k}\|u_0\|_1$,
(iii) $\|u_x(t,\cdot)\|_{L^1} \leqslant \|y(t,\cdot)\|_{L^1}$.
Moreover, if $y_0 \in L^1(\mathbb{S})$, we obtain

$$\|y(t,\cdot)\|_{L^1} \leqslant e^{k\|u_0\|_1^2 t}\|y_0\|_{L^1},$$

where $k \in \left(1, \dfrac{13}{12}\right]$.

Proof Let the initial data $u_0 \in H^s(\mathbb{S})$, $s > \dfrac{3}{2}$, and $T > 0$ be the maximal existence time of the solution u to Eq.(2.2) with initial data u_0. If $y_0 \geqslant 0$, then by Lemma 3.6, we have $y(t, \cdot) \geqslant 0$ for all $t \in [0, \infty)$. Note that $u = G * y$ and the positivity of G, we deduce that $u(t, \cdot) \geqslant 0$ for all $t \geqslant 0$. Due to

$$u(t, x) = \frac{e^{-x+\frac{1}{2}}}{4 \sinh \frac{1}{2}} \int_0^x e^{\xi} y(t, \xi) d\xi + \frac{e^{x-\frac{1}{2}}}{4 \sinh \frac{1}{2}} \int_0^x e^{-\xi} y(t, \xi) d\xi$$
$$+ \frac{e^{-x-\frac{1}{2}}}{4 \sinh \frac{1}{2}} \int_x^1 e^{\xi} y(t, \xi) d\xi + \frac{e^{x+\frac{1}{2}}}{4 \sinh \frac{1}{2}} \int_x^1 e^{-\xi} y(t, \xi) d\xi \qquad (3.10)$$

and

$$u_x(t, x) = -\frac{e^{-x+\frac{1}{2}}}{4 \sinh \frac{1}{2}} \int_0^x e^{\xi} y(t, \xi) d\xi + \frac{e^{x-\frac{1}{2}}}{4 \sinh \frac{1}{2}} \int_0^x e^{-\xi} y(t, \xi) d\xi$$
$$- \frac{e^{-x-\frac{1}{2}}}{4 \sinh \frac{1}{2}} \int_x^1 e^{\xi} y(t, \xi) d\xi + \frac{e^{x+\frac{1}{2}}}{4 \sinh \frac{1}{2}} \int_x^1 e^{-\xi} y(t, \xi) d\xi. \qquad (3.11)$$

In view of (3.10)-(3.11) and $y \geqslant 0$, we deduce that

$$u(t, x) + u_x(t, x) = \frac{e^{x-\frac{1}{2}}}{2 \sinh \frac{1}{2}} \int_0^x e^{-\xi} y(t, \xi) d\xi$$
$$+ \frac{e^{x+\frac{1}{2}}}{2 \sinh \frac{1}{2}} \int_x^1 e^{-\xi} y(t, \xi) d\xi \geqslant 0,$$

and

$$u(t, x) - u_x(t, x) = \frac{e^{-x+\frac{1}{2}}}{2 \sinh \frac{1}{2}} \int_0^x e^{\xi} y(t, \xi) d\xi$$
$$+ \frac{e^{-x-\frac{1}{2}}}{4 \sinh \frac{1}{2}} \int_x^1 e^{\xi} y(t, \xi) d\xi \geqslant 0.$$

In view of Theorem 3.1, we obtain that $E(u)$ and $F(u)$ are conservation laws. By Lemma 3.5, it follows that

$$|u_x(t, x)| \leqslant u(t, x) \leqslant \sqrt{k} \|u_0\|_1, \quad \forall (t, x) \in \mathbb{R}_+ \times \mathbb{S}, \ k \in \left(1, \frac{13}{12}\right].$$

By Theorem 3.2, we obtain that the solution u to Eq.(2.2) exists globally in time with the initial u_0.

Due to $y(t,x) = u(t,x) - u_{xx}(t,x)$, it follows that $u = G*y$ and $u_x = G_x * y$. Thanks to Young's inequality. Note that $\|G_x\|_{L^1} \leq \|G\|_{L^1} = 1$, one can easily check (i)-(iii).

On the other hand, using Eq.(2.1), we have

$$\frac{d}{dt}\int_\mathbb{S} y(t,x)dx = \int_\mathbb{S} y_t(t,x)dx = \int_\mathbb{S} -(u^2 y_x + 3uu_x y)dx$$
$$= \int_\mathbb{S} -(u^2 y)_x - uu_x y dx \leq \|uu_x\|_{L^\infty}\int_\mathbb{S} y(t,x)dx$$
$$\leq k\|u_0\|_1^2 \int_\mathbb{S} y(t,x)dx.$$

Since $y_0 \in L^1(\mathbb{S})$, Applying Gronwall's lemma to the above inequality yields

$$\|y(t,\cdot)\|_{L^1} \leq e^{k\|u_0\|_1^2 t}\|y_0\|_{L^1},$$

where $k \in \left(1, \frac{13}{12}\right]$.

This completes the proof of Theorem 3.4. $\square$

By the similar method to the proof of Theorem 3.4, we obtain

Theorem 3.5 Let $u_0 \in H^s(\mathbb{S})$, $s > \frac{3}{2}$. Assume $y_0 = (u_0 - u_{0,xx})$ is non-positive, and $y_0^{1/3} \in L^2(\mathbb{S})$. Then Eq.(1.1) (or Eq.(2.2)) has a unique global strong solution

$$u = u(\cdot, u_0) \in C([0,\infty); H^s(\mathbb{S})) \cap C^1([0,\infty); H^{s-1}(\mathbb{S})).$$

Define $y(t,\cdot) := u(t,\cdot) - u_{xx}(t,\cdot)$. Then $E(u) = \int_\mathbb{S} u^2 + u_x^2\, dx$ and $F(u) = \int_\mathbb{S} y^{\frac{2}{3}}\, dx$ are two conservation laws and we have for all $t \in \mathbb{R}_+$,

(i) $y(t,\cdot) \leq 0$, $u(t,\cdot) \leq 0$, and $|u_x(t,\cdot)| \leq -u(t,\cdot)$ on $\mathbb{S}$,
(ii) $\|u(t,\cdot)\|_{L^1} \leq \|y(t,\cdot)\|_{L^1}$, $\|u(t,\cdot)\|_{L^\infty} \leq \sqrt{k}\|u(t,\cdot)\|_1 = \sqrt{k}\|u_0\|_1$,
(iii) $\|u_x(t,\cdot)\|_{L^1} \leq \|y(t,\cdot)\|_{L^1}$.

Furthermore, if $y_0 \in L^1(\mathbb{S})$ it follows that

$$\|y(t,\cdot)\|_{L^1} \leq e^{k\|u_0\|_1^2 t}\|y_0\|_{L^1},$$

where $k \in \left(1, \frac{13}{12}\right]$.

4 Global weak solutions

In this subsection, we shall prove that there exists a unique global weak solution to Eq.(2.2), provided the initial potential y_0 satisfies certain sign conditions.

First, let us recall the definition of strong and weak solutions.

Definition 4.1 [30] *If $u(t,x) \in \mathcal{C}([0,T); H^s(\mathbb{S})) \cap \mathcal{C}^1([0,T); H^{s-1}(\mathbb{S}))$ with $s > \dfrac{3}{2}$ is a solution to Eq.(2.2), then $u(t,x)$ is called a strong solution to Eq.(2.2) (or (1.1)). If $u(t,x)$ is a strong solution on $[0,T)$, $\forall T > 0$, then it is called global strong solution to Eq.(2.2) (or (1.1)).*

Note that Eq.(1.1) has the soliton waves with corner at their peaks. Obviously, these solitons are not strong solutions to Eq.(2.2). In order to provide a mathematical framework for the study of these solitons, we first give a definition of weak solutions to Eq.(2.2). Letting

$$F(u) = u^2 u_x + G * \left(3uu_x u_{xx} + 2u_x^3 + 3u^2 u_x\right).$$

Then Eq.(2.2) can be written as follows

$$u_t + F(u) = 0, \quad u(0,x) = u_0. \tag{4.1}$$

Definition 4.2 [30] *Assume the initial data $u_0 \in H^s(\mathbb{S})$, $s \in \left[0, \dfrac{3}{2}\right]$. If $u(t,x) \in L^\infty_{loc}([0,T);$
$H^s(\mathbb{S}))$ and satisfies*

$$\int_0^T \int_{\mathbb{S}} u\varphi_t - F(u)\varphi \, dx dt + \int_{\mathbb{S}} u_0 \varphi(0,x) dx = 0$$

for all $\varphi \in \mathcal{C}_c^\infty([0,T) \times \mathbb{S})$, where $\mathcal{C}_c^\infty([0,T) \times \mathbb{S})$ denotes the space of all functions on $[0,T) \times \mathbb{S}$, which is restricted to $[0,T) \times \mathbb{S}$ is a smooth function on $\mathbb{R}^2$ with compact support contained in $(-T,T) \times \mathbb{S}$. Then $u(t,x)$ is called a weak solution to Eq.(2.2). If function $u(t,x)$ is a weak solution on $[0,T)$ for every $T > 0$, then it is called a global weak solution to Eq.(2.2).

Proposition 4.1 [27, 29, 30] (i) *Every strong solution is a weak solution.*

(ii) *If u is a weak solution and $u \in \mathcal{C}([0,T); H^s(\mathbb{S})) \cap \mathcal{C}^1([0,T); H^{s-1}(\mathbb{S}))$ with $s > \dfrac{3}{2}$, then it is a strong solution.*

(iii) *The peaked solutions are global weak solutions of Eq.(1.1).*

Next, we recall a partial integration result for Bochner spaces.

Lemma 4.1 [22] Let $T > 0$. If
$$f, g \in L^2((0,T); H^1(\mathbb{S})) \quad \text{and} \quad \frac{df}{dt}, \frac{dg}{dt} \in L^2((0,T); H^{-1}(\mathbb{S})),$$
then f, g are a.e. equal to a function continuous from $[0,T]$ into $L^2(\mathbb{S})$ and
$$\langle f(t), g(t) \rangle - \langle f(s), g(s) \rangle = \int_s^t \left\langle \frac{df(\tau)}{d\tau}, g(\tau) \right\rangle d\tau + \int_s^t \left\langle \frac{dg(\tau)}{d\tau}, f(\tau) \right\rangle d\tau$$
for all $s, t \in [0, T]$.

Throughout the paper, let $\{\rho_n\}_{n \geq 1}$ denote the mollifiers
$$\rho_n(x) = \left(\int_{\mathbb{S}} \rho(\xi) d\xi \right)^{-1} n\rho(nx), \quad x \in \mathbb{R}, \ n \geq 1,$$
where $\rho \in C_c^\infty(\mathbb{R})$ is defined by
$$\rho(x) := \begin{cases} e^{\frac{1}{x^2-1}}, & \text{for } |x| < 1, \\ 0, & \text{for } |x| \geq 1. \end{cases}$$

Lemma 4.2 [8] Given $\mu \in M^+(\mathbb{S})$, there exists a sequence $f_n = \rho_n * \mu \in C^\infty(\mathbb{S})$ such that
$$f_n \geq 0, \quad \|f_n\|_{L^1} \leq \|\mu\|_{M(\mathbb{S})}, \quad \text{and} \ f_n \to \mu \ \text{in } D'(\mathbb{S}).$$

Let us now present the existence and uniqueness of the weak solutions.

Theorem 4.1 Let $u_0(x) \in H^1(\mathbb{S})$. Assume $y_0 = (u_0 - u_{0,xx}) \in M^+(\mathbb{S})$. Then Eq.(2.2) has a weak solution
$$u \in W^{1,\infty}(\mathbb{R}_+ \times \mathbb{S}) \cap L^\infty(\mathbb{R}_+; H^1(\mathbb{S}))$$
with initial data $u(0,x) = u_0$ and such that
$$y(t,x) = (u(t,x) - u_{xx}(t,x)) \geq 0, \quad \forall (t,x) \in [0,T) \times \mathbb{S},$$
and for any fixed $T > 0$, $y(t,x)$ is bounded on $[0,T)$.

Proof Since $u_0 \in H^1(\mathbb{S})$, in view of Lemma 3.5, it follows that
$$\|u_0\|_{L^1} \leq \|u_0\|_{L^\infty} \leq \sqrt{\frac{13}{12}} \|u_0\|_1. \tag{4.2}$$

We first prove that there exists a solution u with initial data u_0, which belongs to $H^1_{loc}(\mathbb{R}_+ \times \mathbb{S}) \cap L^\infty(\mathbb{R}_+, H^1(\mathbb{S}))$, satisfying Eq.(2.2) in the sense of distributions.

If we define $u_0^n := \rho_n * u_0 \in H^\infty(\mathbb{S})$ for $n \geqslant 1$. Obviously, we obtain

$$u_0^n \longrightarrow u_0 \quad \text{in } H^1(\mathbb{S}), \qquad n \to \infty. \tag{4.3}$$

In view of Young's inequality and Lemma 4.2, for all $n \geqslant 1$, we have that

$$\begin{aligned} \|u_0^n\|_1 &= \|\rho_n * u_0\|_1 \leqslant \|u_0\|_1, \\ \|u_0^n\|_{L^1} &= \|\rho_n * u_0\|_{L^1} \leqslant \|u_0\|_{L^1}, \\ \|y_0^n\|_{L^1} &= \|\rho_n * y_0\|_{L^1} = \|y_0\|_{M(\mathbb{S})}. \end{aligned} \tag{4.4}$$

Note that for all $n \geqslant 1$,

$$y_0^n = u_0^n - u_{0,xx}^n = \rho_n * (y_0) \geqslant 0.$$

Applying Theorem 3.4, it follows that there exists a global strong solution

$$u^n = u^n(\cdot, u_0^n) \in C([0,\infty); H^s(\mathbb{S})) \cap C^1([0,\infty); H^{s-1}(\mathbb{S})), \quad \forall s > \frac{3}{2},$$

and $u^n(t,x) - u_{xx}^n(t,x) \geqslant 0$ for all $(t,x) \in \mathbb{R}_+ \times \mathbb{S}$.

In view of Theorem 3.4, Lemma 3.5 and (4.4), it follows that

$$\begin{aligned} \|u_x^n(t,\cdot)\|_{L^\infty}^2 &\leqslant \|u^n(t,\cdot)\|_{L^\infty}^2 \leqslant \frac{13}{12}\|u^n(t,\cdot)\|_1^2 \\ &= \frac{13}{12}\|u_0^n(\cdot)\|_1^2 \leqslant \frac{13}{12}\|u_0(\cdot)\|_1^2. \end{aligned} \tag{4.5}$$

By the above inequality, we have that

$$\begin{aligned} \|(u^n(t,\cdot))^2 u_x^n(t,\cdot)\|_{L^2} &\leqslant \|u^n(t,\cdot)\|_{L^\infty}^2 \|u_x^n(t,\cdot)\|_{L^2} \\ &\leqslant \frac{13}{12}\|u^n(t,\cdot)\|_1^3 \leqslant \frac{13}{12}\|u_0\|_1^3. \end{aligned} \tag{4.6}$$

By Young's inequality and (4.5), for all $t \geqslant 0$ and $n \geqslant 1$, we deduce

$$\begin{aligned} &\left\| G * \partial_x \left[\frac{3}{2} u^n (u_x^n)^2 + (u^n)^3\right] + G * \frac{1}{2}(u_x^n)^3 \right\|_{L^2} \\ &\leqslant \|G_x\|_{L^2} \left(\frac{3}{2} \|u^n\|_{L^\infty} \|(u_x^n)^2\|_{L^1} + \|u^n\|_{L^\infty} \|(u^n)^2\|_{L^1} \right) \\ &\quad + \frac{1}{2}\|G\|_{L^2} \|u_x^n\|_{L^\infty} \|(u_x^n)^2\|_{L^1} \\ &\leqslant \|G_x\|_{L^2} \|u^n\|_{L^\infty} \left(\frac{3}{2}\|u_x^n\|_{L^2}^2 + \|u^n\|_{L^2}^2 \right) + \frac{1}{2}\|G\|_{L^2}\|u^n\|_{L^\infty}\|u_x^n\|_{L^2}^2 \end{aligned}$$

$$\leq \frac{1}{2}\sqrt{\frac{13}{12}}\|u_0\|_1(3\|G_x\|_{L^2}\|u^n\|_1^2 + \|G\|_{L^2}\|u^n\|_1^2)$$
$$= \frac{1}{2}\sqrt{\frac{13}{12}}(3\|G_x\|_{L^2} + \|G\|_{L^2})\|u_0\|_1^3. \qquad (4.7)$$

Combining (4.6)-(4.7) with Eq.(2.2), for all $t \geq 0$ and $n \geq 1$, we find

$$\|\frac{d}{dt}u^n(t,\cdot)\|_{L^2} \leq \frac{1}{2}\sqrt{\frac{13}{12}}\left(\sqrt{\frac{13}{3}} + \|G\|_{L^2} + 3\|G_x\|_{L^2}\right)\|u_0\|_1^3. \qquad (4.8)$$

For fixed $T > 0$, by (4.5) and (4.8), we have

$$\int_0^T \int_{\mathbb{R}} ([u^n(t,x)]^2 + [u_x^n(t,x)]^2 + [u_t^n(t,x)]^2) dx dt \leq K(T), \qquad (4.9)$$

where $K(T) = \frac{13}{48}\left(\sqrt{\frac{13}{3}} + \|G\|_{L^2} + 3\|G_x\|_{L^2}\right)^2 \|u_0\|_1^6 T + \|u_0\|_1^2 T$.

It follows that the sequence $\{u^n\}_{n\geq 1}$ is uniformly bounded in the space $H^1((0,T) \times \mathbb{S})$. Thus we can extract a subsequence such that

$$u^{n_k} \rightharpoonup u \quad \text{weakly in } H^1((0,T) \times \mathbb{S}) \text{ for } n_k \to \infty, \qquad (4.10)$$

and

$$u^{n_k} \longrightarrow u \quad \text{a.e. on } (0,T) \times \mathbb{S} \text{ for } n_k \to \infty, \qquad (4.11)$$

for some $u \in H^1((0,T) \times \mathbb{S})$. By Theorem 3.4 and (4.4), we have that for fixed $t \in (0,T)$, the sequence $u_x^{n_k}(t,\cdot) \in BV(\mathbb{S})$ satisfies

$$\mathbb{V}[u_x^{n_k}(t,\cdot)] = \|u_{xx}^{n_k}(t,\cdot)\|_{L^1} \leq \|u^{n_k}(t,\cdot)\|_{L^1} + \|y^{n_k}(t,\cdot)\|_{L^1}$$
$$\leq 2\|y^{n_k}(t,\cdot)\|_{L^1} \leq 2\|y_0^{n_k}\|_{L^1} \exp\left(\frac{13}{12}\|u_0^{n_k}\|_1^2 t\right)$$
$$\leq 2\|y_0\|_{M(\mathbb{S})} \exp\left(\frac{13}{12}\|u_0\|_1^2 t\right)$$

and

$$\|u_x^{n_k}(t,\cdot)\|_{L^\infty} \leq \sqrt{\frac{13}{12}}\|u_x^{n_k}(t,\cdot)\|_1 = \sqrt{\frac{13}{12}}\|u_0^{n_k}\|_1 \leq \sqrt{\frac{13}{12}}\|u_0\|_1.$$

By virtue of Helly's theorem cf. [24], we obtain that there exists a subsequence, denoted again by $\{u_x^{n_k}(t,\cdot)\}$, which converges at every point to some function $v(t,\cdot)$ of finite variation with

$$\mathbb{V}(v(t,\cdot)) \leq 2\|y_0\|_{M(\mathbb{S})} \exp\left(\frac{13}{12}\|u_0\|_1^2 t\right).$$

Since for almost all $t \in (0,T)$, $u_x^{n_k}(t, \cdot) \to u_x(t, \cdot)$ in $D'(\mathbb{S})$ in view of (4.11), it follows that $v(t, \cdot) = u_x(t, \cdot)$ for a.e. $t \in (0,T)$. Therefore, we have

$$u_x^{n_k}(t, \cdot) \longrightarrow u_x(t, \cdot) \quad \text{a.e. on } (0,T) \times \mathbb{S} \text{ for } n_k \to \infty, \qquad (4.12)$$

and for a.e. $t \in (0,T)$,

$$\mathbb{V}[u_x(t, \cdot)] = \|u_{xx}(t, \cdot)\|_{M(\mathbb{S})} \leqslant 2\|y_0\|_{M(\mathbb{S})} \exp\left(\frac{13}{12}\|u_0\|_1^2 t\right). \qquad (4.13)$$

By Theorem 3.4 and (4.5), we have

$$\left\|\frac{3}{2}u^n(u_x^n)^2 + (u^n)^3 + \frac{1}{2}(u_x^n)^3\right\|_{L^2}$$

$$\leqslant \frac{3}{2}\|u^n\|_{L^\infty}^2\|u_x^n\|_{L^2} + \|u^n\|_{L^\infty}^2\|u^n\|_{L^2} + \frac{1}{2}\|u_x^n\|_{L^\infty}^2\|u_x^n\|_{L^2}$$

$$\leqslant \frac{13}{12}\|u_0\|_1^2(2\|u_x^n\|_{L^2} + \|u^n\|_{L^2})$$

$$\leqslant \frac{13}{3}\|u_0\|_1^2(\|u_x^n\|_{L^2}^2 + \|u^n\|_{L^2}^2)^{\frac{1}{2}}$$

$$= \frac{13}{3}\|u_0\|_1^3.$$

For fixed time $t \in (0,T)$, we derive that the sequence

$$\left\{\frac{3}{2}u^n(u_x^n)^2 + (u^n)^3 + \frac{1}{2}(u_x^n)^3\right\}$$

is uniformly bounded in $L^2(\mathbb{S})$. Therefore, it has a subsequence, denoted again by $\left\{\frac{3}{2}u^n(u_x^n)^2 + (u^n)^3 + \frac{1}{2}(u_x^n)^3\right\}$, converging weakly in $L^2(\mathbb{S})$. By (4.11), we deduce that the weak $L^2(\mathbb{S})$-limit is $\frac{3}{2}uu_x^2 + u^3 + \frac{1}{2}u_x^3$.

Note that G, $G_x \in L^\infty(\mathbb{S})$. As $n \to \infty$, it follows that

$$G * \partial_x \left[\frac{3}{2}u^n(u_x^n)^2 + (u^n)^3\right] + G * \frac{1}{2}(u_x^n)^3$$

$$\longrightarrow G * \partial_x \left[\frac{3}{2}uu_x^2 + u^3\right] + G * \frac{1}{2}u_x^3. \qquad (4.14)$$

Combining (4.11)-(4.12) with (4.14), we have that u satisfies Eq.(2.2) in $D'((0,T) \times \mathbb{S})$.

Since for all $t \in \mathbb{R}_+$ and all $n \geqslant 1$, $u_t^{n_k}(t, \cdot)$ is uniformly bounded in $L^2(\mathbb{S})$ and $\|u^{n_k}(t, \cdot)\|_1$ has an uniform bound. therefore, the family $t \mapsto u^{n_k}(t, \cdot) \in H^1(\mathbb{S})$ is weakly equicontinuous on $[0,T]$ for any $T > 0$. An application of the Arzela-Ascoli theorem yields that $\{u^{n_k}\}$ contains a subsequence, denoted again by $\{u^{n_k}\}$, which converges weakly in $H^1(\mathbb{S})$, uniformly in $t \in [0,T]$. The limit function is u. Due to T is arbitrary, we have that u is locally and weakly continuous from $[0,\infty)$ into $H^1(\mathbb{S})$, i.e.

$$u \in C_{w,loc}(\mathbb{R}_+; H^1(\mathbb{S})).$$

For a.e. $t \in \mathbb{R}_+$, since $u^{n_k}(t,\cdot) \rightharpoonup u(t,\cdot)$ weakly in $H^1(\mathbb{S})$, in view of Theorem 3.4 and (4.4), we obtain

$$\begin{aligned}
\|u(t,\cdot)\|_{L^\infty} &\leqslant \sqrt{\frac{13}{12}}\|u(t,\cdot)\|_1 \leqslant \sqrt{\frac{13}{12}} \liminf_{n_k \to \infty} \|u^{n_k}(t,\cdot)\|_1 \\
&= \sqrt{\frac{13}{12}} \liminf_{n_k \to \infty} \|u_0^{n_k}(t,\cdot)\|_1 \\
&\leqslant \sqrt{\frac{13}{12}}\|u_0\|_1,
\end{aligned} \qquad (4.15)$$

for a.e. $t \in \mathbb{R}_+$.

Consequently, one can easily check that

$$u \in L^\infty(\mathbb{R}_+ \times \mathbb{S}) \cap L^\infty(\mathbb{R}_+; H^1(\mathbb{S})).$$

Note that by Theorem 3.4 and (4.5), we have

$$\begin{aligned}
\|u_x^n(t,\cdot)\|_{L^\infty} &\leqslant \|u^n(t,\cdot)\|_{L^\infty} \leqslant \sqrt{\frac{13}{12}}\|u^n(t,\cdot)\|_1 \\
&\leqslant \sqrt{\frac{13}{12}}\|u_0^n(t,\cdot)\|_1 \leqslant \sqrt{\frac{13}{12}}\|u_0\|_1.
\end{aligned} \qquad (4.16)$$

Combining (4.16) with (4.12), we deduce that

$$u_x \in L^\infty(\mathbb{R}_+ \times \mathbb{S}).$$

This shows that

$$u \in W^{1,\infty}(\mathbb{R}_+ \times \mathbb{S}) \cap L^\infty(\mathbb{R}_+; H^1(\mathbb{S})).$$

Note that $y_0 \in M^+(\mathbb{S})$, in view of (4.13) and Theorem 3.4, it follows that

$$\begin{aligned}
\|u(t,\cdot) - u_{xx}(t,\cdot)\|_{M(\mathbb{S})} &\leqslant \|u(t,\cdot)\|_{L^1} + \|u_{xx}(t,\cdot)\|_{M(\mathbb{S})} \\
&\leqslant \|u(t,\cdot)\|_{L^\infty} + 2\|y_0\|_{M(\mathbb{S})} \exp\left(\frac{13}{12}\|u_0\|_1^2 t\right) \\
&\leqslant \sqrt{\frac{13}{12}}\|u_0\|_1 + 2\|y_0\|_{M(\mathbb{S})} \exp\left(\frac{13}{12}\|u_0\|_1^2 t\right).
\end{aligned} \qquad (4.17)$$

For any fixed T, $\forall t \in [0,T]$, we prove that

$$(u(t,\cdot) - u_{xx}(t,\cdot)) \in M(\mathbb{S}).$$

Therefore, by virtue of (4.11)-(4.12), we obtain that for all $t \in [0,T]$, as $n \to \infty$,

$$(u^{n_k}(t,\cdot) - u^{n_k}_{xx}(t,\cdot)) \to (u(t,\cdot) - u_{xx}(t,\cdot)) \quad \text{in } D'(\mathbb{S}).$$

Since $u^{n_k}(t,\cdot) - u^{n_k}_{xx}(t,\cdot) \geq 0$, for all $(t,x) \in \mathbb{R}_+ \times \mathbb{S}$, it follows that for a.e. $t \in [0,T]$

$$(u(t,\cdot) - u_{xx}(t,\cdot)) \in M^+(\mathbb{S}).$$

This completes the proof of Theorem 4.1. □

Theorem 4.2 *Given $u_0 \in H^1(\mathbb{S})$ and $y_0 \in M^+(\mathbb{S})$. If $u, v \in W^{1,\infty}(\mathbb{R}_+ \times \mathbb{S}) \cap L^\infty(\mathbb{R}_+; H^1(\mathbb{S}))$ are the weak solutions of Eq.(2.2) with initial data u_0. Then $u = v$ for a.e. $(t,x) \in (\mathbb{R}_+, \mathbb{S})$. Moreover, $E(u) = \int_{\mathbb{S}}(u^2 + u_x^2)dx$ is a conservation law and $u \in \mathcal{C}(\mathbb{R}; H^1(\mathbb{R})) \cap \mathcal{C}^1(\mathbb{R}; L^2(\mathbb{R}))$.*

Proof First, by a regularization technique, we will show that $E(u)$ is a conservation law.

In view of Theorem 4.1, we have that u solves Eq.(2.2) in the sense of distributions, it follows that for a.e. $t \in \mathbb{R}_+$,

$$\rho_n * u_t + \rho_n * u^2 u_x + \rho_n * p_x * \left(\frac{3}{2}uu_x^2 + u^3\right) + \rho_n * p * \frac{1}{2}u_x^3 = 0. \tag{4.18}$$

Multiplying (4.18) by $\rho_n * u$, integration and in view of Lemma 4.1, we obtain that for a.e. $t \in \mathbb{R}_+$,

$$\frac{1}{2}\frac{d}{dt}\int_{\mathbb{S}}(\rho_n * u)^2 dx + \int_{\mathbb{S}}(\rho_n * u)(\rho_n * u^2 u_x)dx$$
$$+ \int_{\mathbb{S}}(\rho_n * u)\left(\rho_n * p_x * (\frac{3}{2}uu_x^2 + u^3) + \rho_n * p * \frac{1}{2}u_x^3\right)dx = 0. \tag{4.19}$$

Differentiating (4.18), we obtain a relation which multiplied by $(\rho_n * u_x)$, after integration and in view of $\partial_x^2(G * f) = G * f - f$. Then for a.e. $t \in \mathbb{R}_+$, we obtain

$$\frac{1}{2}\frac{d}{dt}\int_{\mathbb{S}}(\rho_n * u_x)^2 dx + \int_{\mathbb{S}}(\rho_n * u_x)\left(\rho_n * p * (\frac{3}{2}uu_x^2 + u^3)\right)dx$$
$$+ \int_{\mathbb{S}}(\rho_n * u_x)(\rho_{n,x} * u^2 u_x) - (\rho_n * u_x)\left(\rho_n * (\frac{3}{2}uu_x^2 + u^3)\right)dx$$

$$+ \int_S (\rho_n * u_x)\left(\rho_n * p_x * \frac{1}{2}u_x^3\right) dx = 0. \tag{4.20}$$

Similar to the arguments given on pages 11-12 in [28], we infer that for all fixed $t \in \mathbb{R}_+$, $E(u)$ is conservation law.

As $u \in C_{w,loc}(\mathbb{R}; H^1(\mathbb{R}))$ and $E(u)$ is a conservation law, we deduce

$$\|u(t,\cdot) - u(s,\cdot)\|_1^2 = \|u(t,\cdot)\|_1^2 - 2(u(t,\cdot), u(s,\cdot))_1 + \|u(t,\cdot)\|_1^2$$
$$= 2\|u_0(\cdot)\|_1^2 - 2(u(t,\cdot), u(s,\cdot))_1$$
$$\longrightarrow 0, \quad \text{as } t \to s.$$

In virtue of Eq.(2.2), we obtain

$$u \in C(\mathbb{R}_+; H^1(\mathbb{R})) \cap C^1(\mathbb{R}_+; L^2(\mathbb{R})).$$

Finally, similar to the argument given on pages 13-15 in [28], we show the uniqueness of the weak solutions to Eq.(2.2) with the initial u_0.
This completes the proof of Theorem 4.2. □

Example 4.1 Assume that the initial data

$$u_0 = \frac{\cosh\left(x - [x] - \frac{1}{2}\right)}{\sinh\left(\frac{1}{2}\right)}, \quad x \in \mathbb{R}.$$

A easy computation shows that $u_0 - u_{0,xx} \in M^+(\mathbb{S})$, and as one can check, Eq.(1.1) with u_0 has the unique global weak solution

$$u(t,x) = \frac{\cosh\left(x - t - [x - t] - \frac{1}{2}\right)}{\sinh\left(\frac{1}{2}\right)}, \quad x \in \mathbb{R}, \ t \geq 0.$$

This weak solution is a peakon periodic solution.

It is interesting to note that the Novikov equation on the line admits the peakon solutions [27]

$$u(t,x) = \pm\sqrt{c}\,e^{-|x-ct|}, \quad c > 0, \ x \in \mathbb{R}, \ t \geq 0.$$

The most nature extension to the periodic solution would be to consider [8]

$$\sum_{n=-\infty}^{\infty} \pm\sqrt{c}\,e^{-|x-ct|} = \pm\sqrt{c}\,\frac{\cosh\left(x - ct - [x - ct] - \frac{1}{2}\right)}{\sinh\left(\frac{1}{2}\right)}, \quad c > 0, \ x \in \mathbb{R}, \ t \geq 0.$$

The peakon periodic solution, although not a classical solution of Eq. (1.1), is a solution in a considerably stronger sense than the more traditionally studied weak solutions for conservation laws (such as shock waves). Indeed, it is piecewise analytic and satisfies Eq.(1.1) pointwise away from the singularities.

Following the proof of theorems 4.1-4.2, in view of Theorem 3.5, we can obtain

Theorem 4.3 Let $u_0(x) \in H^1(\mathbb{S})$. Assume $y_0 = (u_{0,xx} - u_0) \in M^+(\mathbb{S})$. Then Eq.(2.2) has a unique and global solution

$$u \in W^{1,\infty}(\mathbb{R}_+ \times \mathbb{S}) \cap L^\infty\left(\mathbb{R}_+; H^1(\mathbb{S})\right)$$

with initial data $u(0,x) = u_0$, and $(u_{xx}(t,\cdot) - u(t,\cdot)) \in M^+(\mathbb{S})$ is bounded on $[0,T]$, for any fixed $T > 0$. Moreover, $E(u) = \int_{\mathbb{R}} (u^2 + u_x^2)dx$ is a conservation law and the solution $u \in \mathcal{C}_{loc}(\mathbb{R}_+; H^1(\mathbb{S})) \cap \mathcal{C}^1_{loc}(\mathbb{R}_+; L^2(\mathbb{S}))$.

Acknowledgments This work was partially supported by CPSF (Grant No.: 2013T60086) and NSFC (Grant No.: 11401122). The authors thank the references for their valuable comments and constructive suggestions.

References

[1] H. Amann, Ordinary Differential Equation, W. de Gruyter, Berlin, 1990.

[2] R. Camassa and D. Holm, An integrable shallow water equation with peaked solitons, *Phys. Rev. Letters* **71** (1993) 1661-1664.

[3] R. Camassa, D. Holm and J. Hyman, An integrable shallow water equation, *Adv. Appl. Mech.* **31** (1994) 1-33.

[4] G.M. Coclite, K.H. Karlsen and N.H. Risebro, Numberical schemes for computing discontinuous solutions of the Degasperis-Procesi equation, *IMA J. Numer. Anal.* **28** (2008) 80-105.

[5] A. Constantin, On the scattering problem for the Camassa-Holm equation, *Proc. R. Soc. London A* **457** (2001) 953-970.

[6] A. Constantin, Finite propagation speed for the Camassa-Holm equation, *J. Math. Phys.* **46** (2005) 023506, 4 pp.

[7] A. Constantin and J. Escher, Wave breaking for nonlinear nonlocal shallow water equation, *Acta Math.* **181** (1998) 229-243.

[8] A. Constantin and J. Escher, Well-posedness, global existence and blowup phenomena for a periodic quasi-linear hyperbolic equation, *Comm. Pure Appl. Math.* **51** (1998) 475-504.

[9] A. Constantin, J. Escher, On the blow-up rate and the blow-up set of breaking waves for a shallow water equation, *Math. Z.* **233** (2000) 75-91.

[10] A. Constantin and L. Molinet, Global weak solutions for a shallow water equation, *Comm. Math. Phys.* **211** (2000) 45-61.

[11] A. Constantin and L. Molinet, The initial value problem for a generalized Boussinesq equation, *Differential and Integral equations* **15** (2002) 1061-1072.

[12] A. Degasperis, D.D. Holm and A.N.W. Hone, A new integrable equation with peakon solution, *Theoret. and Math. Phys* **133** (2002) 1463-1474.

[13] A. Degasperis and M. Procesi, Asymptotic integrability, in: A. Degasperis, G. Gaeta (Eds.), *Symmetry and Perturbation Theory, World Scientific* (1999) 23-37.

[14] J. Escher, Y. Liu and Z. Yin, Shock waves and blow–up phenomena for the periodic Degasperis–Procesi equation, *Indiana Univ. Math. J.* **56** (2007) 87-117.

[15] A. Fokas and B. Fuchssteiner, Symplectic structures, their Bäcklund transformation and hereditary symmetries, *Physica D* **4** (1981), 47-66.

[16] D.D. Holm and M.F. Staley, Wave structure and nonlinear balances in a family of evolutionary PDEs, *SIAM J. Appl. Dyn. Syst.* **2** (2003) 323-380.

[17] A.N.W. Hone, H. Lundmark and J. Szmigielski, Explicit multipeakon solutions of Novikov's cubically nonlinear integrable Camassa–Holm type equation, *Dynamics of Partial Differential Equations* **6** (2009) 253-289.

[18] A.N.W. Hone and J. Wang, Integrable peakon equations with cubic nonlinearity, *J. Phys. A: Math. Theor.* **41** (2008) 372002.

[19] T. Kato, Quasi-linear equation of evolution, with applications to partical differential equations, in: Spectral Theorey and Differential Equation, in: Lecture Notes in Math., vol. 488, Spring-Verlag, Berlin, 1975, pp. 25-70.

[20] T. Kato and G. Ponce, Commutator estimation and the Euler and Navier–Stokes Equation, *Comm. Pure Appl. Math.* **41** (1998) 891-907.

[21] H. Lundmark, Formation and dynamics of shock waves in the Degasperis-Procesi equation, *J. Nonlinear Sci.* **17** (2007) 169-198.

[22] J. Malek, J. Necas, M. Rokyta and M. Ruzicka, Weak and Measure-valued Solutions to Evolutionary PDEs, London: Chapman & Hall, 1996.

[23] V.S. Mikhailov and V.S. Novikov, Perturbative symmetry approach, *J. Phys. A: Math. Gen.* **35** (2002) 4775-4790.

[24] I.P. Natanson, Theory of Functions of a Real Variable, New York: F. Ungar Publ. Co., 1998.

[25] V.S. Novikov, Generalizations of the Camassa-Holm equation, *J. Phys. A: Math. Theor.* **42** (2009) 342002.

[26] F. Tiglay, The periodic Cauchy problem for Novikov's equation, *Int. Math. Res. Notices* (2010).

[27] X. Wu and Z. Yin, Well-posedness and global existence for the Novikov equation, *Annali Sc. Norm. Sup. Pisa,* XI (2012) 707-727.

[28] X. Wu and Z. Yin, Global weak solutions for the Novikov equation, *J. Phys. A:Math. Theor.* **44** (2011) 055202 (17pp).

[29] Z. Yin, On the Cauchy problem for an integrable equation with peakon solutions, *Illinois J. Math.* **47** (2003) 649-666.

[30] Z. Yin, Global weak solutions to a new periodic integrable equation with peakon solutions, *J. Funct. Anal.* **212** (2004) 182-194.

Quasineutral Limit of the Pressureless Euler-Poisson Equation for Ions*

Pu Xueke (蒲学科) and Guo Boling (郭柏灵)

Abstract In this paper, we consider the quasineutral limit of the Euler-Poisson equation for cold ions when the Debye length tends to zero. In the cold ion case the Euler-Poisson equation is pressureless and hence fails to be Friedrich symmetrisable, excluding the application of the PsDO energy estimates method of Grenier to obtain uniform estimates independent of ε. We use ε-weighted norms to overcome this difficulty, which combine energy estimates in different levels with weights depending on ε. Finally, that the quasineutral regimes are the compressible Euler equations is proven for well prepared initial data. As a natural extension, we also obtain the zero temperature limit of the Euler-Poisson equation.

Keywords Euler-Poisson equation; Quasineutral limit; compressible Euler equation

1 Introduction

In this paper, we consider the Euler-Poisson equation for cold ions in plasma

$$(EP) \begin{cases} \partial_t n + div(n\mathbf{u}) = \mathbf{0}, & (1.1a) \\ \partial_t \mathbf{u} + \mathbf{u} \cdot \nabla \mathbf{u} = -\nabla \phi, & (1.1b) \\ \varepsilon \Delta \phi = e^\phi - n, & (1.1c) \end{cases}$$

where n is the density of the ions, $\mathbf{u} = (u_1, \cdots, u_d)$ is the velocity field and ϕ is the electric potential at time $t \in \mathbb{R}^+$ and position $x \in \mathbb{R}^d$, $d \leqslant 3$. Here e^ϕ is the rescaled electron density by the famous Boltzmann relation and $\varepsilon \ll 1$ is a small parameter representing the squared scaled Debye length $\varepsilon = \lambda_D^2/L^2 = \epsilon_0 k_B T_i/N_i e^2 L^2$, where λ_D is the Debye length, L is the characteristic observation length, ϵ_0 is the vacuum

---

* Quarterly of Applied Mathematics, 2016, LXXIV(2): 245–273. DOI:https://doi.org/10.1090/qam/1424

permittivity, κ_B is the universal Boltzmann constant, T_i and N_i are respectively the average temperature and density of ions and e is the fundamental electric charge. For typical plasma applications, the Debye length is very small compared to the characteristic length of physical interest and it is therefore necessary to consider the limiting system when $\varepsilon \to 0$. For more physical background of the Euler-Poisson equation or the ion-acoustic plasma, one may refer to [14].

Formally, by letting $\varepsilon \to 0$, we obtain from the third equation in (1.1) that $\phi = \ln n$, and hence the following compressible Euler system

$$(EQ) \quad \begin{cases} \partial_t n + div(n\mathbf{u}) = 0, \\ \partial_t \mathbf{u} + \mathbf{u} \cdot \nabla \mathbf{u} + \nabla \ln n = 0. \end{cases} \quad (1.2)$$

This limit system (1.2) is an hyperbolic symmetrisable system, whose classical result for the existence and uniqueness of sufficiently smooth solutions in small time interval is available in [17]. The system (1.2) have to be supplemented by suitable initial conditions. We shall assume that the plasma is uniform and electrically neutral near infinity, i.e., $n \to n^{\pm}$ and $\mathbf{u} \to 0$ as $x \to \pm\infty$. More precisely, let $\tilde{n}$ be a smooth strictly positive function, constant outside $x \in [-1, +1]$, going to $n^{\pm}$ as $x \to \infty$. We assume that the initial conditions $(n_0^0, \mathbf{u}_0^0)$ satisfy

$$(n_0^0 - \tilde{n}) \in H^s(\mathbb{R}^d), \quad \mathbf{u}_0^0 \in \mathbf{H}^s(\mathbb{R}^d), \quad n_0^0 \geqslant \sigma > 0, \quad (1.3)$$

for some $s > 3/2$ and some constant $\sigma > 0$.

Theorem 1.1 *Let $(n_0^0, \mathbf{u}_0^0) \in H^{s'} \times H^{s'}$ be initial data with $s' > \dfrac{d}{2} + 1$ and satisfy (1.3). Then there exists $T > 0$, maximal time of existence and a solution $(n^0, \mathbf{u}^0)$ of (1.2) on $0 \leqslant t < T$ with initial data $(n_0^0, \mathbf{u}_0^0)$ such that for every $T' < T$,*

$$(n^0 - \tilde{n}, \mathbf{u}^0) \in (C([0,T]; H^{s'}) \times C([0,T]; \mathbf{H}^{s'})) \cap (C^1([0,T]; H^{s'-1}) \times C^1([0,T]; \mathbf{H}^{s'-1}))$$

and T depends only on $\|(n_0^0 - \tilde{n}, \mathbf{u}_0^0)\|_{H^{s'} \times \mathbf{H}^{s'}}$.

Remark 1.1 *For the maximal existence time T, either $T = \infty$ in the case of global existence, or $T < \infty$ and the solution blows up when $t \to T$:*

$$\limsup_{t \to T} \left(\|\mathbf{u}^0\|_{L^\infty} + \|n^0\|_{L^\infty} + \left\|\frac{1}{n^0}\right\|_{L^\infty} + \int_0^t (\|\partial_x \mathbf{u}^0\|_{L^\infty} + \|\partial_x n^0\|_{L^\infty}) d\tau \right) = \infty. \quad (1.4)$$

Therefore, we will work on a time interval $[0, T']$ for $T' < T$ (but arbitrary close to T) in order to insure $0 < \sigma' < n^0(t,x) < \sigma''$ for all (t,x), for some constants $\sigma', \sigma'' > 0$. Here, σ' may approach to 0 as T' goes to T.

Let us define $\phi^0 = \ln n^0$. The main result in this paper is the following

Theorem 1.2 *Let $s' \in \mathbb{N}$ with $s' > \left[\frac{d}{2}\right] + 2$ be sufficiently large. Let $(n_0^0, \mathbf{u}_0^0) \in H^{s'} \times \mathbf{H}^{s'}$ and $(n^0, \mathbf{u}^0)$ be the solution of the limit system (1.2) on $[0, T)$ with initial data $(n_0^0, \mathbf{u}_0^0)$, given in Theorem 1.1. Then there exists solutions $(n^\varepsilon(t), \mathbf{u}^\varepsilon(t))$ of (1.1) with the same initial data on $[0, T^\varepsilon)$ with $\liminf_{\varepsilon \to 0} T^\varepsilon \geq T$. Moreover, for every $T' < T$ and for every ε small enough, $\varepsilon^{-1}(n^\varepsilon - n^0)$ and $\varepsilon^{-1}(\mathbf{u}^\varepsilon - \mathbf{u}^0)$ are bounded in $L^\infty([0, T']; H^s)$ and $L^\infty([0, T']; \mathbf{H}^s)$, respectively, for some $s < s'$.*

This theorem will be proved in Section 2. Without essential difficulties, we can show that the same result holds on the torus $\mathbb{T}^d$ following the method in the present paper. The details are omitted.

Before proving this theorem, we make some points that stimulated our work in the present paper. The more general isothermal Euler-Poisson equation for ions has the following form

$$\begin{cases} \partial_t n + div(n\mathbf{u}) = 0, & (1.5a) \\ \partial_t \mathbf{u} + \mathbf{u} \cdot \nabla \mathbf{u} + \frac{T_i}{n}\nabla n = -\nabla \phi, & (1.5b) \\ \varepsilon \Delta \phi = e^\phi - n, & (1.5c) \end{cases}$$

where $T_i > 0$ is the ion temperature. When $T_i = 0$ (compared with the electron temperature), this equation reduces to (1.1) for the cold ions. When $T_i > 0$, the pressure term introduces a smoothing effect that enables Cordier and Grenier [2] to prove the quasineutral limit as $\varepsilon \to 0$ by using the pseudodifferential energy estimates method of [6]. It is shown that, under suitable conditions, the solution of (1.5) converges to the following Euler equation as $\varepsilon \to 0$

$$\begin{cases} \partial_t n + div(n\mathbf{u}) = 0, \\ \partial_t \mathbf{u} + \mathbf{u} \cdot \nabla \mathbf{u} + (T_i + 1)\nabla \ln n = 0. \end{cases} \quad (1.6)$$

But their method cannot be applied to the case when $T_i = 0$, where such a smoothing effect vanishes. The main difference between (1.1) and (1.5) is that (1.5) has the pressure term $T_i \nabla \ln n$, which is crucial in proving the quasineutral result of the Euler-Poisson equation (1.5). With this term, the hyperbolic part of (1.5) is Friedrich symmetrisable and the general framework of pseudodifferential operator energy estimates methods of Grenier [6] can be applied. One may refer to [2] for more details of application of this method in treating the quasineutral limit of (1.5). But without the pressure term, as is the case in the present paper, the pseudodifferential energy method cannot apply and no quasineutral limit can be drawn without introducing new techniques, since the hyperbolic part (the equations (1.1a) and (1.1b)) of (1.1) is not symmetrisable.

Some historic results follow. For the Euler-Poisson equation (1.5), Guo and Pausader [8] constructed global smooth irrotational solutions with small amplitude for this equation with fixed $\varepsilon > 0$ and $T_i > 0$. Very recently, Guo and the first author of the present paper [9] derived the KdV equation from (1.5) for the full range of $T_i \geqslant 0$, which inspired the first author [19] to derive the Kadomtsev-Petviashvili II equation and the Zakharov-Kuznetsov equation (ZKE) via the Gardner-Morikawa type transformations. Lannes *et al* independently studied the ZKE limit of the Euler-Poisson equation under long wavelength limit [15]. Very recently, Han-Kwan also studied the KdV and ZKE limit from the kinetic Vlasov-Poisson system under long wavelength limit. Guo *et al* [7] made a breakthrough for the Euler-maxwell two-fluid system in 3D and proved that irrotational, smooth and localized perturbations of a constant background with small amplitude lead to global smooth solutions. The longwavelength limit of Euler-Poisson system in [15, 19] could be interpreted as a quasineutral limit of the Euler-Poisson equation to the constant solution $(1, 0, 0)$. However, the quasineutral limit of the present paper concerns of solutions $(n, \mathbf{u}, \phi)$ to a general compressible Euler equation (1.6) is new.

For the quasineutral limit of the Euler-Poisson equation, as pointed above, Cordier and Grenier proved the quasineutral limit from the isothermal Euler-Poisson equation (1.5) to the Euler equation (1.6) in [2]. Wang [24] studied the quasineutral limit of the Euler-Poisson system with and without viscosity. We also would like to remark that Loeper [16] proved quasineutral limit results for the electron Euler-Poisson equation without pressure term and derived the incompressible Euler equation and hence is irrelevant to the present result. Han-Kwan [10] studied the quasineutral limit of the Vlasov-Poisson system with massless electrons, where the author derived the classical isothermal Euler equation for cold ions using the relative entropy method. Very recently, Gerard-Varet *et al* studied the quasineutral limit of the system (1.5) in $\mathbb{R}^3_+$ and boundary layers were constructed. For numerical studies for the pressureless Euler-Poisson equation (1.1), the reader may refer to a recent paper of Degond *et al* [3], which analyzes various schemes for the Euler-Poisson-Boltzmann equation. For more results on the quasi-neutral limit results of the Euler-Poisson equation and related models, one may refer to various recent papers and the references therein, see [1,4,5,12,18,20–22,24,25] to list only a few. The difference between [20] and the present paper is that [20] shortly discusses the weak convergence of weak solutions, while the present paper considers the strong convergence of smooth solutions which requires subtle estimates.

Now, we make a remark on Cordier and Grenier's results on the quasineutral limit for the Euler-Poisson equation for $T_i > 0$ fixed. The presence of the pressure term $T_i \nabla \ln n$ in (1.5) enables them to derive the Euler equation as Debye length goes to zero. But their estimates are not uniform in $T_i > 0$, because the smoothing effects of the pressure term vanish as $T_i \to 0$. One may refer to Guo-Pu [9] for a clear study why this smoothing effects vanish. But by similar method we use to prove Theorem 1.2 in Section 2, we can indeed give uniform in $T_i > 0$ estimates and hence improve their results. This would yield the zero temperature limit as $T_i \to 0$ stated in Theorem 1.3 below. Let $(n^{\varepsilon,T_i}, \mathbf{u}^{\varepsilon,T_i}, \phi^{\varepsilon,T_i})$ be a solution of (1.5) and $(n^{0,T_i}, \mathbf{u}^{0,T_i}, \phi^{0,T_i})$ be a solution of (1.6) with the same initial data. Let

$$n^{\varepsilon,T_i} = n^{0,T_i} + \varepsilon n^{1,T_i}, \quad \mathbf{u}^{\varepsilon,T_i} = \mathbf{u}^{0,T_i} + \varepsilon \mathbf{u}^{1,T_i}, \quad \phi^{\varepsilon,T_i} = \phi^{0,T_i} + \varepsilon \phi^{1,T_i}.$$

Here ϕ^{ε,T_i} and n^{ε,T_i} satisfy the Poisson equation (1.5c) and indeed ϕ^{ε,T_i} can be solved via $\phi^{\varepsilon,T_i} = \phi^{\varepsilon,T_i}[n^{\varepsilon,T_i}]$ and $\phi^{0,T_i} = \ln n^{0,T_i}$. We will prove the following

Theorem 1.3 *Let $(n^{0,T_i}, \mathbf{u}^{0,T_i}, \phi^{0,T_i})$ be a solution of (1.6) on $[0,T)$ with initial data $(n_0, \mathbf{u}_0) \in H^{s'}$ with s' sufficiently large. There exists solutions $(n^{\varepsilon,T_i}, \mathbf{u}^{\varepsilon,T_i}, \phi^{\varepsilon,T_i})$ of (1.5) with the same initial data on $[0, T^{\varepsilon,T_i})$ with $\liminf_{\varepsilon, T_i \to 0} T^{\varepsilon,T_i} \geq T$. Moreover, for every $T' < T$ and for ε, T_i small enough, $(n^{1,T_i}, \mathbf{u}^{1,T_i})$ are bounded in $L^\infty([0,T']; H^s)$ for some $s < s'$.*

By such a result, we can indeed derive the Euler equation (1.2) from (1.5) by first letting $\varepsilon \to 0$ and then letting $T_i \to 0$. The limit $T_i \to 0$ is usually known as the cold ion limit. Furthermore, $T_i > 0$ and $\varepsilon > 0$ don't depend on each other. They can go to zero independently.

The next section is devoted to the proof of Theorem 1.2. For this purpose, we write the solution of (1.1) as $n^\varepsilon = n^0 + \varepsilon n^1$ and $\mathbf{u}^\varepsilon = \mathbf{u}^0 + \varepsilon \mathbf{u}^1$ and consider the remainder system (R_ε) of n^1 and $\mathbf{u}^1$. The main idea is then to show that $(n^1, \mathbf{u}^1)$ is uniformly bounded in $H^s \times \mathbf{H}^s$ when $\varepsilon \to 0$. To overcome the difficulty of non-symmetrisability of (1.1), we introduce some triple norm $|||\cdot|||_{\varepsilon,s}$

$$|||\mathbf{u}^1|||^2_{\varepsilon,s} = \|\mathbf{u}^1\|^2_{H^s} + \varepsilon\|\nabla\mathbf{u}^1\|^2_{H^s},$$
$$|||\phi^1|||^2_{\varepsilon,s} = \|\phi^1\|^2_{H^s} + \varepsilon\|\phi^1\|^2_{H^s} + \varepsilon^2\|\Delta\phi^1\|^2_{H^s},$$
(1.7)

and then show that $|||(\mathbf{u}^1,\phi^1)|||_{\varepsilon,s}$ is uniformly bounded on some time interval independent of ε. The main novelty of the proof is then to combine the s-order energy estimates with the $(s+1)$-order energy estimates with weights 1 and ε. By such a combination, we obtain some Gronwall type inequality for $|||(\mathbf{u}^1,\phi^1)|||_{\varepsilon,s}$, which enables us to obtain uniform estimates independent of ε. The estimates of $\|n^1\|_{H^s}$ is obtained through some elliptic estimates from the Poisson equation (1.1c) in Section 2.2. This method was successfully employed in our previous paper to handle the long wavelength limit of the Euler-Poisson system to the KdV equation in [9], and could be potentially useful in treating the quasineutral limit for the pressureless electron Euler-Poisson equations.

Theorem 1.3 is proven in Section 3 in the same spirit of the proof of Theorem 1.2.

We introduce several notations. We let L^p denote the usual Lebesgue space of p-th integrable functions normed by $\|\cdot\|_{L^p}$. When $p=2$, we usually use $\|\cdot\|$ instead of $\|\cdot\|_{L^2}$. The Sobolev space H^s, $s\in\mathbb{Z}$, $s\geqslant 0$ is defined as $H^s(\mathbb{R}^d) = \{f(x): \sum_{|\alpha|\leqslant s}\|\partial^\alpha f\|^2 < \infty\}$, where $\alpha = (\alpha_1,\cdots,\alpha_d)$ is a multi-index, $|\alpha| = \sum\alpha_i$ and $\partial^\alpha = \frac{\partial^{|\alpha|}}{\partial^{\alpha_1}_{x_1}\cdots\partial^{\alpha_d}_{x_d}}$. H^s is a Banach space with norm $\|f\|_{H^s} = (\sum_{|\alpha|\leqslant s}\|\partial^\alpha f\|^2)^{1/2}$. For definiteness, we will restrict ourselves to the physical space dimensions $d\leqslant 3$ in this paper.

2 Proof of Theorem 1.2

The purpose of this section is to prove Theorem 1.2. Let $(n^\varepsilon,\mathbf{u}^\varepsilon,\phi^\varepsilon)$ satisfy the Euler-Poisson equation (1.1), and $(n^0,\mathbf{u}^0,\phi^0)$ be a sufficiently smooth solution of the Euler equation (1.2). We let

$$n^\varepsilon = n^0 + \varepsilon n^1, \quad \mathbf{u}^\varepsilon = \mathbf{u}^0 + \varepsilon\mathbf{u}^1, \quad \phi^\varepsilon = \phi^0 + \varepsilon\phi^1. \tag{2.1}$$

Here ϕ^ε and n^ε satisfy the Poisson equation (1.1c) and indeed ϕ^ε can be solved via $\phi^\varepsilon = \phi^\varepsilon[n^\varepsilon]$ and $\phi^0 = \ln n^0$. Then $(n^1,\mathbf{u}^1,\phi^1)$ satisfy the remainder system (R_ε):

$$(R_\varepsilon)\begin{cases}\partial_t n^1 + \nabla\cdot(n^0\mathbf{u}^1 + \mathbf{u}^0 n^1) + \varepsilon\nabla\cdot(n^1\mathbf{u}^1) = 0, & (2.2\mathrm{a})\\ \partial_t\mathbf{u}^1 + \mathbf{u}^0\cdot\nabla\mathbf{u}^1 + \mathbf{u}^1\cdot\nabla\mathbf{u}^0 + \varepsilon\mathbf{u}^1\cdot\nabla\mathbf{u}^1 = -\nabla\phi^1, & (2.2\mathrm{b})\\ -\varepsilon\Delta\phi^1 = \Delta\phi^0 + n^1 - n^0\phi^1 + \sqrt{\varepsilon}R^1, & (2.2\mathrm{c})\end{cases}$$

where
$$R^1 = \varepsilon^{-3/2}(n^0 + \varepsilon n^0 \phi^1 - e^{\phi^0 + \varepsilon \phi^1}). \tag{2.3}$$

To prove Theorem 1.2, we need only to derive some uniform bound for the remainder equation (2.2). To slightly simplify the presentation, we assume that (2.2) has smooth solutions in a small time T_ε dependent on ε. Let $\tilde{C}$ be a constant to be determined later, much larger than the bound of $\|(n_0^1, \mathbf{u}_0^1)\|_s$, such that on $[0, T_\varepsilon]$,

$$\sup_{[0, T_\varepsilon]} \|(n^1, \mathbf{u}^1, \phi^1)\|_{H^s} \leqslant \tilde{C}. \tag{2.4}$$

We will prove that $T_\varepsilon > T$ as $\varepsilon \to 0$ for some $T > 0$. Recalling the expressions for n and $\mathbf{u}$ in (2.1), we immediately know that there exists some $\varepsilon_1 = \varepsilon_1(\tilde{C}) > 0$ such that on $[0, T_\varepsilon]$,

$$\sigma'/2 < n^\varepsilon < 2\sigma'', \quad |\mathbf{u}^\varepsilon| \leqslant 1/2, \quad \text{for all} \quad 0 < \varepsilon < \varepsilon_1. \tag{2.5}$$

2.1 Estimates for R_1

We first bound R^1 in terms of ϕ^1. More precisely, we have the following:

Lemma 2.1 *Let $(n^0, \mathbf{u}^0, \phi^0)$ be a sufficiently smooth solution of (1.2) by Theorem 1.1. Then for the remainder term (2.3), we have on $[0, T_\varepsilon]$,*

$$\|R^1\|_{H^k} \leqslant C(\sqrt{\varepsilon}\tilde{C})\|\phi^1\|_{H^k}, \quad \text{and}$$
$$\|\partial_t R^1\|_{H^k} \leqslant C(\sqrt{\varepsilon}\tilde{C})(\|\phi^1\|_{H^k} + \|\partial_t \phi^1\|_{H^k}), \quad \forall k \geqslant 0. \tag{2.6}$$

In particular, there exists some $\varepsilon_1 > 0$ and $C_1 = C(1)$ such that

$$\|R^1\|_{H^k} \leqslant C_1 \|\phi^1\|_{H^k}, \quad \text{and}$$
$$\|\partial_t R^1\|_{H^k} \leqslant C_1(\|\phi^1\|_{H^k} + \|\partial_t \phi^1\|_{H^k}), \quad \forall k \geqslant 0, \tag{2.7}$$

for all $0 < \varepsilon < \varepsilon_1$ and $t \in [0, T_\varepsilon]$.

Proof From the Taylor expansion in the integral form, we have

$$R^1 = \varepsilon^{1/2} e^{\phi^0} \int_0^1 e^{\theta \varepsilon \phi^1}(1-\theta) d\theta (\phi^1)^2.$$

By taking L^2 norm, we have

$$\|R^1\| \leqslant \sqrt{\varepsilon}\|e^{\phi^0}\|_{L^\infty} e^{\varepsilon \|\phi^1\|_{L^\infty}} \|\phi^1\|_{L^2} \|\phi^1\|_{L^\infty}.$$

From the continuity assumption (2.4), we have $\|\phi^1\|_{L^\infty} \leqslant C\tilde{C}$ on $[0, T_\varepsilon]$ and hence

$$\|R^1\|_{L^2} \leqslant C(\sqrt{\varepsilon}\tilde{C})\|\phi^1\|_{L^2}, \quad \forall t \in [0, T_\varepsilon].$$

By applying ∂^α with $|\alpha| = k$, $k \geqslant 1$ integers, similar estimates yield

$$\|R^1\|_{H^k} \leqslant C(\sqrt{\varepsilon}\tilde{C})\|\phi^1\|_{H^k}.$$

Taking ∂_t to R^1 and then taking the H^k norm, we obtain

$$\|\partial_t R^1\|_{H^k} \leqslant C(\sqrt{\varepsilon}\tilde{C})(\|\phi^1\|_{H^k} + \|\partial_t\phi^1\|_{H^k}).$$

Finally, choosing $\varepsilon_1 = (1/\tilde{C})^2$ yields (2.7). $\square$

2.2 Elliptical estimates

The following lemmas provide useful estimates between n^1, $\mathbf{u}^1$ and ϕ^1. These will be used widely in the uniform estimates in the next subsection.

Lemma 2.2 *Let $(n^1, \mathbf{u}^1, \phi^1)$ be a smooth solution for the remainder system (R_ε), and α be a multiindex. There exist ε_1 and C such that for any $0 < \varepsilon < \varepsilon_1$ and any multiindices α with $|\alpha| = k \geqslant 0$, there hold*

$$\|\partial_x^\alpha n^1\|^2 \leqslant C + C\|\phi^1\|_{H^k}^2 + C\varepsilon^2\|\Delta\phi^1\|_{H^k}^2, \quad \text{and}$$
$$\|\partial_x^\alpha \phi^1\|^2 + \varepsilon\|\partial_x^\alpha \nabla\phi^1\| + \varepsilon^2\|\Delta\partial_x^\alpha\phi^1\|^2 \leqslant C + C\|\partial_x^\alpha n^1\|^2,$$

on the interval $[0, T_\varepsilon]$.

Proof Taking the L^2 inner product of (2.2c) with ϕ^1 and then integrating by parts yield

$$\varepsilon\|\nabla\phi^1\|^2 + \int n^0|\phi^1|^2 = \int \phi^1\Delta\phi^0 + \int n^1\phi^1 + \sqrt{\varepsilon}\int \phi^1 R^1.$$

Hereafter, $\int = \int_{\mathbb{R}^d} \cdots dx$. As $n^0 > \sigma' > 0$ for $t \leqslant T' < T$, we obtain by Young's inequality

$$\varepsilon\|\nabla\phi^1\|^2 + \sigma'\|\phi^1\|^2 \leqslant \frac{\sigma'}{4}\|\phi^1\|^2 + \frac{4C}{\sigma'}(\|\Delta\phi^0\|^2 + \|n^1\|^2 + \varepsilon\|R^1\|^2).$$

From Lemma 2.1, there exists $\varepsilon_1 > 0$ such that $\|R^1\| \leqslant C_1\|\phi^1\|$. Then by choosing a new smaller ε_1 such that $\varepsilon_1 \leqslant \dfrac{\sigma'^2}{16CC_1}$, we then have for any $0 < \varepsilon < \varepsilon_1$ that

$$\varepsilon\|\nabla\phi^1\|^2 + \sigma'\|\phi^1\|^2 \leqslant \frac{C}{\sigma'}(1 + \|n^1\|^2). \tag{2.8}$$

Similarly, by taking the L^2 inner product of (2.2c) with $\varepsilon\Delta\phi^1$ and integrating by parts, we obtain that

$$\varepsilon^2 \|\Delta\phi^1\|^2 + \varepsilon \int n^0 |\nabla\phi^1|^2 = -\varepsilon \int \Delta\phi^0 \Delta\phi^1 + \varepsilon \int n^1 \Delta\phi^1$$

$$- \varepsilon \int \nabla n^0 \phi^1 \nabla\phi^1 + \varepsilon^{3/2} \int R^1 \phi^1$$

$$\leqslant \frac{\varepsilon^2}{2} \|\Delta\phi^1\|^2 + 4(\|\Delta\phi^0\|^2 + \|n^1\|^2 + \varepsilon\|R^1\|^2 + \varepsilon\|\phi^1\|^2 + \varepsilon\|\nabla\phi^1\|^2),$$

which yields for any $0 < \varepsilon < \varepsilon_1$ for some $\varepsilon_1 > 0$ that

$$\varepsilon^2 \|\Delta\phi^1\|^2 + \varepsilon\sigma' \|\nabla\phi^1\|^2 \leqslant C(1 + \|n^1\|^2), \qquad (2.9)$$

as $n^0 > \sigma' > 0$ for $t \leqslant T' < T$, where the constant C depends on σ'. By combining (2.8) and (2.9) together, we easily obtain that for any $0 < \varepsilon < \varepsilon_1$:

$$\varepsilon^2 \|\Delta\phi^1\|^2 + \varepsilon\sigma' \|\nabla\phi^1\|^2 + \|\phi^1\|^2 \leqslant C(1 + \|n^1\|^2),$$

for some constant C depending on σ'. On the other hand, by taking the L^2 norm of (2.2c), we obtain that for any $0 < \varepsilon < \varepsilon_1$:

$$\|n^1\|^2 \leqslant \varepsilon^2 \|\Delta\phi^1\|^2 + \|n^0\|_{L^\infty}^2 \|\phi^1\|^2 + \|\Delta\phi^0\|^2 + \varepsilon\|R^1\|^2$$

$$\leqslant C(1 + \|\phi^1\|^2 + \varepsilon^2 \|\Delta\phi^1\|^2),$$

thanks again to Lemma 2.1, where C depends on σ'' and C_1. Therefore, we finishes the proof when $k=0$. Higher order estimates can be handled similarly, and we omit further details. $\square$

Lemma 2.3 *Let $(n^1, \mathbf{u}^1, \phi^1)$ be a smooth solution for the remainder system (R_ε), and α be an integer. There exist ε_1 and C such that for any $0 < \varepsilon < \varepsilon_1$ and any multiindices α with $|\alpha| = k$, there holds*

$$\|\partial^\alpha \partial_t n^1\|^2 \leqslant C(1 + \|\mathbf{u}^1\|_{\mathbf{H}^{k+1}}^2 + \|\phi^1\|_{H^{k+1}}^2 + \varepsilon^2 \|\Delta\phi^1\|_{H^{k+1}}^2), \qquad (2.10)$$

on the time interval $[0, T_\varepsilon]$.

Proof We take the L^2 norm of (2.2a) to obtain

$$\|\partial_t n^1\|^2 \leqslant C(\|\mathbf{u}^1\|_{\mathbf{H}^1}^2 + \|n^1\|_{H^1}^2) + \varepsilon^2(\|n^1\|_{L^\infty}^2 \|\nabla\mathbf{u}^1\|^2 + \|\mathbf{u}^1\|_{L^\infty}^2 \|\nabla n^1\|^2)$$

$$\leqslant C(1 + \varepsilon^2(\|n^1\|_{L^\infty}^2 + \|\mathbf{u}^1\|_{L^\infty}^2))(\|\mathbf{u}^1\|_{\mathbf{H}^1}^2 + \|n^1\|_{H^1}^2),$$

for some constant C depending on $(n^0, \mathbf{u}^0)$. By the continuity assumption (2.4) and Lemma 2.2, we have

$$\|\partial_t n^1\|^2 \leqslant C(1+\varepsilon^2\tilde{C})(1+\|\mathbf{u}^1\|_{\mathbf{H}^1}^2 + \|\phi^1\|_{H^1}^2 + \varepsilon^2\|\Delta\phi^1\|_{H^1}^2),$$
$$\leqslant C(1+\|\mathbf{u}^1\|_{\mathbf{H}^1}^2 + \|\phi^1\|_{H^1}^2 + \varepsilon^2\|\Delta\phi^1\|_{H^1}^2),$$

for any $0 < \varepsilon < \varepsilon_1$, for some $\varepsilon_1 > 0$.

Higher order inequalities are proved similarly. Taking ∂^α with $|\alpha| = k \geqslant 1$ to the equation (2.2a), and then taking the L^2 norm to obtain

$$\|\partial^\alpha \partial_t n^1\|^2 \leqslant C(\mathbf{u}^1\|_{\mathbf{H}^{k+1}}^2 + \|n^1\|_{H^{k+1}}^2) + \varepsilon^2(\|n^1\|_{L^\infty}^2 \|\partial^\alpha \nabla \mathbf{u}^1\|^2 + \|\mathbf{u}^1\|_{L^\infty}^2 \|\partial^\alpha \nabla n^1\|^2)$$
$$\leqslant C(1+\varepsilon^2(\|n^1\|_{L^\infty}^2 + \|\mathbf{u}^1\|_{L^\infty}^2))(\|\mathbf{u}^1\|_{\mathbf{H}^{k+1}}^2 + \|n^1\|_{H^{k+1}}^2)$$
$$\leqslant C(1+\|\mathbf{u}^1\|_{\mathbf{H}^{k+1}}^2 + \|\phi^1\|_{H^{k+1}}^2 + \varepsilon^2\|\Delta\phi^1\|_{H^{k+1}}^2),$$

for any $0 < \varepsilon < \varepsilon_1$, for some $\varepsilon_1 > 0$, where we have used the multiplicative estimates in Lemma A.1. $\square$

Lemma 2.4 *Let $(n^1, \mathbf{u}^1, \phi^1)$ be a smooth solution for the remainder system (R_ε), and α be an integer. There exist ε_1 and C such that for any $0 < \varepsilon < \varepsilon_1$ and any multiindices α with $|\alpha| = k$,*

$$\|\partial^\alpha \partial_t \phi^1\|^2 + \varepsilon\|\partial^\alpha \nabla \partial_t \phi^1\|^2 + \varepsilon^2\|\partial^\alpha \Delta \partial_t \phi^1\|^2 \leqslant C(1+\|\partial^\alpha \partial_t n^1\|^2 + \|\phi^1\|_{H^k}^2),$$

on the time interval $[0, T_\varepsilon]$.

Proof Taking ∂_t of (2.2c), we obtain

$$-\varepsilon \Delta \partial_t \phi^1 = \Delta \partial_t \phi^0 + \partial_t n^1 - \partial_t(n^0 \phi^1) + \sqrt{\varepsilon}\partial_t R^1.$$

Taking L^2 inner product with $\partial_t \phi^1$ and then integrating by parts, we obtain

$$\varepsilon\|\nabla \partial_t \phi^1\|^2 + \int n^0 |\partial_t \phi^1|^2 = \int \partial_t \phi^1 (\Delta \partial_t \phi^0 - \partial_t n^0 \phi^1 + \partial_t n^1 + \sqrt{\varepsilon}\partial_t R^1).$$

As $n^0 > \sigma' > 0$ for $t \leqslant T' < T$, we have by Hölder inequality

$$\varepsilon\|\nabla \partial_t \phi^1\|^2 + \sigma'\|\partial_t \phi^1\|^2 \leqslant \frac{\sigma'}{4}\|\partial_t \phi^1\|^2$$
$$+ \frac{4}{\sigma'}(\|\Delta \partial_t \phi^0\|^2 + \|\partial_t n^0\|_{L^\infty}^2 \|\phi^1\|^2 + \|\partial_t n^1\|^2 + \varepsilon\|\partial_t R^1\|^2)$$
$$\leqslant \frac{\sigma'}{4}\|\partial_t \phi^1\|^2 + \frac{4}{\sigma'}(C + C\|\phi^1\|^2 + \|\partial_t n^1\|^2 + \varepsilon\|\partial_t R^1\|^2).$$

By choosing a small $\varepsilon_1 > 0$ such that $16C_1\varepsilon_1 \leqslant \sigma'^2$, we then have for any $0 < \varepsilon < \varepsilon_1$ that

$$\varepsilon\|\nabla\partial_t\phi^1\|^2 + \sigma'\|\partial_t\phi^1\|^2 \leqslant \frac{C}{\sigma'}(1 + \|\phi^1\|^2 + \|\partial_t n^1\|^2),$$

thanks to Lemma 2.1. Similarly, by taking inner product with $\varepsilon\Delta\partial_t\phi^1$, we obtain

$$\varepsilon^2\|\Delta\partial_t\phi^1\|^2 + \sigma'\varepsilon\|\partial_t\phi^1\|^2 \leqslant \frac{C}{\sigma'}(1 + \|\phi^1\|^2 + \|\partial_t n^1\|^2).$$

Adding them together, we obtain that for and any $0 < \varepsilon < \varepsilon_1$ for some $\varepsilon_1 > 0$,

$$\|\partial_t\phi^1\|^2 + \varepsilon\|\nabla\partial_t\phi\|^2 + \varepsilon^2\|\Delta\partial_t\phi^1\|^2 \leqslant C(1 + \|\partial_t n^1\|_{H^k}^2 + \|\phi^1\|_{H^k}^2),$$

for some constant C depending on σ'. Higher order estimates can be treated similarly and we obtain for any α with $|\alpha| = k$ that

$$\|\partial_t\partial^\alpha\phi^1\|^2 + \varepsilon\|\partial_t\partial^\alpha\nabla\phi\|^2 + \varepsilon^2\|\partial_t\partial^\alpha\Delta\phi^1\|^2 \leqslant C(1 + \|\partial_t n^1\|_{H^k}^2 + \|\phi^1\|_{H^k}^2),$$

for some constant C depending on σ'. $\square$

By recalling Lemma 2.3, we have the following:

Corollary 2.1 *Let $(n^1, \mathbf{u}^1, \phi^1)$ be a smooth solution for the remainder system (R_ε), and α be an integer. There exist ε_1 and C such that for any $0 < \varepsilon < \varepsilon_1$ and any multiindices α with $|\alpha| = k$,*

$$\|\partial^\alpha\partial_t\phi^1\|^2 + \varepsilon\|\partial^\alpha\nabla\partial_t\phi^1\|^2 + \varepsilon^2\|\partial^\alpha\Delta\partial_t\phi^1\|^2$$
$$\leqslant C(1 + \|\mathbf{u}^1\|_{\mathbf{H}^{k+1}}^2 + \|\phi^1\|_{H^{k+1}}^2 + \varepsilon^2\|\Delta\phi^1\|_{H^{k+1}}^2),$$

on the time interval $[0, T_\varepsilon]$.

As a direct consequence of Lemma 2.1, Lemmas 2.4 and 2.3, we also have

Corollary 2.2 *Let $(n^1, \mathbf{u}^1, \phi^1)$ be a smooth solution for the remainder system (R_ε), and α be an integer. There exist ε_1 and C such that*

$$\|\partial_t R^1\|_{H^k} \leqslant C(1 + \|\|(\mathbf{u}^1, \phi^1)\|\|_{\varepsilon,k+1}),$$

for any $0 < \varepsilon < \varepsilon_1$ and any multiindices α with $|\alpha| = k$.

2.3 Estimates of the s order

In this subsection, we give several estimates at the s order. However, the H^s-norm of the solutions depends on the H^{s+1}-norm and hence cannot be closed until the next subsection. The main result in this subsection is Proposition 2.1. In the following, $\gamma \geqslant 0$ will always denote a multiindex with $|\gamma| = s$.

Lemma 2.5 Let $\gamma \geqslant 0$ be a multiindex with $|\gamma| = s$, $(n^1, \mathbf{u}^1, \phi^1)$ be a smooth solution for the system (2.2). There exists $\varepsilon_1 > 0$ and $C > 0$ such that

$$\left\{\frac{1}{2}\frac{d}{dt}\|\partial^\gamma \mathbf{u}^1\|_{L^2}^2 + \frac{1}{2}\frac{d}{dt}\int \frac{n^0}{n^0 + \varepsilon n^1}|\partial^\gamma \phi^1|^2 + \frac{\varepsilon}{2}\frac{d}{dt}\int \frac{1}{n^0 + \varepsilon n^1}|\partial^\gamma \nabla \phi^1|^2\right\}$$
$$\leqslant C(1 + \varepsilon^3 |||(\mathbf{u}^1, \phi^1)|||_{\varepsilon,5}^3)(1 + |||(\mathbf{u}^1, \phi^1)|||_{\varepsilon,s\vee 3}^2), \tag{2.11}$$

for any $0 < \varepsilon < \varepsilon_1$, where $s \vee 3 = \max\{s, 3\}$.

Proof Let γ be a multiindex with $|\gamma| = s \geqslant 0$. Taking ∂^γ to (2.2b), we obtain

$$\partial_t \partial^\gamma \mathbf{u}^1 + \partial^\gamma(\mathbf{u}^0 \cdot \nabla \mathbf{u}^1) + \partial^\gamma(\mathbf{u}^1 \cdot \nabla \mathbf{u}^0) + \varepsilon \partial^\gamma(\mathbf{u}^1 \cdot \nabla \mathbf{u}^1) = -\partial^\gamma \nabla \phi^1.$$

Taking L^2 inner product with $\partial^\gamma \mathbf{u}^1$, we obtain

$$\int \partial_t \partial^\gamma \mathbf{u}^1 \partial^\gamma \mathbf{u}^1 = -\int \partial^\gamma \nabla \phi^1 \partial^\gamma \mathbf{u}^1 - \varepsilon \int \partial^\gamma(\mathbf{u}^1 \cdot \nabla \mathbf{u}^1) \partial^\gamma \mathbf{u}^1$$
$$-\int \partial^\gamma(\mathbf{u}^0 \cdot \nabla \mathbf{u}^1) \partial^\gamma \mathbf{u}^1 - \int \partial^\gamma(\mathbf{u}^1 \cdot \nabla \mathbf{u}^0) \partial^\gamma \mathbf{u}^1$$
$$=: I + II + III + IV. \tag{2.12}$$

- *Estimate of the fourth term IV.*

The term IV can be bounded by

$$IV \leqslant C\|\partial^\gamma(\mathbf{u}^1 \cdot \nabla \mathbf{u}^0)\|_{L^2}\|\partial^\gamma \mathbf{u}^1\|_{L^2}$$
$$\leqslant C(\|\mathbf{u}^1\|_{H^s}\|\nabla \mathbf{u}^0\|_{L^\infty} + \|\nabla \mathbf{u}^0\|_{H^s}\|\mathbf{u}^1\|_{L^\infty})\|\mathbf{u}^1\|_{H^s}$$
$$\leqslant C(\|\mathbf{u}^1\|_{H^s}^2 + \|\mathbf{u}^1\|_{H^2}^2), \tag{2.13}$$

where we have used the commutator estimates (A.1), the Sobolev embedding $H^2 \hookrightarrow L^\infty$ when $d \leqslant 3$ and the fact that $(n^0, \mathbf{u}^0)$ is a known smooth solution of the Euler equation (1.2) by Theorem 1.1.

- *Estimate of the third term III.*

By integration by parts, the third term III can be rewritten as

$$III = -\int \mathbf{u}^0 \cdot \nabla \partial^\gamma \mathbf{u}^1 \partial^\gamma \mathbf{u}^1 - \int [\partial^\gamma, \mathbf{u}^0] \cdot \nabla \mathbf{u}^1 \partial^\gamma \mathbf{u}^1$$
$$= \frac{1}{2}\int \nabla \cdot \mathbf{u}^0 |\partial^\gamma \mathbf{u}^1|^2 - \int [\partial^\gamma, \mathbf{u}^0] \cdot \nabla \mathbf{u}^1 \partial^\gamma \mathbf{u}^1. \tag{2.14}$$

By using commutator estimates (A.1), we obtain

$$\left|\int [\partial^\gamma, \mathbf{u}^0] \cdot \nabla \mathbf{u}^1 \partial^\gamma \mathbf{u}^1\right| \leqslant C \|[\partial^\gamma, \mathbf{u}^0] \cdot \nabla \mathbf{u}^1\|_{L^2} \|\partial^\gamma \mathbf{u}^1\|_{L^2}$$
$$\leqslant C(\|\nabla \mathbf{u}^0\|_{L^\infty} \|\nabla \mathbf{u}^1\|_{H^{s-1}} + \|\mathbf{u}^0\|_{H^s} \|\nabla \mathbf{u}^1\|_{L^\infty}) \|\partial^\gamma \mathbf{u}^1\|_{L^2}$$
$$\leqslant C(\|\mathbf{u}^1\|_{H^3}^2 + \|\mathbf{u}^1\|_{H^s}^2),$$

where we have used the Hölder inequality and the Sobolev embedding $H^2 \hookrightarrow L^\infty$. This yields the estimate

$$III \leqslant C(\|\mathbf{u}^1\|_{H^3}^2 + \|\mathbf{u}^1\|_{H^s}^2), \tag{2.15}$$

since the first term on the RHS of (2.14) is bounded by $C\|\mathbf{u}^1\|_{H^s}^2$.

- *Estimate of the second term II.*

Similar to the estimate of III, we have by integration by parts that

$$II = \frac{\varepsilon}{2}\int \nabla \cdot \mathbf{u}^1 \partial^\gamma \mathbf{u}^1 \partial^\gamma \mathbf{u}^1 - \varepsilon \int [\partial^\gamma, \mathbf{u}^1] \cdot \nabla \mathbf{u}^1 \partial^\gamma \mathbf{u}^1$$
$$\leqslant C\varepsilon \|\nabla \cdot \mathbf{u}^1\|_{L^\infty} \|\partial^\gamma \mathbf{u}^1\|_{L^2}^2$$
$$\quad + C\varepsilon(\|\nabla \mathbf{u}^1\|_{L^\infty} \|\nabla \mathbf{u}^1\|_{H^{s-1}} + \|\nabla \mathbf{u}^1\|_{L^\infty} \|\mathbf{u}^1\|_{H^s}) \|\partial^\gamma \mathbf{u}^1\|_{L^2}$$
$$\leqslant C\varepsilon \|\mathbf{u}^1\|_{H^3} \|\mathbf{u}^1\|_{H^s}^2. \tag{2.16}$$

- *Estimate of the first term I.*

By integration by parts, the term I in (2.12) is rewritten as

$$I = \int \partial^\gamma \phi^1 \partial^\gamma \nabla \cdot \mathbf{u}^1.$$

To handle this term, we note that from the remainder equation (2.2a),

$$(n^0 + \varepsilon n^1)\partial^\gamma \nabla \cdot \mathbf{u}^1 + [\partial^\gamma, n^0 + \varepsilon n^1]\nabla \cdot \mathbf{u}^1 + \partial_t \partial^\gamma n^1$$
$$+ \partial^\gamma((\mathbf{u}^0 + \varepsilon \mathbf{u}^1) \cdot \nabla n^1 + \mathbf{u}^1 \cdot \nabla n^0) + \partial^\gamma(n^1 \nabla \cdot \mathbf{u}^0) = 0. \tag{2.17}$$

Inserting this into I, we obtain

$$I = -\int \frac{\partial^\gamma \phi^1}{n^0 + \varepsilon n^1} \partial_t \partial^\gamma n^1 - \int \frac{\partial^\gamma \phi^1}{n^0 + \varepsilon n^1} \partial^\gamma ((\mathbf{u}^0 + \varepsilon \mathbf{u}^1) \cdot \nabla n^1)$$
$$- \int \frac{\partial^\gamma \phi^1}{n^0 + \varepsilon n^1} [\partial^\gamma, n^0 + \varepsilon n^1] \nabla \cdot \mathbf{u}^1 - \int \frac{\partial^\gamma \phi^1}{n^0 + \varepsilon n^1} \partial^\gamma (\mathbf{u}^1 \cdot \nabla n^0)$$
$$- \int \frac{\partial^\gamma \phi^1}{n^0 + \varepsilon n^1} \partial^\gamma (n^1 \cdot \nabla \mathbf{u}^0) =: \sum_{i=1}^5 I_i. \tag{2.18}$$

In the following, we estimate $I_3 \sim I_5$ while leaving the estimates of I_1 and I_2 to the next lemmas.

For I_3, we have

$$I_3 \leqslant C\|\partial^\gamma \phi^1\|_{L^2}(\|\nabla(n^0 + \varepsilon n^1)\|_{L^\infty}\|\nabla \cdot \mathbf{u}^1\|_{H^{s-1}} + \|n^0 + \varepsilon n^1\|_{H^s}\|\nabla \cdot \mathbf{u}^1\|_{L^\infty})$$
$$\leqslant C\|\phi^1\|_{H^s}(\|\mathbf{u}^1\|_{H^s} + \|\mathbf{u}^1\|_{H^3}) + C\varepsilon(\|n^1\|_{H^3} + \|\mathbf{u}^1\|_{H^3})\|\phi^1\|_{H^s}(\|n^1\|_{H^s} + \|\mathbf{u}^1\|_{H^s})$$
$$\leqslant C(1 + \varepsilon\|(n^1, \mathbf{u}^1)\|_{H^3})\|(n^1, \mathbf{u}^1, \phi^1)\|_{H^s}^2 + C\|\mathbf{u}^1\|_{H^3}^2,$$

where we have used the Hölder inequality, commutator estimates (A.1) and the fact that n^0 and $n^0 + \varepsilon n^1$ are bounded from above and below by positive numbers when $\varepsilon < \varepsilon_1$ is small enough in (2.5).

For I_4, directly applying the Hölder inequality and Lemma A.1 yields

$$I_{24} \leqslant C\|\phi^1\|_{H^s}(\|\mathbf{u}^1\|_{L^\infty}\|\nabla n^0\|_{H^s} + \|\mathbf{u}^1\|_{H^s}\|\nabla n^0\|_{L^\infty})$$
$$\leqslant C\|\phi^1\|_{H^s}^2 + C\|\mathbf{u}^1\|_{H^s}^2 + C\|\mathbf{u}^1\|_{H^2}^2.$$

Similarly, I_5 can be bounded by

$$I_{25} \leqslant C\|\phi^1\|_{H^s}^2 + C\|n^1\|_{H^s}^2 + C\|n^1\|_{H^2}^2.$$

Summarizing, we have that

$$I \leqslant I_1 + I_2 + C\|(n^1, \mathbf{u}^1)\|_{H^3}^2 + C(1 + \varepsilon^2\|(n^1, \mathbf{u}^1)\|_{H^3}^2)\|(n^1, \mathbf{u}^1, \phi^1)\|_{H^s}^2. \tag{2.19}$$

To end the proof of Lemma 2.5, we need to get suitable estimates for I_1 and I_2. However, this is not straightforward and to make it easier to read, we leave the proof to the next two lemmas. □

Lemma 2.6 *The term of I_2 in (2.18) is bounded by*

$$I_{21} \leqslant C(1 + \varepsilon^3 |||(\mathbf{u}^1, \phi^1)|||_{\varepsilon,4}^3)(1 + |||(\mathbf{u}^1, \phi^1)|||_{\varepsilon,s\vee 3}^2), \tag{2.20}$$

for some constant $C > 0$ and for all $0 < \varepsilon < \varepsilon_1$.

Proof First, we observe that I_2 in (2.18) can be decomposed into

$$I_2 = -\int \frac{\partial^\gamma \phi^1(\mathbf{u}^0 + \varepsilon \mathbf{u}^1)}{n^0 + \varepsilon n^1} \cdot \nabla \partial^\gamma n^1 - \int \frac{\partial^\gamma \phi^1}{n^0 + \varepsilon n^1} [\partial^\gamma, \mathbf{u}^0 + \varepsilon \mathbf{u}^1] \cdot \nabla n^1 =: I_{21} + I_{22}.$$

By commutator estimate (A.1),

$$I_{22} \leqslant C\|\partial^\gamma \phi^1\|(\|\nabla \mathbf{u}^0 + \varepsilon \nabla \mathbf{u}^1\|_{L^\infty}\|\nabla n^1\|_{H^{s-1}} + \|\mathbf{u}^0 + \varepsilon \mathbf{u}^1\|_{H^s}\|\nabla n^1\|_{L^\infty})$$
$$\leqslant C\|\phi^1\|_{H^s}(\|n^1\|_{H^s} + \|n^1\|_{H^3}) + C\varepsilon(\|(n^1, \mathbf{u}^1)\|_{H^3})\|\phi^1\|_{H^s}(\|(n^1, \mathbf{u}^1)\|_{H^s})$$
$$\leqslant C\|n^1\|_{H^3}^2 + C(1 + \varepsilon\|(n^1, \mathbf{u}^1)\|_{H^3})\|(n^1, \mathbf{u}^1, \phi^1)\|_{H^s}^2. \quad (2.21)$$

To treat I_{21}, we first note that from the remainder equation (2.2c),

$$\partial^\gamma \nabla n^1 = \nabla \partial^\gamma (n^0 \phi^1) - \varepsilon \partial^\gamma \nabla \Delta \phi^1 - \partial^\gamma \nabla \Delta \phi^0 - \sqrt{\varepsilon} \partial^\gamma \nabla R^1. \quad (2.22)$$

Hence I_{21} is accordingly divided into

$$I_{21} = -\int \frac{\partial^\gamma \phi^1(\mathbf{u}^0 + \varepsilon \mathbf{u}^1)}{n^0 + \varepsilon n^1} \cdot \nabla \partial^\gamma (n^0 \phi^1) + \varepsilon \int \frac{\partial^\gamma \phi^1(\mathbf{u}^0 + \varepsilon \mathbf{u}^1)}{n^0 + \varepsilon n^1} \cdot \partial^\gamma \nabla \Delta \phi^1$$
$$+ \int \frac{\partial^\gamma \phi^1(\mathbf{u}^0 + \varepsilon \mathbf{u}^1)}{n^0 + \varepsilon n^1} \cdot \partial^\gamma \nabla \Delta \phi^0 + \sqrt{\varepsilon} \int \frac{\partial^\gamma \phi^1(\mathbf{u}^0 + \varepsilon \mathbf{u}^1)}{n^0 + \varepsilon n^1} \cdot \partial^\gamma \nabla R^1$$
$$=: \sum_{i=1}^{4} I_{21i}. \quad (2.23)$$

For the first term I_{211}, we have

$$I_{211} = -\int \frac{\partial^\gamma \phi^1(\mathbf{u}^0 + \varepsilon \mathbf{u}^1)}{n^0 + \varepsilon n^1} \cdot n^0 \nabla \partial^\gamma \phi^1 - \int \frac{\partial^\gamma \phi^1(\mathbf{u}^0 + \varepsilon \mathbf{u}^1)}{n^0 + \varepsilon n^1} \cdot [\partial^\gamma, n^0] \nabla \phi^1$$
$$= \frac{1}{2} \int \nabla \cdot \left(\frac{n^0(\mathbf{u}^0 + \varepsilon \mathbf{u}^1)}{n^0 + \varepsilon n^1}\right) |\partial^\gamma \phi^1|^2 - \int \frac{\partial^\gamma \phi^1(\mathbf{u}^0 + \varepsilon \mathbf{u}^1)}{n^0 + \varepsilon n^1} \cdot [\partial^\gamma, n^0] \nabla \phi^1.$$

By direct computation and Sobolev embedding, we have

$$\left\|\nabla \cdot \left(\frac{n^0(\mathbf{u}^0 + \varepsilon \mathbf{u}^1)}{n^0 + \varepsilon n^1}\right)\right\|_{L^\infty} \leqslant C + C\varepsilon^2(\|n^1\|_{H^3}^2 + \|\mathbf{u}^1\|_{H^3}^2),$$

which yields

$$I_{211} \leqslant C\|\phi^1\|_{H^s}^2 + C\varepsilon^2(\|n^1\|_{H^3}^2 + \|\mathbf{u}^1\|_{H^3}^2)\|\phi^1\|_{H^s}^2 + C\|\partial^\gamma \phi^1\|_{L^2}^2.$$
$$\cdot (1 + \varepsilon\|\mathbf{u}^1\|_{L^\infty})(\|\nabla n^0\|_{L^\infty}\|\nabla \phi^1\|_{H^{s-1}} + \|n^0\|_{H^s}\|\nabla \phi^1\|_{L^\infty})$$
$$\leqslant C(1 + \varepsilon^2(\|n^1\|_{H^3}^2 + \|\mathbf{u}^1\|_{H^3}^2))\|\phi^1\|_{H^s}^2 + C\|\phi^1\|_{H^3}^2, \quad (2.24)$$

thanks to the commutator estimates (A.1). For I_{212}, by integration by parts, we obtain

$$I_{212} = \varepsilon \int \frac{(\mathbf{u}^0 + \varepsilon \mathbf{u}^1)\partial^\gamma \phi^1}{n^0 + \varepsilon n^1} \cdot \partial^\gamma \nabla \Delta \phi^1$$

$$= -\varepsilon \int \frac{(\mathbf{u}^0+\varepsilon\mathbf{u}^1)}{n^0+\varepsilon n^1} \cdot \partial^\gamma \nabla\phi^1 \cdot \partial^\gamma \nabla\nabla\phi^1 - \varepsilon \int \nabla\left(\frac{(\mathbf{u}^0+\varepsilon\mathbf{u}^1)}{n^0+\varepsilon n^1}\right) \partial^\gamma \phi^1 \cdot \partial^\gamma \nabla\nabla\phi^1$$
$$= \frac{3\varepsilon}{2}\int \nabla \cdot \left(\frac{(\mathbf{u}^0+\varepsilon\mathbf{u}^1)}{n^0+\varepsilon n^1}\right)|\partial^\gamma \nabla\phi^1|^2 + \varepsilon \int \nabla^2\left(\frac{(\mathbf{u}^0+\varepsilon\mathbf{u}^1)}{n^0+\varepsilon n^1}\right)\partial^\gamma\phi^1\partial^\gamma\nabla\phi^1.$$

By direct computation and Sobolev embedding $H^2 \hookrightarrow L^\infty$, we have

$$\|\nabla\cdot\left(\frac{(\mathbf{u}^0+\varepsilon\mathbf{u}^1)}{n^0+\varepsilon n^1}\right)\|_{L^\infty} \leqslant C+C\varepsilon^2(\|n^1\|_{H^3}^2+\|\mathbf{u}^1\|_{H^3}^2),$$
$$\|\nabla^2\left(\frac{(\mathbf{u}^0+\varepsilon\mathbf{u}^1)}{n^0+\varepsilon n^1}\right)\|_{L^\infty} \leqslant C+C\varepsilon^3(\|n^1\|_{H^4}^3+\|\mathbf{u}^1\|_{H^4}^3),$$

which yield

$$I_{212} \leqslant C\left(1+\varepsilon^3(\|(n^1,\mathbf{u}^1)\|_{H^4}^3)\right)\left(\|\phi^1\|_{H^s}^2+\varepsilon\|\nabla\phi^1\|_{H^s}^2\right). \tag{2.25}$$

For I_{213}, by Hölder inequality, we obtain

$$I_{213} \leqslant C\left(1+\|\mathbf{u}^1\|_{L^2}^2+\|\phi^1\|_{H^s}^2\right). \tag{2.26}$$

For I_{214}, by integration by parts,

$$I_{214} = -\sqrt{\varepsilon}\int \nabla\cdot\left(\frac{(\mathbf{u}^0+\varepsilon\mathbf{u}^1)}{n^0+\varepsilon n^1}\right)\partial^\gamma\phi^1\partial^\gamma R^1 - \sqrt{\varepsilon}\int \frac{(\mathbf{u}^0+\varepsilon\mathbf{u}^1)}{n^0+\varepsilon n^1} \cdot \nabla\partial^\gamma\phi^1\partial^\gamma R^1.$$

By using Lemma 2.1, we obtain

$$I_{214} \leqslant C(1+\varepsilon^2\|(n^1,\mathbf{u}^1)\|_{H^3}^2)\|\phi^1\|_{H^s}^2$$
$$+C(1+\varepsilon\|\mathbf{u}^1\|_{L^\infty})(\|\phi^1\|_{H^s}^2+\varepsilon\|\nabla\phi^1\|_{H^s}^2)$$
$$\leqslant C(1+\varepsilon\|(\mathbf{u}^1,n^1)\|_{H^3})(\|\phi^1\|_{H^s}^2+\varepsilon\|\nabla\phi^1\|_{H^s}^2). \tag{2.27}$$

By (2.23) and putting (2.24)-(2.27) together, we obtain

$$I_{21} \leqslant C+C\|(n^1,\mathbf{u}^1,\phi^1)\|_{H^3}^2$$
$$+C(1+\varepsilon^3\|(n^1,\mathbf{u}^1)\|_{H^4}^3)(\|(n^1,\mathbf{u}^1,\phi^1)\|_{H^s}^2+\varepsilon\|\nabla\phi^1\|_{H^s}^2)$$
$$\leqslant C(1+\varepsilon^3|||(\mathbf{u}^1,\phi^1)|||_{\varepsilon,4}^3)(1+|||(\mathbf{u}^1,\phi^1)|||_{\varepsilon,s\vee 3}^2), \tag{2.28}$$

where we have used the definition of the norm $|||\cdot|||_{\varepsilon,s}$ in (1.7) and Lemma 2.2 to replace the norms of n^1 with the norms of ϕ^1, By putting (2.21) and (2.28) together and using Lemma 2.2, we obtain (2.20). This ends the proof of Lemma 2.6. □

Lemma 2.7 *For I_1 in (2.18), we have the estimate*

$$I_1 \leqslant -\frac{1}{2}\frac{d}{dt}\int \frac{n^0}{n^\varepsilon}|\partial^\gamma\phi^1|^2 - \frac{\varepsilon}{2}\frac{d}{dt}\int \frac{1}{n^\varepsilon}|\partial^\gamma\nabla\phi^1|^2$$
$$+C(1+\varepsilon^3|||(\mathbf{u}^1,\phi^1)|||_{\varepsilon,5}^3)(1+|||(\mathbf{u}^1,\phi^1)|||_{\varepsilon,s\vee 3}^2).$$

Proof From the remainder equation (2.2c), we obtain

$$\partial^\gamma \partial_t n^1 = \partial^\gamma \partial_t(n^0 \phi^1) - \varepsilon \partial^\gamma \partial_t \Delta \phi^1 - \partial^\gamma \partial_t \Delta \phi^0 - \sqrt{\varepsilon}\partial^\gamma \partial_t R^1. \qquad (2.29)$$

In this way, we can divide I_1 in (2.18) into the following

$$I_1 = \varepsilon \int \frac{\partial^\gamma \phi^1}{n^0 + \varepsilon n^1} \partial^\gamma \partial_t \Delta \phi^1 - \int \frac{\partial^\gamma \phi^1}{n^0 + \varepsilon n^1} \partial^\gamma \partial_t(n^0 \phi^1)$$

$$- \int \frac{\partial^\gamma \phi^1}{n^0 + \varepsilon n^1} \partial^\gamma \partial_t \Delta \phi^0 - \sqrt{\varepsilon} \int \frac{\partial^\gamma \phi^1}{n^0 + \varepsilon n^1} \partial^\gamma \partial_t R^1 = \sum_{i=1}^4 I_{1i}. \qquad (2.30)$$

In the following, we treat the RHS terms of (2.30) one by one.

- *Estimate of I_{12}.*

For the term I_{12}, we have

$$I_{12} = -\int \frac{\partial^\gamma \phi^1}{n^0 + \varepsilon n^1} \partial^\gamma(\phi^1 \partial_t n^0) - \int \frac{\partial^\gamma \phi^1}{n^0 + \varepsilon n^1} \partial^\gamma(n^0 \partial_t \phi^1)$$

$$= -\int \frac{\partial^\gamma \phi^1}{n^0 + \varepsilon n^1} \partial^\gamma(\phi^1 \partial_t n^0) - \int \frac{n^0 \partial^\gamma \phi^1}{n^0 + \varepsilon n^1} \partial_t \partial^\gamma \phi^1 - \int \frac{\partial^\gamma \phi^1}{n^0 + \varepsilon n^1}[\partial^\gamma, n^0]\partial_t \phi^1$$

$$=: \sum_{i=1}^3 I_{12i}.$$

For the first term I_{121}, since n^0 is known and is assumed to be smooth in Theorem 1.1, we have

$$I_{121} \leqslant C\|\phi^1\|_{H^s}^2,$$

thanks to the multiplicative estimate in Lemma A.1. For the second term I_{122}, by integration by parts and Lemma 2.3, we have

$$I_{122} = -\frac{1}{2}\frac{d}{dt}\int \frac{n^0}{n^0 + \varepsilon n^1}|\partial^\gamma \phi^1|^2 + \frac{1}{2}\int \partial_t\left(\frac{n^0}{n^0 + \varepsilon n^1}\right)|\partial^\gamma \phi^1|^2$$

$$\leqslant -\frac{1}{2}\frac{d}{dt}\int \frac{n^0}{n^0 + \varepsilon n^1}|\partial^\gamma \phi^1|^2 + C(1 + \varepsilon\|\partial_t n^1\|_{L^\infty})\|\partial^\gamma \phi^1\|_{L^2}^2$$

$$\leqslant -\frac{1}{2}\frac{d}{dt}\int \frac{n^0}{n^0 + \varepsilon n^1}|\partial^\gamma \phi^1|^2 + C(1 + \varepsilon^2\|\|(\mathbf{u}^1, \phi^1)\|\|_{\varepsilon,3}^2)\|\phi^1\|_{H^s}^2,$$

where we have used the Sobolev embedding $H^2 \hookrightarrow L^\infty$ for $d \leqslant 3$. For the third term I_{123}, by commutator estimates in Lemma A.1 and Lemma 2.3, we have

$$I_{123} \leqslant C\|\partial^\gamma \phi^1\|_{L^2}(\|\nabla n^0\|_{L^\infty}\|\partial_t \phi^1\|_{H^{s-1}} + \|n^0\|_{H^s}\|\partial_t \phi^1\|_{L^\infty})$$

$$\leqslant C(1 + \|\|(\mathbf{u}^1, \phi^1)\|\|_{\varepsilon,s}^2 + \|\|(\mathbf{u}^1, \phi^1)\|\|_{\varepsilon,3}^2).$$

Summarizing, we have

$$I_{12} \leqslant -\frac{1}{2}\frac{d}{dt}\int \frac{n^0}{n^0+\varepsilon n^1}|\partial^\gamma \phi^1|^2$$
$$+ C(1+\varepsilon^2|||(\mathbf{u}^1,\phi^1)|||^2_{\varepsilon,3})(1+|||(\mathbf{u}^1,\phi^1)|||^2_{\varepsilon,s}+|||(\mathbf{u}^1,\phi^1)|||^2_{\varepsilon,3}). \qquad (2.31)$$

- *Estimate of I_{13}.*

For the term I_{13}, since ϕ^0 is known and smooth, it is easy to obtain

$$I_{13} \leqslant C(1+\|\phi^1\|^2_{H^s}). \qquad (2.32)$$

- *Estimate of I_{14}.*

For the term I_{14}, by integration by parts, we have

$$I_{14} = \sqrt{\varepsilon}\int \frac{\partial^{\gamma+\gamma_1}\phi^1}{n^0+\varepsilon n^1}\partial^{\gamma-\gamma_1}\partial_t R^1 + \sqrt{\varepsilon}\int \partial^{\gamma_1}\left(\frac{1}{n^0+\varepsilon n^1}\right)\partial^\gamma \phi^1 \partial^{\gamma-\gamma_1}\partial_t R^1,$$

where $\gamma_1 \leqslant \gamma$ is a multiindex with $|\gamma_1|=1$. By Lemma 2.2, Corollary 2.2 and the definition of the triple norm $|||\cdot|||_{\varepsilon,s}$, we have the bound

$$I_{14} \leqslant C\sqrt{\varepsilon}\|\partial^{\gamma+\gamma_1}\phi^1\|_{L^2}\|\partial_t R^1\|_{H^{s-1}} + C(1+\varepsilon\|\partial^{\gamma_1}n^1\|_{L^\infty})\|\partial^\gamma \phi^1\|_{L^2}\|\partial_t R^1\|_{H^{s-1}}$$
$$\leqslant C\|\partial_t R^1\|^2_{H^{s-1}} + C\varepsilon\|\nabla\phi^1\|^2_{H^s} + C(1+\varepsilon^2\|n^1\|^2_{H^3})\|\phi^1\|^2_{H^s}$$
$$\leqslant C(1+\varepsilon^2|||(\mathbf{u}^1,\phi^1)|||^2_{\varepsilon,3})(1+|||(\mathbf{u}^1,\phi^1)|||^2_{\varepsilon,s}). \qquad (2.33)$$

- *Estimate of I_{11}.*

We next deal with the term I_{11} in (2.30). By integration by parts, we have

$$I_{11} = -\varepsilon\int \frac{1}{n^0+\varepsilon n^1}\partial^\gamma\nabla\phi^1 \cdot \partial_t\partial^\gamma\nabla\phi^1 - \varepsilon\int \nabla\left(\frac{1}{n^0+\varepsilon n^1}\right)\partial^\gamma\phi^1 \cdot \partial_t\partial^\gamma\nabla\phi^1$$
$$= -\frac{\varepsilon}{2}\frac{d}{dt}\int \frac{1}{n^0+\varepsilon n^1}|\partial^\gamma\nabla\phi^1|^2 + \frac{\varepsilon}{2}\int \partial_t\left(\frac{1}{n^0+\varepsilon n^1}\right)|\partial^\gamma\nabla\phi^1|^2$$
$$- \varepsilon\int \nabla\left(\frac{1}{n^0+\varepsilon n^1}\right)\partial^\gamma\phi^1 \cdot \partial_t\partial^\gamma\nabla\phi^1 =: I_{111} + I_{112} + I_{113}.$$

Since from Sobolev embedding and Lemma 2.3, we have

$$\|\partial_t\left(\frac{n^0}{n^0+\varepsilon n^1}\right)\|_{L^\infty} \leqslant C(1+\varepsilon\|\partial_t n^1\|_{L^\infty}) \leqslant C(1+\varepsilon^2|||(\mathbf{u}^1,\phi^1)|||^2_{\varepsilon,3}),$$

it is immediate that

$$I_{112} \leqslant C(1+\varepsilon^2|||(\mathbf{u}^1,\phi^1)|||^2_{\varepsilon,3})(\varepsilon\|\partial^\gamma\nabla\phi^1\|^2).$$

For the term I_{113}, we have by integration by parts,

$$
\begin{aligned}
I_{113} &= \varepsilon \int \Delta \left(\frac{1}{n^0 + \varepsilon n^1}\right) \partial^\gamma \phi^1 \partial_t \partial^\gamma \phi^1 + \varepsilon \int \nabla \left(\frac{1}{n^0 + \varepsilon n^1}\right) \partial^\gamma \nabla \phi^1 \partial_t \partial^\gamma \phi^1 \\
&= -\varepsilon \int \partial^{\gamma_1} \Delta \left(\frac{1}{n^\varepsilon}\right) \partial^\gamma \phi^1 \partial_t \partial^{\gamma-\gamma_1} \phi^1 - \varepsilon \int \Delta \left(\frac{1}{n^\varepsilon}\right) \partial^{\gamma+\gamma_1} \phi^1 \partial_t \partial^{\gamma-\gamma_1} \phi^1 \\
&\quad -\varepsilon \int \partial^{\gamma_1} \nabla \left(\frac{1}{n^\varepsilon}\right) \nabla \partial^\gamma \phi^1 \partial_t \partial^{\gamma-\gamma_1} \phi^1 - \varepsilon \int \nabla \left(\frac{1}{n^\varepsilon}\right) \nabla \partial^{\gamma+\gamma_1} \phi^1 \partial_t \partial^{\gamma-\gamma_1} \phi^1,
\end{aligned}
\tag{2.34}
$$

where $\gamma_1 \leqslant \gamma$ is a multiindex with $|\gamma_1| = 1$. By direct computation, Hölder inequality and Sobolev embedding $H^2 \hookrightarrow L^\infty$, it is easy to obtain

$$
\begin{aligned}
\left\|\partial^\alpha \left(\frac{1}{n^\varepsilon}\right)\right\|_{L^\infty} &\leqslant C(1 + \varepsilon^{|\alpha|}\|n^1\|_{H^{2+|\alpha|}}^{|\alpha|}) \\
&\leqslant C(1 + \varepsilon^{|\alpha|}|||(\mathbf{u}^1, \phi^1)|||_{\varepsilon, 2+|\alpha|}^{|\alpha|}),
\end{aligned}
\tag{2.35}
$$

for any smooth function n^0 and any multiindex α, thanks to Lemma 2.2. On the other hand, from Lemma 2.4 and Lemma 2.3, we obtain

$$
\begin{aligned}
\|\partial_t \partial^{\gamma-\gamma_1} \phi^1\|_{L^2}^2 &\leqslant C(1 + \|\phi^1\|_{H^{s-1}}^2 + \|\partial_t n^1\|_{H^{s-1}}^2) \\
&\leqslant C(1 + |||(\mathbf{u}^1, \phi^1)|||_{\varepsilon, s}^2).
\end{aligned}
\tag{2.36}
$$

Since the order of the derivatives on $n^0/(n^0 + \varepsilon n^1)$ in (2.34) does not exceed 3, by using Hölder inequality, (2.35) and (2.36), I_{113} can be bounded by

$$
\begin{aligned}
I_{113} &\leqslant C(1 + \varepsilon^3 |||(\mathbf{u}^1, \phi^1)|||_{\varepsilon, 5}^3) \cdot \varepsilon (\|\partial^\gamma \phi^1\|_{L^2} + \|\partial^{\gamma+\gamma_1} \phi^1\|_{L^2} \\
&\quad + \|\nabla \partial^\gamma \phi^1\|_{L^2} + \|\nabla \partial^{\gamma+\gamma_1} \phi^1\|_{L^2}) \cdot \|\partial_t \partial^{\gamma-\gamma_1} \phi^1\|_{L^2} \\
&\leqslant C(1 + \varepsilon^3 |||(\mathbf{u}^1, \phi^1)|||_{\varepsilon, 5}^3) \\
&\quad \times (\|\partial^\gamma \phi^1\|_{L^2}^2 + \varepsilon \|\partial^\gamma \nabla \phi^1\|_{L^2}^2 + \varepsilon^2 \|\partial^\gamma \Delta \phi^1\|_{L^2}^2 + \|\partial_t \partial^{\gamma-\gamma_1} \phi^1\|_{L^2}^2) \\
&\leqslant C(1 + \varepsilon^3 |||(\mathbf{u}^1, \phi^1)|||_{\varepsilon, 5}^3)(1 + |||(\mathbf{u}^1, \phi^1)|||_{\varepsilon, s}^2),
\end{aligned}
$$

where we have used the definition of $||| \cdot |||_{\varepsilon, s}$ in (1.7) and the L^2 boundedness of the Riesz operator [23]. To be more precise, there exists some constant $C > 0$ such that $\|\partial_i \partial_j \phi^1\| \leqslant C \|\Delta \phi^1\|$ since $\partial_i \partial_j = -R_i R_j \Delta$, where R_i is the ith Riesz operator. In particular, we have $\|\nabla \partial^{\gamma+\gamma_1} \phi^1\| \leqslant C \|\partial^\gamma \Delta \phi\|$.

Summarizing, we have

$$
I_{11} = -\frac{\varepsilon}{2} \frac{d}{dt} \int \frac{1}{n^\varepsilon} |\partial^\gamma \nabla \phi^1|^2 + C(1 + \varepsilon^3 |||(\mathbf{u}^1, \phi^1)|||_{\varepsilon, 5}^3)(1 + |||(\mathbf{u}^1, \phi^1)|||_{\varepsilon, s}^2).
\tag{2.37}
$$

By (2.30) and the estimates of (2.31), (2.32), (2.33) and (2.37), we have

$$I_1 \leqslant -\frac{1}{2}\frac{d}{dt}\int \frac{n^0}{n^\varepsilon}|\partial^\gamma \phi^1|^2 - \frac{\varepsilon}{2}\frac{d}{dt}\int \frac{1}{n^\varepsilon}|\partial^\gamma \nabla\phi^1|^2$$
$$+ C(1+\varepsilon^3|||(\mathbf{u}^1,\phi^1)|||^3_{\varepsilon,5})(1+|||(\mathbf{u}^1,\phi^1)|||^2_{\varepsilon,s\vee 3}).$$

This ends the proof of Lemma 2.7. □

Now, we can end the proof of Lemma 2.5.

End of proof of Lemma 2.5 The proof of Lemma 2.5 is closed by (2.12), (2.13), (2.15), (2.16), (2.19) and Lemmas 2.6 and 2.7. □

Proposition 2.1 Let $s \geqslant 0$ be a positive integer, $(n^1, \mathbf{u}^1, \phi^1)$ be a smooth solution for the system (2.2). There exists $\varepsilon_1 > 0$ and $C, C' > 0$ such that for any $0 < \varepsilon < \varepsilon_1$, there holds

$$\|\mathbf{u}^1(t)\|^2_{H^s} + \|\phi^1(t)\|^2_{H^s} + \varepsilon\|\nabla\phi^1(t)\|^2_{H^s}$$
$$\leqslant C'(\|\mathbf{u}^1(0)\|^2_{H^s} + \|\phi^1(0)\|^2_{H^s} + \varepsilon\|\nabla\phi^1(0)\|^2_{H^s})$$
$$+ C\int_0^t (1+\varepsilon^3|||(\mathbf{u}^1,\phi^1)|||^3_{\varepsilon,5})(1+|||(\mathbf{u}^1,\phi^1)|||^2_{\varepsilon,s\vee 3})d\tau, \quad (2.38)$$

where C' depends only on σ' and σ''.

Proof This is shown by integrating (2.11) over $[0,t]$ and summing them up for $|\gamma| \leqslant s$, and then using $\sigma' < n^0 < \sigma''$ and $\frac{\sigma'}{2} < n^\varepsilon < 2\sigma''$ for any $t \in [0, T_\varepsilon]$ in (1.4) and (2.5) for $0 < \varepsilon < \varepsilon_1$. □

However, this Gronwall inequality is not closed since the right hand side of (2.38) depends on $\varepsilon^2\|\Delta\phi^1\|^2_{H^s}$, which does not appear on the left hand side. This will be treated in the next subsection.

2.4 Weighted $s+1$ order estimates

In the following, we also let γ be a multiindex with $|\gamma| = s$. The main result in this subsection is Proposition 2.2.

Lemma 2.8 Let $s \geqslant 0$ be a positive integer, $(n^1, \mathbf{u}^1, \phi^1)$ be a smooth solution for the system (2.2). There exists $\varepsilon_1 > 0$ and $C, C' > 0$ such that

$$\frac{\varepsilon}{2}\frac{d}{dt}\int |\partial^\gamma \nabla\mathbf{u}^1|^2 \leqslant -\frac{\varepsilon}{2}\frac{d}{dt}\int \frac{n^0}{n^\varepsilon}|\partial^\gamma \nabla\phi^1|^2 - \frac{\varepsilon^2}{2}\frac{d}{dt}\int \frac{1}{n^\varepsilon}|\partial^\gamma \Delta\phi^1|^2$$

$$+ C(1+\varepsilon^2|||(\mathbf{u}^1,\phi^1)|||_{\varepsilon,4}^2)(1+|||(\mathbf{u}^1,\phi^1)|||_{\varepsilon,s\vee 3}^2), \quad (2.39)$$

for any $0 < \varepsilon < \varepsilon_1$.

Proof Let γ be a multiindex with $|\gamma| = s \geqslant 0$. Taking ∂^γ in the second equation of (2.2), we obtain

$$\partial_t \partial^\gamma \mathbf{u}^1 + \partial^\gamma(\mathbf{u}^0 \cdot \nabla \mathbf{u}^1) + \partial^\gamma(\mathbf{u}^1 \cdot \nabla \mathbf{u}^0) + \varepsilon \partial^\gamma(\mathbf{u}^1 \cdot \nabla \mathbf{u}^1) = -\partial^\gamma \nabla \phi^1.$$

Taking L^2 inner product with $-\varepsilon \partial^\gamma \Delta \mathbf{u}^1$, we obtain

$$\frac{\varepsilon}{2}\frac{d}{dt}\int |\partial^\gamma \nabla \mathbf{u}^1|^2 = -\varepsilon \int \partial_t \partial^\gamma \mathbf{u}^1 \partial^\gamma \Delta \mathbf{u}^1$$

$$= \varepsilon \int \partial^\gamma \nabla \phi^1 \partial^\gamma \Delta \mathbf{u}^1 + \varepsilon^2 \int \partial^\gamma(\mathbf{u}^1 \cdot \nabla \mathbf{u}^1) \partial^\gamma \Delta \mathbf{u}^1$$

$$+ \varepsilon \int \partial^\gamma(\mathbf{u}^0 \cdot \nabla \mathbf{u}^1) \partial^\gamma \Delta \mathbf{u}^1 + \varepsilon \int \partial^\gamma(\mathbf{u}^1 \cdot \nabla \mathbf{u}^0) \partial^\gamma \Delta \mathbf{u}^1$$

$$=: I^\varepsilon + II^\varepsilon + III^\varepsilon + IV^\varepsilon. \quad (2.40)$$

- *Estimate of IV^ε.*

The term IV^ε can be bounded by

$$IV^\varepsilon = -\varepsilon \int \partial^\gamma \nabla(\mathbf{u}^1 \cdot \nabla \mathbf{u}^0) \partial^\gamma \nabla \mathbf{u}^1$$

$$= -\varepsilon \int \left(\partial^\gamma(\nabla \mathbf{u}^1 \cdot \nabla \mathbf{u}^0) + \partial^\gamma(\mathbf{u}^1 \cdot \nabla^2 \mathbf{u}^0)\right) \partial^\gamma \nabla \mathbf{u}^1$$

$$\leqslant C\varepsilon(\|\partial^\gamma(\nabla \mathbf{u}^1 \cdot \nabla \mathbf{u}^0)\|_{L^2}^2 + \|\partial^\gamma(\mathbf{u}^1 \cdot \nabla^2 \mathbf{u}^0)\|_{L^2}^2) + C\varepsilon\|\partial^\gamma \nabla \mathbf{u}^1\|_{L^2}^2$$

$$\leqslant C(\|\mathbf{u}^1\|_{H^s}^2 + \varepsilon\|\nabla \mathbf{u}^1\|_{H^s}^2 + \|\mathbf{u}^1\|_{H^3}^2), \quad (2.41)$$

where we have used the commutator estimates (A.1), the Sobolev embedding $H^2 \hookrightarrow L^\infty$ when $d \leqslant 3$ and the fact that n^0 and $\mathbf{u}^0$ are known smooth solutions of the Euler equation (1.2) by Theorem 1.1.

- *Estimate of III^ε.*

By integration by parts, the third term III^ε can be rewritten as

$$III^\varepsilon = \varepsilon \int \mathbf{u}^0 \cdot \nabla \partial^\gamma \mathbf{u}^1 \partial^\gamma \Delta \mathbf{u}^1 + \varepsilon \int [\partial^\gamma, \mathbf{u}^0] \cdot \nabla \mathbf{u}^1 \partial^\gamma \Delta \mathbf{u}^1$$

$$= \varepsilon \int \sum_{i,j,k} u_i^0 \partial_i \partial^\gamma u_k^1 \partial^\gamma \partial_j \partial_j u_k^1 + \varepsilon \int \sum_{\beta=1}^{\gamma} C_\gamma^\beta \partial^\beta \mathbf{u}^0 \cdot \partial^{\gamma-\beta} \nabla \mathbf{u}^1 \cdot \partial^\gamma \Delta \mathbf{u}^1$$

$$= -2\varepsilon \int \sum_{i,j,k} \partial_j u_i^0 \partial_i \partial^\gamma u_k^1 \partial^\gamma \partial_j u_k^1 + \varepsilon \int \sum_{i,j,k} \partial_i u_i^0 \partial^\gamma \partial_j u_k^1 \partial^\gamma \partial_j u_k^1$$

$$- \varepsilon \int \sum_{\beta=1}^\gamma C_\gamma^\beta \partial^\beta \mathbf{u}^0 \cdot \partial^{\gamma-\beta}\nabla^2\mathbf{u}^1 \cdot \partial^\gamma \nabla \mathbf{u}^1$$

$$- \varepsilon \int \sum_{\beta=1}^\gamma C_\gamma^\beta \partial^\beta \nabla\mathbf{u}^0 \cdot \partial^{\gamma-\beta}\nabla \mathbf{u}^1 \cdot \partial^\gamma \nabla \mathbf{u}^1$$

$$=: III_1^\varepsilon + III_2^\varepsilon + III_3^\varepsilon + III_4^\varepsilon,$$

where $\mathbf{u}^0 = (u_1^0, \cdots, u_d^0)$ and $\mathbf{u}^1 = (u_1^1, \cdots, u_d^1)$ and we have used integration by parts twice in the third equality. For III_1^ε, III_2^ε and III_3^ε, we easily obtain

$$|III_1^\varepsilon, III_2^\varepsilon, III_3^\varepsilon| \leqslant C\varepsilon \|\nabla \mathbf{u}^1\|_{H^s}^2.$$

For III_4^ε, by Hölder inequality, we obtain

$$|III_4^\varepsilon| \leqslant C\varepsilon \|\mathbf{u}^1\|_{H^s} \|\nabla \mathbf{u}^1\|_{H^s} \leqslant C(\|\mathbf{u}^1\|_{H^s}^2 + \varepsilon \|\nabla \mathbf{u}^1\|_{H^s}^2).$$

Summing them up, we obtain

$$|III^\varepsilon| \leqslant C(\|\mathbf{u}^1\|_{H^s}^2 + \varepsilon \|\nabla \mathbf{u}^1\|_{H^s}^2). \qquad (2.42)$$

- *Estimate of II^ε.*

For the second term II^ε, by integration by parts, we have

$$II^\varepsilon = \varepsilon^2 \int \mathbf{u}^1 \cdot \partial^\gamma \nabla \mathbf{u}^1 \partial^\gamma \Delta \mathbf{u}^1 + \varepsilon^2 \int [\partial^\gamma, \mathbf{u}^1] \cdot \nabla \mathbf{u}^1 \partial^\gamma \Delta \mathbf{u}^1$$

$$= -\varepsilon^2 \int \sum_{i,j,k} \partial_j u_i^1 \partial^\gamma \partial_i u_k^1 \partial^\gamma \partial_j u_k^1 + \frac{\varepsilon^2}{2} \int \sum_{i,j,k} \partial_i u_i^1 \partial^\gamma \partial_j u_k^1 \partial^\gamma \partial_j u_k^1$$

$$- \varepsilon^2 \int [\partial^\gamma, \nabla \mathbf{u}^1] \cdot \nabla \mathbf{u}^1 \cdot \partial^\gamma \nabla \mathbf{u}^1 - \varepsilon^2 \int [\partial^\gamma, \mathbf{u}^1] \cdot \nabla^2 \mathbf{u}^1 \cdot \partial^\gamma \nabla \mathbf{u}^1 =: \sum_{i=1}^4 II_i^\varepsilon.$$

By Hölder inequality, II_1^ε and II_2^ε are bounded by

$$|II_1^\varepsilon, II_2^\varepsilon| \leqslant C(\varepsilon \|\nabla \mathbf{u}^1\|_{L^\infty})(\varepsilon \|\nabla \mathbf{u}^1\|_{H^s}^2)$$
$$\leqslant C(\varepsilon \|\mathbf{u}^1\|_{H^3})(\varepsilon \|\nabla \mathbf{u}^1\|_{H^s}^2).$$

By commutator estimates,

$$\|[\partial^\gamma, \nabla \mathbf{u}^1] \cdot \nabla \mathbf{u}^1\|_{L^2} \leqslant C(\|\nabla^2 \mathbf{u}^1\|_{L^\infty} \|\nabla \mathbf{u}^1\|_{H^{s-1}} + \|\nabla \mathbf{u}^1\|_{H^s} \|\nabla \mathbf{u}^1\|_{L^\infty})$$

$$\leqslant C\|\mathbf{u}^1\|_{H^4}(\|\mathbf{u}^1\|_{H^s}+\|\nabla\mathbf{u}^1\|_{H^s}).$$

Hence
$$|II_3^\varepsilon|\leqslant C\varepsilon\|\mathbf{u}^1\|_{H^4}(\|\mathbf{u}^1\|_{H^s}^2+\varepsilon\|\nabla\mathbf{u}^1\|_{H^s}^2).$$

Similarly, by commutator estimates
$$|II_4^\varepsilon|\leqslant C\varepsilon^2(\|\nabla\mathbf{u}^1\|_{L^\infty}\|\nabla^2\mathbf{u}^1\|_{H^{s-1}}+\|\mathbf{u}^1\|_{H^s}\|\nabla^2\mathbf{u}^1\|_{L^\infty})\|\nabla\mathbf{u}^1\|_{H^s}$$
$$\leqslant C\varepsilon\|\mathbf{u}^1\|_{H^4}(\|\mathbf{u}^1\|_{H^s}^2+\varepsilon\|\nabla\mathbf{u}^1\|_{H^s}^2).$$

Summing them up, we obtain
$$|II^\varepsilon|\leqslant C\varepsilon\|\mathbf{u}^1\|_{H^4}(\|\mathbf{u}^1\|_{H^s}^2+\varepsilon\|\nabla\mathbf{u}^1\|_{H^s}^2). \tag{2.43}$$

- *Estimate of I^ε.*

In the following, we treat the first term I^ε in (2.40). By integration by parts thrice,
$$I^\varepsilon=-\varepsilon\int \partial^\gamma\Delta\phi^1\partial^\gamma\nabla\cdot\mathbf{u}^1.$$

By using (2.17), we obtain
$$I^\varepsilon=\int\frac{\varepsilon\partial^\gamma\Delta\phi^1}{n^\varepsilon}\partial_t\partial^\gamma n^1+\int\frac{\varepsilon\partial^\gamma\Delta\phi^1}{n^\varepsilon}\partial^\gamma(\mathbf{u}^\varepsilon\cdot\nabla n^1)$$
$$+\int\frac{\varepsilon\partial^\gamma\Delta\phi^1}{n^\varepsilon}[\partial^\gamma,n^\varepsilon]\nabla\cdot\mathbf{u}^1+\int\frac{\varepsilon\partial^\gamma\Delta\phi^1}{n^\varepsilon}\partial^\gamma(\mathbf{u}^1\cdot\nabla n^0)$$
$$+\int\frac{\varepsilon\partial^\gamma\Delta\phi^1}{n^\varepsilon}\partial^\gamma(n^1\cdot\nabla\mathbf{u}^0)=:\sum_{i=1}^5 I_i^\varepsilon. \tag{2.44}$$

In the following, we estimate $I_3^\varepsilon\sim I_5^\varepsilon$ while leaving the estimates of I_1^ε and I_2^ε to the next two lemmas. For I_3^ε, we have

$$I_3^\varepsilon\leqslant C\varepsilon\|\partial^\gamma\Delta\phi^1\|_{L^2}(\|\nabla(n^0+\varepsilon n^1)\|_{L^\infty}\|\nabla\cdot\mathbf{u}^1\|_{H^{s-1}}+\|n^0+\varepsilon n^1\|_{H^s}\|\nabla\cdot\mathbf{u}^1\|_{L^\infty})$$
$$\leqslant C\varepsilon^2\|\Delta\phi^1\|_{H^s}^2+\|\mathbf{u}^1\|_{H^s}^2+\|\mathbf{u}^1\|_{H^3}^2+C\varepsilon^2(\|n^1\|_{H^3}^2+\|\mathbf{u}^1\|_{H^3}^2)(\|n^1\|_{H^s}^2+\|\mathbf{u}^1\|_{H^s}^2),$$

where we have used the commutator estimates (A.1) and the fact that $n^0+\varepsilon n^1$ are bounded from above and below by positive numbers when $\varepsilon<\varepsilon_1$ is small enough in (2.5). Recalling Lemma 2.2 and the definition of the triple norm (1.7), we obtain

$$I_3^\varepsilon\leqslant C(1+\varepsilon^2\||(\mathbf{u}^1,\phi)\||_{\varepsilon,3}^2)\||(\mathbf{u}^1,\phi)\||_{\varepsilon,s}^2+C\|\mathbf{u}^1\|_{H^3}^2.$$

For I_4^ε, we have

$$I_4^\varepsilon \leqslant C\varepsilon \|\Delta\phi^1\|_{H^s}(\|\mathbf{u}^1\|_{L^\infty}\|\nabla n^0\|_{H^s} + \|\mathbf{u}^1\|_{H^s}\|\nabla n^0\|_{L^\infty})$$
$$\leqslant C\varepsilon^2\|\Delta\phi^1\|_{H^s}^2 + C\|\mathbf{u}^1\|_{H^s}^2 + C\|\mathbf{u}^1\|_{H^2}^2.$$

Similarly, I_5^ε can be bounded by

$$I_5^\varepsilon \leqslant C\varepsilon \|\Delta\phi^1\|_{H^s}^2 + C\|n^1\|_{H^s}^2 + C\|n^1\|_{H^2}^2$$
$$\leqslant C(1+|||\phi|||_{\varepsilon,s}^2) + C\|n^1\|_{H^2}^2,$$

thanks to Lemma 2.2. Summarizing, we obtain from (2.44)

$$I^\varepsilon \leqslant I_1^\varepsilon + I_2^\varepsilon + C\|\mathbf{u}^1\|_{H^3}^2 + C\|n^1\|_{H^2}^2$$
$$+ C(1+\varepsilon^2|||(\mathbf{u}^1,\phi)|||_{\varepsilon,3}^2)(1+|||(\mathbf{u}^1,\phi)|||_{\varepsilon,s}^2), \tag{2.45}$$

where I_1^ε and I_2^ε will be treated in the next two lemmas. $\square$

Lemma 2.9 *The term I_2^ε in (2.44) can be estimated as*

$$I_2^\varepsilon \leqslant C(1+\varepsilon^2|||(\mathbf{u}^1,\phi^1)|||_{\varepsilon,3}^2)(1+|||(\mathbf{u}^1,\phi^1)|||_{\varepsilon,s}^2) + |||\phi^1|||_{\varepsilon,3}^2,$$

for all $0 < \varepsilon < \varepsilon_1$ for some $\varepsilon_1 > 0$.

Proof Recall that I_2^ε is given by (2.44). By integration by parts, it can be rewritten as

$$I_2^\varepsilon = \int \frac{\varepsilon \partial^\gamma \Delta\phi^1}{n^\varepsilon}\mathbf{u}^\varepsilon \cdot \nabla \partial^\gamma n^1 + \int \frac{\varepsilon \partial^\gamma \Delta\phi^1}{n^\varepsilon}[\partial^\gamma, \mathbf{u}^\varepsilon] \cdot \nabla n^1 =: I_{21}^\varepsilon + I_{22}^\varepsilon.$$

- *Estimate of I_{22}^ε.*

For the second term I_{22}^ε, we obtain

$$I_{22}^\varepsilon \leqslant C\varepsilon \|\partial^\gamma \Delta\phi^1\|_{L^2}\|[\partial^\gamma, \mathbf{u}^\varepsilon]\cdot \nabla n^1\|_{L^2}$$
$$\leqslant C\varepsilon \|\partial^\gamma \Delta\phi^1\|_{L^2}(\|\nabla \mathbf{u}^\varepsilon\|_{L^\infty}\|\nabla n^1\|_{H^{s-1}} + \|\mathbf{u}^\varepsilon\|_{H^s}\|\nabla n^1\|_{L^\infty})$$
$$\leqslant C\varepsilon \|\Delta\phi^1\|_{H^s}(\|n^1\|_{H^s} + \|n^1\|_{H^3} + \varepsilon\|\mathbf{u}^1\|_{H^3}\|n^1\|_{H^s} + \varepsilon\|n^1\|_{H^3}\|\mathbf{u}^1\|_{H^s})$$
$$\leqslant C\varepsilon^2\|\Delta\phi^1\|_{H^s}^2 + C(\|n^1\|_{H^s}^2 + \|n^1\|_{H^3}^2 + \varepsilon^2\|(\mathbf{u}^1,n^1)\|_{H^3}^2(\|(n^1,\mathbf{u}^1)\|_{H^s}^2))$$
$$\leqslant C(1+\varepsilon^2|||(\mathbf{u}^1,\phi^1)|||_{\varepsilon,3}^2)(1+|||(\mathbf{u}^1,\phi^1)|||_{\varepsilon,s}^2) + |||\phi^1|||_{\varepsilon,3}^2, \tag{2.46}$$

thanks to the commutator estimates (A.1) in the second step, Lemma 2.2 in the last step and the definition of the triple norm in (1.7).

- *Estimate of I_{21}^ε.*

In the rest of this lemma, we focus us on the treatment of I_{21}^ε. Recalling the remainder equation (2.2c) (see also (2.22)), I_{21}^ε can be divided into

$$I_{21}^\varepsilon = \int \frac{\varepsilon \partial^\gamma \Delta \phi^1}{n^\varepsilon} \mathbf{u}^\varepsilon \cdot \nabla \partial^\gamma (n^0 \phi^1) - \varepsilon^2 \int \frac{\partial^\gamma \Delta \phi^1}{n^\varepsilon} \mathbf{u}^\varepsilon \cdot \partial^\gamma \nabla \Delta \phi^1$$

$$- \int \frac{\varepsilon \partial^\gamma \Delta \phi^1}{n^\varepsilon} \mathbf{u}^\varepsilon \cdot \partial^\gamma \nabla \Delta \phi^0 - \sqrt{\varepsilon} \int \frac{\varepsilon \partial^\gamma \Delta \phi^1}{n^\varepsilon} \mathbf{u}^\varepsilon \cdot \partial^\gamma \nabla R^1 =: \sum_{i=1}^4 I_{21i}^\varepsilon. \quad (2.47)$$

In the following, we estimate the four terms on the RHS of (2.47). For the first term I_{211}^ε, we have

$$I_{211}^\varepsilon = \int \frac{\varepsilon n^0 \partial^\gamma \Delta \phi^1}{n^\varepsilon} \mathbf{u}^\varepsilon \cdot \nabla \partial^\gamma \phi^1 + \int \frac{\varepsilon \partial^\gamma \Delta \phi^1}{n^\varepsilon} \mathbf{u}^\varepsilon \cdot [\partial^\gamma, n^0] \nabla \phi^1$$

$$= \frac{\varepsilon}{2} \int \nabla \cdot \left(\frac{n^0 \mathbf{u}^\varepsilon}{n^\varepsilon}\right) |\partial^\gamma \nabla \phi^1|^2 - \varepsilon \int \sum_{i,j} \partial_i \left(\frac{n^0 u_j^\varepsilon}{n^\varepsilon}\right) \partial^\gamma \partial_i \phi^1 \partial^\gamma \partial_j \phi^1$$

$$+ \int \frac{\varepsilon \partial^\gamma \Delta \phi^1}{n^\varepsilon} \mathbf{u}^\varepsilon \cdot [\partial^\gamma, n^0] \nabla \phi^1 =: I_{2111}^\varepsilon + I_{2112}^\varepsilon + I_{2113}^\varepsilon, \quad (2.48)$$

where $u_j^\varepsilon = (\mathbf{u}^\varepsilon)_j = u_j^0 + \varepsilon u_j^1$. By using Sobolev embedding $H^2 \hookrightarrow L^\infty$,

$$\left\|\nabla \left(\frac{n^0 \mathbf{u}^\varepsilon}{n^\varepsilon}\right)\right\|_{L^\infty} \leqslant C + C\varepsilon^2 (\|n^1\|_{H^3}^2 + \|\mathbf{u}^1\|_{H^3}^2),$$

which yields for the first two terms on the RHS of (2.48)

$$|I_{2111}^\varepsilon, I_{2112}^\varepsilon| \leqslant C\varepsilon (1 + \varepsilon^2 (\|n^1\|_{H^3}^2 + \|\mathbf{u}^1\|_{H^3}^2)) \|\nabla \phi^1\|_{H^s}^2$$

$$\leqslant C(1 + \varepsilon^2 \|\|(\mathbf{u}^1, \phi^1)\|\|_{\varepsilon,3}^2) \|\|\phi^1\|\|_{\varepsilon,s}^2.$$

For the term I_{2113}^ε, by the commutator estimates (A.1), we have

$$\|[\partial^\gamma, n^0]\nabla \phi^1\|_{L^2} \leqslant C(\|\nabla n^0\|_{L^\infty} \|\nabla \phi^1\|_{H^{s-1}} + \|n^0\|_{H^s} \|\nabla \phi^1\|_{L^\infty})$$

$$\leqslant C(\|\phi^1\|_{H^3} + \|\phi^1\|_{H^s}),$$

which implies that by Hölder inequality

$$I_{2113}^\varepsilon \leqslant C \left\|\frac{1}{n^0 + \varepsilon n^1}\right\|_{L^\infty} \|\varepsilon \partial^\gamma \Delta \phi^1 (\mathbf{u}^0 + \varepsilon \mathbf{u}^1)\|_{L^2} \|[\partial^\gamma, n^0]\nabla \phi^1\|_{L^2}$$

$$\leqslant C(\|\phi^1\|_{H^3}^2 + \|\phi^1\|_{H^s}^2) + C(1 + \varepsilon^2 \|\mathbf{u}^1\|_{L^\infty}^2)(\varepsilon^2 \|\partial^\gamma \Delta \phi^1\|_{L^2}^2)$$

$$\leqslant C(1 + \varepsilon^2 \|\mathbf{u}^1\|_{H^2}^2) \|\|\phi^1\|\|_{\varepsilon,s}^2 + C\|\phi^1\|_{H^3}^2,$$

thanks to (2.5), the Sobolev embedding $H^2 \hookrightarrow L^\infty$ and the definition of the triple norm (1.7).

For the first term I_{212}, by integration by parts and Lemma 2.2, we have

$$|I_{212}^\varepsilon| = |\frac{\varepsilon^2}{2}\int \nabla \cdot \left(\frac{\mathbf{u}^0 + \varepsilon \mathbf{u}^1}{n^0 + \varepsilon n^1}\right)|\partial^\gamma \Delta \phi^1|^2|$$
$$\leqslant C(1+\varepsilon^2\|(n^1,\mathbf{u}^1)\|_{H^3}^2)\cdot \varepsilon^2\|\partial^\gamma\Delta\phi^1\|_{L^2}^2$$
$$\leqslant C(1+\varepsilon^2\||(\mathbf{u}^1,\phi^1)\||_{\varepsilon,3}^2)\||\phi^1\||_{\varepsilon,s}^2.$$

For the third term I_{213}^ε, by Hölder inequality and Sobolev embedding, we have

$$|I_{213}^\varepsilon| \leqslant \|\frac{\varepsilon\partial^\gamma\Delta\phi^1(\mathbf{u}^0+\varepsilon\mathbf{u}^1)}{n^0+\varepsilon n^1}\|_{L^2}\|\partial^\gamma\nabla\Delta\phi^0\|_{L^2}$$
$$\leqslant \|\partial^\gamma\nabla\Delta\phi^0\|_{L^2}^2 + C(1+\varepsilon^2\|\mathbf{u}^1\|_{L^\infty}^2)(\varepsilon^2\|\partial^\gamma\Delta\phi^1\|_{L^2}^2)$$
$$\leqslant C + C(1+\varepsilon^2\|\mathbf{u}^1\|_{H^2}^2)\||\phi^1\||_{\varepsilon,s}^2.$$

For the fourth term I_{214}^ε, we have by Hölder inequality,

$$I_{214}^\varepsilon \leqslant C\|\sqrt{\varepsilon}\partial^\gamma\nabla R^1\|_{L^2}^2 + C(1+\varepsilon^2\|\mathbf{u}^1\|_{L^\infty}^2)(\varepsilon^2\|\partial^\gamma\Delta\phi^1\|_{H^s}^2)$$
$$\leqslant C\varepsilon(\|\nabla\phi^1\|_{H^s}^2+\|\phi^1\|^2) + C(1+\varepsilon^2\|\mathbf{u}^1\|_{H^2}^2)(\varepsilon^2\|\partial^\gamma\Delta\phi^1\|_{L^2}^2)$$
$$\leqslant C(1+\varepsilon^2\|\mathbf{u}^1\|_{H^2}^2)\||\phi^1\||_{\varepsilon,s}^2,$$

where we have used Lemma 2.1 with $|\alpha| = s+1$ there and the fact that $\|\phi^1\|_{H^{s+1}} \approx \|\nabla\phi^1\|_{H^s} + \|\phi^1\|_{H^s}$. Summarizing, we obtain

$$I_{21}^\varepsilon \leqslant C + C\|\phi^1\|_{H^3}^2 + C(1+\varepsilon^2\||(\mathbf{u}^1,\phi^1)\||_{\varepsilon,3}^2)\||\phi^1\||_{\varepsilon,s}^2. \tag{2.49}$$

Putting (2.46) and (2.49) together, we end the proof of Lemma 2.9. □

Lemma 2.10 *The term I_1^ε in (2.44) is bounded by*

$$I_1^\varepsilon \leqslant -\frac{\varepsilon}{2}\frac{d}{dt}\int\left(\frac{n^0}{n^\varepsilon}\right)|\partial^\gamma\nabla\phi^1|^2 - \frac{\varepsilon^2}{2}\frac{d}{dt}\int\frac{1}{n^\varepsilon}|\partial^\gamma\Delta\phi^1|^2$$
$$+ C(1+\varepsilon^2\||(\mathbf{u}^1,\phi^1)\||_{\varepsilon,4}^2)(1+\||(\mathbf{u}^1,\phi^1)\||_{\varepsilon,s\vee 3}^2), \tag{2.50}$$

for all $0 < \varepsilon < \varepsilon_1$, for some $\varepsilon_1 > 0$.

Proof Recall that I_1^ε is given in (2.44). From the remainder equation (2.2c) (see also (2.29)), I_1^ε can be divided into the following

$$I_1^\varepsilon = -\varepsilon\int\frac{\varepsilon\partial^\gamma\Delta\phi^1}{n^0+\varepsilon n^1}\partial^\gamma\partial_t\Delta\phi^1 + \int\frac{\varepsilon\partial^\gamma\Delta\phi^1}{n^0+\varepsilon n^1}\partial^\gamma\partial_t(n^0\phi^1)$$

$$+ \int \frac{\varepsilon \partial^\gamma \Delta \phi^1}{n^0 + \varepsilon n^1} \partial^\gamma \partial_t \Delta \phi^0 + \sqrt{\varepsilon} \int \frac{\varepsilon \partial^\gamma \Delta \phi^1}{n^0 + \varepsilon n^1} \partial^\gamma \partial_t R^1 =: \sum_{i=1}^{4} I_{1i}^\varepsilon. \qquad (2.51)$$

In the following, we will estimate the RHS terms one by one.

- *Estimate of I_{11}^ε.*

For the first term I_{11}^ε, we have

$$I_{11}^\varepsilon = -\frac{\varepsilon^2}{2} \frac{d}{dt} \int \frac{1}{n^0 + \varepsilon n^1} |\partial^\gamma \Delta \phi^1|^2 + \frac{\varepsilon^2}{2} \int \partial_t \left(\frac{1}{n^0 + \varepsilon n^1}\right) |\partial^\gamma \Delta \phi^1|^2.$$

Using Lemma 2.3 and Sobolev embedding, we have

$$\left\| \partial_t \left(\frac{1}{n^0 + \varepsilon n^1}\right) \right\|_{L^\infty} \leqslant C(1 + \varepsilon \|\partial_t n^1\|_{L^\infty}) \leqslant C(1 + \varepsilon^2 \|\|(\mathbf{u}^1, \phi^1)\|\|_{\varepsilon,3}^2),$$

which yields that

$$I_{11}^\varepsilon \leqslant -\frac{\varepsilon^2}{2} \frac{d}{dt} \int \frac{1}{n^0 + \varepsilon n^1} |\partial^\gamma \Delta \phi^1|^2 + C(1 + \varepsilon^2 \|\|(\mathbf{u}^1, \phi^1)\|\|_{\varepsilon,3}^2) \|\|\phi^1\|\|_{\varepsilon,s}^2. \qquad (2.52)$$

- *Estimate of I_{12}^ε.*

For the second term I_{12}^ε, we have

$$I_{12}^\varepsilon = \int \frac{\varepsilon \partial^\gamma \Delta \phi^1}{n^0 + \varepsilon n^1} \partial^\gamma (\partial_t n^0 \phi^1) + \int \frac{\varepsilon \partial^\gamma \Delta \phi^1}{n^0 + \varepsilon n^1} \partial^\gamma (n^0 \partial_t \phi^1)$$

$$= \int \frac{\varepsilon \partial^\gamma \Delta \phi^1}{n^0 + \varepsilon n^1} \partial^\gamma (\partial_t n^0 \phi^1) + \int \frac{\varepsilon \partial^\gamma \Delta \phi^1}{n^0 + \varepsilon n^1} n^0 \partial_t \partial^\gamma \phi^1 + \int \frac{\varepsilon \partial^\gamma \Delta \phi^1}{n^0 + \varepsilon n^1} [\partial^\gamma, n^0] \partial_t \phi^1$$

$$=: \sum_{i=1}^{3} I_{12i}^\varepsilon.$$

For the first term I_{121}^ε, by the multiplicative estimates (A.1), we have

$$\|\partial^\gamma (\partial_t n^0 \phi^1)\|_{L^2} \leqslant C(\|\partial_t n^0\|_{L^\infty} \|\phi^1\|_{H^s} + \|\partial_t n^0\|_{H^s} \|\phi^1\|_{L^\infty})$$
$$\leqslant C(\|\phi^1\|_{H^3} + \|\phi^1\|_{H^s}),$$

which yields

$$I_{121}^\varepsilon \leqslant C(\|\phi^1\|_{H^3} + \|\phi^1\|_{H^s}) \|\varepsilon \Delta \phi^1\|_{H^s}$$
$$\leqslant C(\|\phi^1\|_{H^3}^2 + \|\|\phi^1\|\|_{\varepsilon,s}^2).$$

For the second term I_{122}^ε, we have by integration by parts,

$$I_{122}^\varepsilon = -\int \nabla \left(\frac{n^0}{n^0 + \varepsilon n^1}\right) \varepsilon \partial^\gamma \nabla \phi^1 \partial_t \partial^\gamma \phi^1 - \int \left(\frac{n^0}{n^0 + \varepsilon n^1}\right) \varepsilon \partial^\gamma \nabla \phi^1 \partial_t \nabla \partial^\gamma \phi^1$$

$$= \int \nabla\left(\frac{n^0}{n^\varepsilon}\right) \varepsilon \partial^{\gamma+\gamma_1} \nabla \phi^1 \partial_t \partial^{\gamma-\gamma_1} \phi^1 + \int \nabla \partial^{\gamma_1}\left(\frac{n^0}{n^\varepsilon}\right) \varepsilon \partial^\gamma \nabla \phi^1 \partial_t \partial^{\gamma-\gamma_1} \phi^1$$
$$- \frac{\varepsilon}{2}\frac{d}{dt}\int \frac{n^0}{n^\varepsilon}|\partial^\gamma \nabla \phi^1|^2 + \frac{\varepsilon}{2}\int \partial_t\left(\frac{n^0}{n^\varepsilon}\right)|\partial^\gamma \nabla \phi^1|^2 =: I^\varepsilon_{122i},$$

where $\gamma_1 \leqslant \gamma$ is a multiindex with $|\gamma_1| = 1$. By Lemma 2.4 and Lemma 2.3, we have

$$\|\partial_t \partial^{\gamma-\gamma_1}\phi^1\|^2_{L^2} \leqslant C(\|\partial_t \partial^{\gamma-\gamma_1}n^1\|^2 + \|\phi^1\|^2_{H^{s-1}})$$
$$\leqslant C(1 + \||(\mathbf{u}^1,\phi^1)\||^2_{\varepsilon,s}),$$

where $\gamma_1 \leqslant \gamma$, $|\gamma_1| = 1$ and hence $|\gamma - \gamma_1| = s - 1$. On the other hand, by direct computation, we have

$$\|\nabla\left(\frac{n^0}{n^\varepsilon}\right)\|_{L^\infty} \leqslant C(1 + \varepsilon\|\nabla n^1\|_{L^\infty}) \leqslant C(1 + \varepsilon^2\||\phi^1\||^2_{\varepsilon,3}),$$

$$\|\nabla \partial^{\gamma_1}\left(\frac{n^0}{n^\varepsilon}\right)\|_{L^\infty} \leqslant C(1 + \varepsilon^2\|\nabla n^1\|^2_{L^\infty} + \varepsilon\|\nabla \partial^{\gamma_1}\phi^1\|_{L^\infty})$$
$$\leqslant C(1 + \varepsilon^2\||\phi^1\||^2_{\varepsilon,4}),$$

where we have used Lemma 2.2 and Sobolev embedding. Hence, by Hölder inequality, we obtain

$$I^\varepsilon_{1221}, I^\varepsilon_{1222} \leqslant C\|\nabla\left(\frac{n^0}{n^\varepsilon}\right)\|_{L^\infty}\|\varepsilon \partial^{\gamma+\gamma_1}\nabla\phi^1\|_{L^2}\|\partial_t \partial^{\gamma-\gamma_1}\phi^1\|_{L^2}$$
$$+ C\|\nabla \partial^{\gamma_1}\left(\frac{n^0}{n^\varepsilon}\right)\|_{L^\infty}\|\varepsilon \partial^\gamma \nabla \phi^1\|_{L^2}\|\partial_t \partial^{\gamma-\gamma_1}\phi^1\|_{L^2}$$
$$\leqslant C(1 + \varepsilon^2\||\phi^1\||^2_{\varepsilon,4})(1 + \||(\mathbf{u}^1,\phi^1)\||^2_{\varepsilon,s}),$$

where we have used the boundedness of the Riesz operator. Similarly,

$$\|\partial_t\left(\frac{n^0}{n^\varepsilon}\right)\|_{L^\infty} \leqslant C(1 + \varepsilon^2\|\partial_t n^1\|^2_{L^\infty}) \leqslant C(1 + \varepsilon^2\||(\mathbf{u}^1,\phi^1)\||^2_{\varepsilon,3}),$$

which yields that

$$I^\varepsilon_{1224} \leqslant C(1 + \varepsilon^2\||(\mathbf{u}^1,\phi^1)\||^2_{\varepsilon,3})\||\phi^1\||^2_{\varepsilon,s}.$$

Therefore, I^ε_{122} is bounded by

$$I^\varepsilon_{122} \leqslant -\frac{\varepsilon}{2}\frac{d}{dt}\int \frac{n^0}{n^\varepsilon}|\partial^\gamma \nabla \phi^1|^2 + C(1 + \varepsilon^2\||(\mathbf{u}^1,\phi^1)\||^2_{\varepsilon,4})(1 + \||(\mathbf{u}^1,\phi^1)\||^2_{\varepsilon,s}).$$

For the third term I^ε_{123}, we have

$$I^\varepsilon_{123} \leqslant C\varepsilon^2\|\Delta \phi^1\|^2_{H^s} + C\|[\partial^\gamma, n^0]\partial_t \phi^1\|^2_{L^2}$$

$$\leqslant C\varepsilon^2 \|\Delta\phi^1\|_{H^s}^2 + C(\|\nabla n^0\|_{L^\infty}^2 \|\partial_t\phi^1\|_{H^{s-1}}^2 + \|\partial_t\phi^1\|_{L^\infty}^2 \|n^0\|_{H^s}^2)$$
$$\leqslant C(1 + |||(\mathbf{u}^1,\phi^1)|||_{\varepsilon,s}^2 + |||(\mathbf{u}^1,\phi^1)|||_{\varepsilon,3}^2),$$

thanks to Lemma A.1 in the second inequality and Lemma 2.4 and Lemma 2.3 in the last inequality. In summary, I_{12}^ε can be bounded by

$$I_{12}^\varepsilon \leqslant -\frac{\varepsilon}{2}\frac{d}{dt}\int \frac{n^0}{n^\varepsilon}|\partial^\gamma \nabla\phi^1|^2$$
$$+ C(1+\varepsilon^2 |||(\mathbf{u}^1,\phi^1)|||_{\varepsilon,4}^2)(1+|||(\mathbf{u}^1,\phi^1)|||_{\varepsilon,s}^2 + |||(\mathbf{u}^1,\phi^1)|||_{\varepsilon,3}^2). \quad (2.53)$$

- *Estimate of I_{13}^ε.*

For the third term I_{13}^ε, it is straightforward that

$$I_{13}^\varepsilon \leqslant C(1+\varepsilon^2 \|\Delta\phi^1\|_{H^s}^2). \quad (2.54)$$

- *Estimate of I_{14}^ε.*

For the fourth term I_{14}^ε, we have

$$|I_{14}^\varepsilon| \leqslant C\varepsilon^2 \|\Delta\phi^1\|_{H^s}^2 + C\varepsilon \|\partial_t R^1\|_{H^s}^2$$
$$\leqslant C\varepsilon^2 \|\Delta\phi^1\|_{H^s}^2 + C_1(\|\phi^1\|_{H^s}^2 + \varepsilon\|\partial_t\phi^1\|_{H^s}^2)$$
$$\leqslant C\varepsilon^2 \|\Delta\phi^1\|_{H^s}^2 + C_1(\|\phi^1\|_{H^s}^2 + \|\partial_t n^1\|_{H^{s-1}}^2)$$
$$\leqslant C(1+|||(\mathbf{u}^1,\phi^1)|||_{\varepsilon,s}^2), \quad (2.55)$$

where we have used Hölder inequality in the first inequality, Lemma 2.1 in the second inequality, Lemma 2.4 in the third inequality and Lemma 2.3 in the last inequality. Here, we also have used the fact that $\|\partial_t\phi^1\|_{H^s} \approx \|\partial_t\phi^1\|_{H^{s-1}} + \|\partial_t\partial^\alpha \nabla\phi^1\|_{L^2}$ with $|\gamma| = s-1$ and $\|\phi^1\|_{H^{s-1}} \leqslant \|\phi^1\|_{H^s}$ for all integers $s > 0$.

By (2.51), using (2.52), (2.53), (2.54) and (2.55), we obtain the estimate (2.50) for I^ε. □

Now, we can end the proof of Lemma 2.8.

End of proof of Lemma 2.8 By using (2.40), the estimates of (2.41), (2.42), (2.43) and (2.45), and Lemma 2.9 and 2.10, we close the proof of Lemma 2.8. □

Summarizing these lemmas, we obtain the following

Proposition 2.2 Let $s \geqslant 0$ be a positive integer, $(n^1, \mathbf{u}^1, \phi^1)$ be a smooth solution for the system (2.2). There exists $\varepsilon_1 > 0$ and $C, C' > 0$ such that for any $0 < \varepsilon < \varepsilon_1$, there holds

$$\varepsilon \|\nabla \mathbf{u}^1(t)\|_{H^s}^2 + \varepsilon \|\nabla \phi^1(t)\|_{H^s}^2 + \varepsilon^2 \|\Delta \phi^1(t)\|_{H^s}^2$$
$$\leqslant C'(\varepsilon \|\nabla \mathbf{u}^1(0)\|_{H^s}^2 + \varepsilon \|\nabla \phi^1(0)\|_{H^s}^2 + \varepsilon^2 \|\Delta \phi^1(0)\|_{H^s}^2)$$
$$+ C \int_0^t (1 + \varepsilon^2 |||(\mathbf{u}^1, \phi^1)|||_{\varepsilon,4}^2)(1 + |||(\mathbf{u}^1, \phi^1)|||_{\varepsilon,s\vee 3}^2) d\tau, \tag{2.56}$$

where C' depends only on σ' and σ''.

Proof This is shown by integrating (2.39) over $[0, t]$ and summing them up for $|\gamma| \leqslant s$, and then using $\sigma' < n^0 < \sigma''$ and $\sigma'/2 < n^\varepsilon < 2\sigma''$ for any $t \in [0, T_\varepsilon]$ in (1.4) and (2.5) for $0 < \varepsilon < \varepsilon_1$. □

2.5 End of proof of Theorem 1.2

Now, we are in a good position to end the proof of Theorem 1.2.

Proof Let $s \geqslant 7$ be an integer. By Proposition 2.1 and Proposition 2.2 and recalling the definition of the norm (1.7), we obtain the following Gronwall type inequality

$$|||(\mathbf{u}^1, \phi^1)(t)|||_{\varepsilon,s}^2 \leqslant C'C_\varepsilon(0)$$
$$+ C \int_0^t (1 + \varepsilon^3 |||(\mathbf{u}^1, \phi^1)|||_{\varepsilon,5}^3)(1 + |||(\mathbf{u}^1, \phi^1)|||_{\varepsilon,s}^2) d\tau, \tag{2.57}$$

where $C_\varepsilon(0) = |||(\mathbf{u}^1, \phi^1)(0)|||_{\varepsilon,s}^2$ and we have used the fact that $||| \cdot |||_{\varepsilon,3} \leqslant ||| \cdot |||_{\varepsilon,s}$ for $s \geqslant 3$. From (2.4), there exists $\varepsilon_1 > 0$ such that for any $0 < \varepsilon < \varepsilon_1$, $\varepsilon^3 |||(\mathbf{u}^1, \phi^1)|||_{\varepsilon,5}^3 \leqslant 1$, and hence (2.57) yields

$$|||(\mathbf{u}^1, \phi^1)(t)|||_{\varepsilon,s}^2 \leqslant C_2 C_\varepsilon(0) + C_2 \int_0^t (1 + |||(\mathbf{u}^1, \phi^1)|||_{\varepsilon,s}^2) d\tau, \tag{2.58}$$

where $C_2 = \max\{C', 2C\}$. On the other hand, from Lemma 2.2, there exists constant $C_3 > 1$ such that

$$\|n^1(t)\|_{H^s}^2 \leqslant C_3(1 + |||\phi^1|||_{\varepsilon,s}^2). \tag{2.59}$$

Let $C_0 = \sup_{0 < \varepsilon < 1} C_\varepsilon(0)$. We let $\tilde{C}$ in (2.4) satisfy $\tilde{C} \geqslant 2C_3(1 + C_2 C_0)e^{C_2 T_\varepsilon}$, then from (2.58)

$$|||(\mathbf{u}^1, \phi^1)|||_{\varepsilon,s}^2 \leqslant (1 + C_2 C_0)e^{C_2 T_\varepsilon} \leqslant \tilde{C},$$

and hence from (2.59)

$$\|n^1(t)\|_{H^s}^2 \leq C_3(1 + (1 + C_2C_0)e^{C_2T_\varepsilon}) \leq \tilde{C}.$$

Then by the continuity principle, it is standard to get the uniform in ε estimates for $|||(n^1, \mathbf{u}^1, \phi^1)|||_{\varepsilon,s}$. In particular, for every $T' < T$, $\varepsilon^{-1}(n^\varepsilon - n^0)$ and $\varepsilon^{-1}(\mathbf{u}^\varepsilon - \mathbf{u}^0)$ are bounded in $L^\infty([0, T']; H^s)$ and $L^\infty([0, T']; \mathbf{H}^s)$, respectively, uniformly in ε for ε small enough for some $s < s'$. □

3 Proof of Theorem 1.3

This section is devoted to the proof of Theorem 1.3. Let $(n^{\varepsilon,T_i}, \mathbf{u}^{\varepsilon,T_i}, \phi^{\varepsilon,T_i})$ be a solution of (1.5) and $(n^{0,T_i}, \mathbf{u}^{0,T_i}, \phi^{0,T_i})$ be a solution of (1.6) with the same initial data. Let

$$n^{\varepsilon,T_i} = n^{0,T_i} + \varepsilon n^{1,T_i}, \quad \mathbf{u}^{\varepsilon,T_i} = \mathbf{u}^{0,T_i} + \varepsilon \mathbf{u}^{1,T_i}, \quad \phi^{\varepsilon,T_i} = \phi^{0,T_i} + \varepsilon \phi^{1,T_i}. \quad (3.1)$$

Then $(n^{1,T_i}, \mathbf{u}^{1,T_i}, \phi^{1,T_i})$ satisfy the remainder system:

$$(R_\varepsilon) \begin{cases} \partial_t n^{1,T_i} + \nabla \cdot (n^{0,T_i}\mathbf{u}^{1,T_i} + \mathbf{u}^{0,T_i}n^{1,T_i}) + \varepsilon \nabla \cdot (n^{1,T_i}\mathbf{u}^{1,T_i}) = 0, & (3.2a) \\ \partial_t \mathbf{u}^{1,T_i} + \mathbf{u}^{0,T_i} \cdot \nabla \mathbf{u}^{1,T_i} + \mathbf{u}^{1,T_i} \cdot \nabla \mathbf{u}^{0,T_i} + \varepsilon \mathbf{u}^{1,T_i} \cdot \nabla \mathbf{u}^{1,T_i} \\ \quad + \dfrac{T_i \nabla n^{1,T_i}}{n^{0,T_i} + \varepsilon n^{1,T_i}} - \dfrac{T_i n^{1,T_i} \nabla n^{0,T_i}}{n^{0,T_i}(n^{0,T_i} + \varepsilon n^{1,T_i})} = -\nabla \phi^{1,T_i}, & (3.2b) \\ -\varepsilon \Delta \phi^{1,T_i} = \Delta \phi^{0,T_i} + n^{1,T_i} - n^{0,T_i}\phi^{1,T_i} + \sqrt{\varepsilon} R^1, & (3.2c) \end{cases}$$

where R^1 is given in (2.3).

We need only to show uniform in ε and T_i estimates for $(n^{1,T_i}, \mathbf{u}^{1,T_i})$. We also let $\tilde{C}$ to be a constant to be determined later, much larger than the bound of $\|(n_0, \mathbf{u}_0)\|_{H^s}$, such that on $[0, T^{\varepsilon,T_i}]$

$$\sup_{[0,T^{\varepsilon,T_i}]} \|(n^{1,T_i}, \mathbf{u}^{1,T_i}, \phi^{1,T_i})\|_{H^s} \leq \tilde{C}. \quad (3.3)$$

From the expression (), there exists some $\varepsilon_1 = \varepsilon_1(\tilde{C}) > 0$ and $T_{i1} > 0$ such that $\sigma'/2 < n^{\varepsilon,T_i} < 2\sigma'$ and $|\mathbf{u}^{\varepsilon,T_i}| \leq 1/2$ on $[0, T^{\varepsilon,T_i}]$ for all $0 < \varepsilon < \varepsilon_1$ and $0 < T_i < T_{i1}$.

Let γ be a multiindex with $|\gamma| = s$. By taking ∂^γ to (3.2b), and then taking L^2 inner product with $\partial^\gamma \mathbf{u}^{1,T_i}$, we obtain an equality similar to (2.12) but with the following two more terms on the RHS:

$$V + VI := -T_i \int \partial^\gamma \mathbf{u}^{1,T_i} \partial^\gamma \left(\frac{\nabla n^{1,T_i}}{n^{0,T_i} + \varepsilon n^{1,T_i}} \right)$$

$$+ T_i \int \partial^\gamma \mathbf{u}^{1,T_i} \partial^\gamma \left(\frac{\nabla n^{0,T_i}}{n^{0,T_i}} \frac{n^1}{n^{0,T_i} + \varepsilon n^{1,T_i}} \right). \tag{3.4}$$

- *Estimate of the term VI.*

For this term, we obtain by using Leibnitz formula and multiplicative estimates

$$VI \leqslant T_i C(\varepsilon \tilde{C}) \|(\partial^\gamma \mathbf{u}^{1,T_i}, \partial^\gamma n^{1,T_i})\|^2,$$

where $\tilde{C}$ is given in (2.4).

- *Estimate of the term V.*

For this term, we rewrite

$$V = -T_i \int \frac{\partial^\gamma \mathbf{u}^{1,T_i} \cdot \nabla \partial^\gamma n^{1,T_i}}{n^{0,T_i} + \varepsilon n^{1,T_i}} - T_i \int \partial^\gamma \mathbf{u}^{1,T_i} \cdot [\partial^\gamma, \frac{1}{n^{0,T_i} + \varepsilon n^{1,T_i}}] \nabla n^{1,T_i}$$
$$= T_i \int \frac{\partial^\gamma \nabla \cdot \mathbf{u}^{1,T_i} \partial^\gamma n^{1,T_i}}{n^{0,T_i} + \varepsilon n^{1,T_i}} + T_i \int \partial^\gamma \mathbf{u}^{1,T_i} \partial^\gamma n^{1,T_i} \cdot \nabla(\frac{1}{n^{0,T_i} + \varepsilon n^{1,T_i}})$$
$$- T_i \int \partial^\gamma \mathbf{u}^{1,T_i} \cdot [\partial^\gamma, \frac{1}{n^{0,T_i} + \varepsilon n^{1,T_i}}] \nabla n^{1,T_i} =: V_1 + V_2 + V_3.$$

For V_2 and V_3, we obtain by using commutator estimates (A.1) that

$$V_2 + V_3 \leqslant T_i C(\varepsilon \tilde{C}) \|(\partial^\gamma \mathbf{u}^{1,T_i}, \partial^\gamma n^{1,T_i})\|^2.$$

Recalling decomposition (2.17), we decompose V_1 into

$$V_1 = -T_i \int \frac{\partial_t \partial^\gamma n^{1,T_i} \partial^\gamma n^{1,T_i}}{(n^{0,T_i} + \varepsilon n^{1,T_i})^2} - T_i \int \frac{[\partial^\gamma, n^{0,T_i} + \varepsilon n^{1,T_i}] \nabla \cdot \mathbf{u}^{1,T_i} \partial^\gamma n^{1,T_i}}{(n^{0,T_i} + \varepsilon n^{1,T_i})^2}$$
$$- T_i \int \frac{\partial^\gamma((\mathbf{u}^{0,T_i} + \varepsilon \mathbf{u}^{1,T_i}) \cdot \nabla n^{1,T_i}) \partial^\gamma n^{1,T_i}}{(n^{0,T_i} + \varepsilon n^{1,T_i})^2} - T_i \int \frac{\partial^\gamma(\mathbf{u}^{1,T_i} \cdot \nabla n^{0,T_i}) \partial^\gamma n^{1,T_i}}{(n^{0,T_i} + \varepsilon n^{1,T_i})^2}$$
$$- T_i \int \frac{\partial^\gamma(n^{1,T_i} \nabla \cdot \mathbf{u}^{0,T_i}) \partial^\gamma n^{1,T_i}}{(n^{0,T_i} + \varepsilon n^{1,T_i})^2} =: V_{11} + \cdots + V_{15}.$$

The term V_{11} can be estimated as

$$V_{11} = -\frac{T_i}{2} \frac{d}{dt} \int \frac{|\partial^\gamma n^{1,T_i}|^2}{(n^{0,T_i} + \varepsilon n^{1,T_i})^2}.$$

The estimate of term V_{12} is similar to I_3 in (2.18) and the estimate of term V_{13} is similar to II in (2.12). After integration by parts and commutator estimates, they can be estimated as

$$V_{12} + V_{13} \leqslant T_i C(\varepsilon \tilde{C}) \|(n^{1,T_i}, \mathbf{u}^{1,T_i})\|_{H^s}^2.$$

The estimates of the last two terms V_{14} and V_{15} are direct since the derivative on n^{1,T_i} and $\mathbf{u}^{1,T_i}$ is no greater than $|\gamma|$ and we have

$$V_{14} + V_{15} \leqslant T_i C(\varepsilon \tilde{C}) \|(n^{1,T_i}, \mathbf{u}^{1,T_i})\|_{H^s}^2.$$

These estimates lead to the following

$$V + VI \leqslant -\frac{T_i}{2} \frac{d}{dt} \int \frac{|\partial^\gamma n^{1,T_i}|^2}{(n^{0,T_i} + \varepsilon n^{1,T_i})^2} + T_i C(\varepsilon \tilde{C}) \|(n^{1,T_i}, \mathbf{u}^{1,T_i})\|_{H^s}^2. \quad (3.5)$$

Therefore, similar to Proposition 2.1, there exist $\varepsilon_1 > 0$ and $T_{i1} > 0$ and constants $C', C > 0$ such that

$$T_i \|n^{1,T_i}(t)\|_{H^s}^2 + \|\mathbf{u}^{1,T_i}(t)\|_{H^s}^2 + \|\phi^{1,T_i}(t)\|_{H^s}^2 + \varepsilon \|\nabla \phi^{1,T_i}(t)\|_{H^s}^2$$
$$\leqslant C'(T_i \|n^{1,T_i}(0)\|_{H^s}^2 + \|\mathbf{u}^{1,T_i}(0)\|_{H^s}^2 + \|\phi^{1,T_i}(0)\|_{H^s}^2 + \varepsilon \|\nabla \phi^{1,T_i}(0)\|_{H^s}^2)$$
$$+ C \int_0^t T_i C(\varepsilon \tilde{C}) \|(n^{1,T_i}, \mathbf{u}^{1,T_i})\|_{H^s}^2$$
$$+ (1 + \varepsilon^3 |||(\mathbf{u}^{1,T_i}, \phi^{1,T_i})|||_{\varepsilon,5}^3)(1 + |||(\mathbf{u}^{1,T_i}, \phi^{1,T_i})|||_{\varepsilon,s\vee 3}^2) d\tau, \quad (3.6)$$

where C' depends only on σ' and σ''.

Likewise, at the weighted $(s+1)$-order, there holds

$$T_i \varepsilon \|\nabla n^{1,T_i}(t)\|_{H^s}^2 + \varepsilon \|\nabla \mathbf{u}^{1,T_i}(t)\|_{H^s}^2 + \varepsilon \|\nabla \phi^{1,T_i}(t)\|_{H^s}^2 + \varepsilon^2 \|\Delta \phi^{1,T_i}(t)\|_{H^s}^2$$
$$\leqslant C'(T_i \varepsilon \|\nabla n^{1,T_i}(0)\|_{H^s}^2 + \varepsilon \|\nabla \mathbf{u}^{1,T_i}(0)\|_{H^s}^2 + \varepsilon \|\nabla \phi^{1,T_i}(0)\|_{H^s}^2 + \varepsilon^2 \|\Delta \phi^{1,T_i}(0)\|_{H^s}^2)$$
$$+ C \int_0^t T_i \varepsilon C(\varepsilon \tilde{C}) \|(n^{1,T_i}, \mathbf{u}^{1,T_i})\|_{H^{s+1}}^2$$
$$+ (1 + \varepsilon^2 |||(\mathbf{u}^{1,T_i}, \phi^{1,T_i})|||_{\varepsilon,4}^2)(1 + |||(\mathbf{u}^{1,T_i}, \phi^{1,T_i})|||_{\varepsilon,s\vee 3}^2) d\tau, \quad (3.7)$$

where C' depends only on σ' and σ''.

Let now $|||n^{1,T_i}|||_{\varepsilon,s}^2 = \|n^{1,T_i}\|_{H^s}^2 + \varepsilon \|\nabla n^{1,T_i}\|_{H^s}^2$. Putting (3.6) and (3.7) together, we then obtain the following Gronwall type inequality

$$|||(\sqrt{T_i} n^{1,T_i}, \mathbf{u}^{1,T_i}, \phi^{1,T_i})(t)|||_{\varepsilon,s}^2 \leqslant C' C_\varepsilon(0)$$
$$+ \int_0^t C(\varepsilon \tilde{C})(1 + \varepsilon^3 |||(\mathbf{u}^{1,T_i}, \phi^{1,T_i})|||_{\varepsilon,5}^3)(1 + |||(\sqrt{T_i} n^{1,T_i}, \mathbf{u}^{1,T_i}, \phi^{1,T_i})|||_{\varepsilon,s}^2) d\tau, \quad (3.8)$$

where $C_\varepsilon(0) = |||(\mathbf{u}^{1,T_i}, \phi^{1,T_i})(0)|||_{\varepsilon,s}^2$ and $s \vee 3 = s$ for $s \geqslant 3$. From (3.3), there exists $\varepsilon_1 > 0$ such that for any $0 < \varepsilon < \varepsilon_1$, $\varepsilon^3 |||(\mathbf{u}^{1,T_i}, \phi^{1,T_i})|||_{\varepsilon,5}^3 \leqslant 1$ and $C(\varepsilon \tilde{C}) \leqslant C(1)$, and hence (3.8) yields

$$|||(\sqrt{T_i} n^{1,T_i}, \mathbf{u}^{1,T_i}, \phi^{1,T_i})(t)|||_{\varepsilon,s}^2 \leqslant C_2 C_\varepsilon(0)$$

$$+ C_2 \int_0^t (1 + |||(\sqrt{T_i}n^{1,T_i}, \mathbf{u}^{1,T_i}, \phi^{1,T_i}))|||_{\varepsilon,s}^2)d\tau,$$

where $C_2 = \max\{C', 2C(1)\}$. Then invoking Lemma 2.2 and following the proof of Theorem 1.2 in Section 2.5, we obtain the uniform both in ε and T_i estimates for $\|(n^{1,T_i}, \mathbf{u}^{1,T_i}, \phi^{1,T_i})\|_{H^s}$. In deed, we obtain that $\|n^{1,T_i}\|_{H^s}^2 + T_i\varepsilon\|\nabla n^{1,T_i}\|_{H^s}^2$ is uniformly bounded. In particular, for every $T' < T$, $\varepsilon^{-1}(n^\varepsilon - n^0)$ and $\varepsilon^{-1}(\mathbf{u}^\varepsilon - \mathbf{u}^0)$ are bounded in $L^\infty([0,T']; H^s)$ and $L^\infty([0,T']; \mathbf{H}^s)$, respectively, uniformly in ε and T_i for ε small enough for some $s < s'$. This completes the proof.

A Commutator estimates

For reader's convenience, we give two important inequalities that are widely used throughout this paper [13, Lemma X1 and Lemma X4].

Lemma A.1 *Let α be any multi-index with $|\alpha| = k$ and $p \in (1, \infty)$. Then there exists some constant $C > 0$ such that*

$$\begin{aligned}\|\partial_x^\alpha(fg)\|_{L^p} &\leqslant C\{\|f\|_{L^{p_1}}\|g\|_{\dot{H}^{k,p_2}} + \|f\|_{\dot{H}^{k,p_3}}\|g\|_{L^{p_4}}\}, \\ \|[\partial_x^\alpha, f]g\|_{L^p} &\leqslant C\{\|\nabla f\|_{L^{p_1}}\|g\|_{\dot{H}^{k-1,p_2}} + \|f\|_{\dot{H}^{k,p_3}}\|g\|_{L^{p_4}}\},\end{aligned} \quad (\text{A.1})$$

where $f, g \in \mathcal{S}$, the Schwartz class and $p_2, p_3 \in (1, +\infty)$ such that

$$\frac{1}{p} = \frac{1}{p_1} + \frac{1}{p_2} = \frac{1}{p_3} + \frac{1}{p_4}.$$

References

[1] Y. Brenier. Convergence of the Vlasov-Poisson system to the incompressible Euler equations. *Comm. Partial Differential Equations* 25(3-4), (2000)737-754.

[2] S. Cordier and E. Grenier. Quasineutral limit of an Euler-Poisson system arising from plasma physics, *Comm. Partial Differential Equations*, 35(5&6), (2000)1099-1113.

[3] P. Degond, H. Liu, D. Savelief and M-H. Vignal, Numerical approximation of the Euler-Poisson-Boltzmann model in the quasineutral limit. *J. Sci. Comput.*, 51, (2012)59-86.

[4] D. Gerard-Varet, D. Han-kwan and F. Rousset. Quasineutral limit of the Euler-Poisson system for ions in a domain with boundaries. *Indiana Univ. Math. J.*, 62(2), (2013)359-402.

[5] E. Grenier. Oscillatory perturbations of the Navier-Stokes equation. *J. Math. Pures Appl.* 76, (1997)477-498.

[6] E. Grenier. Pseudo-differential energy estimates of singular perturbations, *Comm. Pure Appl. Math.*, 50(9), (1997)821-865.

[7] Y. Guo, A.D. Ionescu and B. Pausader. Global solutions of the Euler-Maxwell two-fluid system in 3D. arXiv: 1303.1060v1.

[8] Y. Guo and B. Pausader. Global smooth ion dynamics in the Euler-Poisson system, *Commun. Math. Phys.*, 303,(2011)89-125.

[9] Y. Guo and X. Pu. KdV limit of the Euler-Poisson system. *Arch. Rational Mech. Anal.*, 211, (2014)673-710.

[10] D. Han-Kwan. Quasineutral limit of the Valsov-Poisson system with massless electrons. *Comm. Partial Differential Equations*, 36, (2011)1385-1425.

[11] D. Han-Kwan. From Vlasov-Poisson to Korteweg-de Vries and Zakharov-Kuznetsov. *Commun. Math. Phys.*, 324(3), (2013)961-993.

[12] S. Jiang, Q. Ju, H. Li and Y. Li. Quasi-neutral limit of the full bipolar Euler-Poisson system. *Sci. China Math.* 53(2010), no. 12, 3099-3114.

[13] T. Kato and G. Ponce. Commutator estimates and the Euler and navier-Stokes equations, *Comm. Pure Appl. Math.*, 41, (1988)891-907.

[14] N. Krall and A. Trivelpiece. *Principles of plasma physics*, San Francisco Press, 1986.

[15] D. Lannes, F. Linares and J.-C. Saut. The Cauchy problem for the Euler-Poisson system and derivation of the Zakharov-Kuznetsov equation. *Studies in Phase Space Analysis with Aplications to PDEs*, in Series: *Progress in Nonlinear Differential Equations and Applications*, vol. 84, M. Cicognani, F. Colombini, D. Del Santo Eds., Birkhaüser, (2013)183-215.

[16] G. Loeper. Quasi-neutral limit of the Euler-Poisson and Euler-Monge-Ampere systems, *Comm. Partial Differential Equations*, 30, (2005)1141-1167.

[17] A. Majda. *Compressible fluid flow and systems of conservation laws in several space variables*, Applied Mathematial Sciences, 53, Springer-Verlag, New York-Berlin, 1984.

[18] Y. Peng, S. Wang. Convergence of compressible Euler-Maxwell equations to incompressible Euler equations, *Comm. Partial Differential Equations*, 33, (2008)349-376.

[19] X. Pu. Dispersive limit of the Euler-Poisson system in higher dimensions, *SIAM J. Math. Anal.*, 45(2), (2013)834-878.

[20] X. Pu. Quasineutral limit of the pressureless Euler-Poisson equation, *Appl. Math. Lett.*, 30, (2014)33-37.

[21] M. Slemrod and N. Sternberg. Quasi-neutral limit for the Euler-Poisson system, *J. Nonlinear Sciences*, 11, (2001)193-209.

[22] E. Shlomo, H. Liu and E. Tadmor. Critical thresholds in Euler-Poisson equations. *Indiana Univ. Math. J.* 50, (2001)109-157.

[23] E.M. Stein. *Singular integrals and differentiability properties of functions*, Princeton University Press, Princeton, New Jersey, 1970.

[24] S. Wang. Quasineutral limit of Euler-Poisson system with and without viscosity. *Comm. Partial Differential Equations*, 29(2004), no. 3-4, 419-456.

[25] S. Wang and S. Jiang. The convergence of the Navier-Stokes-Poisson system to the incompressible Euler equations. *Comm. Partial Differential Equations*, 31(2006), no. 4, 571-591.

Well-posedness and Blow-up Scenario for a New Integrable Four-component System with Peakon Solutions*

Mi Yongsheng (米永生), Guo Boling (郭柏灵) and Mu Chunlai (穆春来)

Abstract In this paper, we are concerned with the Cauchy problem of the new integrable four-component system with cubic nonlinearity. We establish the local well-posedness in a range of the Besov spaces. Then the precise blow-up scenario for strong solutions to the system is derived.

Keywords Besov spaces; peakon solutions; local well-posedness

1 Introduction

In this paper, we consider the following Cauchy problem of the new integrable four-component system with cubic nonlinearity

$$\begin{cases} m_{1,t} + \dfrac{1}{9}(m_1(f_1g_1 + f_2g_2))_x - \dfrac{1}{9}m_1f_1g_1 - \dfrac{1}{9}m_2f_1g_2 = 0, & t > 0, x \in \mathbb{R}, \\ m_{2,t} + \dfrac{1}{9}(m_2(f_1g_1 + f_2g_2))_x - \dfrac{1}{9}m_2f_2g_1 - \dfrac{1}{9}m_2f_2g_2 = 0, & t > 0, x \in \mathbb{R}, \\ m_{3,t} + \dfrac{1}{9}(m_3(f_1g_1 + f_2g_2))_x + \dfrac{1}{9}m_3f_1g_1 + \dfrac{1}{9}m_4f_2g_1 = 0, & t > 0, x \in \mathbb{R}, \\ m_{4,t} + \dfrac{1}{9}(m_4(f_1g_1 + f_2g_2))_x + \dfrac{1}{9}m_4f_1g_2 + \dfrac{1}{9}m_4f_2g_2 = 0, & t > 0, x \in \mathbb{R}, \\ m_i = u_i - u_{i,xx}, \quad i = 1,2,3,4, & t > 0, x \in \mathbb{R}, \\ u_i(0,x) = u_{i,0}(x), \quad i = 1,2,3,4, & x \in \mathbb{R}, \end{cases} \quad (1.1)$$

where $f_1 = u_1 - u_{1,x}, f_2 = u_2 - u_{2,x}, g_1 = u_3 + u_{3,x}, g_2 = u_4 + u_{4,x}$.

In [1], Xia and Qiao prove that the system (1) is completely integrable in the sense of Lax pair, Hamiltonian structure, and conservation laws. They also show that the system (1) admits peaked soliton (peakon) and multi-peakon solutions.

---
* Discrete Contin. Dyn. Syst., 2016, 36(4): 2171–2191. DOI: 10.3934/dcds.2016.36.2171

The original Camassa-Holm equation

$$n_t + vn_x + 2v_x n = 0, n = v - v_{xx},$$

can itself be derived from the Korteweg-deVries equation by tri-Hamiltonian duality. The Camassa-Holm equation was originally proposed as a model for surface water waves in the shallow water regime (see [2, 3]) and has been studied extensively in the last twenty years because of its many remarkable properties. Camassa-Holm equation has traveling wave solutions of the form $ce^{-|x-ct|}$, called peakons, which capture the main feature of the exact traveling wave solutions of greatest height of the governing equations (see [4–6]). Moreover, the shape of some peakons is stable under small perturbations, making these waves recognizable physically (see [7, 8]). It was shown in [9–11] that the inverse spectral or scattering approach was a powerful tool to handle Camassa-Holm equation and the Camassa-Holm equation is a completely integrable. The geometric formulations [12–15], well-posedness and breaking waves, meaning solutions that remain bounded while its slope becomes unbounded in finite time [16–20] have been discucssed. Moreover, the Camassa-Holm equation has global conservative solutions [21, 22] and dissipative solutions [23, 24].

Note that the nonlinearity in the Camassa-Holm equation is quadratic. Two integrable Camassa-Holm-type equations with cubic nonlinearity have been discovered: One is the Novikov equation, and the second is the following equation

$$\begin{cases} m_t + (u^2 - u_x^2)m_x + 2u_x m^2 = 0, & t > 0, x \in \mathbb{R}, \\ m = u - u_{xx}, & t > 0, x \in \mathbb{R}, \\ u(x, 0) = u_0(x), & x \in \mathbb{R}, \end{cases} \quad (1.2)$$

which was found independently by Fokas [25], by Fuchssteiner [26] and by Olver and Rosenau [27] by applying the general method of tri-Hamiltonian duality to the bi-Hamiltonian representation of the modified Korteweg-deVries equation. Later, it was obtained by Qiao [28] from the two-dimensional Euler equations, where the variables $u(t, x)$ and $m(t, x)$ represent, respectively, the velocity of the fluid and its potential density. In [29] it was shown that equation (2) admits a Lax pair, and hence can be solved by the method of inverse scattering. In [30], authors establish the local well-posedness and derive the blow-up scenario. With analytic initial data, they then show that its solutions are analytic in both variables, globally in space and locally in time. They also give geometric descriptions to this integrable equation (2).

Motivated by the references cited above, the goal of this paper is to establish the local well-posedness of System (1) in the nonhomogeneous Besov spaces and derive the precise blow-up scenario of strong solutions to the system. The proof of the local well-posedness is inspired by the argument of approximate solutions by Danchin [33] in the study of the local wellposedness to the Camassa-Holm equation. However, one problematics issue is that we here deal with two-component system with a higher order nonlinearity in the Besov spaces, making the proof of several required nonlinear estimates somewhat delicate. These difficulties are nevertheless overcome by carefully estimates for each iterative approximation of solutions to (1).

Now we are in the position to state the local existence result and Blow-up criteria, where the definition of Besov spaces $B_{p,r}^s$ and $E_{p,r}^s(T)$ will be given in Section 2.

Theorem 1.1 Let $p, r \in [1, \infty]$ and $s > \max\left\{\frac{5}{2}, 2+\frac{1}{p}\right\}$. Assume that $(u_{1,0}, u_{2,0}, u_{3,0}, u_{4,0}) \in B_{p,r}^s \times B_{p,r}^s \times B_{p,r}^s \times B_{p,r}^s$. There exists a time $T > 0$ such that the initial-value problem (1) has a unique solution $(u_1, u_2, u_3, u_4) \in E_{p,r}^s(T) \times E_{p,r}^s(T) \times E_{p,r}^s(T) \times E_{p,r}^s(T)$ and the map $(u_{1,0}, u_{2,0}, u_{3,0}, u_{4,0}) \mapsto (u_1, u_2, u_3, u_4)$ is continuous from a neighborhood of $(u_{1,0}, u_{2,0}, u_{3,0}, u_{4,0})$ in $B_{p,r}^s \times B_{p,r}^s \times B_{p,r}^s \times B_{p,r}^s$ into $C([0,T]; B_{p,r}^{s'}) \cap C^1([0,T]; B_{p,r}^{s'-1}) \times C([0,T]; B_{p,r}^{s'}) \cap C^1([0,T]; B_{p,r}^{s'-1}) \times C([0,T]; B_{p,r}^{s'}) \cap C^1([0,T]; B_{p,r}^{s'-1}) \times C([0,T]; B_{p,r}^{s'}) \cap C^1([0,T]; B_{p,r}^{s'-1})$ for every $s' < s$ when $r = \infty$ and $s' = s$ whereas $r < \infty$.

From Theorem 1.1, the following local well-posedness result was obtained (with a slight modification).

Corollary 1.1 Let $(m_{1,0}, m_{2,0}, m_{3,0}, m_{4,0}) = ((1-\partial_x^2)u_{1,0}, (1-\partial_x^2)u_{2,0}, (1-\partial_x^2)u_{3,0}, (1-\partial_x^2)u_{4,0}) \in H^s(\mathbb{R}) \times H^s(\mathbb{R}) \times H^s(\mathbb{R}) \times H^s(\mathbb{R})$ with $s > \frac{1}{2}$. Then there exists a time $T > 0$ such that the initial-value problem (1) has a unique strong solution $(m_1, m_2, m_3, m_4, m_1) \in C([0,T]; H^s) \cap C^1([0,T]; H^{s-1}) \times C([0,T]; H^s) \cap C^1([0,T]; H^{s-1}) \times C([0,T]; H^s) \cap C^1([0,T]; H^{s-1}) \times C([0,T]; H^s) \cap C^1([0,T]; H^{s-1})$ and the map $(m_{1,0}, m_{2,0}, m_{3,0}, m_{4,0}) \to (m_1, m_2, m_3, m_4, m_1)$ is continuous from a neighborhood of $(m_{1,0}, m_{1,0}, m_{1,0}, m_{1,0})$ in $H^s \times H^s \times H^s \times H^s \times H^s$ into $C([0,T]; H^s) \cap C^1([0,T]; H^{s-1}) \times C([0,T]; H^s) \cap C^1([0,T]; H^{s-1}) \times C([0,T]; H^s) \cap C^1([0,T]; H^{s-1}) \times C([0,T]; H^s) \cap C^1([0,T]; H^{s-1})$.

We state a blow-up criterion for the four-component system (1).

Theorem 1.2 Let $(m_{1,0}, m_{2,0}, m_{3,0}, m_{4,0}) = ((1-\partial_x^2)u_{1,0}, (1-\partial_x^2)u_{2,0}, (1-$

$\partial_x^2)u_{3,0}, (1-\partial_x^2)u_{4,0}) \in H^s(\mathbb{R}) \times H^s(\mathbb{R}) \times H^s(\mathbb{R}) \times H^s(\mathbb{R})$ with $s > \frac{1}{2}$ be as in Corollary 3.1. with $s > \frac{1}{2}$. Let (m_1, m_2, m_3, m_4) be the corresponding solution to (1). Assume $T > 0$ is the maximum time of existence. Then

$$T < \infty \Rightarrow \int_0^T \sum_{i=1}^4 \|m_i(\tau)\|_{L^\infty}^2 d\tau = \infty. \tag{1.3}$$

Making use of Sobolev's embedding theorem and Theorem 1.2, one can get the blow-up criterion as follows.

Corollary 1.2 Let $(m_{1,0}, m_{2,0}, m_{3,0}, m_{4,0}) \in H^s(\mathbb{R}) \times H^s(\mathbb{R}) \times H^s(\mathbb{R}) \times H^s(\mathbb{R})$ with $s > \frac{1}{2}$ and $T > 0$ be the maximal existence time of the corresponding solution (m_1, m_2, m_3, m_4) to System (1). Then the solution $(m_1, m_2, m_3, m_4, m_1)$ blows up in finite time if and only if

$$\limsup_{t \to T} \|m_1(t)\|_{H^s} = \infty \text{ or } \limsup_{t \to T} \|m_2(t)\|_{H^s} = \infty \text{ or }$$
$$\limsup_{t \to T} \|m_3(t)\|_{H^s} = \infty \text{ or } \limsup_{t \to T} \|m_4(t)\|_{H^s} = \infty. \tag{1.4}$$

The rest of this paper is organized as follows. In Section 2, we prove the local well-posedness of the initial value problem (1) in the Besov space. In Section 3, we derive the precise blow-up scenario for strong solutions to the system.

2 Local well-posedness in the Besov spaces

In this section, we shall establish local well-posedness of the initial value problem (1) in the Besov spaces. We can rewrite initial value problem (1) as follows

$$m_{1,t} + \frac{1}{9}(f_1 g_1 + f_2 g_2)m_{1,x} = -\frac{1}{9}(m_1 g_1 - m_3 f_1 + m_2 g_2 - m_4 f_2 - f_1 g_1)m_1$$
$$+ \frac{1}{9}m_2 f_1 g_2, \quad t > 0, x \in \mathbb{R},$$
$$m_{2,t} + \frac{1}{9}(f_1 g_1 + f_2 g_2)m_{2,x} = -\frac{1}{9}(m_1 g_1 - m_3 f_1 + m_2 g_2 - m_4 f_2)m_2$$
$$+ \frac{1}{9}f_2 g_1 m_1 + \frac{1}{9}m_2 f_2 g_2, \quad t > 0, x \in \mathbb{R},$$
$$m_{3,t} + \frac{1}{9}(f_1 g_1 + f_2 g_2)m_{3,x} = -\frac{1}{9}(m_1 g_1 - m_3 f_1 + m_2 g_2 - m_4 f_2 + f_1 g_1)m_3$$
$$- \frac{1}{9}m_4 f_2 g_1, \quad t > 0, x \in \mathbb{R},$$
$$m_{4,t} + \frac{1}{9}(f_1 g_1 + f_2 g_2)m_{4,x} = -\frac{1}{9}(m_1 g_1 - m_3 f_1 + m_2 g_2 - m_4 f_2)m_4$$

$$-\frac{1}{9}f_1g_2m_3 - \frac{1}{9}m_4f_2g_2, \quad t>0, x\in\mathbb{R},$$
$$m_i = u_i - u_{i,xx}, \quad i=1,2,3,4, \ t>0, x\in\mathbb{R},$$
$$u_i(0,x) = u_{i,0}(x), \quad i=1,2,3,4, \ x\in\mathbb{R}, \tag{2.1}$$

where $f_1 = u_1 - u_{1,x}, f_2 = u_2 - u_{2,x}, g_1 = u_3 + u_{3,x}, g_2 = u_4 + u_{4,x}$.

First, for the convenience of the readers, we recall some facts on the Littlewood-Paley decomposition and some useful lemmas.

Notations $\mathcal{S}$ stands for the Schwartz space of smooth functions over $\mathbb{R}^d$ whose derivatives of all order decay at infinity. The set $\mathcal{S}'$ of temperate distributions is the dual set of $\mathcal{S}$ for the usual pairing. We denote the norm of the Lebesgue space $L^p(\mathbb{R})$ by $\|\cdot\|_{L^p}$ with $1 \leq p \leq \infty$, and the norm in the Sobolev space $H^s(\mathbb{R})$ with $s\in\mathbb{R}$ by $\|\cdot\|_{H^s}$.

Proposition 2.1 (Littlewood-Paley decomposition [32]) Let $\mathcal{B} \doteq \left\{\xi\in\mathbb{R}^d, |\xi|\leq\frac{4}{3}\right\}$ and $\mathcal{C} \doteq \left\{\xi\in\mathbb{R}^d, \frac{4}{3}\leq|\xi|\leq\frac{8}{3}\right\}$. There exist two radial functions $\chi\in C_c^\infty(\mathcal{B})$ and $\varphi\in C_c^\infty(\mathcal{C})$ such that

$$\chi(\xi) + \sum_{q\geq 0}\varphi(2^{-q}\xi) = 1, \quad \forall \xi\in\mathbb{R}^d,$$

$$|q-q'|\geq 2 \Rightarrow \mathrm{Supp}\varphi(2^{-q}\cdot)\cap\mathrm{Supp}\varphi(2^{-q'}\cdot) = \varnothing,$$

$$q\geq 1 \Rightarrow \mathrm{Supp}\chi(\cdot)\cap\mathrm{Supp}\varphi(2^{-q'}\cdot) = \varnothing,$$

$$\frac{1}{3}\leq \chi(\xi)^2 + \sum_{q\geq 0}\varphi(2^{-q}\xi)^2 \leq 1, \quad \forall\xi\in\mathbb{R}^d.$$

Furthermore, let $h\doteq\mathcal{F}^{-1}\varphi$ and $\tilde{h}\doteq\mathcal{F}^{-1}\chi$. Then for all $f\in\mathcal{S}'(\mathbb{R}^d)$, the dyadic operators Δ_q and S_q can be defined as follows

$$\Delta_q f \doteq \varphi(2^{-q}D)f = 2^{qd}\int_{\mathbb{R}^d} h(2^q y)f(x-y)dy \text{ for } q\geq 0,$$

$$S_q f \doteq \chi(2^{-q}D)f = \sum_{-1\leq k\leq q-1}\Delta_k = 2^{qd}\int_{\mathbb{R}^d}\tilde{h}(2^q y)f(x-y)dy,$$

$$\Delta_{-1}f \doteq S_0 f \text{ and } \Delta_q f \doteq 0 \text{ for } q\leq -2.$$

Hence,

$$f = \sum_{q\geq 0}\Delta_q f \text{ in } \mathcal{S}'(\mathbb{R}^d),$$

where the right-hand side is called the nonhomogeneous Littlewood-Paley decomposition of f.

Lemma 2.1 (Bernstein's inequality [34])   Let $\mathcal{B}$ be a ball with center 0 in $\mathbb{R}^d$ and $\mathcal{C}$ a ring with center 0 in $\mathbb{R}^d$. A constant C exists so that, for any positive real number λ, any non negative integer k, any smooth homogeneous function σ of degree m and any couple of real numbers (a,b) with $b \geqslant a \geqslant 1$, there hold

$$\text{Supp}\,\hat{u} \subset \lambda\mathcal{B} \Rightarrow \sup_{|\alpha|=k} \|\partial^\alpha u\|_{L^a} \leqslant C^{k+1}\lambda^{k+d(\frac{1}{a}-\frac{1}{b})}\|u\|_{L^a},$$

$$\text{Supp}\,\hat{u} \subset \lambda\mathcal{C} \Rightarrow C^{-k-1}\lambda^k\|u\|_{L^a} \leqslant \sup_{|\alpha|=k} \|\partial^\alpha u\|_{L^a} \leqslant C^{k+1}\lambda^k\|u\|_{L^a},$$

$$\text{Supp}\,\hat{u} \subset \lambda\mathcal{C} \Rightarrow \|\sigma(D)u\|_{L^b} \leqslant C_{\sigma,m}\lambda^{m+d(\frac{1}{a}-\frac{1}{b})}\|u\|_{L^a},$$

for any function $u \in L^a$.

Definition 2.1 (Besov space)   Let $s \in \mathbb{R}, 1 \leqslant p, r \leqslant \infty$. The inhomogenous Besov space $B_{p,r}^s(\mathbb{R}^d)$ ($B_{p,r}^s$ for short) is defined by

$$B_{p,r}^s \doteq \{f \in \mathcal{S}'(\mathbb{R}^d); \|f\|_{B_{p,r}^s} < \infty\},$$

where

$$\|f\|_{B_{p,r}^s} \doteq \begin{cases} \left(\sum_{q \in \mathbb{Z}} 2^{qsr}\|\Delta_q f\|_{L_p}^r\right)^{\frac{1}{r}}, & \text{for } r < \infty, \\ \sup_{q \in \mathbb{Z}} 2^{qs}\|\Delta_q f\|_{L_p}, & \text{for } r = \infty. \end{cases}$$

If $s = \infty, B_{p,r}^\infty \doteq \cap_{s \in \mathbb{R}} B_{p,r}^s$.

Proposition 2.2 (see [34])   Suppose that $s \in \mathbb{R}, 1 \leqslant p, r, p_i, r_i \leqslant \infty (i=1,2)$. We have

(1) Topological properties: $B_{p,r}^s$ is a Banach space which is continuously embedded in $\mathcal{S}'$.

(2) Density: C_c^∞ is dense in $B_{p,r}^s \Leftrightarrow 1 \leqslant p, r \leqslant \infty$.

(3) Embedding: $B_{p_1,r_1}^s \hookrightarrow B_{p_2,r_2}^{s-n(\frac{1}{p_1})-\frac{1}{p_2}}$, if $p_1 \leqslant p_2$ and $r_1 \leqslant r_2$. $B_{p,r_2}^{s_2} \hookrightarrow B_{p,r_1}^{s_1}$ locally compact, if $s_1 < s_2$.

(4) Algebraic properties: $\forall s > 0$, $B_{p,r}^s \cap L^\infty$ is an algebra. Moreover, $B_{p,r}^s$ is an algebra, provided that $s > \dfrac{n}{p}$ or $s \geqslant \dfrac{n}{p}$ and $r = 1$.

(5) Complex interpolation:

$$\|u\|_{B_{p,r}^{\theta s_1 + (1-\theta)s_2}} \leqslant C\|u\|_{B_{p,r}^{s_1}}^\theta \|u\|_{B_{p,r}^{s_2}}^{1-\theta}, \quad \forall u \in B_{p,r}^{s_1} \cap B_{p,r}^{s_2}, \quad \forall \theta \in [0,1].$$

(6) *Fatou lemma:* If $(u_n)_{n\in\mathbb{N}}$ is bounded in $B^s_{p,r}$ and $u_n \to u$ in $\mathcal{S}'$, then $u \in B^s_{p,r}$ and
$$\|u\|_{B^s_{p,r}} \leqslant \liminf_{n\to\infty} \|u_n\|_{B^s_{p,r}}.$$

(7) Let $m \in \mathbb{R}$ and f be an S^m-multiplier (i.e., $f : \mathbb{R}^d \to \mathbb{R}$ is smooth and satisfies that $\forall \alpha \in \mathbb{N}^d$, there exists a constant C_α, s.t. $|\partial^\alpha f(\xi)| \leqslant C_\alpha(1+|\xi|^{m-|\alpha|})$ for all $\xi \in \mathbb{R}^d$). Then the operator $f(D)$ is continuous from $B^s_{p,r}$ to $B^{s-m}_{p,r}$.

Now we state some useful results in the transport equation theory, which are crucial to the proofs of our main theorems later.

Lemma 2.2 (see [33, 34]) *Suppose that $(p,r) \in [1,+\infty]^2$ and $s > -\frac{d}{p}$. Let v be a vector field such that ∇v belongs to $L^1([0,T]; B^{s-1}_{p,r})$ if $s > 1 + \frac{d}{p}$ or to $L^1([0,T]; B^{\frac{d}{p}}_{p,r} \cap L^\infty)$ otherwise. Suppose also that $f_0 \in B^s_{p,r}$, $F \in L^1([0,T]; B^s_{p,r})$ and that $f \in L^\infty(L^1([0,T]; B^s_{p,r}) \cap C([0,T]; \mathcal{S}')$ solves the d-dimensional linear transport equations*

$$(T) \quad \begin{cases} \partial_t f + v \cdot \nabla f = F, \\ f|_{t=0} = f_0. \end{cases}$$

Then there exists a constant C depending only on s, p and d such that the following statements hold:

(1) *If $r = 1$ or $s \neq 1 + \frac{d}{p}$, then*

$$\|f\|_{B^s_{p,r}} \leqslant \|f_0\|_{B^s_{p,r}} + \int_0^t \|F(\tau)\|_{B^s_{p,r}} d\tau + C\int_0^t V'(\tau)\|f(\tau)\|_{B^s_{p,r}} d\tau,$$

or

$$\|f\|_{B^s_{p,r}} \leqslant e^{CV(t)} C\left(\|f_0\|_{B^s_{p,r}} + \int_0^t e^{-CV(\tau)} \|F(\tau)\|_{B^s_{p,r}} d\tau\right) \quad (2.2)$$

hold, where $V(t) = \int_0^t \|\nabla v(\tau)\|_{B^{\frac{d}{p}}_{p,r} \cap L^\infty} d\tau$ if $s < 1 + \frac{d}{p}$ and $V(t) = \int_0^t \|\nabla v(\tau)\|_{B^{s-1}_{p,r}} d\tau$ else.

(2) *If $s \leqslant 1 + \frac{d}{p}$ and $\nabla f_0 \in L^\infty$, $\nabla f \in L^\infty([0,T] \times \mathbb{R}^d)$ and $\nabla F \in L^1([0,T]; L^\infty)$, then*

$$\|f\|_{B^s_{p,r}} + \|\nabla f\|_{L^\infty}$$
$$\leqslant e^{CV(t)}\left(\|f_0\|_{B^s_{p,r}} + \|\nabla f_0\|_{L^\infty} + \int_0^t e^{-CV(\tau)}\|F(\tau)\|_{B^s_{p,r}} + \|\nabla F(\tau)\|_{L^\infty} d\tau\right)$$

with $V(t) = \int_0^t \|\nabla v(\tau)\|_{B_{p,r}^{\frac{d}{p}} \cap L^\infty} d\tau.$

(3) If $f = v$, then for all $s > 0$, the estimate (6) holds with

$$V(t) = \int_0^t \|\nabla v(\tau)\|_{B_{p,r}^{s-1}} d\tau.$$

(4) If $r < +\infty$, then $f \in C([0,T]; B_{p,r}^s)$. If $r = +\infty$, then $f \in C([0,T]; B_{p,r}^{s'})$ for all $s' < s$.

Lemma 2.3 (existence and uniqueness see [33, 34]) Let $(p, p_1, r) \in [1, +\infty]^3$ and $s > -d \min\left\{\frac{1}{p_1}, \frac{1}{p'}\right\}$ with $p' \doteq \left(1 - \frac{1}{p}\right)^{-1}$. Assume that $f_0 \in B_{p,r}^s, F \in L^1([0,T]; B_{p,r}^s)$. Let v be a time dependent vector field such that $v \in L^\rho([0,T]; B_{\infty,\infty}^{-M})$ for some $\rho > 1, M > 0$ and $\nabla v \in L^1([0,T]; B_{p,r}^{\frac{d}{p}} \cap L^\infty)$ if $s < 1 + \frac{d}{p_1}$ and $\nabla v \in L^1([0,T]; B_{p_1,r}^{s-1})$ if $s > 1 + \frac{d}{p}$ or $s = 1 + \frac{d}{p_1}$ and $r = 1$. Then the transport equations (T) has a unique solution $f \in L^\infty([0,T]; B_{p,r}^s) \cap (\cap_{s' < s} C[0,T]; B_{p,1}^{s'})$ and the inequalities in Lemma 2.2 hold true. Moreover, $r < \infty$, then we have $f \in C[0,T]; B_{p,1}^s)$.

Lemma 2.4 (1-D Morse-type estimates [33, 34]) Assume that $1 \leqslant p, r \leqslant +\infty$, the following estimates hold:

(i) For $s > 0$,

$$\|fg\|_{B_{p,r}^s} \leqslant C(\|f\|_{B_{p,r}^s}\|g\|_{L^\infty} + \|g\|_{B_{p,r}^s}\|f\|_{L^\infty});$$

(ii) $\forall s_1 \leqslant \frac{1}{p} < s_2 \left(s_2 \geqslant \frac{1}{p} \text{ if } r = 1\right)$ and $s_1 + s_2 > 0$, we have

$$\|fg\|_{B_{p,r}^{s_1}} \leqslant C\|f\|_{B_{p,r}^{s_1}}\|g\|_{B_{p,r}^{s_2}};$$

(iii) In Sobolev spaces $H^s = B_{2,2}^s$, we have for $s > 0$,

$$\|f\partial_x g\|_{H^s} \leqslant C(\|f\|_{H^{s+1}}\|g\|_{L^\infty} + \|\partial_x g\|_{H^s}\|f\|_{L^\infty}),$$

where C is a positive constant independent of f and g.

Definition 2.2 For $T > 0, s \in \mathbb{R}$ and $1 \leqslant p \leqslant +\infty$, we set

$$E_{p,r}^s(T) \equiv C([0,T]; B_{p,r}^s) \cap C^1([0,T]; B_{p,r}^{s-1}), \quad \text{if } r \leqslant +\infty,$$

$$E_{p,\infty}^s(T) \equiv L^\infty([0,T]; B_{p,\infty}^s) \cap lip^1([0,T]; B_{p,\infty}^{s-1})$$

and $E_{p,r}^s \equiv \cap_{T>0} E_{p,r}^s(T).$

In the following, we denote $C > 0$ a generic constant only depending on p, r, s. Uniqueness and continuity with respect to the initial data are an immediate consequence of the following result.

Proposition 2.3 *Let $1 \leqslant p, r \leqslant +\infty$ and $s > \max\left\{2 + \dfrac{1}{p}, \dfrac{5}{2}\right\}$. Suppose that $(u_1^{(i)}, u_2^{(i)}, u_3^{(i)}, u_4^{(i)}) \in \{L^\infty([0, T]; B_{p,r}^s) \cap C([0, T]; \mathcal{S}')\}^4 (i = 1, 2)$ be two given solutions of the initial-value problem (5) with the initial data $(u_{1,0}^{(i)}, u_{2,0}^{(i)}, u_{3,0}^{(i)}, u_{4,0}^{(i)}) \in B_{p,r}^s \times B_{p,r}^s \times B_{p,r}^s \times B_{p,r}^s (i = 1, 2)$. Then for every $t \in [0; T]$, we have*

$$\sum_{i=1}^{4} \|u_i^{(1)}(t) - u_i^{(2)}(t)\|_{B_{p,r}^{s-1}}$$
$$\leqslant \sum_{i=1}^{4} \|u_{i,0}^{(1)} - u_{i,0}^{(2)}\|_{B_{p,r}^{s-1}} \times \exp\left\{C \int_0^t \sum_{i=1}^{4} \left(\|u_i^{(1)}(\tau)\|_{B_{p,r}^s}^2 + \|u_i^{(2)}(\tau)\|_{B_{p,r}^s}^2\right) d\tau\right\}.$$
(2.3)

Proof Denote $u_i^{(12)} = u_i^{(2)} - u_i^{(1)}$, $m_i^{(12)} = m_i^{(2)} - m_i^{(1)}$, $i = 1, 2, 3, 4$. It is obvious that
$$u_i^{(12)} \in L^\infty([0, T]; B_{p,r}^s) \cap C([0, T]; \mathcal{S}'), \quad i = 1, 2, 3, 4,$$
which implies that $u_i^{(12)} \in C([0, T]; B_{p,r}^{s-1}), i = 1, 2, 3, 4$, and $(u_1^{(12)}, u_2^{(12)}, u_3^{(12)}, u_4^{(12)}, m_1^{(12)}, m_2^{(12)}, m_3^{(12)}, m_4^{(12)})$ solves the transport equations

$$\begin{cases} m_{1,t}^{(12)} + \dfrac{1}{9}[f_1^{(1)} g_1^{(1)} + f_2^{(1)} g_2^{(1)}] m_{1,x}^{(12)} = F_1, \\ m_{2,t}^{(12)} + \dfrac{1}{9}[f_1^{(1)} g_1^{(1)} + f_2^{(1)} g_2^{(1)}] m_{2,x}^{(12)} = F_2, \\ m_{2,t}^{(12)} + \dfrac{1}{9}[f_1^{(1)} g_1^{(1)} + f_2^{(1)} g_2^{(1)}] m_{3,x}^{(12)} = F_3, \\ m_{4,t}^{(12)} + \dfrac{1}{9}[f_1^{(1)} g_1^{(1)} + f_2^{(1)} g_2^{(1)}] m_{4,x}^{(12)} = F_4, \\ m_i^{(12)}|_{t=0} = m_{i,0}^{(12)} = m_{i,0}^{(2)} - m_{i,0}^{(1)}, \quad i = 1, 2, 3, 4, \end{cases}$$
(2.4)

with

$$F_1 = (f_1^{(2)} g_1^{(12)} + f_1^{(12)} g_1^{(1)} + f_2^{(2)} g_2^{(12)} + f_2^{(12)} g_2^{(1)}) m_{1,x}^{(2)}$$
$$- \frac{1}{9}\left[(m_1^{(2)} g_1^{(2)} - m_3 f_1^{(2)} + m_2^{(2)} g_2^{(2)} - m_4^{(2)} f_2^{(2)} - f_1^{(2)} g_1^{(2)}) m_1^{(12)}\right.$$
$$+ (m_1^{(2)} g_1^{(12)} + m_1^{(12)} g_1^{(1)}) m_1^{(1)} - (f_1^{(2)} m_3^{(12)} + f_1^{(12)} m_3^{(1)}) m_1^{(1)}$$
$$+ (m_2^{(2)} g_2^{(12)} + m_2^{(12)} g_2^{(1)}) m_1^{(1)}$$

$$+ (f_2^{(2)}m_4^{(12)} + f_2^{(12)}m_4^{(1)})m_1^{(1)} - (f_1^{(2)}g_1^{(12)} + f_1^{(12)}g_1^{(1)})m_1^{(1)}\Big]$$

$$+ \frac{1}{9}\Big[(f_1^{(2)}g_2^{(12)} + f_1^{(12)}g_2^{(2)})m_2^{(2)} + m_2^{(12)}f_1^{(1)}g_2^{(1)}\Big],$$

$$F_2 = (f_1^{(2)}g_1^{(12)} + f_1^{(12)}g_1^{(1)} + f_2^{(2)}g_2^{(12)} + f_2^{(12)}g_2^{(1)})m_{2,x}^{(2)}$$

$$- \frac{1}{9}\Big[(m_1^{(2)}g_1^{(2)} - m_3 f_1^{(2)} + m_2^{(2)}g_2^{(2)} - m_4^{(2)}f_2^{(2)})m_2^{(12)} - f_2^{(2)}g_1^{(2)}m_1^{(12)}$$

$$+ (m_1^{(2)}g_1^{(12)} + m_1^{(12)}g_1^{(1)})m_2^{(1)} - (f_1^{(2)}m_3^{(12)} + f_1^{(12)}m_3^{(1)})m_2^{(1)}$$

$$+ (m_2^{(2)}g_2^{(12)} + m_2^{(12)}g_2^{(1)})m_2^{(1)}$$

$$+ (f_2^{(2)}m_4^{(12)} + f_2^{(12)}m_4^{(1)})m_2^{(1)} - (f_2^{(2)}g_1^{(12)} + f_2^{(12)}g_1^{(1)})m_1^{(1)}\Big]$$

$$+ \frac{1}{9}\Big[(f_2^{(2)}g_2^{(12)} + f_2^{(12)}g_1^{(1)})m_2^{(2)} + m_2^{(12)}f_2^{(1)}g_2^{(1)}\Big],$$

$$F_3 = (f_1^{(2)}g_1^{(12)} + f_1^{(12)}g_1^{(1)} + f_2^{(2)}g_2^{(12)} + f_2^{(12)}g_2^{(1)})m_{3,x}^{(2)}$$

$$- \frac{1}{9}\Big[(m_1^{(2)}g_1^{(2)} - m_3 f_1^{(2)} + m_2^{(2)}g_2^{(2)} - m_4^{(2)}f_2^{(2)} - f_1^{(2)}g_1^{(2)})m_3^{(12)}$$

$$+ (m_1^{(2)}g_1^{(12)} + m_1^{(12)}g_1^{(1)})m_3^{(1)} - (f_1^{(2)}m_3^{(12)} + f_1^{(12)}m_3^{(1)})m_3^{(1)} + (m_2^{(2)}g_2^{(12)}$$

$$+ m_2^{(12)}g_2^{(1)})m_3^{(1)} + (f_2^{(2)}m_4^{(12)} + f_2^{(12)}m_4^{(1)})m_3^{(1)} + (f_1^{(2)}g_1^{(12)} + f_1^{(12)}g_1^{(1)})m_3^{(1)}\Big]$$

$$- \frac{1}{9}\Big[(f_2^{(2)}g_1^{(12)} + f_2^{(12)}g_1^{(2)})m_4^{(2)} + m_4^{(12)}f_2^{(1)}g_1^{(1)}\Big],$$

$$F_4 = (f_1^{(2)}g_1^{(12)} + f_1^{(12)}g_1^{(1)} + f_2^{(2)}g_2^{(12)} + f_2^{(12)}g_2^{(1)})m_{4,x}^{(2)}$$

$$- \frac{1}{9}\Big[(m_1^{(2)}g_1^{(2)} - m_3 f_1^{(2)} + m_2^{(2)}g_2^{(2)} - m_4^{(2)}f_2^{(2)})m_4^{(12)} + f_1^{(2)}g_2^{(2)}m_3^{(12)}$$

$$+ (m_1^{(2)}g_1^{(12)} + m_1^{(12)}g_1^{(1)})m_4^{(1)} - (f_1^{(2)}m_3^{(12)} + f_1^{(12)}m_3^{(1)})m_4^{(1)} + (m_2^{(2)}g_2^{(12)}$$

$$+ m_2^{(12)}g_2^{(1)})m_4^{(1)} + (f_2^{(2)}m_4^{(12)} + f_2^{(12)}m_4^{(1)})m_4^{(1)} + (f_1^{(2)}g_2^{(12)} + f_1^{(12)}g_2^{(1)})m_3^{(1)}\Big]$$

$$- \frac{1}{9}\Big[(f_2^{(2)}g_2^{(12)} + f_2^{(12)}g_1^{(2)})m_2^{(4)} + m_4^{(12)}f_2^{(1)}g_2^{(1)}\Big].$$

According to Lemma 2.2, we have

$$e^{-C\int_0^t \|\partial_x[\frac{1}{9}[f_1^{(1)}g_1^{(1)}+f_2^{(1)}g_2^{(1)}]](\tau')\|_{B_{p,r}^{s-2}}d\tau'}\|m_1^{(12)}(t)\|_{B_{p,r}^{s-3}}$$

$$\leqslant \|m_{1,0}^{(12)}\|_{B_{p,r}^{s-3}} + C\int_0^t e^{-C\int_0^\tau \|\partial_x[\frac{1}{9}[f_1^{(1)}g_1^{(1)}+f_2^{(1)}g_2^{(1)}]](\tau')\|_{B_{p,r}^{s-2}}d\tau'}\|F_1\|_{B_{p,r}^{s-3}}d\tau,$$

$$e^{-C\int_0^t \|\partial_x[\frac{1}{9}[f_1^{(1)}g_1^{(1)}+f_2^{(1)}g_2^{(1)}]](\tau')\|_{B_{p,r}^{s-2}}d\tau'}\|m_3^{(12)}(t)\|_{B_{p,r}^{s-3}}$$

$$\leqslant \|m_{2,0}^{(12)}\|_{B_{p,r}^{s-3}} + C\int_0^t e^{-C\int_0^\tau \|\partial_x[\frac{1}{9}[f_1^{(1)}g_1^{(1)}+f_2^{(1)}g_2^{(1)}]](\tau')\|_{B_{p,r}^{s-2}} d\tau'} \|F_2\|_{B_{p,r}^{s-3}} d\tau,$$

$$e^{-C\int_0^t \|\partial_x[\frac{1}{9}[f_1^{(1)}g_1^{(1)}+f_2^{(1)}g_2^{(1)}]](\tau')\|_{B_{p,r}^{s-2}} d\tau'} \|m_3^{(12)}(t)\|_{B_{p,r}^{s-3}}$$

$$\leqslant \|m_{3,0}^{(12)}\|_{B_{p,r}^{s-3}} + C\int_0^t e^{-C\int_0^\tau \|\partial_x[\frac{1}{9}[f_1^{(1)}g_1^{(1)}+f_2^{(1)}g_2^{(1)}]](\tau')\|_{B_{p,r}^{s-2}} d\tau'} \|F_3\|_{B_{p,r}^{s-3}} d\tau, \quad (2.5)$$

and

$$e^{-C\int_0^t \|\partial_x[\frac{1}{9}[f_1^{(1)}g_1^{(1)}+f_2^{(1)}g_2^{(1)}]](\tau')\|_{B_{p,r}^{s-2}} d\tau'} \|m_4^{(12)}(t)\|_{B_{p,r}^{s-3}}$$

$$\leqslant \|m_{4,0}^{(12)}\|_{B_{p,r}^{s-3}} + C\int_0^t e^{-C\int_0^\tau \|\partial_x[\frac{1}{9}[f_1^{(1)}g_1^{(1)}+f_2^{(1)}g_2^{(1)}]](\tau')\|_{B_{p,r}^{s-2}} d\tau'} \|F_4\|_{B_{p,r}^{s-3}} d\tau. \quad (2.6)$$

For $s > \max\left\{2+\frac{1}{p}, \frac{5}{2}\right\}$, by Lemma 2.4 and the definitions of f_1, f_2, g_1, g_2, we have

$$\|F_i\|_{B_{p,r}^{s-3}} \leqslant C \sum_{j=1}^4 \|u_j^{(12)}\|_{B_{p,r}^{s-1}} \sum_{j=1}^4 \left(\|u_j^{(1)}\|_{B_{p,r}^s}^2 + \|u_j^{(2)}\|_{B_{p,r}^s}^2\right), \quad i = 1, 2, 3, 4.$$

Therefore, inserting the above estimates to (9)-(10) we obtain

$$e^{-C\int_0^t \|\partial_x[\frac{1}{9}[f_1^{(1)}g_1^{(1)}+f_2^{(1)}g_2^{(1)}]](\tau')\|_{B_{p,r}^{s-2}} d\tau'} \sum_{j=1}^{j=4} \|m_j^{(12)}(t)\|_{B_{p,r}^{s-3}}$$

$$\leqslant \sum_{j=1}^{j=4} \|m_{j,0}^{(12)}\|_{B_{p,r}^{s-3}} + C\int_0^t e^{-C\int_0^\tau \|\partial_x[\frac{1}{9}[f_1^{(1)}g_1^{(1)}+f_2^{(1)}g_2^{(1)}]](\tau')\|_{B_{p,r}^{s-2}} d\tau'}$$

$$\times \sum_{j=1}^4 \|u_j^{(12)}\|_{B_{p,r}^{s-1}} \sum_{j=1}^4 \left(\|u_j^{(1)}\|_{B_{p,r}^s}^2 + \|u_j^{(2)}\|_{B_{p,r}^s}^2\right) d\tau. \quad (2.7)$$

According to Proposition 2.2, we obtain for all $s \in \mathbb{R}$ and $i = 1, 2, 12, j = 1, 2, 3, 4$,

$$\|u_j^{(i)}\|_{B_{p,r}^{s+2}} \approx \|m_j^{(i)}\|_{B_{p,r}^s}. \quad (2.8)$$

Hence, thanks to

$$\left\|\partial_x\left[\frac{1}{9}[f_1^{(1)}g_1^{(1)} + f_2^{(1)}g_2^{(1)}]\right]\right\|_{B_{p,r}^{s-2}} \leqslant C \sum_{j=1}^4 \|u_j^{(1)}\|_{B_{p,r}^s}^2,$$

(12) and then applying the Gronwall's inequality, we reach (7). □

Definition 2.3 For $T > 0, s \in \mathbb{R}$ and $1 \leqslant p \leqslant +\infty$, we set

$$E_{p,r}^s(T) \doteq C([0,T]; B_{p,r}^s) \cap C^1([0,T]; B_{p,r}^{s-1}) \text{ if } r < +\infty,$$

$$E_{p,\infty}^s(T) \doteq L^\infty([0,T]; B_{p,\infty}^s) \cap lip^1([0,T]; B_{p,\infty}^{s-1})$$

and $E_{p,r}^s \doteq \cap_{T>0} E_{p,r}^s(T)$.

Now let us start the proof of Theorem 1.1, which is motivated by the proof of local existence theorem about the Camassa-Holm equation in [33]. Firstly, we shall use the classical Friedrichs regularization method to construct the approximate solutions to the Cauchy problem problem (5).

Lemma 2.5 *Assume that $u_j^{(0)} = 0, j = 1,2,3,4$. Let $1 \leqslant p, r \leqslant +\infty, s > \max\left\{\dfrac{5}{2}, 2+\dfrac{1}{p}\right\}$ and $u_{i,0} \in B_{p,r}^s, i = 1,2,3,4$. Then there exists a sequence of smooth functions $(u_1^{(i)}, u_2^{(i)}, u_3^{(i)}, u_4^{(i)})_{i \in \mathbb{N}} \in C(\mathbb{R}^+; B_{p,r}^\infty) \times C(\mathbb{R}^+; B_{p,r}^\infty) \times C(\mathbb{R}^+; B_{p,r}^\infty) \times C(\mathbb{R}^+; B_{p,r}^\infty)$ solving the following linear transport equation by induction*

$$\begin{cases} \left(\partial_t + \dfrac{1}{9}[f_1^{(i)} g_1^{(i)} + f_2^{(i)} g_2^{(i)}]\partial_x\right) m_1^{(i+1)} = -\dfrac{1}{9}(m_1^{(i)} g_1^{(i)} - m_3^{(i)} f_1^{(i)} + m_2^{(i)} g_2^{(i)} \\ \qquad - m_4^{(i)} f_2^{(i)} - f_1^{(i)} g_1^{(i)}) m_1^{(i)} + \dfrac{1}{9} m_2^{(i)} f_1^{(i)} g_2^{(i)}, \\ \left(\partial_t + \dfrac{1}{9}[f_1^{(i)} g_1^{(i)} + f_2^{(i)} g_2^{(i)}]\partial_x\right) m_2^{(i+1)} = -\dfrac{1}{9}(m_1^{(i)} g_1^{(i)} - m_3^{(i)} f_1^{(i)} + m_2^{(i)} g_2^{(i)} \\ \qquad - m_4^{(i)} f_2) m_2^{(i)} + \dfrac{1}{9} f_2^{(i)} g_1^{(i)} m_1^{(i)} + \dfrac{1}{9} m_2^{(i)} f_2^{(i)} g_2^{(i)}, \\ \left(\partial_t + \dfrac{1}{9}[f_1^{(i)} g_1^{(i)} + f_2^{(i)} g_2^{(i)}]\partial_x\right) m_3^{(i+1)} = -\dfrac{1}{9}(m_1^{(i)} g_1^{(i)} - m_3^{(i)} f_1^{(i)} + m_2^{(i)} g_2^{(i)} \\ \qquad - m_4^{(i)} f_2^{(i)} + f_1^{(i)} g_1^{(i)}) m_3^{(i)} - \dfrac{1}{9} m_4^{(i)} f_2^{(i)} g_1^{(i)}, \\ \left(\partial_t + \dfrac{1}{9}[f_1^{(i)} g_1^{(i)} + f_2^{(i)} g_2^{(i)}]\partial_x\right) m_4^{(i+1)} = -\dfrac{1}{9}(m_1^{(i)} g_1^{(i)} - m_3^{(i)} f_1^{(i)} + m_2^{(i)} g_2^{(i)} \\ \qquad - m_4^{(i)} f_2^{(i)}) m_4^{(i)} - \dfrac{1}{9} f_1^{(i)} g_2^{(i)} m_3^{(i)} - \dfrac{1}{9} m_4^{(i)} f_2^{(i)} g_2^{(i)}, \\ u_j^{(i+1)}(x, 0) = u_{j,0}^{(i+1)}(x) = S_{i+1} u_{j,0}, \quad j = 1,2,3,4. \end{cases}$$
(2.9)

Moreover, there is a positive T such that the solutions satisfy the following properties

(i) $(u_1^{(i)}, u_2^{(i)}, u_3^{(i)}, u_4^{(i)})_{i \in \mathbb{N}}$ *is uniformly bounded in* $E_{p,r}^s(T) \times E_{p,r}^s(T) \times E_{p,r}^s(T) \times E_{p,r}^s(T)$.

(ii) $(u_1^{(i)}, u_2^{(i)}, u_3^{(i)}, u_4^{(i)})_{i \in \mathbb{N}}$ *is a Cauchy sequence in* $C([0,T]; B_{p,r}^{s-1}) \times C([0,T]; B_{p,r}^{s-1}) \times C([0,T]; B_{p,r}^{s-1}) \times C([0,T]; B_{p,r}^{s-1})$.

Proof Since all the data $S_{n+1} u_{j,0}$ $(j = 1,2,3,4)$ belong to $B_{p,r}^\infty$, Lemma 2.3 enables us to show by induction that for all $i \in \mathbb{N}$, the equation (13) has a global solution which belongs to $C(\mathbb{R}^+; B_{p,r}^\infty) \times C(\mathbb{R}^+; B_{p,r}^\infty) \times C(\mathbb{R}^+; B_{p,r}^\infty) \times C(\mathbb{R}^+; B_{p,r}^\infty)$.

Thanks to $s > \max\left\{2 + \dfrac{1}{p}, \dfrac{5}{2}\right\}$, we find $B_{p,r}^{s-2}$ is an algebra. From this, (13) and the definitions of $f_1^{(i)}, f_2^{(i)}, g_1^{(i)}, g_2^{(i)}$, one obtains

$$\left\|-\frac{1}{9}(m_1^{(i)}g_1^{(i)} - m_3^{(i)}f_1^{(i)} + m_2^{(i)}g_2^{(i)} - m_4^{(i)}f_2^{(i)} - f_1^{(i)}g_1^{(i)})m_1^{(i)} + \frac{1}{9}m_2^{(i)}f_1^{(i)}g_2^{(i)}\right\|_{B_{p,r}^{s-2}}$$

$$\leqslant C \left(\sum_{j=1}^4 \|u_j^{(i)}\|_{B_{p,r}^s}\right)^3,$$

$$\left\|-\frac{1}{9}(m_1^{(i)}g_1^{(i)} - m_3^{(i)}f_1^{(i)} + m_2^{(i)}g_2^{(i)} - m_4^{(i)}f_2)m_2^{(i)}\right.$$
$$\left. + \frac{1}{9}f_2^{(i)}g_1^{(i)}m_1^{(i)} + \frac{1}{9}m_2^{(i)}f_2^{(i)}g_2^{(i)}\right\|_{B_{p,r}^{s-2}}$$

$$\leqslant C \left(\sum_{j=1}^4 \|u_j^{(i)}\|_{B_{p,r}^s}\right)^3,$$

$$\left\|-\frac{1}{9}(m_1^{(i)}g_1^{(i)} - m_3^{(i)}f_1^{(i)} + m_2^{(i)}g_2^{(i)} - m_4^{(i)}f_2^{(i)} + f_1^{(i)}g_1^{(i)})m_3^{(i)} - \frac{1}{9}m_4^{(i)}f_2^{(i)}g_1^{(i)}\right\|_{B_{p,r}^{s-2}}$$

$$\leqslant C \left(\sum_{j=1}^4 \|u_j^{(i)}\|_{B_{p,r}^s}\right)^3,$$

and

$$\left\|-\frac{1}{9}(m_1^{(i)}g_1^{(i)} - m_3^{(i)}f_1^{(i)} + m_2^{(i)}g_2^{(i)} - m_4^{(i)}f_2^{(i)})m_4^{(i)}\right.$$
$$\left. - \frac{1}{9}f_1^{(i)}g_2^{(i)}m_3^{(i)} - \frac{1}{9}m_4^{(i)}f_2^{(i)}g_2^{(i)}\right\|_{B_{p,r}^{s-2}}$$

$$\leqslant C \left(\sum_{j=1}^4 \|u_j^{(i)}\|_{B_{p,r}^s}\right)^3.$$

Thanks to Lemma 2.2, the proof of Proposition 2.3 and the above inequality, we have the following inequality for all $i \in \mathbb{N}$

$$e^{-C\int_0^t \|\partial_x \frac{1}{9}[f_1^{(i)}g_1^{(i)} + f_2^{(i)}g_2^{(i)}](\tau')\|_{B_{p,r}^{s-2}} d\tau'} \|u_1^{(i+1)}(t)\|_{B_{p,r}^s}$$

$$\leqslant \|u_{1,0}^{(i)}\|_{B_{p,r}^s} + C\int_0^t e^{-C\int_0^\tau \|\partial_x \frac{1}{9}[f_1^{(i)}g_1^{(i)} + f_2^{(i)}g_2^{(i)}](\tau')\|_{B_{p,r}^{s-1}} d\tau'} \left(\sum_{j=1}^4 \|u_j^{(i)}\|_{B_{p,r}^s}\right)^3 d\tau,$$

$$e^{-C\int_0^t \|\partial_x \frac{1}{9}[f_1^{(i)}g_1^{(i)}+f_2^{(i)}g_2^{(i)}](\tau')\|_{B_{p,r}^{s-2}}d\tau'} \|u_2^{(i+1)}(t)\|_{B_{p,r}^s}$$

$$\leqslant \|u_{2,0}^{(i)}\|_{B_{p,r}^s} + C\int_0^t e^{-C\int_0^\tau \|\partial_x \frac{1}{9}[f_1^{(i)}g_1^{(i)}+f_2^{(i)}g_2^{(i)}](\tau')\|_{B_{p,r}^{s-1}}d\tau'} \left(\sum_{j=1}^4 \|u_j^{(i)}\|_{B_{p,r}^s}\right)^3 d\tau,$$

$$e^{-C\int_0^t \|\partial_x \frac{1}{9}[f_1^{(i)}g_1^{(i)}+f_2^{(i)}g_2^{(i)}](\tau')\|_{B_{p,r}^{s-2}}d\tau'} \|u_3^{(i+1)}(t)\|_{B_{p,r}^s}$$

$$\leqslant \|u_{3,0}^{(i)}\|_{B_{p,r}^s} + C\int_0^t e^{-C\int_0^\tau \|\partial_x \frac{1}{9}[f_1^{(i)}g_1^{(i)}+f_2^{(i)}g_2^{(i)}](\tau')\|_{B_{p,r}^{s-1}}d\tau'} \left(\sum_{j=1}^4 \|u_j^{(i)}\|_{B_{p,r}^s}\right)^3 d\tau,$$

and

$$e^{-C\int_0^t \|\partial_x \frac{1}{9}[f_1^{(i)}g_1^{(i)}+f_2^{(i)}g_2^{(i)}](\tau')\|_{B_{p,r}^{s-2}}d\tau'} \|u_4^{(i+1)}(t)\|_{B_{p,r}^s}$$

$$\leqslant \|u_{4,0}^{(i)}\|_{B_{p,r}^s} + C\int_0^t e^{-C\int_0^\tau \|\partial_x \frac{1}{9}[f_1^{(i)}g_1^{(i)}+f_2^{(i)}g_2^{(i)}](\tau')\|_{B_{p,r}^{s-1}}d\tau'} \left(\sum_{j=1}^4 \|u_j^{(i)}\|_{B_{p,r}^s}\right)^3 d\tau.$$

Hence, we have

$$e^{-C\int_0^t \|\partial_x \frac{1}{9}[f_1^{(i)}g_1^{(i)}+f_2^{(i)}g_2^{(i)}](\tau')\|_{B_{p,r}^{s-2}}d\tau'} \sum_{j=1}^4 \|u_j^{(i+1)}(t)\|_{B_{p,r}^s}$$

$$\leqslant \sum_{j=1}^4 \|u_{j,0}^{(i)}\|_{B_{p,r}^s} + C\int_0^t e^{-C\int_0^\tau \|\partial_x \frac{1}{9}[f_1^{(i)}g_1^{(i)}+f_2^{(i)}g_2^{(i)}](\tau')\|_{B_{p,r}^{s-1}}d\tau'} \left(\sum_{j=1}^4 \|u_j^{(i)}\|_{B_{p,r}^s}\right)^3 d\tau. \tag{2.10}$$

Let us choose a $T > 0$ such that $4C\left(\sum_{j=1}^4 \|u_{j,0}\|_{B_{p,r}^s}\right)^2 T < 1$, and suppose by induction that for all $t \in [0,T]$,

$$\sum_{j=1}^4 \|u_j^{(i)}(t)\|_{B_{p,r}^s} \leqslant \frac{\sum_{j=1}^4 \|u_{j,0}\|_{B_{p,r}^s}}{\left(1 - 4C\left(\sum_{j=1}^4 \|u_{j,0}\|_{B_{p,r}^s}\right)^2 t\right)^{\frac{1}{2}}}. \tag{2.11}$$

Indeed, since $B_{p,r}^{s-1}$ is an algebra, one obtains from (15) and the definitions of $f_1^{(i)}, f_2^{(i)}, g_1^{(i)}, g_2^{(i)}$ that

$$C\int_0^t \|\partial_x \frac{1}{9}[f_1^{(i)}g_1^{(i)}+f_2^{(i)}g_2^{(i)}](\tau')\|_{B_{p,r}^{s-2}}d\tau'$$

$$\leqslant C \int_\tau^t \left(\sum_{j=1}^4 \|u_j^{(i)}(t)\|_{B_{p,r}^s} \right)^2 d\tau' \leqslant C \int_\tau^t \frac{\left(\sum_{j=1}^4 \|u_{j,0}\|_{B_{p,r}^s} \right)^2}{1 - 4C \left(\sum_{j=1}^4 \|u_{j,0}\|_{B_{p,r}^s} \right)^2 t} d\tau'$$

$$= \frac{1}{4} \ln \left(1 - 4C \left(\sum_{j=1}^4 \|u_{j,0}\|_{B_{p,r}^s} \right)^2 \tau \right) - \frac{1}{4} \ln \left(1 - 4C \left(\sum_{j=1}^4 \|u_{j,0}\|_{B_{p,r}^s} \right)^2 t \right).$$

And then inserting the above inequality and (15) into (14) leads to

$$\sum_{j=1}^4 \|u^{(i+1)}(t)\|_{B_{p,r}^s}$$

$$\leqslant \frac{\sum_{j=1}^4 \|u_{j,0}\|_{B_{p,r}^s}}{\left(1 - 4C \left(\sum_{j=1}^4 \|u_{j,0}\|_{B_{p,r}^s} \right)^2 t \right)^{\frac{1}{4}}} + \frac{C}{\left(1 - 4C \left(\sum_{j=1}^4 \|u_{j,0}\|_{B_{p,r}^s} \right)^2 t \right)^{\frac{1}{4}}}$$

$$\times \int_0^t \left(1 - 4C \left(\sum_{j=1}^4 \|u_{j,0}\|_{B_{p,r}^s} \right)^2 \tau \right)^{\frac{1}{4}} \frac{\left(\sum_{j=1}^4 \|u_{j,0}\|_{B_{p,r}^s} \right)^2}{\left(1 - 4C \left(\sum_{j=1}^4 \|u_{j,0}\|_{B_{p,r}^s} \right)^2 \tau \right)^{\frac{3}{2}}} d\tau$$

$$\leqslant \left(1 + C \int_0^t \frac{\left(\sum_{j=1}^4 \|u_{j,0}\|_{B_{p,r}^s} \right)^2}{\left(1 - 4C \left(\sum_{j=1}^4 \|u_{j,0}\|_{B_{p,r}^s} \right)^2 t \right)^{\frac{5}{4}}} d\tau \right)$$

$$\times \frac{\sum_{j=1}^4 \|u_{j,0}\|_{B_{p,r}^s}}{\left(1 - 4C \left(\sum_{j=1}^4 \|u_{j,0}\|_{B_{p,r}^s} \right)^2 t \right)^{\frac{1}{4}}}$$

$$= \frac{\sum_{j=1}^{4} \|u_{j,0}\|_{B_{p,r}^s}}{\left(1 - 4C\left(\sum_{j=1}^{4} \|u_{j,0}\|_{B_{p,r}^s}\right)^2 t\right)^{\frac{1}{2}}}. \tag{2.12}$$

Hence, one can see that

$$\sum_{j=1}^{4} \|u_j^{(i+1)}(t)\|_{B_{p,r}^s} \leqslant \frac{\sum_{j=1}^{4} \|u_{j,0}\|_{B_{p,r}^s}}{\left(1 - 4C\left(\sum_{j=1}^{4} \|u_{j,0}\|_{B_{p,r}^s}\right)^2 t\right)^{\frac{1}{2}}},$$

which implies that $(u_1^{(i)}, u_2^{(i)}, u_3^{(i)}, u_4^{(i)})_{i \in \mathbb{N}}$ is uniformly bounded in $C([0;T]; B_{p,r}^s) \times C([0;T]; B_{p,r}^s) \times C([0;T]; B_{p,r}^s) \times C([0;T]; B_{p,r}^s)$. Using the equation (13) and the similar argument in the proof Proposition 2.3, one can easily prove that $(\partial_t u_1^{(i)}, \partial_t u_2^{(i)}, \partial_t u_3^{(i)}, \partial_t u_4^{(i)})_{i \in \mathbb{N}}$ is uniformly bounded in $C([0;T]; B_{p,r}^{s-1}) \times C([0;T]; B_{p,r}^{s-1}) \times C([0;T]; B_{p,r}^{s-1}) \times C([0;T]; B_{p,r}^{s-1})$. Hence, $(u_1^{(i)}, u_2^{(i)}, u_3^{(i)}, u_4^{(i)})_{i \in \mathbb{N}}$ is uniformly bounded in $E_{p,r}^s(T) \times E_{p,r}^s(T) \times E_{p,r}^s(T) \times E_{p,r}^s(T)$.

Now, it suffices to show that $(u_1^{(i)}, u_2^{(i)}, u_3^{(i)}, u_4^{(i)})_{i \in \mathbb{N}}$ is a Cauchy sequence in $C([0;T]; B_{p,r}^{s-1}) \times C([0;T]; B_{p,r}^{s-1}) \times C([0;T]; B_{p,r}^{s-1}) \times C([0;T]; B_{p,r}^{s-1})$. In fact, for all $i, k \in \mathbb{N}$, from (13), we have

$$\left(\partial_t + \frac{1}{9}[f_1^{(i+k)} g_1^{(i+k)} + f_2^{(i+k)} g_2^{(i+k)}]\partial_x\right)(m_1^{(i+k+1)} - m_1^{(i+1)}) = \widetilde{F}_1,$$

$$\left(\partial_t + \frac{1}{9}[f_1^{(i+k)} g_1^{(i+k)} + f_2^{(i+k)} g_2^{(i+k)}]\partial_x\right)(m_2^{(i+k+1)} - m_2^{(i+1)}) = \widetilde{F}_2,$$

$$\left(\partial_t + \frac{1}{9}[f_1^{(i+k)} g_1^{(i+k)} + f_2^{(i+k)} g_2^{(i+k)}]\partial_x\right)(m_3^{(i+k+1)} - m_3^{(i+1)}) = \widetilde{F}_3,$$

$$\left(\partial_t + \frac{1}{9}[f_1^{(i+k)} g_1^{(i+k)} + f_2^{(i+k)} g_2^{(i+k)}]\partial_x\right)(m_4^{(i+k+1)} - m_4^{(i+1)}) = \widetilde{F}_4,$$

with

$$\widetilde{F}_1 = (f_1^{(i+k)}(g_1^{(k+i)} - g_1^{(i)}) + (f_1^{(k+i)} - f_1^{(i)})g_1^{(i)} + f_2^{(i+k)}(g_2^{(k+i)} - g_2^{(i)})$$
$$+ (f_2^{(k+i)} - f_2^{(i)})g_2^{(i)})m_{1,x}^{(i+1)} - \frac{1}{9}\Big[(m_1^{(i+k)} g_1^{(i+k2)} - m_3^{(i+k)} f_1^{(i+k)} + m_2^{(i+k)} g_2^{(i+k)}$$

$$
\begin{aligned}
&\quad - m_4^{(i+k)} f_2^{(i+k)} - f_1^{(i+k)} g_1^{(i+k)})(m_1^{(k+i)} - m_1^{(i)}) + (m_1^{(i+k)}(g_1^{(k+i)} - g_1^{(i)}) \\
&\quad + (m_1^{(k+i)} - m_1^{(i)}) g_1^{(i)}) m_1^{(i)} - (f_1^{(i+k)}(m_3^{(k+i)} - m_3^{(i)}) + (f_1^{(k+i)} \\
&\quad - f_1^{(i)}) m_3^{(i)}) m_1^{(i)} + (m_2^{(i+k)}(g_2^{(k+i)} - g_2^{(i)}) + (m_2^{(k+i)} - m_2^{(i)}) g_2^{(i)}) m_1^{(i)} \\
&\quad + (f_2^{(i+k)}(m_4^{(k+i)} - m_4^{(i)}) \\
&\quad + (f_2^{(k+i)} - f_2^{(i)}) m_4^{(i)}) m_1^{(i)} - (f_1^{(i+k)}(g_1^{(k+i)} - g_1^{(i)}) + (f_1^{(k+i)} - f_1^{(i)}) g_1^{(i)}) m_1^{(i)} \Big] \\
&\quad + \frac{1}{9} \Big[(f_1^{(i+k)}(g_2^{(k+i)} - g_2^{(i)}) + (f_1^{(k+i)} - f_1^{(i)}) g_2^{(i+k)}) m_2^{(i+k)} \\
&\quad + (m_2^{(k+i)} - m_2^{(i)}) f_1^{(i)} g_2^{(i)} \Big],
\end{aligned}
$$

$$
\begin{aligned}
\widetilde{F}_2 &= (f_1^{(i+k)}(g_1^{(k+i)} - g_1^{(i)}) + (f_1^{(k+i)} - f_1^{(i)}) g_1^{(i)} + f_2^{(i+k)}(g_2^{(k+i)} - g_2^{(i)}) \\
&\quad + (f_2^{(k+i)} - f_2^{(i)}) g_2^{(i)}) m_{2,x}^{(i+1)} - \frac{1}{9} \Big[(m_1^{(i+k)} g_1^{(i+k)} - m_3^{i+k} f_1^{(i+k)} + m_2^{(i+k)} g_2^{(i+k)} \\
&\quad - m_4^{(i+k)} f_2^{(i+k)})(m_2^{(k+i)} - m_2^{(i)}) - f_2^{(i+k)} g_1^{(i+k)}(m_1^{(k+i)} - m_1^{(i)}) \\
&\quad + (m_1^{(i+k)}(g_1^{(k+i)} - g_1^{(i)}) + (m_1^{(k+i)} - m_1^{(i)}) g_1^{(i)}) m_2^{(i)} - (f_1^{(i+k)}(m_3^{(k+i)} - m_3^{(i)}) \\
&\quad + (f_1^{(k+i)} - f_1^{(i)}) m_3^{(i)}) m_2^{(i)} + (m_2^{(i+k)}(g_2^{(k+i)} - g_2^{(i)}) + (m_2^{(k+i)} \\
&\quad - m_2^{(i)}) g_2^{(i)}) m_2^{(i)} + (f_2^{(i+k)}(m_4^{(k+i)} - m_4^{(i)}) + (f_2^{(k+i)} - f_2^{(i)}) m_4^{(i)}) m_2^{(i)} \\
&\quad - (f_2^{(i+k)}(g_1^{(k+i)} - g_1^{(i)}) + (f_2^{(k+i)} - f_2^{(i)}) g_1^{(i)}) m_1^{(i)} \Big] \\
&\quad + \frac{1}{9} \Big[(f_2^{(i+k)}(g_2^{(k+i)} - g_2^{(i)}) + (f_2^{(k+i)} - f_2^{(i)}) g_1^{(i+k)}) m_2^{(i+k)} \\
&\quad + (m_2^{(k+i)} - m_2^{(i)}) f_2^{(i)} g_2^{(i)} \Big],
\end{aligned}
$$

$$
\begin{aligned}
\widetilde{F}_3 &= (f_1^{(i+k)}(g_1^{(k+i)} - g_1^{(i)}) + (f_1^{(k+i)} - f_1^{(i)}) g_1^{(i)} + f_2^{(i+k)}(g_2^{(k+i)} - g_2^{(i)}) \\
&\quad + (f_2^{(k+i)} - f_2^{(i)}) g_2^{(i)}) m_{3,x}^{(i+1)} - \frac{1}{9} \Big[(m_1^{(i+k)} g_1^{(i+k)} - m_3^{(i+k)} f_1^{(i+k)} + m_2^{(i+k)} g_2^{(i+k)} \\
&\quad - m_4^{(i+k)} f_2^{(i+k)} - f_1^{(i+k)} g_1^{(i+k)})(m_3^{(k+i)} - m_3^{(i)}) + (m_1^{(i+k)}(g_1^{(k+i)} - g_1^{(i)}) \\
&\quad + (m_1^{(k+i)} - m_1^{(i)}) g_1^{(i)}) m_3^{(i)} - (f_1^{(i+k)}(m_3^{(k+i)} - m_3^{(i)}) + (f_1^{(k+i)} - f_1^{(i)}) m_3^{(i)}) m_3^{(i)} \\
&\quad + (m_2^{(i+k)}(g_2^{(k+i)} - g_2^{(i)}) + (m_2^{(k+i)} - m_2^{(i)}) g_2^{(i)}) m_3^{(i)} + (f_2^{(i+k)}(m_4^{(k+i)} - m_4^{(i)}) \\
&\quad + (f_2^{(k+i)} - f_2^{(i)}) m_4^{(i)}) m_3^{(i)} + (f_1^{(i+k)}(g_1^{(k+i)} - g_1^{(i)}) + (f_1^{(k+i)} - f_1^{(i)}) g_1^{(i)}) m_3^{(i)} \Big] \\
&\quad - \frac{1}{9} \Big[(f_2^{(i+k)}(g_1^{(k+i)} - g_1^{(i)})
\end{aligned}
$$

$$+ (f_2^{(k+i)} - f_2^{(i)})g_1^{(i+k)})m_4^{(i+k)} + (m_4^{(k+i)} - m_4^{(i)})f_2^{(i)}g_1^{(i)}\Big],$$

$$\widetilde{F}_4 = (f_1^{(i+k)}(g_1^{(k+i)} - g_1^{(i)}) + (f_1^{(k+i)} - f_1^{(i)})g_1^{(i)} + f_2^{(i+k)}(g_2^{(k+i)} - g_2^{(i)})$$
$$+ (f_2^{(k+i)} - f_2^{(i)})g_2^{(i)})m_{4,x}^{(i+1)} - \frac{1}{9}\Big[(m_1^{(i+k)}g_1^{(i+k)} - m_3^{(i+k)}f_1^{(i+k)}$$
$$+ m_2^{(i+k)}g_2^{(i+k)} - m_4^{(i+k)}f_2^{(k+i)})(m_4^{(k+i)} - m_4^{(i)}) + f_1^{(i+k)}g_2^{(k)}(m_3^{(k+i)} - m_3^{(i)})$$
$$+ (m_1^{(i+k)}(g_1^{(k+i)} - g_1^{(i)}) + (m_1^{(k+i)} - m_1^{(i)})g_1^{(i)})m_4^{(i)}$$
$$- (f_1^{(i+k)}(m_3^{(k+i)} - m_3^{(i)}) + (f_1^{(i+k)} - f_1^{(i)})m_3^{(i)})m_4^{(i)} + (m_2^{(i+k)}(g_2^{(k+i)} - g_2^{(i)})$$
$$+ (m_2^{(k+i)} - m_2^{(i)})g_2^{(i)})m_4^{(i)} + (f_2^{(i+k)}(m_4^{(k+i)} - m_4^{(i)}) + (f_2^{(k+i)} - f_2^{(i)})m_4^{(i)})m_4^{(i)}$$
$$+ (f_1^{(i+k)}(g_2^{(k+i)} - g_2^{(i)}) + (f_1^{(i+k)} - f_1^{(i)})g_2^{(i)})m_3^{(i)}\Big] - \frac{1}{9}\Big[(f_2^{(i+k)}(g_2^{(k+i)} - g_2^{(i)})$$
$$+ (f_2^{(k+i)} - f_2^{(i)})g_1^{(i+k)})m_4^{(i+k)} + (m_4^{(k+i)} - m_4^{(i)})f_2^{(i)}g_2^{(i)}\Big].$$

Applying Lemma 2.2 again, then for every $t \in [0,T]$, $j=1,2,3,4$, we obtain

$$e^{-C\int_0^t \|\partial_x \frac{1}{9}[f_1^{(i+k)}g_1^{(i+k)} + f_2^{(i+k)}g_2^{(i+k)}](\tau')\|_{B_{p,r}^{s-1}}d\tau'} \left(\|m_j^{(i+k+1)}(t) - m_j^{(i+1)}(t)\|_{B_{p,r}^{s-3}}\right)$$
$$\leqslant \|m_{j,0}^{(i+k+1)}(t) - m_{j,0}^{(i+1)}(t)\|_{B_{p,r}^{s-3}}$$
$$+ C\int_0^t e^{-C\int_0^\tau \|\partial_x \frac{1}{9}[f_1^{(i+k)}g_1^{(i+k)} + f_2^{(i+k)}g_2^{(i+k)}](\tau')\|_{B_{p,r}^{s-1}}d\tau'} \|\widetilde{F}_j\|_{B_{p,r}^{s-3}}d\tau.$$

Similar to the proof of Proposition 2.3, in the case of $s > \max\left\{2 + \dfrac{1}{p}, \dfrac{5}{2}\right\}$, one can deduce that

$$\|\widetilde{F}_l\|_{B_{p,r}^{s-3}} \leqslant C\sum_{j=1}^{4}\left(\|u_j^{(i+k)} - u_j^{(i)}\|_{B_{p,r}^{s-1}}\right)\sum_{j=1}^{4}\left(\|u_j^{(i)}\|_{B_{p,r}^{s}}^2 + \|u_j^{(i+k)}\|_{B_{p,r}^{s}}^2 + \|u_j^{(i+1)}\|_{B_{p,r}^{s}}^2\right).$$

Since $(u_1^{(i)}, u_2^{(i)}, u_3^{(i)}, u_4^{(i)})_{i \in \mathbb{N}}$ is uniformly bounded in $E_{p,r}^s(T) \times E_{p,r}^s(T) \times E_{p,r}^s(T) \times E_{p,r}^s(T)$ and

$$u_{j,0}^{(i+k+1)} - u_{j,0}^{(i+1)} = S_{k+i+1}u_{j,0} - S_{i+1}u_{j,0} = \sum_{q=i+1}^{i+k}\Delta_q u_{j,0}, \quad j=1,2,3,4,$$

we get a constant C_T independent of i, k such that for all $t \in [0,T]$,

$$\sum_{j=1}^{4}\|(u_j^{(k+i+1)} - u_j^{(i+1)})(t)\|_{B_{p,r}^{s-1}} \leqslant C_T\left(2^{-i} + \int_0^t \sum_{j=1}^{4}\|(u^{(k+i)} - u_j^{(i)})(\tau)\|_{B_{p,r}^{s-1}}d\tau\right).$$

Arguing by induction with respect to the index i, one can easily prove that

$$\sum_{j=1}^{4}\|(u_j^{(k+i+1)} - u_j^{(i+1)})(t)\|_{L_T^\infty(B_{p,r}^{s-1})} \leq \frac{(TC_T)^{i+1}}{(i+1)!}\sum_{j=1}^{4}\|u_j^{(k)}\|_{L_T^\infty(B_{p,r}^{s-1})}$$

$$+ C_T \sum_{q=0}^{i} 2^{q-i}\frac{(TC_T)^q}{q!}.$$

As $\|u_j^{(k)}\|_{L_T^\infty(B_{p,r}^{s-1})}, j = 1, 2, 3, 4$, and C are bounded independently of k, there exists constant C_T' independent of i, k such that

$$\sum_{j=1}^{4}\|(u_j^{(k+i+1)} - u_j^{(i+1)})(t)\|_{L_T^\infty(B_{p,r}^{s-1})} \leq C_T' 2^{-i}.$$

Thus $(u_1^{(i)}, u_2^{(i)}, u_3^{(i)}, u_4^{(i)})_{i\in\mathbb{N}}$ is a Cauchy sequence in $C([0,T]; B_{p,r}^{s-1}) \times ([0,T]; B_{p,r}^{s-1}) \times ([0,T]; B_{p,r}^{s-1})$. □

Proof of Theorem 1.1 Thanks to Lemma 2.5, we obtain that $(u_1^{(i)}, u_2^{(i)}, u_3^{(i)}, u_4^{(i)})_{i\in\mathbb{N}}$ is a Cauchy sequence in $C([0,T]; B_{p,r}^{s-1}) \times C([0,T]; B_{p,r}^{s-1}) \times C([0,T]; B_{p,r}^{s-1}) \times C([0,T]; B_{p,r}^{s-1})$, so it converges to some function $(u_1, u_2, u_3, u_4) \in C([0,T]; B_{p,r}^{s-1}) \times C([0,T]; B_{p,r}^{s-1}) \times C([0,T]; B_{p,r}^{s-1}) \times C([0,T]; B_{p,r}^{s-1})$. We now have to check that (u_1, u_2, u_3, u_4) belongs to $E_{p,r}^s(T) \times E_{p,r}^s(T) \times E_{p,r}^s(T) \times E_{p,r}^s(T)$ and solves the Cauchy problem (5). Since $(u_1^{(i)}, u_2^{(i)}, u_3^{(i)}, u_4^{(i)})_{i\in\mathbb{N}}$ is uniformly bounded in $L^\infty([0,T]; B_{p,r}^s) \times L^\infty([0,T]; B_{p,r}^s) \times L^\infty([0,T]; B_{p,r}^s) \times L^\infty([0,T]; B_{p,r}^s)$ according to Lemma 2.5, the Fatou property for the Besov spaces (Proposition 2.2) guarantees that (u_1, u_2, u_3, u_4) also belongs to $L^\infty([0,T]; B_{p,r}^s) \times L^\infty([0,T]; B_{p,r}^s) \times L^\infty([0,T]; B_{p,r}^s) \times L^\infty([0,T]; B_{p,r}^s)$.

On the other hand, as $(u_1^{(i)}, u_2^{(i)}, u_3^{(i)}, u_4^{(i)})_{i\in\mathbb{N}}$ converges to (u_1, u_2, u_3, u_4) in $C([0,T]; B_{p,r}^{s-1}) \times C([0,T]; B_{p,r}^{s-1}) \times C([0,T]; B_{p,r}^{s-1}) \times C([0,T]; B_{p,r}^{s-1})$, an interpolation argument ensures that the convergence holds in $C([0,T]; B_{p,r}^{s'}) \times C([0,T]; B_{p,r}^{s'}) \times C([0,T]; B_{p,r}^{s'}) \times C([0,T]; B_{p,r}^{s'})$, for any $s' < s$. It is then easy to pass to the limit in the equation (2.9) and to conclude that (u_1, u_2, u_3, u_4) is indeed a solution to the Cauchy problem (2.1). Thanks to the fact that (u_1, u_2, u_3, u_4) belongs to $L^\infty([0,T]; B_{p,r}^s) \times L^\infty([0,T]; B_{p,r}^s) \times L^\infty([0,T]; B_{p,r}^s) \times L^\infty([0,T]; B_{p,r}^s)$, the right-hand side of the equation (5) belongs to $L^\infty([0,T]; C([0,T]; B_{p,r}^s))$. In particular, for the case $r < \infty$, Lemma 2.3 enables us to conclude that $(u_1, u_2, u_3, u_4) \in C([0,T]; B_{p,r}^{s'}) \times C([0,T]; B_{p,r}^{s'}) \times C([0,T]; B_{p,r}^{s'}) \times C([0,T]; B_{p,r}^{s'})$ for any $s' < s$. Finally, using the equation again, we see that $(\partial_t u_1, \partial_t u_2, \partial_t u_3, \partial_t u_4) \in C([0,T]; B_{p,r}^{s'}) \times C([0,T]; B_{p,r}^{s'}) \times C([0,T]; B_{p,r}^{s'}) \times C([0,T]; B_{p,r}^{s'})$ if $r < \infty$, and in $L^\infty([0,T]; B_{p,r}^{s-1}) \times L^\infty([0,T]; B_{p,r}^{s-1}) \times L^\infty([0,T]; B_{p,r}^{s-1}) \times$

$L^\infty([0,T]; B_{p,r}^{s-1})$ otherwise. Therefore, (u_1, u_2, u_3, u_4) belongs to $E_{p,r}^s(T) \times E_{p,r}^s(T) \times E_{p,r}^s(T) \times E_{p,r}^s(T)$. Moreover, a standard use of a sequence of viscosity approximate solutions $(u_{1,\varepsilon}, u_{2,\varepsilon}, u_{3,\varepsilon}, u_{4,\varepsilon},)_{\varepsilon>0}$ for the Cauchy problem (5) which converges uniformly in $C([0,T]; B_{p,r}^s) \cap C^1([0,T]; B_{p,r}^{s-1}) \times C([0,T]; B_{p,r}^s) \cap C^1([0,T]; B_{p,r}^{s-1}) \times C([0,T]; B_{p,r}^s) \cap C^1([0,T]; B_{p,r}^{s-1}) \times C([0,T]; B_{p,r}^s) \cap C^1([0,T]; B_{p,r}^{s-1})$ gives the continuity of the solution (u_1, u_2, u_3, u_4) in $E_{p,r}^s \times E_{p,r}^s \times E_{p,r}^s \times E_{p,r}^s$. The proof of Theorem is complete. □

3　Blow-up criteria

Proof of Theorem 1.2　We will prove the theorem by induction with respect to the regular index $s\left(s > \dfrac{1}{2}\right)$ as follows.

Step 1　For $s \in \left(\dfrac{1}{2}, 1\right)$, by Lemma 2.3 and System (2.1), we have

$$\|m_1\|_{H^s} \leqslant \|m_{1,0}\|_{H^s} + C\int_0^t \|(f_1g_1 + f_2g_2)_x\|_{L^\infty} \|m_1(\tau)\|_{H^s} d\tau$$
$$+ C\int_0^t \|-(m_1g_1 - m_3f_1 + m_2g_2 - m_4f_2 - f_1g_1)m_1 + m_2f_1g_2\|_{H^s} d\tau,$$

$$\|m_2\|_{H^s} \leqslant \|m_{2,0}\|_{H^s} + C\int_0^t \|(f_1g_1 + f_2g_2)_x\|_{L^\infty} \|m_2(\tau)\|_{H^s} d\tau$$
$$+ C\int_0^t \|-(m_1g_1 - m_3f_1 + m_2g_2 - m_4f_2)m_2 + f_2g_1m_1 + m_2f_2g_2\|_{H^s} d\tau,$$

$$\|m_3\|_{H^s} \leqslant \|m_{3,0}\|_{H^s} + C\int_0^t \|(f_1g_1 + f_2g_2)_x\|_{L^\infty} \|m_3(\tau)\|_{H^s} d\tau$$
$$+ C\int_0^t \|-(m_1g_1 - m_3f_1 + m_2g_2 - m_4f_2 + f_1g_1)m_3 - m_4f_2g_1\|_{H^s} d\tau,$$

$$\|m_4\|_{H^s} \leqslant \|m_{4,0}\|_{H^s} + C\int_0^t \|(f_1g_1 + f_2g_2)_x\|_{L^\infty} \|m_4(\tau)\|_{H^s} d\tau$$
$$+ C\int_0^t \|-(m_1g_1 - m_3f_1 + m_2g_2 - m_4f_2)m_4 - f_1g_2m_3 - m_4f_2g_2\|_{H^s} d\tau,$$
(3.1)

for all $0 < t < T$.

Noting that $u_i = (1-\partial_x^2)^{-1}m_i = p * m_i$ with $p(x) = \dfrac{1}{2}e^{-|x|}(x \in \mathbb{R})$, $u_{i,x} = \partial_x p * m_i$, $u_{i,xx} = u_i - m_i$, $i = 1,2,3,4$, together with the Young inequality implies that for all $s \in \mathbb{R}$,

$$\|u_i\|_{L^\infty}, \|u_{i,x}\|_{L^\infty}, \|u_{i,xx}\|_{L^\infty} \leqslant C\|m_i\|_{L^\infty}, \quad i=1,2,3,4,$$
$$\|u_i\|_{H^s}, \|u_{i,x}\|_{H^s}, \|u_{i,xx}\|_{H^s} \leqslant C\|m_i\|_{H^s}, \quad i=1,2,3,4. \tag{3.2}$$

From (17)-(18) and the definitions of $f_1^{(i)}, f_2^{(i)}, g_1^{(i)}, g_2^{(i)}$, we have

$$\|(f_1 g_1 + f_2 g_2)_x\|_{L^\infty} \leqslant C \sum_{j=1}^{4} \|m_j\|_{L^\infty}^2. \tag{3.3}$$

Thanks to Lemma 2.4, (17)-(18) and the definitions of f_1, f_2, g_1, g_2, we obtain

$$\big\| -(m_1 g_1 - m_3 f_1 + m_2 g_2 - m_4 f_2 - f_1 g_1)m_1 + m_2 f_1 g_2 \big\|_{H^s}$$
$$\leqslant C \sum_{j=1}^{4} \|m_j\|_{H^s}^2 \sum_{j=1}^{4} \|m_j\|_{L^\infty}^2,$$
$$\big\| -(m_1 g_1 - m_3 f_1 + m_2 g_2 - m_4 f_2)m_2 + f_2 g_1 m_1 + m_2 f_2 g_2 \big\|_{H^s}$$
$$\leqslant C \sum_{j=1}^{4} \|m_j\|_{H^s}^2 \sum_{j=1}^{4} \|m_j\|_{L^\infty}^2,$$
$$\big\| -(m_1 g_1 - m_3 f_1 + m_2 g_2 - m_4 f_2 + f_1 g_1)m_3 - m_4 f_2 g_1 \big\|_{H^s}$$
$$\leqslant C \sum_{j=1}^{4} \|m_j\|_{H^s}^2 \sum_{j=1}^{4} \|m_j\|_{L^\infty}^2,$$
$$\big\| -(m_1 g_1 - m_3 f_1 + m_2 g_2 - m_4 f_2)m_4 - f_1 g_2 m_3 - m_4 f_2 g_2 \big\|_{H^s}$$
$$\leqslant C \sum_{j=1}^{4} \|m_j\|_{H^s}^2 \sum_{j=1}^{4} \|m_j\|_{L^\infty}^2. \tag{3.4}$$

Plugging (19)-(20) into the inequality in (17) leads to

$$\|m_i\|_{H^s} \leqslant \|m_{i,0}\|_{H^s} + C \int_0^t \sum_{j=1}^{4} \|m_j\|_{L^\infty}^2 \sum_{j=1}^{4} \|m_j(\tau)\|_{H^s}^2 d\tau, \quad i=1,2,3,4. \tag{3.5}$$

Thus, we have

$$\sum_{i=1}^{4} \|m_i\|_{H^s} \leqslant \sum_{i=1}^{4} \|m_{i,0}\|_{H^s} + C \int_0^t \sum_{j=1}^{4} \|m_j\|_{L^\infty}^2 \sum_{j=1}^{4} \|m_j(\tau)\|_{H^s}^2 d\tau. \tag{3.6}$$

Taking advantage of Gronwall's inequality, one gets

$$\sum_{i=1}^{4} \|m_i\|_{H^s} \leqslant \sum_{i=1}^{4} \|m_{i,0}\|_{H^s} e^{C \int_0^t \sum_{j=1}^{4} \|m_i\|_{L^\infty}^2 d\tau}. \tag{3.7}$$

Therefore, if the maximal existence time $T < \infty$ satisfies

$$\int_0^T \sum_{j=1}^4 \|m_i\|_{L^\infty}^2 d\tau < \infty,$$

the inequality (23) implies that

$$\limsup_{t \to T} \sum_{i=1}^4 \|m_i(t)\|_{H^s} < \infty, \tag{3.8}$$

which contradicts the assumption that $T < \infty$ is the maximal existence time. This completes the proof of the theorem for $s \in \left(\frac{1}{2}, 1\right)$.

Step 2 For $s \in (1, 2)$, by differentiating the first equation in (5) with respect to x, we have

$$\partial_x m_t + \frac{1}{9}[f_1 g_1 + f_2 g_2]\partial_x m_x = -\frac{1}{9}[f_1 g_1 + f_2 g_2]_x m_x + \left(-\frac{1}{9}(m_1 g_1 - m_3 f_1 + m_2 g_2\right.$$
$$\left. - m_4 f_2 - f_1 g_1)m_1 + \frac{1}{9}m_2 f_1 g_2\right)_x. \tag{3.9}$$

Applying Lemma 2.3 to (29) yields

$$\|\partial_x m_1(t, \cdot)\|_{H^{s-1}} \leqslant \|\partial_x m_{1,0}\|_{H^{s-1}} + C \int_0^t \|(f_1 g_1 + f_2 g_2)_x\|_{L^\infty} \|\partial_x m\|_{H^{s-1}} d\tau$$
$$+ C \int_0^t \left\| -\frac{1}{9}[f_1 g_1 + f_2 g_2]_x m_x + \left(-\frac{1}{9}(m_1 g_1 - m_3 f_1 + m_2 g_2\right.\right.$$
$$\left.\left. - m_4 f_2 - f_1 g_1)m_1 + \frac{1}{9}m_2 f_1 g_2\right)_x \right\|_{H^{s-1}} d\tau. \tag{3.10}$$

Thanks to Lemma 2.4, (17)-(18) and the definitions of f_1, f_2, g_1, g_2, we obtain

$$\left\| -\frac{1}{9}[f_1 g_1 + f_2 g_2]_x m_x + \left(-\frac{1}{9}(m_1 g_1 - m_3 f_1 + m_2 g_2\right.\right.$$
$$\left.\left. - m_4 f_2 - f_1 g_1)m_1 + \frac{1}{9}m_2 f_1 g_2\right)_x \right\|_{H^{s-1}}$$
$$\leqslant C \sum_{j=1}^4 \|m_j\|_{H^s}^2 \sum_{j=1}^4 \|m_j\|_{L^\infty}^2,$$

$$\|(f_1 g_1 + f_2 g_2)_x\|_{L^\infty} \leqslant \sum_{j=1}^4 \|m_j\|_{L^\infty}^2, \tag{3.11}$$

which together with (26) yields

$$\|\partial_x m_1\|_{H^{s-1}} \lesssim \|m_{1,0}\|_{H^s} + C\int_0^t \sum_{j=1}^4 \|m_j\|_{L^\infty}^2 \sum_{j=1}^4 \|m_j\|_{H^s}^2(\tau)d\tau. \tag{3.12}$$

Similarly, we get

$$\|\partial_x m_i\|_{H^{s-1}} \lesssim \|m_{i,0}\|_{H^s} + C\int_0^t \sum_{j=1}^4 \|m_j\|_{L^\infty}^2 \sum_{j=1}^4 \|m_j\|_{H^s}^2(\tau)d\tau, \quad i=2,3,4. \tag{3.13}$$

Thus, we have

$$\sum_{i=1}^4 \|\partial_x m_i\|_{H^{s-1}} \lesssim \sum_{i=1}^4 \|m_{i,0}\|_{H^s} + C\int_0^t \sum_{j=1}^4 \|m_j\|_{L^\infty}^2 \sum_{j=1}^4 \|m_j\|_{H^s}^2(\tau)d\tau. \tag{3.14}$$

This along with (22) with $s-1$ instead of s ensures

$$\sum_{i=1}^4 \|\partial_x m_i\|_{H^s} \lesssim \sum_{i=1}^4 \|m_{i,0}\|_{H^s} + C\int_0^t \sum_{j=1}^4 \|m_j\|_{L^\infty}^2 \sum_{j=1}^4 \|m_j\|_{H^s}^2(\tau)d\tau. \tag{3.15}$$

Similar to Step 1, we can easily prove the theorem for $s \in [1,2)$.

Step 3 Suppose $2 \leqslant k \in \mathbb{N}$. By induction, we assume that (4) holds when $k-1 \leqslant s < k$, and prove that it holds for $k \leqslant s < k+1$. To this end, we differentiate (1) k times with respect to x, producing

$$\partial_t \partial_x^k m_1 + \frac{1}{9}(f_1 g_1 + f_2 g_2)\partial_x(\partial_x^k m_1)$$
$$= -\frac{1}{9}\sum_{l=0}^{k-1} C_k^l \partial_x^{k-l}[f_1 g_1 + f_2 g_2]\partial_x^{l+1} m_1 + \partial_x^k\bigg(-\frac{1}{9}(m_1 g_1 - m_3 f_1 + m_2 g_2$$
$$- m_4 f_2 - f_1 g_1)m_1 + \frac{1}{9}m_2 f_1 g_2\bigg), \tag{3.16}$$

which together with Lemma 2.3 with $s - k \in (0,1)$, implies that

$$\|\partial_x^k m_1(t)\|_{H^{s-k}} \lesssim \|\partial_x^k m_{1,0}\|_{H^{s-k}} + C\int_0^t \|(f_1 g_1 + f_2 g_2)_x\|_{L^\infty} \|\partial_x^k m_1\|_{H^{s-1}} d\tau$$
$$+ C\int_0^t \bigg\|\sum_{l=0}^{k-1} C_k^l \partial_x^{k-l}(f_1 g_1 + f_2 g_2)\partial_x^{l+1} m + \partial_x^k\bigg(-\frac{1}{9}(m_1 g_1 - m_3 f_1 + m_2 g_2$$
$$- m_4 f_2 - f_1 g_1)m_1 + \frac{1}{9}m_2 f_1 g_2\bigg)\bigg\|_{H^{s-k}} d\tau.$$

Using the estimates in Lemma 2.4, the Sobolev embedding inequality and inequalities (18)-(19), we have

$$\left\| \sum_{l=0}^{k-1} C_k^l \partial_x^{k-l}(f_1 g_1 + f_2 g_2) \partial_x^{l+1} m + \partial_x^k \left(-\frac{1}{9}(m_1 g_1 - m_3 f_1 + m_2 g_2 \right.\right.$$
$$\left.\left. - m_4 f_2 - f_1 g_1) m_1 + \frac{1}{9} m_2 f_1 g_2 \right) \right\|_{H^{s-k}}$$
$$\leqslant C \sum_{j=1}^{4} \|m_j\|_{H^{k-\frac{1}{2}+\varepsilon_0}}^2 \sum_{j=1}^{4} \|m_j\|_{H^s}, \qquad (3.17)$$

where the genius constant $\varepsilon_0 \in \left(0, \frac{1}{4}\right)$ so that $H^{\frac{1}{2}+\varepsilon_0}(\mathbb{R}) \hookrightarrow L^\infty(\mathbb{R})$ holds. Thus, we get

$$\|\partial_x^k m_1\|_{H^{s-k}} \leqslant \|m_{1,0}\|_{H^s} + C \int_0^t \sum_{j=1}^{4} \|m_j\|_{H^{k-\frac{1}{2}+\varepsilon_0}}^2 \sum_{j=1}^{4} \|m_j\|_{H^s}(\tau) d\tau, \qquad (3.18)$$

where we used the Sobolev embedding theorem $H^{k-\frac{1}{2}+\varepsilon_0}(\mathbb{R}) \hookrightarrow L^\infty(\mathbb{R})$ with $k \geqslant 2$. Similarly, we have

$$\|\partial_x^k m_i\|_{H^{s-k}} \leqslant \|m_{i,0}\|_{H^s} + C \int_0^t \sum_{j=1}^{4} \|m_j\|_{H^{k-\frac{1}{2}+\varepsilon_0}}^2 \sum_{j=1}^{4} \|m_j\|_{H^s}(\tau) d\tau, \quad i = 2, 3, 4. \qquad (3.19)$$

Then, we obtain

$$\sum_{i=1}^{4} \|\partial_x^k m_i\|_{H^{s-k}} \leqslant \sum_{i=1}^{4} \|m_{j,0}\|_{H^s}$$
$$+ C \int_0^t \sum_{j=1}^{4} \|m_j\|_{H^{k-\frac{1}{2}+\varepsilon_0}}^2 \sum_{j=1}^{4} \|m_j\|_{H^s} \sum_{j=1}^{4} \|m_j\|_{H^{k-\frac{1}{2}+\varepsilon_0}}^2 (\tau) d\tau, \qquad (3.20)$$

which along with (22) with $s - k \in (0, 1)$ instead of s, ensures that

$$\sum_{i=1}^{4} \|m_i\|_{H^{s-k}} \leqslant \sum_{i=1}^{4} \|m_{j,0}\|_{H^s} + C \int_0^t \sum_{j=1}^{4} \|m_j\|_{H^s} \sum_{j=1}^{4} \|m_j\|_{H^{k-\frac{1}{2}+\varepsilon_0}}^2 (\tau) d\tau. \qquad (3.21)$$

Applying Gronwall's inequality then gives

$$\sum_{i=1}^{4} \|m_i\|_{H^s} \leqslant \sum_{i=1}^{4} \|m_{i,0}\|_{H^s} \times e^{C \int_0^t \sum_{j=1}^{4} \|m_j\|_{H^{k-\frac{1}{2}+\varepsilon_0}}^2 (\tau) d\tau}. \qquad (3.22)$$

In consequence, if the maximal existence time $T < \infty$ satisfies

$$\int_0^T \sum_{j=1} \|m_j\|^2_{H^{k-\frac{1}{2}+\varepsilon_0}}(\tau) d\tau < \infty,$$

thanks to the uniqueness of solution in Corollary 1.1, we then find that $\|m_i\|_{H^{k-\frac{1}{2}+\varepsilon_0}}$ ($i = 1, 2, 3, 4$) are uniformly bounded in $t \in (0, T)$ by the induction assumption, which along with (38) implies

$$\limsup_{t \to T} \sum_{i=1}^4 \|m_i(t)\|_{H^s} < \infty, \tag{3.23}$$

which leads to a contradiction.

Therefore, Steps 1 to 3 complete the proof of Theorem 1.2.

Acknowledgements We would like to thank the referees very much for their valuable comments and suggestions.

References

[1] B. Xia and Z. Qiao. Integrable multi-component Camassa-Holm system. arXiv:1310.0268v1 [nlin.SI] 1 Oct 2013.

[2] R. Camassa and D. Holm. An integrable shallow water equation with peaked solitons. *Phys. Rev. Lett.*, 71 (1993) 1661-1664.

[3] A. Constantin and D. Lannes. The hydrodynamical relevance of the Camassa-Holm and Degasperis-Procesi equations. *Arch. Ration. Mech. Anal.*, 192 (2009) 165-186).

[4] A. Constantin. The trajectories of particles in Stokes waves. *Invent. Math.*, 166 (2006) 523-535.

[5] J.F. Toland. Stokes waves. *Topol. Methods Nonlinear Anal.*, 7 (1996) 1-48.

[6] A. Constantin and J. Escher. Particle trajectories in solitary water waves. *Bull. Amer. Math. Soc.*, 44 (2007), 423-431.

[7] A.Constantin and W. Strauss. Stability of peakons. *Comm. Pure Appl. Math.*, 53 (2000) 603-610.

[8] J. Lenells. A variational approach to the stability of periodic peakons. *J. Nonlinear Math. Phys.*, 11 (2004) 151-163.

[9] A. Constantin and H. P. McKean. A shallow water equation on the circle. *Comm. Pure Appl. Math.*, 52 (1999) 949-982.

[10] A. Constantin, V. Gerdjikov and R. Ivanov. Inverse scattering transform for the Camassa-Holm equation. *Inverse Problems*, 22 (2006) 2197-2207.

[11] A. Constantin. On the inverse spectral problem for the Camassa-Holm equation. *J. Funct. Anal.*, 155 (1998) 352-363.

[12] A. Constantin, T. Kappeler, B. Kolev, and P. Topalov. On geodesic exponential maps of the Virasoro group. *Ann. Global Anal. Geom.*, 31 (2007) 155-180.

[13] A. Constantin and B. Kolev. Geodesic flow on the diffeomorphism group of the circle. *Commentarii Mathematici Helvetici*, 78 (2003) 787-804.

[14] S. Kouranbaeva. The Camassa-Holm equation as a geodesic flow on the diffeomorphism group. *J. Math. Phys.*, 40 (1999) 857-868.

[15] G. Misiolek. A shallow water equation as a geodesic flow on the Bott-Virasoro group. *J. Geom. Phys.*, 24 (1998) 203-208.

[16] A. Constantin. Existence of permanent and breaking waves for a shallow water equation: a geometric approach. *Ann. Inst. Fourier (Grenoble).*, 50 (2000) 321-362.

[17] A. Constantin and J. Escher. Wave breaking for nonlinear nonlocal shallow water equations. *Acta Math.*, 181 (1998), 229-243.

[18] A. Constantin and J. Escher. On the blow-up rate and the blow-up set of breaking waves for a shallow water equation. *Math. Z.*, 233 (2000) 75-91.

[19] A. Constantin and J. Escher. Global existence and blow-up for a shallow water equation. *Ann. Scuola Norm. Sup. Pisa.*, 26 (1998) 303-328.

[20] Y. Li and P. Olver. Well-posedness and blow-up solutions for an integrable nonlinearly dispersive model wave equation. *J. Differential Equations.*, 162 (2000) 27-63.

[21] A. Bressan and A. Constantin. Global conservative solutions of the Camassa-Holm equation. *Arch. Ration. Mech. Anal.*, 183 (2007) 215-239.

[22] H. Holden and X. Raynaud. Global conservative solutions of the Camassa-Holm equations-a Lagrangianpoiny of view. *Comm. Partial Differential Equations*, 32 (2007) 1511-1549.

[23] A. Bressan and A. Constantin. Global dissipative solutions of the Camassa-Holm equation. *Anal. Appl.*, 5 (2007)1-27.

[24] H. Holden and X. Raynaud. Dissipative solutions for the Camassa-Holm equation. *Discrete Contin. Dyn. Syst.*, 24 (2009) 1047-1112.

[25] A. S. Fokas. The Korteweg-de Vries equation and beyond. *Acta Appl. Math.*, 39(1995) 295-305.

[26] B. Fuchssteiner. Some tricks from the symmetry-toolbox for nonlinear equations: generalizations of the Camassa-Holm equation. *Physica D.*, 95 (1996) 229-243.

[27] P. Olver and P. Rosenau. Tri-Hamiltonian duality between solitons and solitary-wave solutions having compact support. *Phys. Rev. E.*, 53 (1996) 1900-1906.

[28] Z. Qiao. A new integrable equation with cuspons and W/M-shape-peaks solitons. *J. Math. Phys.*, 47 (2006) 112701 1-13.

[29] Z. Qiao and X. Li. An integrable equation with nonsmooth solitons. *Theor. Math. Phys.*, 267 (2011) 584-589.

[30] Y. Fu, G. Gu, Y. Liu, and Z. Qu. On the Cauchy problemfor the integrable Camassa-Holm type equation with cubic nonlinearity, arXiv:1108.5368v2, 1-27.

[31] G. Gui and Y. Liu. On the global existence and wave-breaking criteria for the two-component Camassa-Holm system. *J. Funct. Anal.*, 258 (2010) 4251-4278.

[32] J. Chemin. Localization in Fourier space and Navier-Stokes system. Phase Space Analysis of Partial Differential Equations. *Proceedings, CRM series, Pisa*, 2004, 53-136.

[33] R. Danchin. A few remarks on the Camassa-Holm equation, Differential Integral Equations, 14 (2001) 953-988.

[34] R. Danchin. Fourier analysis methods for PDEs. *Lecture Notes*, 14 November, 2003.

Darboux Transformation and Classification of Solution for Mixed Coupled Nonlinear Schrödinger Equations*

Ling Liming (凌黎明), Zhao Li chen (赵立臣) and Guo Boling (郭柏灵)

Abstract We derive generalized localized wave solution formula for mixed coupled nonlinear Schödinger equations (mCNLSE) by performing the unified Darboux transformation. Based on the dynamical behavior of solution, the classification of the localized wave solutions on the nonzero background is given explicitly. Especially, the parameter conditions for breather, dark soliton and rogue wave solution of mCNLSE are given in detail. Moreover, we analyze the interaction between dark soliton solution and breather solution. These results would be helpful for nonlinear localized wave excitations and applications in vector nonlinear systems.

Keywords Darboux transformation; mCNLSE; localized wave; rogue wave; dark soliton.

1 Introduction

Nonlinear Schrödinger equation (NLSE) is an important model in mathematical physics, which can be applied to hydrodynamics [1], plasma physics [2], molecular biology [3] and optics [4]. Recently, the localized waves for NLSE on the plane wave background were observed in experiments in succession. For instance, Kuznetzov-Ma soliton was confirmed in 2012 [3]; Akhmediev breather was verified in numerical experiment [5]; Peregrine soliton was experimentally observed in nonlinear fibre optics system [6], water tank [7,8] and plasma [9]; dark soliton was observed on the surface of water [10]. These experiments illustrate that the localized waves are very interesting and important for the NLSE. Indeed these solutions for NLSE on the plane wave

---

* Commun. Nonlinear Sci. Numer. Simulat., 32, 285-304 (2016).

background were known well long time ago [3]. There exists Akhmediev breather, Kuznetzov-Ma soliton and Peregrine soliton for the focusing NLSE; there exists dark soliton for the defocusing one.

However, the localized wave solutions for coupled nonlinear Schrödinger equations (CNLSE) are more colorful and complex than scalar NLSE. Firstly, the spectral problem of NLSE is 2×2, the coupled NLSE is 3×3. The inverse scattering method of CNLSE on the nonzero background has not been solved completely up to now [11]. Secondly, for the Darboux transformation (DT) method, the Darboux matrix for 3×3 spectral problem is more complexity than 2×2 spectral problem. What's more, the Darboux matrix for the defocusing or mixed coupled NLSE is no longer positive or negative definite. Since the localized wave solutions for CNLSE are very interesting and meaningful, it is important to give the classification for these localized wave solutions.

Previous to introduce our work, we review a brief research history of CNLSE. The integrability of the CNLSE had been shown by Manakov who had also obtained the bright soliton in a focusing medium by applying the inverse scattering method [12]. An interesting fact for CNLSE is the collision of soliton on the vanishing background can be inelastic [13, 14] or even appear the soliton reflection [15]. For the nonzero background, there are lots of works for the focusing CNLSE [16–22]. But there is few result for the defocusing and mixed case [11]. The dark-dark soliton solution, bright-dark soliton solution and breather solution for defocusing CNLSE were given in reference [23] by inverse scattering method. The soliton solutions for the multi-component NLSE were given in reference [16, 24, 25] through DT. Recently, different types of soliton solutions on the nonzero background were obtained by algebraic geometry reduction method [26].

DT is a powerful method to construct the soliton solution for the integrable equations. There are different methods to derive the DT, for instance, operator decomposition method [27], gauge transformation [28, 29], loop group method [30] and Riemann-Hilbert method [31]. The DT for multi-component NLSE was given in reference [16, 24, 32–34] (and reference therein). Recently, combined Darboux dressing and tau function method for dark-dark soliton is given in reference [35]; Tsuchida gave a detailed analysis of different types of solutions by using the Darboux-Bäcklund transformation [25].

In this work, we focus on the localized wave solutions of the mCNLSE

$$iq_{1,t} + \frac{1}{2}q_{1,xx} + (|q_2|^2 - |q_1|^2)q_1 = 0,$$
$$iq_{2,t} + \frac{1}{2}q_{2,xx} + (|q_2|^2 - |q_1|^2)q_2 = 0,$$
(1.1)

on the plane wave background. The classification of localized wave solutions for mCNLSE (1.1) on the nonzero background is not an easy work but be important for nonlinear wave theory and physical applications. Indeed, there is few result about the classification of localized wave solutions for CNLSE, even for the focusing CNLSE. Recently, there are lots of works about focusing CNLSE through DT. For instance, vector rogue wave (type-I and type-II) solution, bright-dark-rogue wave solution [17–20] and high order solution [36] have been given. However, for the mCNLSE or the defocusing CNLSE, the Darboux matrix no longer keeps positive or negative definite. To the best of our knowledge, this problem has never been solved with a proper method in the DT theory. In this work, we use the matrix analysis method to deal with this problem. Through this method, we can obtain the complete classification of nonsingular solution for mCNLSE (1.1). Besides the DT, the inverse scattering method [15] and the bilinear method [13, 37, 38] can be used to derive soliton solution of the CNLSE too.

This paper is organized as following. In section 2, we classify the localized wave solutions of mCNLSE (1.1) as two categories. Denote the wave vector of plane wave background for the i-th component as a_i. The first case is $a_1 = a_2$. In this case, we can obtain the degenerate rogue wave, degenerate breather (type-I and type-II), degenerate dark-dark soliton and bright-dark soliton. The second case is $a_1 \neq a_2$. In this case, we can obtain the general rogue wave solution, general breather (type I and type II) solution and general dark-dark soliton solution. The coexistence of different types solution is given by two different categories. The rogue wave solutions for mCNLSE (1.1) are given with a compact form, which is the natural extension for classical one. The classification of rogue wave solutions is given based on the dynamic behavior. A method for looking for different types of rogue wave is given in detail. In section 3, we give the interaction between different types of soliton solutions. Since the rogue wave solution is the limit of breather (type I) solution, we deem that rogue wave is the same kind with breather solution, although their dynamics behaviors are different. Thus we merely analyze the interaction between breather type solution and dark soliton. Finally, we give some discussions and conclusions.

2 Darboux transformation and classification of solution

As we are well known that, the mCNLSE (1.1) admits the following Lax pair:

$$\Phi_x = U\Phi, \quad U(\lambda, Q) \equiv i(\lambda\sigma_3 + Q), \tag{2.1a}$$

$$\Phi_t = V\Phi, \quad V(\lambda, Q) \equiv i\lambda^2\sigma_3 + i\lambda Q - \frac{1}{2}\sigma_3(iQ^2 - Q_x), \tag{2.1b}$$

where

$$\sigma_3 = \text{diag}(1, -1, -1), \quad Q = \begin{bmatrix} 0 & -\bar{q}_1 & \bar{q}_2 \\ q_1 & 0 & 0 \\ q_2 & 0 & 0 \end{bmatrix},$$

the overbar represents the complex conjugation (similarly hereinafter). The compatibility condition of Lax pair (2.1) gives the mCNLSE (1.1). The unified DT is obtained in reference [39] with an integral form. Here we give another representation with a limit form:

Theorem 2.1 ([39], Ling, Zhao and Guo) *The following unified DT,*

$$\Phi[1] = T_1\Phi, \quad T_1 = I - \frac{P_1}{\lambda - \bar{\lambda}_1}, \tag{2.2}$$

where

$$\text{if } \lambda_1 \notin \mathbb{R}, \quad P_1 = \frac{(\lambda_1 - \bar{\lambda}_1)|y_1\rangle\langle y_1|J}{\langle y_1|J|y_1\rangle}, \tag{2.3a}$$

$$\text{if } \lambda_1 \in \mathbb{R}, \quad P_1 = \lim_{\lambda_1 \to \bar{\lambda}_1} \frac{(\lambda_1 - \bar{\lambda}_1)|y_1\rangle\langle y_1|J}{\langle y_1|J|y_1\rangle}, \tag{2.3b}$$

$J = \text{diag}(1, -1, 1)$, $|y_1\rangle \equiv v_1(x,t)\Phi_1$, $\langle y_1| = |y_1\rangle^\dagger$, $v_1(x,t)$ *is an arbitrarily complex function,* Φ_1 *is a special vector solution for linear system* (2.1) *with* $\lambda = \lambda_1$, *and* $\dagger$ *represents Hermite conjugation (similarly hereinafter), converts the above linear system* (2.1) *into a new linear system*

$$\Phi[1]_x = U[1]\Phi[1], \quad U[1] = U(\lambda, Q[1]), \tag{2.4a}$$

$$\Phi[1]_t = V[1]\Phi[1], \quad V[1] = V(\lambda, Q[1]), \tag{2.4b}$$

and transformation between potential functions is

$$Q[1] = Q + [\sigma_3, P_1], \tag{2.5}$$

where commutator $[A, B] \equiv AB - BA$.

In our previous work [39], we merely use the DT to derive the dark-dark soliton solution. In this work, we derive all of localized wave solutions by DT and give its classification by the matrix analysis method. To keep the results with the inverse scattering method, we restrict the parameter $\lambda_1 \in \{z|\text{Im}(z) \geqslant 0\}$. The expression for P_1 (2.3b) is considered under the limit sense, since there exist special solutions which satisfy $\langle y_1|J|y_1\rangle = 0$ when $\lambda_1 \in \mathbb{R}$.

To obtain the localized wave solutions on the plane wave background, we consider the following plane wave solutions as the seed solutions for DT

$$q_1[0] = c_1 e^{i\theta_1}, \quad \theta_1 = \left[a_1 x - \left(\frac{1}{2}a_1^2 + c_1^2 - c_2^2\right)t\right],$$
$$q_2[0] = c_2 e^{i\theta_2}, \quad \theta_2 = \left[a_2 x - \left(\frac{1}{2}a_2^2 + c_1^2 - c_2^2\right)t\right], \tag{2.6}$$

where a_1, a_2, c_1 and c_2 are real constants. Through above transformation (2.5) and the plane wave seed solution, we can obtain different types of nonlinear wave solutions. If $\lambda_1 \notin \mathbb{R}$, the DT can be used to derive breather solution, rogue wave solution and bright-dark soliton solution. If $\lambda_1 \in \mathbb{R}$, the dark-dark soliton solution can be obtained through this transformation (2.5).

Next we give a method how to choose the special solution Φ_1 to construct the nonsingular exact solution of mCNLSE (1.1). Through above DT (2.2) and (2.5), we can obtain many new nonlinear wave solutions for mCNLSE(1.1), which have not been reported before. To use the transformation (2.2), we firstly need to solve the linear system (2.1) with seed solution (2.6) by the gauge transformation method [17]:

$$\Phi = D\Psi, \quad D = \text{diag}(1, e^{i\theta_1}, e^{i\theta_2}). \tag{2.7}$$

Then matrix Ψ satisfies the following linear system:

$$\Psi_x = iU_0\Psi,$$
$$\Psi_t = i\left(\frac{1}{2}U_0^2 + \lambda U_0 - \frac{1}{2}\lambda^2 + c_1^2 - c_2^2\right)\Psi,$$

where

$$U_0 = \begin{bmatrix} \lambda & -c_1 & c_2 \\ c_1 & -\lambda - a_1 & 0 \\ c_2 & 0 & -\lambda - a_2 \end{bmatrix}.$$

We can obtain the different kinds of solution by choosing different special solutions (2.7). The studies on coupled focusing or defocusing NLSE have shown that the relative wave vector plays important role in determining dynamics of nonlinear waves [19, 20, 40]. Therefore, we classify them as two main cases according to the relative wave vector.

2.1 Case I: When $a_2 = a_1$

In this case, we have the following matrix decomposition

$$U_0 M = M D_0, \quad D_0 = \text{diag}(\chi - \lambda, \mu - \lambda, -(\lambda + a_1)), \quad \chi \neq \mu, \qquad (2.8)$$

where

$$M = \begin{bmatrix} 1 & 1 & 0 \\ \dfrac{c_1}{\chi + a_1} & \dfrac{c_1}{\mu + a_1} & c_2 \\ \dfrac{c_2}{\chi + a_1} & \dfrac{c_2}{\mu + a_1} & c_1 \end{bmatrix},$$

and χ and μ are two different roots of the quadratic equation:

$$\xi^2 + (a_1 - 2\lambda)\xi - 2a_1\lambda + c_1^2 - c_2^2 = 0. \qquad (2.9)$$

Then the fundamental solution (2.7) can be given as $\Phi = DMN$, where $N = \text{diag}(e^{\mathrm{i}A(\chi)}, e^{\mathrm{i}A(\mu)}, e^{\mathrm{i}A(-a_1)})$, and

$$A(\chi) \equiv (\chi - \lambda)x + \left[\frac{1}{2}\chi^2 + (c_1^2 - c_2^2 - \lambda^2)\right]t.$$

By the decomposition (2.8), we can simplify the solution form of mCNLSE (1.1) through using the following identity:

$$\begin{aligned} 2\lambda - \chi + \frac{c_2^2 - c_1^2}{\chi + a_1} &= 0, \\ 2\lambda - \mu + \frac{c_2^2 - c_1^2}{\mu + a_1} &= 0. \end{aligned} \qquad (2.10)$$

Interestingly, a further classification on nonlinear wave solutions can be made through relations between c_1 and c_2.

(a) When $c_1 > c_2$, the system (1.1) reflects the defocusing mechanism. With this case, there exists dark-dark soliton solution. Besides the dark-dark soliton, there exists breather solution. In what follows, we give the explicit construction method for them.

To give the nonsingular solution of (1.1), one must choose the special solution Φ_1 such that $\Phi_1^\dagger J \Phi_1$ is negative or positive definite. To find the special solution Φ_1, it is meaningful to analyze the following matrix:

$$M^\dagger J M = \begin{bmatrix} \dfrac{2(\bar\lambda-\lambda)}{\bar\chi-\chi} & \dfrac{2(\bar\lambda-\lambda)}{\bar\chi-\mu} & 0 \\ \dfrac{2(\bar\lambda-\lambda)}{\bar\mu-\chi} & \dfrac{2(\bar\lambda-\lambda)}{\bar\mu-\mu} & 0 \\ 0 & 0 & c_1^2-c_2^2 \end{bmatrix},$$

which can be obtained through equations (2.10). Indeed, the above matrix can not be positive definite or negative definite, since above matrix is nothing but the congruent matrix of J and the congruent matrix can not change the sign of characteristic roots. However, we can look for some submatrices which could be positive or negative definite. Thus, if we choose the special solution Φ_1 and $v_1(x,t)$ such that

$$|y_1\rangle = D \begin{bmatrix} e^{i(\chi_1 x+\frac{1}{2}\chi_1^2 t)} \\ \dfrac{c_1}{\chi_1+a_1}e^{i(\chi_1 x+\frac{1}{2}\chi_1^2 t)} + c_2\alpha_1 e^{i(-a_1 x+\frac{1}{2}a_1^2 t)} \\ \dfrac{c_2}{\chi_1+a_1}e^{i(\chi_1 x+\frac{1}{2}\chi_1^2 t)} + c_1\alpha_1 e^{i(-a_1 x+\frac{1}{2}a_1^2 t)} \end{bmatrix}, \qquad (2.11)$$

where α_1 is a nonzero complex constant. Setting parameter

$$\alpha_1 = \left(\frac{2\mathrm{Im}(\lambda_1)}{(c_1^2-c_2^2)\mathrm{Im}(\chi_1)}\right)^{1/2} e^{\frac{\beta_1}{2}+i\gamma_1},$$

where β_1 and γ_1 are real constants, $\mathrm{Im}(\cdot)$ represents the imaginary part of complex number $\cdot$ (similarly hereinafter), then we can obtain the following breather solution by Theorem 2.1,

$$q_1[1] = c_1 \left[\frac{C_1+1}{2}+\frac{C_1-1}{2}\tanh\left(\frac{A_1-\beta_1}{2}\right)+\frac{c_2 D_1}{c_1}\mathrm{sech}\left(\frac{A_1-\beta_1}{2}\right) e^{iB_1}\right] e^{i\theta_1},$$

$$q_2[1] = c_2 \left[\frac{C_1+1}{2}+\frac{C_1-1}{2}\tanh\left(\frac{A_1-\beta_1}{2}\right)+\frac{c_1 D_1}{c_2}\mathrm{sech}\left(\frac{A_1-\beta_1}{2}\right) e^{iB_1}\right] e^{i\theta_1},$$

where

$$A_1 = -2\mathrm{Im}(\chi_1)[x+\mathrm{Re}(\chi_1)t],$$
$$B_1 = -(\mathrm{Re}(\chi_1)+a_1)x+\frac{1}{2}\left[a_1^2-\mathrm{Re}(\chi_1^2)\right]t+\gamma_1+\frac{3}{2}\pi,$$
$$C_1 = \frac{\bar\chi_1+a_1}{\chi_1+a_1}, \quad D_1 = \left(\frac{2\mathrm{Im}(\lambda_1)\mathrm{Im}(\chi_1)}{c_1^2-c_2^2}\right)^{1/2},$$

$\mathrm{Re}(\cdot)$ represents the real part of complex number $\cdot$ (similarly hereinafter).

Through above expression of solution, we can see that this kind of breather solution is composed of a dark soliton solution and a bright one. Notably, this type of breather solution can not be used to derive rogue wave solution through the limit method. This breather solution never appear in the scalar NLSE. Thus we call this type of breather as breather-II to distinguish the breather which can be reduced to rogue wave. By choosing special parameters, we can obtain the figure of breather-II (Fig. 1).

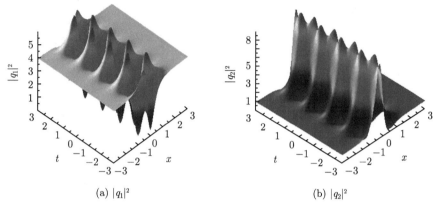

(a) $|q_1|^2$　　　　　　　　(b) $|q_2|^2$

Fig.1　Breather-II type solution: Parameters $a_1 = a_2 = 0$, $c_1 = 2$, $c_2 = 1$, $\lambda_1 = i$, $\chi_1 = 3i$. It is seen that $|q_2|^2$ possesses the feature of bright soliton and breather

If we take $c_2 = 0$, we can obtain so-called bright-dark soliton solution

$$q_1[1] = c_1 \left[\frac{C_1 + 1}{2} + \frac{C_1 - 1}{2} \tanh\left(\frac{A_1 - \beta_1}{2}\right) \right] e^{i\theta_1},$$

$$q_2[1] = D_1 \mathrm{sech}\left(\frac{A_1 - \beta_1}{2}\right) e^{i(B_1 + \theta_1)},$$

where

$$A_1 = -2\mathrm{Im}(\chi_1)[x + \mathrm{Re}(\chi_1)t],$$

$$B_1 = -(\mathrm{Re}(\chi_1) + a_1)x + \frac{1}{2}\left[a_1^2 - \mathrm{Re}(\chi_1^2)\right]t + \gamma_1 + \frac{3}{2}\pi,$$

$$C_1 = \frac{\bar{\chi}_1 + a_1}{\chi_1 + a_1}, \quad D_1 = \sqrt{2\mathrm{Im}(\lambda_1)\mathrm{Im}(\chi_1)}.$$

(b) When $c_1 = c_2$, there is no nontrivial solution of mCNLSE could be constructed through above Theorem 2.1.

(c) When $c_1 < c_2$, the system (1.1) reflects the focusing mechanism. If we choose the special solution Φ_1 and $v_1(x,t)$ such that

$$|y_1\rangle = D \begin{bmatrix} e^{i(\chi_1 x + \frac{1}{2}\chi_1^2 t)} + \alpha_1 e^{i(\mu_1 x + \frac{1}{2}\mu_1^2 t)} \\ \dfrac{c_1}{\chi_1 + a_1} e^{i(\chi_1 x + \frac{1}{2}\chi_1^2 t)} + \dfrac{c_1 \alpha_1}{\mu_1 + a_1} e^{i(\mu_1 x + \frac{1}{2}\mu_1^2 t)} \\ \dfrac{c_2}{\chi_1 + a_1} e^{i(\chi_1 x + \frac{1}{2}\chi_1^2 t)} + \dfrac{c_2 \alpha_1}{\mu_1 + a_1} e^{i(\mu_1 x + \frac{1}{2}\mu_1^2 t)} \end{bmatrix}, \quad (2.12)$$

where α_1 is a complex parameter. Moreover, we introduce the following notation:

$$\lambda_1 + \frac{a_1}{2} = \alpha \sinh(\gamma_1 + i\beta_1), \quad \alpha = \sqrt{c_2^2 - c_1^2}, \quad \beta_1, \gamma_1 \in \mathbb{R},$$

it follows that

$$\sqrt{(\lambda_1 + \frac{a_1}{2})^2 + \alpha^2} = \alpha \cosh(\gamma_1 + i\beta_1), \quad \chi_1 = \alpha e^{\gamma_1 + i\beta_1} - a_1, \quad \mu_1 = -\alpha e^{-\gamma_1 - i\beta_1} - a_1.$$

Setting $\alpha_1 = e^{i(\chi_1 - \mu_1)[x_1 + \frac{1}{2}(\chi_1 + \mu_1)t_1] - \gamma_1 - i\beta_1}$, $x_1, t_1 \in \mathbb{R}$, we can obtain that

$$q_i[1] = c_i K e^{i\theta_1},$$

where

$$K = \frac{\cosh(A + 2i\beta_1)\cosh(\gamma_1) - \sin(\beta_1)\sin(B + 2i\gamma_1)}{\cosh(A)\cosh(\gamma_1) + \sin(\beta_1)\sin(B)},$$

and

$$A = 2\alpha \sin(\beta_1)\left[\sinh(\gamma_1)(x - x_1) + ((2\alpha \cos(\beta_1)\sinh^2(\gamma_1) + 1)\right.$$
$$\left. - \sinh(\gamma_1)a_1)(t - t_1)\right],$$
$$B = 2\alpha \cosh(\gamma_1)\left[\cos(\beta_1)(x - x_1) + (\alpha \sinh(\gamma_1)(2\cos^2(\beta_1) - 1)\right.$$
$$\left. - \cos(\beta_1)a_1)(t - t_1)\right].$$

We can readily see that (suppose $\sin(\beta_1)\sinh(\gamma_1) > 0$)

$$q_i[1] \to c_i e^{2i\beta_1 + i\theta_1}, \quad \text{as } x \to +\infty,$$
$$q_i[1] \to c_i e^{-2i\beta_1 + i\theta_1}, \quad \text{as } x \to -\infty.$$

This kinds of solution can be used to obtain the rogue wave solution through limit technique, thus we call it as breather-I to distinguish the above breather-II.

In this subsection, the classification of localized wave solution can be concluded as follows:

- If $a_1 = a_2$ and $c_1 > c_2$, there exist dark-dark soliton and breather-II soliton. Moreover, if $c_2 = 0$, the breather solution degenerates as bright-dark soliton.
- If $a_1 = a_2$ and $c_1 = c_2$, there is no nontrivial solution.
- If $a_1 = a_2$ and $c_1 < c_2$, there exist breather-I solution and rogue wave solution. But there is no dark-dark soliton.

2.2 Case II: When $a_1 \neq a_2$

In this case, performing the similar way as above subsection, we have the matrix decomposition:

$$U_0 M = M D_0, \quad D_0 = \text{diag}(\chi - \lambda, \mu - \lambda, \nu - \lambda), \tag{2.13}$$

where $\text{Im}(\chi) \geqslant \text{Im}(\mu) \geqslant \text{Im}(\nu)$,

$$M = \begin{bmatrix} 1 & 1 & 1 \\ \dfrac{c_1}{\chi + a_1} & \dfrac{c_1}{\mu + a_1} & \dfrac{c_1}{\nu + a_1} \\ \dfrac{c_2}{\chi + a_2} & \dfrac{c_2}{\mu + a_2} & \dfrac{c_2}{\nu + a_2} \end{bmatrix},$$

and χ, μ and ν are three different roots of the following cubic equation

$$\xi^3 + (a_1 + a_2 - 2\lambda)\xi^2 + [a_1 a_2 + c_1^2 - c_2^2 - 2(a_1 + a_2)\lambda]\xi + a_2 c_1^2 - a_1 c_2^2 - 2 a_1 a_2 \lambda = 0. \tag{2.14}$$

The fundamental solution (2.7) is $\Phi = DMN$, where $N = \text{diag}(e^{\mathrm{i}A(\chi)}, e^{\mathrm{i}A(\mu)}, e^{\mathrm{i}A(\nu)})$,

$$A(\chi) = (\chi - \lambda)x + \left[\frac{1}{2}\chi^2 + (c_1^2 - c_2^2 - \lambda^2)\right] t.$$

Through the matrix decomposition (2.13), we have the following equations:

$$F(\chi) \equiv 2\lambda - \chi - \frac{c_1^2}{\chi + a_1} + \frac{c_2^2}{\chi + a_2}, \quad F(\chi) = F(\mu) = F(\nu) = 0. \tag{2.15}$$

Proposition 2.1 For any $\lambda \in \mathbb{C}_+ = \{z | \text{Im}(z) > 0\}$, there exist two roots $\chi, \mu \in \mathbb{C}_+$, one root $\nu \in \mathbb{C}_- = \{z | \text{Im}(z) < 0\}$ for the cubic equation (2.14).

Proof Since matrix $M^\dagger J M$ is congruent with matrix J. By above equations (2.15), we can obtain

$$M^\dagger J M = \begin{bmatrix} \dfrac{2(\bar\lambda - \lambda)}{\bar\chi - \chi} & \dfrac{2(\bar\lambda - \lambda)}{\bar\chi - \mu} & \dfrac{2(\bar\lambda - \lambda)}{\bar\chi - \nu} \\ \dfrac{2(\bar\lambda - \lambda)}{\bar\mu - \chi} & \dfrac{2(\bar\lambda - \lambda)}{\bar\mu - \mu} & \dfrac{2(\bar\lambda - \lambda)}{\bar\mu - \nu} \\ \dfrac{2(\bar\lambda - \lambda)}{\bar\nu - \chi} & \dfrac{2(\bar\lambda - \lambda)}{\bar\nu - \mu} & \dfrac{2(\bar\lambda - \lambda)}{\bar\nu - \nu} \end{bmatrix},$$

which possesses two positive roots and one negative root. We can arrange roots with the order $\mathrm{Im}(\chi) \geqslant \mathrm{Im}(\mu) \geqslant \mathrm{Im}(\nu)$. We merely need to prove that $\mathrm{Im}(\mu) > 0$. It is evident that if $\lambda \neq \bar{\lambda}$, we can not obtain the real root for equation (2.14). So $\mathrm{Im}(\mu) \neq 0$. We prove it by contradiction. Assuming $\mathrm{Im}(\mu) < 0$, we can know that the matrix

$$M_1 = \begin{bmatrix} \dfrac{2(\bar{\lambda}-\lambda)}{\bar{\mu}-\mu} & \dfrac{2(\bar{\lambda}-\lambda)}{\bar{\mu}-\nu} \\ \dfrac{2(\bar{\lambda}-\lambda)}{\bar{\nu}-\mu} & \dfrac{2(\bar{\lambda}-\lambda)}{\bar{\nu}-\nu} \end{bmatrix}$$

is negative definite. Thus this matrix possesses two negative roots. Together with the knowledge of linear algebra, which implies that $M^\dagger JM$ possesses two negative roots and one positive root. A contradiction emerges. Thus we complete the proof. □

Based on above proposition, we take special solution Φ_1 and $v_1(x,t)$ such that

$$|y_1\rangle = D \begin{bmatrix} \varphi_1 \\ c_1\psi_1 \\ c_2\phi_1 \end{bmatrix}, \quad \begin{bmatrix} \varphi_1 \\ \psi_1 \\ \phi_1 \end{bmatrix} = \begin{bmatrix} 1 & 1 \\ 1 & 1 \\ \dfrac{1}{\chi_1+a_1} & \dfrac{1}{\mu_1+a_1} \\ \dfrac{1}{\chi_1+a_2} & \dfrac{1}{\mu_1+a_2} \end{bmatrix} \begin{bmatrix} e^{iA_1} \\ e^{iB_1} \end{bmatrix}, \quad (2.16)$$

where $A_1 = \chi_1[(x-x_1) + \frac{1}{2}\chi_1(t-t_1)]$, $B_1 = \mu_1[(x-x_1) + \frac{1}{2}\mu_1(t-t_1)]$. For simplicity, we take $x_1 = t_1 = 0$. It follows that the breather solutions of mCNLSE (1.1) are

$$q_i[1]$$
$$= c_i \left(\dfrac{\dfrac{\bar{\chi}_1+a_i}{\chi_1+a_i} + \dfrac{\bar{\mu}_1+a_i}{\mu_1+a_i}\dfrac{\bar{\chi}_1-\chi_1}{\bar{\mu}_1-\mu_1}e^{2\mathrm{Re}(C_1)} + \dfrac{\bar{\chi}_1+a_i}{\mu_1+a_i}\dfrac{\bar{\chi}_1-\chi_1}{\bar{\chi}_1-\mu_1}e^{C_1} + \dfrac{\bar{\mu}_1+a_i}{\chi_1+a_i}\dfrac{\bar{\chi}_1-\chi_1}{\bar{\mu}_1-\chi_1}e^{\bar{C}_1}}{1 + \dfrac{\bar{\chi}_1-\chi_1}{\bar{\mu}_1-\mu_1}e^{2\mathrm{Re}(C_1)} + \dfrac{\bar{\chi}_1-\chi_1}{\bar{\chi}_1-\mu_1}e^{C_1} + \dfrac{\bar{\chi}_1-\chi_1}{\bar{\mu}_1-\chi_1}e^{\bar{C}_1}} \right) e^{i\theta_i},$$
$$i = 1, 2, \quad (2.17)$$

where $C_1 = \mathrm{i}(B_1 - A_1)$. To analyze the dynamics for breather solutions (2.17), we rewrite them as

$$q_i[1] = E_i \left(\dfrac{\cosh(A + i\varepsilon_i) + \omega \cos(B + i\epsilon_i)}{\cosh(A) + \omega \cos(B)} \right) e^{i\theta_i}, \quad (2.18)$$

where

$$E_i = c_i \sqrt{\dfrac{(\bar{\chi}_1+a_i)(\bar{\mu}_1+a_i)}{(\chi_1+a_i)(\mu_1+a_i)}}, \quad \omega = \dfrac{\sqrt{|\bar{\chi}_1-\chi_1||\bar{\mu}_1-\mu_1|}}{|\bar{\chi}_1-\mu_1|},$$

$$\varepsilon_i = \arg\left(\dfrac{\chi_1+a_i}{\mu_1+a_i}\right), \quad \epsilon_i = -\ln\left|\dfrac{\chi_1+a_i}{\mu_1+a_i}\right|,$$

$$A = (\mathrm{Im}(\chi_1) - \mathrm{Im}(\mu_1))x + (\mathrm{Re}(\chi_1)\mathrm{Im}(\chi_1) - \mathrm{Re}(\mu_1)\mathrm{Im}(\mu_1))t + \frac{1}{2}\ln\left|\frac{\bar\chi_1 - \chi_1}{\bar\mu_1 - \mu_1}\right|,$$

$$B = (\mathrm{Re}(\mu_1) - \mathrm{Re}(\chi_1))x + \frac{1}{2}(\mathrm{Re}(\mu_1^2) - \mathrm{Re}(\chi_1^2))t + \arg\left(\frac{\bar\chi_1 - \chi_1}{\bar\chi_1 - \mu_1}\right).$$

The center of breather solution $|q_i[1]|^2$ is along the line $A = 0$. To analyze the shape for the breather solution (2.18), we merely need to consider the following function

$$F_i \equiv \left|\frac{\cosh(A+i\varepsilon_i) + \omega\cos(B+i\epsilon_i)}{\cosh(A) + \omega\cos(B)}\right|^2, \quad A \in \mathbb{R},\ B \in [0, 2\pi). \tag{2.19}$$

We can find the central points $(A, B) = (0, \pi)$ is a critical point with its value $F_i(0, \pi) \equiv K_i = \dfrac{(\omega\cosh(\epsilon_i) - \cos(\varepsilon_i))^2}{(1-\omega)^2}$. The shape of breather is determined by the following Hessian matrix

$$H_i = \begin{bmatrix}(F_i)_{AA} & (F_i)_{AB} \\ (F_i)_{BA} & (F_i)_{BB}\end{bmatrix}_{(A,B)=(0,\pi)},$$

where

$(F_i)_{AA}$
$$= 2\frac{(\cosh^2(\epsilon_i) - \cos(\varepsilon_i)\cosh(\epsilon_i))\omega^2 + (1 - \cos(\varepsilon_i)\cosh(\epsilon_i))\omega + \cos^2(\varepsilon_i) - 1}{(\omega - 1)^3},$$

$(F_i)_{AB}$
$$= 2\frac{\omega\sin(\varepsilon_i)\sinh(\epsilon_i)}{(\omega - 1)^2},$$

$(F_i)_{BB}$
$$= 2\frac{\omega\left((\cosh^2(\epsilon_i) - 1)\omega^2 + (1 - \cos(\varepsilon_i)\cosh(\epsilon_i))\omega + \cos^2(\varepsilon_i) - \cos(\varepsilon_i)\cosh(\epsilon_i)\right)}{(\omega - 1)^3}.$$

If H_i is a negative definite matrix, the central point is a local maximum. In this case, this type of breather is called as the bright breather solution. If H_i is a negative definite matrix, the central point is a local minimum. In this case, it is called as the dark breather solution. If H_i is an indefinite matrix, the central point is a saddle point. If $K_i \leqslant 1$, the breather is called four petals breather; if $K_i > 1$, the breather is called two peaks breather.

Rogue wave solution

It is well known that rogue wave solution can be reduced from certain type of breather solution. In what follows, we give the calculations (when $\mu_1 \to \chi_1$) to obtain the rogue wave solution. Taking the limit ($\mu_1 \to \chi_1$) for above equations (2.17 with replacing B_1 with $B_1 + \pi$), we can obtain the rogue wave solution

$$q_i[1] = c_i \left(1 - \frac{2ir_1}{\chi_1 + a_i} \frac{(x + p_1 t)^2 + r_1^2 t^2 + \frac{i}{\chi_1 + a_i}(x + p_1 t - ir_1 t)}{(x + p_1 t + \frac{1}{2r_1})^2 + r_1^2 t^2 + \frac{1}{4r_1^2}}\right) e^{i\theta_i}, \quad i = 1, 2,$$

where $p_1 = \text{Re}(\chi_1)$, $r_1 = \text{Im}(\chi_1)$ and χ_1 is a double root.

Furthermore, we can classify the rogue wave solution as four different types by the dynamic behavior. Since $|q_2[1]|^2$ possesses the similar characteristics with $|q_1[1]|^2$, we merely consider $|q_1[1]|^2$. We first solve the following equation

$$(|q_1[1]|^2)_x = 0, \quad (|q_1[1]|^2)_t = 0.$$

Then we have the stationary point

$$(x, t) = \left(-\frac{1}{2r_1}, 0\right), \quad \left(-\frac{A + (2p_1 + a_1)B_1}{2Ar_1}, \frac{B_1}{2Ar_1}\right),$$
$$\left(-\frac{Ar_1 + (a_1 p_1 + p_1^2 - r_1^2)B_2}{2Ar_1^2}, \frac{(p_1 + a_1)B_2}{2Ar_1}\right),$$

where

$$A = (p_1 + a_1)^2 + r_1^2, \quad B_1 = \pm\sqrt{3(p_1 + a_1)^2 - r_1^2}, \quad B_2 = \pm\sqrt{3r_1^2 - (p_1 + a_1)^2}.$$

So there are four extreme points when $\frac{1}{3}r_1^2 < (p_1 + a_1)^2 < 3r_1^2$, or there are three extreme points. Another standard for classification of rogue wave solution is the value of

$$K = |q_1[1]|^2|_{x=-\frac{1}{2r_1}, t=0} = \left[1 - \frac{4r_1^2}{(p_1 + a_1)^2 + r_1^2}\right]^2.$$

When $K > 1$, the central point is higher than the background; or the central point is lower than the background. Thus we can classify the rogue wave to the following four different types:

- If $\dfrac{(p_1 + a_1)^2}{r_1^2} \geqslant 3$, then the rogue wave is called dark rogue wave.

- If $1 \leqslant \dfrac{(p_1 + a_1)^2}{r_1^2} < 3$, then the rogue wave is four petals rogue wave [19].

- If $\dfrac{1}{3} < \dfrac{(p_1 + a_1)^2}{r_1^2} < 1$, then the rogue wave is called two peaks rogue wave.

- If $\dfrac{(p_1+a_1)^2}{r_1^2} \leqslant \dfrac{1}{3}$, then the rogue wave is called bright rogue wave.

It is pointed that similar properties hold for the $|q_2[1]|^2$.
For instance, choosing parameters

$$c_1 = \frac{32}{7}\sqrt{21}, \quad c_2 = \frac{50}{7}\sqrt{7}, \quad a_1 = -a_2 = 1, \quad p_1 = 7, \quad r_1 = 8.$$

It follows that $|q_1[1]|^2$ is four petals type, and $\dfrac{1}{3} < \dfrac{(p_1-a_1)^2}{r_1^2} = \dfrac{9}{16} < 1$. It follows that $|q_2[1]|^2$ is two-peaks type. We can plot the above rogue wave solution by soft Maple (Fig. 2).

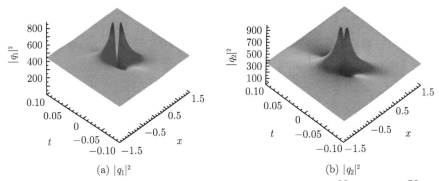

(a) $|q_1|^2$ (b) $|q_2|^2$

Fig.2 Rogue wave solution: Parameters $a_1 = -a_2 = 1$, $c_1 = \dfrac{32}{7}\sqrt{21}$, $c_2 = \dfrac{50}{7}\sqrt{7}$, $\lambda_1 = \dfrac{13}{2} + \dfrac{32}{7}i$, $\chi_1 = 7 + 8i$. It is seen that the solution $|q_1|^2$ possesses four-petals structure and the solution $|q_2|^2$ possesses two-peaks structure

Finally, we consider the method to determine the multiple roots. We set

$$a_1 = \alpha + \beta, \quad a_2 = \alpha - \beta. \tag{2.20}$$

It follows that the matrix $U_0 - \xi$ can be represented as

$$\begin{bmatrix} 2\kappa - \zeta & -c_1 & c_2 \\ c_1 & -\zeta - \beta & 0 \\ c_2 & 0 & -\zeta + \beta \end{bmatrix}, \tag{2.21}$$

where $\zeta = \xi + \alpha$, $\kappa = \lambda_1 + \dfrac{1}{2}\alpha$. Then the determinant of above matrix (2.21) is

$$F = \zeta^3 - 2\kappa\zeta^2 + [c_1^2 - c_2^2 - \beta^2]\zeta + 2\beta^2\kappa - \beta(c_1^2 + c_2^2) = 0. \tag{2.22}$$

The discriminant of above equation (2.22) is

$$D(F) = 64\beta^2 \left(\kappa^4 - \frac{c_1^2 + c_2^2}{2\beta}\kappa^3 + A_2\kappa^2 + A_1\kappa + A_0 \right), \tag{2.23}$$

where

$$A_2 = \left(-\frac{1}{2}\beta^2 + \frac{5}{4}(c_2^2 - c_1^2) + \frac{1}{16}(c_1^2 - c_2^2)^2\beta^{-2} \right),$$

$$A_1 = \frac{9}{16}(c_1^2 + c_2^2)(2\beta + (c_1^2 - c_2^2)\beta^{-1}),$$

$$A_0 = \frac{1}{16}(\beta^2 + c_2^2 - c_1^2)^3\beta^{-2} - \frac{27}{64}(c_1^2 + c_2^2)^2.$$

Proposition 2.2 *The quartic equation $D(F) = 0$ never possesses two pairs of conjugate complex roots.*

Proof We need to analyze the solution of $D(F) = 0$. The discriminant of $D(F) = 0$ is

$$E = -\frac{c_1 c_2}{2^{14}\beta^{10}} \left[(4\beta^2 + c_2^2 - c_1^2)^3 + 27c_1^2 c_2^2(4\beta^2) \right].$$

If $E < 0$, then $D(F) = 0$ possesses two real roots and a pair of complex conjugate root. If $E > 0$, then then $D(F) = 0$ possesses four real roots or two pairs of complex conjugate root. Since $E > 0$, then we can deduce that $c_1 > c_2$ and the solution of $E = 0$ is $\beta^2 = \frac{1}{4}(c_1^{2/3} - c_2^{2/3})$. In what follows, we illustrate the equation $D(F) = 0$ never possesses two pairs of complex conjugate roots. If quartic equation $D(F) = 0$ has two pairs of complex conjugate roots, then we have

$$G(\beta^2) = 16A_0\beta^2 = (\beta^2 + c_2^2 - c_1^2)^3 - \frac{27}{4}\beta^2(c_1^2 + c_2^2)^2$$

$$= \beta^6 + 3(c_2^2 - c_1^2)\beta^4 - 3\left(\frac{5}{2}c_2^2 + \frac{1}{2}c_1^2\right)\left(\frac{5}{2}c_1^2 + \frac{1}{2}c_2^2\right)\beta^2 + (c_2^2 - c_1^2)^3 > 0.$$

On the other hand, we have the discriminant of equation $G(\beta^2) = 0$ is

$$\Delta = \frac{3^9}{4}c_1^2 c_2^2(c_2^2 + c_1^2)^4 > 0,$$

then the equation $G(\beta^2) = 0$ possesses three different real roots β_1^2, β_2^2 and β_3^2. By the Vieta formula, we have

$$\beta_1^2 + \beta_2^2 + \beta_3^2 = 3(c_1^2 - c_2^2) > 0,$$

$$\beta_1^2\beta_2^2 + \beta_2^2\beta_3^2 + \beta_3^2\beta_1^2 = -3\left(\frac{5}{2}c_2^2 + \frac{1}{2}c_1^2\right)\left(\frac{5}{2}c_1^2 + \frac{1}{2}c_2^2\right) < 0,$$

$$\beta_1^2\beta_2^2\beta_3^2 = (c_1^2 - c_2^2)^3 > 0,$$

it follows that the above equation possesses one positive root and two negative roots. Since $\beta_i^2 > 0$, then there is merely one positive root. The following inequality

$$G\left(\frac{1}{4}(c_1^{2/3} - c_2^{2/3})^3\right) = -\frac{27}{16}(c_1^{2/3} - c_2^{2/3})^3 \left(\frac{1}{4}(c_1^{2/3} + c_2^{2/3})^6 + (c_1^2 + c_2^2)^2\right) < 0,$$

illustrate that when $E > 0$, then $G(\beta^2) < 0$. This yields a contradiction. This completes the proof. $\square$

Indeed, the quartic equation $D(F) = 0$ is condition of multiple roots for the cubic equation (2.22). To obtain the rogue wave solution, another condition is the spectral parameters must be nonreal. So we can obtain the existence condition of rogue wave is $E < 0$ i.e. $a_1 \neq a_2$ and $\beta^2 > \frac{1}{4}(c_1^{2/3} - c_2^{2/3})^3$, through above proposition.

For the focusing CNLSE, there is a triple root for the characteristic equation (2.14). However, for mixed case or the defocusing case, there is no triple root for the nonreal spectral parameters. So there is no type-II rogue wave [17]. This fact can be verified by the following elementary fact.

Proposition 2.3 *For any $\kappa \neq \bar{\kappa}$, the characteristic equation (2.14) has no triple root.*

Proof If characteristic equation (2.14) possesses a triple root, then we have

$$\zeta^3 - 2\kappa\zeta^2 + [c_1^2 - c_2^2 - \beta^2]\zeta + 2\beta^2\kappa - \beta(c_1^2 + c_2^2) = \left(\zeta - \frac{2\kappa}{3}\right)^3.$$

It follows that

$$c_1^2 - c_2^2 - \beta^2 - \frac{4}{3}\kappa^2 = 0, \quad -c_2^2\beta - c_1^2\beta + 2\kappa\beta^2 + \frac{8}{27}\kappa^3 = 0.$$

Moreover, we have

$$c_1^2 = \frac{\left(\frac{2}{3}\kappa + \beta\right)^3}{2\beta} > 0, \quad \kappa \neq \bar{\kappa}, \quad c_2^2 = \frac{\left(\frac{2}{3}\kappa - \beta\right)^3}{2\beta} > 0.$$

If $\beta > 0$, we have $\frac{2}{3}\kappa \pm \beta \in \omega\mathbb{R}^+$ or $\omega^2\mathbb{R}^+$. If $\beta < 0$, we have $\frac{2}{3}\kappa \pm \beta \in \omega\mathbb{R}^-$ or $\omega^2\mathbb{R}^-$. Indeed, this is no possible. Thus there is no triple root. $\square$

Since the type-II rogue wave solution is obtained by triple roots, there exists not type-II rogue wave solution for the mCNLSE (1.1) by above theorem.

Homoclinical orbits solution

The homoclinic orbit solution is a kind of space periodical and time exponential decay solution, or called the Akhmediev breather. When time tends to $\pm\infty$, it tends to a plane wave solution with different phase. From the solution expression of (2.17), the homoclinic orbit solution can be obtained through choosing parameters $\text{Im}(\chi_1) = \text{Im}(\mu_1)$.

In the following, we present a way to look for the homoclinic orbit solution for mCNLSE (1.1). Then, it is necessary to analyze the following characteristic equation for matrix (2.21)

$$\zeta^3 - 2\kappa\zeta^2 + (c_1^2 - c_2^2 - \beta^2)\zeta + 2\kappa\beta^2 - (c_1^2 + c_2^2)\beta = 0, \qquad (2.24)$$

where $\kappa = \lambda_1 + \frac{1}{2}\alpha$, $\zeta = \mu_1 + \alpha$, α and β are given in equations (2.20). Suppose another root is $\chi_1 = \mu_1 + \delta$, $\delta \in \mathbb{R}$, then we have

$$3\delta\zeta^2 + (3\delta^2 - 4\delta\kappa)\zeta + \delta^3 - 2\kappa\delta + \delta(c_1^2 - c_2^2 - \beta^2) = 0.$$

It follows that

$$\zeta = -\frac{\delta}{2} + \frac{2\kappa}{3} \pm \frac{1}{6}\sqrt{16\kappa^2 - 3\delta^2 + 12\beta^2 + 12(c_2^2 - c_1^2)},$$

and it also satisfies the following equation

$$(16\kappa^2 - 3\delta^2 - 12c_1^2 + 12c_2^2 + 12\beta^2)\left[3(c_1^2 - c_2^2) + 3(\delta^2 - \beta^2) - 4\kappa^2\right]^2$$
$$= \left[16\kappa^3 + 18(c_2^2 - c_1^2 - 2\beta^2)\kappa + 27\beta(c_1^2 + c_2^2)\right]^2. \qquad (2.25)$$

If we obtain a pair of conjugate complex roots, exact values of δ, β, c_1 and c_2 are substituted into above equation (2.25). Then we can obtain the holoclinic orbit solution through substituting the above mentioned parameters into solutions (2.17).

Then, we give an explicit example to illustrate the method. Suppose $\alpha = 0$, $\beta = 1$, $\delta = 1$, $c_1 = 1$ and $c_2 = 2$, substituting these parameters into (2.25) and solve it about κ, we can obtain that $\kappa = \lambda_1 \approx .633263953 + 1.812212393i$. And then substituting above parameters into characteristic equation (2.24), we have

$$\chi_1 = 1.625455953 + 1.933383832i,$$

$$\mu_1 = 0.625455953 + 1.933383832i.$$

We give the explicit figure for homoclinic orbit solution by choosing special parameters (Fig. 3).

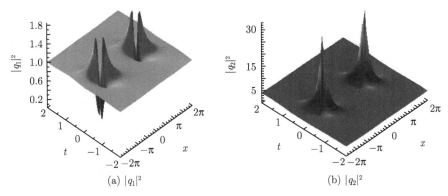

Fig.3 Homoclinic orbit solution: Parameters $a_1 = -a_2 = 1$, $c_1 = 1$, $c_2 = 2$, $\lambda_1 = 0.633263953 + 1.812212393i$, $\chi_1 = 1.625455953 + 1.933383832i$, $\mu_1 = .6254559529 + 1.933383832i$. It is seen that the solution $|q_1|^2$ and $|q_2|^2$ possess the periodical behavior in space

Dark-dark soliton solution

If $\lambda_1 \in \mathbb{R}$, we can obtain the dark-dark soliton solution. Through **proposition** 2.2, the quartic equation (2.23) admits at least two real roots. So there always exists λ_1 such that $D(F) < 0$, it follows that there is a pair of complex roots for cubic equation (2.22). Thus the dark-dark solitons always exist. The details for the dark-dark soliton through DT refer to reference [39]. Similar as above, we choose the special solution Φ_1 and $v_1(x,t)$ such that

$$|y_1\rangle = D \begin{bmatrix} 1 & 1 \\ \dfrac{c_1}{\chi_1 + a_1} & \dfrac{c_1}{\bar{\chi}_1 + a_1} \\ \dfrac{c_2}{\chi_1 + a_2} & \dfrac{c_2}{\bar{\chi}_1 + a_2} \end{bmatrix} \begin{bmatrix} e^{i\chi_1(x+\frac{1}{2}\chi_1 t)} \\ \alpha_1(\bar{\lambda}_1 - \lambda_1)e^{i\bar{\chi}_1(x+\frac{1}{2}\bar{\chi}_1 t)} \end{bmatrix}, \quad (2.26)$$

where $\text{Im}(\chi_1) > 0$. Substituting above special solution into (2.5), we have the dark-dark soliton solution

$$q_i[1] = \frac{c_i}{2}\left[1 + \frac{\bar{\chi}_1 + a_i}{\chi_1 + a_i} + \left(\frac{\chi_1 - \bar{\chi}_1}{\chi_1 + a_i}\right)\tanh(A_1)\right]e^{i\theta_i}, \quad i = 1, 2,$$

where $A_1 = \text{Im}(\chi_1)[x + \text{Re}(\chi_1)t] + \frac{1}{2}\ln\beta_1$ and

$$\beta_1 = -\text{Im}(\chi_1)\text{Im}\left[\alpha_1\left(1 - \frac{c_1^2}{(\bar{\chi}_1 + a_1)^2} + \frac{c_2^2}{(\bar{\chi}_1 + a_2)^2}\right)\right] > 0.$$

In summary, the localized wave solution in this subsection can be concluded as follows:

- If $a_1 \neq a_2$ and $\beta^2 > \frac{1}{4}(c_1^{2/3} - c_2^{2/3})^3$, there exist rogue wave, breather-I solution and dark-dark soliton solution.

- If $a_1 \neq a_2$ and $\beta^2 \leqslant \frac{1}{4}(c_1^{2/3} - c_2^{2/3})^3$, there exist breather-II solution and dark-dark soliton solution.

3 General localized wave and their interactions

The general Darboux matrix for above linear system (2.1) is given in reference [39]. We can summarize as the following theorem:

Theorem 3.1 *Suppose we have N different vector solutions Φ_i for linear system (2.1) with $\lambda = \lambda_i$ ($i = 1, 2, \cdots, N$), denote $|y_i\rangle = v_i\Phi_i$, where v_i is an appropriate function of x and t, then the N-fold DT is*

$$T_N = I - YM^{-1}(\lambda - S)^{-1}Y^\dagger J,$$

where

$$Y = [|y_1\rangle, |y_2\rangle, \cdots, |y_N\rangle],$$
$$S = \text{diag}(\lambda_1^*, \lambda_2^*, \cdots, \lambda_N^*),$$
$$M = \left(\frac{\langle y_i|J|y_j\rangle}{\lambda_j - \bar{\lambda}_i}\right)_{N\times N},$$

and $\langle y_i| = |y_i\rangle^\dagger$, the element $\langle y_i|J|y_i\rangle/(\lambda_i - \bar{\lambda}_i)$ is considered as the meaning of limit $\lambda_i \to \bar{\lambda}_i$. The transformation between potential functions is

$$Q[N] = Q + [\sigma_3, P], \quad P = YM^{-1}Y^\dagger J.$$

In what follows, we consider an explicit application of the above theorem. We choose the seed solution as the plane wave solution (2.6), then there are three kinds of special solutions for $|y_i\rangle$ (equation (2.11), (2.16) and (2.26)), which can be used to construct different types of localized wave solution. Denote

$$|y_i\rangle = \begin{bmatrix} \varphi_i \\ c_1 e^{i\theta_1}\psi_i \\ c_2 e^{i\theta_2}\phi_i \end{bmatrix}.$$

Through above theorem, we can obtain that the general localized wave solution formula for mCNLS (1.1) on the nonzero background:

$$q_1[N] = c_1 e^{i\theta_1}\left(1 - 2\psi M^{-1}\varphi^\dagger\right),$$
$$q_2[N] = c_2 e^{i\theta_2}\left(1 - 2\phi M^{-1}\varphi^\dagger\right),$$

where

$$\varphi = [\varphi_1, \varphi_2, \cdots, \varphi_N],$$
$$\psi = [\psi_1, \psi_2, \cdots, \psi_N],$$
$$\phi = [\phi_1, \phi_2, \cdots, \phi_N].$$

Furthermore, by simple linear algebra formula, setting $\widehat{M} = -\frac{1}{2}M$, we have determinant representation for above general localized wave solution formula:

$$q_1[N] = c_1 \left(\frac{\det(\widehat{M}+X_1)}{\det(\widehat{M})}\right) e^{i\theta_1}, \quad X_1 = \varphi^\dagger \psi,$$

$$q_2[N] = c_2 \left(\frac{\det(\widehat{M}+X_2)}{\det(\widehat{M})}\right) e^{i\theta_2}, \quad X_2 = \varphi^\dagger \phi.$$

In what follows, we prove that the above general localized wave solutions are nonsingular. We can establish the following theorem:

Theorem 3.2 (Dubrovin et.al [45]) *The matrix $\widetilde{M} = i\widehat{M}$ is negative definite.*

Proof We merely give the case when $a_1 \neq a_2$, since the case $a_1 = a_2$ is similar. We prove this fact by analyzing the elements of matrix $\widetilde{M}$. The elements $|y_i\rangle$ possesses two different choices (here we do not consider the limit case).

- $|y_i\rangle = D \begin{bmatrix} 1 & 1 \\ \frac{c_1}{\chi_i + a_1} & \frac{c_1}{\mu_i + a_1} \\ \frac{c_2}{\chi_i + a_2} & \frac{c_2}{\mu_i + a_2} \end{bmatrix} \begin{bmatrix} e^{i\chi_i[x+x_i+\frac{1}{2}\chi_i(t+t_i)]} \\ e^{i\mu_i[x+x_i+\frac{1}{2}\mu_i(t+t_i)]} \end{bmatrix}, \quad i=1,2,\cdots,N_1, \operatorname{Im}(\chi_i) > 0, \operatorname{Im}(\mu_i) > 0, x_i, t_i \in \mathbb{R},$

- $|y_j\rangle = D \begin{bmatrix} 1 & 1 \\ \dfrac{c_1}{\chi_j+a_1} & \dfrac{c_1}{\bar{\chi}_j+a_1} \\ \dfrac{c_2}{\chi_j+a_2} & \dfrac{c_2}{\bar{\chi}_j+a_2} \end{bmatrix} \begin{bmatrix} e^{i\chi_j(x+\frac{1}{2}\chi_j t)} \\ \alpha_i(\bar{\lambda}_j-\lambda_j)e^{i\bar{\chi}_j(x+\frac{1}{2}\bar{\chi}_j t)} \end{bmatrix}, \quad j = N_1+1, N_1+2,\cdots, N_1+N_2,$ $N_1+N_2 = N$, $\mathrm{Im}(\chi_j) > 0$, $\lambda_j \in \mathbb{R}$,

$$\beta_j = -\mathrm{Im}(\chi_j)\mathrm{Im}\left[\alpha_j\left(1 - \frac{c_1^2}{(\bar{\chi}_j+a_1)^2} + \frac{c_2^2}{(\bar{\chi}_j+a_2)^2}\right)\right] > 0.$$

Then we have the following results: If $i \neq j$, we have (here we merely consider the case $1 \leqslant i \leqslant N_1$, $N_1+1 \leqslant j \leqslant N$, the other case can be obtained with a parallel way)

$$\frac{\langle y_i|J|y_j\rangle}{2\mathrm{i}(\bar{\lambda}_i-\lambda_j)}$$

$$= \left[e^{-\mathrm{i}\bar{\chi}_i[x+x_i+\frac{1}{2}\bar{\chi}_i(t+t_i)]},\ e^{-\mathrm{i}\bar{\mu}_i[x+x_i+\frac{1}{2}\bar{\mu}_i(t+t_i)]}\right] \begin{bmatrix} \dfrac{1}{\mathrm{i}(\chi_j-\bar{\chi}_i)} \\ \dfrac{1}{\mathrm{i}(\mu_j-\bar{\mu}_i)} \end{bmatrix} e^{\mathrm{i}\chi_j(x+\frac{1}{2}\chi_j t)}$$

$$= -\int_x^{+\infty}\left(e^{-\mathrm{i}\bar{\chi}_i[s+x_i+\frac{1}{2}\bar{\chi}_i(t+t_i)]+\mathrm{i}\chi_j(s+\frac{1}{2}\chi_j t)} + e^{-\mathrm{i}\bar{\mu}_i[s+x_i+\frac{1}{2}\bar{\mu}_i(t+t_i)]+\mathrm{i}\chi_j(s+\frac{1}{2}\chi_j t)}\right)ds.$$

Similarly, we have

$$\frac{\langle y_i|J|y_i\rangle}{2\mathrm{i}(\bar{\lambda}_i-\lambda_i)} = -\int_x^{+\infty}\left|e^{\mathrm{i}\chi_i[s+x_i+\frac{1}{2}\chi_i(t+t_i)]} + e^{\mathrm{i}\mu_i[s+x_i+\frac{1}{2}\mu_i(t+t_i)]}\right|^2 ds,$$

$$\frac{\langle y_j|J|y_j\rangle}{2\mathrm{i}(\bar{\lambda}_j-\lambda_j)} = -\int_x^{+\infty}\left|e^{\mathrm{i}\chi_j(s+\frac{1}{2}\chi_j t)}\right|^2 ds - \frac{\beta_j}{2\mathrm{Im}(\chi_j)}.$$

It follows that, for any nonzero vector $v = (v_1, v_2, \cdots, v_N)^\mathrm{T}$, we have

$$v^\dagger M v = -\int_x^{+\infty}\left|\sum_{i=1}^{N_1} v_i\left(e^{\mathrm{i}\chi_i[s+x_i+\frac{1}{2}\chi_i(t+t_i)]} + e^{\mathrm{i}\mu_i[s+x_i+\frac{1}{2}\mu_i(t+t_i)]}\right)\right.$$
$$\left. + \sum_{j=N_1+1}^N v_j\left(e^{\mathrm{i}\chi_j(s+\frac{1}{2}\chi_j t)}\right)\right|^2 ds - \left|\sum_{j=N_1+1}^N \frac{v_j\beta_j}{2\mathrm{Im}(\chi_j)}\right|^2 < 0.$$

Thus we complete the proof. □

Remark 3.1 *The theorem 3.2 was first derived in ref. [45] by the general finite gap method. Here we prove it by the elementary method-DT.*

3.1 Interaction between localized waves

In this subsection, we consider the cases that different types of localized waves coexist and interact with each other. Firstly, we consider the case $a_1 \neq a_2$ and $c_1 c_2 \neq 0$.

Taking

$$\begin{bmatrix} \varphi_1 \\ \psi_1 \\ \phi_1 \end{bmatrix} = \begin{bmatrix} 1 & 1 \\ \dfrac{1}{\chi_1 + a_1} & \dfrac{1}{\mu_1 + a_1} \\ \dfrac{1}{\chi_1 + a_2} & \dfrac{1}{\mu_1 + a_2} \end{bmatrix} \begin{bmatrix} e^{i\chi_1[(x+x_1)+\frac{1}{2}\chi_1(t+t_1)]} \\ e^{i\mu_1[(x+x_1)+\frac{1}{2}\mu_1(t+t_1)]} \end{bmatrix},$$

and

$$\begin{bmatrix} \varphi_2 \\ \psi_2 \\ \phi_2 \end{bmatrix} = \begin{bmatrix} 1 & 1 \\ \dfrac{1}{\chi_2 + a_1} & \dfrac{1}{\bar{\chi}_2 + a_1} \\ \dfrac{1}{\chi_2 + a_2} & \dfrac{1}{\bar{\chi}_2 + a_2} \end{bmatrix} \begin{bmatrix} e^{i\chi_2[x+\frac{1}{2}\chi_2 t]} \\ \alpha(\bar{\lambda}_2 - \lambda_2) e^{i\bar{\chi}_2[x+\frac{1}{2}\bar{\chi}_2 t]} \end{bmatrix}, \quad (3.1)$$

where $\beta = -\text{Im}(\chi_2)\text{Im}\left[\alpha\left(1 - \dfrac{c_1^2}{(\bar{\chi}_2 + a_1)^2} + \dfrac{c_2^2}{(\bar{\chi}_2 + a_2)^2}\right)\right] > 0$, x_1 and t_1 are real constants (for simplicity, we set $x_1 = t_1 = 0$), we can obtain the general formulas between breather solution and dark-dark soliton solution:

$$q_i[2] = c_i \left(\frac{M_i}{M}\right) e^{i\theta_i}, \quad i = 1, 2 \quad (3.2)$$

where

$$M = \frac{(e^{C_2 + \bar{C}_2} + \beta)}{\bar{\chi}_2 - \chi_2}\left(\frac{1}{\bar{\chi}_1 - \chi_1} + \frac{e^{2\text{Re}(C_1)}}{\bar{\mu}_1 - \mu_1} + \frac{e^{C_1}}{\bar{\chi}_1 - \mu_1} + \frac{e^{\bar{C}_1}}{\bar{\mu}_1 - \chi_1}\right)$$

$$- e^{C_2 + \bar{C}_2}\left(\frac{1}{\bar{\chi}_1 - \chi_2} + \frac{e^{\bar{C}_1}}{\bar{\mu}_1 - \chi_2}\right)\left(\frac{1}{\bar{\chi}_2 - \chi_1} + \frac{e^{C_1}}{\bar{\chi}_2 - \mu_1}\right),$$

$$M_i = \frac{\left(\dfrac{\bar{\chi}_2 + a_i}{\chi_2 + a_i} e^{C_2 + \bar{C}_2} + \beta\right)}{\bar{\chi}_2 - \chi_2}\left(\frac{\bar{\chi}_1 + a_i}{\chi_1 + a_i}\frac{1}{\bar{\chi}_1 - \chi_1} + \frac{\bar{\mu}_1 + a_i}{\mu_1 + a_i}\frac{e^{2\text{Re}(C_1)}}{\bar{\mu}_1 - \mu_1}\right.$$
$$\left. + \frac{\bar{\chi}_1 + a_i}{\mu_1 + a_i}\frac{e^{C_1}}{\bar{\chi}_1 - \mu_1} + \frac{\bar{\mu}_1 + a_i}{\chi_1 + a_i}\frac{e^{\bar{C}_1}}{\bar{\mu}_1 - \chi_1}\right)$$

$$- e^{C_2 + \bar{C}_2}\left(\frac{\bar{\chi}_1 + a_i}{\chi_2 + a_i}\frac{1}{\bar{\chi}_1 - \chi_2} + \frac{\bar{\mu}_1 + a_i}{\chi_2 + a_i}\frac{e^{\bar{C}_1}}{\bar{\mu}_1 - \chi_2}\right)\left(\frac{\bar{\chi}_2 + a_i}{\chi_1 + a_i}\frac{1}{\bar{\chi}_2 - \chi_1}\right.$$
$$\left. + \frac{\bar{\chi}_2 + a_i}{\mu_1 + a_i}\frac{e^{C_1}}{\bar{\chi}_2 - \mu_1}\right)$$

and
$$C_1 = i(\mu_1 - \chi_1)[x + \frac{1}{2}(\mu_1 + \chi_1)t], \quad C_2 = i\chi_2(x + \frac{1}{2}\chi_2 t).$$

3.1.1 Interaction between dark-dark soliton and breather

From above section, we know that $\text{Im}(\chi_1) \geqslant \text{Im}(\mu_1) > 0$ and $\text{Im}(\chi_2) > 0$. And the velocity of dark-dark soliton is $v_d = -\text{Re}(\chi_2)$, the velocity of breather solution is

$$v_b = \frac{\text{Im}(\mu_1)\text{Re}(\mu_1) - \text{Im}(\chi_1)\text{Re}(\chi_1)}{\text{Im}(\chi_1) - \text{Im}(\mu_1)}, \quad \text{if } \text{Im}(\chi_1) > \text{Im}(\mu_1).$$

We do not consider they possess the same velocity $v_d = v_b$. Without the generality, we can assume $v_b < v_d$. To analyze the interaction between dark-dark soliton and breather, we use the standard method of asymptotic analysis.

After a tedious and elementary calculation, we can obtain the following asymptotical analysis.

(a) Before the interaction($t \to -\infty$), we have

$$q_i[2] \to \frac{c_i}{2}\left(\frac{\bar{\chi}_1 + a_i}{\chi_1 + a_i}\right)\left[1 + \frac{\bar{\chi}_2 + a_i}{\chi_2 + a_i} + \left(\frac{\chi_2 - \bar{\chi}_2}{\chi_2 + a_i}\right)\tanh(C_-)\right]e^{i\theta_i} - c_i e^{i(\theta_i + \varphi_{i,-})}$$

$$+ E_{-,i}\left(\frac{\cosh\left(A_- + i\arg\left(\frac{\bar{\chi}_1 + a_i}{\mu_1 + a_i}\right)\right) + \omega\cos\left(B_- - i\ln\left|\frac{\bar{\chi}_1 + a_i}{\mu_1 + a_i}\right|\right)}{\cosh(A_-) + \omega\cos(B_-)}\right)e^{i\theta_i},$$

where

$$E_{-,i} = c_i\sqrt{\frac{(\bar{\chi}_1 + a_i)(\bar{\mu}_1 + a_i)}{(\chi_1 + a_i)(\mu_1 + a_i)}}, \quad \omega = \frac{\sqrt{|\bar{\chi}_1 - \chi_1||\bar{\mu}_1 - \mu_1|}}{|\bar{\chi}_1 - \mu_1|},$$

$$A_- = (\text{Im}(\chi_1) - \text{Im}(\mu_1))x + (\text{Re}(\chi_1)\text{Im}(\chi_1) - \text{Re}(\mu_1)\text{Im}(\mu_1))t + \frac{1}{2}\ln\left|\frac{\bar{\chi}_1 - \chi_1}{\bar{\mu}_1 - \mu_1}\right|,$$

$$B_- = (\text{Re}(\mu_1) - \text{Re}(\chi_1))x + \frac{1}{2}(\text{Re}(\mu_1^2) - \text{Re}(\chi_1^2))t + \arg\left(\frac{\bar{\chi}_1 - \chi_1}{\bar{\chi}_1 - \mu_1}\right),$$

$$C_- = \text{Im}(\chi_2)\left[x + \text{Re}(\chi_2)t + \frac{1}{Im(\chi_2)}\left(\ln\left|\frac{\chi_1 - \bar{\chi}_2}{\chi_1 - \chi_2}\right| + \frac{1}{2}\ln\beta\right)\right],$$

$$i\varphi_{i,-} = \ln\left(\frac{\bar{\chi}_1 + a_i}{\chi_1 + a_i}\right).$$

(b) After the interaction ($t \to +\infty$), it follows that

$$q_i[2] \to \frac{c_i}{2}\left(\frac{\bar{\mu}_1 + a_i}{\mu_1 + a_i}\right)\left[1 + \frac{\bar{\chi}_2 + a_i}{\chi_2 + a_i} + \left(\frac{\chi_2 - \bar{\chi}_2}{\chi_2 + a_i}\right)\tanh(C_+)\right]e^{i\theta_i} - c_i e^{i(\theta_i + \varphi_{i,+})}$$

$$+ E_{+,i} \left(\frac{\cosh\left(A_+ + i\arg\left(\frac{\chi_1 + a_i}{\mu_1 + a_i}\right)\right) + \omega\cos\left(B_+ - i\ln\left|\frac{\chi_1 + a_i}{\mu_1 + a_i}\right|\right)}{\cosh(A_+) + \omega\cos(B_+)} \right) e^{i\theta_i},$$

where

$$E_{+,i} = \left(\frac{\bar{\chi}_2 + a_i}{\chi_2 + a_i}\right) E_{i,-}, \quad A_+ = A_- + \ln\left|\frac{\chi_1 - \bar{\chi}_2}{\chi_1 - \chi_2}\frac{\mu_1 - \chi_2}{\mu_1 - \bar{\chi}_2}\right|,$$

$$B_+ = B_- + \arg\left(\frac{\bar{\chi}_1 - \bar{\chi}_2}{\bar{\chi}_1 - \chi_2}\frac{\chi_2 - \mu_1}{\bar{\chi}_2 - \mu_1}\right), \quad C_+ = C_- + \ln\left|\frac{\chi_1 - \bar{\chi}_2}{\chi_1 - \bar{\chi}_2}\frac{\mu_1 - \bar{\chi}_2}{\mu_1 - \chi_2}\right|,$$

$$i\varphi_{i,+} = \ln\left(\frac{\bar{\mu}_1 + a_i}{\mu_1 + a_i}\frac{\bar{\chi}_2 + a_i}{\chi_2 + a_i}\right).$$

To demonstrate the dynamic behavior of this type of solution explicitly, we plot some figures with setting related parameters by Maple. In Fig 4, we show the interaction between one breather-II type solution and dark-dark solution. The above asymptotical analysis is valid for this case. In Fig 5, we show the interaction between one homoclinic orbit solution and dark-dark solution. Since homoclinic orbit solution is temporal periodical solution, the asymptotical analysis in the above is not valid for this case. But we could establish a similar analysis on them.

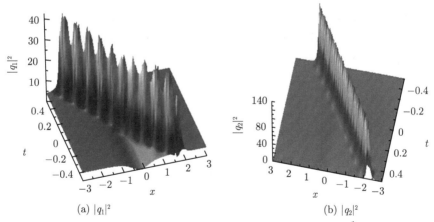

Fig.4 Dark-dark-breather-II type solution. Parameters: $a_1 = -a_2 = \frac{1}{20}$, $c_1 = 2$, $c_2 = 1$, $\lambda_1 = 2 + 5i$, $\lambda_2 = 0$, $\chi_1 = 3.901220880 + 10.25418804i$, $\mu_1 = 0.07993030051 + 0.01536696563i$, $\chi_2 = -0.0416053126 + 1.7328280i$, $\beta = 1$. It is seen that there is breather-II type solution collision with dark-dark soliton, and their interaction is elasticity

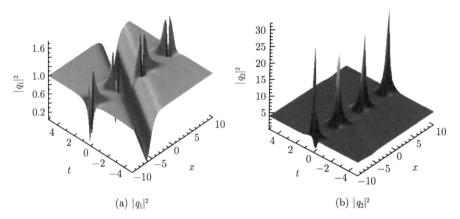

(a) $|q_1|^2$　　　　　　　　　　　(b) $|q_2|^2$

Fig.5　Dark-dark-breather-I type solution. Parameters $a_1 = -a_2 = 1$, $c_1 = 1$, $c_2 = 2$, $\lambda_1 = .633263953301716 + 1.81221239300044\mathrm{i}$, $\lambda_2 = 0$, $\chi_1 = 1.625455953 + 1.933383832\mathrm{i}$, $\mu_1 = .6254559529 + 1.933383832\mathrm{i}$, $\chi_2 = -1.228339172 + 0.7255696805\mathrm{i}$, $\beta = 1$. It is seen that there is homoclinic orbit solution collision with dark-dark soliton, and their interaction is elasticity too

3.1.2　Dark-dark-rogue solution

In this paragraph, we analyze the interaction between dark-dark soliton and rogue wave solution. To derive the dark-dark-rogue solution, we use the limit technique $\mu_1 \to \chi_1$. Denote $\epsilon = \mu_1 - \chi_1$, we can have the following expression

$$\frac{1}{\bar{\chi}_1 - \chi_1 - \epsilon} = \frac{1}{\bar{\chi}_1 - \chi_1} + \frac{\epsilon}{(\bar{\chi}_1 - \chi_1)^2} + o(\epsilon^2),$$

$$\frac{1}{\bar{\epsilon} + \bar{\chi}_1 - \chi_1} = \frac{1}{\bar{\chi}_1 - \chi_1} - \frac{\bar{\epsilon}}{(\bar{\chi}_1 - \chi_1)^2} + o(\bar{\epsilon}^2),$$

$$\frac{1}{\bar{\chi}_1 - \chi_1 - \epsilon + \bar{\epsilon}} = \frac{1}{\bar{\chi}_1 - \chi_1} + \frac{\epsilon - \bar{\epsilon}}{(\bar{\chi}_1 - \chi_1)^2} - \frac{2\epsilon\bar{\epsilon}}{(\bar{\chi}_1 - \chi_1)^3} + o(\epsilon^2, \bar{\epsilon}^2),$$

$$\frac{1}{\bar{\epsilon} + \bar{\chi}_1 - \chi_2} = \frac{1}{\bar{\chi}_1 - \chi_2} - \frac{\bar{\epsilon}}{(\bar{\chi}_1 - \chi_2)^2} + o(\bar{\epsilon}^2),$$

$$\frac{1}{\bar{\chi}_2 - \chi_1 - \epsilon} = \frac{1}{\bar{\chi}_2 - \mu_1} + \frac{\epsilon}{(\bar{\chi}_2 - \mu_1)^2} + o(\epsilon^2),$$

$$\frac{1}{\epsilon + \chi_1 + a_1} = \frac{1}{\chi_1 + a_1} - \frac{\epsilon}{(\chi_1 + a_1)^2} + o(\epsilon^2),$$

and

$$e^{\mathrm{i}\epsilon(x+(\chi_1+\frac{1}{2}\epsilon)t)} = 1 + \mathrm{i}\epsilon(x + \chi_1 t) + o(\epsilon^2),$$

$$e^{-i\bar{\epsilon}(x+(\bar{\chi}_1+\frac{1}{2}\bar{\epsilon})t)} = 1 - i\bar{\epsilon}(x+\bar{\chi}_1 t) + o(\bar{\epsilon}^2),$$

$$e^{i\epsilon(x+(\chi_1+\frac{1}{2}\epsilon)t)-i\bar{\epsilon}(x+(\bar{\chi}_1+\frac{1}{2}\bar{\epsilon})t)}$$
$$= 1 + i\epsilon(x+\chi_1 t) - i\bar{\epsilon}(x+\bar{\chi}_1 t) + \epsilon\bar{\epsilon}|x+\chi_1 t|^2 + o(\epsilon^2, \bar{\epsilon}^2).$$

By the following expansion and (3.2), we can obtain the dark-dark-rogue solution

$$q_i[2] = c_i \left(\frac{M_i}{M}\right) e^{i\theta_i}, \quad i = 1,2, \tag{3.3}$$

where

$$M = \frac{\beta + e^{-2\mathrm{Im}(\chi_2)[x+\mathrm{Re}(\chi_2)t]}}{(\bar{\chi}_1 - \chi_1)(\bar{\chi}_2 - \chi_2)} \left(|x+\chi_1 t|^2 - \frac{i[(x+\chi_1 t) + (x+\bar{\chi}_1 t)]}{(\bar{\chi}_1 - \chi_1)} - \frac{2}{(\bar{\chi}_1 - \chi_1)^2}\right)$$
$$+ \frac{e^{-2\mathrm{Im}(\chi_2)[x+\mathrm{Re}(\chi_2)t]}}{(\bar{\chi}_1 - \chi_2)(\bar{\chi}_2 - \chi_1)} \left(i(x+\bar{\chi}_1 t) + \frac{1}{\bar{\chi}_1 - \chi_2}\right) \left(i(x+\chi_1 t) + \frac{1}{\bar{\chi}_2 - \chi_1}\right),$$

and

$$M_i = F_i \left(\frac{\beta}{\bar{\chi}_2 - \chi_2} + \frac{(\bar{\chi}_2 + a_i)e^{-2\mathrm{Im}(\chi_2)[x+\mathrm{Re}(\chi_2)t]}}{(\chi_2 + a_i)(\bar{\chi}_2 - \chi_2)}\right)$$
$$+ \left[-\frac{1}{(\bar{\chi}_1 - \chi_2)(\chi_2 + a_i)} + \frac{(\bar{\chi}_1 + a_i)}{(\chi_2 + a_i)} \left(\frac{1}{(\bar{\chi}_1 - \chi_2)^2} + \frac{i(x+\bar{\chi}_1 t)}{(\bar{\chi}_1 - \chi_2)}\right)\right]$$
$$\times \left[-\frac{(\bar{\chi}_2 + a_i)}{(\chi_1 + a_i)^2(\bar{\chi}_2 - \chi_1)} + \frac{(\bar{\chi}_2 + a_i)}{(\chi_1 + a_i)} \left(\frac{1}{(\bar{\chi}_2 - \chi_1)^2}\right.\right.$$
$$\left.\left. + \frac{i(x+\chi_1 t)}{(\bar{\chi}_2 - \chi_1)}\right)\right] e^{-2\mathrm{Im}(\chi_2)[x+\mathrm{Re}(\chi_2)t]},$$

and

$$F_i = \frac{\bar{\chi}_1 + a_i}{(\chi_1 + a_i)(\bar{\chi}_1 - \chi_1)} \left(|x+\chi_1 t|^2 - \frac{i[(x+\chi_1 t) + (x+\bar{\chi}_1 t)]}{(\bar{\chi}_1 - \chi_1)} - \frac{2}{(\bar{\chi}_1 - \chi_1)^2}\right)$$
$$+ \frac{1}{(\bar{\chi}_1 - \chi_1)(\chi_1 + a_i)} \left[i(x+\chi_1 t) + i(x+\bar{\chi}_1 t)\frac{\bar{\chi}_1 + a_i}{\chi_1 + a_i} + \frac{2}{\bar{\chi}_1 - \chi_1}\right].$$

The asymptotical analysis of the above subsubsection is still valid for the dark-dark-rogue solution, since the dark-dark-rogue solution is nothing but the limit for solution (3.2). Finally we show the explicit dynamics by plotting figure (Fig. 6).

3.2 Two dark-one bright soliton solution

In this subsection, we consider the special case $a_1 = a_2$. Firstly, we consider $c_1 > c_2 > 0$, choosing the following special solution

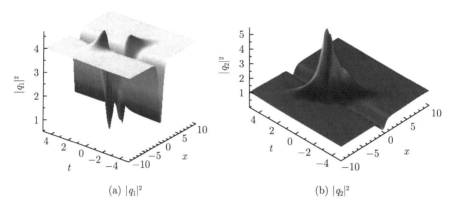

Fig.6 Dark-dark-rogue solution: Parameters $a_1 = -a_2 = 1$, $c_1 = 2$, $c_2 = 1$, $\lambda_1 = 1.24185466772002 + .636002000756738\mathrm{i}$, $\lambda_2 = 0.8333333333$, $\chi_1 = 1.356709486 + 1.087820879\mathrm{i}$, $\chi_2 = 1.414213562\mathrm{i}$, $\beta = 1$. It is seen that there is dark-dark rogue wave solution collision with rogue wave solution

$$|y_1\rangle = \begin{bmatrix} e^{\mathrm{i}(\chi_1 x + \frac{1}{2}\chi_1^2 t)} \\ \dfrac{c_1}{\chi_1 + a_1} e^{\mathrm{i}(\chi_1 x + \frac{1}{2}\chi_1^2 t)} + c_2\alpha_1 e^{\mathrm{i}(-a_1 x + \frac{1}{2}a_1^2 t)} \\ \dfrac{c_2}{\chi_1 + a_1} e^{\mathrm{i}(\chi_1 x + \frac{1}{2}\chi_1^2 t)} + c_1\alpha_1 e^{\mathrm{i}(-a_1 x + \frac{1}{2}a_1^2 t)} \end{bmatrix}$$

and $|y_2\rangle$ (3.1) as above subsection, for simplicity we take $\alpha_1 = 1$, we can obtain the solution (3.2) in above subsection with

$$M = \frac{\left(e^{C_2+\bar{C}_2} + \beta\right)}{\bar{\chi}_2 - \chi_2}\left(\frac{1}{\bar{\chi}_1 - \chi_1} + \frac{(c_1^2 - c_2^2)e^{C_1+\bar{C}_1}}{2(\bar{\lambda}_1 - \lambda_1)}\right) - \frac{e^{C_2+\bar{C}_2}}{(\bar{\chi}_1 - \chi_2)(\bar{\chi}_2 - \chi_1)},$$

$$M_i = \frac{\left(\dfrac{\bar{\chi}_2 + a_i}{\chi_2 + a_i} e^{C_2+\bar{C}_2} + \beta\right)}{\bar{\chi}_2 - \chi_2}\left(\frac{\bar{\chi}_1 + a_i}{\chi_1 + a_i}\frac{1}{\bar{\chi}_1 - \chi_1} + \frac{(c_1^2 - c_2^2)e^{C_1+\bar{C}_1}}{2(\bar{\lambda}_1 - \lambda_1)} + d_i e^{C_1}\right)$$
$$- \frac{\bar{\chi}_1 + a_i}{\chi_2 + a_i}\frac{e^{C_2+\bar{C}_2}}{(\bar{\chi}_1 - \chi_2)}\left(\frac{\bar{\chi}_2 + a_i}{\chi_1 + a_i}\frac{1}{(\bar{\chi}_2 - \chi_1)} + d_i e^{C_1}\right),$$

and

$$C_1 = -\mathrm{i}(\chi_1 + a_1)\left(x + \frac{1}{2}(\chi_1 - a_1)t\right), \quad C_2 = \mathrm{i}\chi_2\left(x + \frac{1}{2}\chi_2 t\right), \quad d_1 = \frac{c_2}{c_1}, \quad d_2 = 1/d_1.$$

We can perform similar asymptotical analysis on them. Choosing special parameters, we can obtain the interaction figure between dark-dark soliton and breather-II solution in the degenerate case. Here we give an example for the dark-dark soliton and breather-II soliton possesses the same velocity (Fig. 7).

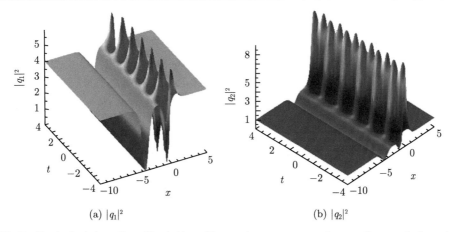

(a) $|q_1|^2$ (b) $|q_2|^2$

Fig.7 Dark-dark-breather-II solution: Parameters $a_1 = a_2 = 0$, $c_1 = 2$, $c_2 = 1$, $\lambda_1 = i$, $\lambda_2 = 0$, $\chi_1 = 3i$, $\chi_2 = 1.732050808i$, $\beta = \exp(10)$. It is seen that there exists dark-dark solution and breather-II solution with the stationary

Secondly, we consider $c_1 > 0$ and $c_2 = 0$. Indeed, the explicit expression of this kind solution can be obtained by choosing parameter $c_2 = 0$ on the above. Thus we can have two dark-one bright soliton solution:

$$q_1[2] = c_1\left(\frac{M_1}{M}\right)e^{i\theta_1}, \quad q_2[2] = c_1\left(\frac{M_2}{M}\right)e^{i\theta_1}, \qquad (3.4)$$

where

$$M = \frac{\left(e^{C_2+\bar{C}_2}+\beta\right)}{\bar{\chi}_2-\chi_2}\left(\frac{1}{\bar{\chi}_1-\chi_1}+\frac{c_1^2 e^{C_1+\bar{C}_1}}{2(\bar{\lambda}_1-\lambda_1)}\right)-\frac{e^{C_2+\bar{C}_2}}{(\bar{\chi}_1-\chi_2)(\bar{\chi}_2-\chi_1)},$$

$$M_1 = \frac{\left(\frac{\bar{\chi}_2+a_1}{\chi_2+a_1}e^{C_2+\bar{C}_2}+\beta\right)}{\bar{\chi}_2-\chi_2}\left(\frac{\bar{\chi}_1+a_1}{\chi_1+a_1}\frac{1}{\bar{\chi}_1-\chi_1}+\frac{c_1^2 e^{C_1+\bar{C}_1}}{2(\bar{\lambda}_1-\lambda_1)}\right)$$
$$-\frac{\bar{\chi}_1+a_1}{\chi_2+a_1}\frac{\bar{\chi}_2+a_1}{\chi_1+a_1}\frac{e^{C_2+\bar{C}_2}}{(\bar{\chi}_2-\chi_1)(\bar{\chi}_1-\chi_2)},$$

$$M_2 = \frac{\left(\frac{\bar{\chi}_2+a_1}{\chi_2+a_1}e^{C_2+\bar{C}_2}+\beta\right)e^{C_1}}{\bar{\chi}_2-\chi_2}-\frac{(\bar{\chi}_1+a_1)e^{C_1+C_2+\bar{C}_2}}{(\bar{\chi}_1-\chi_2)(\chi_2+a_1)},$$

and

$$C_1 = -i(\chi_1+a_1)\left(x+\frac{1}{2}(\chi_1-a_1)t\right), \quad C_2 = i\chi_2\left(x+\frac{1}{2}\chi_2 t\right).$$

In the following, we give the asymptotical analysis for the two dark-one bright soliton solution, which is nothing but nonlinear superposition for the bright-dark soliton and one dark soliton. It is readily to see that the velocity of bright-dark soliton equals to $v_{bd} = -\text{Re}(\chi_1)$ and the velocity of the dark one equals to $v_d = -\text{Re}(\chi_2)$. Assuming $v_{bd} < v_d$, $\text{Im}(\chi_1) > 0$ and $\text{Im}(\chi_2) > 0$. Through the tedious calculation, we have

(a) Before interaction ($t \to -\infty$), we have

$$q_1[2] \to \frac{c_1}{2}\left[1 + \frac{\bar{\chi}_1 + a_1}{\chi_1 + a_1} + \frac{\chi_1 - \bar{\chi}_1}{\chi_1 + a_1}\tanh(A_-)\right]e^{i\theta_1} - c_1\left(\frac{\bar{\chi}_1 + a_1}{\chi_1 + a_1}\right)e^{i\theta_1}$$
$$+ \frac{c_1}{2}\left(\frac{\bar{\chi}_1 + a_1}{\chi_1 + a_1}\right)\left[1 + \frac{\bar{\chi}_2 + a_1}{\chi_2 + a_1} + \frac{\chi_2 - \bar{\chi}_2}{\chi_2 + a_1}\tanh(B_-)\right]e^{i\theta_1},$$
$$q_2[2] \to \sqrt{2\text{Im}(\lambda_1)\text{Im}(\chi_1)}\,\text{sech}(A_-)e^{i(\theta_1 + C_-)},$$

where

$$A_- = \text{Im}(\chi_1)\left[x + \text{Re}(\chi_1)t + \frac{1}{2\text{Im}(\chi_1)}\ln\left(\frac{c_1^2}{2}\frac{\text{Im}(\chi_1)}{\text{Im}(\lambda_1)}\right)\right],$$
$$B_- = \text{Im}(\chi_2)\left[x + \text{Re}(\chi_2)t + \frac{1}{2\text{Im}(\chi_2)}\left(\ln\beta - 2\ln\left|\frac{\chi_1 - \chi_2}{\bar{\chi}_1 - \chi_2}\right|\right)\right],$$
$$C_- = -(\text{Re}(\chi_1) + a_1)x + \frac{1}{2}\left[a_1^2 - \text{Re}(\chi_1^2)\right]t + \frac{3}{2}\pi.$$

(b) After interaction ($t \to +\infty$), we have

$$q_1[2] \to \frac{c_1}{2}\left(\frac{\bar{\chi}_2 + a_1}{\chi_2 + a_1}\right)\left[1 + \frac{\bar{\chi}_1 + a_1}{\chi_1 + a_1} + \frac{\chi_1 - \bar{\chi}_1}{\chi_1 + a_1}\tanh(A_+)\right]e^{i\theta_1} - c_1\left(\frac{\bar{\chi}_2 + a_1}{\chi_2 + a_1}\right)e^{i\theta_1}$$
$$+ \frac{c_1}{2}\left[1 + \frac{\bar{\chi}_2 + a_1}{\chi_2 + a_1} + \frac{\chi_2 - \bar{\chi}_2}{\chi_2 + a_1}\tanh(B_+)\right]e^{i\theta_1},$$
$$q_2[2] \to \sqrt{2\text{Im}(\lambda_1)\text{Im}(\chi_1)}\,\text{sech}(A_+)e^{i(\theta_1 + C_+)},$$

where

$$A_+ = A_- + \ln\left|\frac{\bar{\chi}_1 - \chi_2}{\chi_1 - \chi_2}\right|,$$
$$B_+ = B_- + \ln\left|\frac{\chi_1 - \chi_2}{\bar{\chi}_1 - \chi_2}\right|,$$
$$C_+ = B_- + \arg\left(\frac{\bar{\chi}_1 - \bar{\chi}_2}{\bar{\chi}_1 - \chi_2}\right).$$

To show its dynamics behavior, we choose two different groups of parameters. The figure 8 and figure 9 show that the component $|q_1[2]|^2$ possesses two dark soliton and

$|q_2[2]|^2$ possesses one bright soliton. In figure 8, the two dark solitons possess different velocities. The interaction can be analyzed by above asymptotical analysis. In figure 9, the two dark solitons possess same velocity with stationary.

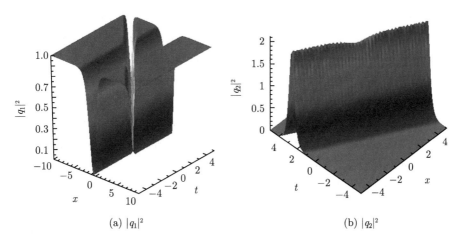

(a) $|q_1|^2$ (b) $|q_2|^2$

Fig.8 Two dark-one bright soliton: Parameters $a_1 = a_2 = 0$, $c_1 = 1$, $c_2 = 0$, $\lambda_1 = 1 + i$, $\lambda_2 = 0$, $\chi_1 = 1.786151378 + 2.272019650i$, $\chi_2 = i$, $\beta = 1$. It is seen that there are two dark soliton in the component $|q_1|^2$ and one bright soliton in the component $|q_2|^2$

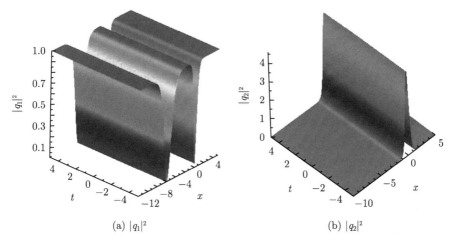

(a) $|q_1|^2$ (b) $|q_2|^2$

Fig.9 Two dark-one bright soliton: Parameters $a_1 = a_2 = 0$, $c_1 = 1$, $c_2 = 0$, $\lambda_1 = i$, $\lambda_2 = 0$, $\chi_1 = 2.414213562i$, $\chi_2 = i$, $\beta = \exp(10)$. It is seen that there are two dark soliton in the component $|q_1|^2$ with stationary and one bright soliton in the component $|q_2|^2$ with stationary

4 Discussions and Conclusions

In this paper, we provide a method to derive nonsingular localized wave solutions of mixed coupled nonlinear Schrödinger equations for which it is essential to deal with indefinite Darboux matrix. Furthermore, we present the classification for nonlinear localized wave solutions of the model through combining Darboux transformation and matrix analysis methods. The explicit conditions and ideal excitation forms for these nonlinear waves are presented in detail, which are meaningful for further physical studies on them. Since the dark soliton and breather type solution could coexist in the special background, the interaction between them is a very interesting topic naturally. Based on the iteration of Darboux transformation, we could construct the explicit solution to describe their interaction.

The high order solution can be obtained by the generalized DT [41–44]. Indeed, high-order solution can be obtained through limit technique from the solution formula in Theorem 3.1 too. We would like to consider the general high order solution with a proper form in the future. The methods here can be extended directly to the defocusing CNLSE or even general multi-component NLSE, the mixed or defocusing Sasa-Satuma system, three wave system, long wave-short wave model and other AKNS reduction system with indefinite Darboux matrix cases.

Recently, a classification on soliton solution for multi-component NLSE was presented in ref. [25], which mainly involving the dark-dark soliton, the bright-dark soliton and the breather solution (called by bright soliton there). The nonlinear wave solutions and the derivation method presented here are distinctive from the results.

Finally, what need mentioned is, when we posted our manuscript on arxiv, Professor Tuschida gave us a comment that the non-singularity condition had been given through the general algebro-geometric scheme in reference [45]. The differences between our work and reference [45] have been given in [46].

Acknowledgements The authors thank the anonymous referee for her/his valuable comments and advisements to improve our manuscript. This work is supported by National Natural Science Foundation of China (Contact No. 11401221, 11405129) and Fundamental Research Funds for the Central Universities (Contact No. 2014ZB0034).

References

[1] V. E. Zakharov, and A. B. Shabat. *Exact theory of two-dimensional self-focusing and one-dimensional self-modulation of waves in nonlinear media.* Zh. Eksp. Teor. Fiz., **61**, 118-134 (1971).

[2] H. Bailung and Y. Nakamura. *Observation of modulational instability in a multicomponent plasma with negative ions.* **50**, 231-242 (1993).

[3] B. Kibler, J. Fatome , C. Finot, et al.. *Observation of Kuznetsov-Ma soliton dynamics in optical fibre.* Scientific Reports, **2**, (2012).

[4] G.P. Agrawal. *Nonlinear Fiber Optics.*. 4th Edition. Boston: Academic Press, 2007.

[5] J.M. Dudley, G. Genty, F. Dias, et al.. *Modulation instability, Akhmediev Breathers and continuous wave supercontinuum generation.* Optics Express, **17**: 21497-21508 (2009).

[6] B. Kibler, J. Fatome, C. Finot, et al.. *The Peregrine soliton in nonlinear fibre optics.* Nature Physics, **6**, 790-795 (2010).

[7] A. Chabchoub, N.P. Hoffmann and N. Akhmediev. *Rogue wave observation in a water wave tank.* Phys. Rev. Lett., **106**: 204502 (2011).

[8] A Chabchoub, N Hoffmann, M Onorato, et al.. *Super rogue waves: observation of a higher-order breather in water waves.* Phys. Rev. X, **2**, 011015 (2012).

[9] H. Bailung, S.K. Sharma and Y. Nakamura. *Observation of Peregrine solitons in a multicomponent plasma with negative ions.* Phys. Rev. Lett., **107**, 255005 (2011).

[10] A. Chabchoub, O. Kimmoun, H. Branger, et al.. *Experimental observation of dark solitons on the surface of water.* Phys. Rev. Lett., **110**, 124101 (2013).

[11] B. Prinari, M.J. Ablowitz and G. Biondini. *Inverse scattering transform for the vector nonlinear Schrödinger equation with nonvanishing boundary conditions.* J. Math. Phys., **47**, 063508 (2006).

[12] S.V. Manakov, Zh. Éksp. Teor. Fiz. 65, 505 1973[Sov. Phys. JETP 38, 248 1974.]

[13] T. Kanna, M. Lakshmanan. *Exact soliton solutions, shape changing collisions, and partially coherent solitons in coupled nonlinear Schrödinger equations.* Phys. Rev. Lett., **86**, 5043 (2001).

[14] T. Kanna, M. Lakshmanan, P. Tchofo Dinda, et al.. *Soliton collisions with shape change by intensity redistribution in mixed coupled nonlinear Schrödinger equations.* Phys. Rev. E, **73**, 026604: 1-15. (2006).

[15] D.S. Wang, D.J. Zhang and J. Yang. *Integrable properties of the general coupled nonlinear Schrödinger equations.* J. Math. Phys., **51**, 023510. (2010).

[16] M.G. Forest, O C. Wright. *An integrable model for stable: unstable wave coupling phenomena.* Physica D: Nonlinear Phenomena, **178**, 173-189 (2003).

[17] B. Guo and L. Ling. *Rogue wave, breathers and bright-dark-rogue solutions for the coupled Schrödinger equations.* Chin. Phys. Lett., **28**, 110202 (2011).

[18] F. Baronio, A. Degasperis, M. Conforti, et al.. *Solutions of the vector nonlinear Schrödinger equations: evidence for deterministic rogue waves.* Phys. Rev. Lett., **109**, 044102 (2012).

[19] L. Zhao, J. Liu. *Localized nonlinear waves in a two-mode nonlinear fiber.* JOSA B, **29**, 3119-3127 (2012).

[20] L. Zhao, J. Liu. *Rogue-wave solutions of a three-component coupled nonlinear Schrödinger equation.* Phys. Rev. E, **87**: 013201 (2013).

[21] X. Wang, Y. Li, Y. Chen. *Generalized Darboux transformation and localized waves in coupled Hirota equations.* Wave Motion, (2014).

[22] J. He, L. Guo, Y. Zhang and A. Chabchoub. *Theoretical and experimental evidence of non-symmetric doubly localized rogue waves.* Accepted by Proceedings of the Royal Society A, (2014).

[23] G. Dean, T. Klotz, B. Prinari, et al.. *Dark-dark and dark-bright soliton interactions in the two-component defocusing nonlinear Schrödinger equation.* Applicable Analysis, **92**, 379-397 (2013).

[24] Q.H. Park and H.J. Shin. *Systematic construction of multicomponent optical solitons.* Phys. Rev. E, **61**, 3093 (2000).

[25] T. Tsuchida. *Exact solutions of multicomponent nonlinear Schrödinger equations under general plane-wave boundary conditions.* arXiv preprint arXiv: 1308.6623, 2013.

[26] C. Kalla. *Breathers and solitons of generalized nonlinear Schrödinger equations as degenerations of algebro-geometric solutions.* J. Phys. A: Math. Theor., **44**, 335210 (2011).

[27] P. Deift and E. Trubowitz. *Inverse scattering on the line.* Comm. Pure and Appl. Math., **32**, 121-251 (1979).

[28] V. B. Matveev and M A. Salle. *Darboux transformations and solitons.* (Berlin: Springer-Verlag, 1991).

[29] C.H. Gu, H.S. Hu, Z. Zhou. *Darboux transformations in integrable systems: theory and their applications to geometry.* (Springer, 2006).

[30] C.-L. Terng and K. Uhlenbeck. *Bäcklund transformations and loop group actions.* Comm. Pure Appl. Math., **53**, 1-75 (2000).

[31] S. P. Novikov, S. V. Manakov, V. E. Zakharov, and L. P. Pitaevskii. *Theory of solitons: the inverse scattering method.* (Springer, 1984).

[32] O.C. Wright and Gregory M. Forest. *On the Bäcklund-gauge transformation and homoclinic orbits of a coupled nonlinear Schrödinger system.*, Physica D: Nonlinear Phenomena, **141**, 104-116 (2000).

[33] A. Degasperis, S. Lombardo. *Multicomponent integrable wave equations: I. Darboux-dressing transformation*. J. Phys. A: Math. Theor., **40**, 961 (2007).

[34] A. Degasperis, S. Lombardo. *Multicomponent integrable wave equations: II. Soliton solutions*. J. Phys. A: Math. Theor., **42**, 385206 (2009).

[35] A. de O Assuncao, H. Blas and M. da Silva. *New derivation of soliton solutions to the $AKNS_2$ system via dressing transformation methods*. J. Phys. A: Math. Theor., **45**, 085205 (2012).

[36] L. Ling, B. Guo, and L. Zhao. *High-order rogue waves in vector nonlinear Schrödinger equations*. Phys. Rev. E, **89**, 041201(R) (2014).

[37] B. Feng. *General N-soliton solution to a vector nonlinear Schrödinger equation*. J. Phys. A: Math. Theor., **47**, 355203 (2014).

[38] D. Zhang, S. Zhao, Y. Sun. and J. Zhou. *Solutions to the modified Korteweg-de Vries equation*. Rev. Math. Phys., **26**, 1430006 (2014).

[39] L. Ling, L. Zhao, B. Guo. *Darboux transformation and multi-dark soliton for N-component coupled nonlinear Schrödinger equations*. Accepted by Nonlinearity.

[40] F. Baronio, M. Conforti, A. Degasperis, et al.. *Rogue wave solutions for coupled defocusing nonlinear Schrödinger equations*. Phys. Rev. Lett., **113**, 034101 (2014).

[41] B. Guo, L. Ling and Q P. Liu. *Nonlinear Schrödinger equation: Generalized Darboux transformation and rogue wave solutions*. Physical Review E, **85**, 026607 (2012).

[42] J. He, H. Zhang, L. Wang and A. Fokas. *Generating mechanism for higher-order rogue waves*. Phys. Rev. E, **87**, 052914 (2013).

[43] B. Guo, L. Ling and Q P. Liu. *High-Order Solutions and Generalized Darboux Transformations of Derivative Nonlinear Schrödinger Equations*. Stud. Appl. Math., **130**, 317-344 (2013).

[44] D. Bian, B. Guo and L. Ling. *High-Order Soliton Solution of Landau-Lifshitz Equation*. Stud. Appl. Math., **134**, 181-214 (2015).

[45] B. A. Dubrovin, T. M. Malanyuk, I. M. Krichever and V. G. Makhankov. *Exact solutions of the time-dependent Schrödinger equation with self-consistent potentials*. Sov. J. Part. Nucl., **19**, 252-269 (1988).

[46] L. Ling, L. Zhao and B. Guo. *Reply to "Comment on "Darboux transformation and classification of solution for mixed coupled nonlinear Schrödinger equations""*. arXiv:1408.2230 (2014).